P9-ELP-429

Biology

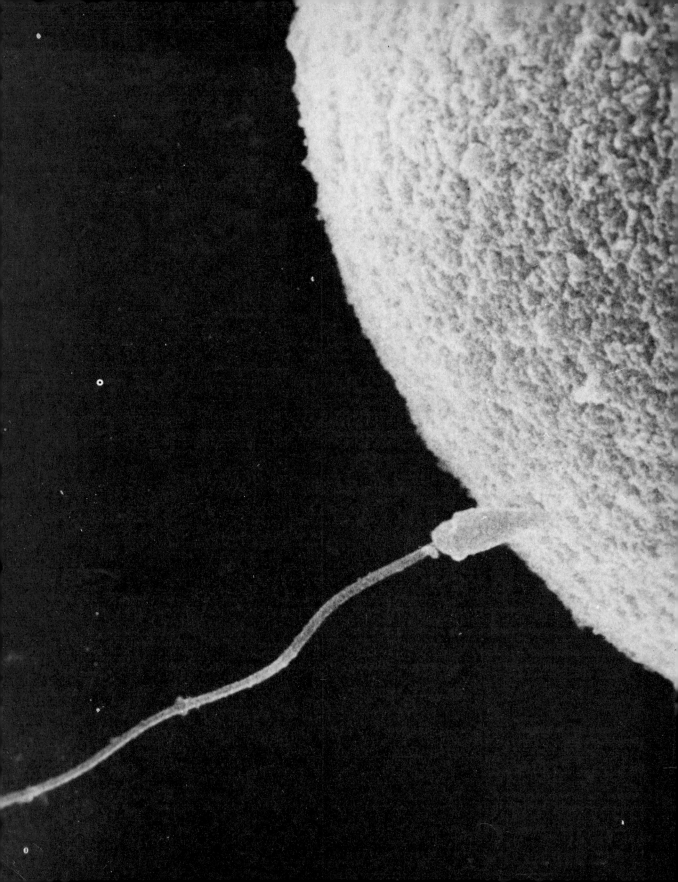

Biology

Ebert, James D.
*Carnegie Institution of Washington
and
Marine Biological Laboratory*

Loewy, Ariel G.
Haverford College

Miller, Richard S.
Yale University

Schneiderman, Howard A.
University of California, Irvine

Holt, Rinehart and Winston, Inc.

*New York Chicago San Francisco Atlanta
Dallas Montreal Toronto London Sydney*

Copyright © 1973 by Holt, Rinehart and Winston, Inc.
All rights reserved
ISBN: **0-03-084863-6**
Library of Congress Catalog Card Number: 72-91237
Printed in the United States of America
3 4 5 6   0 7 1   9 8 7 6 5 4 3 2

Frontispiece: Scanning electron micrograph of sea urchin sperm shortly after contact with the surface of the egg. By courtesy of Don W. Fawcett and Everett Anderson.

Text design by Jerry Tillet.
Photo research by Yvonne Freund.

CREDITS AND ACKNOWLEDGMENTS

Chapter 1. 1-3: Reprinted with permission of Macmillan Publishing Co., Inc. from *Race and Races* by Richard A. Goldsby. Copyright © 1971 by Richard A. Goldsby; 1-4A: Wide World; 1-4B: John Launois, Black Star; 1-5; John Moss, Black Star; 1-6: Metropolitan Museum of Art; 1-7, -8: NASA; 1-9: Werner Bishop, Magnum; 1-10: Los Angeles County Air Pollution Control Division. **Chapter 2.** 2-1: Hooke, *Micrographia*, 1664, Linda Hall Library, London; 2-2, -3, -4, -6A, -7A, -9, -34: From *Cell Structure and Function*, 2/E, by Ariel G. Loewy and Philip Siekevitz. Copyright © 1969 by Holt, Rinehart and Winston. Reprinted by permission of Holt, Rinehart and Winston; 2-6B: M. C. Ledbetter, Brookhaven National Laboratory; 2-7B, -12, -15B, -28A, -30: G. E. Palade, Rockefeller University; 2-11: J. F. Hoffman; 2-13, -17, -22, -27B, -31: D. W. Fawcett, from Fawcett, *The Cell: An Atlas of Fine Structure*, W. B. Saunders, 1966; 2-14A: After A. L. Houwink and P. A. Roelfson, *Acta Botanica Neerlandica*, **3**:285, 1954; 2-14B: After K. Mulhethaler, from *Cells and Organelles* by Alex Novikoff and Eric Holtzman. Copyright © 1970 by Holt, Rinehart and Winston. Reprinted by permission of Holt, Rinehart and Winston; 2-15A: J. D. McNabb and E. Sanborn, *Journal of Cell Biology*, **22**:702, no. 3, 1964. Reproduced by permission of The Rockefeller University; 2-16A: M. Farquhar and G. E. Palade, *Journal of Cell Biology*, **17**:379, no. 2, 1963. Reproduced by permission of The Rockefeller University; 2-19: H. Bonnett, Jr. and E. Newcomb, *Journal of Cell Biology*, **27**:431, no. 2, 1965. Reproduced by permission of The Rockefeller University; 2-20: Daniel S. Friend; 2-21, -25, -28C, -29B: Novikoff/Holtzman; 2-23A: Arnold Schecter, from Fawcett; 2-23B: D. W. Fawcett, from Bloom and Fawcett, *Textbook of Histology*, 9/E, W. B. Saunders, 1968, after E. Roth; 2-24A: Sergei Sorokin, from Fawcett; 2-14B: Jean Andre, from Fawcett; 2-26: I. Gibbons, University of Hawaii; 2-27A: Shinya Inoué, "Polarization Optical Studies of the Mitotic Spindle. I. The Demonstration of Spindle Fibers in Living Cells," *Chromosoma* (Berl.), **5**:491, 1953. Berlin-Gottinger-Heidelberg: Springer; 2-28B: S. Fleischer, B. Fleischer, and W. Stoeckenius, *Journal of Cell*
(continued on p. 773.)

PREFACE

Biology—
Today and Tomorrow

This book arises out of our conviction that, in its recent cataclysmic evolution, the science of biology has come of age. And, as in all the races and societies of man, "coming of age" brings with it new privileges and new responsibilities. By any reckoning, the responsibilities loom large.

As biology comes of age, so must its teachers and students. Biology, like its sister sciences chemistry and physics, which are themselves rapidly becoming part of the warp and woof of biology, has immense potential for both good and evil. The almost overwhelming advances in biomedicine and biomedical engineering—the development of antibiotics and vaccines and the fashioning of artificial organs; and in agriculture—the emergence of hybrid corn and other crops and livestock to feed the world—will soon be dwarfed by advances we are just beginning to perceive. But with these advances come responsibilities for larger decisions that will inevitably shape the lives of future generations.

As this book is being completed, one of the debates in the Congress and in the Courts centers on determining the vital balance between the need for increased power, to be provided by atomic plants, and the protection of the environment from radioactive and thermal pollution. The question is being asked in all seriousness, "Is a little pollution a good thing?" Which organisms are favored by raising the temperature of the ocean? Given the need for more electricity, where in the political process—and on what grounds—do we determine which is the lesser of two possible risks, radioactive and thermal pollution or pollution by the products of burning fossil fuels, products including cancer-causing hydrocarbons?

This question is related to another of great moment: The need for a vastly expanded program of cancer research. How does the public and its elected representatives determine whether the field of biomedicine is ready for a billion dollar a year program? How does it assess scientific evidence, especially when, as it so often happens, it is conflicting? Is there a conflict between "basic" and "relevant" research? Should we be preoccupied with "instant relevance?" How does information coalesce into ideas? How does research evolve into technological innovation? We know that for many innovations of great economic and social importance the "lag time" is long, measured not in months or years, but decades. One example, which we trace in Chapter 15, the oral contraceptive pill, required over sixty years of research in anatomy, physiology, systematic botany, and steroid chemistry, all of which would have come to naught but for the powerful stimulus of Margaret Sanger and other pioneers in the field of population control.

The atomic age was thrust upon a world unprepared for it. It is essential, we believe, that the impact of the biological revolution—its sweeping new concepts, its vast potential for technological advance, and its overwhelming social consequences—fall on "the mind prepared."

Strides are being made in capturing the drama of today's biology, in showing its emergence as an experimental science, and in fostering the inquiring mind. But at the same time the barrier to public judgment and understanding of the subject is increasing as the tide of new facts flows endlessly.

Thus there is a constant need for distillation and synthesis. This book is our distillation of biology. We have tried to provide for the beginning student the *essence* of the subject. Clifford Grobstein put it well when he wrote, in *The Strategy of Life,* that "Like science in general, biology is rooted in the substantial and the material." We have thus sought to convey the substantial and material facts about "life as external reality." But we have tried to convey *more* than selected facts and representative examples; we have tried to bring into focus the ideas of biology —their origin, testing and social impact.

We have paid serious attention to the social implications of biology. However our readers should not expect to find a single "catch-all" chapter on "social biology." We find merit in discussing the social implications of a discovery in the context of other findings to which it is related. Thus aging and "cells out of control"—cancer and congenital defects—are considered together with the mechanisms regulating development; and sterility resulting from "V.D.," and homosexuality and transvestism, are assessed against the background of the hormones in human reproduction. And so it goes: The origin of race and races is taken up as a question in population genetics, and pesticides and pollution are viewed in the context of environmental biology. Population control appears several times in our story, as seen through the eyes first of the reproductive biologist and then of the student of populations.

We have tried to make the book both timely and timeless. It is timely in the sense that new discoveries and insights are integrated into the text in sufficient depth to permit their full understanding. No subject has been ignored because it is distasteful or sensitive. At the same time, no subject is included just because "it always has been." The test is always: Is the idea important to an understanding of today's and tomorrow's world? In this sense the book is timeless.

Although we hope that a large audience will find our book useful to read without relation to a formal course of study, it is intended to be a *textbook for beginning students.* We agree with the sentiment voiced by Loewy and Siekevitz in introducing their *Cell Structure and Function,* that it is the beginner who deserves an initial statement that is fully representative of the contemporary quality and mode of the field. Thus we do not distinguish among beginners. We believe our book will be useful to beginning students no matter what their previous training or academic "track" may be. As we see it, our book can stand alone as a textbook for a one-semester course meeting the requirements of non-majors (but when does one become irrevocably "fixed" as a "non-biologist?"), or used in conjunction with paperbacks and other readings, as the text for a year's course for biology majors, whatever their preprofessional aims.

Our principal concern has not been the "package" or framework within which our readers will be receiving their training in biology. That responsibility rests with the instructor, who knows the background and experience and the motivations of his students. Rather, we have tried to convey the ideas of our subject, its spirit, and the sense of excitement surrounding it.

We have developed each topic gradually, but with sufficient rigor to make it meaningful no matter whether the instructor elects to follow the sequence we have chosen or whether he follows any one of several other equally valid routes to the same end.

For example, we have tried to offer an incisive treatment of cell regulatory mechanisms, leading up to a consideration of development and its controls, both the cell's inner controls and the environmental forces shaping development. We have worked hard to make topics which students traditionally find difficult, for example the electrochemical changes underlying the nerve impulse, clear and easy to remember. These key concepts are presented within the framework of a modern treatment of the nervous system and the physiological basis of animal behavior, including an analysis of the limbic system, motivation, learning, and the physiological effects of drugs.

Plants have been given major emphasis, with plant structure, function, diversity and development being treated extensively. In fact the sections on diversity of both animals and plants, with which the book is concluded, are, we believe, unique, recapturing the interest of natural history in a framework of modern comparative biochemistry, physiology and behavior.

The ideas of biology constitute an increasingly large part of our social heritage. More and more we are contemplating from whence we came. It was our late President, John F. Kennedy, who said: "All of us have in our veins the exact same percentage of salt in our blood that exists in the ocean, and therefore, we have salt in our blood, in our sweat, in our tears. We are tied to the ocean. And when we go back to the sea—

whether it is to sail or to watch it—we are going back from whence we came."

And the President might have added that in our embryonic lives we spend some time in that "other ocean," the fluid of the amnion.

Thus we ponder our origins and our place in the universe. Our understanding of the living world about us is ever changing, ever expanding. If we are to face the future with confidence, we must understand how our ideas are rooted in the past—and nourished and winnowed in the present. It is toward that end that we offer this book.

ACKNOWLEDGMENTS

The preparation of this book was aided by the contributions of Barbara Burkett (University of Miami), Donald Fosket (University of California, Irvine) and William Sistrom (University of Oregon) and by the constructive criticism of many colleagues. Jane Milek (Nassau Community College), Eleanor L. Potorski (Suffolk County Community College), Harry Reasor (Miami-Dade Junior College), Jerome Williams (Eastfield College), and Herbert Wisner (University of Oregon) were frank and helpful critics of the entire book. Parts of the book were "field-tested" in Barbara Burkett's introductory course for non-majors at Miami. Early in the development of this book, Samuel Moffat, a science writer who is based in California, played an important role in his collaboration with three of the present authors.

Others who offered suggestions on the manuscript at several stages in its preparation included Joseph Arditti, Ernest Ball, Peter and Susan Bryant, Peter Dixon, Elizabeth Fosket, George Hunt, Stuart Krassner, Gary Lynch, James McGaugh, Anne M. Schneiderman, David Tepfer, and Marcel Verzeano, all of the University of California, Irvine; John J. Baird (California State Colleges), William Balamuth (University of California, Berkeley), Robert D. Barnes (Gettysburg College), Walter Bock (Columbia University), William E. Boggs, C. R. Botticelli (Boston University), C. Loring Brace (University of Michigan), Fanny-Fern Davis (Okaloosa-Walton Junior College), T. Delevoryas (Yale University), Marie Gilstrap (Highline Community College), Elizabeth

U. Green (Haverford College), E. J. Kormondy (Oberlin College), Gabriel W. Lasker (Wayne State University), Wilbur Mangas, Stanley L. Miller (University of California, San Diego), Todd Newberry (University of California, Santa Cruz), A. E. O'Donnell (Stanford Research Institute), Robert Ornduff and Oscar H. Paris (University of California, Berkeley), Irwin Sherman (University of California, Riverside), E. Shneour (University of Utah), Keith Short (University of Dublin), Sidney Titelbaum (Chicago City College), Paul F. White (Polk Junior College), Anne D. Willoughby (Haverford College), and Stephen Zornetzer (University of Florida).

We are pleased to acknowlege the help of our illustrators, Gaetano diPalma, Eric Hieber and George Kelvin, as well as that of Yvonne Freund, our photo researcher.

A number of suggestions emerged as our manuscript was being transformed into a book by our Editors and Publisher, Ian Baldwin, Jr., Brian Heald, Michael Notter, Dan Serebrakian and Don Schumacher. To them, and to numerous others who helped along the way, we are deeply grateful.

J.D.E., *Baltimore, Maryland and*   December, 1972
        Woods Hole, Massachusetts
A.G.L., *Haverford, Pennsylvania*
R.S.M., *New Haven, Connecticut*
H.A.S., *Irvine, California*

CONTENTS

Biology

CHAPTER 1

The Biological Revolution

Three billion years ago life began on this planet. Whether it arose spontaneously from nonliving matter or was seeded by spores from a distant corner of the universe we do not as yet know. What followed, however, is becoming a well-chronicled drama, a sweep of accelerating events which has transformed our planet from a lifeless rock reflecting the rays of the sun to an intricate, teeming, minutely balanced living world in which the sun's energy and the materials of the earth are continually converted into the multitude of structures and processes of life.

This wonderful and complex planet of ours may not appear so very unusual to a spaceship approaching it from afar—to an astronaut 100,000 miles away the earth is but a small body spinning through space at 66,600 miles an hour, one of a group of nine planets circling in their orbits about the sun. In physical terms the earth is not especially remarkable; it is surrounded by a gaseous envelope, the atmosphere. It has a diameter of approximately 7900 miles and a circumference of about 25,000 miles. Its surface area of nearly 200 million square miles is about one-third land and the rest water. It has an inner solid core and a molten outer core, enclosed in a rocky mantle that extends to the earth's surface. However, what makes the earth unique in our solar system is that everywhere on its surface, on land, in its rivers, in its lakes and oceans, and for large distances into the atmosphere, there is *life*. And within this envelope of life that we call the **biosphere,** man is only one of a remarkable variety of organisms that have adapted to almost the entire range of physical conditions found on earth.

The highest human settlement in the world is in Chile at an altitude of 5700 meters (18,700 feet). Most humans would be starved for oxygen at this altitude and would be dizzy, weak, and unable to work. However, a few people and many species of animals have adapted to these high altitudes. Beetles live in the highest meadows of the Himalayas, earthworms are found at the extremes of cold and altitude at the snow line of the Andes, and eagles and vultures soar as high as 7000 meters.

At another extreme, marine animals have been collected from 10,500 meters below the ocean's surface at pressures of one ton per square centimeter; we will undoubtedly find living organisms at even greater depths when we have devised suitable apparatus for collecting them.

The lowest air temperature ($-88°C$) was recorded in Antarctica in 1960. Temperatures almost this low have also been recorded in the Yukon Territory of Canada on several occasions, and the plants and animals of the Arctic are obviously capable of surviving at temperatures this low and probably lower. More remarkably, the green alga *Chlorella*, a single-celled plant, has been frozen to $-182°C$ and survived, and dry seeds of plants have germinated after three weeks in liquid air at $-190°C$. At the other extreme of temperature blue-green algae and certain bacteria live in hot springs (such as those in Yellowstone Park) at temperatures only 12 degrees centigrade below the boiling point of water.

Life is remarkable for reasons other than adaptability. The sweep of evolution has given rise to living systems which are increasingly able to *control* the conditions under which they live and thus escape the constraints and the capriciousness of their physical surroundings. Thus, for instance, while some organisms developed spores, eggs, or seeds which are able to survive drought or extreme cold, others evolved **warm-bloodedness** and thereby extended their range of life activities beyond that of their ancestors.

Man is a new arrival on this earth (Fig. 1-1);

yet through the coupled evolution of his hand and his brain, the stage is now set for ever-increasing control of the arbitrary constraints of our environment to a degree which is still almost unimaginable. But before attempting to evaluate the magnitude and consequences of this *biological revolution*, we must ask ourselves how evolution—the upward sweep of the history of life on our earth—could have occurred, how living matter (more than other physical matter) is endowed supremely well with this ability to evolve.

LIVING MATTER USES ENERGY AND CREATES ORDER

There is a law of physics, the *second law of thermodynamics*, which asserts that **systems** in isolation "run down." Let us explain what we mean by "isolation" and by "running down," for the second law is an elusive concept unless one understands the meaning of each of its components in very concrete terms.

The most important concept of thermodynamics is energy, which is defined as the ability to do work. The word work has a familiar, everyday ring to it: we all have a feeling for what it is to do work and we all know that one needs energy to perform it. The work with which we are most familiar is mechanical work in which a force moves a mass through a given distance. But there are other forms of work along with their respective forms of energy. Electrical work can be performed by the expenditure of electrical energy, chemical work by chemical energy and so on. Furthermore, these various forms of energy are interconvertible so that chemical energy, for instance, can be transduced into mechanical energy and even vice versa. Living matter is particularly effective in performing energy transductions: our muscles convert chemical energy to mechanical energy that results in movement; when a firefly flashes its signal, it converts chemical energy to light energy; an electric eel converts the chemical energy stored in its "batteries" to electrical energy; the human eye converts light energy to chemical energy and then to electrical energy. Cells, the basic units of which all living organisms are composed, are specialists in converting one form of energy to another.

The important thing to recognize about these various forms of energy is that when they interconvert, they do so quantitatively, that is, without any energy being created or destroyed in the process. This is what the *first law of thermodynamics* states and numerous experiments have demonstrated that this law is obeyed just as scrupulously by living matter as it is by the heat engines investigated by the early students of thermodynamics.

To establish the first law, the systems under investigation had to be isolated from their environment so that no energy flowed between them and their environment. It was soon discovered that such isolated systems "ran down," that is, their ability to perform work decreased as work was being performed by them. The second law of thermodynamics was able to explain this important observation by distinguishing between **free energy** (G) and **enthalpy** (H) and showing how these two properties of matter are related to each other.

Free energy is that portion of the energy of a system which is able to perform work. When an isolated system performs work, the free energy of the system decreases. The enthalpy is the energy given off or absorbed and can be measured directly in an isolated system by the change in temperature which the system experiences. By studying heat engines it was soon discovered that free energy and enthalpy are not necessarily directly related to each other. That is, the loss of free energy during the performance of work by an isolated system is not necessarily equal to the increase in enthalpy associated with the process. It turned out that another term is involved in the equation. This term consists of two other properties of matter, one of which is the **absolute temperature** which is measured in Kelvin units. The other property is the **entropy,** which is related to the state of orderliness or organization of the system. The higher the state of organization, the lower is the entropy. The second law states that when a system performs work it runs down not only because the free energy decreases but because the entropy, its state of disorganization,

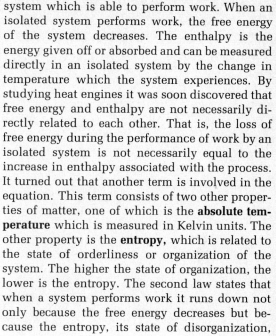

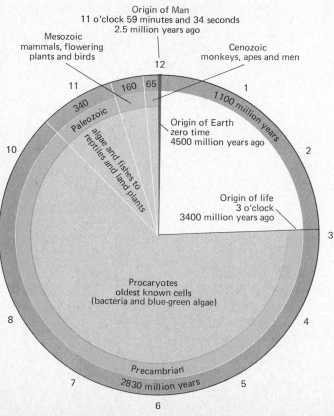

Figure 1-1 Man is a new arrival on the earth. If the age of the earth were 12 hours, then man has been on the scene for only 26 seconds. Notice how the pace of evolution accelerates, how the evolutionary periods become shorter and shorter.

increases. The formal relationship among these four properties of matter (G, H, T, and S) is stated by the second law of thermodynamics as follows:

$$\Delta G = \Delta H - T\,\Delta S$$

where ΔG is the change in *free energy* measured in *calories.* (One calorie is defined as the energy necessary to raise the temperature of one gram of water by one degree centigrade.)

ΔH is the change in *enthalpy* and is also measured in calories.

T is the *absolute temperature* and is measured in degrees Kelvin.

ΔS is the change in *entropy* and is measured in calories per degree Kelvin.

To put the second law of thermodynamics into words, in a given system the change in the amount of energy capable of doing work is equal to the amount of heat given off or absorbed minus the product of the absolute temperature and the change in entropy (disorganization).

As we have said at the beginning of this section, it is a universal law of nature that systems in isolation "run down." We now have a more precise way of describing what running down means. When work is performed free energy (G) is utilized and, according to the above equation, when free energy decreases, the enthalpy (H) must decrease or the entropy (S) must increase or both. Therefore, when a system in isolation performs work, its state of disorganization tends to increase and the corollary of this is that organization of a system can only be restored or increased by supplying free energy to it from the outside.

We are familiar with this principle in our daily existence. We know that machines capable of performing work must be supplied with external free energy to keep them going, that watches must be wound, engines must be fueled, and in short, that perpetual motion machines cannot exist in the real world. We also know, on a more general level, that it takes energy to create and maintain order, whether this order represents a developing embryo, a computer program, or books on a library shelf. The second law of thermodynamics simply formalizes our intuition and states with precision the relationship between energy and order.

G. N. Lewis, one of the great contributors to thermodynamic theory, was known to have been puzzled about biological evolution. While he observed physical systems to run down and become disordered, he noticed that biological systems have the ability to increase their order or complexity whether in **ontogeny,** (development of the fertilized egg to the mature individual), or in **phylogeny,** (evolutionary history of an organism).

Was G. N. Lewis right to think that living matter seemed to defy the laws of thermodynamics? We now realize that he was not; he failed to recognize that living matter cannot be regarded as an isolated system, that an organism, although very individualistic, is an open system as far as the flow of energy and of materials is concerned. Thus although we can recognize ourselves in the mirror every morning as being each day the self-same person, we do not derive this constancy of qualities from being at *equilibrium* with our surroundings, for an equilibrium is the end result of all energy transformations—it is a system at rest, it is death. A living system is a **steady state,** or better, a large number of interlocking steady states, which are maintained at their complex level of organization by a constant flux of free energy and materials moving through the system. The individuality of an organism is similar to the individuality of a candle flame which, although it *appears* the same from minute to minute, does so only through the constant flow of energy and materials supplied by oxidizing wax (Fig. 1-2).

However, the availability of energy and of materials is not sufficient to explain the maintenance of the highly organized steady state we observe in biological organisms. Energy becomes useful if it is absorbed by an organized structure. Thus we come to a fundamental property of a steady state system such as living matter: structure is necessary for the proper *utilization* of energy and energy is necessary for the *elaboration* and *maintenance* of structure. As we shall find in the following chapters, there is a minimal biological organization, transmitted from cell to cell, which forms the "informational" basis for the utilization of energy necessary to power the growth and development of the organism. As organization unfolds in the development of a plant or animal, so does its ability to utilize

Three billion years ago, as life began on our planet, the energy of the sun served to warm the earth, which in turn radiated energy into space. After life began, some of the sun's energy was captured by organisms and used by them to maintain and extend their organization. As we shall see later, the theory of evolution explains how, by a process of selection, there evolved increasingly complex and effective states of biological organization. Larger and larger areas of the globe and more and more **ecological niches** were occupied by species that filled these niches with greater and greater effectiveness. It is a staggering thought that some 2 million biological species exist on our earth today. Most of them have survived millions of years of intense competition to fill a given niche, each living in its own characteristic way. The biotic world on our planet is truly a pluralistic society with 2 million coexisting, highly specialized, closely adaptive ways of life.

Let us return, however, to our discussion of life as an energy-capturing process. We have implied that there is a directional arrow in evolution from simple to complex, from a few sparsely distributed species to many species populating every nook and cranny of our earth. This ascending arrow can be characterized by the amount of energy trapped and processed by living matter through the course of evolution. If one were able to measure such an upward sweep of energy through time, he would find a series of discontinuities or steps, each of which marks a *revolutionary* change brought about by an *evolutionary* adaptation. One such dramatic event was the conquest of land by plants, extending life's activities to the terrestrial third of our planet. Each new evolutionary adaptation brings about a new set of biotic relationships, a revolution characterized by the utilization of greater proportions of energy coming from the sun.

THE BRAIN, THE HAND, AND THE INDUSTRIAL REVOLUTION

Human evolution has passed through a series of revolutions, each cascading over its predecessor with increasing rapidity. The processes of evolu-

---

B2 ••• **The News-Journal papers**

Weather

Showers and thunderstorms from the lachians and Carolinas into New Eng Scattered rain showers and snow shov Rockies, with rain likely along the nort and 50s were predicted from the uppe sippi Valley to the northern and central Northeast through the Appalachians to well as in the southern Rockies and the desert Southwest, the southern half from Florida to the mid Atlantic Coast

Forecast through 7 p.m.

High temperatures

FRONTS: Warm

Forecasts 162.

New Castle County
Today: Mostly cloudy. thundershowers. High erly 15-20 mph.
Tonight: Cloudy wit thundershowers. westerly 10-20 r
Precipitation: percent today

Delmarv
Today:
High
Ton

ructure and each other in that we shall

does all this it ultimately ll see later, al- als, this **food** nts, which in f the sun.

is due to vapors h temperatures. ures are caused dly released

of the flame emperature than cause owing to the of oxygen at the edge, r more rapidly. flame is not only shape of the wick, namics of the rise of rs and the heated air flame. rising adjacent to the d air to rise past the dle thus keeping the ing and thus preventing ten wax in the center

nt wick glows at edge of e,because of the abundance flame is very hot.

hread in wick shortens upon ringing about bending of us wick maintains a constant which determines the unchanging rance of flame.

Outer shell of wax has higher ting point.

Inner shell of wax has lower melting point thus allowing for stable pool of molten wax which flows up wick by capillary action.

6. Wax provides materials and energy.

1-2 A candle flame is an example of a steady system maintaining an "unchanging" individuity through a constant flow of energy and material. As is the case with living matter, a constant flux of energy is required to maintain the organization of this system.

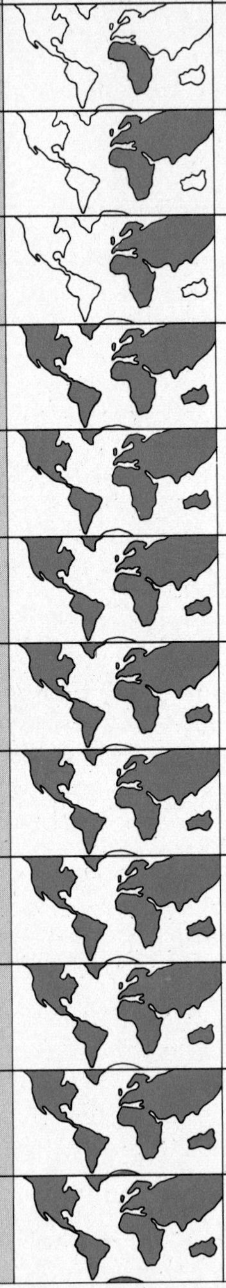

years ago	cultural stage	area populated	assumed density (per square mile)	population (millions)	revolutions
1,000,000	old stone age		0.0112	0.125	Paleolithic
300,000	middle stone age		0.032	1.00	
25,000	new stone age		0.105	3.34	
10,000			0.105	5.32	Neolithic
6,000	village farming and early urban		2.64	86.5	
2,000	village farming and urban		2.76	133	Urban
320	farming and industrial		9.7	545	Industrial
220			12.8	728	
170			16.2	906	
70			28.8	1,610	Scientific
20			43.0	2,400	
2000 A.D.			121.0	6,270	

Figure 1-3 Human population density has passed through a series of rapid escalations. In fact the way one can recognize a revolutionary development is to look for a sudden expansion of the population density. The effect of the industrial and scientific revolutions on the population density of the world is a cause for concern.

tion and of revolution are not conflicting or contradictory. Evolution means unfolding, developing, changing—a process which may proceed slowly by many minute steps. However, at certain infrequent intervals change can occur more rapidly when some new or basic factors come into play. We refer to such periods of rapid change as *revolutions*, a word which in its literal sense means *turning around*. It implies a basic change in relationships, which allows for new opportunities to be seized and therefore for rapid evolutionary change. Survival on dry land, warm-bloodedness, flight, and the opposable thumb[1] are examples of revolutionary changes that were followed by new and extensive evolutionary developments. Revolution is part and parcel of evolution.

In human evolution, revolutions occurred when a new relationship between man and nature unleashed new selective forces; in the first 1 to 2 million years of human evolution these forces brought about a number of important biological changes and in the last 10,000 years a multitude of cultural changes. One important index of these sweeping evolutionary changes in the biology and culture of our ancestors is population density. Figure 1-3 illustrates how human population density passed through a series of escalations, each of which coincided with a revolutionary period in our past. In an ultimate sense population density can be related to the amount of energy a population can capture from its environment, and one can therefore conclude that each revolution in human existence is brought about by the utilization of a larger proportion of the sun's energy.

The first or **paleolithic** revolution began some 2 million years ago when our ancestors started to collect and later manufacture stone tools used for hunting and associated activities (Fig. 1-4). During this long period of human evolution, which represents 99 percent of our **hominid** past, the brain expanded from 400 to 1400 cubic centimeters and, in addition to manufacturing increasingly sophisticated stone tools,

[1] We refer to the opposable thumb of the primates which evolved slowly to become the most important feature of the toolmaking hand of our early human ancestors.

man learned how to control fire, hunt collectively, communicate by speech, think symbolically, engage in art, and organize himself into small societies. Although our ancestors were efficient hunters, the earth was not able to sustain more than a few million paleolithic men.

The second or **neolithic** revolution (Fig. 1-5) began some 10 to 15,000 years ago when men learned some of the basic skills of farming and animal husbandry while leading a seminomadic existence. The control of plant growth, primitive though it was, made available to man a larger portion of the sun's energy and gave rise to a large increase in the human population (Fig. 1-3). By the end of the neolithic period, 6000 years ago, almost 100 million men lived on the earth.

The third or **urban** revolution (Fig. 1-6) which followed saw man settle permanently in fertile valleys. His newly efficient agriculture allowed man to accumulate and store a surplus of grain. This in turn made possible the formation of complex societies in which a large proportion of members could engage in activities other than the primary production of food. Thus men were released for the practice of crafts (artisans), the recording of information (scribes), the mediating of power between neighboring societies (soldiers), the regulation and codification of beliefs (priests), and the establishment and implementation of law (rulers). And so, some 5000 years ago, an urban revolution began in which the fundamental features of modern society were laid down—an outstanding example of how man's relationship to nature brings about changes in man's relationship to man. We conclude that in the last million years toolmaking and other social activities of man provided an opportunity for natural selection, which first brought about a rapid expansion of his intellectual and related biological capabilities. This then enabled man to embark on a road of cultural evolution, which in turn has experienced a number of accelerating revolutions.

Some 350 years ago the **industrial** revolution began when man started to build machines which replaced some of the human and animal muscle power needed to run his urban societies. Again the earth experienced a large increase in population density and again we can conclude that a

Figure 1-4 A few Stone Age survivors of the paleolithic period have been discovered recently. These Tasaday tribesmen belong to a tiny group of 25 cavemen who were discovered on the island of Mindanao (Philippines) in June of 1971. They use stone tools which they fashion, use for a brief period and then discard. They weave very primitive baskets and are believed to communicate with a very simple language. The government of the Philippines has set aside land to protect them from the encroachment of their neolithic, urban, industrial, and scientific fellow men.

Figure 1-5 Neolithic farming methods in Africa.

Figure 1-6 Detail of Sumerian relief showing life in the fertile valleys of Mesopotamia, scene of the first urban revolution.

greater portion of solar energy was used to power the activities of man. This time, however, some of the energy utilized to support human activities was released by machines which use fossil fuels in which solar energy had been fixed by plants many millennia ago.

THE SCIENTIFIC REVOLUTION

In the last 150 years a human activity has begun in earnest whose origins we can trace to the early Greeks and Egyptians. We refer to man's increasingly successful attempts to understand the world in "scientific" terms. It is beyond our purpose here to define the word "scientific" in a unique and entirely satisfactory fashion. We would prefer instead to use the procedure that science generally uses in approaching a problem, in order to clarify the nature of the phenomenon of scientific inquiry itself. Let us begin by arguing that science is a human activity involving two separable but closely interrelated processes.

1. Scientific activity makes us aware that the phenomena and the processes of the universe are interrelated. Therefore scientific activity brings about an increasingly *coherent* vision of the universe. We use this increase in coherence as an index of the truth value of a given view. Science, like any other human activity, does not have any *direct* access to truth, but it does have criteria for judging the likelihood that a certain view is true. *Coherence* is such a criterion and we judge that we have made progress when a certain theory increases the number and diversity of the phenomena it brings together. For instance, an attempt to relate motions of planets to each other resulted in laws of motion that became applicable to the study of movement of other bodies. Concepts such as force, gravity, momentum, and inertia arising from the astronomical studies of Kepler, Galileo, and Newton provided a coherent picture within which many kinds of phenomena, large and small, fast and slow, could be included. Again, Lavoisier's theory of oxidation brought together such diverse events as the rusting of a nail, the burning of a candle, and the breathing of a mouse. Furthermore, the reverse process, reduction, such as the electrolysis of water and the photosynthetic activity of a plant, also became intelligible.

2. Besides building a picture of a world in which an increasing number of phenomena become related to each other, scientific activity leads to another crucially important result. It appears that the increasingly rich and coherent picture of our universe allows man, or the products of his intellect, such as the machines he builds, to exert greater and greater *control* over our surroundings. And so our ability to exert control is used as another criterion of the truth value of a given view of the universe. Thus an understanding of the laws of motion permitted man to build increasingly powerful machines, and the understanding of oxidation allows us to probe into chemical reactions, including those which occur in living cells.

Although the criteria of coherence and control do not provide an *a priori* justification for calling something true, it is nevertheless possible to anchor them to evolution itself. We defined evolution at the beginning of this chapter as a process which endows living matter with increasing control over the capriciousness of the environment. The growth of human intellect and the scientific revolution it has engendered is the latest and most powerful development in the evolutionary fight against environmental capriciousness. We can conclude that the process of natural selection buttresses the evolution of science by perpetuating in our culture those analyses or visions which increase the number of phenomena coming under our scrutiny and enlarge our control of these phenomena.

In recent years, as a result of a number of experiments with the "spontaneous" synthesis of organic compounds from simpler components, biologists have come to believe that, given the right preconditions on a planet, and given sufficient time, biological evolution on that planet becomes a virtual certainty rather than a chance event of low probability. Because we now believe that planet formation is a direct consequence of star formation rather than an infrequent event, astronomers have come to feel that the 10^{18} stars of our universe are centers of as many solar systems. These two views, namely the high probability of evolution and the stupendous number of solar systems, lead us to believe that the universe is teeming with life. And since, as we have just concluded, the scientific revolution is a major consequence of the evolutionary process, we can expect the universe to be honeycombed with technological societies. Our own technological society is only 350 years old, a mere blink of the eye in our galaxy which is 10 to 15 billion years old. Unless technology is an inherently self-destructive product of evolution, one might expect to find technological societies that number their age in many millions of years.

With this in mind we have begun an international effort to detect and hopefully decode radio messages sent from distant civilizations in our universe (Fig. 1-7). Tuning in to this one-way dialogue of the heavens may have incalculable importance for us, because through it we may attain level of information and wisdom obviating millions of years of trial and error on our part. In fact civilizations which have survived for millions of years might provide us with the insights necessary for our own survival. By tuning in to the wisdom of the universe we would become part of a process of natural selection operating on a cosmic level.

Let us return from these flights of fancy to some of the more recent consequences of the scientific revolution.

LIFE TURNS ON ITSELF

During the last 30 years a number of developments in science have begun to spin off a series of new revolutions. Developments in electronics and in the theory of information transfer have initiated a **cybernetic** revolution in which many complex operations formerly performed exclusively by the human mind can now be performed by machines (Fig. 1-8). Thus what the industrial revolution had done in supplementing and expanding the work of muscles, the cybernetic revolution is extending to the work of the human mind. The consequences of the cybernetic revolution are only beginning to be felt, although its immediate effect may be to cause much pain and human dislocation. The ultimate effect of the cybernetic revolution would be to eliminate drudgery from human work.

The **atomic** revolution will decrease our de-

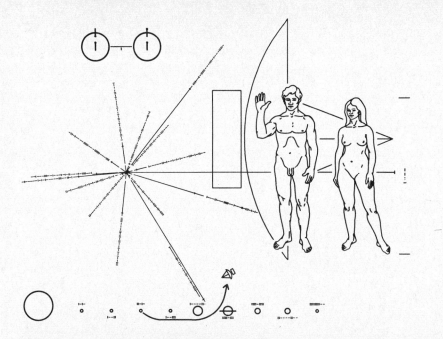

Figure 1-7 Adam and Eve travel into space. When Pioneer F was recently sent into space the picture at left was etched on a plaque attached to the hull of the spacecraft, a first attempt by our species to make a statement about itself which some other intelligent, scientifically educated being in outer space might be able to decode, possibly millions of years from now.

Figure 1-8   Just as the industrial revolution has supplemented the use of the muscle, so the cybernetic revolution has extended the use of the mind. Overall view of the Mission Operations Control Room in the Manned spacecraft center in Houston, Texas. This is the third day of the Apollo 8 lunar orbit mission. Seen on the television monitor is a picture of the earth which was telecast from the spacecraft 176,000 miles away.

pendence on fossil fuels, but it can be shown that ultimately the efficient harnessing of the sun's energy will provide the only large-scale source of energy which can sustain a complex society for long periods of time.

During the last 20 years the most exciting development in science is the ushering in of the **biological** revolution. During this period something has happened in biology that is qualitatively different from what went on before. As a result of developments in cytology, genetics, and biochemistry, a new and unified science has developed which has begun to interpret or explain biological phenomena such as heredity, growth, differentiation, and even evolution in molecular terms. We are discovering that the relationship between structure and function which allows living matter to capture energy and utilize it for increasing organization reaches all the way down to the basic molecules of the living machine.

These **macromolecules** of the cell, the nucleic acids and proteins, are indeed the "moving parts" of a system which operates with a high degree of precision even though its crucial molecular components in any given cell may be few in number. Thus we find that a single molecule of deoxyribonucleic acid, or DNA, in a bacterium can duplicate itself and control the metabolic processes of the bacterium, a fact which we would certainly not have considered possible from our previous knowledge of molecules derived from nonliving systems. The success of the molecular approach to the study of life derives, of course, from the success of these molecules in operating the living machine. The ingenuity of the molecular biologist is a reflection of the ingenuity of the molecules of life!

To demonstrate how central our recent insights into the molecular machinery of life have been one need only point at their universal validity. The *code* which our cells use to store their hereditary information in DNA is universal from the simplest virus to man. The molecular devices our cells use for translating the language of nucleic acids into the language of proteins are also universal. In fact we have come to believe that almost everything the cell does, whether it is the control of the movement of materials through its membrane, the transduction of chemi-

cal energy into mechanical work, or the burning of food to manufacture high energy compounds, follows some broad, general patterns utilized by all cells. However, superimposed on these universal pieces of molecular machinery are the phenomena accounting for the diversity of living things, and the astounding thing about biological systems is that enough variation can overlay the basic similarities to account for the inexhaustible diversity encountered in the world of life. What is fascinating and tantalizing about the machinery of life is that it can account both for the *universality* and the *diversity* of living systems, both these properties finding their explanation on a molecular level.

It is to be expected that with the increasing power of our insights comes an increasing power to control. We can only begin to perceive in very broad terms the full scope of biological control that the biological revolution will make available to us. Control over disease, life expectancy, and population growth are rapidly coming into our grasp. But rapid and deep control over human evolution itself can only be imagined in the dimmest of outlines, not only because we are uncertain about its scientific feasibility but also because we cannot begin to imagine how the human control of human evolution would ever become socially operable.

Whichever way we think of the consequences of the biological revolution, whether we view these developments with hope or despair, it should be abundantly clear that we are witnessing one of the most dramatic events in the long history of biological evolution—the evolution of the human mind and of science, its cultural product, to the point that the mind is able to *turn on itself* and make itself the object of its own scrutiny. What greater goal can there be for man than the goal of self-awareness which will bring the rich diversity of his experience into a single coherent pattern, ranging from the molecules of his cells to the images of his dreams?

KNOWLEDGE AND ACTION

When we dream of a vast expansion of human power and insight, we do so with some trepidation. The experience of the last half century has taught us that man's increasing power over his environment has raised more problems than it has solved. Although our present technological development could, in principle, eradicate hunger and poverty, human society has developed unevenly leaving some societies at the neolithic and even paleolithic levels and a majority of human beings at a preindustrial level. Our vast power to affect our environment can be and has been abused by turning it against the less developed and less powerful societies.

Although most of the benefits of the industrial revolution have reached only one-third of the world's population, some of its consequences, such as the reduction of infant mortality and the increase in life expectancy, have reached a much larger proportion of the world's population. This has increased the rate of population growth in the poorer countries, which in turn, given their powerlessness in world affairs, increases the poverty of the world's poor (Figure 1-9).

Figure 1-9   Starvation in India.

Meanwhile in rich countries, where most of the world's raw materials are manufactured into products, the environment has come under increasing strain. While two-thirds of the world hungers, one-third is poisoning its water, polluting its air (Fig. 1-10), and exhausting its natural resources. For instance, in the technologically advanced countries, the high level of agricultural productivity is maintained by widespread use of persistent pesticides. One such pesticide is DDT, which has reduced many species of hawks and eagles to very low population levels. The peregrine falcon has become extinct throughout most of its European and North American range and it is doubtful that this bird can be saved from total extinction. Our national emblem, the bald eagle, has also become a victim of pesticides and may suffer a similar fate.

The decline of the blue whale is another story documented in recent years. This whale is the largest animal that ever lived on earth. It is 25 feet long at birth and is about 53 feet long by the time it is weaned at an age of seven months. Blue whales may live as long as 40 years and grow to lengths of 100 feet or more. One 83-foot male weighed 242,397 pounds! Since the 1930s, the blue whale population has declined steadily from overfishing, in spite of repeated warnings from biologists that the species may soon become extinct. Although the harvest of whales is supposedly regulated by international agreement, individual nations will not abide by these agreements and so the number of blue whales steadily decreases. Experts now estimate that the species will be extinct within the next decade or two, and that other species of whales may follow very shortly.

What happens when we lose a species or a lake or a forest? Studies of populations and environments have shown us that less desirable species soon take over the roles of the extinct species. Whenever we pollute or destroy parts of an environment, species that need certain foods or require rather special conditions are lost, but weeds and pests remain. Rats and gulls and pigeons will be with us long after the peregrine falcon and the blue whale are gone. We therefore lose some of the quality of our environment by reducing its diversity and removing many of the species that make our environment attractive and interesting.

A more serious question is whether we actually need the enormous diversity of plants and animals found on earth. The peregrine falcon and the blue whale may not be absolutely essential to the economy of the biosphere. We might even follow Benjamin Franklin's original suggestion to use the wild turkey, rather than the bald eagle, as our national emblem. However, we might soon find it necessary to look for a substitute for the wild turkey as well, winding up with a herring gull perched atop a garbage dump.

We do not, in fact, know what the special contribution of each species, with its particular adaptations to its environment, is to the biosphere. We do know, however, that when a lake or a forest or a river begins to lose its diversity of species, it has moved from a healthy to an unhealthy condition in which its future is in doubt. It has lost part of its stability, and is likely to collapse and cease to function as an integrated system. The biosphere, like the human body, must retain all of its critical components in order to survive, for it is a vast living machine of enormous complexity. It includes all the units of life,

from molecules to the complex assemblages of populations and communities which we refer to as **ecosystems,** such as grasslands, forests, lakes, and seas, each with its own special qualities. The biosphere is the most complex and highly integrated of all biological systems. Energy from the sun penetrates the earth's atmosphere where it is received by a huge array of green plants, trees, grasses, and even microscopic algae, all of which have the remarkable ability to use solar energy to convert various inorganic chemicals into organic substances that can be used as the building blocks of life. Food which is manufactured by these plants can then be eaten by animals, which may in turn be eaten by other animals. When these plants and animals die, to be replaced by others of their kind, their bodies decompose and return nutrients to the soils and waters of the earth where they may be used by other plants to repeat the process of producing new supplies of usable energy to run this huge, biological machine, "the spaceship earth." The biosphere is therefore the total life-support system of the earth, run by energy from the sun. Its continued functioning depends on the special mixture of gases and chemicals that occur in

Figure 1-10 Smog engulfs the Los Angeles Civic Center one day in 1956. What in the fifties appeared to be specific for Los Angeles is today a major problem for most large cities. In certain parts of Tokyo, for instance, children walk to school with facemasks designed to protect them from the effects of pollution which, on occasion, causes residents to collapse in the streets.

their proper quantities on this particular planet, and upon the delicate balance of relationships between the various kinds of plants and animals that give it life. Obviously any threat to the biosphere that would seriously alter these balances is a threat to life itself.

Because the biosphere is too vast and huge a level of organization to study conveniently in its totality, we often neglect its significance. This is compounded by our habit of thinking of the natural world of the biosphere in terms of "Mother Nature," infinitely capable of adjusting to manmade changes in the environment and always able to heal the "minor wounds" we inflict on the face of the earth. We are only now beginning to appreciate the enormous impact of man upon his environment and to recognize the importance of the biosphere as a unit of organization.

For the first time we have become aware that many of our actions have global effects. We now have evidence that dust and other kinds of particles distributed into the air by man's activities are changing the temperature of the earth's atmosphere. These changes are relatively subtle ones, a matter of a few degrees in temperature, but how much change the biosphere can stand and still function normally no one yet knows. Virtually every modern technological advance has produced by-products, thousands of new chemicals, which eventually enter our rivers and streams. About 80 percent of the circulating fresh water in the United States is used for diluting industrial wastes. For example, the manufacturing of one ton of nylon pollutes about 5000 cubic meters of water. As we shall see in later chapters there are many critical steps in the transfer of energy and nutrients through ecosystems. The loss of any one of these stages can damage the system beyond repair. We must ask ourselves if any one of the chemicals we are discharging into the environment might not cause irreversible damage to the life-support system of the biosphere.

These few examples of biological imbalance brought about by our expanding technological society show that there is much that we have to study and many actions we will have to take before a balance between society and the environment can be restored. We shall have to learn a great deal about the ecological relations of the multitude of organisms in our biosphere, and also much about ourselves and our own social relations, before effective action is possible. But we shall have to act frequently before the information is all in. For, as we have already pointed out, science is not a one-way process with all knowledge preceding all action. Applicability, or the effectiveness of control, is one of the criteria by which we judge a scientific advance. *Thus knowledge leads to action and action leads to further knowledge.*

Many biologists, including the authors of this book, feel that man today stands at the crossroads of success or failure as a biological species. The exciting and often startling discoveries of modern biology have given us new insights into the meaning of life and the future of man. These discoveries have also shown us that man has the ability and the power to affect not only his own destiny but also the quality and possibly the future of life itself. This book is written for the generation which will decide this future.

We have attempted in this chapter to cover a vast panorama of biological concerns and to suggest why understanding them should be of importance to each and every one of us.

We have viewed evolution as an energy-acquiring phenomenon, constantly expanding its degree of organization. We have also sounded an alarm that our increasing power may prove to be detrimental to our further evolution. We hope that we have convinced the reader that knowledge of all levels of biological organization is important for effective action in human affairs.

This book is organized into sections which follow ascending levels of biological organization:

Level I deals with the cell as the basic common denominator of all living systems. It is here we discuss the relationship between structure and function at the molecular level.

Level II deals with development and takes us from the cellular to the organismic level.

Level III is the world of organisms, delineated by their way of life. Here we show how the functioning of different groups is related to their particular ecological role.

Level IV views the population as the arena for evolutionary change and as the source of biological diversity.

These levels of biological organization can be approached in a variety of orders. One could start with the organism, the most familiar of biological levels of organization, and reach "down" to the cell and "up" to the population. However, one could also start with the population and with evolution, one of its central phenomena, and move to the lower levels of organization. The biosphere is a single, vast system which can be penetrated at any level as long as all the various interconnections are ultimately made.

SUGGESTED READING LIST

Abrahamson, J., et al, *Who Shall Live — Man's Control Over Birth and Death*. Hill and Wang, New York, 1970.

Berrill, N.J., *Inherit the Earth — The Story of Man and His Changing Planet*. Fawcett Publications Inc., Greenwich, Connecticut, 1967.

Childe, V. G., *Man Makes Himself*. The New American Library, New York, 1951.

Ehrenfeld, D.W., *Biological Conservation*. Holt, Rinehart, and Winston, Inc., New York, 1970.

Goldsby, R., *Race and Races*. The Macmillan Company, New York, 1971.

Grobstein, C., *The Strategy of Life*. W.H. Freeman, San Francisco, 1965.

Henshaw, P.S., *This Side of Yesterday — Extinction or Utopia*. John Wiley and Sons, Inc., New York, 1971.

Loewy, A.G. and P. Siekevitz, *Cell Structure and Function*. Holt, Rinehart, and Winston, Inc., New York, 1969.

Pfeiffer, J.E., *The Emergence of Man*. Harper and Row, New York, 1969.

Piel, G., *The Acceleration of History*. Alfred A. Knopf, New York, 1972.

Part One

The Cell

Here we consider the cell
as the basic common denominator
of living systems and discuss
the relations of structure and
function at the molecular level.

CHAPTER 2

The Unit of Life

The body of an adult human being contains 60 trillion cells (60×10^{12}), functioning together in close cooperation. A redwood tree, weighing more than 1000 tons and towering over 300 feet into the air, is made up of 20,000 times that many cells. An amoeba, just as alive, although so small that we can barely see it with the naked eye, is a single cell.

The **cell** is the smallest independent unit of biological activity. Whether cells are part of a redwood tree or belong to the body of a man or make up the entire organism like an amoeba, they conform to a general plan of structure and function which enables us in the following seven chapters to talk about cells and cell processes in a way that is applicable to all cells in all organisms.

This concept of the fundamental unity of the cell is a very recent one, although one might trace its origins all the way back to the investigations of Robert Hooke. In 1665 Hooke examined a thin slice of cork under a microscope of his own design. He described the cork as resembling a honeycomb containing many pores that suggested little boxes. These he named cells, although what he actually saw were dead cell walls from which the living cells had disappeared (Fig. 2-1). Hooke later studied slices of living plant tissue and observed "juices" inside the plant cells. These juices make up the living cell and are described in this chapter.

18

A CELL THEORY

After Hooke, many biologists examined cells of all sorts and drew pictures of what they saw, but it was nearly 150 years before the significance of cellular structure became apparent. Early in the nineteenth century several scientists began laying the groundwork for a general theory about cells. In 1838 the botanist Matthias Schleiden suggested that each plant cell leads a double life. It is in some respects an independent organism, but at the same time it cooperates with the other cells that make up the whole plant.

A year later Theodor Schwann applied this hypothesis to animal cells and came to the conclusion that all organisms are made of cells that are essentially alike. This was followed 19 years later by a further generalization which seemed to complete the picture. At that time Rudolf Virchow, a German pathologist, proposed that cells come only from other cells. He wrote: "Where a cell exists, there must have been a preexisting cell, just as the animal arises only from an animal and the plant only from a plant."

These three men, drawing upon their own observations and those of many others, created an early version of the **cell theory.** This theory *predicted* that all organisms are composed of *similar* units, called cells, each of which lives a partly independent existence and is "separately born and destined separately to die." These cells are the structural and functional unit of life and are able to duplicate themselves. Extensive observations during the rest of the nineteenth century confirmed the predictions of the cell theory so that it ceased to be a *theory* and became an established *fact.* However, only in the last 30 years has it become apparent how similar the basic processes of all cells really are.

By a mere examination of cells, even with a light microscope, we might be led to think that cells differ far more than they resemble each other. They cover an enormous range of sizes. Eggs are single cells; and the largest bird's egg, that of the ostrich, is 20 centimeters in diameter. A sperm cell without the tail may be only one forty-thousandth that size, or about 5 microns. Nerve cells connecting muscles of vertebrates to the spinal cord may be several meters long.

Cells also vary greatly in appearance. Mature plant cells commonly have a stiff cellulose wall and a large cavity called a **vacuole** filled with fluid (Fig. 2-2). Most animal cells lack a cell wall and a vacuole (Fig. 2-3). Some cells move freely (Fig. 2-4); others are fixed in place. A cell in a stem that conducts sugar down to the roots looks very different from a root hair cell that absorbs minerals from the soil. A long thin muscle cell bears little resemblance to a kidney cell.

Beyond this diversity in size and appearance, cells vary greatly in the functions they perform. Some cells are separate living organisms and carry out all the functions of life without establishing partnerships with any other cells. Such one-celled organisms include bacteria and numerous other organisms known as **protists** (Fig. 2-4). Protists that photosynthesize are called **algae** and those that cannot are called **protozoans.** Other cells are highly specialized and are found only as part of many-celled or **multicellular or-**

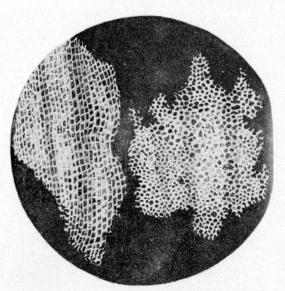

Figure 2-1   Robert Hooke's cork cells. Hooke made a thin slice of cork and looked at it under his microscope. Cork is a dead tissue showing the cell walls clearly but the living cell itself inside the little boxes is, of course, no longer present. Hooke therefore did not see the actual cells but he saw the spaces vacated by them.

ganisms. In multicellular organisms groups of similar cells form a **tissue** such as muscle tissue or nerve tissue. Groups of tissues form an **organ** which carries out a particular function for the organism. The stomach of an animal or the leaf of a plant is an organ. In organs the component cells of each tissue become highly specialized. Some cells secrete juices that digest food, wax that waterproofs skin, or calcium that builds bone. Others are sensitive to light, heat, or touch. Muscle cells are specialized for contraction whereas nerve cells carry electrical impulses. Despite these differences, the cell theory holds that these diversified and specialized cells all share many properties.

As the cells of an organism become specialized they come to depend upon other cells of the organism for their existence, just as a steel worker in our society depends upon the farmer who produces his food and upon the clothing manufacturer who makes his clothes. Yet Schleiden emphasized that even specialized cells lived lives of their own while contributing to the existence of the organism. This suggests that in a suitable environment a single cell of a multicellular organism should be able to survive and even divide, thereby reproducing its own kind. For many years it was not possible to find the right environment for cells other than one-celled organisms. Then early in the twentieth century we learned to grow animal and plant cells in laboratory cultures. **Cell culture,** as the technique is called, has become a valuable tool for a wide variety of investigations. It has shown that cells that rely on other cells for their survival in nature can, under certain conditions, lead independent existences.

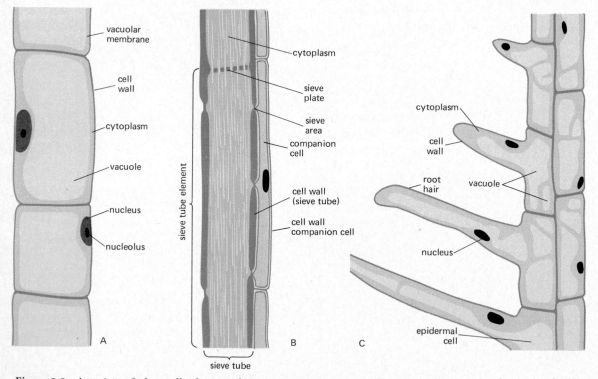

Figure 2-2   A variety of plant cells showing the variation in structure and function which can be observed in higher plants. A. Parenchyma cells, relatively undifferentiated plant cells, still capable of cell division and growth. B. Phloem sieve tubes and companion cells which have the function of conducting food through the stems of plants. C. Root hair cells in various stages of development which have the function of absorbing water and minerals from the soil.

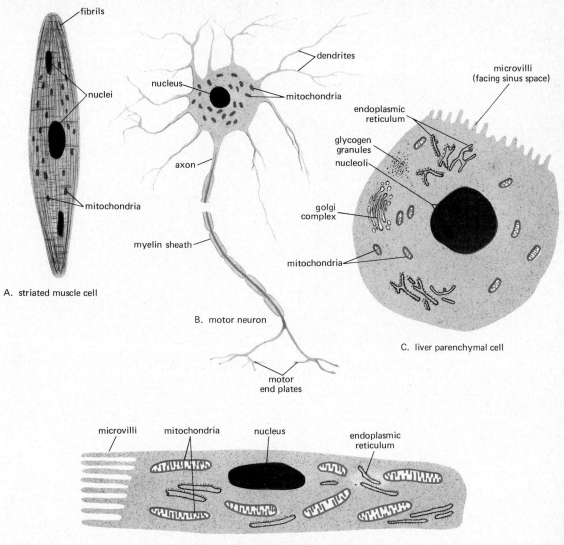

Figure 2-3   A variety of animal cells showing the variation in structure and function which can be observed. A. **Striated muscle cell:** contains many nuclei, rich in fibrils involved in contraction, and in **mitochondria** providing the ATP which power it. B. **Motor neuron:** It has long extensions of cytoplasm, sometimes meters long, which conduct electrical impulses. C. **Liver parenchymal cell:** Can be considered a "typical" cell containing a wide spectrum of structures generally found in cells. D. **Kidney cell:** Narrow, one surface covered with tiny finger-like structures called microvilli which are involved in the exchange of materials, with numerous mitochondria lying close to membranes providing the ATP to power the transport of substances.

THE MODERN CONCEPT

The modern concept of the cell as the "common denominator" of life has gone far beyond the theory of Schleiden and Schwann. We have learned the extent of these similarities because of two important developments. The first is the **electron microscope,** which is capable of magnifying cellular details at least 500 times more than light microscopes. The other is the merging of genetics and biochemistry into a *molecular biology* of great breadth and power.

These advances confirm the cell theory in a way that its founders could not have anticipated. More clearly than ever the cell *is* the fundamental unit of life. We now know that the molecular structure and function of vital parts of the cell—those that control heredity, for instance, or generate energy—are strikingly similar in all living organisms (at least on Earth). Despite the obvious differences among a bacterium, a protozoan, a root cell of a tree, and a nerve cell of a man, all of these cells have in common a number of important pieces of molecular machinery. For ex-

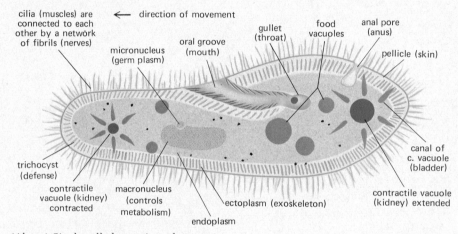

Figure 2-4   (Above) Single-celled organism, the protozoan, *Paramecium*. This tiny creature carries out within a single cell all of the activities and functions necessary for its own survival and the survival of its species. It possesses a variety of small structures—the organelles ("little organs")—which perform functions that are similar to those of certain multicellular animals. Its body is covered with about 2500 tiny hair-like structures called cilia (a name which comes from the Latin word for "eyelid") by which it moves. *Paramecium* is discussed again in Ch. 24. (Right) A single celled plant, the alga, *Chlamydomonas*. This microscopic organism possesses a single large chloroplast and photosynthesizes. It has the typical cellulose wall of a plant cell but it has a number of small vacuoles instead of the single large vacuole found in the mature cells of most higher plants (see Fig. 2-2). Like *Paramecium* it has a variety of organelles, including two whiplike threads called **flagella** by which it moves. It also has a structure called a pyrenoid that is probably involved in making starch from the sugars that are formed in the chloroplast by photosynthesis. It is discussed again in Chapter 23.

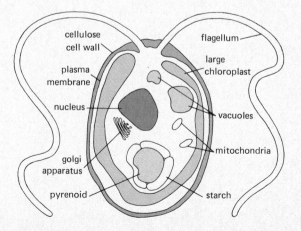

ample, they duplicate genetic material, utilize hereditary instructions in synthesizing proteins, carry out the transfer of energy, convert chemical energy to work, and control the exchange of materials—all in a similar way. In later chapters we shall present evidence that these striking similarities among cells of very different types result from the fact that all cells, regardless of their immediate origin, share a common ancestry and are linked by the long thread of evolutionary history.

The rest of this chapter introduces the structure of the cell, particularly the observations which have been made possible through the use of the electron microscope.

Seeing is believing . . . and vice versa

The process of observation is often more complex than the student initially realizes. An instructor might point to a structure that the student is unable to see; the student will soon discover that seeing requires training and experience that allow the observer, in a sense, to anticipate what he is going to see. When a biologist observes new structures which have never been seen before, he faces two problems.

First, before he can begin to "see" the new structure, he needs to observe a number of examples of it under a variety of conditions. This will build up a *concept* in the observer's mind that will actually determine what he sees every time he looks at one particular example of the structure. Second, as we discuss shortly, whatever the procedures for preparing and observing the object, they invariably affect the object in some way. Therefore the most subtle and complex problem the observer faces is to conclude what the object must have been like before he modified it prior to or during the process of observation.

The traditional stumbling block of the microscopist is the **artifact,** a modification of the object which misleads the microscopist as to its real nature. If, however, we recognize that to varying extents all observations both actively involve the object and modify it in some way, we may turn artifacts to our advantage. The trick is to discover, often through patient experimentation, what a particular modification of an object means in terms of its original properties.

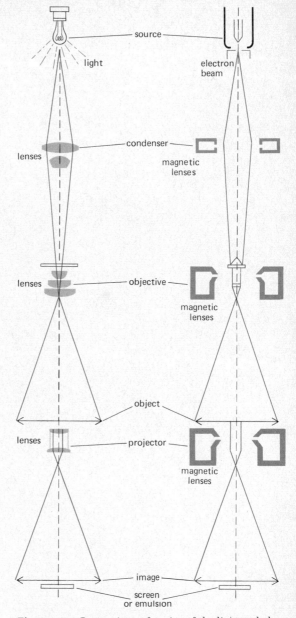

Figure 2-5 Comparison of optics of the light and electron microscopes. The light microscope uses visible light and the eye or a photographic emulsion as the recording instrument. The electron microscope uses an electron beam and a fluorescent screen or a photographic emulsion as the recording instrument. Because electrons carry a charge, electron beams can be focused with magnetic lenses.

The electron microscope

These warnings regarding the complexity of the process of observation apply even more to electron microscopy than to light microscopy.

The unaided human eye can distinguish two objects that are about 1/10 mm or 100 microns apart. We say that it has a **resolving power** of 100 microns. This means that if you look at two dots that are less than 100 microns apart they merge into one blurry dot, or if you look at two lines less than 100 microns apart they appear as one line. Microscopes enable us to separate structures closer than this. A good light microscope has a resolving power about 400 times better than the naked eye and will separate objects that are only 0.25 micron apart. It is not possible to construct a light microscope that will do better than this because it is a fact of physics that the resolving power of a microscope can be no better than approximately one-half the wavelength of the light being used to observe an object. This means that when using white light with an average wavelength of 0.5 micron, the limit of resolution of the light microscope will be 0.25 micron.

It is important to distinguish clearly between *resolving power* and *magnification*. You could enlarge a photograph taken through an excellent light microscope of two dots less than 0.25μ apart until it covered the side of a barn and increase the magnification 10 million or more times. But the two dots would still appear as one blurred dot.

Electron microscopes have greater resolving power than light microscopes. Instead of light the electron microscope utilizes a beam of electrons focused by lenses which are magnetic fields acting on the negatively charged electrons. Because the eye cannot see directly the images produced by the beam, they are recorded on a fluorescent screen or on photographic film (Fig. 2-5). The electron microscope enables us to distinguish objects only 5Å in diameter, 500 times smaller than the smallest structures that can be seen under the light microscope (Table 2-1). This is because the wavelength of an electron beam is much shorter than that of light.

An object can be seen under the electron microscope if it absorbs either more or fewer electrons than its surroundings. Since the scattering of electrons interferes with observation, it is helpful to focus the beam sharply and to use exceedingly thin slices of living material which have previously been killed by treatment with a fixative and then dried. What we observe with the

---

Table 2-1 Size relationships, units of measurement, and instruments for observation of biological objects

Units of measurement	Nanometers (nm)	Objects observed	Instrument for observation
1 centimeter (cm) = $^1/_{100}$ meter = 0.4 inch	10^7	Giant egg cells	Naked eye
1 millimeter (mm)	10^6	Very large cells	
	10^5	Large cells	Light microscope
	10^4	Average eucaryotic cell	
1 micron (μ) or micrometer (μm)	10^3	Average procaryotic cell	
	10^2	Largest virus, smallest bacterium	Electron microscope
	10	Small virus	
1 nanometer (nm) or millimicron (mμ)	1	Macromolecule	X-ray diffraction
1 Angstrom unit (Å)	0.1	Small organic molecule	

electron microscope are thin slices of dry remains of living material that are frequently incinerated by the electron beam so that only their carbon skeleton remains. From these remains we try to reconstruct a picture of what the cell looks like when it is alive and whole.

To accentuate the difference in electron absorption between a given structure and its surroundings, the electron microscopist also uses stains. These are compounds which contain heavy metals and serve the same purpose as the colored stains used for light microscope specimens. A specimen must also be embedded in plastic so that slices as thin as 50 nanometers can be cut. When looking at electron micrographs remember that we are seeing only a thin slice of a cell which has been fixed at one fleeting moment in its existence.

By mentally putting together the thin slices, we try to reconstruct their three-dimensional shape. In addition, by fixing cells at various stages of their development we try to reconstruct the process of development itself. As you become more familiar with the objects and processes in the cell you will no doubt be able to picture the cell as a dynamic, three-dimensional structure.

A dramatic recent application of electron microscopy is the development of the **scanning electron microscope.** This microscope examines specimens by reflecting the electron beam off their surfaces. To improve resolution, a narrowly focused spot is used to scan the field, and an image is produced by summating the results of the scanning motion. The advantage of the scanning electron microscope is that surfaces of great complexity can be visualized in three dimensions at finer resolutions and greater depths of focus than those obtainable with light microscopy. Several excellent scanning electromicrographs—including the frontispiece—are used in this text.

A "TYPICAL" CELL

It has become increasingly clear in recent years that two basic cell types occur in nature: the simple **procaryotic cell** and the more complex **eucaryotic cell.** Procaryotic cells include bacteria and blue-green algae. Eucaryotic cells include the cells of all other organisms. Their name denotes a major difference between them. The Greek word *karyon* means "kernel" or "nucleus." Procaryotic cells are "before the nucleus" whereas eucaryotic cells are "with a true nucleus." Table 2-2 summarizes some of the key differences between procaryotic and eucaryotic cells. The eucaryotic cell has two distinct regions: the **nucleus** which is surrounded by a nuclear membrane or envelope in which the hereditary materials are "packaged," and the **cytoplasm** in which are crowded various structures and membrane-bound compartments. The procaryotic cell has only one distinct region: its nucleus is not set off from the cytoplasm by a nuclear membrane and it lacks the other internal compartments typically found in eucaryotic cells. As we shall see later on, procaryotic organisms appear to have preceded eucaryotes in the course of evolution. In the description of a "typical cell" which follows we shall use a eucaryotic animal or plant cell because such cells include many more kinds of cell structures than do procaryotic cells. We return to procaryotic cells in Chapter 9 when we discuss bacteria.

It is customary, when introducing the cell, to talk about a typical cell, that is, one that contains the structures found in most cells, although we know that there is no such thing as a typical cell. Despite the great similarity in the general plan, each eucaryotic cell type has specific modifications that adapt it to its particular life-style or function. Some cells have large numbers of small bodies called mitochondria, others have an extensive system of internal membranes, still others have specialized parts for manufacturing secretions. Figures 2-6A and 2-7A are generalized drawings of a "typical" plant and animal cell. Compare these drawings with the electron micrographs (Figures 2-6B and 2-7B) in order to supplement these schematized versions with a more realistic picture. Notice that the eucaryotic plant cell contains almost all of the internal structures possessed by animal cells as well as a cell wall, chloroplasts, and one or more large vacuoles.

The membrane systems of the cell

Plasma membrane. We begin our description of the cell by starting with the outside of an ani-

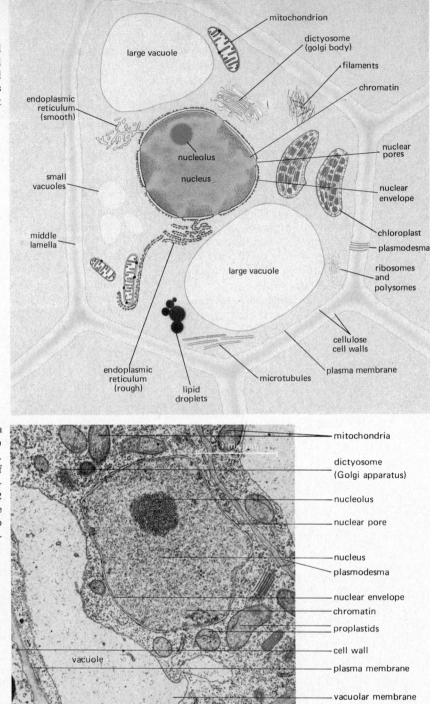

Figure 2-6

A. Drawing of a generalized plant cell. The cellulose cell wall, the chloroplasts and the unusually large vacuoles are characteristic of plant cells.

Labels (top drawing): mitochondrion; dictyosome (golgi body); filaments; chromatin; large vacuole; endoplasmic reticulum (smooth); nucleolus; nucleus; nuclear pores; nuclear envelope; small vacuoles; middle lamella; chloroplast; plasmodesma; large vacuole; ribosomes and polysomes; cellulose cell walls; plasma membrane; endoplasmic reticulum (rough); lipid droplets; microtubules

B. Electron micrograph of a plant cell, from the root tip of *Arabidopsis thaliana*. The short line at the top of the photograph is a *size reference*. This cell is about 2 microns in diameter. The scale of miles on a road map is a similar type of size reference (× 21,000).

Labels (micrograph): mitochondria; dictyosome (Golgi apparatus); nucleolus; nuclear pore; nucleus; plasmodesma; nuclear envelope; chromatin; proplastids; cell wall; plasma membrane; vacuolar membrane; endoplasmic reticulum; vacuole; 1 micron

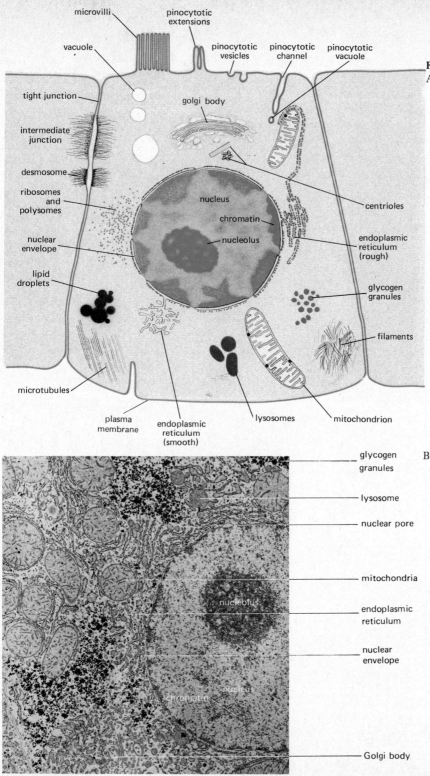

microvilli

pinocytotic
extensions

vacuole

pinocytotic
vesicles

pinocytotic
channel

pinocytotic
vacuole

tight junction

golgi body

intermediate
junction

desmosome

ribosomes
and
polysomes

nuclear
envelope

lipid
droplets

microtubules

plasma
membrane

endoplasmic
reticulum
(smooth)

lysosomes

mitochondrion

nucleus

chromatin

nucleolus

centrioles

endoplasmic
reticulum
(rough)

glycogen
granules

filaments

Figure 2-7

A. Drawing of a generalized animal cell. The cellulose wall found in the plant cell is absent. The surface of the animal cell undergoes a number of modifications such as microvilli, pinocytotic vesicles or extensions, and a variety of cell-to-cell junctions.

glycogen
granules

lysosome

nuclear pore

mitochondria

endoplasmic
reticulum

nuclear
envelope

Golgi body

nucleolus

nucleus

chromatin

B. Electron micrograph of animal cell from the liver of a newborn rat ($\times$ 9,000).

mal cell. The first structure we encounter is the cell's "skin," the plasma membrane or *cell membrane*. It is only 75Å thick, but it has the important task of controlling what enters and leaves the cell. Molecules and ions in solution move from regions of high concentration to regions of lower concentration by **diffusion** (Fig. 2-8). The plasma membrane acts as a barrier to the diffusion of most molecules and ions, thus slowing down the rate at which molecules tend to enter or leave the cell. Furthermore this barrier is *differential* being more rapidly penetrated by some molecules than by others (Fig. 2-9). Water molecules, oxygen, carbon dioxide, and a few other simple molecules diffuse freely across the plasma membrane. Larger molecules and most positively charged atoms or molecules are stopped by the plasma membrane. We refer to the plasma membrane as being *differentially permeable* or *semipermeable*.

Studies of the movement of materials in and out of the cell provided a picture of the structure and composition of the plasma membrane long before it was seen in the electron microscope. The rapid penetration of *fat-soluble* substances indicated that the plasma membrane contains a layer of fatty or oily substances which are called **lipids** (see Chapter 4, pp. 116-119), and the rapid penetration of small *water-soluble* substances showed that this lipid layer is interrupted by small water-filled pores. The physical properties of the cell surface or of purified membrane preparations (Fig. 2-11) showed that **protein** (see Chapter 4, pp. 125-131) is also present in the membrane. It was therefore gratifying when these deductions regarding the structure and properties of the membrane were confirmed directly by the electron microscope, which shows a three-layered plasma membrane composed of two heavily staining, electron dense outer regions presumably containing protein and a more transparent inner region believed to contain the lipid. Figure 2-10 shows a highly enlarged electron micrograph of the plasma membrane along with some of our present ideas regarding the composition and orientation of the materials in it.

In addition to the *diffusion* of substances across the plasma membrane, many substances move across it by **active transport** which is discussed in Chapter 4. In the process of active transport a specific substance can be moved against a diffusion gradient from a region of low concentration of the substance to a region of high concentration. Movement against the gradient requires the expenditure of energy and involves special molecular mechanisms that enable parts of the plasma membrane to pick up specific molecules and escort them into the cell.

Table 2-2 A comparison of procaryotic and eucaryotic cells

Structure	Procaryotic cell	Eucaryotic cell
Chromosomes	Only one chromosome per cell, contain DNA only	Several chromosomes per cell, contain DNA and protein
Size	Small (0.5–5 microns)	Large (10–100 microns)†
Nucleus	No nuclear membrane, no mitotic apparatus	Nuclear membrane, nucleolus, mitotic apparatus
Organelles	Absent	Mitochondria, chloroplasts, centrioles, lysosomes
Internal membrane system	Simple folds (mesosome)	Complex folds (endoplasmic reticulum, Golgi complex)
Flagella	Simple (no internal system of microtubules)	Complex ($9 \times 2 + 2$) system of microtubules
Cytoskeleton	None	Filaments and microtubules

† Typical plant cells are about 25μ in diameter, typical animal cells are about 10μ in diameter.

Modifications of the plasma membrane. Under a light microscope the plasma membrane seems to stretch tightly about the cell, but the electron microscope shows that this is usually not the case. The plasma membrane of animal cells sometimes folds outwards to form finger-like projections called **microvilli** (Fig. 2-15). Regimented arrays of microvilli are commonly found in cells particularly active in the absorption of materials, such as those of the intestinal lining. The microvilli provide an increased surface for absorption and are probably involved in the active transport of sugars and other energy-containing molecules from the food in the intestinal tract.

Sometimes the outward folds in the plasma membrane are very irregular like those in Figure 2-13. These irregular protrusions often form little pockets called **vesicles** which surround droplets of fluid. Vesicles may also be formed by the inward folding of the plasma membrane (Fig. 2-12).

Vesicles are not static entities but can move toward the cell interior. When the opening to the surface is pinched off, the vesicle becomes a fluid-filled sac known as a **vacuole.** By forming vacuoles in this way many cells engulf fluid **(pinocytosis)** and food **(phagocytosis).**

In plant cells the plasma membrane is surrounded by a multilayered cell wall composed

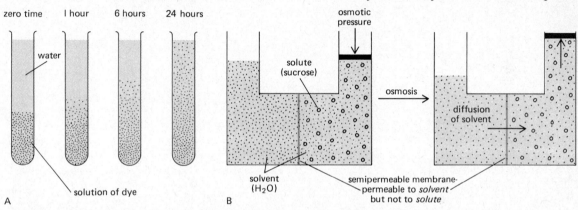

Figure 2-8   Diffusion and osmosis. A. **Diffusion** is the movement of a substance from a region where its concentration is high to another region in which the concentration of that substance is lower. If you put a solution of a colored dye in a test tube and carefully layer over it pure water, the dye molecules will slowly distribute themselves evenly throughout the test tube. This occurs because the concentration of dye molecules in the upper part of the test tube is lower and random motion results in a net transport of dye into the upper part of the tube. If you could observe *individual* dye molecules you would see that each one of them moves in a random direction. As a result, since at the start of the experiment there are more dye molecules in the bottom of the tube, the overall movement of the dye molecules will be from bottom to top.

What happens when all the dye molecules are distributed evenly in the test tube? Since there are just as many dye molecules in the top of the test tube as in the bottom, there is no overall movement of dye molecules in any particular direction. However the individual dye molecules are still moving around just as much as before. The key features of diffusion are that each molecule moves as an individual and its movements are random. The result of diffusion is the even distribution of the diffusing substance.

B. **Osmosis** is a special kind of diffusion. It is the diffusion of water across a membrane which is impermeable to other substances. The plasma membrane is such a differentially permeable or *semipermeable* membrane. Some molecules can pass readily through it whereas others cannot. The diagram illustrates the net movement of water through a semipermeable membrane which separates a solution of some **solute** from pure water. The membrane is permeable to water but impermeable to the solute. Since the concentration of *water* is higher in the left compartment than in the right compartment, there will be a net diffusion of water from the left compartment to the right compartment. This movement of water is called osmosis. It can be prevented by exerting a pressure in the right hand compartment. Osmotic pressure is defined as the pressure necessary to prevent the flow of water into the compartment containing the solute. When the pressure is not applied, water moves through the membrane into the solute-containing compartment.

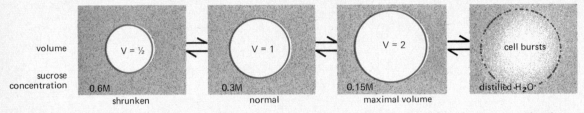

| volume | V = ½ | V = 1 | V = 2 | cell bursts |

sucrose
concentration   0.6M   0.3M   0.15M   distilled H₂O

    shrunken   normal   maximal volume

Figure 2-9   Osmotic behavior of an animal cell. The osmotic movement of water depends on *how many* molecules or ions of solute a solution contains rather than on the *kind* of solute. The same number of molecules of two wholly different substances ordinarily exert the same osmotic pressure. The term *isotonic* or *isosmotic* describes two solutions that have equal osmotic concentrations. In comparing solutions of different concentrations, the solution with the higher concentration of solute and therefore a higher osmotic concentration is called *hypertonic* and the one that has a lower concentration of solute and a lower osmotic concentration is called *hypotonic*. In osmosis, water molecules move through a semipermeable membrane into a hypertonic solution until the osmotic concentrations are equal on both sides of the membrane.

Cell membranes are permeable to water but impermeable to many other substances. They are thus semipermeable. When a cell is placed in pure water, water tends to move into the hypertonic interior of the cell by osmosis and the cell tends to swell. The osmotic concentration of a cell can be measured by placing it in various concentrations of a non-penetrating solute like sucrose and determining the concentration of sucrose that causes no volume change in the cell. The diagrams show the effects of various concentrations of sucrose on an animal cell, in this case a sea urchin egg. Concentrations are measured in moles[1] of sucrose per liter of water.

When a sea urchin egg is placed in a non-penetrating solute like sucrose, it remains at its original size in 0.3 M sucrose which is the isosmotic or isotonic concentration. In hypertonic sucrose the cell shrinks and in hypotonic sucrose the cell expands and finally bursts. Other animal cells behave in a similar way. Plant cells have a cellulose wall which cannot stretch nearly as much as an animal cell.

---

[1] The mole is the basic unit used to denote the quantities of substances participating in physical reactions such as diffusion and in chemical reactions. The number of particles in one mole of any substance (molecules, atoms or ions) is always precisely the same, namely 6.023×10^{23}. Thus one mole of water contains 6.023×10^{23} water molecules, one mole of copper contains 6.023×10^{23} copper atoms and one mole of sodium ions contains 6.023×10^{23} sodium ions.

mainly of cellulose (Fig. 2-14; see also p. 116 and Chapter 10). In mature plants a single large vacuole may take up much of the cell's volume. It is bounded by a membrane and is filled with fluid. The remainder of the cellular material forms a thin layer between the vacuole and the plasma membrane of the cell.

Certain modifications of the plasma membrane called **junctions** bind the cells of multicellular organisms together and may function in cell to cell communication. The cells of multicellular animals are held together by structures called **desmosomes** (*desmos* means "binding" in Greek) which occur in patches on the plasma membrane. Desmosomes are complex modifications of the plasma membrane in which some material is laid down in the intercellular space and on the inside of each membrane (Fig. 2-16). Filaments radiate from the desmosome to the interior of the cell. Desmosomes appear to bind animal cells together in the same way that

spot welds bind two pieces of metal together. In animal cells we also find other junctions which are thought to function in cell to cell communication—such as **tight junctions,** in which the two membranes of adjoining cells are so close that their two outermost layers appear to be fused (Fig. 2-16A). Unlike desmosomes, tight junctions usually occur as a continuous ring around the cell, whereas desmosomes occur in patches.

The cells of multicellular plants are held together by a special layer of glue called the **middle lamella** which binds the walls of two plant cells together. Cell to cell communication involves cytoplasmic strands called **plasmodesmata** which run through openings in the cell wall and connect the interiors of adjacent cells (Fig. 2-16B).

Internal membranes.   Not all the inward folds of the plasma membrane lead to dead ends near the surface. Some form channels that lead deep into the interior. These channels connect to a

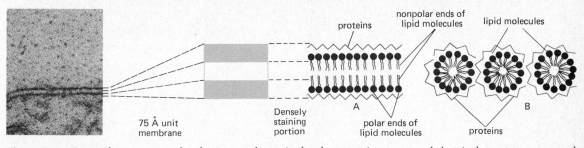

Figure 2-10 Layered appearance of a plasma membrane in the electron microscope and chemical structures proposed to account for its layered appearance. Plasma membranes are composed of lipids and proteins. Lipid molecules, which are discussed in Chapter 4, have two ends: a water soluble or *polar* head-end and a fat soluble or *nonpolar* tail. A common view of the plasma membrane shown in A is that the two outer layers are composed of proteins forming a sandwich, with the "filling" of the sandwich composed of a double layer of lipid molecules. The lipid molecules are oriented with the fat soluble ends facing each other and their water-soluble ends facing the proteins. Recent studies indicate that the proteins and lipids may be arranged in different ways in different cells. One suggestion is illustrated in B and shows the proteins coating clusters of lipid molecules. It also appears likely that plasma membranes are interrupted by numerous specialized pores.

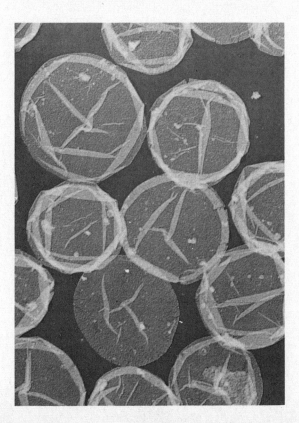

network of flattened membranes extending into many parts of the cell. In electron micrographs the cavities between the sheets of membrane often look like long parallel tubes. Careful study of consecutive slices, or *serial sections*, of the same cell shows that these cavities are interconnected. This system of vesicles makes up the **endoplasmic reticulum** (Figs. 2-6 and 2-7). Its appearance varies in different cell types and in different stages of growth of the same type of cell, but it is found in almost every cell. A key function of the endoplasmic reticulum (ER) is to provide a large network for transporting materials within the cell.

The membrane of the endoplasmic reticulum appears to be similar to the cell's outer membrane. But the side of the ER membrane facing the interior of the cell may be smooth (Fig. 2-17), like the interior of the plasma membrane, or may have particles attached to it that give it a roughened appearance (Fig. 2-18). These particles, 150 to 200Å across, are called **ribosomes.** Ribosomes synthesize all of the proteins of the cell. They

Figure 2-11 Purified membranes of human red blood cells. These membranes are obtained by placing red blood cells in distilled water. This causes water to enter the red cells by osmosis until the cells burst, release their contents and leave their membrane behind. These membranes, often referred to as **red cell ghosts,** have been used to demonstrate that the plasma membrane is composed of both lipid and protein and that it is 75-100 Å thick ($\times$ 4500).

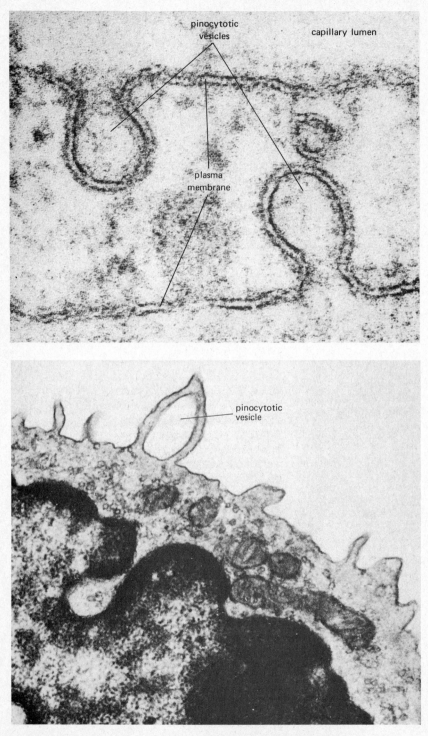

pinocytotic vesicles

capillary lumen

plasma membrane

pinocytotic vesicle

Figure 2-12 Pinocytotic vesicles may form by the infolding of the plasma membrane as in this cell from the wall of a blood capillary. Endothelial cells, as they are called, are very thin, the entire width of the cell being represented by the distance between the two plasma membranes shown in the picture. When a pinocytotic vesicle forms it acts like a ferry, transporting a droplet of material from the fluid surrounding the capillary to the lumen or inside of the capillary. One can imagine that the formation of pinocytotic vesicles by endothelial cells permits the rapid bulk transport of materials across the capillary wall ($\times$ 395,000).

Figure 2-13 The plasma membrane often forms blunt outfoldings which may entrap fluid in tiny cavities called pinocytotic vesicles. The electron micrograph shows the outfoldings of the plasma membrane of a leucocyte or white blood cell of a salamander. In one area the tips of the outfoldings have fused to form a vesicle within which is trapped some of the fluid surrounding the leucocyte. Subsequently this pinocytotic vesicle moves into the cell. By forming vesicles of this sort the leucocyte can transport droplets of fluid from the outside of the cell to the inside. Once the vesicle is inside the cell, the substance may be utilized directly or may be digested to smaller molecules. Pinocytosis is an important method by means of which cells may transport large amounts of materials ($\times$ 31,000).

Figure 2-14   Structure of the plant cell wall.

A. Fibrillar structures of the cellulose cell wall run in different directions in the different layers of this multilayered structure. The fibrils are rendered visible by depositing a thin layer of metal on the specimen from an angle, thus forming long shadows in the electron micrograph (× 38,000).

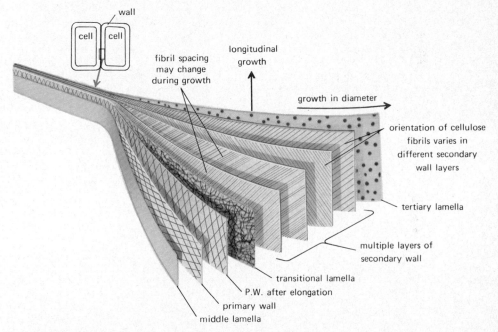

B. Drawings of the various layers of the cell wall of a higher plant showing the complexity of the multilayered structure of a cell wall.

may appear either as free particles or strung together like beads on a string (Fig. 2-19). Ribosomes are particularly abundant in cells that make protein that is exported to other parts of the organism, such as the liver cells which synthesize blood plasma proteins. A typical animal or plant cell will contain between 500,000 and 5 million ribosomes.

The endoplasmic reticulum connects with two other parts of the cell, the **Golgi complex** (often called *dictyosomes* in plants) and the membranes around the nucleus. The Golgi complex is a system of tightly packed, smooth vesicles near the nucleus (Fig. 2-20). They are shaped like a flattened stack of pancakes 1 micron wide. It is believed that the Golgi complex may serve as a

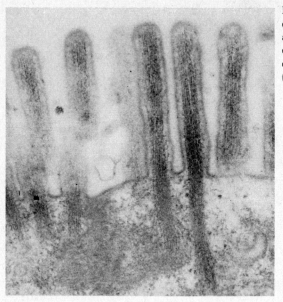

Figure 2-15 Microvilli are regular finger-like outfoldings of the plasma membrane which greatly increase its surface area. In the intestinal cell illustrated here they form a so called "brush border." Each intestinal cell may have 3000 or more individual microvilli. Longitudinal section (left) (× 50,000).

tangential section   cross section

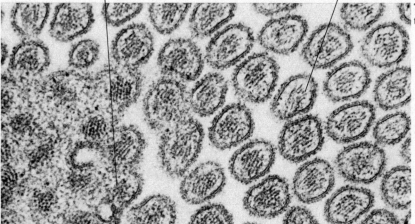

Tangential and cross sections of the microvilli (× 80,000).

storage and packaging depot for proteins and polysaccharides (see Chapter 4, pp. 125-131) that are eventually secreted from the cell.

The transfer of proteins to the Golgi complex from the cytoplasm, where they are synthesized, is believed to occur in the following way: Some ribosomes that are attached to the ER membranes are thought to transfer the proteins which they synthesize into the interior of the ER vesicles; these vesicles then travel to the Golgi complex where they can be observed to fuse with the stacked Golgi sacs. Thus we have a picture of ER membranes becoming incorporated into the Golgi complex. This "flow" is balanced by the Golgi complex forming vesicles through a process of budding and returning these membranes to the endoplasmic reticulum. It is also worth noting that whereas proteins are synthesized on ribosomes at the surface of endoplasmic reticulum membranes and then transported to the Golgi

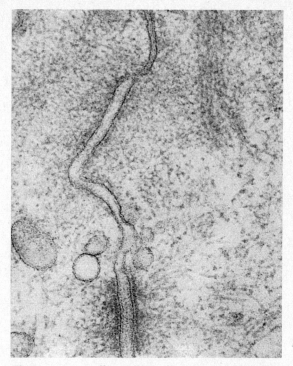

Figure 2-16A Cell to cell attachments in animal cells. The "junctional complex" is composed of a **tight junction, intermediate junction** and **desmosome.** The tight junction forms a continuous ring around the cell and probably prevents the leakage of material between cells. The other cell attachments are probably involved in the exchange of material between cells or in binding cells together (× 96,000).

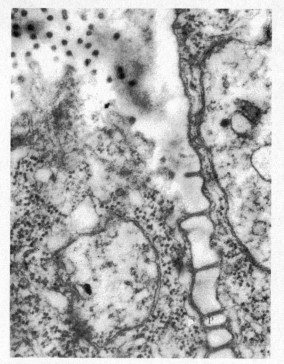

Figure 2-16B Cell to cell communication in plant cells is accomplished by **plasmodesmata** (P). These cytoplasmic strands run through pores in the cell wall and connect the interiors of adjacent cells. This electron micrograph of two adjacent cells from the shoot tip of a club moss shows plasmodesmata penetrating the cell wall. It also shows that the plasma membrane lines the pores in the cell wall and is continuous with the plasma membranes of the cells. The *upper left* shows a surface view of the plasmodesmata in a fold of the cell wall.

complex, some polysaccharides, such as the cellulose of the plant cell wall, seem to be synthesized directly at the Golgi complex.

In addition to leading into the Golgi complex, the endoplasmic reticulum also leads into the space between the two membranes that enclose the nucleus, which we discuss shortly.

These several kinds of vesicular structures exchange material among themselves. Some of this exchange occurs through direct connections, as among the ER, the Golgi complex, and the nuclear envelope. Most of the exchange occurs through a budding of vesicles of one sort which then fuse with vesicles of another sort. In general, we have come to think of these membrane-bound structures as being in constant flux, continuously changing their position, structure, content, and relationship with each other.

In summary, the membrane system of the eucaryotic cell is (1) the site of synthesis and packaging of a variety of compounds and (2) a system of transport within the cell and toward the exterior of excretory cells. In more general terms, we can think of the membrane system as separating the cell into two kinds of compartments: one set of separate compartments enclosed by the membrane system but occasionally and temporarily interconnecting; and the *cell interior*, permanently continuous, in which molecules can diffuse freely from one end of the cell to the other.

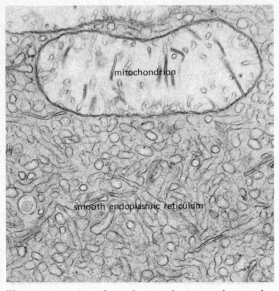

Figure 2-17 Vesicles of smooth (agranular) endoplasmic reticulum in the interstitial cells of the testis (× 32,000).

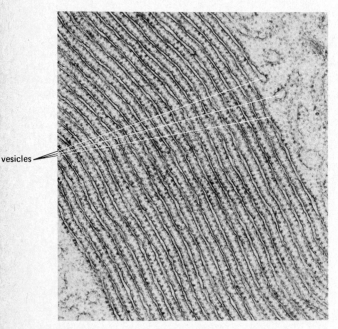

Figure 2-18 Flattened vesicles of rough (granular) endoplasmic reticulum from the **acinar cell** of the **pancreas.** The granules are ribosomes attached to the side of the membrane which is in contact with the cytoplasmic matrix (interior side) (× 82,000).

The cell interior

So far we have explored only those parts of the cell directly connected to its plasma membrane. Now let us cross through the membrane. We enter the cytoplasm, the sometimes fluid, sometimes jelly-like material inside the cell. The cytoplasm contains many kinds of small molecules and charged atoms or *ions* in *solution*. It also contains larger molecules that are in *suspension*, forming what is known as a *colloid*. These fine particles in suspension do not settle out because they are in a state of constant movement, like the particles of dust in the air. The cytoplasm has the unusual property of being able to flow like a thick, viscous fluid, or to take on the *elastic* properties of a gel. On the whole it is more solid near the outer membrane and more fluid in the interior.

Recently, the components of the cytoplasm which endow it with elasticity were discovered. They are elongated rods called **microfilaments** and **microtubules.** These structures provide cells with structural rigidity forming a sort of **cyto-skeleton** and some of them are involved in cell movement.

Microfilaments are thin (40-50Å) elongated structures which are usually difficult to resolve in the electron microscope. However, they can be readily observed when they occur in large quantity and are organized in parallel array (Fig. 2-22). In some cells such as muscle cells (Chapter 11), microfilaments play a role in cell movement and are able to convert the chemical energy of the cell into the mechanical work of cell contraction. In many other cells microfilaments appear to play a role in cytoplasmic streaming or in other movements which occur almost universally in eucaryotic cells (Fig. 2-21).

Microtubules are larger (200-300Å), more easily discernible structures (Fig. 2-23A, B), their cylindrical walls being built from approximately a dozen protein subunits (Fig. 2-23C). They also play a role in movement and are probably responsible for the conversion of chemical energy into work in beating cilia, dividing cells and moving chromosomes. In addition they play a special role in cells which have characteristic shapes, such as the disc-shaped red blood cells of some fish. There they appear to endow the cell

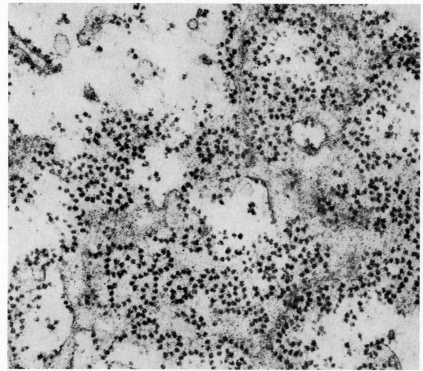

Figure 2-19 Polysomes on the endoplasmic reticulum in plasma cells. Polysomes are groups of ribosomes strung together like beads on a string ($\times$ 100,000).

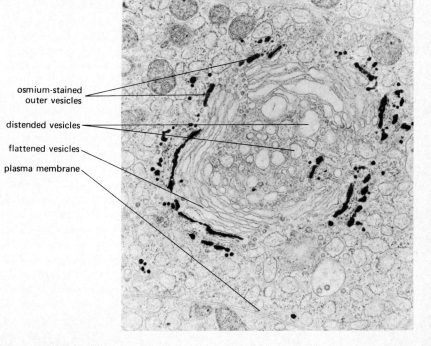

osmium-stained outer vesicles

distended vesicles

flattened vesicles

plasma membrane

Figure 2-20 The Golgi complex in the epididymis of a mouse. Though its existence was deduced from its stainability with osmium, it remained for the electron microscope to reveal the structure of concentric vesicles of the Golgi complex. We now know, as can be seen in this micrograph, that only the outer vesicles stain with osmium, but since this is the only structure in this region of the cell which stains with osmium, there can be no doubt that the Golgi complex was indeed visible before it was studied in the electron microscope.

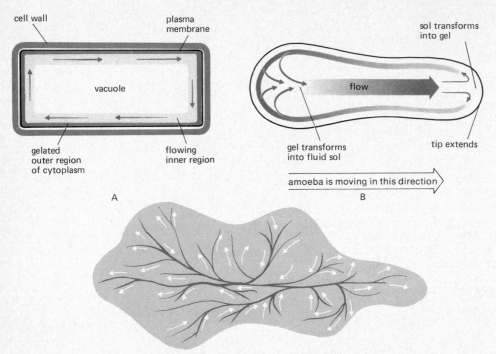

Figure 2-21 Examples of cytoplasmic streaming. A. Cyclosis, the rapid streaming of cytoplasm around a central vacuole in a plant cell. B. Amoeboid movement through the extension of a pseudopod. The movement is accompanied by the conversion of solid (called a gel) to liquid (called a sol) and liquid to solid in different regions of the cytoplasm. C. Multichanneled streaming in the slime mold *Physarum*. The body of this slime mold consists of a flat continuous sheet of cytoplasm containing many nuclei but no cell membranes. In this **plasmodium**, as it is called, the cytoplasmic streaming has a shuttle motion with the direction of streaming reversing every 30 to 60 seconds. Therefore, while one pseudopod expands another one contracts, as the plasmodium is composed of many regions, each with its own rhythm.

with the mechanical rigidity necessary to maintain its nonspherical shape.

Most animal cells and some plant cells have two **centrioles,** cylindrical structures ($0.15\mu \times 0.3$-0.5μ) lying perpendicular to each other, usually between the Golgi complex and the nucleus (Fig. 2-24A). They contain a bundle of nine *triplets* of microtubules arranged in beautiful radial symmetry (Fig. 2-24B). This arrangement of microtubules is denoted as 9×3. Centrioles duplicate during cell division. In some cells they become structures called **kinetosomes** or **basal bodies,** which migrate to the surface of cells and give rise to **cilia** and **flagella,** the hairlike projections of the cell surface, that are capable of organized motion (Fig. 2-25). Basal

bodies have the same 9×3 organization of microtubules as do the centrioles. However cilia and flagella have an outer ring of nine *pairs* of fibers surrounding two additional fibers in the center. This arrangement is denoted $(9 \times 2) + 2$. Interestingly this remarkable piece of bilaterally symmetrical organization is identical in all forms from the alga *Chlamydomonas* to the sperm of man (Fig. 2-26). It is a convincing piece of evidence that all eucaryotes descended from a common ancestral flagellated cell.

Finally, the **spindle,** a structure which forms in the nucleus during cell division, also contains microtubules (Fig. 2-27). As we shall see later, the spindle is involved with the distribution and movement of the chromosomes.

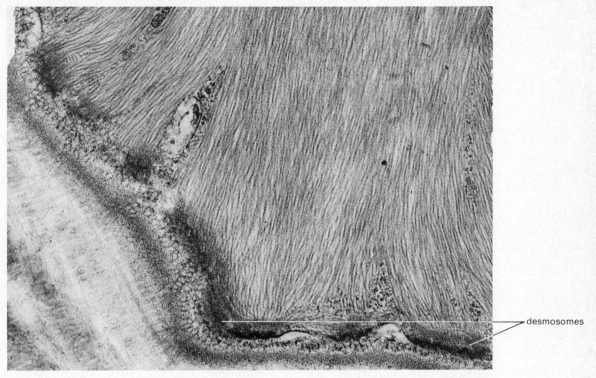

desmosomes

Figure 2-22   Cytoplasmic filaments found in unusual abundance in a basal cell of the **epidermis** (× 88,000).

In summary, the cytoplasm lying outside the vesicles of the membrane system has a complex structure. It is the supporting medium of the nucleus, endoplasmic reticulum, and other cellular structures. It also contains microfilaments and microtubules, some of which seem to give it structural rigidity, while others are involved in the conversion of chemical energy into mechanical work.

Cell organelles

As we move through the cytoplasm we find suspended in it many structures known as **organelles.** These fall in two general categories: those which are encased by two membranes and those encased by a single membrane.

Mitochondria are bounded by two membranes. Generally shaped like sausages, they are perhaps 2 to 10 microns long, and 0.5 to 1.0 micron in diameter. They occur in large numbers and a typical animal or plant cell will contain between 500 and 2500 mitochondria. The outer membrane of a mitochondrion is relatively smooth, but the inner membrane is folded into perpendicular partitions, or **cristae** (Fig. 2-28). The partitions provide a large surface area for the enzymes or natural chemical catalysts, which form groups called *assemblies* which are built directly into the membrane structure. A close look at the surface of the inner membrane reveals small spherical structures supported by stalks. The spheres, the stalks, and the membrane to which they are attached together form the assemblies of enzymes which are responsible for the chemical activities of the mitochondria.

In the mitochondria the foods of the cell, such as carbohydrates and fats, are oxidized to carbon dioxide and water. The energy thus released helps convert other molecules into high energy forms that can be used for work throughout the cell. For this reason the mitochondrion is

Figure 2-23 Cytoplasmic microtubules.

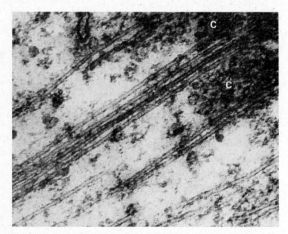

A. Cross section in kidney cell. Arrows point to cross section of microtables (× 86,000).

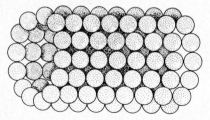

B. Microtubules of mitotic spindle inserting into chromosomes (C). (× 70,000).

C. Microtubules are composed of protein molecules, denoted by spheres in this drawing. These protein subunits assemble themselves into a microtubule. The self-assembly of protein subunits into organized structures is one of the fundamental ways in which cells form large structures made of many molecules.

Figure 2-24 Centrioles.

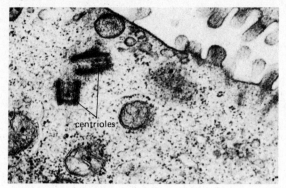

A. General structure showing two centrioles lying in the typical, right-angled position with respect to each other. From the embryonic chick epithelium (× 41,000).

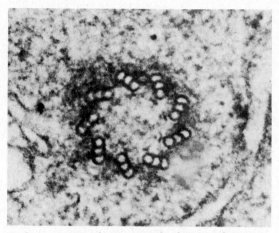

B. Cross section of one centriole showing characteristic microtubular organization of nine triplets (9 × 3). From embryonic, chick pancreas (× 164,000).

known as the *powerhouse of the cell.* Some cells have many mitochondria, some relatively few, depending on their energy requirements. Those few cells that entirely lack mitochondria must depend on less efficient ways of producing energy.

Mitochondria reproduce themselves by dividing and carry their own hereditary information in the form of a circular molecule of deoxyribonucleic acid or DNA, discussed in detail in Chapter 3 (See also Fig. 8-13).

Cells of most plants and many protists contain a number of other organelles called **plastids**

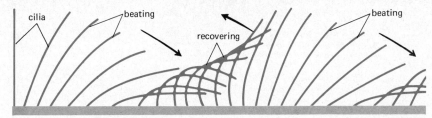

Figure 2-25 Organized motion of cilia bringing about the directional movement of a cell. The cilia on the surface of a cell beat in coordinated waves so that at any given moment some are in the effective or "downstroke" position of the cycle and some are recovering.

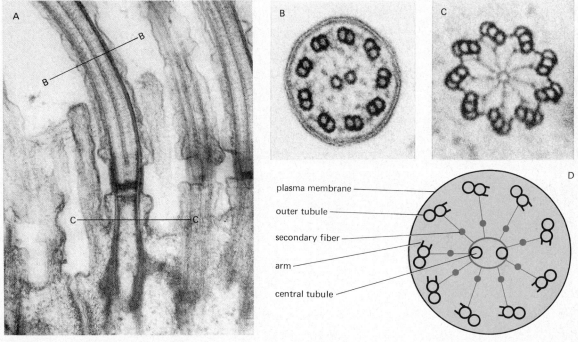

Figure 2-26 Structure of cilia and flagella.
A. Cilia on the gill of the fresh-water mussel, *Anodonta cataracta*, in tangential section. The shaft of the cilium extends outward from the basal body.
B. Cross section of a flagellum from the protozoan, *Trichonympha*, in position B-B of part A (× 127,500).
C. Cross section of a basal body in *Trichonympha*, in position C-C of part A (× 127,500).
D. Diagrammatic representation of the cross-section of a cilium or flagellum. Notice the (9 × 2) + 2 structure of microtubules, the presence of arms and of nine secondary fibers. Compare with part B.

which are also bounded by two membranes. Some are colorless, such as the ones that synthesize starch from sugar in nongreen plant cells; others are colored, such as the ones found in flower petals. The most important plastids are found in green plants and various protists. They are the green structures called **chloroplasts** (Fig. 2-29). They are usually larger than mitochondria and may be up to 20 microns in diameter. A typical plant cell will contain between 50 and 200 chloroplasts. In addition to their two membranes, chloroplasts have other membranes in their interior which form a highly laminated structure. These membranes are associated with

Figure 2-27   The mitotic spindle.

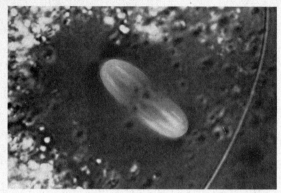

A. The spindle as revealed by the polarizing microscope. The longitudinal organization of spindle
fibers gives the spindle the optical properties that
make it observable in the polarizing microscope.
From the **oocyte** of the marine worm, *Chaetopterus
pergamentaceous.*

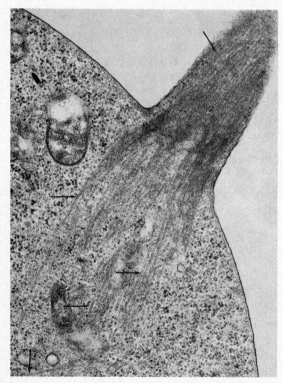

B. Electron micrograph of the longitudinal section
of the spindle of a dividing **erythroblast** in **bone
marrow.** Arrows point to spindle microtubules
(× 46,000).

chlorophyll and other pigments necessary for
photosynthesis, the process by which plants
utilize light energy from the sun and build sugars
from carbon dioxide and water, thereby providing
the rest of the living world with food. The membranes in the chloroplast form sacs which often
stack themselves into cylindrical piles called
grana. The sacs (or **thylakoids**) contain particles
of two different sizes (100Å and 175Å) which are
located in the membrane and which no doubt
play an important role in photosynthesis which
we do not know yet. Chlorophyll is located in
the grana and there the initial steps of the photosynthetic process occur.

Plastids duplicate by division or grow from
simpler self-duplicating bodies called **proplastids.** Like mitochondria, plastids carry some
of their own hereditary material in the form of a
circular molecule of DNA.

In the cytoplasm there are also nearly spherical bodies, encased in a single membrane, which
contain digestive enzymes. These organelles
called **lysosomes** (Fig. 2-30), are the *recycling
centers* of the cell. The digestive enzymes in
the lysosomes are able to break down large
molecules such as proteins, carbohydrates, and
fats, into small molecules which the cell can use.
The proteins of the cell and organelles like mitochondria wear out after continuous use and must
be replaced. Lysosomes fuse with these worn out
molecules and organelles, digest them and return
the small molecules to the cytoplasm for reuse.
Lysosomes also digest substances the cell may
take in by phagocytosis or pinocytosis. It is fortunate that the powerful digestive enzymes of the
lysosomes are contained in the protective envelope of the lysosomal membrane. Otherwise
they would destroy the cell itself. In fact, when
the cell dies, the lysosomes open, releasing enzymes which digest the cell. This final function
has earned lysosomes the name "suicide bags."

Nucleus and cell division

The most prominent structure in the cytoplasm
is the **nucleus** (Fig. 2-31). Like mitochondria and
plastids, it is surrounded by two membranes
which form a **nuclear envelope** which has a
large number of pores. The nucleus performs two

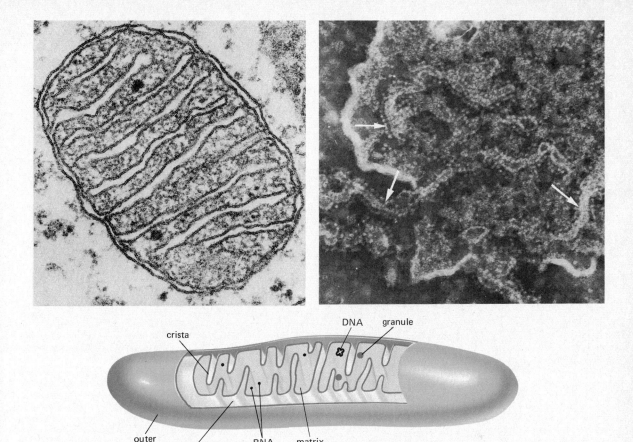

Figure 2-28 The mitochondrion. (Upper left.) Pancreas acinar cell mitochondrion showing outer membrane and inner membranes (× 256,000). (Upper right.) Isolated and disrupted beef heart mitochondria, negatively stained with phosphotungstic acid (× 135,000). The picture clearly shows small spheres (arrows) attached to what we know from other pictures are the **cristae** or inner membranes of the mitochondria. (Below.) Diagrammatic representation of a mitochondrion.

key functions. First, it controls the ongoing activities of a cell by regulating which protein molecules are produced and when, a matter we shall discuss in Chapters 4 to 6. In addition it contains most of the cell's hereditary information, the instructions that determine whether a particular cell will become a *Paramecium*, part of a sequoia tree, or part of a man. It passes this information on to daughter cells in the process of cell division. As we shall prove in Chapter 3, the information is in DNA molecules that are located on long, slender structures called **chromosomes.** Different organisms vary in the number of chromosomes

present in their cells: a fruit fly has 8; a broad bean, 12; and humans, 46.

When a cell is *not* dividing, a stage known as **interphase,** the nucleus shows little evidence of the chromosomes except for a tangle of thread-like material, called **chromatin,** and one or more dark bodies, the **nucleoli** (Fig. 2-31). The threads are actually the chromosomes, which are very spread out. At the time of cell division the nucleus changes its appearance. The thread-like chromosomes coil up and become short and thick rods. In this compact form they take up stain and can be seen in the light microscope (Fig. 2-32).

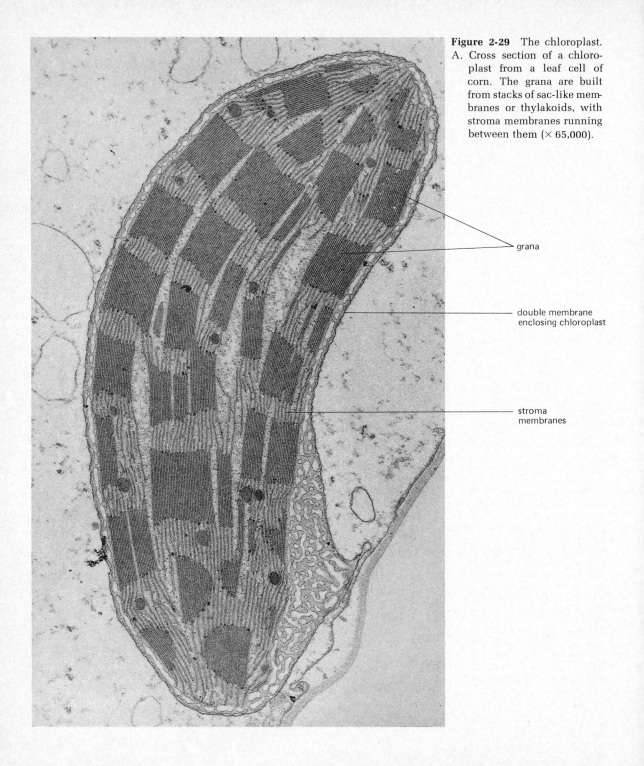

Figure 2-29 The chloroplast.
A. Cross section of a chloroplast from a leaf cell of
corn. The grana are built
from stacks of sac-like membranes or thylakoids, with
stroma membranes running
between them (× 65,000).

grana

double membrane
enclosing chloroplast

stroma
membranes

B. Diagrammatic representation of chloroplast containing grana, thylakoids, stroma membranes, and particles.

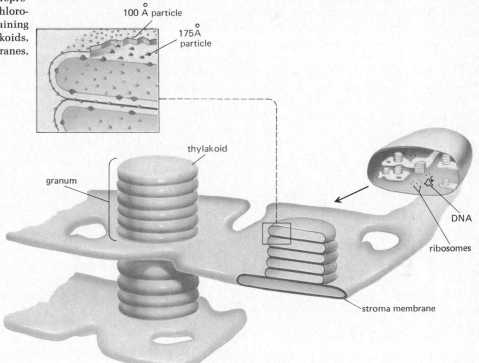

100 Å particle

175Å particle

thylakoid

granum

DNA

ribosomes

stroma membrane

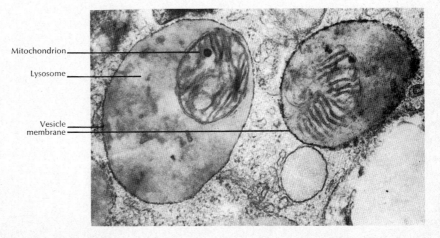

Mitochondrion

Lysosome

Vesicle membrane

Figure 2-30. Lysosomes in a kidney cell. Two of the three lysosomes in this photograph are digesting a mitochondrion. Note that a single membrane surrounds each lysosome whereas a double membrane surrounds the mitochondria they are digesting (× 57,000).

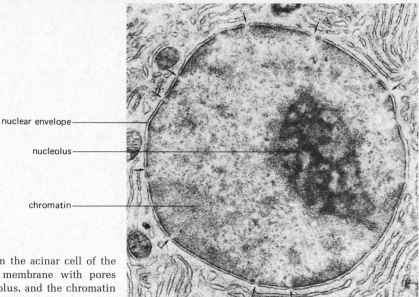

nuclear envelope——

nucleolus——

chromatin——

Figure 2-31 The nucleus from the acinar cell of the pancreas. Notice the double membrane with pores (marked by arrows), the nucleolus, and the chromatin material of the chromosomes (× 22,000).

Cell division involves two closely related events. First the chromosomes divide, producing two identical sets of chromosomes, and each set becomes part of a new nucleus. This process is called **mitosis.** Then the cytoplasm divides and each of the two new daughter cells receives one nucleus with a complete set of the chromosomes that were present in the parent cell. The cytoplasm is not divided as exactly as the chromosomes, but each of the daughter cells receives about half of the various cell organelles. Some of these organelles, such as chloroplasts and mitochondria, reproduce themselves, whereas others such as the ribosomes are synthesized by the new cell. At the end of cell division, the daughter cells are about the same size, have roughly the same number of organelles, and have precisely the same number of chromosomes.

The process of mitosis is a continuous series of events but we divide it into four phases for the convenience of description (Fig. 2-33).

1. Prophase.   The chromosomes become visible and can be seen to be longitudinally divided into identical **chromatids,** but are held together in a particular area, the **centromere.** We now know that the hereditary material (DNA) had been duplicated in the previous interphase; prophase

is the point at which the duplicated chromatids shorten and thicken through a process of coiling and pick up material which can be stained with certain dyes. During prophase, the nucleolus and the nuclear membrane begin to disappear. In animal cells and a few motile plant cells, such as the sperm cells of mosses, it is possible to detect a pair of organelles, the **centrioles,** near the nucleus. As we mentioned earlier, centrioles seem to be identical to the basal bodies of flagella and cilia and appear able to initiate the formation of microtubules. During prophase the centrioles begin to move apart and a new structure called the *spindle* is synthesized between them. The spindle is made up of **spindle fibers,** each of which is a bundle of microtubules. Some spindle fibers extend from one spindle pole to the other. Other spindle fibers are attached to the centromere of each chromosome and appear to control the movements of the chromosomes during mitosis. Although most plant cells seem to lack centrioles, they form a spindle in the same way that animal cells do.

2. Metaphase.   The centrioles are now positioned at opposite poles of the nuclear region, with a spindle stretched between them. Chromosomes are now at their shortest and thickest and have migrated to the **equatorial plane** of the spindle to which they appear to be attached by their centromeres. The nucleolus and nuclear membranes have completely disappeared.

3. Anaphase. The centromeres have split and the daughter chromosomes (the former chromatids) now move in opposite directions toward the two poles. The centromeres appear to be attached to spindle fibers and lead the way, while the arms of the chromosomes drag behind. The mechanism by which spindle fibers move the chromosomes during anaphase was demonstrated by making a small burn in a spindle fiber with a focused beam of ultraviolet light. The burn mark on this spindle fiber moved from the equatorial plane of the cell to the pole and finally off the end of the fiber. This result indicates that spindle fibers do not contract during chromosome movement. Instead it would seem that spindle protein subunits (see Fig. 2-23C) are added at the equator and removed at the pole.

4. Telophase. The chromosomes are now at opposite poles, and the reverse of prophase now begins: chromosomes lengthen and become thinner by uncoiling and the nucleolus and nuclear membrane begin to reappear. In the meantime the whole cell begins to divide. In animals, cells divide by a pinching off or cleavage process in which the cell membrane separating the two daughter cells grows inward and divides them. In most plants, the cells divide by forming a **cell plate.** This is a series of droplets which form in the interior of the cell and fuse to form a disc which grows outward to the surface of the cell, finally dividing it in two. A cellulose wall is then formed in the space between the daughter cells. After telophase, two daughter cells are left, each containing a nucleus in interphase.

We conclude that mitosis is a process which separates daughter chromosomes after they have been **replicated,** into two identical sets, thus ensuring that the daughter cells formed by subsequent cleavage each have an identical set of hereditary material. The shortening and thickening of the chromosomes, as well as their lining up at metaphase, are devices which allow the chromosomes to be transported without becoming entangled.

The nucleolus, which disappears during mitosis and then reappears in each new cell during the telophase stage, contains many particles like those observed lining the endoplasmic reticulum. The electron microscope shows that they resemble ribosomes (Fig. 2-31). These are in fact immature ribosomes in the process of being constructed, after which they presumably pass into the cytoplasm. Electron micrographs also show pores in the nuclear membrane (Fig. 2-31) through which the ribosomes and other substances manufactured in the nucleus may pass.

In summary, the eucaryotic cell nucleus is an organelle surrounded by two membranes, and carrying a number of chromosomes which contain the hereditary material of the cell. After the hereditary material is duplicated, the process of mitosis separates daughter chromosomes into two identical sets. This process is followed by cell division. The end result is two daughter cells with identical heredities.

THE CELL AT WORK

We have stressed in the course of our description of the cell that it is not static and inert. Cells engage in constant activity, both physical and chemical. Materials are being absorbed, processed, and shipped out. Cytoplasm and organelles move about, membranes change their shapes and relationships. To perform all this work, the cell constantly utilizes energy which comes to the cell from outside.

We can look at the cell as composed of a number of interconnected machines[1] that are capable of performing different kinds of work in synchrony. As we shall see, these machines are powered with chemical energy derived from a high-energy compound, called **adenosine triphosphate** (ATP). The chemical energy is used by these machines to perform different kinds of work (Fig. 2-34). Thus some machines perform

---

[1] When we use the word "machine," here we do not mean a mechanical machine. We mean simply a device which converts energy from one form to another regardless of the details of the mechanism. The process of converting energy from one form to another, say from chemical energy to mechanical energy, is called **transducing.**

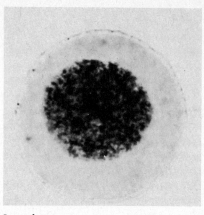

Interphase

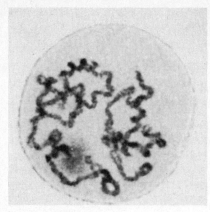

Early Prophase

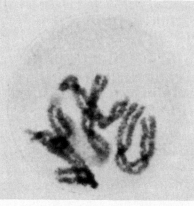

Late Prophase

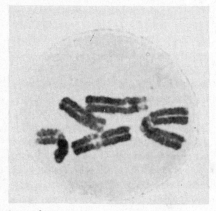

Metaphase

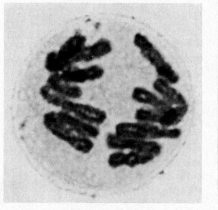

Anaphase

Telophase

Figure 2-32   Micrographs of the process of mitosis in a plant cell. Several stages in the division of the nucleus in the microspores of the plant *Trillium erectum*.

mechanical work, others chemical work, and yet others regulatory work. Let us look at each of these in turn.

All life involves motion, whether it is the contraction of muscle cells, the beating of flagella, the streaming of cytoplasm, or the movement of chromosomes. To produce motion there are special proteins (which we discuss later when studying muscle) that transduce chemical energy into *mechanical work*.

A cell also works to accumulate some substances and to excrete others. This is why certain substances are more concentrated inside the cell than outside, whereas other substances are found in higher concentrations outside the cell. The accumulation or excretion of materials against concentration gradients is called *active transport* or *osmotic work* and requires energy. The presence of a relatively impermeable plasma membrane greatly reduces the energy that the cell must

expend on osmotic work. Once materials are transported into the cell or excreted out of it, the low permeability of the membrane often keeps them there without further energy expenditure.

Sometimes positive and negative ions are accumulated or excreted in different amounts (*electrical work*). This brings about a difference in electrical charge on opposite sides of the plasma membrane. Nerve cells use these electrical charges in conducting nerve impulses.

Equally important is the tremendous amount of *chemical work* to be done. Cells must have energy to maintain their organization, that is, to reduce disorder (entropy). Cells obtain energy in one form; then, by a series of reactions, they transform it into another form so it is readily available for their own use. Plants and animals resemble each other in the way they do this, except that green plants and animals depend on different sources of energy (Fig. 2-34).

A. Interphase: Nuclear membrane present. Chromosomes not visible as compact structures. Nucleolus present. Two sets of centrioles lie close to each other.

B. Early prophase: Chromosomes become visible as long, thin threads. Nucleolus beginning to disappear. Centrioles begin to move apart.

C. Middle prophase: Chromosomes short and thick. Chromosomes appear double (chromatids). Chromosomes held together at centromere. Centrioles have moved farther apart. Spindle is

beginning to be organized.

D. Late prophase: Chromosomes are moving toward equatorial plane of nucleus. Nuclear membrane has almost disappeared. Nucleolus no longer present. Centrioles are nearly at opposite poles of the nucleus. Spindle is nearly complete.

E. Metaphase : Chromosomes are at equatorial plane of nucleus. Nuclear membrane has disappeared. Centrioles at opposite poles of the nucleus. Centromere of each double-stranded chromosome appears attached to a fibril of spindle.

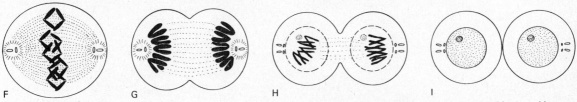

F. Early anaphase: Centromeres have begun to move to opposite poles dragging the chromosomes behind them.

G. Late anaphase: The two sets of chromosomes, now single-stranded, gather at opposite poles. Cell division begins.

H. Telophase: Chromosomes elongate, become thinner and less distinct. Nucleolus reappears. Nuclear membrane begins to form. Centrioles become replicated. Cell division nearly complete.

I. Chromosomes again not visible. Nuclear membrane complete. Cell division complete.

Figure 2-33 Mitosis in an animal cell.

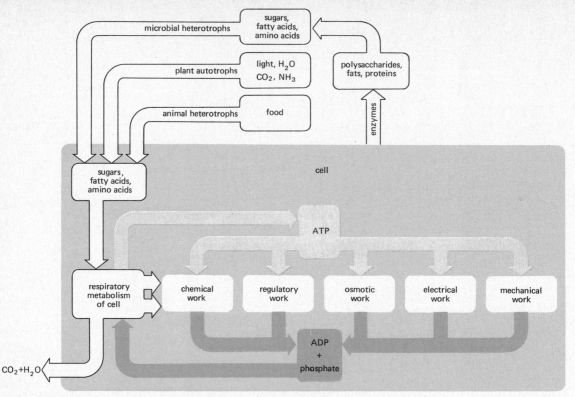

Figure 2-34 The flow of energy through the various machines of the cell.

The plant utilizes light energy to manufacture carbohydrates, fats and proteins. The animal eats these substances as food. Both kinds of organisms then break down these substances, store energy from the breakdown in usable form, and draw on these energy reserves for all the work that goes on in the cells.

Although a cell may resemble a factory in some respects, it is not like the factories we are accustomed to, for the cell is a *self-replicating* factory which produces all the products necessary to make another factory just like itself. Even more remarkable, its parts are superbly miniaturized, with a tiny memory tape which can direct its own duplication while at the same time controlling the manufacture of all the other parts necessary for maintenance, repair, and growth (*regulatory work*).

We emerge with the concept that the cell has a certain structure so that it can make use of the energy available to it to perform work, but we can

just as well say that the cell has to use energy to perform the work which is necessary for the maintenance of its structure. We shall find this two-way relationship between structure and function to be the most significant way one can look at the processes of the cell.

How does the cell use energy to maintain its structure and what kinds of structures does it use to handle its energy transformations? Not by using the clanging machinery we imagine in a factory, nor even with the hot or high pressure vats of a chemical plant, but with a quiet, cool, low-power molecular machinery we can only understand by looking more closely at the chemistry of the cell. We shall do that in Chapters 4 through 6 when we examine the processes by which the cell makes its living. But first let us look at the cell from the point of view of the hereditary material which it stores in its nucleus and which provides the cell with the instructions necessary to carry out all its functions.

SUGGESTED READING LIST

Bloom, W. and D.W. Fawcett, *A Textbook of Histology*, ninth edition. W.B. Saunders, Philadelphia, 1968.

Brachet, J. (ed.), *The Living Cell*. Scientific American, Freeman, San Francisco, 1961.

Brown, W.V. and E.M. Bertke, *Textbook of Cytology*. C.V. Mosby, Saint Louis, 1969.

De Robertis, E.D.P., W.W. Nowinski, and F.A. Saez, *Cell Biology*, fifth edition. W.B. Saunders, Philadelphia, 1970.

Dupraw, E.J., *Cell and Molecular Biology*. Academic Press, New York, 1968.

Fawcett, D.W., *An Atlas of Fine Structure — The Cell*. W.B. Saunders, Philadelphia, 1966.

Jensen, W.A., *The Plant Cell*. Wadsworth, Belmont, California, 1964.

Jensen, W.A. and R.B. Park, *Cell Ultrastructure*. Wadsworth, Belmont, California, 1967.

Ledbetter, M.C. and K.R. Porter, *Introduction to Fine Structure of Plant Cells*. Springer-Verlag, New York, 1970.

Loewy, A.G. and P. Siekevitz, *Cell Structure and Function*, second edition. Holt, Rinehart and Winston, New York, 1969.

Meek, G.A., *Practical Electron Microscopy for Biologists*. Wiley-Interscience, London, 1970.

Novikoff, A.B. and E. Holtzman, *Cells and Organelles*. Holt, Rinehart and Winston, New York, 1970.

Pfeiffer, J., *The Cell*. Life Science Library, Time Inc., New York, 1964.

Ris, H. et al, *Topics in the Study of Life in Cell Biology*. Harper and Row, New York, 1971.

Swanson, C.P., *The Cell*, third edition. Prentice-Hall, Englewood Cliffs, New Jersey, 1969.

Wilson, E.B., *The Cell in Development and Heredity*, reprinted 1953. Macmillan, New York, 1953.

Wolfe, S.L., *Biology of the Cell*. Wadsworth, Belmont, California, 1972.

CHAPTER 3

From Generation to Generation

HEREDITY, THE SOURCE OF LIKENESS AND OF VARIABILITY

Everyone knows that cats have kittens, dogs have puppies, and humans have babies. Everyone also knows that no two puppies, kittens, or babies are exactly alike—not even "identical" twins. We have here two apparently divergent yet fundamental properties of living matter:

1. The ability to produce offspring which resemble the parents in minute detail (faithfulness).
2. The ability of offspring to vary slightly, but in a great many ways, from one another (variability).

As wondrous as these simultaneous properties of **faithfulness** and **variability** of living matter may be, we describe in this chapter something equally miraculous—the machinery of heredity which can account for these apparently paradoxical properties of biological reproduction.

The history of our understanding of the laws of heredity has followed a tortuous and, at times, interrupted path. It was nurtured by three major lines of investigation, all of which began in the middle of the last century. To treat the subject historically, we would have to follow simultaneously (1) the breeding experiments started by Mendel in 1856, (2) the cellular investigations of nuclei and chromosomes begun by Hertwig in 1870, and (3) the study of nucleic acids initiated by Miescher in 1868. Furthermore we would have to demonstrate by constant cross reference how these three avenues of research influenced one another. To do this would be fascinating, but very complicated for the writer and student alike. Hence we have opted in favor of clarity rather than historical chronology and cover in this chapter first, a description of the chromosomal events associated with fertilization and gamete formation, then a description of genetic studies rendered more understandable by our knowledge of the chromosomal events. In the following chapter, we cover sufficient chemistry so that by Chapter 6 we can complete our story on the faithfulness and variability of biological reproduction.

THE BEHAVIOR OF CHROMOSOMES (FERTILIZATION AND MEIOSIS)

We have defined heredity as the machinery responsible for the production of both *likeness* and *variability* in organisms. If inheritance is *machinery*, what do we inherit? Today we know that we inherit a *material* and that this material brings about, through many physical and chemical interactions, the formation of properties which are characteristic of the organism. This important concept now appears self-evident, but our predecessors were far from clear about the distinction between the hereditary material and the biological characters it engenders. Hippocrates, and even Darwin, believed that we inherited something very close to the *characters* themselves, produced by each part of the body, affected by the environment, and somehow collected in the semen. Aristotle, on the other hand, was clear that we inherited a *potentiality* to produce a biological character, rather than the character itself, but he was far from clear that this potentiality had a material substance for its basis. It took the last 100 years of experimentation to show that parents indeed transmit a hereditary material to their offspring, which is a form of *information* in that it represents a potentiality which can, under the proper environmental conditions, both intra- and extracellular, be *translated* into biological structures and activities.

Most of the hereditary machinery is active when a cell divides and its daughter cells grow. But historically it was through studies of activities associated with sexual reproduction that its properties have been clarified. Let us therefore begin our study of heredity by considering biological sexuality.

Sexuality is a process by which the hereditary material of two individuals of the same species can be combined and reassorted. As we show in Chapter 15 the significance of sexuality is evolutionary, rather than physiological, because life—that is, cell division and cell growth—can proceed without sex. Biological evolution, however, is greatly accelerated if the hereditary material of a species can be reshuffled during reproduction.

We begin with the diploid cells of eucaryotic organisms, such as the garden pea, the fruit fly, and man. Toward the end of the chapter we consider bacteria and viruses and show how the concept of the gene has evolved through intensive work with these simpler organisms.

In eucaryotes the most dramatic event of sexuality is fertilization, which occurs when the sperm produced by one individual fuses with the egg of another individual to form a fertilized egg, or **zygote** (Fig. 3-1). The important thing about this fusion (first recognized as late as 1876 by Hertwig, working with the sea urchin egg), was that the nucleus of the sperm and little else enters the egg and fuses with the nucleus of the egg; it took another 26 years before it was demonstrated conclusively that only one sperm was necessary for fertilization. Long ago Leonardo da Vínci (1452–1519) pointed out that in most instances the **direction** (male variant × female normal or male normal × female variant) of a genetic cross ordinarily does not affect the results of the cross. One could therefore conclude from this and from the fact that the sperm contributes little other than nucleus that nuclei carry the hereditary material. How compact the hereditary material must be can be ascertained from the fact that the nucleus from the sperm of a man represents $1/10^{10}$ (one ten-thousand billionth) of his body weight. All the potentialities necessary for the development of all the properties of a man—his height, his intelligence, his smile, his hair color, his gestures, his handwriting, his emotions—reside in this tiny amount of hereditary material. Surely, it is worth studying the properties of this substance and the structures carrying it.

We already know that the nucleus contains chromosomes. They are the most characteristic and invariant content of the nucleus. It is possible to count the number of chromosomes in the sperm and egg of a species and show that they are normally constant and equal in number, and that each chromosome has characteristic proportions, giving it its own recognizable individuality

(Fig. 3-2). We usually say that there are n **(haploid)** chromosomes in the sperm and egg, the male and female gametes. In the pea family n is equal to 7. After the fusion of the egg and sperm nuclei, we find $2n$ chromosomes in the resulting nucleus, and in higher plants and animals the chromosome number remains $2n$ **(diploid)** in all the cells of the organism resulting from the growth and development of the fertilized egg. However, if these $2n$ organisms are to produce offspring with $2n$ chromosomes in each cell, the number of chromosomes found in their gametes must be reduced back to n, and indeed it is!

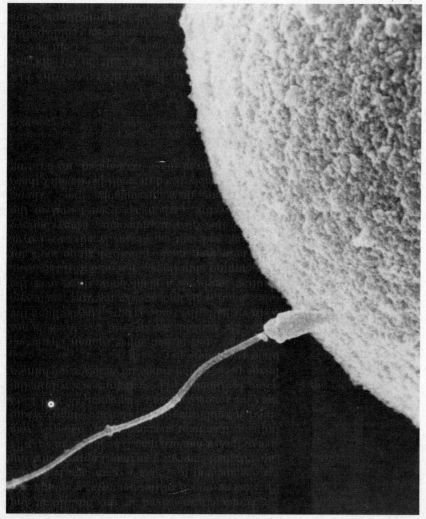

Figure 3-1 Fertilization in the sea urchin. The sperm carries very little else besides the nucleus and the motile machinery in the tail. Yet the amount of its hereditary material is nearly the same as that of the egg.

The process by which the chromosome number is reduced from 2n to n is called **meiosis** and usually occurs during gamete formation. Figure 3-3 illustrates the life cycle of the pea and of man. The main difference between these organisms is that the pea produces male and female gametes (eggs and pollen) on the same plant, whereas in man and other higher animals male and female organisms produce sperms and eggs, respectively. We encounter other variations in the life histories of procaryotes, lower plants, and lower animals throughout the book, especially in Chapter 15.

Meiosis is often thought of as a complicated form of mitosis. This is misleading, for as we have seen mitosis has to do with chromosome *duplication* followed by cell duplication, whereas meiosis has to do not only with *halving* of the chromosome number but also, very importantly, with the reshuffling of the hereditary material to produce gametes differing from each other in their hereditary makeup. This *reshuffling* occurs at two levels of the meiotic process and for purposes of clarity Figure 3-4 shows two schemes in which the two kinds of reshuffling processes are artificially separated from each other. The similarities between meiosis and mitosis lie in their common function, that is, separating the two sets of chromosomes from each other and ensuring that the daughter cells receive complete **complements** of chromosomes. This requires the choreography we encountered in the mitotic process: (1) the shortening and thickening of the long chromosomal threads into the condensed chromosomes during *prophase*, (2) the migration of the chromosomes to the *metaphase* plate, (3) the separation of the

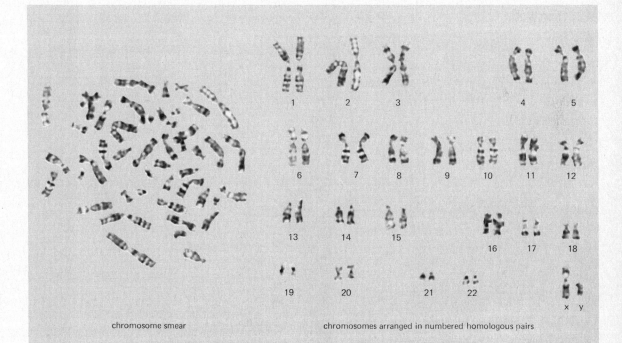

chromosome smear   chromosomes arranged in numbered homologous pairs

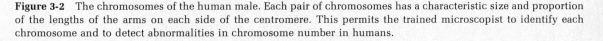

Figure 3-2   The chromosomes of the human male. Each pair of chromosomes has a characteristic size and proportion of the lengths of the arms on each side of the centromere. This permits the trained microscopist to identify each chromosome and to detect abnormalities in chromosome number in humans.

chromosomes along a mitotic spindle in *ana-phase*, and (4) the reversal of the process of chromosome shortening in *telophase*. However, as we shall see, in meiosis the choreography is repeated twice, since two nuclear divisions occur.

We have selected a hypothetical cell in which *n* is only 2. Notice that during fertilization the sperm contributes two chromosomes (indicated by black) and the egg also contributes two chromosomes (brown). Notice also that in the gametes and in the cells of diploid organisms we depict the chromosomes in their condensed state as they would normally appear in metaphase of mitosis. We do this to illustrate that each chromosome in a haploid set has its own characteristic size and shape. The shape of a chromosome can most easily be quantified by measuring the length of each arm of the chromosome on either side of the centromere and expressing them as a ratio. Thus we can characterize the 23 pairs of human chromosomes by measuring the average length and the ratio of the lengths of their arms (Fig. 3-2). Notice that after fertilization, chromosomes in the cells of diploid organisms occur in **homologous** pairs, one member of a pair being derived from each parent.

The first significant event relating to meiosis occurs in the interphase preceding it, when each chromosome duplicates to form two strands or chromatids which, however, do not separate. This temporary lack of separation is depicted by drawing only one centromere for each pair of chromatids. The second significant event finds the homologous maternal and paternal chromosomes coming together to pair in a precise manner, so that every part of the maternal chromosome comes to lie next to the equivalent part of the paternal chromosome. It is generally thought that the sex act in higher organisms involves copulation or fertilization. Fundamentally, however, the sexual process (that is, union and exchange) can be thought to take place here in meiosis, first when homologous chromosomes pair and later when they may exchange segments (crossing over).

Next the alignment of chromosomes on a metaphase plate occurs, just as in mitosis, prior to a separation of homologous chromosomes. Another important thing to recognize is that each chromosome pair or **bivalent** lines up independ-ently from all the others with respect to the direction of separation of its maternal and paternal chromosomes (Fig. 3-4a). This means that although the two daughter cells each get a complete set of chromosomes, their composition with respect to their *maternal* or *paternal* origin is random. In our simplified case, with *n* equal to 2,

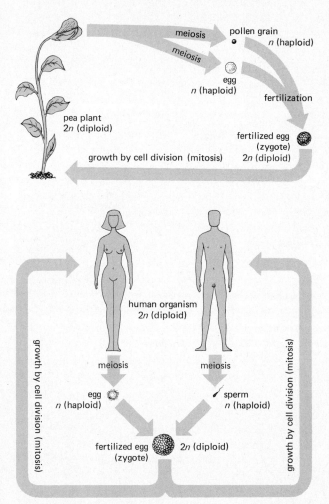

Figure 3-3 The life cycle of the pea and of man. In the pea, though by no means in all higher plants, the same plant produces both pollen (male gamete) and egg (female gamete). In man, one organism (the female) produces eggs and the other (the male) produces sperm. In both man and the pea, meiosis, the process whereby the chromosome number is reduced from 2n to n, occurs immediately prior to gamete formation.

this gives rise to only four different combinations, but with a larger number of chromosomes the number of combinations can become very large (2^n). In human beings, for instance, the number of chromosomal combinations which can be found in the gametes would be 2^{23}, or 8,388,608.

The final significant stage of meiosis is the second division, which separates the chromatids from each other. Notice that we are left with four cells, each containing a haploid set of chromosomes. Thus haploidy is achieved by a single duplication of chromosomes in the interphase before meiosis begins, followed by *two* cell divisions in which *no chromosome duplication occurs*.

In the second division also, the orientation of each chromosome pair is random, which, in this case, would appear not to be a source of variation, since now each chromosome is composed of a pair of chromatids, *both* of which are of either maternal or paternal origin. However—and this brings us to our second scheme of meiosis (Fig. 3-4b)—this is not always the case, since exchanges between segments of *nonsister* (one maternal and one paternal) chromatids may in fact have taken place during the prophase of the first division, so that by the second division two of the four chromatids are of mixed (maternal and paternal) origin. Most commonly, when homologous chromosomes are paired, exchanges occur between one of the maternal and one of the paternal chromatids; however, all four chromatids can participate. These exchanges occur with a finite frequency, but precisely where along the length of the chromatid the exchange takes place varies randomly. The result of this process is a tremendous increase of the hereditary variability of the cells formed by the first meiotic division. Since we now have paternal chromosomes containing chromatids with maternal material and vice versa, random orientation during the second division increases the hereditary variability even more.

In summary, meiosis brings about two important functions of sexuality:

1. It reduces the number of chromosomes from the diploid number to the haploid number, thereby allowing the formation of gametes containing only a single full complement or set of chromosomes.

2. By exchanging chromosomal segments and by reshuffling the chromosomes, meiosis creates a gigantic pool of hereditary variability in the gametes produced by an organism.

As we see in Chapters 18 and 19 this pool of variability provides the basis of evolution through natural selection. It is important to understand meiosis to interpret the results obtained by Mendel and his successors when they crossed higher organisms and observed their offspring.

THE DISCRETE NATURE OF THE GENE

Much painstaking work on the heredity of higher organisms had been done before Mendel by a number of very able and experienced biologists. Their work had led to naught because they operated on the wrong assumptions. They were natural historians concerned with multifaceted and complex problems of evolution. Their appreciation and view of the organisms they observed were equally multifaceted and complex. They recorded many properties in the parental organisms and sought to observe how they were transmitted to the offspring. Furthermore they looked at properties which seemed important to the organism in its fight for survival, or to man as in the case of domesticated animals and plants. These properties, such as coat color in rodents or milk production in cows, were either difficult to quantify or dependent on a multiplicity of hereditary factors. As a result no quantitative correlations could be made and, indeed, no quantitative records were kept. In fact very little effort was made to keep track of the generations, so that results which came from different generations were often collated. Mendel himself sized up the situation succinctly when in 1865 he wrote in his major article on plant hybridization:

Those who survey the work done in this department will arrive at the conviction that among all the numerous experiments made, not one has been carried out to such an extent and in such a way as to make it possible to determine the number of different forms under which the offspring of hybrids appear, or to arrange these forms with certainty according to their separate generations, or definitely to ascertain their statistical relations.

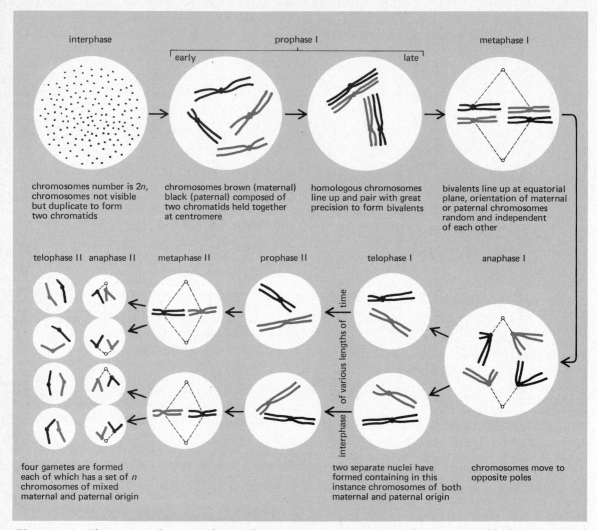

Figure 3-4A The process of meiosis showing how gametes come to contain chromosomes of both maternal and paternal origin.

Ten years before this article was published Mendel had begun a meticulous, purposeful, and prophetic set of breeding experiments with the garden pea, *Pisum,* which led to his formulating the laws of heredity some 35 years before their significance was appreciated by other biologists. Mendel owed his success to the following:

1. The selection of an organism, the garden pea, which (a) was available in a variety of forms with easily observable differences and (b) could readily be cross-pollinated experimen-

tally without much chance of unwanted crosses occurring. That is, Mendel could, if he wished, cross-breed separate plants (note that a single plant produces both male and female gametes and that, in the pea plant, fertilization normally occurs within a single flower), but cross-pollination rarely occurs accidentally.

2. Ensuring that each strain was pure by self-pollinating them for the first two years and selecting those plants that gave offspring all of which resembled their parents.

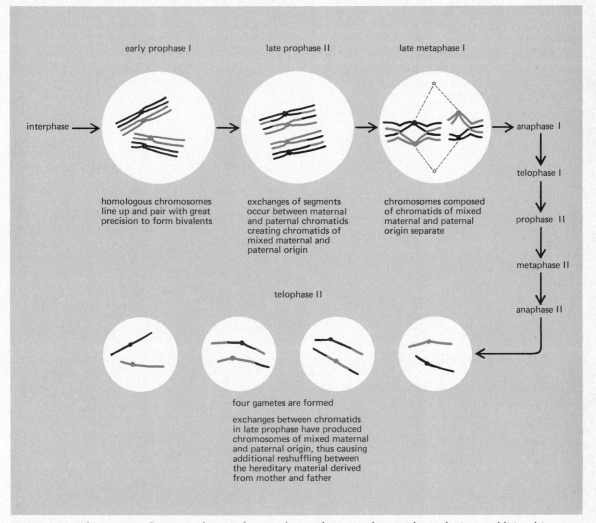

early prophase I

late prophase II

late metaphase I

interphase →

homologous chromosomes
line up and pair with great
precision to form bivalents

exchanges of segments
occur between maternal
and paternal chromatids
creating chromatids of
mixed maternal and
paternal origin

chromosomes composed
of chromatids of mixed
maternal and paternal
origin separate

→ anaphase I

↓

telophase I

↓

prophase II

↓

metaphase II

↓

anaphase II

telophase II

four gametes are formed

exchanges between chromatids
in late prophase have produced
chromosomes of mixed maternal
and paternal origin, thus causing
additional reshuffling between
the hereditary material derived
from mother and father

Figure 3-4B The process of meiosis showing how exchanges between chromatids results in an additional increase in the genetic variability of the gametes.

3. Selecting seven characters for observation that gave *all or none*, rather than intermediate, results. For example, his crosses between plants with purple flowers and plants with white flowers always resulted in purple-flowered plants and white-flowered plants, but never in plants with flowers of an intermediate or blended color.

4. Performing large numbers of crosses and keeping meticulous records of the characters and the generations in which they occurred.

   Important, however, as these methods may

be, Mendel brought the crucial "prejudice" of favoring *numerical simplicity* to his experiments. Expressing the results of large numbers of crosses in terms of ratios of small whole numbers like 3:1 or 1:1, or even 9:3:3:1, was an approach for which there was no *a priori* justification. Mendel knew very little about nuclei and probably nothing about chromosomes. He certainly did not know that chromosomes came in pairs in the cells of higher organisms and were reassorted during the formation of gametes. In short he knew nothing about the physical basis of the hereditary laws he was formulating. He had, however, a prejudice, amply justified in retrospect: that obtaining simple whole numbers in his breeding experiments signified that he was observing a basic piece of hereditary machinery. Interestingly it was because of this very numerical simplicity that his findings were rejected by even the greatest biologists of his time, such as Nägeli, to whom Mendel had sent a copy of his great article. To these biologists the laws of heredity were far from simple and clearly could not be described by numerical relations involving simple whole numbers.

We now present some of the data which Mendel obtained and which demonstrate how a theory of heredity can be formulated. We do this somewhat differently from Mendel because we have the advantage of knowing about the behavior of chromosomes during fertilization and gamete formation. Those who wish to see how Mendel succeeded in his formulation in the absence of all such knowledge are urged to read Mendel's brilliant article, one of the greatest intellectual achievements of modern science.

Let us take one of the seven characters Mendel studied and perform two of the standard crosses that Mendel carried out. Mendel had two strains of garden peas which differed in flower color—one of them purple, the other white. We shall not record the sex of the parents, because Mendel found that the direction of the cross did not influence the results.

First Cross

Parental generation (P)	Purple × white
First filial generation (F₁)	**All purple**

The result of this experiment seems simple and might look convincing to today's observer. To convince his skeptical contemporaries Mendel performed it with all seven of his characters and carried out *in toto* seven trials, involving 70 plants and 287 fertilizations. The following analysis of Mendel's theory attempts to identify the logical structure of its components.

Conclusion 1.   From this experiment emerged the first conclusion of Mendel's laws of hered- ity—that in the hybrid or F_1 generation, one character of a given pair of characters is dominant over the other one, since only that character is expressed.

In the second cross, Mendel interbred members of the hybrid or F_1 generation.

Second Cross

First filial generation (F₁)	Purple × purple (929 plants)	
Second filial generation (F₂)	Purple (705 plants)	White (224 plants)
or	3.15 :	1
or	3/4 :	1/4

Conclusion 2.   The white character, which disappears in the F_1, reappears unchanged in ¼ of the F_2 offspring. This is a most important conclusion. Most students of heredity believed at the time in the **blending** character of heredity, rather than in its **discreteness**. The characters they had chosen to observe, which we now know are dependent on many genetic factors, had behaved in a blending manner; that is, they gave intermediate characters in the F_1 and no reappearance of pure, unchanged parental characters in the F_2.

Mendel's results, however, allowed him to make a clear and unambiguous hypothesis: there is a hereditary factor responsible for the character, white flower, which is transmitted from the white parent *through the* F_1 to the ¼ white offspring in the F_2.

Conclusion 3.   But because the F_1 is purple it must also have a hereditary factor for purple. Therefore we come to the next conclusion.

Conclusion 4.   There must be at least two hereditary factors for flower color carried in the F_1 and therefore, by implication, in all individuals. Although logically the number could be more than two, it is characteristic of the making of scientific inferences to assume the simplest hypothesis until forced to revise it. As we shall see the number two we assume here has stood the test of time remarkably well.

Conclusion 4, like Conclusion 2, is very important. It establishes that hereditary factors come in pairs, a deduction we *now* can understand because we know that chromosomes come in pairs in the cells of higher organisms.

Conclusion 5.   Now that we know that there are two factors, one for purple and one for white, we can think of dominance in the F_1 as being a property of the relationship of the factors. We say that the F_1 has a factor for purple flower which is *dominant over* the factor for white flower, which we say is *recessive*.

We are now ready for some definitions, which in a sense are hidden hypotheses.

Definition 1.   We say that the factors responsible for purple or white flowers are part of the **genotype,** the hereditary material of the individual that is transmitted essentially unchanged from one generation to another. Thus we have learned that the hereditary factors are discrete or separate factors or units of heredity that do not blend or intermingle. They do, however, *interact,* dominance being one form of interaction in which only one factor in a pair is expressed. The character which finally becomes expressed in the organism is part of the **phenotype.** Inherent in this definition, therefore, is the hypothesis that differentiates between the collection of factors now called *genes,* which constitute the genotype, and their expression in the form of characters which constitute the phenotype. The basic problem of cell biology, whether it is molecular genetics, biochemistry, or cytology, is to work out the precise mechanism whereby a genotype determines the formation of a given set of characters or phenotype.

Definition 2.   The concept of pairs of factors or genes has emerged based on (1) the fact that the F_1 must carry a gene for white, since the white character reappears in the F_2 and (2) dominance, which implies dominance over something, namely, over its recessive counterpart. Such pairs of genes are called **alleles.** Inherent in the definition of allele is the hypothesis that allelic genes are in some way related, a notion which was eventually to be explained by their position at equivalent loci on homologous chromosomes and by the essential similarity of the hereditary information they carry. And Mendel in selecting seven pairs of characters all giving him a 3/4:1/4 **segregation** in the F_2, in fact demonstrated that these characters depended on seven pairs of genes or alleles, each pair having a dominant and a recessive gene. The concept of segregation (Mendel called this his **first law of segregation**) means that alleles which are together in one organism can become separated in its offspring. This phenomenon is, of course, accounted for by the formation of gametes during meiosis.

In the two crosses described, we used the terms purple and white to describe the phenotypes. We now express these crosses by symbolizing their genotype or gene content. We use the symbol C for the dominant gene determining the expression of a purple (or colored) flower; the recessive gene we denote by c. Therefore, the purple-flowered parent is CC and the white-flowered parent is cc. We know the purple-flowered parent is CC and not Cc because all the offspring in the F_1 are purple-colored. As we see later, had the parent been Cc, half of the offspring of a cross with cc would be white-flowered.

The F_1, as we have already argued, must be Cc since it must contain c to give white-flowered offspring in the F_1, and it must have C since it is purple-flowered itself.

First Cross

P	CC	$\times$	cc
	Purple		White
F_1		Cc	
		Purple	

What about the genetic composition of the F_2? Here we can make some predictions. But before we do we must examine the kind of gametes the F_1 generation can produce. In doing this, as well as in numerous instances from now on, we shall assume that genes are carried by chromosomes. By making this assumption and demonstrating that this allows us to explain events or make predictions regarding them, we shall continue to strengthen the hypothesis that genes are indeed located on chromosomes. Now, if the genes C and c are carried on homologous chromosomes in the diploid pea plant, then during gamete formation they will be separated from each other to form both eggs and pollen, half of which will carry the gene C and the other half the gene c. When F_1 pea plants producing these gametes are crossed or self-fertilized, we can find out the genetic constitution of their offspring.

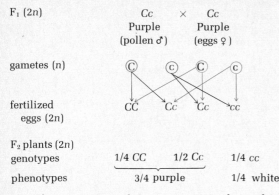

The important things to notice about this cross are the following:

1. The phenotype purple is composed of two genotypes: 1/4 of the total offsprng are **homozygous** CC and 1/2 are **heterozygous** Cc.
2. Together they form the purple phenotype category which bears a 3/4:1/4 ratio to the white phenotype as observed by Mendel.

In summary, by assuming that C and c are carried on homologous chromosomes and separated from each other by meiosis during gamete formation, we have been able to account for Mendel's 3/4:1/4 segregation which he observed with the seven pairs of characters he studied.

But not only should we be able to account for Mendel's observations, we should be able to make predictions.

Prediction 1. The purple-flowered plants in the F_2 generation are not one homogeneous group, but composed of homozygotes and heterozygotes in a 1 to 2 ratio, that is, 1/4 of the F_2 generation is homozygous dominant and 1/2 is heterozygous.

Prediction 2. Since each plant is the result of a single union of one egg and one sperm, all flowers on a given plant will be identical in that they all produce ⓒ gametes if the purple F_2 plant is homozygous, or 1/2 ⓒ and 1/2 ⓒ gametes if the plant is heterozygous.

These predictions should be testable by a crossing or breeding experiment in which the results of the cross will permit us to characterize each purple-flowered F_2 plant either as a homozygote or as a heterozygote. The most convenient cross which can be performed to test these predictions is a cross between the F_2 plant

and a homozygous recessive plant of the parental stock. This so-called **back (or test) cross** can be shown to yield the phenotypes (and, in this case, the genotypes) in a 1/2 to 1/2 ratio if the parent is heterozygous, and all dominant phenotypes if the parent is homozygous.

Back Cross

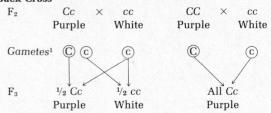

The test cross is convenient in that it takes a relatively low number of offspring to determine whether a given F_2 plant, when crossed with the homozygous recessive parent, gives only a dominant phenotype or a 1/2 dominant to 1/2 recessive phenotype ratio in its offspring.

Mendel himself did not use the back cross, but allowed the F_2 plants to self-fertilize. In this case the homozygous plants also gave offspring all with the dominant phenotype, but the heterozygote of course gave a 3/4 dominant to 1/4 recessive distribution of phenotypes. And, indeed, Mendel was able to demonstrate by performing a large number of crosses with all seven of his character pairs that the F_2 class of dominant phenotypes was composed of 1/3 homozygotes and 2/3 heterozygotes (that is, in the case of purple-flowered F_2 self-fertilizations, 36 plants yielded exclusively purple-flowered offspring, whereas 64 plants yielded some purple and some white offspring). This may not appear like a very good 1/3 to 2/3 ratio, but when Mendel added up the results of all his seven crosses, he obtained a ratio of 1 to 1.93.

The question as to what constitutes proof of a given ratio (that is, is 36:64 a 1/3 to 2/3 ratio or is it not) is one which we can answer today by performing certain statistical tests, which, unfortunately, we cannot describe in an introduc-

[1] In the case of a homozygous organism, it will be our convention to show only one gamete as that is all that is produced. When two different gametes are produced by a heterozygous organism, we shall represent both gametes implying that they are produced in a 1:1 ratio.

tory book such as this one. In Mendel's time such tests had not yet been developed and Mendel circumvented the problem by performing many more crosses than we need to perform today. He must also have exerted a conscious or unconscious selective bias in collecting his data, as was demonstrated by the statistician, Fisher, who calculated that Mendel's ratios were better than could be expected on the average. The dependence of genetic experimentation on statistical techniques is, of course, due to the random nature of the meiotic and fertilization processes, a fact which has disturbed some scientists ever since Mendel published his article. Thus even in the 1940s and 1950s, Lysenko and some of his followers in the Soviet Union argued against the validity of Mendelian genetics, because it is based on a random process requiring statistical methods which Lysenko thought were of dubious validity for the performing of inferences. However, Lysenkoism has all but disappeared from the Soviet Union, and it is likely that in this case the dispute was political rather than scientific. Today an overwhelming majority of the scientific community accepts **the law of gene segregation** enunciated by Mendel and the resultant implication of discreteness of genes and the non-blending nature of inheritance.

GENES COME IN GROUPS (INDEPENDENT ASSORTMENT AND LINKAGE)

We have learned that gene pairs, or alleles, segregate independently from one another and we now understand that this is a natural consequence of (1) the process of meiosis which separates homologous chromosomes, putting them into different gametes and of (2) fertilization, which is a random process recombining homologous chromosomes and, consequently, alleles according to the laws of chance.

The next question which Mendel asked was whether or not the segregation of one gene pair is independent of the segregation of another gene pair. If it is independent, it should be possible to predict the outcome of the results just by combining the frequencies of the segregating gene

pairs themselves. Let us, then, take two pairs of characters of pea seeds studied by Mendel:

Y for *yellow seed,*
   which is dominant to y for *green seed*

R for *round seed,*
   which is dominant to r for *wrinkled seed.*

By now we can predict what happens when a yellow, round, seed-producing parent is crossed with a green, wrinkled seed-producing parent.

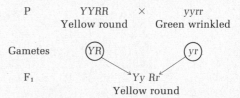

P   YYRR   ×   yyrr
          Yellow round   Green wrinkled

Gametes   YR   yr

F₁   Yy Rr
                       Yellow round

All the F₁ are yellow, round seed producers and they all have the genotype *YyRr.* The two F₁ parents, however, each produce four kinds of gametes in equal frequency and we can predict from their frequency what the frequency of the different classes of offspring should be. Let us first look at the classes of different phenotypes produced. If gene pairs are going to **assort independently,** then if we expect ¾ of the offspring to be yellow and ¾ to be round, we expect the *product* of their frequencies, or ⁹/₁₆, to be both round and yellow. Similarly, if we expect ¾ to be yellow and ¼ to be wrinkled, then we expect ³/₁₆ to be both yellow and wrinkled. The dichotomous diagram in Table 3-1 is a convenient device for stating all the possibilities and multiplying out all the frequencies to obtain the final frequencies of the phenotypes.

Table 3-1   Frequencies of phenotypes in a two factor cross

Frequency of yellow-green phenotypes	Frequency of round-wrinkled phenotypes	Frequency of combined phenotypes
¾ yellow	¾ round	⁹/₁₆ Yellow round
	¼ wrinkled	³/₁₆ Yellow wrinkled
¼ green	¾ round	³/₁₆ Green round
	¼ wrinkled	¹/₁₆ Green wrinkled

Thus we obtain the famous 9:3:3:1 ratios, which in the case of the above-mentioned seed characters, Mendel found to be

315 Yellow round
108 Green round
101 Yellow wrinkled
32 Green wrinkled

The independent assortment of pairs of different genes Mendel found to hold true for all seven of his characters. We now know this to be either a miraculous "fluke" or the result of a deliberate choice of the seven characters by Mendel. However, before going into this any further, let us look at the distribution of *genotypic* classes. Here also the flow sheet we have used before will simplify our labors (Table 3-2).

We see that we obtain nine different genotypic classes (1:2:1:2:4:2:1:2:1) and that they group themselves into the four different phenotypic classes (9:3:3:1). Most textbooks develop this subject by multiplying out all 16 fertilizations involving the four male and four female gametes, producing 16 organisms which then have to be collated into the nine genotypic classes and four phenotypic classes. This is an immensely laborious procedure for general use. However, to give the student some practice we reproduce the **Punnett Square** (Fig. 3-5) with most squares blank and invite the student to fill them in, summarizing his results in Table 3-3.

Let us now think of independent assortment of two pairs of alleles in relation to what we know about meiosis and fertilization. Mendel demonstrated that the seven factor pairs he studied *all* assorted independently from one another. We would expect this to happen if all the seven pairs of characters were situated on different pairs of chromosomes, for we have seen in our description of meiosis that chromosome pairs behave independently of one another, so that a parent $AaBb$ indeed can produce all four kinds of gametes (AB, Ab, aB, and ab) provided genes A and B are situated on different chromosomes. Suppose, however, that the gene pairs A,a and B,b are situated on the same pair of homologous chromosomes. Would we still get independent assortment of the gene pairs?

Let us assume that we have an individual in which the paternal chromosome carries the genes A,B and the maternal chromosome carries the genes a,b. If we assume that no crossing over or chromosomal exchange takes place during meiosis, then this individual will produce only two kinds of gametes with respect to these two genes. We symbolize linked genes by placing them above each other, which allows us to illustrate which two genes are on one (let us say paternal) chromosome and which two genes are on the other, or maternal chromosome:

$$F_1 \qquad \frac{AB}{ab} \qquad \times \qquad \frac{AB}{ab}$$

Table 3-2   Frequencies of genotypes in a two factor cross

Frequencies of Y,y genes	Frequencies of R,r genes	Frequencies of genotypes	Frequencies of phenotypes
$\frac{1}{4}$ YY	$\frac{1}{4}$ RR	$\frac{1}{16}$ YYRR	
	$\frac{2}{4}$ Rr	$\frac{2}{16}$ YYRr	
	$\frac{1}{4}$ rr	$\frac{1}{16}$ YYrr	$\frac{9}{16}$ Yellow round
$\frac{2}{4}$ Yy	$\frac{1}{4}$ RR	$\frac{2}{16}$ YyRR	
	$\frac{2}{4}$ Rr	$\frac{4}{16}$ YyRr	
	$\frac{1}{4}$ rr	$\frac{2}{16}$ Yyrr	$\frac{3}{16}$ Yellow wrinkled
$\frac{1}{4}$ yy	$\frac{1}{4}$ RR	$\frac{1}{16}$ yyRR	$\frac{3}{16}$ Green round
	$\frac{2}{4}$ Rr	$\frac{2}{16}$ yyRr	
	$\frac{1}{4}$ rr	$\frac{1}{16}$ yyrr	$\frac{1}{16}$ Green wrinkled

Gametes   *AB   ab*   *AB   ab*

F_2 $\frac{1}{4}\,\dfrac{AB}{AB}$ $\frac{1}{4}\,\dfrac{AB}{ab}$ $\frac{1}{4}\,\dfrac{ab}{AB}$ $\frac{1}{4}\,\dfrac{ab}{ab}$

Or, since it does not matter whether *AB* is on the maternal or paternal chromosome, we have

$\frac{1}{4}\,\underbrace{\dfrac{AB}{AB}\qquad \frac{2}{4}\,\dfrac{AB}{ab}}_{3/4}\qquad \frac{1}{4}\,\dfrac{ab}{ab}$ Genotypic ratio

$\underbrace{\qquad\qquad}_{3/4}\qquad\qquad \frac{1}{4}$ Phenotypic ratio

Thus if no chromosomal exchange takes place, we would expect gene pairs which are **linked** (that is, situated on the same chromosome) to segregate together and give us simple $3/4{:}1/4$ ratios. In Mendel's study no two pairs of factors turned out to be linked to each other in this way and since, as was found later, the pea has only seven pairs of chromosomes, it is most unlikely that he could have stumbled on these precise seven pairs of genes by chance alone. It is most likely that Mendel had the law of independent assortment in mind and selected for study those seven pairs that did assort independently. This is not surprising given Mendel's strength of conviction regarding the regularity of the laws of hered-

ity and his need to overcome the resistance of his contemporaries against the simple numerical ratios he championed.

We conclude that the law of independent assortment as stated by Mendel is not a general law, because it applies only to gene pairs situated on *different* chromosome pairs. However, this law played a very important role in developing the idea, since it was necessary to understand the special case of independent assortment before one could *generalize* on it by introducing the ideas of linkage and crossing over.

The law of independent assortment started breaking down as a general principle almost as soon as Mendel's laws were rediscovered in 1900 when Correns, one of the three biologists who rediscovered Mendel's article, observed two factor pairs which tended to stay linked to each other in the F_2. Another study in 1902 by Bateson and Punnett involving two gene pairs in the sweet pea (*P* for purple flower, *p* for red flower; and *L* for long pollen grain, *l* for round pollen grain) showed that the F_2 gave instead of the customary 9:3:3:1 ratios the following:

Purple long	Purple round	Red long	Red round
1528	106	117	381

The original cross had been made between purple long *PL/PL* and red round *pl/pl* parents. The results therefore give an excess of **parental** types and an insufficiency of **recombinants** (purple round and red long). These results make

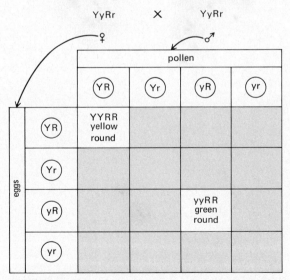

Figure 3-5 Punnett Square of two factor cross to be filled in by student who should add up the frequencies in the different squares, fill in Table 3-3, and confirm the frequencies of genotypes and phenotypes listed in Table 3-2.

Table 3-3   Summary of results gathered from Punnett Square (to be filled in by student)

Genotypes	Frequency	Phenotypes	Frequency
YYRR	1/16	} Yellow round	————
YYRr	————		
YyRR	————		
YyRr	————		
YYrr	————	} Yellow wrinkled	————
Yyrr	————		
yyRR	————	} Green round	————
yyRr	————		
yyrr	————	} Green wrinkled	————

it hard to analyze mathematically precisely what proportion of the offspring remained linked to each other and what proportion recombined. This is because in the F_1 *both* parents produced gametes which were the results of recombination. The situation can be simplified if the F_1 is *back crossed* to the homozygous recessive in which, if recombination occurs between identical gene pairs, no effect on the results will be observed. Let us, for instance, take two factor pairs in corn (see diagram below).

Examination of this back cross shows why the frequency of the progeny of the back cross is equal to the frequency of the gametes produced by the F_1 parent, since the homozygous recessive parent produces only *cs* gametes. This therefore allows the calculation of recombination frequency during the meiosis that produces the gametes in the F_1.

T. H. Morgan, the great American geneticist, was the first to analyze results such as these and conclude that the excess of parental types is due to two particular genes being located on the same chromosome. He suggested that the appearance of recombinants was due to a chromosomal exchange that occurs when the site of exchange or chromosomal break develops somewhere *between* the two particular genes being studied. Remember that when we discussed meiosis we pointed out that chromosome breakage followed by exchange occurs in two or four chromatids or **tetrads.** Usually the exchange between two nonsister chromatids results in two reciprocally recombinant chromosomes. This event is depicted in Figure 3-6.

Morgan's important hypothesis that linkage is due to genes being situated on the same chromosome and that recombination is due to cross-

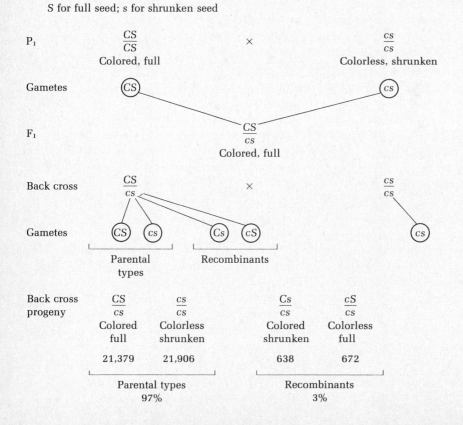

C for colored seeds; *c* for colorless seeds
S for full seed; *s* for shrunken seed

ing over (or chromosome exchange) has been tested in many different ways.

1. Quantitative studies of recombination frequencies (using genetic data) and crossing over frequencies (using cytological data) were performed and were found to exhibit the 1:2 frequency we would predict from the preceding analysis.
2. Very ingenious experiments were performed, using special strains of the fruit fly, *Drosophila melanogaster,* in which chromosomes with particular abnormalities can be found. These abnormalities are such that they can be seen under the microscope, thus allowing homologous chromosomes to be distinguished and a

crossover event to be recognized. Therefore, by using the proper genes or **genetic markers,** it can be demonstrated that in an organism in which the genes have recombined, crossing over will also have taken place between the homologous chromosomes. These experiments are somewhat lengthy to describe, but students who are interested may want to look up in genetics textbooks Stern's experiments with the fruit fly, or Creighton and McClintock's experiments with corn (see *Genetics,* by R. P. Levine).
3. By collecting sufficient genes and performing sufficient back crosses between any two genes, it is possible to arrange them into **linkage groups.** It turns out that these groups are: (a) mutually exclusive (that is, if A is linked with B and C is linked with D, but A is not linked with C, then A will not be linked with

Figure 3-6   Crossing over and recombination during meiosis.

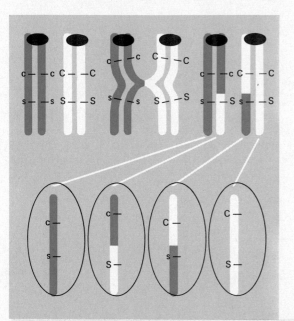

A. Crossing over (exchange) of chromosomal material and recombinations of genes occurring during meiosis of the F_1 parent in a test cross. This F_1 corn plant (*CS/cs*) was produced by crossing a *CS/CS* plant with a *cs/cs* plant, resulting in an organism which carries in all its cells a CS chromosome from one parent and a cs chromosome from the other parent. If, in a particular meiotic division, recombination occurs between the *C* and *S* loci, then two out of the four gametes produced will be recombinants.

B. Homologous chromosomes during late prophase of the first meiotic division of a salamander spermatocyte. It is clear that only one chromatid from each chromosome is involved in the chiasma at the right.

D); (b) finite in number; and, (c) equal in number to the number n, where n is the number of chromosomes in a haploid set (Table 3-4).

Table 3-4 Correspondence between number of linkage groups and number of chromosomes in a variety of organisms

Organism	Number of linkage groups	Number of chromosomes in haploid set
Garden pea	7	7
Drosophila melanogaster	4	4
Drosophila pseudoobscura	5	5
Corn	10	10

We shall encounter many more demonstrations of the chromosome theory of inheritance, but before we do we look more closely at the question of how genes interact with one another to produce the characters we observe when we perform the crosses.

GENES INTERACT WITH EACH OTHER

Mendel's work defined a gene pair or allelic genes as factors which, when crossed, segregate to form only *two* phenotypes in a simple 3:1 ratio. If four phenotypes are produced in a 9:3:3:1 ratio, Mendel concluded that he was looking at two sets of genes. We have just shown that two sets of genes on the same chromosome may still give us four phenotypes, but in different ratios. It soon turned out that simple relationships between gene and character were the exceptions rather than the rule. Fortunately, Mendel and his early followers did select the simple cases since, as it happens frequently in the development of science, the difficult cases are special modifications of the simpler ones and cannot be understood without first understanding the simpler ones. In retrospect, now that we know a great deal regarding the action of genes and the complexity of metabolism and development, it is surprising that so many simple relationships

exist between genes and complex characters, such as plant size, seed color, and fruit shape. Most genes (and in higher organisms there may be as many as 100,000 in each gamete) have a complicated effect on the phenotypic characters that they affect. The following list of relationships is simply illustrative of the types of modifications of the Mendelian factors one can observe.

Manifold effects of genes

It was already clear to Mendel that a single gene pair may affect more than one character. The gene pair affecting flower color, for instance, also affected seed color and the color of a region in the axil of a leaf. In the fruit fly, for instance, the vestigial gene, vg, not only reduces the wing size but also alters certain bristles and decreases the fly's fecundity and life expectancy.

Incomplete dominance

In many instances the effect of two dominant genes is more pronounced or, in any case, different from the heterozygous state. In the case of a number of plants with red flowers, the heterozygous state is pink and the homozygous recessive white. In man, for instance, a certain recessive gene, when homozygous, is responsible for the fatal anemia, thalassemia major, whereas in the heterozygote a less fatal anemia (thalassemia minor), recognizable by the shape of the red blood cells, is produced.

Furthermore dominance can be modified in a variety of ways as, for instance, the gene for baldness in man is dominant in males and recessive in females.

Multiple alleles

It soon became evident after the rediscovery of Mendel's article that there can be more than two alleles for a gene at a given **locus.** In diploid cells, of course, there can only be two alleles (one on each chromosome) in any given cell, but there can be more than two kinds of genes at a certain locus occurring in a population of a certain species. In man, for instance, there are the three alleles, I^A, I^B, and i, each of which produce a certain kind of antigen on the surface of red blood cells. (**Antigens** elicit the production of **antibodies** when injected into the blood of a vertebrate. These antibodies, once produced, will precipitate the antigens which had elicited their production. Antibodies formed against antigens present on the surface of human red blood cells will **agglutinate** them, that is, cause them to clump together.)

An antigen normally found in a mammalian cell will not elicit antibody production from the organism to which the cell belongs, which is to be expected because if this occurred, the organism would agglutinate its own red blood cells. In the case of the blood groups, a man with a genotype $I^A I^A$ does, however, have antibodies which agglutinate the red blood cells of a person with genotype $I^B I^B$, but do not agglutinate red cells of a person having genotype ii. The whole set of relationships among genotype, antigen, antibody, and agglutination reactions is shown in Table 3-5. These relationships are of considerable practical importance, because they determine what kind of blood an individual can receive in a blood transfusion; the general rule is that one must avoid the agglutination of the donor's cells by the antibodies of the recipient. The antibodies of the donor do not have any significant agglutinating effect on the red cells of the recipient, since the donor's blood gets diluted considerably.

Notice that an individual in the AB blood group will not agglutinate the blood of any donor (universal receiver), whereas the red blood cells of an individual in the O blood group will not be agglutinated by any receiver (universal donor).

Other blood group genes exist in addition to those of the ABO system. One of importance is the Rh system, in which the serum of the mother may agglutinate the red cells of the fetus she carries—an abnormality which may be treated by exchanging the baby's blood immediately after birth and by other methods (p. 383).

Because of the presence of a wide variety of blood group systems, it has become possible to determine the most *probable* father of a baby from a group of *possible* fathers.

We have looked at the ABO system in detail, because not only does it exemplify the phenomenon of multiple alleles but also because it is of general and practical importance. Many other series of multiple alleles have been observed, in some cases as many as a dozen.

Interaction between alleles

Mendel's work demonstrated that genes are inherited as discrete or separate entities. This does not mean, however, that they do not interact in the formation of the phenotype. We have already encountered incomplete dominance, which is one kind of interaction. There are other kinds of interactions between alleles, such as **lethals.**

Table 3-5 Relationship between genotype, phenotype, and agglutination reaction in human blood groups—a case of multiple alleles

Genotype	Antigen on red cells	Phenotype (blood group)	Antibody in blood serum	Agglutinate cells of individual with blood type
$I^A I^A$	A	A	Anti-B	B, AB
$I^A i$	A	A	Anti-B	B, AB
$I^B I^B$	B	B	Anti-A	A, AB
$I^B i$	B	B	Anti-A	A, AB
$I^A I^B$	A, B	AB	None	None
ii	None	O	Anti-A and Anti-B	A, B, AB

A certain strain of yellow mice when bred gives ⅔ yellow and ⅓ black offspring. The black mice will breed true, but all the yellow offspring will always give the 2 yellow to 1 black ratio. This result seemed to be completely at odds with the Mendelian ratios, until it was discovered that some of the mice had died at an early stage during embryonic development and that these dead embryos represented one-quarter of the progeny of a yellow × yellow cross. We can therefore explain the results as follows:

A^y is the dominant gene for yellow.

a is the recessive gene for nonyellow or black.

Parents	$A^y a$	×	$A^y a$	
	Yellow		Yellow	
Progeny	¼ $A^y A^y$	½ $A^y a$		¼ aa
	Lethal	Yellow		Black
	(die before birth)	Live		Live

We have here, then, an interaction between alleles in which the homozygous dominant class is lethal and the homozygous recessive is black, but the heterozygote is yellow, an entirely new effect, which is the result of the interaction between the dominant and the recessive alleles.

Interaction between sets of different alleles

Any complex biological character, such as height, shape of nose, resistance to a certain disease, or intelligence, must be the result of the interaction between the environment and a large number of genes. The precise genetic basis of such interactions has, of course, not yet been determined, but several examples involving a few gene pairs have been worked out. A classical example involving only two gene pairs is that of the shape of the comb in fowl, which is determined by two genes R and P. R ( ) gives a "rose" comb and P ( ) produces a "pea" comb, but we can also have "walnut" and "single" combs, depending on the combination of genes. In the following cross, empty parentheses ( ) are used to symbolize that either a dominant or recessive gene could be involved.

P_1	$RRpp$		$rrPP$
	Rose		Pea
F_1		$RrPp$	
		Walnut	

F_2

¾ R ( )
— ¾ P ( ) → ⁹⁄₁₆ R ( ) P ( ) walnut
— ¼ pp → ³⁄₁₆ R ( ) pp rose

¼ rr
— ¾ P ( ) → ³⁄₁₆ rr P ( ) pea
— ¼ pp → ¹⁄₁₆ rr pp single

In this case we still have the expected number of four phenotypic classes generally encountered in a two-factor cross. Many cases are known in which two or more phenotypic classes look the same, so that ratios such as 9:6:1, or 15:1, or 12:3:1, or 12:4 are obtained. The interesting thing about these ratios is that they would have been unexplainable before the Mendelian ratios of 9:3:3:1 had been established. Again we see how important it is that Mendel chose to study the simple cases first.

Not all cases are as simple as those used here, and the methods of inheritance of most complex characters in higher organisms have yet to be studied and understood. We give the following example only to illustrate one of the more complex characters, such as the inheritance of coat color in mice in which the role of six interacting gene pairs has been worked out (Table 3-6).

Modifying genes

At times, instead of having a number of genes influencing a certain character to a similar extent, we have a certain gene pair having the major influence and a large, sometimes undetermined number of genes having a smaller or modifying effect. Thus in mice, for instance, the W gene is lethal when homozygous. When heterozygous, it produces white spotting. The amount of white spotting, however, depends on a number of modifier genes that determine the amount of white spotting in the heterozygote (Ww), ranging from a black mouse with a few white spots all the way to an almost completely white mouse.

Quantitative inheritance

In pre-Mendelian times students of heredity studied characters with a *continuous* form of heredity in which the phenotypes blended into each other. They did so because they considered them to be biologically or commercially important. In retrospect we now know that they were too complicated to study genetically at the time. Since then, however, it has been possible to analyze examples of so-called continuous or blending inheritance and in the simpler cases to specify the number of gene pairs involved.

One of the simpler cases studied was that of the color of the kernels of wheat. When a red kerneled strain was crossed with a white kerneled one, an F_1 of intermediate color was obtained. In the F_2 a very large amount of almost *continuous* variation was obtained that ranged from the red parent on one hand to the white parent on the other. The variability was so continuous that it was almost impossible to divide

Table 3-6   Interaction of genes for coat color in mice

Genes					Gametic formula	Phenotype
C	A	B	D	P	S	CABDPS	Wild-type agouti
					s	CABDPs	Spotted agouti
				p	S	CABDpS	Pink-eyed agouti
					s	CABDps	Pink-eyed, spotted agouti
			d	P	S	CABdPS	Dilute agouti
					s	CABdPs	Spotted, dilute agouti
				p	S	CABdpS	Pink-eyed, dilute agouti
					s	CABdps	Pink-eyed, spotted, dilute agouti
		b	D	P	S	CAbDPS	Cinnamon
					s	CAbDPs	Spotted cinnamon
				p	S	CAbDpS	Pink-eyed cinnamon
					s	CAbDps	Pink-eyed, spotted cinnamon
			d	P	S	CAbdPS	Dilute cinnamon
					s	CAbdPs	Spotted, dilute cinnamon
				p	S	CAbdpS	Pink-eyed, dilute cinnamon
					s	CAbdps	Pink-eyed, spotted, dilute cinnamon
	a	B	D	P	S	CaBDPS	Black
					s	CaBDPs	Spotted black
				p	S	CaBDpS	Pink-eyed black
					s	CaBDps	Pink-eyed, spotted black
			d	P	S	CaBdPS	Dilute black
					s	CaBdPs	Spotted, dilute black
				p	S	CaBdpS	Pink-eyed, dilute black
					s	CaBdps	Pink-eyed, spotted, dilute black
		b	D	P	S	CabDPS	Brown
					s	CabDPs	Spotted brown
				p	S	CabDpS	Pink-eyed brown
					s	CabDps	Pink-eyed, spotted brown
			d	P	S	CabdPS	Dilute brown
					s	CabdPs	Spotted, dilute brown
				p	S	CabdpS	Pink-eyed, dilute brown
					s	Cabdps	Pink-eyed, spotted, dilute brown
c with any other genes						c	Albino, may be of any of 32 genotypes above

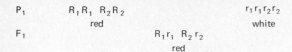

<table>
</table>

	R_1R_2	R_1r_2	r_1R_2	r_1r_2
R_1R_2	$R_1R_1R_2R_2$ red	$R_1R_1R_2r_2$ red	$R_1r_1R_2R_2$ red	$R_1r_1R_2r_2$ red
R_1r_2	$R_1R_1R_2r_2$ red	$R_1R_1r_2r_2$ red	$R_1r_1R_2r_2$ red	$R_1r_1r_2r_2$ red
r_1R_2	$R_1r_1R_2R_2$ red	$R_1r_1R_2r_2$ red	$r_1r_1R_2R_2$ red	$r_1r_1R_2r_2$ red
r_1r_2	$R_1r_1R_2r_2$ red	$R_1r_1r_2r_2$ red	$r_1r_1R_2r_2$ red	$r_1r_1r_2r_2$ white

Figure 3-7 Quantitative inheritance of multiple factors in wheat. Two different alleles (R_1 and R_2) cause the wheat kernel to be red. As the above diagram shows, the depth of color is related to the number of dominant genes present, $R_1R_1R_2R_2$ being deepest red while $R_1r_1r_2r_2$ or $r_1r_1R_2R_2$ are lightest. The homozygous recessive ($r_1r_1r_2r_2$) is white.

the offspring into classes. However, by performing large numbers of crosses, it was possible to show that the white and red parental phenotypes occurred in $1/64$ of the total offspring in the F_2. This shows that *three* independent genes determining kernel color are involved. A simpler case involving only two of these three genes affecting the kernel color of wheat is illustrated in Fig. 3-7. It also illustrates how complicated it becomes numerically to analyze cases of *quantitative* inheritance involving larger numbers of genes. In recent years statistical techniques have been developed to analyze such cases and estimate the number of genes affecting a certain quantitative character. These estimates are based on the fact that as larger and larger numbers of genes are involved, the distribution of the F_2 becomes narrower and narrower; that is, a smaller proportion of parental types appears (Table 3-7).

However, quantitative inheritance can be recognized easily, even when large numbers of genes are involved. Notice in the previous cross that the F_1, although intermediate between the parents, shows little variation, whereas in the F_2, variation occurs from one parental type all the way to the other. This phenomenon had been observed before Mendel's time, but could, of course, not be explained in a simple manner. Note also that the variation in the F_1 is an indication of the effect of the environment on the quantitative character, since all the individuals in the F_1 are presumably of the same genotype.

Table 3-7 Frequency distributions expected in F_2 populations from crosses of parents differing in a dimensional trait by 12 units when the difference in the parents is determined by one, two, three, and six pairs of independent genes, with equal additive effects.

Gene pairs	P₁	Class centers F₁												P₂	n
	101	102	103	104	105	106	107	108	109	110	111	112	113		
1	1,024						2,048						1,024	4,096
2	256			1,024			1,536			1,024			256	4,096
3	64		384		960		1,280		960		384		64	4,096
6	1	12	66	220	495	792	924	792	495	220	66	12	1	4,096

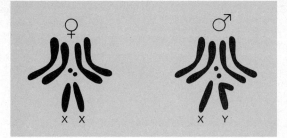

Figure 3-8 The chromosome complement of the *Drosophila* female and male. Notice that the only difference is in the X and Y chromosomes, the female being XX and the male XY.

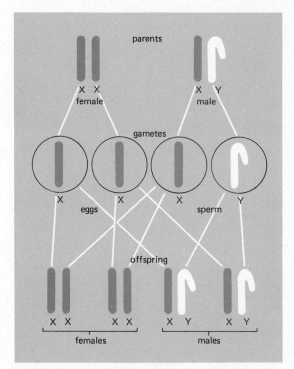

Figure 3-9 Sex determination in the fruit fly, *Drosophila melanogaster*. Notice that the male offspring receive the Y chromosome of the male parent and the female offspring receive the X chromosome of the male parent. Of course, the female parent can provide only an X chromosome and, therefore, it is the contribution of the male parent which determines the sex of the offspring.

CHROMOSOMES DETERMINE SEX

If we look at the chromosomes of a male fruit fly and compare them with those of the female fly, we discover that there is a definite difference in the chromosomes (Fig. 3-8). The male chromosome, called the Y chromosome, has a kink in it whereas the female chromosome, called the X chromosome, is straight. An examination of the eggs and sperm of the fruit fly shows that the egg carries an X chromosome, and the sperm carries either an X or a Y chromosome. It would appear therefore that sex determination depends on whether an X-containing sperm or a Y-containing sperm fertilizes the X-containing egg (Fig. 3-9).

It is clear from the diagram in Figure 3-9 why the sex ratio should be approximately 1:1. Sex determination appears to be a function of the balance between genes present on the X chromosome and the other chromosomes or **autosomes.** Thus, for instance, if a fly has the Y missing accidentally (XO), it is a male, and if a fly happens to by XXY, it is a female. A fly which is XXX is called a *superfemale*, because its female secondary sexual characteristics are accentuated, although its fertility is impaired.

The XY system of sex determination is very widespread, occurring in many invertebrates, some fish, most plants, and mammals. However, the question as to which chromosomes are involved in the sex-determining balance can vary in XY systems. In man it is the presence of the Y chromosome which determines maleness and its absence, femaleness. Thus the infrequently occurring XO human is a female and the XXYs are males.

In other organisms the female is XY and the male XX, and in yet others, such as the domestic fowls, the female is XO and the male XX. This interesting cytological phenomenon of sex determination was combined with the results of genetic experiments performed by T. H. Morgan and his co-workers to yield fascinating results which led to a coherent theory of the relationship between genes and chromosomes.

Mendel had pointed out that the results of his breeding experiments were unaffected by the direction of the cross. It did not matter whether the pollen of a white-flowered plant was placed on the stigma of a red-flowered plant or vice versa. But in 1910, Morgan, Sturtevant, Muller, and Bridges reported that when a red-eyed female was mated to a white-eyed male different results were obtained than if the cross was performed in the opposite direction. Furthermore within each cross different results were obtained in the males than in the females. Let us use the word *red* for the red-eyed flies and the word *white* for the white-eyed flies. Then

Cross I

P	Red ♀	×	White ♂
F₁	Red ♀ and red ♂		
F₂	½ Red ♀	¼ Red ♂	¼ White ♂

Cross II

P	White ♀	×	Red ♂
F₁	½ Red ♀		½ White ♂
F₂	¼ White ♀	¼ Red ♀	¼ White ♂ ¼ Red ♂

Morgan and his co-workers explained the results of these crosses by assuming:

1. That the gene pair for red- or white-eye color was situated on the X chromosome, but
2. Was absent from the Y chromosome. (Further studies revealed that there are very few genes present on the Y chromosome of *Drosophila*.)

These assumptions explain the results of both crosses shown in Figure 3-10. White eye is recessive so that *WW* and *Ww* are red-eyed and *ww* is white-eyed.

This important experiment did not, of course, *prove* that genes are located on the chromosomes. What the experiment did show was that by assuming that certain genes are located on particular chromosomes, it is possible to explain the results in terms of Mendelian theory, even though the results at first appear to be at variance with that theory.

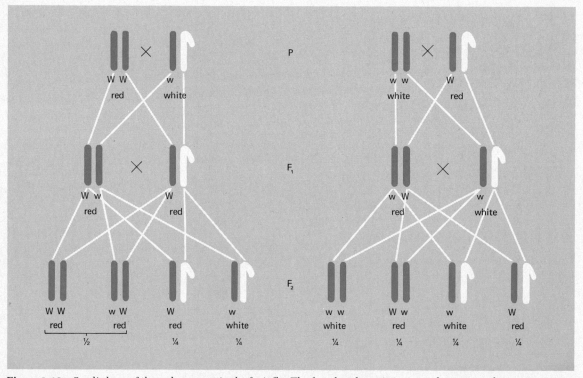

Figure 3-10   Sex linkage of the red eye gene in the fruit fly. The fact that the ratios are not the same in the two sexes can be explained by assuming that the W,w alleles are located on the X chromosome but missing from the Y chromosome.

Many other sex-linked genes have since been found in the fruit fly and other organisms. They are easily detected because of the difference in ratios found between the sexes. In man some 40 sex-linked genes are known, including the genes for color blindness and hemophilia—both of which are recessives. A gene for hemophilia, which can be traced to Queen Victoria, found its way to many royal houses in Europe, including that of the last Russian czar, and it can be argued that this gene had an important influence on modern European history.

There is a misconception about sex-linked abnormalities in man, which is that the male shows the abnormality and the female is the carrier. This appears to be the case because the allele causing the abnormality is infrequent in the population. Therefore the probability of the female having both genes is low, occurring only if a female carrier and a male with the disease have offspring.

GENE GROUPS ARE ARRANGED IN LINEAR ORDER (MAPPING)

We have already learned that there is a quantitative relationship between crossover frequency as determined by direct microscopic observation and recombination frequency as measured by genetic experiments. There must therefore be a physical meaning for recombination frequency or, to put it in the form of a question, "What could the physical reason be that genes A and B, for instance, have a high recombination frequency and genes B and C a low recombination frequency?" Morgan, as early as 1911, suggested that the recombination frequency between two genes must be related to the *distance between the genes on the chromosome;* the lower the recombination frequency, the closer they are to each other. Sturtevant, who was at the time an undergraduate student of Morgan's, concluded that if this is correct, it should be possible to construct genetic maps indicating the relative position of genes on their respective chromosomes.

The procedure developed for genetic mapping is simple. We take three genes known to be linked and carry out a back-cross. Recombination frequencies are calculated and, using these as an estimate of the physical distance, the genes are arranged in linear order. The exciting thing about this procedure is that once this is done no qualitative discrepancies are known to occur. That is, if A and B are close, and B and C are far apart, then we can construct two possible maps:

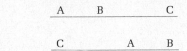

or

But a further measurement of recombination frequency between A and C will be able to distinguish between these possibilities and, if it turns out that A and C are farther apart than B and C, then the map

becomes *uniquely* established. By uniquely we mean that when crossover frequencies to other genes are measured, the results are not at variance with the sequence A, B, and C we had previously established. By studying additional genes (or "markers") we simply add more points on our *linear* map. This result may seem self-evident, but to the scientific workers at the beginning of the century it must have been an exciting revelation that these discrete units of action we call genes are in fact arranged in a definite and linear relationship to one another. The linear order of genes has remained an important generalization of biological structure. In recent years biologists have discovered bacterial chromosomes which are circular, rather than open-ended. Although the linear order theory applies here as well, it should be noted that with such chromosomes recombination frequencies are obtained which may seem anomalous unless we take the circularity into account. What it amounts to is that in a circular chromosome any pair of genes has two distances between them—a long distance and a short distance (unless they are exactly opposite one another on the circle), and we have to be clear which distance we are measuring.

The following is an example of a typical mapping experiment. Three genes in the fruit fly, *Drosophila melanogaster,* are used. All three are recessive, sex-linked mutations.

$$+ = \text{normal}, cv = \text{cross veinless}$$
$$+ = \text{normal}, ec = \text{echinus}$$
$$+ = \text{normal}, ct = \text{cut}$$

In order to perform a back cross the proper heterozygous females had to be created by crossing a cross veinless fly with an echinus cut fly. Note that since these genes are located on the X chromosomes, the male flies are haploid for them.

P $\dfrac{cv ++}{cv ++}$ × $+ ec\ ct$

 ♀   ♂

cross veinless   echinus, cut

F₁ $\dfrac{cv + +}{+ ec\ ct}$ $cv\ ec\ ct$

♀   ♂

cross veinless,
echinus, cut

The $\dfrac{cv + +}{+ ec\ ct}$ female was back crossed to a $cv\ ec\ ct$ male. The male produces only one kind of gamete, but the female produces eight possible gametes, with the frequencies depending on the recombination events occurring among the three pairs of genes. The nice thing about doing a back cross with the *homozygous* recessive male (carrying only one X chromosome) is that it will make any recessive gene carried by the female homozygous in some of the offspring and therefore recognizable phenotypically. The following are the results of such an experiment. For the purposes of brevity, the genotypes of the maternal gametes (for example, $cv + +$) are noted rather than of the diploid cells $\left(\text{for example, } \dfrac{cv + +}{cv\ ec\ ct}\right)$.

The results of the back cross experiment are

Parental types

cross veinless (cv ++)	2,207	
echinus, cut (+ ec ct)	2,125	81.5%

Recombinants

cross veinless, echinus (cv ec +)	273	
cut (++ ct)	265	10.1%
echinus (+ ec +)	217	
cross veinless, cut (cv + ct)	223	8.3%
normal (+++)	5	
cross veinless, echinus, cut (cv, ec, ct)	3	0.1%

Let us examine the above results:

1. The parental or noncrossover types can be recognized easily because they are the largest class of flies. Furthermore we can recognize them phenotypically, since we know that the maternal flies were $\dfrac{cv + +}{+ ec\ ct}$.

2. Double crossovers can be identified by the fact that they are lowest in frequency since the frequency of two infrequent events occurring together is lower than each separate event. This also allows determination of the order of the genes because, in a double crossover, it is only the *middle gene which changes in position.* By comparing the double crossovers (+++ and $cv\ ec\ ct$) with the parental types ($cv ++$ and $+ ec\ ct$) we can see that cv changes in position, but ec and ct remain together which places cv between ec and ct. Therefore, the maternal flies can now be represented as $\dfrac{+ cv +}{ec + ct}$. Furthermore, we can now determine the relative distance between the genes by placing the genes in the four remaining classes of recombinants in the proper order and noting where and how frequently the crossovers occur.

3. Because we know the gene order, we know that $cv\ ec\ +$ is really $ec\ cv\ +$ and $++ ct$ is $++ ct$. By comparing these two sets of recombinants with the maternal flies, it is clear that $ec\ cv\ +$ and $++ ct$ flies represent crossovers between ec and cv (10.1%).

4. In a similar manner, $+ ec\ +$ is really $ec ++$ and $cv + ct$ is $+ cv\ ct$ when the genes are in the proper order. Therefore, by comparison with the maternal flies, $ec ++$ and $+ cv\ ct$ represent crossovers between cv and ct (8.3%).

5. Crossovers between genes ec and ct occur in all four classes; $cv\ ec\ +$, $++ ct$, $+ ec\ +$, and $cv + ct$ (18.4%).

6. Finally, let us look at the quantitative relationships. Crossover distances are expressed in terms of recombination frequencies. The frequencies, including the double crossovers, are

Between ec and cv	10.1 + 0.1 = 10.2
Between cv and ct	8.3 + 0.1 = 8.4
Between ec and ct	10.1 + 8.3 = 18.4

This gives us the following linkage map:

$$\underset{\longleftarrow\!-\!-\!-\!-\!18.4\!-\!-\!-\!-\!\longrightarrow}{\overset{ec \quad 10.2 \quad cv \quad 8.4 \quad ct}{}}$$

Notice that the distance between *ec* and *ct* is 18.4 when the recombinations are counted directly, but 18.6 when the distances *ec–cv* and *cv–ct* are added up. The reason for this is that whenever we get a double crossover between any two genes, we count them as nonrecombinants. Thus the number of recombinations observed are less than the number of crossovers, and as this discrepancy gets larger, the distance between given genes increases. We therefore obtain a more accurate measure of distance between genes if we add up a number of short distances than if we measure directly the recombination frequencies between two genes placed far apart from one another. It was also found, however, that genes which are *very* close to each other recombine less frequently than predicted by their distance from one another. This phenomenon, called **interference,** simply means that the occurrence of a crossover at a certain point of a chromosome reduces the frequency of another crossover occurring close by. There are probably good geometrical reasons involving the structure of the chromosome for the phenomenon of interference.

We now know how to construct maps of the linear order of genes on a chromosome. Is it possible to identify the chromosome on which a particular group of linked genes is situated? In the case described this is easy, since the genes we examined are sex-linked and the appearance of the X chromosome, as we have seen, is known through the phenomenon of sex determination. Connecting linkage groups to other specific (microscopically identifiable) chromosomes has been a more arduous but entirely feasible task.

The main tools for genetic identification of chromosomes have been the use of chromosomal abnormalities and, in the case of insects, the use of the giant salivary gland chromosomes.

By irradiating organisms with X-ray, or by allowing mutations to occur spontaneously, a number of different abnormalities of the chromosomes can be obtained. Major abnormalities, such as the loss or gain of one or more chromosomes, are not very useful for genetic studies, since they are either lethal or cause sterility in the organism. Among the minor changes we have:

1. **Deletions**—when a portion of a chromosome is missing.
2. **Duplications**—when a given portion of a chromosome appears more than once.
3. **Translocations**—when a portion of a chromosome appears on another or nonhomologous chromosome.
4. **Inversions**—when a portion of a chromosome runs in the opposite direction than it normally does.

All four cases can be studied genetically by constructing linkage maps. Thus in *deletion*, certain genes will be missing and those on either side of the deletion will be closer together. In *duplication*, a certain sequence will occur twice; in *inversion*, the gene order will be inverted, and in *translocation*, certain genes will now belong to a different linkage group. But the exciting thing is that the chromosomes in which these abnormal-

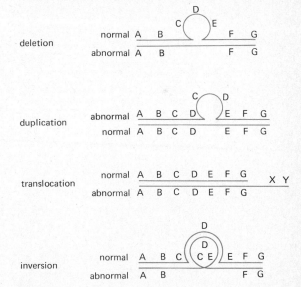

Figure 3-11   The relationship between certain chromosomal abnormalities and the appearance of chromosomes in the pairing phase (prophase) of meiosis. Because of the great precision of the pairing, the presence of unpaired regions can be detected with ease. This phenomenon can also be used for assigning a linkage group identified genetically to a particular chromosome identified microscopically.

ities appear can also be identified microscopically. This is done most easily during the first part of meiosis when homologous chromosomes pair. In the offspring of a normal parent and a parent in which a chromosomal abnormality has occurred, the homologous chromosomes, in their attempt to pair, will form a variety of interesting patterns which are characteristic of the particular chromosomal abnormality. Figure 3-11 shows some of these patterns.

Correlating linkage groups obtained by genetic crosses with chromosomal abnormalities detected microscopically is a tedious process, but the encouraging thing is that it needs to be done only once for a given chromosome. For instance, once an inversion has been detected genetically by constructing a linkage map and a figure characteristic of an inversion has been found in a given chromosome during chromosome pairing in the prophase of meiosis, the chromosome has been identified and can now be labeled as to the particular group of genes it carries. This process of chromosome identification has been carried out with varying degrees of success, depending on the amount of genetic information we have for a given organism and the number of chromosomes present. In *Drosophila,* where we know of hundreds of genes and there are only four chromosomes, the identification of all four chromosomes was easy. In man, however, only a few linkage groups have been identified with a particular chromosome and considerably more such exploration is needed before each chromosome is mapped for a particular group of genes.

We now briefly discuss salivary gland chromosomes and the elegant techniques which have been developed with them. In the salivary glands of insects there are giant nuclei containing giant chromosomes which appear as if they were in the prophase of meiosis. These chromosomes are uncoiled or fully extended so that they are very long. They are the result of many chromosome divisions, forming hundreds (possibly thousands) of chromosome threads which lie side by side so that these chromosomes are not only very long but also very thick. When stained, the salivary gland chromosomes show an intricate and highly characteristic pattern of bands (Fig. 3-12), so much so that every portion of each chromosome can be identified by direct observation. It follows that any abnormality in the chromosome can also be identified microscopically (Fig. 3-13). Furthermore, by studying smaller and smaller abnormalities it has become possible for the *Drosophila* geneticist to localize particular genes down to the nearest band. The culmination of this painstaking research has been to compare genetic maps with such cytological maps (Fig. 3-14).

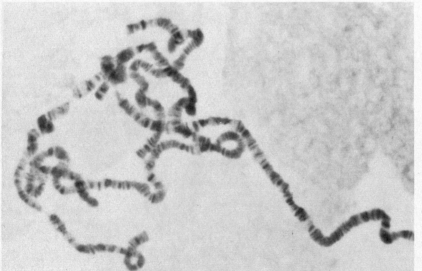

Figure 3-12 The giant chromosomes from a salivary gland cell of the fruit fly. The band pattern is characteristic for each chromosome and this allows each chromosome to be identified microscopically. The chromosomes are stuck together at their respective centromeres and therefore the small fourth chromosome is sometimes difficult to see.

Figure 3-13 The effects of chromosomal abnormalities on the microscopic appearance of salivary gland chromosomes of the fruit fly. In addition to the characteristic figures, the band pattern provides additional evidence as to the particular type of chromosomal abnormality. Notice, for instance, how readily duplications and inversions can be identified.

A. **Deletion.** a, b, and c show how deletion is brought about through a double break in the chromosome and the loss of the piece.

B. **Duplication** and double duplication shown clearly by repeat of characteristic band pattern in the 16A region of the chromosome.

C. **Inversion** can be detected genetically by an alteration of gene order in the genetic map, but it can also be observed microscopically by observing the reversal of the loop of one of the paired chromosomes.

D. **Translocations** are easily identifiable through the pairing of two nonhomologous chromosomes. Here the effects of a reciprocal translocation is observed, that is, when two nonhomologous chromosomes have exchanged parts pair with two normal chromosomes.

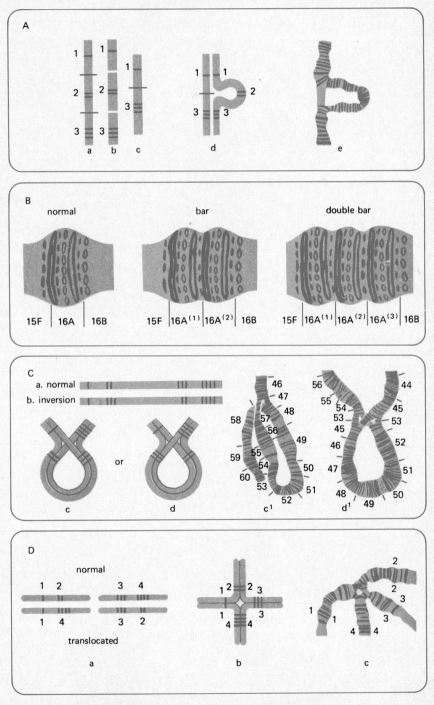

The order of the genes, as determined by this cytological method, turned out to be the same as the order of genes worked out by purely genetic methods of analysis. This result is truly a triumph of the work initiated by Morgan and his students at the beginning of this century. However, the genetic *distances* did not turn out to be perfectly correlated to the cytological distances (Fig. 3-14). This simply means that crossover frequency is not uniform along the length of the chromosome and, given the fact that chromosomes are highly differentiated, complex structures, there is no real reason why it should be.

We have come to the end of one of the most astounding stories of human intellectual history. It started in Mendel's garden and ended in the "fly rooms" of T. H. Morgan and his students. The work which was performed permitted us to identify the discrete units of hereditary activity and localize them on the microscopically visible chromosomes. It was not until some 30 years later, when the chemistry of the hereditary material had been worked out and its message had been decoded, that such grandeur of scientific experimentation and inference reappeared.

WHAT IS A GENE?

Just as organisms have evolved since the beginning of life on earth, so have our ideas regarding them evolved ever since we began to think about them. The concept of the gene is an interesting example of how the meaning of a word in the language of science undergoes modifications, sometimes slowly and continuously (evolution), at other times rapidly and discontinuously (revolution), but at all times a thread or connection is maintained, thus justifying the continued use of the same word.

The gene of Mendel and Morgan was a unit of activity as defined by a number of genetic experiments. Its main qualities were that (1) it was able to appear in more than one form, (2) these various forms of the gene were positioned at a precise place on a given chromosome, and (3) it could exchange its position at that precise place on the chromosome with an equivalent position on a homologous chromosome.

There was something paradoxical about this early concept of the gene. Although it was seen as a unit of activity, in fact it was the *change in activity*, that is, the difference in activity between the gene and its allele that was really being observed. One can imagine, for instance, that the character called "red eye" or "white eye" is controlled by many genes, but it is only by the existence of *two forms of the gene W that our attention is drawn to a hereditary unit on the X chromosome of the fruit fly which contributes to the formation of red eyes.* Thus the genetics of

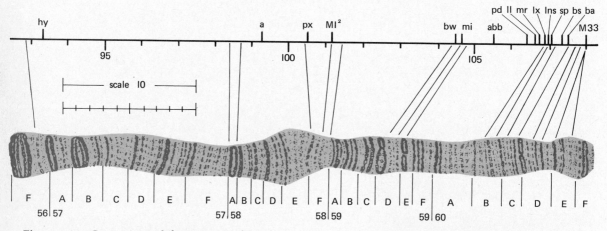

Figure 3-14 Comparison of the genetic and cytological maps of a portion of a salivary gland chromosome in the fruit fly. Notice that the order of the genes in the two maps is identical, though some differences in the relative distances do occur. (From *Principles of Genetics*, 5th ed. by E. W. Sinnot, L. D. Dunn, and T. Dobzhansky. Copyright 1958 by McGraw-Hill Book Company, Inc. Used with permission of McGraw-Hill Book Company.)

Mendel and Morgan tells us nothing about the activity of the gene itself; rather it draws our attention to some kind of activity at a particular locus on a chromosome through the observation of a change of activity produced by an allele of one gene. All this has changed in the last 30 years, for today we not only know what the gene is *structurally* but also (and this is a necessary part of a complete definition in biology) what it does *functionally*. The scientific revolution in molecular biology of the last 30 years has provided many important answers to questions concerning the machinery of living things, most of them concentrated in this area of gene structure and function. We describe some of these answers in nonchemical terms in the rest of this chapter, leaving other answers requiring the knowledge of chemistry to succeeding chapters.

Mutation

If our concept of the gene so far is based on its ability to exist in more than one form, we might at this point wonder about the origin and nature of this variability. Darwin already had recognized that changes in the characteristics of an organism sometimes occur and he called them **sports.** Evolutionists, as will be seen in Chapter 18, discussed at great length whether evolution consisted of the selection of these sports in the form of many small changes or fewer large changes. However, the method of inheritance of these sports was not understood, a fact which plagued Darwin greatly in his later years. De Vries, as early as 1890, studied variegation in plants that consisted of changes in pigmentation in leaves or flowers, often in patches. The interesting thing about variegation is that it can be shown to occur suddenly in a mature plant, affecting only a portion of it. However, if that portion produced a flower and seed, one could demonstrate that the effect was inheritable. De Vries coined the word **mutation** for the sudden occurrence of inherited changes such as variegation, but applied it to the change in the *character* or phenotype. When Mendel was rediscovered, the meaning of the word mutation was easily shifted to refer to the change in the genetic material that must occur when a certain gene changes to an allele. That one

allele turns into another by a single sudden, highly infrequent event soon became a controversial assumption. It was not until 1909, when a white eye mutation appeared in a true breeding stock of red eye (wild-type) *Drosophila*, that the first mutation observed in a laboratory of a known Mendelian gene was documented by the appropriate breeding experiments. After that a number of other mutational events were recorded in populations of laboratory flies, and the reader might try to explain why these were most frequently dominants or sex-linked recessives.

When laboratory mutations began to turn up which involved some of the alleles also found in nature in so-called natural populations, the following view began to emerge. Each genetic locus has a "wild-type" gene which occurs with the highest frequency of all the alleles at that particular locus. Each wild-type gene can mutate in one or more ways with a low, but characteristic, frequency. The frequency with which a certain allele occurs in a certain population depends not only on its mutation rate but also on the selection rate, as well as a number of other factors discussed in Chapter 18. It is sufficient here for us to understand that the change in allele frequencies in populations is what evolution is about. One interesting fact to emerge from these early studies on mutations is that the wild-type allele is usually dominant to the other alleles.

It became of interest in the 1930s to find out something about the rate of mutations in different organisms and different cells of the same organism. Genetic techniques were worked out in the 1920s and 1930s whereby the rate of lethal mutations could be studied quantitatively in *Drosophila*; in 1927 Muller, using these techniques, made the important discovery that X-rays increase the rate at which mutations occur. Since then we have learned a great deal more about the artificial *induction* of mutations, especially by using microorganisms as the experimental material. The results of these studies can be summarized as follows.

1. Mutations can be induced by forms of radiation of *low wavelength and high-energy yield per quantum,* starting with relatively low-energy

ultraviolet radiation and going all the way up to the "hard" high-energy X-rays.

2. In the case of ultraviolet rays it can be shown that the wavelengths most strongly absorbed by **nucleic acids** are the most efficient at bringing about mutations. With other forms of radiation it appears that the mechanism involved is the bringing about of ionization of atoms, which secondarily causes chemical changes in the hereditary material.

3. The mutations induced by radiation are, on the whole, the same mutations which have been observed to occur spontaneously. Radiation therefore simply increases the *rate* at which mutations occur. Furthermore both natural and induced mutations are frequently reversible.

4. There is a linear relationship between dosage and induction of mutations. However, the straight line does not extrapolate to zero. Furthermore the "spontaneous" mutation frequency is greater than can be accounted for by the presence of naturally occurring ionizing radiation.

5. The presence of a spontaneous mutation rate, not completely accounted for by radiation, suggested that other agents besides radiation might be mutagenic; in fact, during World War II the British discovered that mustard gas was mutagenic. Since then many other substances including simple compounds such as nitrous acid were found to be mutagenic.

6. Mutation frequencies vary considerably from gene locus to gene locus, tissue to tissue, and organism to organism. In fact a given gene may change its mutation rate when moved from its normal position on a certain chromosome to another position. In man the sex-linked recessive gene for hemophilia occurs in 1 per 50,000 X chromosomes per generation. If such a rate is typical, it is possible to calculate that any human egg or sperm has a 0.5 percent chance to have a new mutant in it. This spontaneous mutation rate would account for the relatively low frequency of inheritable defects in the human population; but since there is a linear relationship between dosage and mutation rate, any increase in the amount of radiation experienced by the human population will naturally induce a proportional increase in genetic abnormalities. This is why biologists in recent years have been very concerned about the increase in radiation exposure caused by our modern technology and by the testing of atomic weapons. The problem is not only that human suffering is increased by increasing the proportion of genetic defects in our population but it should also be emphasized that nobody knows as yet how much of an increase in genetic variability our species can take before the normal regulatory processes of evolution break down and an irreversible deterioration of our genetic constitution occurs. In the absence of any definitive knowledge we are well advised to treat assurances from military or industrial sources with the greatest skepticism.

7. Finally, radiation frequently induces chromosomal abnormalities or rearrangements which genetically may behave like single gene mutations until the appropriate mapping studies or microscopic observations demonstrate that, in fact, a chromosomal alteration has taken place.

Thus, in summary, one answer to the question regarding the nature of the gene is that it is made of the kind of substance that is capable of reproducing itself, of mutating, and of reproducing the mutant form.

One gene, one enzyme

The question of what the gene does may sound naive to the uninitiated, for in its final effect on the phenotype the activity of the gene is as manifold and varied as the biological world itself. However, if one stops to think for a moment about this collection of genes on the chromosome having a chemistry that could not be so immensely varied, one might come to the conclusion that the *primary* action of the gene may indeed follow a single basic pattern. This is what Beadle and Tatum must have had in mind when they first formulated experiments designed to discover the primary action of the gene.

Beadle and Tatum selected for their studies the mold *Neurospora*. This organism had several advantages.

1. It is haploid, so that the effects of recessive mutations can be observed readily.
2. Genetic techniques with this organism had already been worked out.
3. Its nutritional requirements are simple for the wild-type; that is, it can be grown on a *carbon source*, such as glucose, small amounts of inorganic *salts*, and *biotin*, the only organic compound it needs that it cannot manufacture from its carbon source.

After irradiating *Neurospora* with X-rays and UV radiation, they obtained mutant strains which could not grow on the above **minimal medium.** However, Beadle and Tatum developed a **complete medium** by adding malt and yeast extracts to the minimal medium, which did allow the mutant strains to grow. The rest was merely a matter of trial and error and simple chemistry. By adding various known substances, such as amino acids and vitamins, to the minimal medium, they found what each of the mutant strains required. It turned out, no doubt to their delight, that each strain required only *one substance*. The substances were pyridoxine (vitamin B_6), thiamine (vitamin B_1), and para-aminobenzoic acid. Genetic analysis showed that each strain was characterized by a single recessive mutation that brought about a *nutritional lesion*, that resulted in an inability of the organism to synthesize a certain organic compound required for its growth. They concluded that the X-rays and UV radiation had induced a mutation in a gene which in its wild-type form was involved in the synthesis of a particular compound. The mutation brought about the loss of that ability. Since, as we see in the following chapter, the synthesis of organic compounds in the cell is brought about by enzymes, Beadle and Tatum concluded that genes control the synthesis of enzymes. This hypothesis was in fact demonstrated to be true by showing that each mutant strain lacked the enzyme normally responsible for the synthesis of the compound which was missing in the respective strain (Fig. 3-15).

Further work discovered many additional mutants (X-ray or ultraviolet light induced or even spontaneous), each of which lacked the ability to produce some organic compound necessary for the growth of the organism. There were, of course, mutations which could not grow even on the complete medium, but those had to be ignored because they could not be studied by these *nutritional* methods.

From this work, started by Beadle and Tatum and then extended in numerous other laboratories, the first general hypothesis developed regarding gene action—the *one gene–one enzyme* hypothesis. According to this hypothesis each gene is responsible for the synthesis of a certain enzyme. A mutation brings about the loss of this ability.

This hypothesis was of great importance at the time because it focused the attention of biologists on the possibility that **proteins** are the primary products of genes[1]. However, the hypothesis itself has since been modified considerably and we discuss this matter in some detail in Chapter 6.

The chemistry of the gene

We have already pointed out that the concept that the gene has one basic primary action goes together with the concept that it is one particular kind of compound. To biological workers in the late 1930s and early 1940s, it seemed reasonable that it had to be a large molecule, capable of great variability in its structure. There are only two candidates meeting these requirements: the *proteins* and the *nucleic acids*. Although much of the early evidence pointed to nucleic acids, as late as 1941 many biologists believed that only proteins could have the variability necessary to be the genetic material. It was not that nucleic acids were newcomers on the scene, for they had been discovered by Miescher in the same period that Mendel was publishing the results of his work with peas. Furthermore, while Morgan and his students performed the many interesting experiments which linked genes to chromosomes, biochemists and cytologists working with nucleic acids clearly demonstrated that deoxyribonucleic acid (DNA) was part of the chromosome, and ribonucleic acid (RNA) was in some way related to protein synthesis. However, it was also known

---

[1] As we shall see in the following chapter all enzymes are proteins.

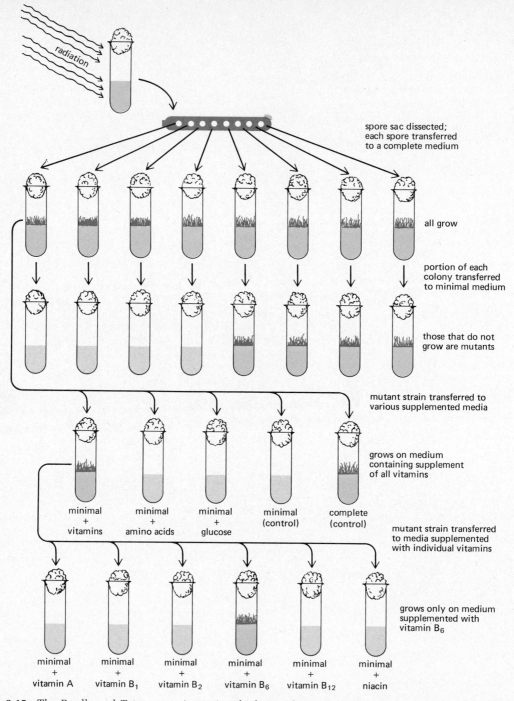

Figure 3-15   The Beadle and Tatum experiment in which specific compounds were identified as growth requirements in a variety of mutants of the bread mold, *Neurospora*. In this particular experiment a growth requirement for vitamin B₆ was detected.

that chromosomes contained some protein and therefore what was needed was some definitive proof, that is, experiments which demonstrated unambiguously which compound carried the *hereditary information.* And two such decisive experiments were soon performed.

One of these consisted of a series of experiments with the pneumonia-producing organism, *Diplococcus pneumoniae,* culminating in 1944 in a clear-cut demonstration that DNA is capable of carrying genetic information.

It had been known since 1920 that some strains of this organism caused the disease (virulent) and others did not (avirulent). It was also known that heat-killed virulent bacteria did not cause the disease, but, to everybody's great surprise, a mixture of avirulent living bacteria and virulent heat-killed bacteria caused pneumonia when injected into mice. Furthermore it was possible to isolate living virulent bacteria from the sick mice. It was therefore possible to conclude that the presence of heat-killed virulent bacteria *transformed* the *living avirulent strains into virulent strains* (Fig. 3-16A). This was not,

however, the only conclusion one could make. One could, for instance, suppose that the mice somehow resurrected the dead virulent bacteria, an explanation not generally resorted to in scientific writings. However the first hypothesis lent itself nicely to additional experimentation, which Avery, MacLeod and McCarty in fact carried out. The first question they asked was whether or not one needed a live mouse to bring about the transformation. It turned out that one did not! The mouse is simply a convenient selective growth medium; that is, if the transformation occurred in only a very few cells, the effect would be greatly amplified in the mouse, because the tissue of the mouse would allow the disease-producing bacteria to multiply rapidly and thus outgrow the avirulent cells that might at first have been present in much greater numbers. With this in mind, Avery, MacLeod, and McCarty devised several *in vitro*[1] experiments in which they added relatively large numbers of heat-treated

[1] *In vivo* means within the living system while *in vitro* literally means in glass — that is, in the test tube.

Figure 3-16   The genetic transformation of the pneumococcus bacterium.

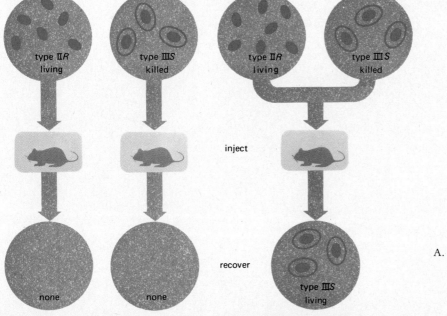

A. In the mouse. Heat-killed preparations of type IIIS virulent transform type IIR non-virulent into type IIIS virulent bacteria.

cells to increase the probability of the transformation and devised techniques for selecting the transformed bacteria (Fig. 3-16B).

In using these methods they were able to show that the transformation they had observed in mice could also occur in the test tube. Once they had done this the way was clear to experiment with fragments of the dead virulent cells and even with purified compounds. Thus eventually they demonstrated that highly purified DNA from virulent Diplococcus cells transformed in the test tube some of the avirulent cells to virulent cells (Fig. 3-16C).

Since the work of Avery, MacLeod, and McCarty, numerous cases have been reported of transformation of a variety of bacteria *in vitro* with DNA from different sources. Here is a typical experiment. In pneumococcus we know of some strains that are sensitive to the antibiotic streptomycin, and of others which are streptomycin resistant. This is a hereditary difference, resistance or sensitivity being passed on from one cell to another at cell division. If a large culture of resistant cells are grown and DNA purified from them, it is possible to demonstrate that such a preparation contains no living cells and no protein impurities. This purified DNA from resistant cells is then mixed with a growing population of streptomycin-sensitive cells and allowed to incubate for some time. Samples are then taken from this mixture and spread on petri plates containing streptomycin in the growth medium. Under such conditions sensitive cells will not grow, but resistant cells will form discrete circular colonies on the agar that can be counted with ease. The total cell population can, of course, be counted by **plating** the cells on streptomycin-free agar. When this is done, it can be shown that a fairly high proportion (often as high as 60 percent) of the sensitive cells are converted from streptomycin-sensitive to streptomycin-resistant cells. These experiments prove that DNA carries the information which is capable of transforming a bacterium carrying a given inherited character

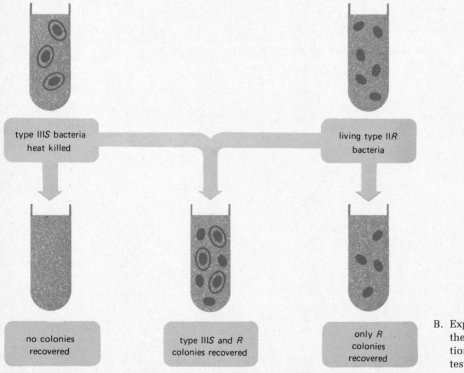

type IIIS bacteria heat killed

living type IIR bacteria

no colonies recovered

type IIIS and R colonies recovered

only R colonies recovered

B. Experiment showing that the genetic transformation also can occur in the test tube.

into a bacterium with a different inherited character, the new inherited character being derived from the strain from which the DNA had been purified. This is sexuality at the molecular level and we discuss it more fully in Chapter 9.

As convincing as these experiments strike us today, the biological world apparently needed one more decisive experiment, which came in 1952 when Hershey and Chase reported their now famous Waring blender experiment with the bacterium, *Escherichia coli*, and the bacterial virus or bacteriophage, T2. The experiment we are about to describe, although ingenious, is deceptively straightforward and simple. It must be recognized, however, that it relied on a number of techniques which had been developed by a small but active group of "phage workers" over the decade preceding this experiment.

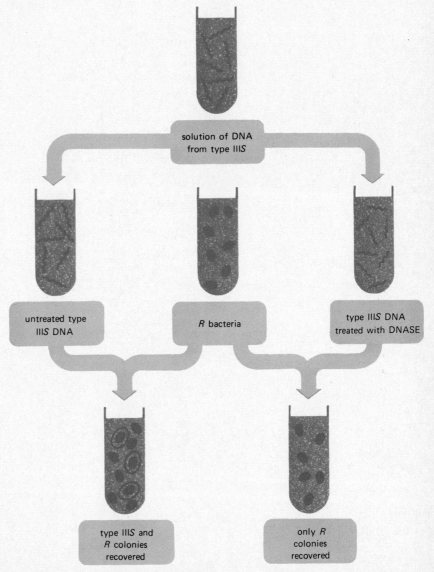

C. Experiment showing that the transforming agent is, in fact, DNA. DNASE is an enzyme which specifically breaks down DNA.

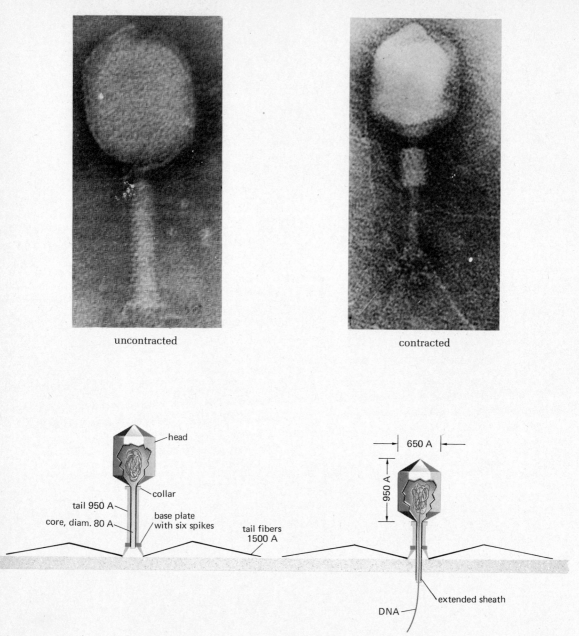

uncontracted   contracted

Figure 3-17 T2 bacteriophage in its normal and contracted forms. This virus is one of the most complex and differentiated forms known, being made of a number of structures each built of one or more structural proteins. The contraction of the tail contributes to the process whereby the phage DNA is injected into the *E. coli* cell. The tailplate appears to contain some lysozyme that has the function of "chewing" a hole into the mucopolysaccharide wall of the bacterium.

T2 bacteriophage is a submicroscopic particle or virus which under the electron microscope can be seen to consist of a capsid, a tail, and a few other structures (Fig. 3-17). These structures are made of a number of proteins and one gigantic molecule of DNA which is packed tightly inside the capsid. When phage particles are mixed with *E. coli* bacteria, they can be observed under the electron microscope to attach themselves by their tails to the outer coat of the bacteria. After some 20 to 30 minutes phage particles appear in the bacteria, after which the bacteria break open or lyse, releasing some 100 to 200 new phage particles. This rapid and efficient process of phage replication can occur only inside a living bacterium. The question therefore as to whether phages or other viruses are living organisms is debatable. They are endowed with one of the properties of living systems, namely the ability to replicate, but they lack other properties, such as an energy-producing metabolism, and therefore rely on living host cells to furnish these. However, because of the simplicity of the process of phage reproduction, it has become one of the most closely studied and most fruitful model systems for investigating the transfer of hereditary information in biological systems.

Hershey and Chase (Fig. 3-18) labeled one culture of phage with radioactive phosphorus (^{32}P) and another culture with radioactive sulfur (^{35}S). This can be done readily by growing phages on *E. coli* bacteria which had been previously labeled by growing them in media containing these radioactive materials. As we shall see in the following chapter, proteins contain sulfur but no phosphorus and nucleic acids contain phosphorus but no sulfur. Therefore the presence of radioactive phosphorus in a certain preparation shows that there are nucleic acids in it, and

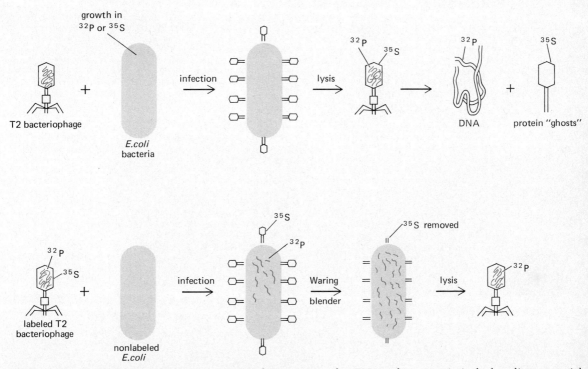

Figure 3-18 The Hershey and Chase experiment demonstrating that DNA and not protein is the hereditary material for the replication of bacteriophage in the bacterium *E. coli*.

the presence of radioactive sulfur shows that we are dealing with proteins. Hershey and Chase then performed two experiments. In one tube they mixed *E. coli* bacteria with ^{35}S-labeled phages and in another tube they mixed bacteria with ^{32}P-labeled phages. After 20 minutes they centrifuged the bacteria to the bottom of the tube and resuspended them; in both cases they found, from the presence of radioactivity, that phages had attached themselves to the bacteria and had therefore been centrifuged with them (Fig. 3-18). At this point they subjected these preparations to a short treatment with the Waring blender. The intense agitation brought about by the blender stripped the attached phage particles from the surface of the bacteria. The phage particles were then separated from the bacteria by centrifugation at moderate speed. This left the small phages in the supernatant, but brought the bacteria down to the bottom of the tube. Now they were able to test the supernatant and the precipitate for radioactivity. In the ^{32}P experiment, most of the radioactivity was in the *E. coli* cells. In the ^{35}S experiment most of the radioactivity remained in the phage particles. From this result Hershey and Chase concluded that the DNA (^{32}P-label) had entered the bacteria, while the protein in the form of empty (DNA-free) phage particles had remained outside. Should this conclusion be correct, it should be possible to demonstrate that the bacteria thus labeled with most of the ^{32}P and with very little of the ^{35}S could produce entirely new phages. And this, in fact, they did. Upon lysis, phage particles composed of DNA still labeled with ^{32}P and new unlabeled protein were released. Thus Hershey and Chase proposed that during phage infection, DNA is injected into the bacterium, leaving the empty protein coat of the phage particles outside. The DNA which has been injected obviously carries the information necessary for the manufacture of both the DNA and the proteins of the new phage particles. DNA is therefore the hereditary material! How DNA does the wonderful things that are required to make more DNA while also providing the necessary information for the manufacture of proteins, we consider in the following two chapters. There is, however, one property of the gene which we have not considered in detail and that is concerned with the recombination of genetic material.

Haploid organisms and gene recombination

Classical Mendelian experiments have led us to think that recombination occurs at selected points *between* genes, that the gene in fact is an indivisible unit. Recent experiments with bacteria and viruses have shown, however, that recombination takes place even *within* a gene. But to have genetic recombination sex is necessary. Do viruses and bacteria have sex? Thirty years ago we would have said No. Now, however, because of the work of Lederberg and his successors, the sexuality of bacteria and viruses is even better understood than that of higher organisms.

The reason bacteria had been thought to be asexual organisms is that nobody had seen bacteria mate, but Lederberg devised genetic methods which were able to determine whether they *had* mated even if very infrequently. We discuss the sexuality of bacteria and viruses in chapter 9. It is sufficient to recognize here that, because of the large number of microorganisms one can handle at any one time and their short generation time, it is possible to detect very infrequent recombinational events. It was therefore possible to show that recombination even occurred *within* genes. Thus our early concept of the gene as a minimal *indivisible* unit of heredity has faded; instead there has developed, as we have seen, a much more concrete concept of the gene as a definite chemical entity, an idea which we will be able to develop further when we learn about the chemistry of DNA in the following chapter.

GENETICS AND MAN

The laws of heredity apply to man just as they do to a bacterium, a fruit fly, or a garden pea. However, because man is the most complex organism on earth, the complexity of the interaction between hereditary and environmental factors is at its greatest in man. It can be argued that *any* trait of man has its hereditary basis. Even the most patent environmentally determined traits might be correlated with a wide range of hereditarily influenced factors. Conversely, the most clear-cut examples of genetically controlled characters may be modified by environmental manipulation. There is nevertheless a great deal of difference of opinion among students of man as to the relative importance of the hereditary and environmental factors which influence complex characters such as behavior and intelligence. This is not only because the problems are complex and the information is difficult to obtain, but also because man, being in this case both the student and the subject of his studies, faces problems of objectivity that in our present divided world inhibit the communication and cooperation necessary for scientific progress.

The nature versus nurture argument is an old one. In Shakespeare's *The Tempest*, Prospero says of Caliban:

A devil, a born devil, on whose nature
Nurture can never stick; on whom my pains,
Humanely taken, all, all lost, quite lost.

Dickens on the other hand structures many of his novels around the theme of human degradation as caused by the environmental ravages of early industrial society. In this century, because of the work of the cultural anthropologists and the theorizing of some members of the psychoanalytic school, the newborn child had increasingly been considered to be a blank sheet on which the environment wrote a continuous and frequently indelible message. More recently, this trend has, however, begun to reverse due to the studies of the behavior of animals in nature (ethology), neurological and behavioral studies of a variety of laboratory animals, and some careful studies on human populations.

The work on animal behavior in nature is discussed in Chapter 14. Suffice it to say here that studies of complex behavior, such as nest building or bird song, have shown that in some animals these activities are entirely unlearned and controlled by hereditary factors. On the other hand, in other animals similar behavior is directly learned from the parents. What these studies prove is that both "nature" and "nurture" are important, but they *also* prove that there *can* be such a thing as hereditary information transmitted from generation to generation determining complex behavior patterns.

In laboratory animals such as rats and mice, elaborate tests for performance or temperamental traits have been devised and strains have been developed which differ markedly in their performance in these tests. In rats, for instance, "maze-bright" and "maze-dull" strains have been bred, showing great differences in their ability to learn this type of performance. However, the importance of the environment has also emerged from these studies, because in certain strains poorly performing animals may change into well-performing animals when subjected to different environmental conditions.

In human populations in which the environment cannot be easily controlled, the student of human behavior has to depend so far on a very few experimental situations yielding very sparse data. One extremely useful approach is the comparison of twins derived from the same fertilized egg (identical) with twins derived from different fertilized eggs (fraternal). As Table 3-8 demonstrates, the **concordance** of identical twins with the mental disease, schizophrenia, is 59 percent, while that of fraternal twins is only 11 percent. However, a concordance of 59 percent means

Table 3-8 Concordance with respect to schizophrenia in male and female monozygotic and dizygotic twins. (*Rosenthal and Kety* (eds.), *The Transmission of Schizophrenia. London: Pergamon Press, 1968*)

		Monozygotic		Dizygotic Same Sex	
		Female (%)	Male (%)	Female (%)	Male (%)
Luxenburger	(1928)	88	67	0	0
Rosanoff	(1934)	78	42	14	9
Essen-Möller	(1941)	75	67	18	9
Slater	(1957)	73	45	16	10
Mitsuda	(1957)	50	50	—	—
Inouye	(1961)	62	58	—	—
Gottesman and Shields	(1966)	45	38	12	6
Kringlen	(1967)	29	23	9	6
		33	42	—	—
Mean Concordance		59%	48%	11%	6%

that in 41 percent of the cases only one of a pair of identical twins developed the mental disease. This shows that even though there is a strong hereditary basis for schizophrenia, there are some environmental conditions under which the disease does not develop. Therefore studies on the role of heredity may also yield useful results on the role of the environment, for by examining intensively the histories of those identical twins in which only one member of the pair developed the syndrome, it might be possible to identify important environmental components which protect humans from it.

Another approach has been to compare adopted and natural children in homes which are carefully matched for what are considered to be relevant conditions.

Using such methods many traits, both normal and abnormal, physical and behavioral, have been studied. The difficulties encountered, however, are legion. In some cases, such as the schizophrenia studies, the data are sparse, since schizophrenic identical twins represent a very small percentage of the population. In other instances the problem is that of clear evaluation of the trait. A notorious example of this is the measurement of intelligence. One would like to have a test that is independent of the cultural experience of the subjects, so that comparison can be made of different socioeconomic classes, ethnic and racial groups. No such test has so far been devised. Furthermore there is the more subtle question as to what the test really measures. Is intelligence really a unitary thing, or are there components which vary independently, each affected by nature and nurture in different ways? We have as yet no clear answers to these questions. Nevertheless one can argue that genetic studies may in fact sort out some of the separate components of intelligence, which in turn will lead to better tests. Thus better genetic studies will lead to improved tests, just as better tests will improve our genetic insights. Such an approach may appear circular, but it is really *helical* in that at every turn of the circle new insights are achieved and therefore such investigations are not fundamentally different from other branches of science.

We are only at the very beginning of the study of human genetics, but some progress has already been made in identifying the genetic basis of both normal and abnormal human traits. We must ask ourselves, however, whether such new insights are useless or valuable, harmless or dangerous. To consider these questions, we must focus our sights not only on the individual human organism, but on all of human society which, since the human species can freely interbreed, represents the total pool of alleles from which the individual organism draws his genotype.

GENETICS AND HUMAN SOCIETY

That genetics applies to human society cannot be doubted. The controversy, however, is in *how much* and *how* it applies, which is not surprising because, as we have shown in the preceding section, there is much we still need to find out before we understand the hereditary basis of human behavioral traits. In addition to the present insufficiency of our knowledge, there is a more subtle and complex problem which is related to all the other basic, existential problems affecting humanity. We refer to the relationship between fact and value or, in more traditional terms, between science and ethics. This relationship is at the root of all the major human questions, such as the optimum size of human populations or the rational use of the earth's resources, and is pertinent to the balance between humanity and the earth, its environment.

As we shall see when we study evolution, its mechanism consists of all the factors affecting the frequency of alleles in a population. It is important to recognize that evolution has not stopped in human populations. Evolution is due to differential survival of different genotypes. Human technology has simply brought about new conditions which must have altered in many ways the pattern of differential survival of human genotypes. One valid and relatively uncontroversial activity for geneticists may be to study the new selective forces in modern society and see what influence they have on gene frequencies in human populations. It appears, for instance, that the lower income groups in technological societies have a higher birth rate than upper income groups. This may have a selective influence on gene frequencies in the population, but precisely what genes are affected and how rapidly is at present far beyond our ability to establish. Even more difficult would be for us to decide whether the effect produced by differential survival of members of different economic classes should disturb us or please us. Although this is a difficult question, it is one we should want to, and indeed must, face.

Traditionally, scientists have felt that questions of value or ethics lie outside the field of scientific endeavor. This may have been a useful hypothesis while it lasted! Today we recognize that questions of *value* and questions of *fact* are delicately intertwined so that it becomes not just difficult but in fact self-defeating to try to separate them. We find that no question of fact is free of assumptions regarding value and no question of value is independent of questions of fact. By accepting a close relationship between fact and value, we are harvesting the fruits of our earlier biological tradition which proclaimed the unity of mind and body. Thus if we can find out what is good for the growth of the organism's body, should we not at some point be able to find out what is good for the growth of his mind or spirit? One might argue that what is good for a human's mind may depend on many other factors. But this is not surprising, for this is precisely what we find to be the case for *all* phenotypic characters, namely that they are the result of a delicate interplay between nature and nurture. Even in simpler problems such as human nutrition, we find that what may be a good diet for an Eskimo in a cold climate may be a very dangerous diet for a New Yorker living in a centrally heated apartment.

The notion that we may find out through study and experiment what is good for the human spirit may sound hopelessly naive or idealistic to us today, since we are confronted perpetually with mankind's division into groups or factions which perceive their interests to be mutually irreconcilable. Arriving at a universal or rational ethic based on the biological reality of man and the environmental reality in which he lives is almost inconceivable to us in our present strife-torn world. But the reverse is probably also true — reducing strife without enlarging humanity's perception of the universality of ethics is also not possible. So, if we are to have a goal as human beings and scientists, it would have to be to chip away at both ends of this spectrum. Creating a better and more rational world will allow for a better science of society and finding out more about man and society will help free us from self-destructive social relations.

If seen from this point of view, facing slowly and reasonably, and with humility, the reality that, as man increasingly understands hereditary and evolutionary principles governing human society, he will also want to enlarge his control of them, is no different than facing other major human problems.

Evolution is the oldest, most successful *laissez-faire* system devised by nature. But when the industrial revolution started man on a road which has led him toward an enlarged understanding and control of the world around him, it has also led him toward deeper insights regarding his own nature. For the first time in biological evolution on this particular planet, matter has evolved to the point that it has begun to *turn back on itself*, and understand itself sufficiently to begin to control its own biological destiny. We have tasted the fruit of self-knowledge and it is our fate as humans to struggle and bring about the conditions necessary for a benign and successful utilization of our insights.

We have been at this game for only 100 years, since Mendel first began to breed his garden peas. If the evolution of the mind should turn out to be a successful biological trait, rather than a device accidentally built to self-destruct, we have aeons of human and social evolution ahead of us. Each generation has reason to think that it stands at a crossroad. Our generation certainly has good cause to feel that way, for what happens in the very near future in the relationship between the development of science and the evolution of society will affect fundamentally whether there is to be a long-range future for man on this earth.

SUGGESTED READING LIST

Hayes, W., *The Genetics of Bacteria and their Viruses.* Basil, Blackwell, and Mott, Oxford, England, 1964.

Levine, L., *Papers on Genetics, A Book of Readings.* C.V. Mosby, St. Louis, 1971.

Levine, R.P., *Genetics,* second edition. Holt, Rinehart, and Winston, New York, 1968.

Mendel, G., "Experiments in Plant Hybridization," an English translation of Mendel's original article reprinted in J.A. Peters (ed.), *Classic Papers in Genetics.* Prentice-Hall, Englewood Cliffs, New Jersey, 1959.

Morgan, T.H., "Sex-limited Inheritance in Drosophila," reprinted in J.A. Peters (ed.), *Classic Papers in Genetics.* Prentice-Hall, Englewood Cliffs, New Jersey, 1959.

Sinnott, E.W., L.C. Dunn, and T. Dobzhansky, *Principles of Genetics.* McGraw-Hill, New York, 1950.

Srb, A.M., R.D. Owen, and R.S. Edgar, *General Genetics.* W.H. Freeman, San Francisco, California, 1965.

Stent, G.S. (ed.), *Papers on Bacterial Viruses.* Little, Brown, and Co., Boston, Mass., 1960.

Stern, C., *Principles of Human Genetics,* second edition. W.H. Freeman, San Francisco, California, 1960.

Strickberger, M.W., *Genetics.* Macmillan, New York, 1968.

Sturtevant, A.H., *A History of Genetics.* Harper and Row, New York, 1965.

Sturtevant, A.H. and G.W. Beadle, *An Introduction to Genetics.* Dover, New York, 1962.

Sutton, H.E., *An Introduction to Human Genetics.* Holt, Rinehart, and Winston, New York, 1965.

Watson, J.D., *The Molecular Biology of the Gene,* second edition. W.A. Benjamin, New York, 1969.

Whitehouse, H.L.K., *Toward an Understanding of the Mechanism of Heredity.* Edward Arnold, London, 1965.

Zubay, G.L. (ed.), *Papers in Biochemical Genetics.* Holt, Rinehart, and Winston, New York, 1968.

CHAPTER 4

The Molecules of Life

The world is made of 92 natural elements[1] which combine with one another in different ways to form the multitude of chemical compounds around us (Table 4-1). Living matter is special in many different ways, one of the most striking being its *specificity*. Thus of all the 92 elements available in the earth's crust and atmosphere only four — oxygen, carbon, hydrogen, and nitrogen — are major components of and have been shown to be necessary for living matter (see Table 4-2). Some twelve other elements also necessary for sustaining life are found in much smaller quantities. In the case of these **trace** or **ultratrace** elements, the amounts of a given element are so small that it has been exceedingly difficult to prove that they are indeed required. And so it is entirely probable that our list of 16 elements utilized in living matter is as yet incomplete.

To return to our point regarding the specificity of living matter, the remarkable thing we want to emphasize is that life is endowed with the ability to discriminate among the chemical elements in its surroundings, concentrating some (for example, hydrogen), diluting others (for ex-

[1] We will not define the word **"element"** at the outset. Its meaning will become clearer with use, making definition easier later in the chapter.

Table 4-1 The periodic table

Legend:

6	→ atomic number
C	→ symbol for element (here, Carbon)
12.011	→ atomic weight

Elements essential for living organisms

IA	IIA	IIIB	IVB	VB	VIB	VIIB	VIII	VIII	VIII	IB	IIB	IIIA	IVA	VA	VIA	VIIA	0
1 H 1.00797																	2 He 4.0026
3 Li 6.939	4 Be 9.012											5 B 10.811	6 C 12.011	7 N 14.007	8 O 15.9994	9 F 18.998	10 Ne 20.183
11 Na 22.990	12 Mg 24.312											13 Al 26.98	14 Si 28.086	15 P 30.97	16 S 32.064	17 Cl 35.453	18 Ar 39.95
19 K 39.102	20 Ca 40.08	21 Sc 44.96	22 Ti 47.90	23 V 50.94	24 Cr 52.00	25 Mn 54.94	26 Fe 55.85	27 Co 58.93	28 Ni 58.71	29 Cu 63.54	30 Zn 65.37	31 Ga 69.72	32 Ge 72.59	33 As 74.92	34 Se 78.96	35 Br 79.91	36 Kr 83.80
37 Rb 85.47	38 Sr 87.62	39 Y 88.91	40 Zr 91.22	41 Nb 92.91	42 Mo 95.94	43 Tc 99	44 Ru 101.07	45 Rh 102.91	46 Pd 106.4	47 Ag 107.87	48 Cd 112.40	49 In 114.82	50 Sn 118.69	51 Sb 121.75	52 Te 127.60	53 I 126.90	54 Xe 131.30
55 Cs 132.90	56 Ba 137.34	57–71 La series*	72 Hf 178.49	73 Ta 180.95	74 W 183.85	75 Re 186.2	76 Os 190.2	77 Ir 192.2	78 Pt 195.1	79 Au 196.97	80 Hg 200.59	81 Tl 204.37	82 Pb 207.19	83 Bi 208.98	84 Po 210	85 At 210	86 Rn 222
87 Fr 223	88 Ra 226	89– Ac series†															

*Lanthanide series	57 La 138.91	58 Ce 140.12	59 Pr 140.91	60 Nd 144.24	61 Pm 147	62 Sm 150.35	63 Eu 151.96	64 Gd 157.25	65 Tb 158.92	66 Dy 162.50	67 Ho 164.93	68 Er 167.26	69 Tm 168.93	70 Yb 173.04	71 Lu 174.97
†Actinide series	89 Ac 227	90 Th 232.04	91 Pa 231	92 U 238.03	93 Np 237	94 Pu 239	95 Am 241	96 Cm 242	97 Bk 249	98 Cf 252	99 Es 254	100 Fm 253	101 Md	102 No	103 Lw

The periodic table is a systematic and orderly arrangement of all known elements which, at a glance, tells of the relationship of one element to another. For example, in addition to being arranged in order of increasing atomic number, the elements are grouped according to the number of electrons in the outermost shell (**valence electrons**)—that is, all the elements in a vertical column have the same number of valence electrons. Because many chemical properties are determined by the number of electrons in the outermost shell, grouping together elements with the same number of valence electrons has the effect of gathering together elements of chemical similarity. Column 0, for instance, contains the noble or inert gases, all of which are extremely unreactive because each has a full outer electron shell. Column VIIA contains the halogens, all of which have seven valence electrons and react in similar ways.

The Lanthanide and Actinide series are separated from the rest of the table, as they have many chemical characteristics in common because of the special electronic configuration that they all possess.

ample, sodium), and excluding others (for example, aluminum) entirely.

In this chapter we discuss the chemical elements which are important for the world of life and some of the many different **compounds** they form. Because biology is, among other things, a special kind of chemistry, the world of life can only be understood by understanding the chemical building blocks which make up the living machine.

ATOMS, IONS, AND MOLECULES

An important property of the universe in which we live is its discontinuity. Matter is not thinly and evenly spread throughout the universe but comes in galaxies, which in turn are composed of stars and planets. Planets are composed of chemical compounds, and it is truly remarkable that despite the size and diversity of the universe and the many chemical compounds occurring in it, there are only 92 naturally occurring elements responsible for this great diversity of materials.

Even more remarkably, all 92 elements are composed of only three basic units, the **proton,** the **electron,** and the **neutron.** Protons and neutrons make up the very dense **nucleus** of each atom and the electrons move in a much larger volume around the nucleus. The proton carries a **unit positive charge,** the electron a **unit negative charge,** and the neutron is uncharged. In recent years physicists have been taking these three basic units apart and are discovering a bewildering array of even smaller **elementary particles,** some of which serve to hold the nucleus together. Fortunately this need not concern us since it would seem that the phenomena of life are not directly involved with the forces that hold an atomic nucleus together.

The simplest atom in our list is hydrogen (H), which is composed of just one electron and one proton (Fig. 4-1). The single proton is the nucleus or center of this atom and is 1800 times heavier than the electron, which moves in a spherical *orbit* around the proton. The electron is so light that it is impossible to say precisely where in space the electron exists at a particular

Table 4-2 Elements generally required for life[1]

Element	Symbol	Atomic number	Relative abundance in atoms (percent)	
			Human body	Earth's crust
Hydrogen	H	1	60.3 ⎫	Trace
Oxygen	O	8	25.5 ⎬ Major	62.55
Carbon	C	6	10.5	Trace
Nitrogen	N	7	2.42 ⎭	Trace
Sodium	Na	11	0.730 ⎫	2.64
Calcium	Ca	20	0.226	1.95
Phosphorus	P	15	0.134	Trace
Sulfur	S	16	0.132 ⎬ Trace	Trace
Potassium	K	19	0.036	1.42
Chlorine	Cl	17	0.032	Trace
Magnesium	Mg	12	0.010	Trace
Manganese	Mn	25	Trace ⎭	Trace
Iron	Fe	26	Ultratrace	1.92
Copper	Cu	29	Ultratrace	Trace
Zinc	Zn	30	Ultratrace	Trace
Cobalt	Co	27	Ultratrace	Trace

[1] *Trace* elements are required in concentrations of one atom in 10^3 to 10^4 atoms; *ultratrace* elements in less than one in 10^5 atoms.

instant in time. This has nothing to do with our inability to "see" the electron, but rather is the result of a basic physical principle which states that there will always be this uncertainty connected with the location of atomic particles. This uncertainty forces us to view the electron's negative charge as being distributed in a spherical cloud around the proton. This cloud is most dense at the proton nucleus and thins out to nothing at large distances from the nucleus. It is possible to calculate the *density* of the electron cloud around a nucleus and draw a line around the region within which the electron will be present, for instance, 90 percent of the time.

Hydrogen is the simplest of the elements, having an atomic weight of 1.008; this means simply that it is about $\frac{1}{12}$ as heavy as an atom of carbon which has its atomic weight set arbitrarily at 12. The next atom, with an atomic number of 2, is helium (He) which contains two protons, two electrons, and one neutron. The neutron carries no charge, has almost the same mass as the proton, and is located with the proton at the center of the atom. Protons and neutrons, although a very small part of the total volume of an atom, nevertheless are responsible for most of its mass. As we go to higher elements with greater **atomic numbers** the numbers of electrons, protons, and neutrons increase in a systematic manner. The atomic number is equal to the number of positive charges, or protons, in a given nucleus. We can therefore now define an **element:** It is a kind of matter whose atoms all have the same atomic number.

Each successive element carries one additional electron and one additional proton, thus ensuring electrical neutrality of the atom; for example, lithium has three protons and three electrons, and beryllium four electrons and four protons and so forth. Neutrons on the other hand do not increase in as logical a manner. Lithium has three neutrons and three protons, and neon has ten protons and ten neutrons, while argon has 18 protons and 22 neutrons, and krypton 36 protons and 48 neutrons. Furthermore the number of neutrons in a given atom can also vary. Hydrogen always has one proton and one electron, but besides the usual form it may also have one

Figure 4-1   The hydrogen atom.

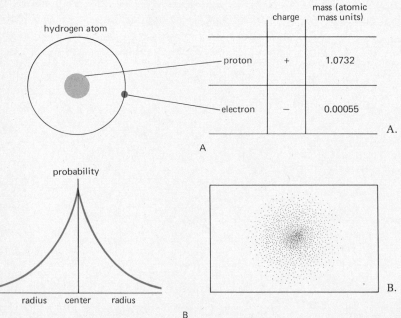

	charge	mass (atomic mass units)
proton	+	1.0732
electron	−	0.00055

A. Classical picture of hydrogen atom with electron circling around proton.

B. "Picture" of hydrogen atom showing probable density of electron around the proton.

neutron, in which case it is the **isotope** *deuterium* (^{2}H), or two neutrons, in which case it is the isotope *tritium* (^{3}H). This number placed above and to the left of the symbol of the element is the **mass number,** which is equal to the sum of the number of protons and neutrons.

Deuterium can be separated from hydrogen because it is twice as heavy. Tritium can be detected easily because it is unstable and loses a neutron at a finite rate to form a deuterium atom, thereby giving off radiation. Such an isotope is called a **radioactive** isotope. It can be detected readily with devices capable of measuring various forms of radiation produced by radioactive elements and has therefore been very useful in many kinds of biological research to which we refer later. Oxygen, besides the usual ^{16}O, has two additional isotopes, ^{17}O and ^{18}O, both of which are stable. Carbon has a number of isotopes, one of which, ^{14}C, is frequently used for biological research because it is radioactive; ^{32}P and ^{35}S are two other radioactive isotopes which are frequently utilized.

The superscript number describing the particular isotope can be used to work out the composition of that atom, for example, ^{18}O. Oxygen has an atomic number of 8. It therefore has 8 electrons and 8 protons and ^{18}O must therefore have 10 neutrons. The most common isotope, ^{16}O, has, of course, 8 neutrons in its nucleus. We now return to electrons, which endow elements with their individual characteristics.

The third element in the periodic table (Table 4-1) is lithium (atomic number 3). It has 3 electrons, but the third electron does not fit into the inner orbit where the first two are distributed but distributes itself in a 2nd orbit. This new orbit can hold a maximum of 8 electrons and is characteristic of the second row of elements listed in the periodic table. The third row in the periodic table contains elements with one to eight electrons in a third orbit. The fourth row, however, contains 18 elements. These elements add electrons to both their third and fourth orbits. The fourth row ends with krypton, which has 18 electrons in its third orbit and 8 electrons in its fourth orbit.

The important thing to recognize is that the *chemical* properties of the elements, that is, the properties responsible for the interaction between different elements to form *chemical compounds,* depend almost entirely on the *outer* orbits, which contain a maximum of two electrons for the first row of the periodic table and a maximum of 8 electrons for all subsequent rows. The vertical column at the extreme right of the periodic table (helium, neon, argon, krypton, etc.) consists of elements with the maximum number of electrons (2 for helium and 8 for the remaining elements) in their outer orbit (Fig. 4-2). It is possible to generalize and state that elements will react with one another in such a way as to form as closely as possible the electron configuration of these *inert gases,* and thereby form a "stable" outer orbit of eight or nearly eight electrons.

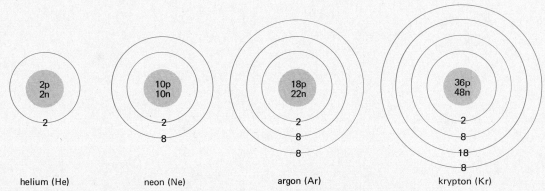

Figure 4-2   The inert gases, showing a maximum number of electrons in the outer orbit.

Thus at the far left of the periodic table we have a vertical column of elements (hydrogen, lithium, sodium, potassium, rubidium, etc.) which have a single electron in their outer orbit. These elements have a tendency to lose that electron without changing their number of protons,

Figure 4-3   Cations and anions.

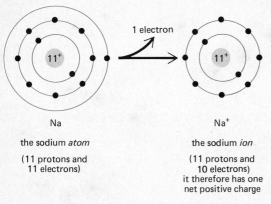

Na

the sodium *atom*

(11 protons and 11 electrons)

Na$^+$

the sodium *ion*

(11 protons and 10 electrons) it therefore has one net positive charge

A

A. Na$^+$, a cation, is produced when the sodium atom loses one electron, thus leaving the particle with a net positive charge.

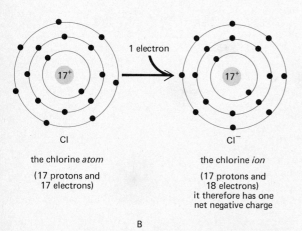

Cl

the chlorine *atom*

(17 protons and 17 electrons)

Cl$^-$

the chlorine *ion*

(17 protons and 18 electrons) it therefore has one net negative charge

B

B. Cl$^-$, an anion, is produced when the chlorine atom picks up one electron, thus producing a particle with a net negative charge.

thus forming a positively charged ion or **cation** (Fig. 4-3). Similarly, the elements in the next vertical column (beryllium, magnesium, calcium, strontium, etc.) have two electrons in their outer orbit and are capable of losing both these electrons, thus forming cations with two positive charges.

At the other extreme of the periodic table, next to the inert gases, there is a column of elements (fluorine, chlorine, bromine, iodine, etc.) which have seven electrons in their outer orbit. These can *stabilize* this orbit by *adding* an electron, thus forming **anions** with single negative charges (Fig. 4-3).

The way in which an atom gains or loses electrons is by taking them away from or giving them to other atoms. Thus sodium can react with chlorine, the former giving an electron to the latter. This reaction brings about the formation of a compound composed of cations and anions called a **salt**—in this case *sodium chloride*, or table salt (Fig. 4-4). Many salts dissolve readily in water, allowing the ions to move freely with respect to one another. We refer to such a state as being **fully dissociated.** However, the strong attraction between the positive and negative ions will keep these ions within a definite average distance from one another, an interaction called an **ionic bond.** It is important to recognize that the salt, sodium chloride, is a chemical compound having its own definite physical and chemical properties that bear little resemblance to the properties of the elements sodium and chlorine. Sodium is a soft, highly reactive solid and chlorine a highly reactive gas. However, the salt, sodium chloride, is a hard, extremely stable crystalline solid of very high melting point, its solubility in water being an important characteristic for the world of life.

We have listed hydrogen among the elements capable of losing one electron and becoming **monovalent** (singly charged) cations. Hydrogen, however, is very special. When it reacts with chlorine, for instance, to form hydrogen chloride, it produces a compound with distinct chemical characteristics called an **acid** (Fig. 4-5). An acid is a substance which releases protons (or hydrogen ions); its properties are the result of the presence of these positively charged ions. As we shall see

Figure 4-4 Sodium chloride, a salt. The positive and negative charges of Na⁺ and Cl⁻ attract each other and form an ionic bond.

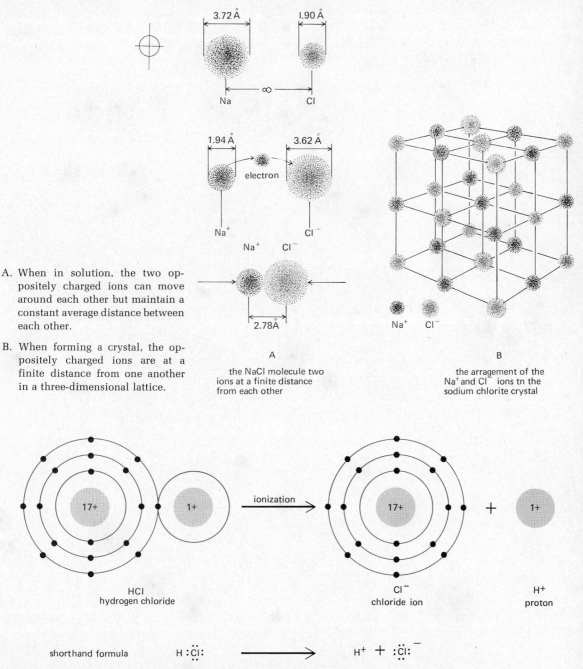

A. When in solution, the two oppositely charged ions can move around each other but maintain a constant average distance between each other.

B. When forming a crystal, the oppositely charged ions are at a finite distance from one another in a three-dimensional lattice.

A

the NaCl molecule two ions at a finite distance from each other

B

the arragement of the Na⁺ and Cl⁻ ions tn the sodium chlorite crystal

HCl
hydrogen chloride

ionization

Cl⁻
chloride ion

H⁺
proton

shorthand formula

$H : \overset{..}{\underset{..}{Cl}} :$ ⟶ $H^+ + : \overset{..}{\underset{..}{Cl}} : {}^-$

Figure 4-5 The ionization of an acid. Hydrogen chloride (hydrochloric acid) is an acid because, upon dissociation, it produces a proton (H⁺). The shorthand formula shows only the outer shell of electrons.

later, the hydrogen ion or proton (H^+) plays an extremely important role in the interactions of compounds found in living matter.

Giving or taking electrons is, however, not the only way in which elements can interact with each other. Electrons can also be *shared*; when two or more atoms come together and share electrons they form chemical or **covalent** bonds. Let us examine the nature of the covalent bond and a few chemical compounds held together by covalent bonds.

The electron around the proton in hydrogen is distributed in a spherical orbit called an *s orbital*. The two electrons around helium also form *s* orbitals. However, electrons at a greater distance from the nucleus can distribute them-

selves differently; thus they may form figure eight patterns called *p orbitals*. Even other distributions are possible (Fig. 4-6A).

In the periodic table some special elements are located centrally in the second and third horizontal rows. These elements have neither the tendency to lose electrons like those in the first or second column, nor to gain them like those in the seventh column. Instead they react with one another and with hydrogen to share electrons. Let us take carbon—the basic structural element of the compounds found in living matter. It has 6 electrons around its nucleus, the 2 closest being paired in an *s* orbital resembling that of helium. The four other electrons are distributed in four orbitals which are a mixture of *s* and *p*

Figure 4-6   Electron orbitals and the formation of covalent bonds.

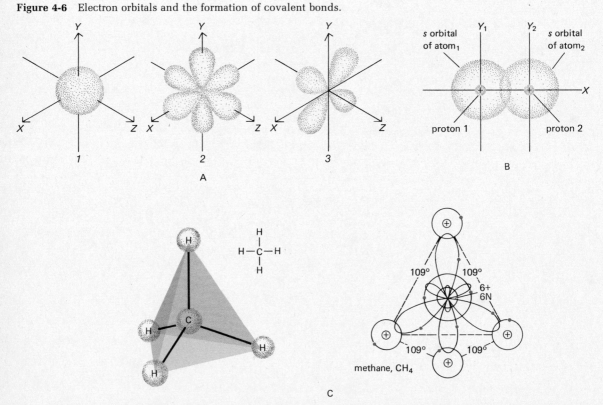

A. A variety of orbitals. 1. The *s* orbital; 2. the *p* orbital; 3. the *d* orbital. B. The formation of the hydrogen molecule (H_2), a molecule held together by a covalent bond. C. Methane (CH_4), a molecule in which the four electrons of four hydrogen atoms (H) and four electrons in the outer shell of one carbon (C) form an outer shell of eight electrons resembling those of the inert gas, neon.

and are called **hybrid orbitals.** Their important property is their ability to overlap with the orbital of an electron from another atom. The overlapping pair of electrons, each belonging to a different atom, holds the atoms together by a covalent bond to form a molecule (Fig. 4-6B).

Since carbon has four valence electrons, each distributed in a hybrid orbital, these electrons can pair with four unshared electrons from other atoms. We say therefore that the carbon has a **valence** of 4. Hydrogen has a valence of 1, because it has a single electron in an *s* orbital; thus one carbon atom can react with four hydrogen atoms to form *methane*—a compound in which the four unpaired electrons of carbon in hybrid orbitals form *pairs* with the four electrons of hydrogen in *s* orbitals (Fig. 4-6C). This produces a molecule containing few valence electron pairs which resemble in their distribution the inert gas, neon. It should come as no surprise that methane is a relatively inert gas.

We can now generalize and say that reactions between atoms that form covalent compounds bring about the sharing of orbitals between unpaired electrons. This resembles the four paired electron distribution of inert gases such as neon and argon.

The convention which we follow in writing the *structural formulas* of covalent compounds is to draw a single line between respective atoms representing the shared electron pair. In the following structural formulas we write:

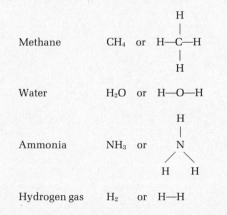

Methane	CH_4	or	H—C—H (with H above and below)
Water	H_2O	or	H—O—H
Ammonia	NH_3	or	N (with H below left and right)
Hydrogen gas	H_2	or	H—H

There are times, however, when an atom shares more than one electron pair with another atom. Oxygen gas, for instance, is composed of two oxygen molecules sharing two electron pairs. We say it is *divalent* and write it with a double bond: O_2 or $O{=}O$.

Nitrogen is *trivalent* because it shares three electron pairs. It reacts with another nitrogen atom to form nitrogen gas and is written with a triple bond: N_2 or $N{\equiv}N$.

Having briefly reviewed a few facts regarding atoms, ions, and molecules, we are now ready to describe the ions and molecules which are important to life.

Let us look again in Table 4-2 at the list of elements required for life. Of these we return to H, O, C, N, P, and S later, for they are the constituents of the molecules of the living machine. The remaining ten elements are mostly present in the cells as ions, with Na^+, K^+, Ca^{++}, Mg^{++} and Cl^- being both the most abundant and most important of the ten. There is much we still do not understand about the role of these ions. We know that they interact in a variety of ways with the giant molecules (macromolecules) of the cell and we also know that living matter has been most conservative about the relative concentrations of these ions it maintains in the cell. Thus Table 4-3 shows the relative concentrations of the five ions in seawater and then follows them up the evolutionary scale to man. It is not too much of an oversimplification to say that we still carry in ourselves the seawater in which life first began.

WATER AND LIFE

Water is the *dispersion medium* in which the molecules of life are distributed. It is not only the universal solvent without which life cannot exist, but it is also the most abundant substance found in the cell (Table 4-4, p. 108). It is not surprising therefore that water has many interesting properties, all of which depend on its special molecular structure.

Oxygen has eight electrons. Two of these are in the stable *s* orbitals, leaving six electrons in the outer orbitals, four of which are paired, leaving two unpaired. Each of these unpaired electrons can share an orbital with the single electron of two hydrogen atoms. This results in the formation of the compound, water, H_2O or

Notice, however, that the most stable way of sharing these electrons is for the molecule to establish a special geometry in which the atoms form definite angles with respect to one another. We shall have numerous occasions to refer to these **bond angles** and also to **bond distances** (distances between the centers of atoms) of covalent compounds, for they are very important in determining the structure and specificity of the molecules of life.

Because both hydrogen atoms are situated on one side of the water molecule, they exert a strong pull on the four electrons that they share with the oxygen atom. However, this leaves four electrons which are not involved in covalent bonding and, since the oxygen nucleus is so much larger, it exerts a strong pull on the remaining four electrons, which will distribute themselves in a preferred direction, giving the water molecule a property called **polarity** (Fig. 4-7). The polarity of the water molecule is such that if the oxygen atom is visualized at the center of a tetrahedron, then two of the corners will be occupied by hydrogen atoms. The other two corners will represent directions in which the remaining four electrons are most frequently situated, thus bringing about two regions of high negative charge. The directions in which hydrogens are located represent, of course, two regions of positive charge (Fig. 4-7). Thus we find not only that water is polar but that its polarity has a finite geometry with two sets of negative charges and two sets of positive charges which are projecting out from the oxygen nucleus, much like the four corners of a tetrahedron.

This important property brings about an interesting interaction among neighboring water molecules. Each water molecule, because it has two negative and two positive regions, can inter-

Table 4-3   Ionic composition of seawater and the body fluids of several species

	Na^+	K^+	Ca^{2+}	Mg^{2+}	Cl^-	SO_4^{2-}
Vertebrates						
Man	145	5.1	2.5	1.2	103	2.5
(mammal)	100	3.5	1.7	.83	71	1.7
Rat	145	6.2	3.1	1.6	116	
(mammal)	100	4.2	2.1	1.1	80	
Frog	103	2.5	2.0	1.2	74	
(amphibian)	100	2.4	1.9	1.2	72	
Lophius	228	6.4	2.3	3.7	164	
(fish)	100	2.8	1.0	1.6	72	
Invertebrates						
Hydrophilus	119	13	1.1	20	40	0.14
(insect)	100	11	.93	17	34	0.13
Lobster	465	8.6	10.5	4.8	498	10
(arthropod)	100	1.9	2.3	1.0	110	2.2
Venus	438	7.4	9.5	25	514	26
(mollusk)	100	1.7	2.2	5.7	120	5.9
Sea cucumber	420	9.7	9.3	50	487	30
(echinoderm)	100	2.3	2.2	12	120	7.2
Seawater	417	9.1	9.4	50	483	30
	100	2.2	2.3	12	120	7.2

Black numbers are expressed in millimoles (mM) per liter; numbers on color background are relative, expressed in terms of 100 units of Na^+.

act with four other water molecules. In liquid water these interactions are temporary, *being made and being broken* continuously, but when water freezes to form ice, the water molecules arrange themselves in a permanent **crystal lattice,** such as that depicted in Figure 4-8. The remarkable thing about this arrangement in the ice crystal lattice is that, unlike most crystals, it is very loosely packed; that is, it has large holes, which explains why ice at zero degrees is lighter than water at the same temperature. In a freezing mixture ice will therefore float to the top of water, a fact with important ecological consequences, for when bodies of water freeze, they freeze from the top; thus the layer of ice insulates the remaining water from rapid heat loss and allows living matter in ponds and streams to survive harsh winters.

The attraction between the electronegative oxygen of one water molecule and the electropositive hydrogen of another molecule is called a **hydrogen bond.**

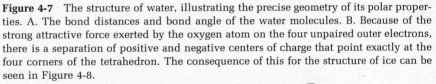

(one Angstrom unit, $1\text{Å} = 10^{-1} \text{ nm} = 10^{-4} \mu = 10^{-7} \text{ mm}$)

This type of bond occurs not only in water but also in other compounds containing hydrogen and oxygen or nitrogen atoms. It is a weak bond (notice its much greater length compared with the covalent H—O bond *within* the water molecule). Because of the ease with which it may be broken, the hydrogen bond is a most important form of interaction in many biological molecules which depend for their proper functioning on weak, easily changing interactions.

A number of other properties of water derive from its unique geometry and its capacity to form hydrogen bonds. One of these properties, called the high **latent heat of evaporation,** enables organisms to regulate their body temperature through sweating.

Possibly the most important property of water is derived from its **solvent power.** Since water is a polar substance, it will interact with other polar substances, thus becoming a solvent for them. As mentioned before, salts which are fully dissociated into ions will ordinarily dissolve in water easily. However, other substances which are not normally dissociated will dissociate when

Figure 4-7   The structure of water, illustrating the precise geometry of its polar properties. A. The bond distances and bond angle of the water molecules. B. Because of the strong attractive force exerted by the oxygen atom on the four unpaired outer electrons, there is a separation of positive and negative centers of charge that point exactly at the four corners of the tetrahedron. The consequence of this for the structure of ice can be seen in Figure 4-8.

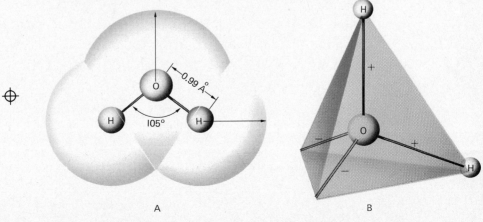

dissolved in water. Still other substances which are only slightly polar have their polarity increased by dissolving in water. In a slightly polar substance the centers of gravity of positive and negative charges are slightly separated. Water has a tendency to increase the separation of charges, thus increasing the polarity of the substance and making it water-soluble.

Water not only brings about dissociation of molecules into ions in other substances, it can also have this effect on itself as well. In a very small percentage of all the water molecules, one of the protons leaves one of the water molecules and associates with another one, thus bringing about the formation of a negatively charged **hydroxyl ion** (OH^-) and a positively charged **hydronium ion** (H_3O^+).

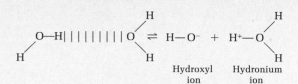

Hydroxyl   Hydronium
ion   ion

Normally for purposes of shorthand we do not write the whole hydronium ion, but merely indicate that the proton (H^+) comes off the water molecule, by writing $H_2O \rightleftharpoons H^+ + OH^-$. This convention is acceptable as long as it is remembered that protons do not normally float around freely in water.

We have already pointed out that an acid is a substance which upon dissociating gives off a proton (H^+). Water is therefore an acid, but it is a *weak acid*, for the dissociation is only very slight

Figure 4-8   The structure of ice.

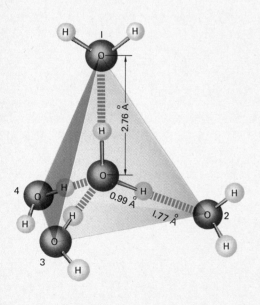

A

A. Tetrahedral coordination of water molecules in ice. Molecules 1 and 2 as well as the central H_2O molecule lie entirely in the plane of the paper. Molecule 3 lies above this plane, and 4 below, so that oxygens 1, 2, 3, and 4 lie at the corners of a regular tetrahedron.

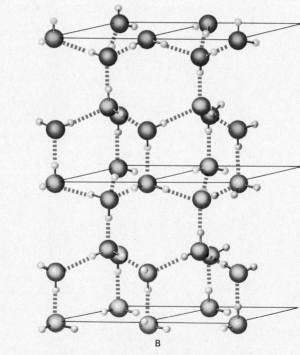

B

B. The arrangement of water molecules in ice. This diagram shows how the water molecules are arranged in a loose network of adjacent tetrahedrons, forming a structure which is actually less tightly packed than water in the liquid state. (From *General Chemistry*, 2nd ed., by Linus Pauling. W. H. Freeman and Company. Copyright © 1953.)

and thus the concentration of protons in water is very low indeed.

A **base** is a substance which picks up protons. Therefore water is also a *weak base,* for as one water molecule gives off a proton another water molecule picks it up, leaving free hydroxyl ions in solution.

$$H_2O + H_2O \rightleftharpoons OH^- + H_3O^+$$

We can formalize the above description of an acid and base in the following way:

$$HA \quad \rightleftharpoons \quad A^- \quad + \quad H^+$$
Acid   Conjugate   Proton
        base

Water participates in the dissociation of acids as follows:

$$HA + H_2O \rightleftharpoons A^- + H_3O^+$$

which in our shorthand version is written as:

$$HA \rightleftharpoons A^- + H^+$$

In the case of a base, we have:

$$A \quad + \quad H_2O \quad \rightleftharpoons \quad AH^+ \quad + \quad OH^-$$
Base   Acid   Hydroxyl
          product   ion

or the shorthand version:

$$A + H^+ \rightleftharpoons HA^+$$

The degree to which acids and bases dissociate varies a great deal. Hydrochloric acid (HCl) in water will dissociate almost completely, releasing a large concentration of protons; it is therefore called a **strong acid.** Carbonic acid, H_2CO_3, (luckily for our teeth since it is present in soft drinks) releases only a low concentration of protons; it is therefore called a **weak acid.** The same is true for bases; a strong base is one which picks up protons with great avidity, thus leaving a large concentration of hydroxyl ions in solution.

Since the concentration of H^+ and OH^- are inversely related (when one of them is high, the other one is low), the strength of acids and bases can be described just by measuring the concentration of one of them. And since H^+ is measured easily, we use it for our scale of acid and base strength. Since, however, the concentration of H^+ in a strong acid can be 100,000,000,000,000 (10^{14}) times greater than in a strong base, we simply use a logarithmic scale (pH) which is defined as follows:

$$pH = \frac{1}{\log H^+ \text{ concentration}} = -\log [H^+]$$

In this scale a low pH represents a high hydrogen ion concentration and vice versa (Fig. 4-9). Before leaving this discussion of water and its relationship to acids and bases we must mention **buffers,** which are of considerable importance to biological systems.

When a weak acid is added to a strong base or a strong acid to a weak base, a salt is obtained which in solution forms an equilibrium with the acid and base producing a mixture with some in-

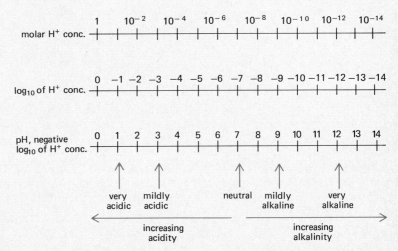

Figure 4-9 The logarithmic nature of the pH scale; the relationship between the hydrogen ion (H^+) concentration and pH.

teresting properties. To illustrate these properties let us take the weak acid, acetic acid, (CH_3COOH) and start adding increasing quantities of sodium hydroxide (NaOH).

$$CH_3COOH + NaOH \leftrightarrows CH_3COO^- + Na^+ + H_2O$$

| Acetic acid (weak acid) | Sodium hydroxide (strong base) | Sodium acetate (salt) |

This process is called **titration** and can be followed by measuring the pH of the solution with an appropriate device called a **pH meter.** Figure 4-10 illustrates the process. The **titration curve** shows a remarkable property. As we begin adding base to the acid the pH will change rapidly, but then toward the middle of the curve it will *change more slowly* and at the end will change rapidly again. Thus at the middle of the titration curve there is a resistance of the solution to changing pH. This resistance is called **buffering,** and a solution of acid, base, and salt that finds itself in a pH range at which this resistance to the pH change occurs is called a **buffered solution** or a **buffer.**

Living matter is tremendously sensitive to pH. This is particularly true for the huge macromolecules of the cell which are greatly affected by pH, and it therefore becomes crucial that the cell protect itself against pH change by *regulating* the pH most carefully. Our blood plasma, for instance, which is in equilibrium with the fluid which bathes every cell in our body, maintains its pH at 7.2 to 7.3 by several very sensitive regulatory processes. Decreasing the pH of our blood

Table 4-4 Compounds found in a typical bacterial cell

	Percent of total weight
Water	70
Carbohydrates	3
Lipids	2
Proteins	15
DNA	1
RNA	6
Other	3

by only one-tenth of a pH unit causes us to pant, a process which accelerates the loss of carbon dioxide from our blood through our lungs. This increases the pH of our blood because the carbon dioxide in it is in equilibrium with carbonic acid. Therefore as the carbon dioxide concentration decreases, so does the carbonic acid concentration. Panting therefore causes an increase in pH (decrease in acidity) of the blood.

Water has many other interesting properties which render it suitable for the crucial role it plays in nature. Let us go on, however, and examine some of the compounds which are dispersed in water to form the complex machinery of the living cell. It is customary to divide the compounds of the cell into a number of categories distinguished from one another by their chemical structure or properties. Table 4-4 shows how much of each of these compounds is found in an average cell. We now discuss them in turn.

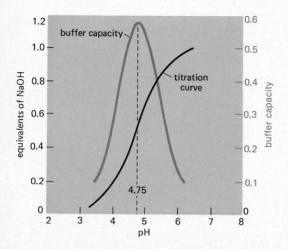

Figure 4-10 (Left) Titration curve of a weak acid (acetic acid) and a strong base, sodium hydroxide. This curve can easily be established by adding known quantities of NaOH to a known quantity of acetic acid and measuring the pH values obtained. The buffer capacity can be obtained by computing the derivative of this curve, that is, the rate of NaOH added per rate of pH change. This curve shows a maximum at the midpoint of the titration curve, that is, the point at which the addition of a given amount of base has the least effect on pH. The pH at which this occurs is numerically equal to the pK_A of the acid. It is also the pH at which the buffer has maximum buffering action.

THE CARBOHYDRATES—READY ENERGY AND PROTECTION

The compounds we shall be discussing from now on all have carbon (C) as their main structural element. They are called **organic compounds,** and it was first thought that they could only be synthesized in living cells. But in 1828 Friedrich Wöhler synthesized urea and since then we have learned to synthesize over 1,000,000 different organic compounds, the majority of which are not even found in living systems.

Carbon has unique properties which render it especially suitable to be the basic structural element of the molecules of life.

First, the periodic table (Table 4-1) shows that carbon is centrally located in the second horizontal row. It is therefore able to react with the four **electropositive atoms** of hydrogen, each of which contribute an electron to be shared in a covalent bond. However, carbon can also react with **electronegative atoms** such as N, O, P, S, and Cl on the other side of the periodic table. In this case carbon contributes electrons to form a variety of different covalent bonds.

Second, carbon can form covalent bonds with other carbon atoms, forming *single* bonds, *double* bonds, or *triple* bonds.

Since carbon can react with one, two, three, or four other carbon atoms, this allows for a versatility of structure which is unique in nature. Furthermore carbon can form open chains or molecules which are closed rings.

Third, an oxide of carbon, carbon dioxide, is a gas which can dissolve in water to become hydrated and form carbonic acid. Carbon dioxide from the air is the source of all carbon in living systems. This ability to distribute itself around the planet in gaseous form, and yet enter the cell by solution, and then become incorporated into the organic compounds of the cell is unique to oxide of carbon. Silicon, however, one row below carbon in the periodic table, and in some properties reminiscent of carbon, forms an oxide (SiO_2) commonly known as quartz, a material hardly capable of sustaining life. Clearly carbon is

uniquely endowed to be the fundamental structural element of living matter.

What then are the various kinds of organic compounds which carbon can form with H, O, N, S, and P? Rather than listing particular compounds we shall list the **functional groups** which make up the organic compounds. Table 4-5 lists some of these groups and shows how they are structurally related to one another. Each functional group represents the basic structure for a series of possible compounds. Let us take, for instance, the hydrocarbon group.

If R' is H and R is H, CH_3, C_2H_5, and C_3H_7, we obtain respectively the following straight chain hydrocarbon series

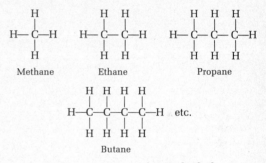

If we do the same thing with an alcohol,

then we obtain the series methanol, ethanol, propanol, butanol, etc.

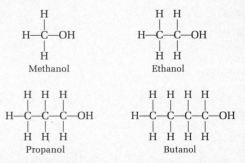

Table 4-5a   Major aliphatic groups

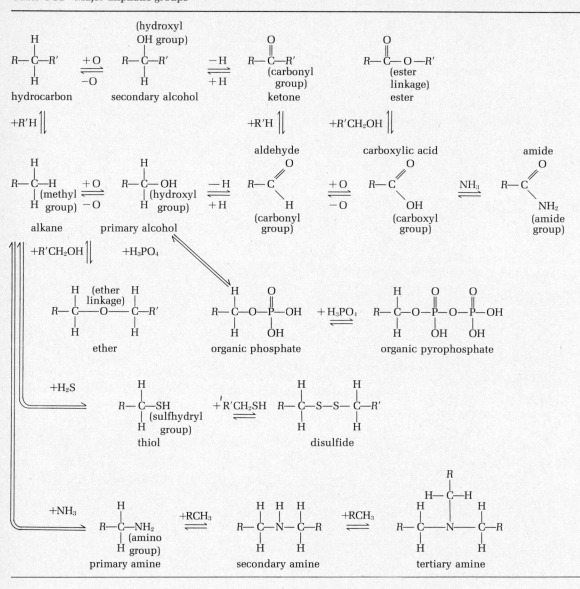

Table 4-5b A number of functional groups and representative compounds containing them

Structural formula	Shorthand form	General name	Compound	Common name
H | R— C—H | H	RCH_3	Alkane	CH_4 C_2H_6 C_3H_8	Methane Ethane Propane
H | R— C— OH | H	RCH_2OH	Alcohol	CH_3OH C_2H_5OH C_3H_7OH	Methanol Ethanol Propanol
O ‖ R— C | H	RCHO	Aldehyde	HCHO CH_3CHO	Formaldehyde Acetaldehyde
O ‖ R— C—R′	RCOR′	Ketone	CH_3COCH_3	Acetone
O ‖ R— C \ OH	RCOOH	Carboxylic acid	HCOOH CH_3COOH	Formic acid Acetic acid
O ‖ R— C— O—R′	RCOOR′	Ester	$CH_3COOC_2H_5$	Ethyl acetate
O ‖ R— C—NH₂	$RCONH_2$	Amide	CH_3CONH_2	Acetamide
H H | | R— C— O— C—R′ | | H H	RCH_2OCH_2R'	Ether	$C_2H_5OC_2H_5$	Diethyl ether
H | R— C—SH | H	RCH_2SH	Sulfhydryl	C_2H_5SH	Ethanethiol
H | R— C—NH₂ | H	RCH_2NH_2	Amine	$C_2H_5NH_2$	Ethylamine

The organic structures mentioned so far are located on chains but can also be part of ring structures, for carbon can form five- and six-membered rings with ease.

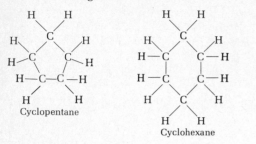

Cyclopentane

Cyclohexane

There is, however, a different kind of six-membered ring structure that carbon can form that has very different properties from cyclohexane. This structure, called **benzene**, has a series of three **conjugated double bonds**, that is, double bonds alternating with single bonds.

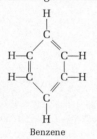

Benzene

Double bonds usually make a chemical compound more reactive. But with benzene this is

not the case. Its great stability is due to the fact that the double bonds in it are in a dynamic state endowing all six bonds with a partially *double bond character*, thus stabilizing the whole structure. Benzene is the basis of an entirely different set of organic structures—the so-called **aromatic** compounds (Table 4-6). All other compounds (that is, those not containing a *benzene ring*) are called **aliphatic.** As we proceed further with our discussion of organic compounds we shall be using a variety of abbreviated or shorthand methods of symbolizing them which should be learned step by step as they are introduced.

Figure 4-11 The tetrahedral structure of carbon.

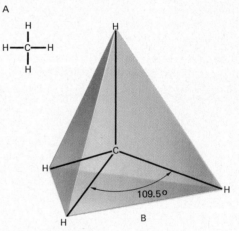

A. Conventional formula.
B. Tetrahedral arrangement in space.

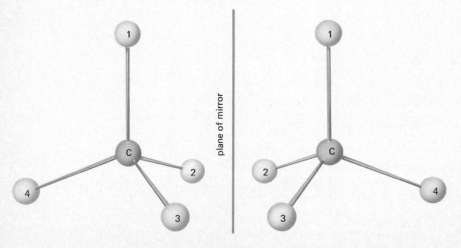

plane of mirror

Figure 4-12 Optical isomers are mirror images of each other. This happens when a carbon atom has four different atoms or chemical groupings attached to it.

One important structural property of carbon which plays a crucial role in the structure of biologically active molecules is the **tetrahedral** arrangement of its bonds in space (Fig. 4-11). Another important property is the ability of carbon compounds to form **optical isomers,** that is, two compounds with identical chemical formulas but which are *mirror images* of each other. This happens when a given carbon atom has four different groups attached to it. Such a carbon atom is called *asymmetric.* Figure 4-12 shows how when this happens, two different structures can be built which are related to each other as an object is to the image in its mirror. Although optical isomers are chemically very similar to one another, they can be distinguished physically because they turn a *plane of polarized light* in opposite directions. A remarkable property of macromolecules found in living systems is that they distinguish between optical isomers.

Table 4-6   Some aromatic compounds

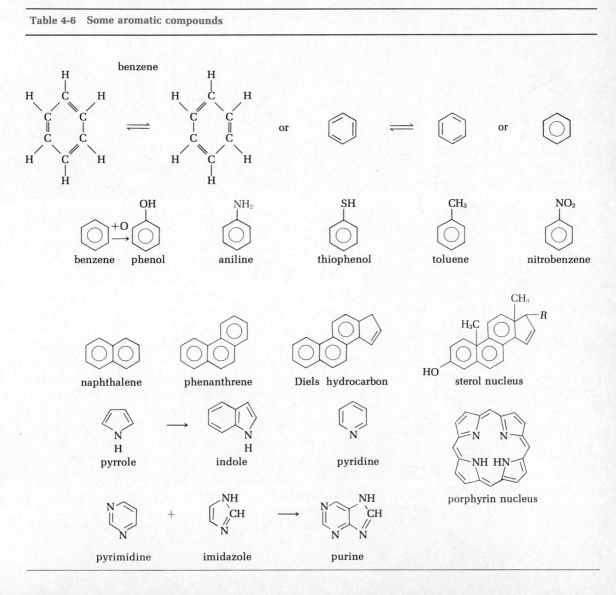

One remaining property of some organic compounds which we must mention is the ability of some to dissociate as acids or bases which we shall call **electrolytes.** The weak electrolytes shown at right are most frequently encountered in organic compounds of living systems. Notice that phosphoric acid can release two protons, so that

$$R—O—\overset{\overset{\displaystyle O}{\|}}{\underset{\underset{\displaystyle OH}{|}}{P}}—O^-$$

is the conjugate base in the first dissociation and, at the same time, the acid in the second dissociation.

The acids and their conjugate bases listed are ordered in increasing value of their pk_A. This number indicates the pH value at which half of the compound is found in its acid form and the other in its conjugate base form. Thus the pk_A gives us an estimate of the strength of the acid or conjugate base. A pk_A of 2 indicates that the phosphoric acid group is a fairly strong acid and the

$$R—O—\overset{\overset{\displaystyle O}{\|}}{\underset{\underset{\displaystyle OH}{|}}{P}}—O^-$$

ion an extremely weak base, while a pk_A of 9 to 10 shows that the ammonium ion is an extremely weak acid and the amino group a fairly strong base.

But enough of organic chemistry! Let us now apply our knowledge to some of the **carbohy-**

Table 4-7 Some weak electrolytes

Acid	pk_A	Conjugate base and proton
$R—O—\overset{\overset{O}{\|}}{\underset{\underset{OH}{\|}}{P}}—OH$ Organic phosphoric acid	2	$R—O—\overset{\overset{O}{\|}}{\underset{\underset{OH}{\|}}{P}}—O^- + H^+$
$R—\overset{\overset{O}{\|\|}}{\underset{\underset{OH}{}}{C}}$ Carboxyl group	3-4	$R—\overset{\overset{O}{\|\|}}{\underset{\underset{O^-}{}}{C}}\ + H^+$ Carboxylate ion
$R—O—\overset{\overset{O}{\|}}{\underset{\underset{OH}{\|}}{P}}—O^-$	7	$R—O—\overset{\overset{O}{\|}}{\underset{\underset{O^-}{\|}}{P}}—O^- + H^+$
$R—NH_3^+$ Ammonium ion	9-10	$R—NH_2 + H^+$ Amino group

drates we find in cells. Carbohydrates have the general formula $(CH_2O)_n$, the ratio of hydrogen to oxygen being 2 to 1 as in water.

D-*glucose* is the living world's most widespread 6-carbon sugar or *hexose*. Five of its carbons have alcohol (—OH) groups, while the sixth terminal carbon forms an aldehyde (—CHO). In solution the straight chain form of glucose is in equilibrium with two ring forms which are **isomers** of each other.

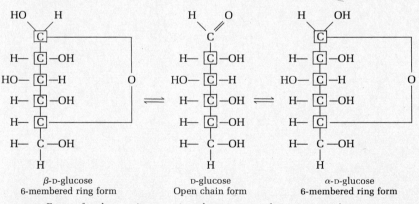

β-D-glucose
6-membered ring form

D-glucose
Open chain form

α-D-glucose
6-membered ring form

Forms of D-glucose. Asymmetric carbon atoms are shown in square boxes.

We shall use the following shorthand representation for the ring form of glucose:

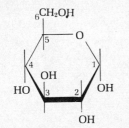

Simplified formula of α-D-glucose

The carbon atoms are represented by the corners of the hexagon and the hydrogen atoms by the end of the vertical lines. The thickened portion of the ring lies in the foreground.

Since glucose has 4 asymmetric carbon atoms (see straight chain formula above), 16 different optical isomers of glucose are possible. However, only three of them are found in nature.

Glucose and fructose are **monosaccharides,** that is, they are made of just one 6-carbon structure (hexose). Fructose, a hexose forming a 5-membered ring, combines with glucose to form the **disaccharide** *sucrose,* common table sugar.

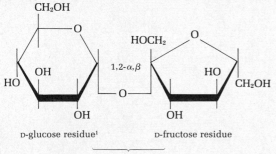

D-glucose residue[1]   D-fructose residue

Sucrose

But it is also possible to have **pentoses** which form five-membered rings. The most important one, *ribose,* we shall describe when discussing nucleic acids.

The bond joining two monosaccharides is created by a reaction which eliminates water. When the bond is broken by the addition of water, the reverse process (**hydrolysis**) takes place.

[1] The word **residue** refers to that portion of the hexose molecule that is left when it has reacted with another hexose and formed an —O— bond by the elimination of an —H from one hexose and an —OH from the other.

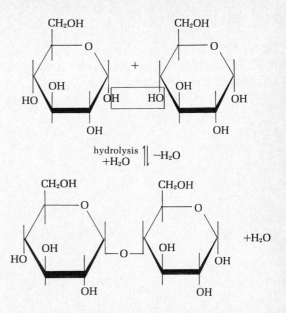

Monosaccharides, such as glucose and fructose, and disaccharides, such as sucrose, are the basic "coins" of energy exchange in nature. They are synthesized in green plants by the process of photosynthesis from carbon dioxide and water utilizing visible light as the energy source. The cell then releases the chemical energy stored in them by the stepwise breakdown of these carbohydrates all the way back to carbon dioxide and water (**respiration**). Thus the carbohydrates provide the cell with a readily available source of energy to power the many energy-requiring reactions necessary to maintain life.

When a number of monosaccharides are attached to each other, **polysaccharides** such as **starch, glycogen,** and **cellulose** are formed. Glycogen is a huge molecule consisting of some 30,000 glucose residues arranged in a branched pattern (Fig. 4-13). Glycogen is found in animal cells (for example, the liver) and is an energy storage device. By synthesizing such a huge molecule from many small molecules the cell effectively takes these small molecules out of solution. When energy is needed, these huge carbohydrate storage molecules can be broken down to di- and monosaccharides. Starch is the carbohydrate food storage product of plant cells. This molecule is made up of about 300 glucose residues forming unbranched chains in a helix.

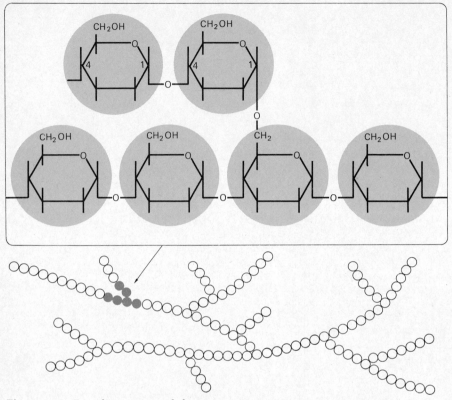

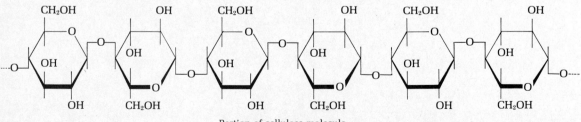

Figure 4-13   Branching pattern of glycogen.

Cellulose or chemical derivatives of it are the *structural* materials of plant cells. Cellulose forms long unbranched molecules made of over 1000 glucose residues which lie next to each other to form the very strong structural material that makes the cell wall surrounding plant cells. It is the major component of wood, cotton, flax, hemp, and many other materials used by man because of their tensile strength.

THE LIPIDS—HIGH ENERGY CONTENT AND SURFACE ACTIVITY

Let us look for a moment at a hydrocarbon like methane (CH_4). This compound is **saturated,** that is, it contains as much hydrogen as it possibly can. In carbon dioxide (CO_2), however, the carbon contains as much oxygen as it possibly can. When methane is heated in the presence of oxy-

Portion of cellulose molecule

gen it burns; that is, it becomes **oxidized** in stages until it ends up as CO_2 and H_2O and gives off energy. Since carbon in methane contains a maximum of hydrogen and no oxygen it is in its most **reduced** state and in that condition it contains the greatest amount of chemical energy. Carbohydrates with the general formula $(CH_2O)_n$ are in a more oxidized state than pure hydrocarbons and therefore contain less energy than they do. Cells cannot normally synthesize and store pure hydrocarbons, but in lipids they come close to doing so, for the ratio of hydrogen to oxygen in some lipids may be as high as 18:1. Lipids are therefore the most concentrated source of biological energy available to the cell.

Unlike carbohydrates, lipids are not necessarily related to one another in their chemical structure. What they do have in common is their solubility in **nonpolar organic solvents.** We have already pointed out that many substances dissolve in water; these are called *polar* substances, because, like water, they have some degree of separation of their centers of positive and negative electric charge. Nonpolar substances such as methane, for instance, have a very symmetrical charge distribution and will therefore not dissolve in polar solvents such as water but *will* dissolve in nonpolar solvents such as benzene or ether. All cellular substances which dissolve in these nonpolar solvents are defined as lipids. Fats are composed of two distinct types of compounds, glycerol and fatty acids, and are formed by the combination of one glycerol molecule with three frequently different fatty acids.

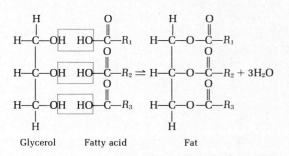

| Glycerol | Fatty acid | Fat |

The symbol R in fatty acids represents a saturated or unsaturated hydrocarbon chain frequently 16 or 18 carbons long, an example of which is capric acid.

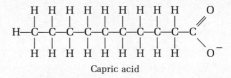

Capric acid

As shown below, using a shorthand notation, a saturated fatty acid has no double carbon bonds ($C{=}C$), while an unsaturated fatty acid will have one or more double bonds.

Saturated fatty acids

$CH_3(CH_2)_{14}COO^-$ palmitic acid
$CH_3(CH_2)_{16}COO^-$ stearic acid

Unsaturated fatty acids

$CH_3(CH_2)_7CH{=}CH(CH_2)_7COO^-$ oleic acid
$CH_3(CH_2)_4CH{=}CH(CH_2)CH{=}CH(CH_2)_7COO^-$ linoleic acid

The melting point of a fat depends on the degree of saturation of its fatty acids. If highly unsaturated, it is a liquid at room temperature and is then referred to as an **oil.** Vegetable oils can be **hydrogenated** and converted into fats. This is how margarine is manufactured. Fats, because of their energy-rich, long hydrocarbon chains and their insolubility in water, are excellent energy storage compounds in both animal and plants.

Fatty acids, however, have another important function. When examining a fatty acid molecule, it can be recognized that it has a split personality in that it consists of both an extremely polar, water-soluble end and an extremely nonpolar, water-insoluble portion. The "schizophrenia" of fatty acid molecules is beautifully demonstrated by their interaction with water—that is, they both do and do not dissolve in water (see below). They

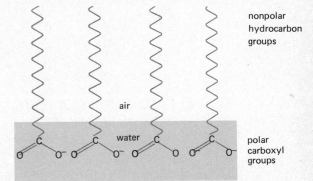

achieve this by forming a *monomolecular* layer at the surface of water, with the carboxyl groups sticking into the water and the hydrocarbon chains waving above the surface.

Because of these **surface-active** properties, fatty acids and lipids play an important role in the structure and function of biological membranes, about which we shall have more to say in subsequent chapters.

In addition to fats and fatty acids, a number of other lipids play important roles in the cell. We shall restrict ourselves to mentioning three examples.

Lecithin, shown below, is an important component of biological membranes.

The molecule has two long hydrophobic (fat soluble) carbon chains at one end and one hydrophilic (water soluble) end containing a negative and a positive charge. It is therefore so strongly surface-active that when added to water it forms a *bimolecular layer* with the nonpolar portions of the molecules on the inside and the polar portions toward the water. This brings about the tendency for lecithin to produce more and more surface, forming a convoluted maximum-area structure (myelin figures) which is very reminiscent of some biological membranes (Fig. 4-14). The bimolecular layer, with some modifications, represents our present concept of membrane structure (Chaps. 2 and 5).

Another class of lipids, which are totally unrelated in their molecular structure to the ones discussed so far, are the **steroids,** of which cholesterol is a well-known example.

lecithin

cholesterol

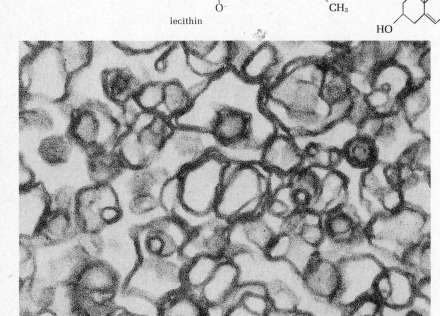

Figure 4-14 Phospholipid microdispersions in water. The phospholipids can be oriented to form vesicles (closed bags) in water. The vesicles which behave as osmometers are studied as model membranes. Some vesicles are multilayered. The interpretation is that each layer is a bimolecular leaflet of phospholipids with their polar groups pointed toward the water and their nonpolar groups pointed toward one another (×100,000).

Steroids occur in cell membranes, but they are also **hormones** noted for the regulatory role they play in multicellular animals (see Chapter 15).

Finally, we must mention **carotenoids** which are synthesized by plants, as being one of the pigments necessary for the process of photosynthesis. Animals cannot manufacture carotenoids, but since carotenoids are necessary for vision, animals must rely on plants for their supply. Carotene, shown at the bottom of this page, is one of the best known examples of this class of compounds.

And now we leave the lipids to discuss two groups of biological molecules, both of them huge, or *macromolecules* that more than any other group of molecules endow the cell with the properties which are characteristic of life.

NUCLEIC ACIDS — BIOLOGICAL INFORMATION STORAGE, RETRIEVAL, AND TRANSFER

When in 1868 Friedrich Miescher discovered in nuclei of pus cells a macromolecular substance containing phosphorus, he was only dimly aware that this important material was involved in the heredity of the cell. He called this substance *nuclein*, and for the remainder of his life he and his students made important contributions to our understanding of its chemistry. As early as 1884, Oskar Hertwig, who demonstrated previously that fertilization of the sea urchin egg involved the fusion of two nuclei, one from the egg and the other one from the sperm, had stated that *"Nuclein is the substance that is responsible not only for fertilization but also for the trans-*

mission of hereditary characteristics. . . ." This most fundamental biological generalization fell into oblivion, only to be resurrected again in the 1940s, when it became the central theme around which modern molecular biology has since developed with explosive rapidity.

What happened in the 1940s to convince biologists that nuclein, now called DNA, was the hereditary material?

First, there was the interesting fact demonstrated by many *cytological* investigations that DNA was located on the chromosomes, which as we have seen are the carriers of the genes. One particular chemical reaction called the Feulgen reaction produces a purple color with DNA specifically and was found to be an excellent cytological stain for chromosomes from both animal and plant cells. A second nucleic acid, RNA (ribonucleic acid), was found both in the nucleus, particularly in the nucleolus, and in the cytoplasm.

Second, the purification of viruses such as *tobacco mosaic virus* (TMV) by Stanley soon proved that DNA or RNA was always associated with these self-reproductive "supermolecules."

Third, as we pointed out in Chapter 3, Avery, MacLeod, and McCarty made the dramatic discovery that extracts from certain strains of pneumococcus bacteria could transform avirulent strains into virulent strains, and that this transformation was transmitted hereditarily. On purification of the extracts they found that the active material was DNA of the bacterial strain from which the extract was prepared.

We also discussed the experiment in which Hershey and Chase, using radioactive isotopes, demonstrated that when bacterial viruses enter their bacterial hosts and reproduce several hun-

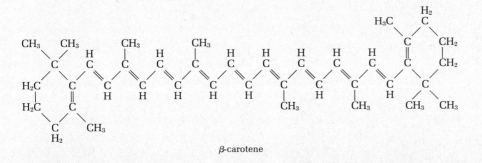

β-carotene

dred more viruses, only the DNA of the virus enters the cells, while the protein coat of the virus is left outside.

This last experiment dealt the death blow to a widely held belief that in fact proteins were the carriers of hereditary information. Today such a belief seems strangely quaint, but as late as 1941 many biologists felt that nucleic acids with their four bases were far too simple to carry the large variety of genetic information which the cell must be able to store.

With the concept in mind that nucleic acids must be the chemical compounds which store and transmit hereditary information, let us begin to describe the chemical properties of these interesting compounds.

First, it is important to remind the reader that from the point of view of organic structure and biological function there are two kinds of nucleic acids, DNA and RNA. As Figure 4-15 shows, the structural units making up DNA and RNA differ in two respects:

1. DNA uses deoxyribose. RNA uses ribose.
2. DNA uses thymine. RNA uses uracil.

The formulas used in Figure 4-15 are shorthand representations, which omit the carbon at the corners of the ring structures. The nitrogen bases bear closer examination, since they have not been encountered here before and since, as we shall see, they are the basic alphabet of the hereditary material. Notice that DNA and RNA have two kinds of bases, the smaller **pyrimidines** and the larger **purines.** Notice also that one of the purines and one pyrimidine contain a carbonyl group ($-CO$), while the other purine and pyrimidine contain an amino group ($-NH_2$) in the same position. We shall return to this most important observation shortly.

How are these subunits consisting of phosphate, pentose, and nitrogen bases linked together in DNA and RNA? The answer turns out to be extremely simple. First, we have a *backbone* consisting of alternate sequences of phosphate–sugar-phosphate-sugar, etc. Second, connected to each ribose is a nitrogen base. There are no chemical restrictions as to which base is to be connected. Any of the four bases can fit a particular ribose molecule (Fig. 4-16). It is important to study precisely how these subunits are interconnected, how the phosphates connect to the third and fifth carbon atom of the pentose molecule, how a nitrogen of the nitrogen bases connects to the first carbon atom of the pentoses, and

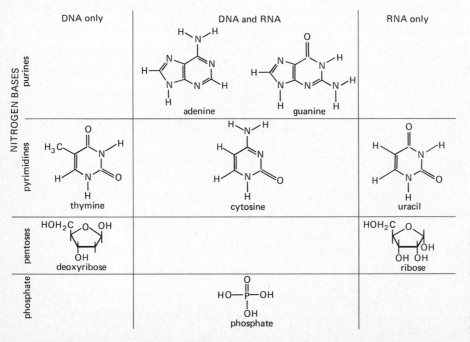

Figure 4-15
The building blocks of DNA and RNA. Memorize the structure of these important units from which the language of heredity is constructed.

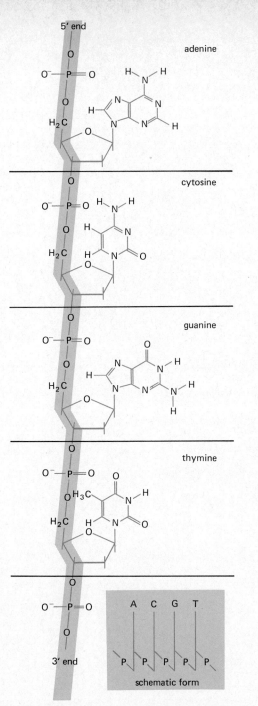

Figure 4-16 How the building blocks of DNA are connected to each other to form the DNA polynucleotide. In the language of chemistry: the deoxyribose nucleotides are connected with 3′, 5′-phosphodiester bridges. The RNA polynucleotide has the same structure, except that ribose and uracil are used instead of deoxyribose and thymine.

how the whole structure can be summarized in schematic form (Fig. 4-16, lower right).

The strand of DNA just described can be extremely long; in the bacterium *Escherichia coli*, for instance, it is almost 1 millimeter long and has a molecular weight of 2×10^9 daltons.[1] How can such a gigantically long thread be synthesized in the cell and how is it "packaged," that is, how is it arranged in space? It turns out, surprisingly perhaps, that the answer to the second question provided a key for the answer to the first.

In 1953 Watson and Crick published an article in the journal *Nature* in which they proposed a three-dimensional structure for DNA. The proposal was based on (1) the observation by Chargaff that in DNA the amount of adenine equals the amount of thymine and the amount of guanine equals the amount of cytosine ($A/T = G/C = 1$); and (2) the conclusions by Franklin and Wilkins, using the technique of **x-ray crystallography**, that DNA is a long thin rod 20Å in diameter and that there is a *repeating structure* along this rod every 34Å.

Watson and Crick proposed that the long thin rod of DNA is composed of two strands twisting around one another like the edges of a spiral staircase, with the twists 34Å apart, thus accounting for the 34Å repeats observed by x-ray crystallographers. They suggested that the nitrogen bases point inward toward each other and that they are arranged in pairs, like the steps of a spiral staircase (Fig. 4-17a), with A always opposite T and G always opposite C. This *pairing of bases* not only fits the size requirements of the proposed double helix but, dramatically, the proper *base pairs* when placed next to one another show a complementarity of chemical structure which permits the formation of two hydrogen bonds in the case of A and T and three hydrogen bonds in the case of G and C (Fig. 4-17b). Thus the DNA structure proposed by Watson and Crick accounted perfectly for the observations on DNA structure made by Chargaff, Franklin, Wilkins, and others. But more importantly, the structure provided a clue for the synthesis of

[1] One dalton is the mass of one hydrogen atom or 1.67×10^{-24} gram. Thus one molecule of water weighs 18 daltons. One gram is about 6×10^{23} daltons.

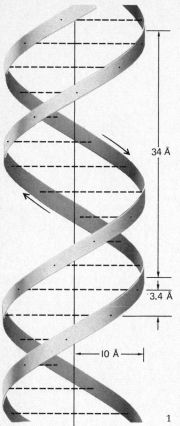

34 Å

3.4 Å

10 Å

1

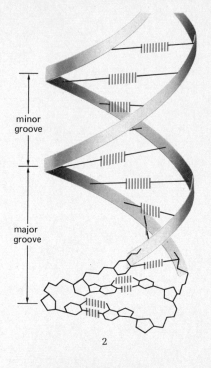

minor groove

major groove

2

Figure 4-17
A. The spiral staircase (double-stranded helix) arrangement of DNA.

1 and 2: open models

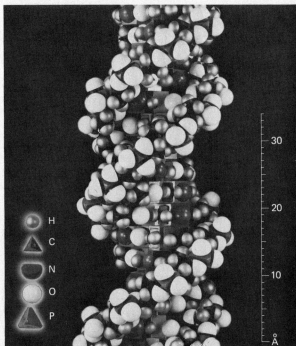

H
C
N
O
P

3: space-filling models

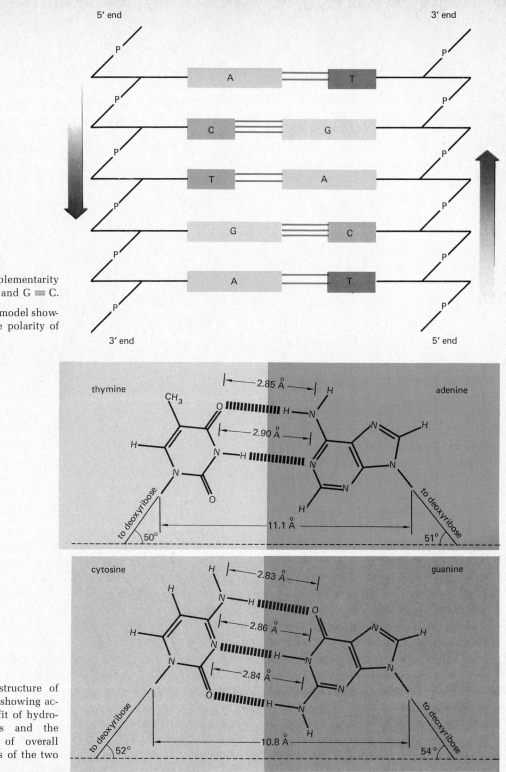

5' end

3' end

B. The complementarity of A = T and G ≡ C.

1. shorthand model showing reverse polarity of strands.

3' end

5' end

thymine

adenine

|— 2.85 Å —|

|— 2.90 Å —|

|——— 11.1 Å ———|

50°

51°

to deoxyribose

to deoxyribose

cytosine

guanine

|— 2.83 Å —|

|— 2.86 Å —|

|— 2.84 Å —|

|——— 10.8 Å ———|

52°

54°

to deoxyribose

to deoxyribose

2. chemical structure of base pairs showing accuracy of fit of hydrogen bonds and the similarity of overall dimensions of the two base pairs.

DNA in the cell and thus for the perpetuation of the hereditary material, a phenomenon which is so fundamental to life.

It is important to recognize that any theory for the synthesis of the hereditary material must provide a mechanism whereby it can be involved in its own duplication. Thus if DNA is the hereditary material, it must have the property of *self-duplication*. The structure of DNA which Watson and Crick proposed did indeed have such a possibility. One simply had to imagine that through base pairing each strand could specify the synthesis of a *complementary* strand, possibly by starting at one end, untwisting itself, and twisting up two new strands each composed of one parent strand and one *replicated* strand (Fig. 4-18). Thus the DNA structure proposed by Watson and Crick provided a plausible model

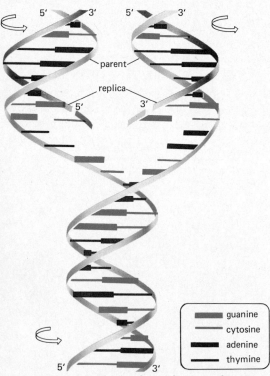

5'   3'   5'   3'

parent

replica
5'   3'

guanine
cytosine
adenine
thymine

5'   3'

Figure 4-18   A model for the replication of DNA. As the new strands are laid down, causing the parent strand to unwind, the daughter strands must also rotate. Some assumptions in this model are still unproven.

for the self-reproduction of DNA, while at the same time offering an explanation for Chargaff's observation of *base equality* (A = T and G = C).

The DNA model provided us with an insight into how the DNA molecule was replicated. But these questions arose: first, as to how the hereditary information was in fact stored in the molecule and second, how the information was transmitted to other cellular processes where it could have a controlling effect on them. In 1953 not enough regarding the role of RNA was known to answer these questions. However, Watson and Crick did point out that the four types of nitrogen bases could be considered to form some sort of **linear code** spelling out in special biological language the orders that control the cellular processes. We now know many of the details of this process and we shall discuss them in Chapter 6. It is sufficient to say here that very soon after the structure of DNA was enunciated, the general concept began to emerge that RNA was involved in the process of information transfer by taking the information from the DNA and in some way controlling the synthesis of proteins.

Let us therefore briefly look at RNA to see whether its structure tells us anything about its important function.

First, RNA was found to be single-stranded.

Second, it was found to occur in various places in the cell, in a wide range of sizes. The fact that one form of RNA was observed to occur in both the nucleus and cytoplasm led to the prediction and ultimate proof that it played a role in conveying information from DNA to the cytoplasm, hence its name, messenger RNA (mRNA). Other RNAs were an integral part of the structure of ribosomes (ribosomal RNAs). Yet a third class, whose role we consider in Chapter 6, included a number of low molecular weight transfer RNAs (tRNAs).

In the years that followed the publication of the Watson-Crick hypothesis, the structure they proposed was amply confirmed by numerous observations. Today it stands, along with the theory of evolution and Mendel's laws, as one of the most firmly established "facts of life," as well as one of the most fruitful insights into the nature of living matter.

PROTEINS—SPECIFICITY OF STRUCTURE AND CONTROL

Proteins are the basic structural and functional compounds of the cell, for not only do they constitute the major portion of the *dry weight* of an actively growing cell but they also are the agents which *control* or *regulate* most of the processes which occur in the cell.

In recent years we have learned a great deal about the structure of proteins and this has given us important insights into how they function.

Every cell contains a very large number of different proteins. The smallest bacterium probably contains a few thousand, while the cells of a mammal probably have 100,000 different types of protein molecules. These proteins differ in composition and size, having molecular weights ranging from thousands to millions of daltons. Proteins differ widely from each other in their specific properties and therefore it is possible to separate them from each other and to prepare them in pure form.

All proteins are built from, at the most, 20 different building blocks which occur universally in living matter. These building blocks, called **amino acids,** have at one end of their molecule a common structure—a carboxyl and an amino group attached to an asymmetric carbon atom— and at the other end a group (**R group**) which is characteristic of each amino acid (Fig. 4-19). Amino acids can be separated from each other by a technique called **chromatography** in which they are made to move along the surface of a solid, being transported by a pair of solvents that are miscible in each other but in which the different amino acids are soluble to different extents. The solid may be in the form of a flat sheet, in which case it is **paper** or **thin layer chromatography,** or it may be finely divided and packed into a cylindrical tube, in which case we refer to **column chromatography** (Fig. 4-20).

Amino acids have at least two **dissociable groups,** an α-carboxyl group, which has a **pK$_a$** of 2–3, and an α-amino group with a pK$_a$ of 9–10. There are a few amino acids, such as glutamic acid and lysine, which have dissociable **side chains** or **R groups.**

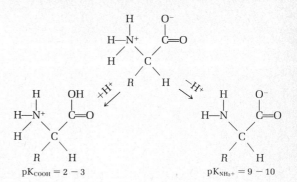

The above equation shows the dissociation of amino acids. At intermediate pH an amino acid is both positively and negatively charged. Adding protons suppresses the dissociation of the —COO$^-$ group; removing protons brings about the dissociation of the —NH$_3^+$ group.

With the exception of glycine, amino acids have an asymmetric carbon atom. They can therefore exist in the form of two optical isomers, but it is a remarkable fact that only *l-amino* acids are found in naturally occurring proteins.

Amino acids are released when proteins are heated in concentrated solutions of strong acids. The great chemist Emil Fischer proved that the bond which is broken by this procedure is the **peptide bond,** formed between α-amino and α-carboxyl groups, and shown below.

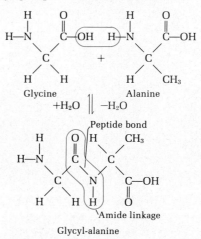

Two amino acids react to form a dipeptide held together by a **peptide bond.**

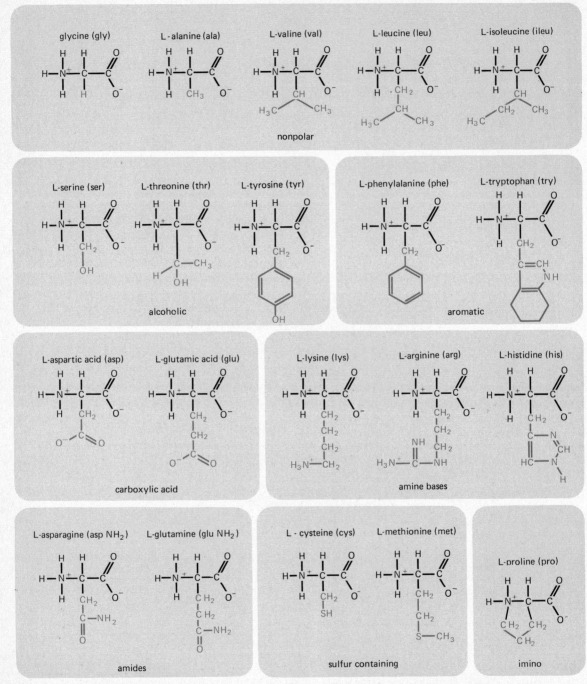

Figure 4-19 The amino acids, the alphabet of protein structure. The **R groups** are colored and consist of that portion of the amino acid which is characteristic of each amino acid. The student should memorize this 20-letter alphabet of protein structure and function. To aid the student in this task we have divided the amino acids into eight groups with similar chemical properties.

Figure 4-20 Chromatography, an important technique used for separating amino acids and other compounds.

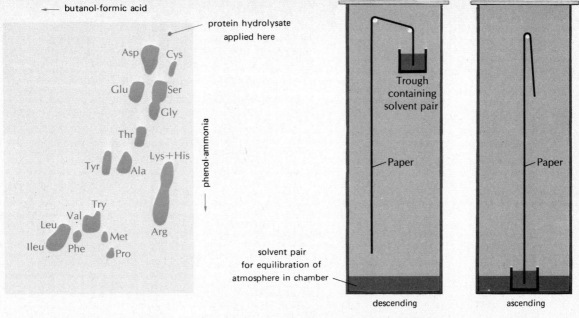

two-dimensional chromatogram

two types of chromatography apparatus

A. **Paper chromatography.** The amino acid mixture is applied in one corner, the paper is inserted in the chromatography chamber, and a given pair of miscible solvents either ascends by capillary action or descends by siphon action. As the solvents move through the filter paper, the amino acids are partially separated along one dimension. Further separation is achieved by turning the paper through 90° and repeating the operation with a different solvent pair. The position of each amino acid is then determined by spraying the filter paper with ninhydrin, which produces a purple color with amino acids. The position of radioactive amino acids can be determined by placing the chromatogram on a sheet of X-ray photographic paper, which after development will show the amino acids as darkened spots. Glutamine and asparagine are deaminated by acid hydrolysis to glutamic and aspartic acids, respectively, and therefore do not appear on the chromatogram.

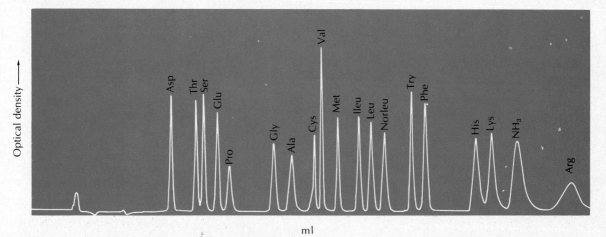

B. **Column chromatography.** Chromatogram of a mixture of amino acids obtained by using an automated amino acid analyzer. This run took 2 hours and utilized 0.05 μmole of each amino acid. Norleucine is not a naturally occurring amino acid but is used here as an "internal standard." Asparagine and glutamine are missing because they are deaminated during hydrolysis of the protein with hydrochloric acid.

The breaking of the peptide bond is called hydrolysis, because a molecule of H_2O is added to the *amino acid residues* to form the free amino acid.

It was also demonstrated that the amino acids in proteins form long unbranched chains or **polymers,** which we call **polypeptides.** This is a deceptively simple generalization. In fact, it is one of the major laws of nature. Let us state it again for emphasis. *Proteins are made of one or more linear unbranched chains composed of up to twenty different amino acids in varying numbers and sequences.*

The next question one might ask is how are these amino acids arranged? Is the arrangement statistical as indeed in many polymers synthesized in the laboratory it turns out to be, or is the sequence finite for each polypeptide chain in a given protein molecule? As Fred Sanger showed in an epoch-making study in the 1940s, proteins are finite, although complex, chemical compounds in which the sequence of amino acids is definite and unique for each polypeptide chain. Figure 4-21 shows the structure of the protein, *insulin*, as worked out by Sanger. Notice that insulin happens to be made of two different poly-

peptide chains held together by **disulfide bonds,** which are formed between the side chains of two cysteine residues. Another disulfide bond occurs between two cysteines within the same chain. Since Sanger's work, the amino acid sequences in a number of other proteins have been determined, and it is now evident that proteins can be made of a single chain or of as many as six chains. The next question which arises is how are these chains arranged in space?

From a variety of physical studies ranging from x-ray diffraction to electron microscopy it has become evident that proteins are usually *globular*, that is, nearly spherical in structure. Since polypeptide chains are long thin molecules, it is clear that, in order to form a globular, compact molecule, the polypeptide must be *folded* in some way. Using the powerful technique of x-ray crystallography, Kendrew and his co-workers in the late 1950s and early 1960s worked out the structure of *myoglobin* and showed that its single polypeptide chain formed a globular molecule by twisting itself into a helix which is interrupted in various places to allow the helices to double back on each other to form a compact although irregular structure (Fig. 4-22).

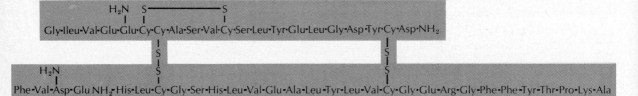

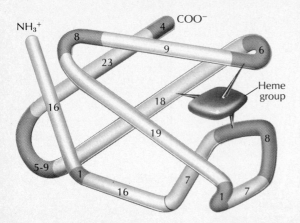

Figure 4-21 (Above). The chemical structure of the protein insulin. This protein consists of two polypeptide chains. Besides the peptide bonds which are present between amino acids, there are also disulfide (—S—S—) bonds, one of which occurs within one of the chains and two others which hold the two chains together.

Figure 4-22 (Left). The three-dimensional structure of myoglobin, a globular protein. This protein is composed of eight straight helixes interrupted by nonhelical portions (colored segments) which form corners of various degrees of tightness and allow the formation of a compact, globular protein molecule. The numbers indicate how many amino acid residues are found within the indicated segment.

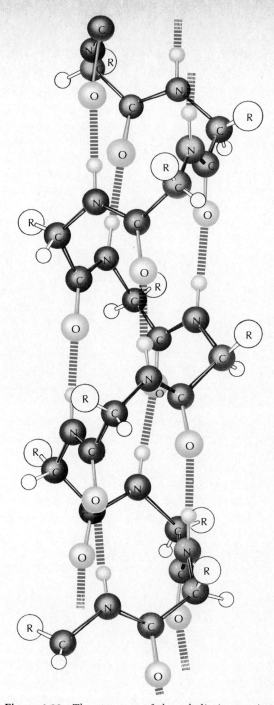

The helix, which Kendrew discovered by direct observation, had in fact been predicted a decade earlier by Pauling and Corey. This so-called *α-helix* is a beautifully ingenious device of nature in which all ⟍N—H and ⟍C═O groups on the polypeptide backbone can form hydrogen bonds with each other (Fig. 4-23).

If hydrogen bonds stabilize the α-helix, what brings about the formation of the folding of the polypeptide chain to form the compact globular protein molecule? It turns out that the side chains of the amino acids are largely responsible for this, since a number of different polar and nonpolar interactions are involved in the formation of the final folded structure.

Recent studies of the *fine structure* of other protein molecules show, as might be expected from the specificity of function of each protein molecule, that no single pattern of protein structure exists. We now believe that the specific structure of each kind of protein molecule is derived from the sequence of amino acids in its polypeptide chains. Since the sequence of amino acids in each type of chain is unique, the final folding pattern will also be unique for each type of protein molecule.

The perceptive student might at this point already see the emergence of a picture of breathtaking generality yet remarkable simplicity. It goes as follows: At one end of the scale we have the proteins, which control the processes of the cell. At the other end we have DNA, which stores the hereditary information. How can the information locked in DNA be used to affect the processes of the cell? The answer turns out to be surprisingly simple. *DNA controls the structure and hence the function of proteins.* The details of how this happens are discussed in Chapter 6, but it would be unfair to leave the student without some clues.

Let us start with the compact, folded protein molecule and work backwards. Its molecular structure allows it to control a given process in the cells. Its structure is determined by the linear arrangement of amino acids in the polypeptide chain. This linear order has in turn been determined by a linear code based on the sequence of

Figure 4-23 The structure of the α-helix in proteins. Every ⟍C═O group and every ⟍N—H group can be united by a hydrogen bond. The R groups represent the side chains of any one of the 20 amino acids and point away from the helix.

nitrogen bases in certain messenger RNA molecules and in the cytoplasm. The sequence of nitrogen bases in this messenger RNA had previously been determined by the sequence of nitrogen bases in the DNA of the nucleus. So, conversely, starting with a linear sequence of nitrogen bases in DNA we end up with a linear sequence of amino acids in the protein, which in turn determines the three-dimensional organization of the protein and thus finally determines the biological function of the protein.

Thus the unique information in the DNA of

	human	monkey	pig, bovine, sheep	horse	dog	rabbit	kangaroo	chicken, turkey	duck	rattlesnake	turtle	tuna fish	moth	neurospora	candida	yeast
human	0															
monkey	1	0														
pig, bovine, sheep	10	9	0													
horse	12	11	3	0												
dog	11	10	3	6	0											
rabbit	9	8	4	6	5	0										
kangaroo	10	11	6	7	7	6	0									
chicken, turkey	13	12	9	11	10	8	12	0								
duck	11	10	8	10	8	6	10	3	0							
rattlesnake	14	15	20	22	21	18	21	19	17	0						
turtle	15	14	9	11	9	9	11	8	7	22	0					
tuna fish	21	21	17	19	18	17	18	17	17	26	18	0				
moth	31	30	27	29	25	26	28	28	27	31	28	32	0			
Neurospora	48	47	46	46	46	46	49	47	46	47	49	48	47	0		
Candida	51	51	50	51	49	50	51	51	51	51	53	48	47	42	0	
yeast	45	45	45	46	45	45	46	46	46	47	49	47	47	41	27	0

Figure 4-24   Differences in amino acid composition of the protein cytochrome *c* obtained from a variety of organisms on the evolutionary scale. A comparison of the number of amino acid differences between the enzyme cytochrome *c* of man and of other organisms. Notice that the more remote the kinship of two organisms, the greater are the differences in amino acid composition. By a detailed study of the pattern of substitutions, it is possible to deduce the kinships of organisms—who is related to whom—and construct a genealogy called a phylogenetic tree (Chapter 21). Kinship relations deduced from biochemistry closely resemble those based on generally accepted morphological grounds. (From the *Atlas of Protein Sequence and Structure* 1967-68, Margaret O. Dayhoff and Richard V. Eck, National Biomedical Research Foundation, Washington, D.C., 1968.)

living cells becomes *transcribed* (DNA to RNA) and *translated* (RNA to protein) into the structure of many kinds of proteins, each of which has its own special structure and function. In fact, not only does *myoglobin*, for instance, differ from *cytochrome c* in a given cell, but the cytochrome *c* of one species differs slightly from the cytochrome *c* of a closely related species. The difference between two kinds of cytochrome *c* becomes greater as more distantly related species are compared, and so we find that we can use the amino acid sequence of proteins to study evolution (Fig. 4-24). And this, after all, should not surprise us, for we might expect that evolution, which brings about a change in the genetic information of a species, must also bring about a change in the amino acid sequence and hence in the protein structure of that species. Thus by studying evolution through changes in protein structure, we come much closer to the fundamental genetic changes in the DNA than we do if we look at grosser characteristics such as the shape or behavior of the organism.

We have come to the end of our discussion of the molecules of the cell and this completes the descriptive phase — optical (Chapter 2), genetic (Chapter 3), and chemical (Chapter 4) — of the cell. In the following two chapters, we combine these three approaches to see how the cell handles the flow of molecules, of energy, and of information, and how it manages to regulate these with respect to one another to produce the smoothly functioning cellular activity which is the basis of all biological phenomena.

SUGGESTED READING LIST

Anfinsen, C.B., *The Molecular Basis of Evolution.* John Wiley and Sons, New York, 1963.

Baker, J.J.W. and G.E. Allen, *Matter, Energy and Life,* second edition. Addison-Wesley, Reading, Massachusetts, 1970.

Barker, R., *Organic Chemistry of Biological Compounds.* Prentice-Hall, Englewood Cliffs, New Jersey, 1971.

Beadle, G., "The genes of men and molds," *Scientific American* offprints, W.H. Freeman, San Francisco, 1948.

Dickerson, R.E. and I. Geis, *The Structure and Action of Proteins.* Harper and Row, New York, 1969.

Gamow, G., *Mr. Tompkins Explores the Atom.* Cornell University Press, Ithaca, New York, 1960.

Hardwick, E.R. and C.M. Knobler, *Chemistry, Man and Matter.* Ginn and Co., Waltham, Massachusetts, 1970.

Henderson, L.J., *The Fitness of the Environment.* The Beacon Press, Boston, Massachusetts, 1958.

Kendrew, J.C., "Myoglobin and the structure of protein," *Science,* 139, 1259 (1963). (Bobbs-Merrill Reprint Series, Howard W. Sams and Co., Indianapolis, Ind.)

Lehninger, A.L., *Biochemistry.* Worth, New York, 1970.

Loewy, A.G. and P. Siekevitz, *Cell Structure and Function,* second edition. Holt, Rinehart and Winston, New York, 1969.

Sanger, F. and L.F. Smith, "The structure of insulin," *Endeavour,* 16, 48 (1957). (Bobbs-Merrill Reprint Series, Howard W. Sams and Co., Indianapolis, Ind.)

Watson, J.D., *The Double Helix.* Atheneum, New York, 1968.

Watson, J.D., *The Molecular Biology of the Gene,* second edition. W.A. Benjamin, New York, 1969.

Watson, J.D. and F.H.C. Crick, "Genetic implications of the structure of deoxyribonucleic acid," *Nature,* 171, 964 (1953). (Bobbs-Merrill Reprint Series, Howard W. Sams and Co., Indianapolis, Ind.)

CHAPTER 5

The Flow of Molecules and of Energy

Most tropical fish enthusiasts are familiar with a common water plant called *Elodea* (Fig. 5-1) which can be purchased at any pet shop. If a sprig of *Elodea* is cut through its stem with a razor blade and placed upside down in a beaker of water, an interesting experiment can be performed. If we look at the cut end of the stem in extremely weak light, we shall probably fail to see any activity. Suppose we now expose the plant to bright light; we soon see a stream of bubbles rising from the cut stem. If the rate at which these bubbles are produced is measured and then the plant is shaded slightly, we can observe a decrease in the rate of bubble production. Adding a pinch of baking soda (sodium bicarbonate), which increases the carbon dioxide concentration of the water, increases the rate of bubble production. The bubbles which *Elodea* produces can be collected in a test tube (Fig. 5-1), and by performing the well-known glowing splint experiment (a glowing splint bursts into flame in the presence of oxygen), we are able to demonstrate that the gas bubbles consist of oxygen which is the end product of a complex **metabolic process** which starts with the absorption of carbon dioxide and water by the plant and utilizes light energy to convert these two substances into *carbohydrates* and oxygen.

We shall discover over and over again in the following chapters that living organisms are highly organized systems that perform work. To build up this organization as well as to maintain it, and to perform work of various kinds, living organisms must utilize energy. The primary source of energy for almost all organisms on earth is **photosynthesis,** a process which, as we saw in *Elodea,* ends in the production of oxygen. It is directly responsible for providing the energy needed by plants, but it is also indirectly responsible for the life of animals and microorganisms, for although some animals feed on other animals, you will encounter an animal that derives its energy from eating plants if you follow this **food chain** back far enough.

WHAT IS METABOLISM?

Between the time sodium bicarbonate was added to the water and the time an increase in rate of oxygen production was observed a large number of physical and chemical reactions took place. These reactions involved various complex cellular structures and many different molecules. The study of **metabolism** attempts to describe the myriad of molecular processes involved in the life of the cell. The synthesis of carbohydrate and accompanying release of oxygen in photosynthesis are only the beginning of the metabolic flow of the cell's molecules and energy. The carbohydrates are broken down again to carbon dioxide and water (respiration) and the energy thus released is used to synthesize the high-energy compound, adenosine triphosphate (ATP), which "drives" the various energy-requiring, work-producing processes of the cell (Fig. 5-2). Many of these energy-requiring processes such as movement are very familiar because everyone knows energy is necessary for mechanical work. But as we have seen in Chapter 2, there are also more subtle and less well-known processes such as: the accumulation of molecules inside the cell (osmotic work), the production of electricity (electrical work), the synthesis of large and complex molecules (chemical work), and the control and interweaving of cellular processes (regulational work).

Thus we can see that the **respiratory metabolism** of the cell produces energy, which is then utilized by the cell in a number of processes involved in one or another form of work. Clearly the energy produced by the respiratory metabolism must be made available in some sort of usable way to the various machines (chemical, mechanical, electrical, or osmotic) which carry out the work. Some sort of intermediary compound must exist, and in fact it does. As we have already remarked, this intermediary, which is the main "coin" of energy exchange in the cell, is ATP. This nucleotide composed of adenine, ribose, and

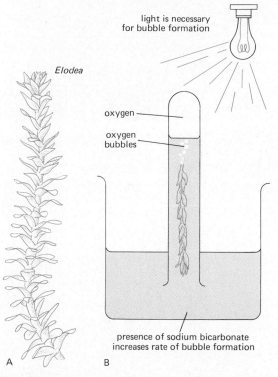

light is necessary
for bubble formation

Elodea

oxygen

oxygen
bubbles

presence of sodium bicarbonate
increases rate of bubble formation

A   B

Figure 5-1.

A. The water plant *Elodea.*

B. Photosynthesis in *Elodea* produces oxygen bubbles. The rate of bubble production can be used to measure the rate of photosynthesis which is affected by factors such as light intensity, carbon dioxide concentration, temperature, etc.

three phosphate groups is called a **high-energy phosphate compound,** because when the last two phosphates are split off by hydrolysis, as much as 9000 cal/mole[1] of ATP are released.

Other organic phosphate compounds, some of which we shall encounter later, are also high-energy phosphate compounds; ATP can be synthesized from ADP by using the energy resulting from the breakdown of such a compound. Conversely, ATP can supply the energy necessary for the synthesis of a high-energy compound by breaking down to ADP. Such processes by which energy is transferred to and from ATP are described as being **energetically coupled** and are symbolized as follows:

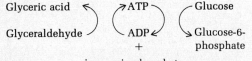

Glyceric acid   ATP   Glucose

Glyceraldehyde   ADP   Glucose-6-phosphate

\+

inorganic phosphate

[1] One cal (or calorie) is the amount of energy required to raise the temperature of 1 gram of water from 14.5 to 15.5°C.

One mole is the number of molecules needed to give the weight in grams which is numerically equal to the number obtained when the atomic weights of all the atoms in a molecule are added together.

In this scheme the oxidation of glyceraldehyde to glyceric acid is coupled through ATP to the synthesis of glucose-6-phosphate from glucose. The net effect is that some of the energy in glyceraldehyde is transferred to glucose-6-phosphate.

The important thing to recognize about high-energy phosphates is that the release of 9000 cal/mole represents a quantity or package of energy which appears to be of just the right magnitude to "drive" chemical reactions to higher energy states without at the same time releasing too much energy and thus threatening the chemical integrity of the machine in which the reactions take place. Remember that the cell is a machine that itself is made up of the kind of molecules it consumes or synthesizes, and thus the problem for the cell is to carry out the process in sufficiently small steps so as not to burn itself up. The 9000 cal/mole are of the right magnitude to fill this need, and it is therefore hardly surprising that ATP is the universal molecule for energy exchange. Another advantage of using many small steps in the metabolism of the cell rather than a few large ones is that, according to the laws of thermodynamics, this allows for greater efficiency and therefore less waste of energy.

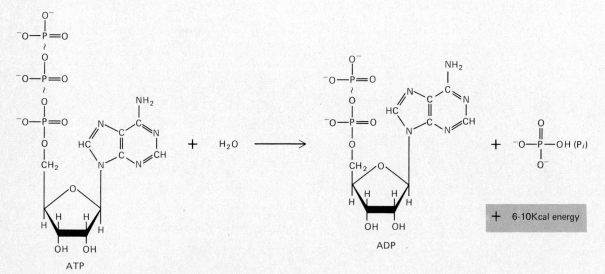

Figure 5-2. The hydrolysis of ATP yields six to ten kcal. of energy per mole. This "quantum" of energy is used by living cells to drive their various energy-requiring processes.

Respiration provides not only the energy for numerous important cellular processes; it also provides, at various stages of the breakdown, the building blocks which the cell uses as raw materials for the synthesis of its various parts.

In this chapter we describe some of the molecular details of the processes we have just mentioned. We begin with carbon dioxide molecules and light rays outside the plant cell and follow the flow of molecules and energy as it takes us through some of the major metabolic processes of the cell. Before we do this, however, we must become acquainted with the properties of **enzymes,** the protein molecules which through their catalytic properties control the processes to be described.

ENZYMES — CATALYSIS AND CONTROL

Organic compounds, based on the covalent chemistry of carbon, generally react very slowly with one another. For instance, glucose, if exposed to the oxygen of the air, will react so very slowly that it would take many millions of years before it all becomes converted to carbon dioxide and water. All that is necessary to speed up the reaction is to heat the mixture of glucose and oxygen to the point at which it *ignites.* At this point the reaction proceeds with sufficient speed, releasing enough energy to maintain the temperature of the mixture at its ignition temperature. We call this process *burning.* Thus we learn that glucose can oxidize rapidly if its temperature is raised sufficiently.

In the cell, however, glucose can be oxidized quite rapidly at much lower temperatures, such as 10 to 20°C. This is possible because enzymes, the protein catalysts of the cell, speed up in very specific ways the many steps the cell uses in converting glucose to carbon dioxide and water.

Enzymes play a dual role in the cell: they can *speed up* the rates of chemical reactions and by so doing they can *control* or *regulate* the rates of reactions so as to integrate or intermesh the processes occurring in the cell. Thus we conclude that enzymes are not only **catalysts** but also **regulators.** The question is: how do they manage to do this?

The reason for the slowness of many reactions involving organic compounds despite the large amounts of energy released by the reactions is shown in Figures 5-3, 5-4, and 5-5.

Figure 5-3 shows an energy profile of a typical organic reaction. Notice that the energy of the reactants is higher than that of the products so that energy is released by the reaction. Notice, however, that for the reaction to proceed it must first go through an **activated state,** which has a higher energy content than the reactants have to begin with. If there are very few molecules at room temperature with enough energy to form the activated state, the reaction will proceed very slowly.

Figure 5-4 shows why raising the temperature will greatly increase the rate of such a reaction. Notice that at 20°C the *energy distribution* of the molecules is such that very few have as much as,

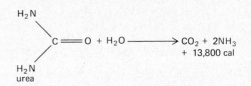

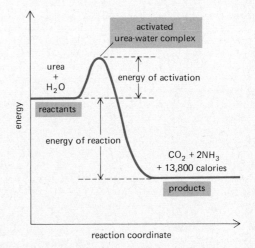

Figure 5-3. Energy diagram of the hydrolysis of urea showing that in order to react, urea and water must have sufficient energy to form activated complex. Since at room temperature very few molecules have sufficient energy to form the activated complex, the rate of the reaction is exceedingly slow.

or more than, the activation energy. Notice, however, as the temperature is raised to 100°C, the distribution of energies of the molecules shifts in such a way as to appreciably increase the proportion of molecules with energies as great as or greater than the activation energy. Therefore at

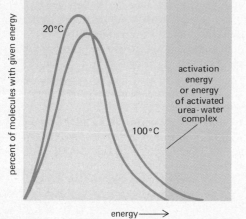

Figure 5-4. Energy distribution of urea in water at two temperatures in relation to activation energy necessary for hydrolysis of urea. At 100°C a much higher proportion of molecules have energies equal to or greater than the activation energy; this explains the considerable effect of temperature on the rate of a chemical reaction.

Figure 5-5. How enzymes speed up a chemical reaction.

100°C the reaction will be able to proceed considerably faster than at 20°C.

Figure 5-5A explains why the enzyme manages to speed up reactions at 20°C which ordinarily can be speeded up by heating the reactants. The diagram shows that the enzyme brings about the formation of an activated state at a much lower energy level than that of the uncatalyzed activated state. Figure 5-5B shows that, as a result of lowering the energy of activation, the enzyme allows a much higher proportion of molecules to react and therefore permits the reaction to proceed faster.

The next question which arises is: how does the enzyme succeed in lowering the energy of activation of a given reaction? Before we answer this question, let us first list a few properties of enzymes in order to become familiar with these interesting molecules.

1. All enzymes are proteins, although they may sometimes require cations or other organic compounds for their activity. If these compounds are permanently attached to the enzyme, they are called **prosthetic groups** (for example, the *heme* group in *hemoglobin*). If the attachment is very brief, however, we call them **coenzymes** (for example, *acetyl coenzyme A*).

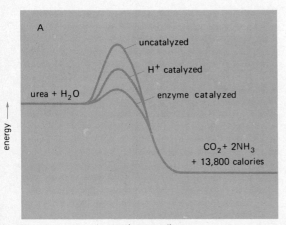

A. Urease lowers the **activation energy** by forming an **enzyme-substrate complex.**

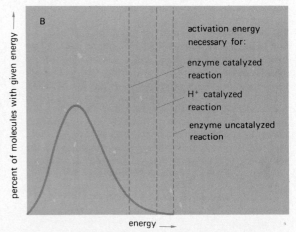

B. By lowering the activation energy a larger proportion of urea molecules can react.

2. Enzymes are highly specific for the particular type of reaction they catalyze. However, the degree of specificity varies from enzymes that can catalyze the reaction of only one compound to some which can catalyze the reactions of a given group no matter what else the remaining part of the molecule happens to be. Most enzymes are specific for only one of the two optical isomers. The compounds whose reactions are catalyzed by enzymes are called **substrates.**

3. Enzymes are highly dependent on pH, having a rather sharp maximum velocity at a pH value which is characteristic for a given enzyme.

4. Like other proteins, whose structure is easily altered at temperatures of 40 to 70°C, enzymes are sensitive to elevated temperatures, although to varying degrees.

5. Enzymes are active at extremely low concentrations. This means that, like other catalysts, their reaction with a given substrate molecule is temporary and the enzyme is released for repeated cycles of activity.

6. Because enzymes are active at low concentrations, it is possible for compounds which inactivate enzymes (**enzyme poisons**) by reacting with them also to be active in very low concentrations.

7. Finally, and most important, enzymes react with the substrate to form a short-lived **enzyme-substrate complex.**

We now return to the question of how enzymes act. The picture which emerges is that the enzyme has a special region on its macromolecular surface that has a special configuration to *fit* closely with the substrate. This region, called the **active site,** is composed of chemically active side chains of the amino acid residues. Thus there may be charged dissociated groups such as the carboxyl group of glutamic acid, and the amino group of lysine, or there may be a histidine side chain or possibly a sulfhydryl group. Such groups are believed to be placed in a special arrangement at the surface so as to fit the substrate closely. Furthermore we have evidence that some of the groups may react temporarily with the substrate. As a result of this multiple interaction between enzyme and substrate, the reaction of the sub-

strate which normally requires a high energy of activation can occur at the much lower energy of activation of the *enzyme-substrate complex* (Fig. 5-5).

The mechanism of enzyme action just described may sound rather abstract, but to document it with specific examples is beyond the scope of this book. However, Figure 5-6 shows the enzyme lysozyme reacting with its substrate, a polysaccharide chain. Notice how the chain fits on a groove in the enzyme surface and how the chain interacts with a number of side chains of amino acid residues. Notice also the arrow indicating where the chain will break as a result of the enzyme's action; the two halves of the substrate chain will then separate from the enzyme and a new chain will attach itself with great precision for another round of reactions leading to the splitting of the new chain.

There is even more recent evidence suggesting that the enzyme's active site changes in configuration when the substrate moves into position on the active site. Such highly specific alterations in the enzyme active site might in fact provide us with the explanation for the tremendous specificity and efficiency of enzyme catalysis.

ON ENTERING THE CELL

The carbon dioxide outside the cell enters the cell by diffusion. This occurs because light shines on the plant cell, using up the carbon dioxide inside the cell and making the CO_2 concentration within the cell lower than that outside. But unlike molecules of perfume that diffuse across a room when the bottle top is removed, the molecules of carbon dioxide must diffuse across a barrier to enter the cell. This barrier, which restricts the movement into and out of the cell, is the plasma membrane (see Chapter 2).

This structure isolates the cell from its surroundings and controls the passage of materials into and out of the cell. It therefore plays a crucial role in the regulation of the cell's contact with its environment, and thus in the determination of what substances are to be in the cell and what substances are not.

It is possible to study the rate at which various substances can enter the cell. When this was

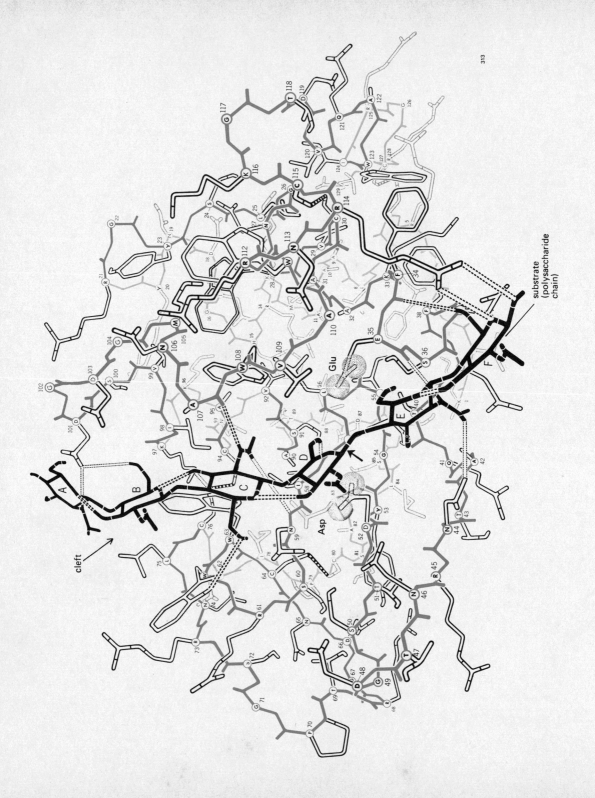

first done, some general rules emerged which helped to develop some ideas concerning the properties and possible structure of the plasma membrane. These general rules and some of the conclusions we derive from them are as follows:

1. The more water soluble a molecule is, the less rapidly it penetrates the cell; the more oil soluble it is, the more rapidly it is capable of penetrating a cell. We therefore conclude that at least part of the plasma membrane is lipid (composed of oil) in nature and molecules must generally penetrate or dissolve through this oily layer to enter the cell.

2. *Small* water-soluble molecules, such as oxygen, ethyl alcohol, carbon dioxide, or water itself, can penetrate the cell much faster than would be predicted from their solubility in oil. Therefore we conclude that the lipid layer covering the cell is not continuous but must have very small pores or cracks through which these small water-soluble molecules can pass.

3. Uncharged molecules can pass through the plasma membrane much more rapidly than charged molecules or *ions* of the same size. On the whole, positively charged ions (cations) can pass through the membrane more quickly than negatively charged ions (anions). We presume, therefore, that the material (probably protein) around the pores in the membrane is negatively charged itself, so that it tends to repel anions, but allows cations to pass through more rapidly. From these and many other observations of the plasma membrane we have made the following conclusions: it is a thin (75Å) layer composed of lipid and of protein arranged in such a manner that the lipid extends almost over the entire area of the membrane, leaving tiny pores through which small, uncharged molecules such as oxygen, water, and carbon dioxide can pass as well as cations.

The structure in Figure 2-10 is a hypothetical model of the plasma membrane that includes the properties just described. Next to it is a high-resolution electron micrograph of the plasma membrane showing the three-layered "leaflet" structure that the model proposes. To date, resolving power has not sufficed to determine whether or not the pores which we have deduced from diffusion experiments in fact exist.

So far we have discussed the entry (and exit) of molecules via diffusion. This process, of course, stops as soon as the concentration of a given type of molecule is the same inside and outside the cell. But if we examine the contents of cells, we find that there are many kinds of molecules which are in fact much more abundant inside the cell than in the surrounding medium, and vice versa. For example, most animal cells contain more potassium than sodium in the cytoplasm, but live in an environment richer in sodium. Human red blood cells have 20 times as much potassium as sodium, but the blood plasma surrounding the red cells is 20 times richer in sodium. The cell membrane is somewhat permeable to both ions. To prevent the loss of potassium the cell is constantly accumulating it, and to prevent the build-up of sodium the cell regularly disposes of it. Plants take in mineral ions from the soil even when they are not absorbing water and even when the ion concentration is much greater in the roots than in the soil. Thus the cell can actually accumulate materials *against* concentration gradients and it can also do the opposite—*excrete* some materials as, for instance, kidney cells do, so that the cell interior contains less of a given substance than its surroundings. These processes of accumulation and excretion require energy which must be expended by the cell. This is called **active transport** or osmotic work.

We do not know as yet what the precise molecular mechanism is by which active transport

Figure 5-6. (Facing page). Lysozyme, an example of the *fit* between an enzyme and the substrate which it catalyzes. This three-dimensional diagram of the enzyme shows the precise location of the substrate on the active site of the enzyme. The amino acid side chains involved in the interaction of the substrate with the enzyme are shown as well as the position of the four disulfide bonds. The four spheres represent the oxygen atoms belonging to the carboxyl groups of a glutamyl and an aspartyl residue, which we believe are involved in the immediate mechanism of the hydrolytic reaction catalyzed by lysozyme. Arrow indicates where substrate chain is broken by the action of the enzyme, (Irving Geis, from *Atlas of Protein Sequence and Structure*, 1967-68, Margaret O. Dayhoff and Richard V. Eck, National Biomedical Research Foundation, Washington, D.C., 1968.)

is brought about. We know that energy is required and, as we shall see later, we know where the energy comes from. We suspect that the membrane contains numerous specific molecular **carriers** which can bind specific molecules or ions, transport them across the membrane, and release them inside (Fig. 5-7). Somehow the carriers can utilize energy supplied by ATP to perform this operation. But although many hypothetical schemes have been suggested regarding the molecular details, no direct evidence in favor of any one scheme has as yet been obtained. There is, however, one form of active transport which can be observed visually, the process of **pinocytosis.** As already described in Chapter 2, it is possible to see with the electron microscope various steps of a process consisting of the infolding of the surface membrane and its pinching off to form spherical vacuoles which finally disintegrate to release their contents into the cytoplasm. This process permits the passage of materials into the cell in bulk form without requiring the passage of any molecules through the plasma membrane.

Pinocytosis is a widely distributed phenomenon in the living world. It was first discovered in tissue cultures of animal cells, which carried out the process so vigorously that it could actually be observed under the light microscope. Amoebae seem to take in much of their food by pinocytosis. Variations of the mechanism are found in other cells. The thin cells lining blood capillaries literally ferry materials from one membrane to the other in small vesicles (Fig. 5-8). Another example is seen in cells of the pancreas which manufacture enzymes necessary for digestion. Packets of these enzymes, or zymogen granules, move to the cell membrane when the pancreas is stimulated to secrete digestive juices. The zymogen granules fuse with the membrane, a fissure develops, and the enzymes spill out of the cell (Fig. 5-9).

The specific molecular mechanism by which the membrane can utilize energy and perform the mechanical work that results in this form of bulk transport has as yet to be elucidated.

We have seen how carbon dioxide and water as well as other substances might enter the cell, and how others might leave it. We stay inside the cell now and follow the carbon dioxide and water molecules as they are utilized by the *Elodea* cell as raw materials for manufacturing all the other substances of which it is composed.

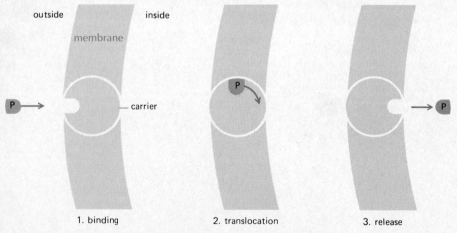

Figure 5-7. A hypothetical carrier transports a specific penetrating molecule across the membrane. The scheme involves three steps: (1) binding of penetrant to carrier, (2) translocation of penetrant across membrane, and (3) release of penetrant on the other side of the membrane. This scheme in its simplest form does not involve an energy source. For active transport, which involves the accumulation or excretion of penetrants against concentration gradients, an energy source supplied by the metabolism of the cell must be coupled to it.

FROM RADIANT ENERGY
TO LIVING MATTER

The leaf of an *Elodea* plant is composed of two layers of cells. If we focus on one of the layers with a light microscope, we see a series of box-shaped cells surrounded by thin cellulose walls, each cell containing a large vacuole. All of the living matter of this plant cell (cytoplasm and nucleus) is found in a thin layer lining the cellulose wall (Fig. 5-10). To the naked eye the green material of the plant (chlorophyll) appears to be distributed evenly throughout the leaf, but the light microscope reveals that it is really localized in cytoplasmic organelles, the chloroplasts.

Examination with the electron microscope reveals that the chloroplasts have a complex in-ternal structure (Fig. 2-29). A double membrane surrounds this organelle, which is composed of large numbers of membranes (**lamellae**) in parallel array. At various points, interconnecting these lamellae, are stacks of coin-shaped membranes forming the **grana,** containing the various pigments (chlorophylls, xanthophylls, carotenoids) that absorb light energy necessary for photosynthesis. Recently, very high resolution studies of the grana show that they in turn are made up of little bodies called **quantosomes,** which are believed to be the minimum photosynthetic units capable of carrying out the early stages of the photosynthetic process. We believe that the quantosome is made up of a highly organized arrangement of lipid, protein, and pigments. Its ability to absorb the energy of light and convert

Figure 5-8.   Pinocytosis, a process whereby the cell transports materials across membranes without their ever passing through the membrane.

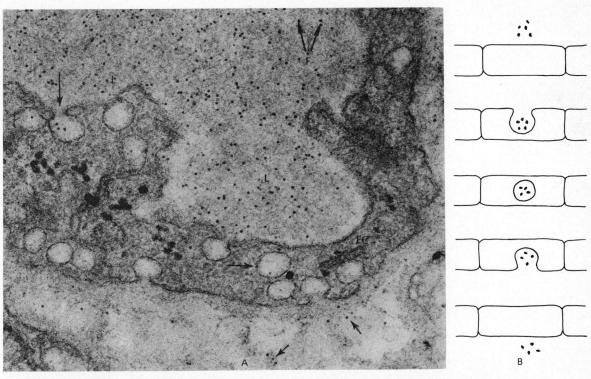

A. Endothelial cell of blood capillary showing how ferritin particles are picked up from the lumen of the blood capillary by pinocytotic vesicles, how these vesicles move across the narrow cell and discharge their contents outside the blood capillary.

B. Diagrammatic representation of pinocytotic process.

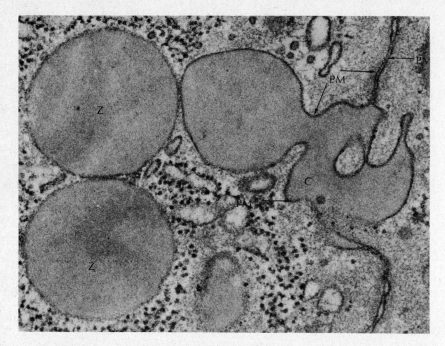

Figure 5-9. Another example of pinocytosis. The membrane of zymogen granules (Z) fuses with the plasma membrane (PM) and their contents (C) are discharged into the lumen.

it into chemical energy appears to be dependent on its highly organized molecular structure.

It is, then, through these tiny quantosomes that the major flow of energy supplying all of living matter on earth is funneled. It has been estimated that plants on earth manufacture more than 500 billion tons of organic material each year. We not only make use of some of this material but also utilize energy trapped in the **fossil fuels** (oil and coal) by the photosynthetic process millions of years ago. Because of this and because the molecular machinery of photosynthesis is interesting as such, we devote time to studying it in some detail.

Light is a form of energy defined by our ability to detect it with our eyes. Its wavelengths range roughly from 400 to 700 microns (μ).[1] Below this range we find the shorter ultraviolet rays and the even shorter gamma and x-rays. Above this range is the infrared region, and above that the huge radio waves. White light is a mixture of light of various colors, which can be shown readily by passing white light through a prism which separates it into the colors of the rainbow. It is a law of photochemistry that only the light which is *absorbed* by a chemical sub-

stance can be used in a chemical process. The light which passes through the system (*transmitted*) or is *reflected* or *scattered* cannot contribute to a chemical reaction. We can therefore immediately predict that green light, which is in part reflected from the leaves of plants, is utilized to a much lesser extent than red and blue

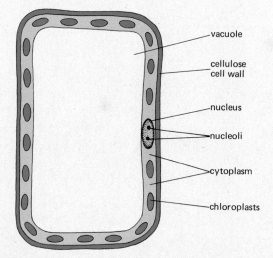

vacuole

cellulose cell wall

nucleus

nucleoli

cytoplasm

chloroplasts

Figure 5-10. A cell from the leaf of the water plant *Elodea*, showing the large vacuole and the thin layer of cytoplasm with the chloroplasts embedded in it.

[1] Remember that $1mm = 10^3\mu = 10^6nm = 10^7\text{Å}$.

light, which is absorbed by the photosynthetic pigment. We can test this experimentally by constructing an **action spectrum** of photosynthesis. This consists of illuminating a plant with light of different wavelengths and determining how efficiently photosynthesis is carried out at each of these wavelengths by measuring the rate of oxygen production. Figure 5-11 shows an action spectrum of photosynthesis; it can be seen that the highest rates of photosynthesis are indeed obtained with red and blue light.

We can now make a further prediction. Knowing that only absorbed light can provide energy for a chemical process and that red and blue light are most efficient in providing energy for photosynthesis, we can predict that one or more pigments must be found in plants which absorb light in the red and blue region. And this, indeed, is what we find to occur. In Figure 5-12 we see the **absorption spectrum** of **chlorophyll a,** one of the plant pigments. The difference between the shape of the absorption spectrum of chlorophyll *a* and the action spectrum of photosynthesis is due in part to the fact that other pigments, such as chlorophyll *b*, the carotenes, and the xanthophylls, are also involved in light absorption.

We have now reached the point at which carbon dioxide has diffused from outside the plant cell into the chloroplast, and visible light (mostly red and blue) has been absorbed by the

pigments which are present in a highly organized state in the quantosomes. Before we proceed any further we must state the overall significance of the photosynthetic reaction.

The carbon in carbon dioxide is found in its most fully oxidized state. The carbon in living organisms is found in a less oxidized (more reduced) state, as can be seen by examining the empirical formula $(CH_2O)_n$ of a carbohydrate. This more reduced state of carbon contains more

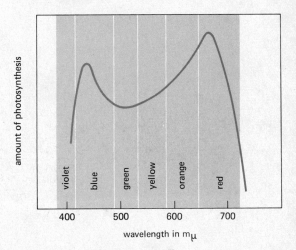

Figure 5-11. The action spectrum of photosynthesis, showing that the maximum rate of photosynthesis occurs in the blue and red regions of the visible spectrum.

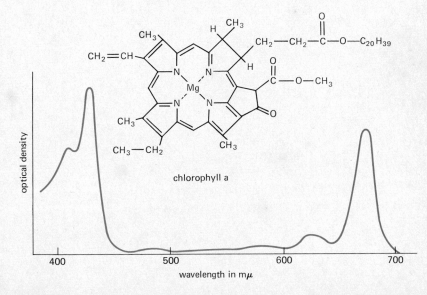

Figure 5-12. The absorption spectrum of chlorophyll a, showing absorption maxima in the blue and red region of the visible spectrum. This accounts for the maxima seen in the action spectrum (Fig. 5-11).

chemical energy, which is what is released in the form of heat when we burn (oxidize) organic compounds such as those found in wood, coal, or oil. Thus, in order to build organic compounds from carbon dioxide, the plant must reduce the carbon dioxide, a process which requires energy obtained from light. However, in order to reduce carbon dioxide the plant must obtain its hydrogen from somewhere. The ultimate donor of the hydrogen is water, which in turn becomes oxidized during the process of photosynthesis. Although in the early stages of organic chemistry oxidation was thought of as the gain of oxygen or the loss of hydrogen and reduction as the reverse of this process, we now know that a more general way of looking at these processes is by considering the loss or gain of electrons. Thus oxidation is the loss of electrons and reduction the gain of electrons. We should therefore think of photosynthesis as the use of light energy to move electrons from water to carbon dioxide (CO_2) and thus form a more reduced organic compound, such as a carbohydrate. Thus as glucose is synthesized we have the following overall reaction:

$$6CO_2 + 12H_2O + \text{light energy}$$
$$\longrightarrow C_6H_{12}O_6 + 6H_2O + 6O_2$$

This equation is the modern version of the "balance sheet" of the photosynthetic process. It is written in this manner because **tracer**[1] studies using the isotope ^{18}O have shown that the oxygen given off is derived not from CO_2 but only from water. Since 12 oxygen atoms are involved we need 12 molecules of H_2O on the left side of the equation. The oxygen of the six water molecules on the right-hand side of the equation is derived from the CO_2 resulting from the reduction of CO_2 to glucose.

Thus we have here an **oxidation-reduction** reaction in which H_2O is oxidized to oxygen and CO_2 is reduced to carbohydrate with light energy required to "drive" the reaction.

A great deal more is known about this process which involves a large number of chemical reactions. We are satisfied here to sketch the bare outline of this fascinating subject.

[1] Remember that radioactive as well as stable isotopes can be used to follow the path of organic reactions in the cell.

We can distinguish between two classes of photosynthetic reactions:

1. The light reactions, which, as the name implies, require the energy of light to drive them.

2. The dark reactions which follow the light reactions and which do not use light but rather require the products of the light reactions.

Evidence that both these classes of reactions occur was obtained early in this century when it was found that at high light intensity, intermittent light is more efficient (per unit of light) for photosynthesis than continuous light. It was correctly concluded at that time that there must be one or more *rate-limiting* dark reactions following one or more light reactions. At high light intensity, the dark periods following the light periods allow the dark reactions to catch up with the light reactions.

When a quantum[2] of light is absorbed by the chlorophyll of the leaf, some electrons become displaced by being raised to a higher energy level. If there is no other substance available to pick up the displaced electrons, they will eventually return to their **ground state,** giving off heat and light (fluorescence) in the process. But in photosynthesis the excited electrons are picked up by a special molecule which acts as an electron acceptor, after which the chlorophyll receives the electrons which have been displaced from an electron donor. We now know that this process occurs in two separate photochemical reactions, the former using light of short wavelength and the other light of long wavelength and probably using a different pigment. We also know that the former reaction involves the oxidation of H_2O and the latter reaction the reduction of the **cofactor** or NADP (nicotinamide adenine dinucleotide phosphate) to NADPH. (Reduced NADP).

[2] Light or any other kind of electromagnetic radiation can be described by its wavelength or by the amount of energy in the little "packets" or *quanta* making up the radiation. The relationship of these quantities is defined by Planck's equation:

$$E = h\nu$$

where E is the energy per quantum

h is 6.624×10^{-27} erg-sec (Planck's constant)

ν is the frequency of the radiation.

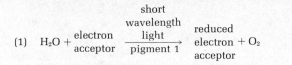

(1) $H_2O +$ electron acceptor $\xrightarrow[\text{pigment 1}]{\begin{array}{c}\text{short}\\\text{wavelength}\\\text{light}\end{array}}$ reduced electron + O_2 acceptor

(2)   Donor + NADP $\xrightarrow[\text{pigment 2}]{\begin{array}{c}\text{long}\\\text{wavelength}\\\text{light}\end{array}}$ NADPH + oxidized donor

The reduced electron acceptor in reaction (1), which now is in a state of higher energy, passes its electrons down an **electron transport system,** the molecular details of which are not as yet understood, but we do know that as this happens the energy of the electrons is utilized to synthesize ATP from ADP and inorganic phosphate. The electrons then reduce a second pigment (reaction 2) which upon irradiation raises the electrons to another high-energy state, where they are now capable of reducing NADP (Fig. 5-13). Thus it appears that the function of the two light reactions is to oxidize H_2O and produce ATP and NADPH. These two compounds are now available for the next steps in photosynthesis, which do not require light.

Our next problem is to work out how carbon dioxide is incorporated into the organic compounds of the cell (carbon dioxide assimilation). Great progress has been made in this area during the last few years by the use of radioisotopes, chromatography, and autoradiography (Fig. 5-14).

By using radioactive carbon dioxide and very short exposure times to light it was possible to discover the first stable organic compound into which the carbon dioxide is incorporated. It turned out to be 3-phosphoglyceric acid (3-PGA), and further work proved that the acceptor molecule for the carbon dioxide was ribulose 1,5-diphosphate (RuDp). We now know many more details of these reactions, including the precise points at which NADPH and ATP are utilized. For our purpose it is important to understand that the phosphoglyceric acid is the starting point of many different kinds of syntheses in the cell, including the synthesis not only of ribulose diphosphate (to complete the photosynthetic cycle) but also of other carbohydrates, such as glucose, and of other classes of compounds, such as amino acids, nitrogen bases, and lipids.

You have had a glimpse of the photosynthetic machinery of the plant cell. We have left out a great number of details and there are also many aspects of this process which are not as yet understood. We do not, for instance, understand the exact structure of the quantosome, which permits electrons to "flow" from one pigment to another, and we certainly do not understand how this flow of electrons is "coupled" to the synthesis of ATP. But we do at long last have a general outline of the process, which we expect will be filled in during the next decade of research.

FROM FOOD TO FUEL

We have seen how the plant cell utilizes light energy to build glucose or other molecules from carbon dioxide and water. We can think of molecules such as glucose as being the food of the plant cell and indeed of the entire world, for as we have already pointed out, the photosynthetic process turns out to be the food supplier of the entire biotic world. The importance of this food to both plant and animal cells is twofold: it provides the cell with energy and it supplies it with the molecular pieces it requires for constructing the variety of cellular materials we call living matter. We begin with the supply of energy.

When we want to use the energy of a fuel like glucose, we heat it up to the temperature at which it burns. This self-sustaining reaction is one of the earliest and most important technological inventions of our ancestors, although they did not burn glucose but rather cellulose, which is a polymer of glucose. The equation summarizing the burning of glucose is

$$C_6H_{12}O_6 + 6O_2 \longrightarrow 6CO_2 + 6H_2O + 690,000 \text{ cal}$$

The cell also oxidizes glucose and the overall reaction of this process may also be written as above, but there are a number of important differences. The cell does not release the energy of glucose in a sudden burst of heat. Instead it utilizes a large number of intermediate chemical steps, many of which are coupled to energy-consuming reactions leading to the synthesis of a special compound which is the *fuel* of the cell.

This compound is none other than ATP. Thus the burning of glucose can be compared to the uncontrolled rush of Niagara Falls, whereas the *respiration* of glucose in the cell is like a series of smaller stepwise locks which are coupled to separate "generators" capable of producing electricity. We now know that the cell is capable of synthesizing 38 molecules of ATP (from ADP and inorganic phosphate) by respiring (oxidizing) one molecule of glucose to CO_2 and H_2O. The overall equation for the respiration of glucose in the cell can be written as

$$C_6H_{12}O_6 + 6O_2 + 38ADP + 38P_i[1]$$
$$\longrightarrow 6CO_2 + 6H_2O + 38ATP + 234,000 \text{ cal}$$

Notice that the cell manages to trap 456,000 calories in the form of the fuel ATP, and only 234,000, or 33 percent, of the energy available in the glucose molecule is lost by the cell as heat. (In warm-blooded animals this heat is of some biological value since it is used, in part, to maintain the body temperature of the organism.) An energy conversion efficiency of 67 percent is really remarkably high when we consider that modern steam generators are able to convert only

[1] P_i is used to symbolize inorganic phosphate ($HPO_4^=$).

30 percent of the energy of coal into usable electrical energy.

How then does the cell perform this remarkable feat? Let us describe the process by following a molecule of glucose through its respiratory degradation and see how ATP is synthesized by the cell.

It should first be pointed out that the cell does not store a large amount of its food supply in the form of glucose. Large quantities of glucose would bring about an abnormally high molecular concentration of the substance in the internal medium of the cell. As we already know the cell overcomes this problem by polymerizing the glucose into higher molecular weight polysaccharide molecules, such as *glycogen* in animals and *starch* in plants. The synthesis of starch from glucose is an energy-requiring reaction and consumes ATP (Fig. 5-15). When the cell is ready to utilize its stored polysaccharide food for the purpose of ATP synthesis, it breaks down these compounds not by hydrolysis to glucose but by **phosphorolysis** to glucose-1-phosphate (Fig. 5-16). This conserves the energy which had been invested by ATP during the synthesis of the polysaccharide and prepares the glucose molecule for further breakdown.

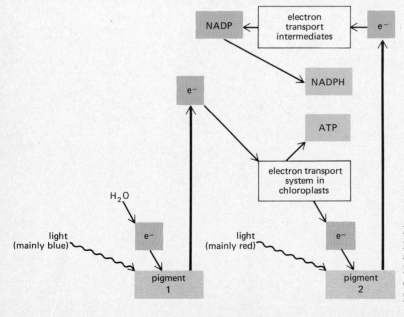

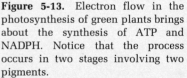

Figure 5-13. Electron flow in the photosynthesis of green plants brings about the synthesis of ATP and NADPH. Notice that the process occurs in two stages involving two pigments.

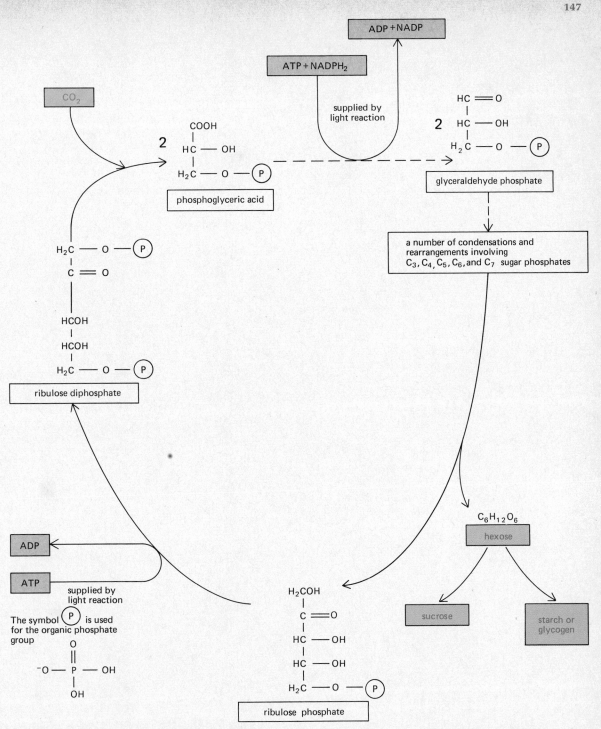

Figure 5-14. Carbon dioxide assimilation—that is, how the carbon in CO_2 is reduced and used to build glucose in the photosynthesis of green plants. Notice that the ATP and NADPH produced by the photochemical reactions (Fig. 5-13) are utilized to supply the energy for the energy-requiring reactions of the above process.

The many small steps by which glucose is broken down can, for purposes of convenience, be grouped into three major stages.

Stage 1 (Fig. 5-17) occurs in the ground substance of the cell and does not require oxygen. It involves the breakdown of the 6-carbon sugar (glucose or glucose-1-phosphate) into the 3-carbon compound, pyruvic acid, and results in the net synthesis of two or three molecules of ATP, depending on whether glucose or a polysaccharide is the starting material. In the presence of oxygen, pyruvic acid is utilized for the next stage of the process. When oxygen is absent, as is the case in many situations in which microorganisms live or in rapidly exercising muscle, the pyruvic acid is broken down further.

In the case of muscle, pyruvic acid is reduced to lactic acid, which is taken away from the muscle by the bloodstream and brought to the liver, where it is resynthesized to glycogen. The overall degradation of glycogen to lactic acid by muscle is called **glycolysis.** It permits the animal organism to rapidly release large amounts of muscular energy, by far exceeding the rate at which its body can supply oxygen to the muscles. As we shall see, glycolysis releases only a small portion of the energy found in glycogen; more important, however, is the fact that this process allows the muscle to release this energy rapidly since it is not dependent on the body's rather sluggish ability to supply oxygen to the muscle cells.

Many microorganisms, on the other hand, live under conditions of minimal or no oxygen. They modify the pyruvic acid in a number of different ways in a process called **fermentation.** Alcoholic fermentation is one of the better known and salubrious examples.

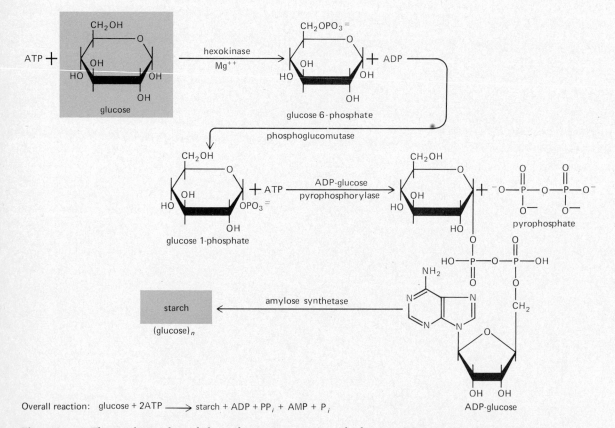

Overall reaction: glucose + 2ATP $\longrightarrow$ starch + ADP + PP$_i$ + AMP + P$_i$

Figure 5-15.   The synthesis of starch from glucose is a reaction which uses ATP as its energy source.

A few important points are illustrated by the scheme in Figure 5-17. If we start with the 6-carbon sugar glucose-6-phosphate, an additional ATP must be invested to produce the fructose-1,6-diphosphate. After this we obtain a series of 3-carbon compounds, but before we can obtain any energy from them we must carry out an oxidation by investing NAD (nicotinamide-adenine dinucleotide) which becomes reduced to NADH. In the absence of oxygen the reduced NADH thus formed must be reoxidized in some way; this is the significance of the last steps in glycolysis and fermentation. In the presence of oxygen, on the other hand, the NADH can be oxidized and pyruvic acid can be picked up by the next stage of the process which eventually leads to the formation of carbon dioxide and water.

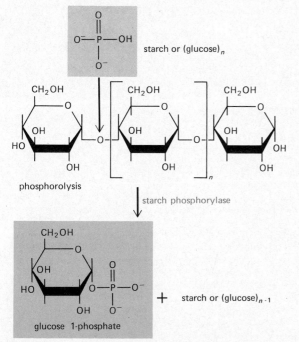

phosphorolysis

starch phosphorylase

glucose 1-phosphate

Figure 5-16. Breakdown of starch to glucose by phosphorolysis. Starch can be hydrolyzed to glucose by heating in weak acid. In nature, however, starch is broken down by phosphorolysis rather than by hydrolysis, thus producing glucose-1-phosphate which still retains some of the energy of starch. This obviates the need to utilize ATP for the phosphorylation of glucose, normally a first step in the glycolytic breakdown of glucose (Fig. 5-17).

The scheme described is the result of two decades (1920–1940) of work by great biochemists including Meyerhof and Emden. The ingenuity of their experiments can be appreciated only when it is remembered that they did not have available, at the time, the technique of radioactive tagging.

The Meyerhof-Emden scheme and the Krebs tricarboxylic acid cycle, which we shall presently discuss, are milestones in the progress of biochemical science, and a great tribute to human ingenuity.

So far we have discussed Stage 1 of the metabolic breakdown of glucose. **Stages 2 and 3** both occur in the mitochondria. Since it is in the mitochondria that 36 of the 38 ATP molecules are produced for every molecule of glucose consumed, this organelle has been appropriately called the "powerhouse of the cell." Electron microscopy has underscored this idea by demonstrating the spatial proximity in numerous cell types between the mitochondria and the site of utilization of ATP. Thus in an osmotically active cell the mitochondria are located in close proximity to the cell surface, which is active in the pumping process; in muscle the mitochondria are situated near the fibers responsible for muscle contraction, and so on.

Stage 2 involves the entry of pyruvic acid into the mitochondria, where it is oxidized by NAD, decarboxylated to a 2-carbon compound, and combined with coenzyme A to form the all-important compound *acetyl coenzyme A* (Fig. 5-18). This compound is important not only in glucose metabolism but also in other metabolic processes such as fat metabolism. Acetyl coenzyme A then enters the cycle by combining with oxalacetic acid, a 4-carbon compound, to form citric acid, a 6-carbon compound. As the cycle churns once around, two CO_2 molecules are "knocked off," thus eventually producing again the 4-carbon oxalacetic acid. The cycle repeats a second and third time, releasing two more CO_2 molecules, and each time forming a total of 6 CO_2 molecules from a single glucose molecule. Thus, to return to the overall balance sheet

$$C_6H_{12}O_6 + 6O_2 \longrightarrow 6CO_2 + 6H_2O + 690,000 \text{ cal.}$$

We have now explained the release of the 6CO_2.

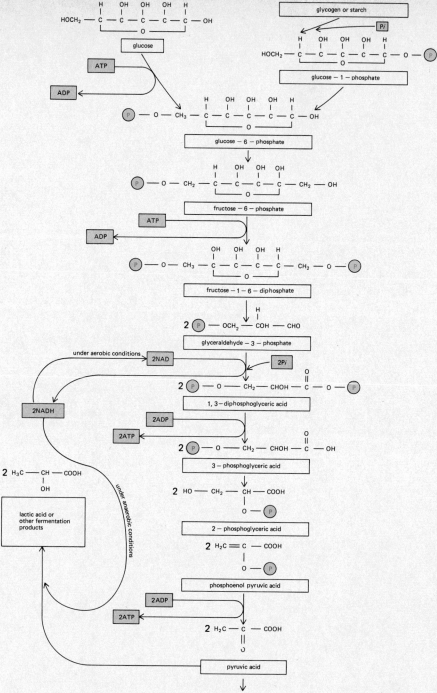

Figure 5-17. Glycolysis, the breakdown of glucose in the absence of oxygen. This process produces only two ATP molecules per molecule of glucose and stops when a three-carbon compound such as lactic acid is produced. Glycolysis or fermentation, therefore, yields very much less energy than the aerobic breakdown of glucose to CO_2 and H_2O (Fig. 5-18). Furthermore, the accumulation of lactic acid or other fermentation products presents a variety of problems for the organism.

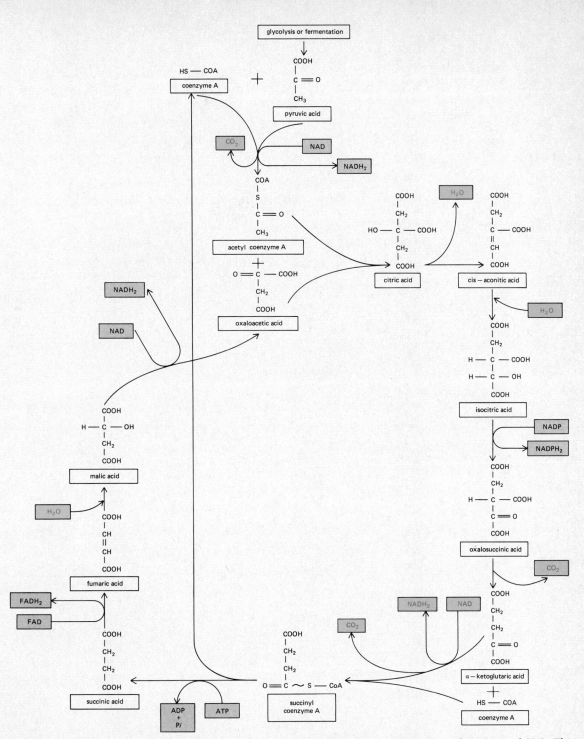

Figure 5-18. The Krebs (tricarboxylic acid) cycle—the aerobic breakdown of pyruvic acid to CO_2 and H_2O. This process produces 36 molecules of ATP for every two molecules of pyruvic acid—a much higher energy yield than the glycolytic process.

(Two CO_2 molecules are released from pyruvic acid, two from oxalosuccinic acid, and, finally, two from α-ketoglutaric acid.) Notice also that at several points in this scheme NAD is reduced to NADH, and in one case another cofactor FAD (flavin adenine dinucleotide) is reduced to $FADH_2$. This, in effect, means that as the carboxylic acid cycle completes one turn there are five steps where an intermediate is oxidized by giving up a pair of electrons to the NAD or FAD cofactors. Before we consider what happens to these electrons, we must examine another important feature of the Krebs cycle.

ATP production, as we shall see, is one of the important functions of the oxidative metabolism we are discussing. But the cell not only requires energy; it also requires *materials* which it utilizes to synthesize all the various molecules it needs for growth. The Krebs cycle is one of the major sources of these materials. Thus the cell is able to withdraw any one of the intermediates and use it as a starting point of synthesis of compounds, such as amino acids, fatty acids, steroids, porphyrins, carbohydrates, and nitrogen bases. Thus it is understandable that we consider the Krebs cycle, occurring in the mitochondria, as the metabolic hub of the cell.

Stage 3 begins when the electrons are removed from the various substrates in the cycle and donated to NAD or FAD. When this happens, a great deal of energy has been transferred to the cofactors. It can be shown, for instance, that when NADH is oxidized, some 53,000 cal are released.

$$NADH + 1/2\ O_2 + H^+ \longrightarrow NAD + H_2O + 53,000\ cal$$

This is far more energy than is needed to synthe-

size one ATP molecule from $ADP + P_i$ (inorganic phosphate). In fact it can be shown that only some 8000 to 12,000 calories are necessary. The cell solves this efficiency problem by allowing the electrons to move down a chain of electron-transporting substances, an arrangement which breaks the process of oxidation into several steps allowing three ATP molecules to be synthesized for every NADH molecule oxidized (Fig. 5-19) and two molecules of ATP for every $FADH_2$ molecule oxidized. This process, which *couples* the oxidation of the coenzymes to the phosphorylation of ADP, is called **oxidative phosphorylation.** The exact molecular mechanism by which this is achieved is one of the most important unsolved problems of modern cell biology. We know that the internal membranes of the mitochondria are composed of repeating units which we call respiratory assemblies (Fig. 5-20). These assemblies are composed of the various intermediates of the electron transport chain and are no doubt in close proximity to the enzymes involved in the phosphorylation process. In some way this close proximity of these various components allows energy to be transmitted from one system to the other, with the important result that some 30 ATP molecules are synthesized for every two molecules of pyruvic acid metabolized. Remember that in Stage 1 two molecules of ATP are synthesized as well as two molecules of NADH, the latter under aerobic conditions allowing the synthesis of six or more ATP molecules. Thus the overall yield of ATP from the oxidation of glucose is 38 molecules.

If we compare the yield of two molecules of ATP from the anaerobic breakdown of glucose to

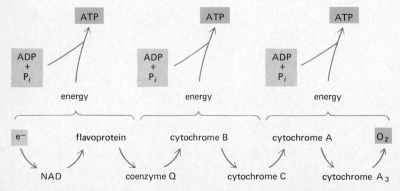

Figure 5-19 The respiratory chain, the transport of electrons in mitochondria, and the synthesis of ATP. The precise mechanism whereby the energy of the electrons is utilized to convert ADP and P_i into ATP has so far eluded our understanding, although a number of outstanding scientists have devoted much time to working on this problem.

that of 38 molecules from the oxidative break-down of glucose, it is clear that oxidative metabolism is much more efficient. The development of oxidative metabolism early in evolution constituted an important step in the evolution of life. This step had to be preceded by the development of photosynthesis, since it was the activity of photosynthetic organisms which supplied the earth with oxygen in the first place. Thus the evolution of photosynthesis not only made the vast energy of the sun available to the early life forms, but, by making oxygen available, it paved the way for the evolution of the highly efficient oxidative metabolism just described.

HOW IS METABOLISM STUDIED?

We have discussed the dramatic discoveries that have contributed to our understanding of the processes by which light energy converts CO_2 and H_2O into food and food in turn is converted into the fuel ATP.

There are many other metabolic schemes which the cell utilizes for the synthesis and breakdown of many of its compounds. The cell must synthesize lipids for its membranes, polysaccharides for its outer wall (if it is a microbial or plant cell), nitrogen bases and pentoses for DNA and RNA, the information storage and information transfer systems, and, most importantly, amino acids for proteins which make up the structural and catalytic systems. Many of these metabolic schemes are by now well understood. We might estimate that at present some 500 to 1000 enzyme-catalyzed metabolic steps are known. A great deal still needs to be done, but it is clear that by now the major outline of the "flow of energy and of small molecules" in the cell has been established.

How have these important advances been made? A large variety of techniques have been used and applied to each problem in a great many different combinations. They become part of the repertoire of the cell biologist as he is trained to use them, first as a student and later as a young researcher. Most practicing scientists contribute to the storehouse of methods available to the scientific community, which explains why knowledge in cell biology is developing at an exponential rate. The following are some of the most widely used methods for the study of metabolic sequences.

1. **The use of enzyme specificity.** As we have pointed out enzymes catalyze reactions in a

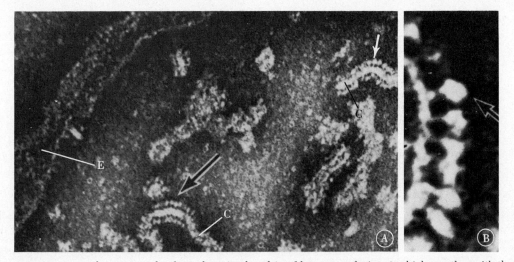

Figure 5-20. Inner membrane particles from the mitochondria of heart muscle (cow) which together with the membrane to which they are attached constitute the respiratory chain shown in Figure 5-19. A is a view ($\times$ 80,000) showing part of the edge of a mitochondrion (E) and several cristae (C). Along the cristae surfaces are arrays of spheres attached to the cristae by stalks (arrows). B is at higher magnification ($\times$600,000) and shows several spheres. It is thought that the membrane contains the electron transport chain and the spherical particles couple the flow of electrons to the phosphorylation of ADP to ATP.

highly specific manner. Thus it is possible to identify an enzyme in a particular cell extract by adding a known substrate and demonstrating that a particular reaction takes place. If an enzyme for a particular reaction is present in the cell extract, it is likely that the reaction also occurs in the cell. Thus, for instance, if there is an enzyme in a cell extract capable of reducing isocitric acid to oxalosuccinic acid *in vitro* (Fig. 5-18), it is likely that this reaction can also operate *in vivo*. Whether it does or not can be ascertained by performing experiments *in vivo*.

2. **The use of isotopic tracers.** One type of *in vivo* experiment is to use compounds containing isotopic tracers. Carbon-14, phosphorus-32, hydrogen-3, and sulfur-35 are radioactive isotopes which can be measured readily with appropriate electronic devices. Oxygen-18, nitrogen-14, and hydrogen-2 are not radioactive, but can be measured with a **mass spectrometer,** a device which can distinguish between atoms of different mass. By feeding living cells isotopically labeled compounds and, after various time intervals, killing the cells and isolating various compounds, we can follow the progress of the labeled atoms through these compounds and thus confirm whether certain reactions indeed occur *in vivo*.

3. **The use of genetic mutations.** As we have seen in Chapter 3, Beadle and Tatum developed a method using genetic mutants of the fungus *Neurospora,* in which they were able to show that certain genes controlled the synthesis of certain enzymes. This ingenious experimental approach was also of great value to the study of metabolism, since it helped to identify many **metabolic intermediates** which had hitherto been unrecognized. Furthermore, in addition to discovering new intermediates, the genetic method served to confirm the presence *in vivo* of many metabolic sequences which had been discovered *in vitro*.

Just as in *Neurospora,* many higher organisms are congenitally incapable of making certain metabolic intermediates or certain enzyme cofactors. These materials must be supplied in the food of the organism if it is to maintain its health and normal growth. We call these substances **vitamins.** What may be a vitamin for a human may not be a vitamin for a rat, if the latter organism has a hereditary makeup capable of producing the proper enzyme, which in turn manufactures the proper metabolic intermediate or cofactor.

4. **The use of metabolic poisons.** Many enzymes are specifically inhibited by certain substances. Because enzymes are present in very low concentrations, the inhibitors also can act at very low concentrations and many of them were known as *poisons* long before the study of metabolism. Thus, for instance, *arsenite* inhibits the enzyme α-ketoglutaric dehydrogenase, which is responsible for the conversion of α-ketoglutaric acid to succinyl coenzyme A; *cyanide* inhibits the electron transport chain; *malonic acid* inhibits succinic dehydrogenase, which converts succinic acid to fumaric acid. When a metabolic poison is added to a cell or a cell extract, it stops a certain metabolic sequence and causes the accumulation or "piling up" of the intermediate preceding the enzyme that is being inhibited. This provides information as to whether certain reactions occur in the cell, and by using several poisons we can also obtain information as to the sequence of reactions. This somewhat laborious approach was the main one available to workers such as Meyerhof and Krebs, who obtained their important results before radioactive tracers or genetic techniques were available.

THE CELL IS NOT A BAG OF ENZYMES

In Chapter 2 we discussed the various structures of the cell. In this chapter, although we did refer to chloroplasts and mitochondria briefly, the student might be left with the impression that grinding up cells, making extracts, and identifying metabolic steps is the only aim of cell biology. Nothing could be further from the truth! The last 15 years of research have shown how important the knowledge of structure is in the understanding of function. In fact two major areas of cell metabolism which we still do not understand (photosynthetic phosphorylation and oxidative phosphorylation), elude our comprehension because we do not as yet have an understanding of the structural aspects involved.

During the last 15 or 20 years a great variety of techniques have been developed which are used to relate a particular enzymatic activity to a particular structure. These can be divided roughly into two major categories.

1. **Cytological methods.** These consist of identifying the biochemical properties of a particular structure while that structure is still present in its normal cellular surroundings. This can be done by feeding cells with intermediates labeled with tritium (^{3}H). It is possible to detect where the intermediate is incorporated by **fixing** (inactivating the proteins) and **sectioning** (slicing into thin sections) the cell and then laying the slices against a photographic film which is sensitive to the radiation produced by tritium. In the following chapter we shall see how this technique, which is called **autoradiography,** has been utilized in studies of chromosomes. Techniques have been developed whereby autoradiography has been utilized even at the electron microscope level.
2. **Centrifugal methods.** These consist of separating cellular structures from one another. If large centrifugal forces are used, even molecules can be separated. A centrifuge is a device which spins the biological material and thus increases the gravitational field acting on the particles or molecules in the sample. This causes the particles or molecules to settle and, depending on their size, shape, and specific gravity, they will settle at different rates. Thus it is possible to separate cell components from each other and study their biochemical properties independently of each other.

Figure 5-22 illustrates an experiment which separates different cell particles. The smaller particles left in the supernatant after a relatively slow centrifugation can be separated from one another by centrifuging at higher velocities. By utilizing a medium of increasing concentration between the top and the bottom of the tube (a sucrose gradient for particles containing proteins or a cesium chloride gradient for nucleic acids), it is possible to improve greatly the resolution of the separation (Fig. 5-21).

Thus we conclude our discussion of the flow of molecules and of energy through the cell, remembering that the cell is a complex maze of membranes and compartments with enzymes localized in special places in this multicomponent system. We have followed carbon dioxide and water molecules into the cell and have seen how light energy is utilized to synthesize glucose from them. We have also seen how glucose is oxidized to carbon dioxide and water and how the energy released is used to synthesize ATP.

In the following chapter we shall see how ATP provides energy for the *flow of information* which controls the processes of the cell.

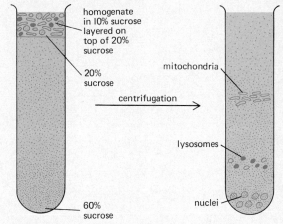

Figure 5-21. Sucrose gradient centrifugation, a powerful new method of separating organelles, particles, and even large molecules. A sucrose gradient is first established in a centrifuge tube by adding simultaneously two concentrations of sucrose in changing amounts. The **homogenate** is then layered at the top of the sucrose gradient and the material is centrifuged. The changing concentration of sucrose exerts a differential frictional resistance and a differential buoyancy on the different particles, thus bringing about better separation than a medium of constant composition. At the end of the centrifugation, a hole is punched at the bottom of the tube and a number of constant volume fractions are collected.

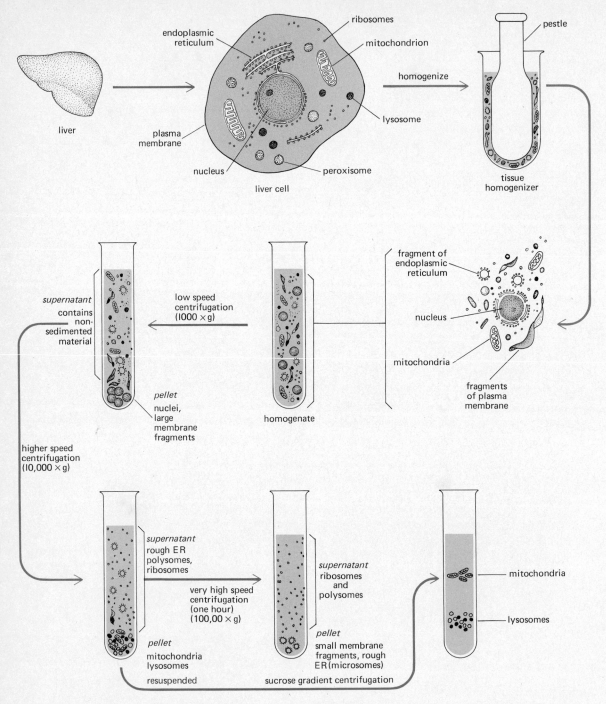

Figure 5-22. Separation of different cell particles or organelles from liver cells. New and better techniques of separation have been developed in recent years and we can expect a great deal of progress to be achieved in the future.

SUGGESTED READINGS

Allen, J.M. (ed.), *Molecular Organization and Biological Function.* Harper and Row, New York, 1967.

Baker, J.J.W. and G.E. Allen, *Matter, Energy and Life.* Addison-Wesley, Reading, Massachusetts, 1970.

Baldwin, E., *Dynamic Aspects of Biochemistry,* fourth edition. Cambridge University Press, New York, 1964.

Baldwin, E., *The Nature of Biochemistry,* second edition. Cambridge University Press, New York, 1967.

Bernard, S., *The Structure and Function of Enzymes.* W.A. Benjamin, Inc., New York, 1968.

Cohen, G.N., *Biosynthesis of Small Molecules.* Harper and Row, New York, 1967.

Goldsby, R.A., *Cells and Energy.* Macmillan, New York, 1967.

Karlson, P., *Introduction to Modern Biochemistry.* Academic Press, New York, 1968.

Lehninger, A.L., *Biochemistry.* Worth, New York, 1970.

Lehninger, A.L., *Bioenergetics.* W.A. Benjamin, New York, 1965.

Loewy, A.G. and P. Siekevitz, *Cell Structure and Function,* second edition. Holt, Rinehart and Winston, New York, 1969.

Van Holde, K.E., *Physical Biochemistry.* Prentice-Hall, Englewood Cliffs, New Jersey, 1971.

Wold, F., *Macromolecules: Structure and Function.* Prentice-Hall, Englewood Cliffs, New Jersey, 1971.

CHAPTER 6

The Flow of Information

If we place a single cell of a bacterium such as *Escherichia coli* in a suitable growth medium, the cell will grow and divide within a 20-minute period and so will its two daughters and four granddaughters. This *exponential growth rate* will be maintained until the decreasing food supply limits it or until the growing bacteria cause a deleterious change in the composition of the culture medium. The mathematical potentialities of an exponential growth curve are, of course, well known. It takes but 9 hours before we have 1 billion cells and, if the growth rate were to continue like this and the food supply were inexhaustible, we would have a culture that weighs as much as our earth in less than two days. What is less well recognized than the *fecundity* of exponential reproduction is the *precision* of the reproductive process in cells. The probability that a cell under normal growth conditions will grow and divide to form two cells is very close to one or, to put it differently, the chances that some mistake in the metabolism of the cell will occur which will interfere with the process of growth and duplication is probably less than one in a million. And this despite the fact that, even in as simple a cell as the bacterium *E. coli*, a minimum of 2000 separate chemical reactions are going on simultaneously. How are all these chemical reactions synchronized to produce a finite cellular structure which exhibits the many structural details described in Chapter 2 and which is capable of growing and dividing with almost perfect precision?

In the preceding chapter we have demonstrated how the process of respiration supplies chemical building blocks and ATP (energy) for the maintenance, growth, and duplication of the cell. In this chapter we attempt to show how this *flow of energy and materials* is synchronized or regulated by a *flow of information* that brings about the integrated behavior of cell processes. However, we must warn the reader that, since this is the youngest area of cell biology, the answers provided are often tentative and incomplete. If we knew more about this area of *cellular regulation*, we would probably be further along the way to understanding aging and many of our most dreaded diseases (Chapter 8).

ENZYME SPECIFICITY—THE KEY TO METABOLIC REGULATION

The first point to make regarding cell regulation is so obvious that we almost hesitate to make it. If we regard the cell as a complex mixture of chemical compounds, then clearly the way these compounds become related to one another is through enzyme action. Since enzymes have considerable specificity—that is, they will catalyze one particular reaction but not another—the presence of certain enzymes will determine whether a particular synthetic process will proceed or not. We give two examples of such **biosynthetic** reactions, showing how bringing together the proper enzymes and substrates results in the synthesis of a cellular compound.

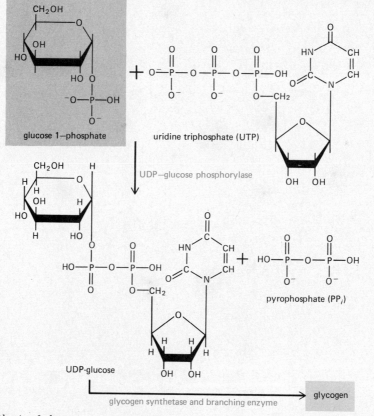

Figure 6-1 The synthesis of glycogen, an example of the role of enzyme specificity in the regulation of a synthetic pathway. Given the proper raw materials (substrate), the necessary catalysts (enzymes), and a source of energy (UTP), the cell will synthesize a specific product—in this case, glycogen, a storage product of animal cells.

1. If we put together uridine triphosphate (UTP), glucose-1-phosphate (G-1-P), and the two enzymes, glycogen synthetase and "branching" enzyme, we synthesize glycogen (Fig. 6-1).
2. If we put together acetyl coenzyme A, ATP, CO_2, reduced nicotinamide adenine dinucleotide (NADPH), and reduced flavin mononucleotide ($FADH_2$) as well as the appropriate enzymes which can be extracted from living tissues, we synthesize a fat (Fig. 6-2).

These are but two of the many synthetic processes which occur in the cell. Such processes will occur if the necessary energy in the form of ATP or some other equivalent high-energy phosphate compound, the necessary substrates, and the proper enzymes are made available. Since both ATP and the substrates depend for their formation on other enzymes, the question of how metabolism is regulated comes down to how enzymes are synthesized and how their concentration and their activity are regulated. Thus we start at the very beginning, with the hereditary information which the cell stores, and find out how it manages to duplicate the information and how it is able to use this information to synthesize and control the enzymes of the cell.

HOW THE INFORMATION IS DUPLICATED

Long before we knew about the details of DNA structure, it had been clear to geneticists that the genetic material had to have two functions.

1. It had to have a way of replicating itself, a reaction that was often termed **autocatalytic,** which simply means that there must be some of this substance present to make more of it.
2. For genetic material to influence the development of the cell it must have a way of controlling cellular processes.

This formulation of the early geneticists can be translated into modern language as follows.

DNA (the cell's genetic material) must have a double role.

1. It must in some way be involved in its own synthesis.
2. It must in some way control the synthesis and possibly the activity of enzymes and probably of other proteins.

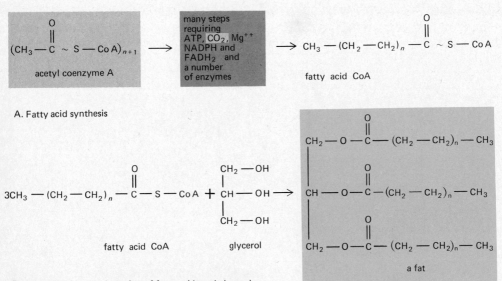

A. Fatty acid synthesis

B. Fat synthesis – condensation of fatty acids and glycerol

Figure 6-2 The synthesis of a fat. Another example of how enzymes control the synthesis of a product.

We begin with the first role. *How is DNA involved in its own production?*

Remember that in our discussion of the structure of DNA we remarked that the double-stranded structure held together with complementary base pairs (A–T and G–C) suggests how DNA might replicate itself. During the last 15 years additional evidence has been accumulated to confirm the original belief of Watson and Crick that DNA is duplicated by a mechanism in which each strand of the DNA molecule becomes the **template** for the synthesis of a complementary strand. Some of this evidence is as follows.

1. One of the most dramatic series of studies was performed by Arthur Kornberg and his co-workers, who purified an enzyme **DNA polymerase.** If this enzyme is *incubated* with the four proper nucleotide triphosphates, deoxyadenosine triphosphate (dAPPP), deoxythymidine triphosphate (dTPPP), deoxyguanosine triphosphate (dGPPP), and deoxycytidine triphosphate (dCPPP), *nothing will happen.* However, if a little bit of DNA is added to the incubation mixture, then something will definitely happen—some DNA will be synthesized (Fig. 6-3). This experiment suggests that the presence of some DNA is necessary for the synthesis of more DNA.

   But how do we know that the DNA which has been synthesized is just like the *primer* or *template* DNA which had been added to the incubation mixture? A simple approach is to determine the **base ratios** of the template DNA and see if the synthesized DNA has the same base ratios. As Table 6-1 shows, the answer, within limits of experimental error, is yes. The synthetic DNA does closely resemble the primer DNA in its base ratios.

2. Kornberg developed an even more sensitive method for demonstrating the similarity between the primer DNA and the synthetic DNA. Although it is not possible to discuss this method in detail, it is sufficient to say that it is possible to measure the frequency with which any given base has each of the four bases as nearest neighbor. Kornberg was able to show that, even when the much more stringent criterion of **nearest-neighbor frequency** is used, the primer DNA and the synthetic DNA are alike (Table 6-2).

   An exciting consequence of the nearest-neighbor study was the confirmation of the prediction of the Watson-Crick model that the strands run in opposite directions (Fig. 6-4).

3. So far the evidence cited comes from test tube (*in vitro*) experiments. Cell biologists, however, are never satisfied unless they also prove that what they observe in the test tube also occurs in the cell.

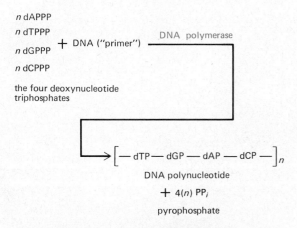

Figure 6-3 The synthesis of DNA using the enzyme DNA polymerase, the four nucleotide triphosphates, and a small amount of DNA as a primer.

Table 6-1 Base composition of templates and products of DNA synthesized *in vitro*

DNA	A	T	G	C
Mycobacterium phlei				
Template	0.65	0.66	1.35	1.34
Product	0.65	0.65	1.34	1.37
E. coli				
Template	1.00	0.97	0.98	1.05
Product	1.04	1.00	0.97	0.98
Calf thymus				
Template	1.14	1.05	0.90	0.85
Product	1.12	1.08	0.85	0.85
T_2 bacteriophage				
Template	1.31	1.32	0.67	0.70
Product	1.33	1.29	0.69	0.70
AT copolymer	1.99	1.93	<0.05	<0.05

If both strands of DNA act as templates for the synthesis of new strands, then the method of DNA replication ought to be **semiconservative.** That is, a parent **duplex** ought to give rise to two daughter duplexes, each containing one of the parent strands, and each of these daughters should give rise to two duplexes, only one of which would have the original parent strand in it (Fig. 6-5). Such experiments were indeed performed with bacterial cells, in which the DNA is free, and with cells in higher organisms, in which the DNA is packaged in some as yet unknown manner in the chromosomes. In these experiments the parent cells were grown in the presence of a radioactive precursor of DNA, a procedure which labels the DNA. Then the label was withdrawn and the cells were allowed to replicate twice in the presence of a "cold" (unradioactive) precursor. After the first division all the DNA molecules (or

chromosomes) were labeled, but at half intensity, and after the second division only half the DNA molecules (or chromosomes) were labeled (Fig. 6-5). This proved that DNA replication in the living cell is semiconservative—a property which was predicted by considering the structure of DNA.

There are, however, many problems regarding the biosynthesis of DNA which remain to be solved. It is not even absolutely certain whether DNA polymerase is the actual enzyme which replicates DNA in the cell or whether it is a **repair enzyme** which helps correct **lesions** in the DNA that occur during the everyday wear and tear of cellular activity. But although much remains to be done, it is clear that we can now discern the basic outlines of the mechanisms by which the cell duplicates its hereditary material.

In this connection one of the most interesting developments in recent years is the discovery that in eucaryotic cells the nucleus is not the only repository of DNA. We now know that mitochondria and chloroplasts contain circular molecules of double-stranded DNA (Fig. 8-13)—a fact which

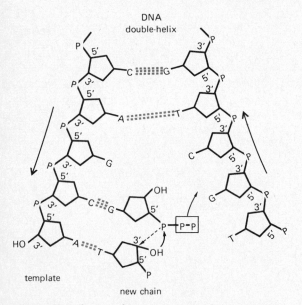

DNA
double-helix

template

new chain

Figure 6-4 The strands run in opposite directions. This figure shows the mechanism of enzymatic replication of DNA as suggested by Kornberg. Notice that, because the strands run in opposite directions, the DNA polymerase can only replicate one strand at a time. A variety of suggestions have been made as to how the other strand is replicated but our understanding of this important process is, in fact, still incomplete.

Table 6-2   Nearest-neighbor base freqencies of DNAs synthesized from templates of native and enzymatically formed calf thymus DNA

	Template	
Neighbors	Native DNA	Enzymatically synthesized DNA
pApA	0.089	0.088
pApG	0.072	0.074
pApC	0.052	0.051
pApT	0.073	0.075
pTpA	0.053	0.059
pTpG	0.076	0.076
pTpC	0.067	0.064
pTpT	0.087	0.083
pGpA	0.064	0.063
pGpG	0.050	0.057
pGpC	0.044	0.042
pGpT	0.056	0.056
pCpA	0.064	0.078
pCpG	0.016	0.011
pCpC	0.054	0.055
pCpT	0.067	0.068

confirms very nicely the earlier hypothesis that these cellular organelles are *self-reproductive* (Chapter 8).

HOW THE MESSAGE IS TRANSCRIBED

The concept that the DNA in the nucleus contains hereditary information which is capable of self-duplication has been widely accepted for some time. We have also known for several years that the ribosomes in the cytoplasm of the cell are involved in the synthesis of proteins. However, we did not know until relatively recently how the information in the nucleus became transferred to the cytoplasm, where it presumably initiates the synthesis of specific proteins. From the use of ^{32}P radioactive label experiments it became clear that RNA was synthesized in the

nucleus and then moved into the cytoplasm. But it was not clear whether ribosomes specific for the synthesis of particular proteins are synthesized in the nucleus or whether all ribosomes are the same but could be modified for a specific protein-synthesizing function by a specific *messenger RNA* (mRNA) synthesized in the nucleus. We now know that the latter is the case. Here again we have obtained evidence from both *in vitro* and *in vivo* experiments:

1. A *short-lived* form of RNA is synthesized in the nucleus and transported into the cytoplasm. This short-lived mRNA has the interesting property of forming **polysomes** which are a number of ribosomes together like beads on a string (Fig. 2-19). It can be shown, for instance, by using centrifugal separation methods, that in freshly ^{32}P labeled cells the newly

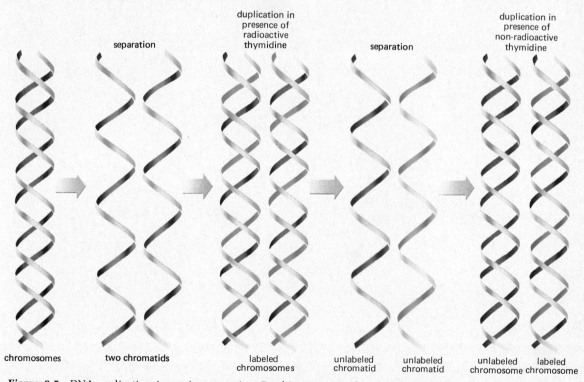

separation — duplication in presence of radioactive thymidine — separation — duplication in presence of non-radioactive thymidine

chromosomes   two chromatids   labeled chromosomes   unlabeled chromatid   unlabeled chromatid   unlabeled chromosome   labeled chromosome

Figure 6-5 DNA replication is semiconservative. By this we mean that during DNA replication, each of the two daughter molecules contains one of the parental strands and one newly synthesized strand. The above diagram shows what would happen during chromosome duplication if the eucaryotic chromosome carried a single, uninterrupted DNA molecule—an assumption which seems to be supported by the evidence at present.

labeled RNA is found in the polysome fraction and not in the single ribosomes (Fig. 6-6).

2. Using *in vitro* experiments we have found that an enzyme (RNA polymerase) which can be purified from the cell can synthesize RNA from the four nucleotides, APPP, UPPP, CPPP, and GPPP. Not surprisingly, but nevertheless to the great delight of biochemists, this RNA-synthesizing system requires DNA (not RNA) as a primer (Fig. 6-7). It turns out that the primer is single-stranded DNA, that is, DNA

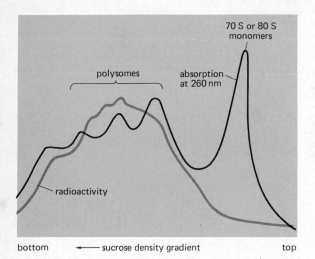

bottom ◄——— sucrose density gradient   top

Figure 6-6 The newly labeled, freshly synthesized RNA appears in the polysome fractions. This suggests that the messenger RNA carries the label, a conclusion which has been confirmed by numerous other experiments. In the above experiment, the ribosomes were separated from the polysome fractions by sucrose gradient centrifugation. The radioactive label had previously been introduced into the RNA by introducing a ^{32}P-labeled precursor of RNA as a short **pulse** into a growing culture of cells. By a pulse, we mean exposing the cells for a short time to radioactivity after which the radioactivity is removed by diluting with a large amount of "cold" precursor and often also by centrifuging the cells down and resuspending them in "cold" growth medium.

which has been uncoiled into single strands by heating. Also in the *in vitro* system both strands of the DNA seem to be transcribed, so that it is possible to show, in this case, that the base ratios of the RNA are the same as those of the primer DNA.

3. Not only is the base ratio the same, but we can show with the **molecular hybridization technique** that the sequence of bases of the DNA primer and the RNA are the same, or at least very similar. In this technique, we heat a mixture of DNA and RNA and then allow it to cool slowly. Using a system of double radioactive labels (for example, ^{3}H for DNA and ^{32}P for RNA), we are able to show that double-stranded molecules consisting of RNA and DNA strands can form. We interpret this to mean that the base sequence of one strand of primer DNA is transcribed into a complementary base sequence of the synthesized RNA which is then able to form a double helix with the primer strand.

Hence we now believe that, under conditions that we do not as yet understand in great detail, one of the two strands of DNA in the nucleus is transcribed by the enzyme RNA polymerase into a single strand of mRNA, which then moves into the cytoplasm and interacts with a number of ribosomes to form a polysome which then initiates protein synthesis.

HOW THE TRANSCRIBED MESSAGE IS TRANSLATED

We already pointed out that the language of the nucleic acids is a *four letter language* written in linear fashion and that the language of the proteins is a *twenty letter language* also written linearly. Clearly, then, there must be a process in the cell whereby the four letter language of the

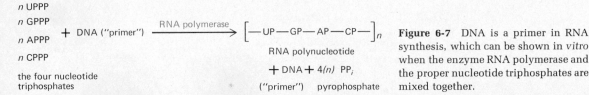

Figure 6-7 DNA is a primer in RNA synthesis, which can be shown in *vitro* when the enzyme RNA polymerase and the proper nucleotide triphosphates are mixed together.

nucleic acids is **translated** into the twenty letter language of the proteins. It is a mark of the genius of Francis Crick that he predicted that an **adapter molecule** of RNA must exist that at one end would interact with the mRNA and at the other end with the amino acid molecules forming the polypeptide chain. If the adapter hypothesis is correct, there must be at least twenty different adapter molecules, that is, at least one having specificity for each of the twenty amino acids. Crick's prediction, in fact, turned out to be true. We now know that there are *twenty classes* of *transfer RNA* (tRNA) molecules. Compared with other RNA molecules their molecular weight of about 23,000 is fairly low. For each class of these

tRNA molecules there is a specific *enzyme* capable of linking it to a given amino acid forming a chemical bond between the amino acid and its specific tRNA (Fig. 6-8). One side of the tRNA molecule, through the action of a specific enzyme, becomes tied to a given amino acid. The other side must in some way become specifically attached to the messenger. But how is this accomplished? Again there are many lines of evidence. They lead towards the following.

We already know that the mRNA and the ribosome become attached. Now we learn that the *activated* tRNA (that is, the RNA which has a given amino acid attached to it) also becomes attached to the ribosome and in such a way that

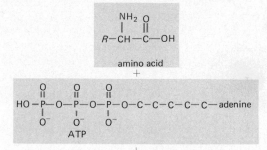

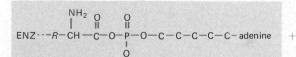

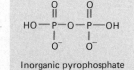

Figure 6-8 A given amino acid is linked with a chemical bond to a particular transfer RNA by a specific enzyme. We show here only one short piece of the end of the tRNA. The ACC sequence at this end is universal for all transfer RNAs (Figure 6-10).

its other end comes into contact with a small portion of the mRNA. Now let us focus on this point of contact between the activated tRNA and the mRNA and ask about the extent and the nature of this contact.

Ever since the structure of DNA had been elucidated, it became relevant to speculate how a language of four letters could be translated into a language of twenty letters. The cosmologist, George Gamow, and later Francis Crick, proposed **codes** whereby three nitrogen bases in the nucleic acid could specify one amino acid. Their respective codes imposed different restrictions on this translation process, but in both cases it was possible to obtain exactly twenty different **triplets** each presumably coding for a given amino acid. In the last few years, by using a variety of genetic and biochemical approaches, the actual code has indeed been worked out by Nirenberg, Ochoa, and others (Fig. 6-9). The code has several features.

1. It is indeed a *triplet* code, in which a sequence of three nitrogen bases codes for a given amino acid.
2. The code, in the language of information theory, is **degenerate,** that is, there can be more than one triplet coding for a given amino acid.
3. Some triplets do not code for an amino acid, but seem to code for initiating and terminating the translation—that is, they act as punctuation. It would appear that one molecule of mRNA can carry more than one message, and therefore can synthesize more than one kind of polypeptide. The function therefore of the initiator and terminator triplets or **codons** is to mark where the translation into a given polypeptide sequence must begin and where it must end.
4. If you have a language of four letters and ask how many three letter words can be written with such a language, assuming no restrictions

Figure 6-9   The genetic code. Notice that a number of amino acids are coded for by more than one trinucleotide codon; it is usually the third trinucleotide that is changed.

1ST ↓	2ND →	U	C	A	G	↓ 3RD
		PHE	SER	TYR	CYS	U
U		PHE	SER	TYR	CYS	C
		LEU	SER	Terminator††	Terminator††	A
		LEU	SER	Terminator††	TRP	G
		LEU	PRO	HIS	ARG	U
C		LEU	PRO	HIS	ARG	C
		LEU	PRO	GLUN	ARG	A
		LEU	PRO	GLUN	ARG	G
		ILEU	THR	ASPN	SER	U
		ILEU	THR	ASPN	SER	C
A		ILEU	THR	LYS	ARG	A
		MET, Initiator†	THR	LYS	ARG	G
		VAL	ALA	ASP	GLY	U
		VAL	ALA	ASP	GLY	C
G		VAL	ALA	GLU	GLY	A
		Initiator†	ALA	GLU	GLY	G

(Table modified from F. C. Crick.)

† In bacteria, the initiator amino acid seems to be formyl methionine.

†† These "nonsense" codons do not code for any amino acid. The mechanism for chain termination is not yet understood.

whatsoever, you can easily calculate that 64 three letter words are possible. The genetic code in Figure 6-9 shows that all 64 words are in fact used, which had not been anticipated by Gamow or by Crick.

5. *The code seems to be universal for all organisms on this planet!*

The decoding of the translational process used by nature is one of the triumphs of modern molecular biology. It was the fruition of two decades of work, started on the one hand by Sanger, when he showed that the polypeptides in the protein insulin had an exact sequence of amino acids, and on the other hand by Watson and Crick,

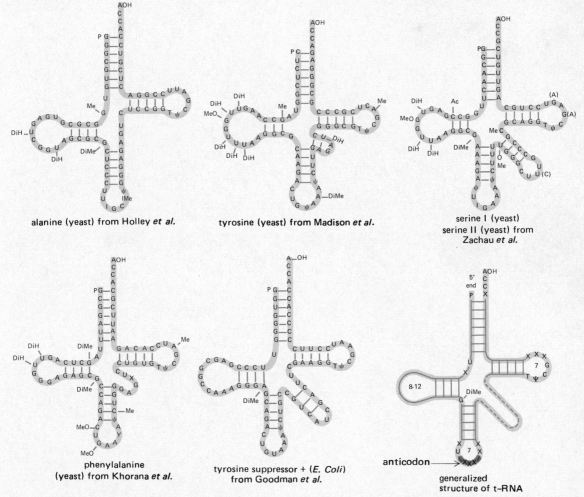

alanine (yeast) from Holley *et al.*

tyrosine (yeast) from Madison *et al.*

serine I (yeast)
serine II (yeast) from
Zachau *et al.*

phenylalanine
(yeast) from Khorana *et al.*

tyrosine suppressor + (*E. Coli*)
from Goodman *et al.*

generalized
structure of t-RNA

Figure 6-10 Cloverleaf pattern of five different tRNA molecules. The fact that the same pattern of folding can be generated though base sequences differ greatly is a dramatic confirmation of the hypothesis that the cloverleaf pattern is, in fact, a universal structure of the 64 different tRNA molecules. The second dramatic feature of the cloverleaf pattern is that it positions a triplet of bases at the center of the middle loop which is consistent with the anticodon (Figure 6-11) predicted from the code (Figure 6-9). One characteristic feature of tRNA is that, in addition to the four usual bases found in RNA (C, A, U, G), there are a number of special bases which are slightly modified forms of the usual bases. The student need not concern himself with these details except to note, for the purposes of comparing the anticodons with the codons, that the bases denoted as "I" and "ψ" are equivalent to G and U, respectively.

when they showed that the double-stranded structure of DNA was stabilized by the complementarity of base sequences along the strands. Of course, we must not forget the contributions of others before them, such as Sumner, who showed that enzymes can be crystallized and are made of protein; Avery, MacLeod, and McCarty, and Hershey and Chase who showed that DNA was the hereditary material; and Beadle and Tatum, who showed that genes are in some way responsible for the synthesis of enzymes; and many others, some of whom we have mentioned in the last five chapters.

Let us go back to the ribosomes and the interaction between the mRNA and tRNA. We have just learned that the mRNA arranges its message in the form of 64 different *codons* composed of given triplets of nitrogen bases. We would therefore predict that this message would be read by 64 different tRNA molecules, each of which would have at one end an **anticodon site** with bases that are complementary to the codon triplets. Thus if a codon for phenylalanine reads UUC (see Fig. 6-9), then a tRNA molecule must exist which can combine with phenylalanine and has an *anticodon* TTG.

This prediction has also been dramatically confirmed in recent work in which the exact structures of a number of tRNA molecules have been determined (Fig. 6-10). In all cases studied so far the anticodon sequence of a particular tRNA is complementary to one of the codons assigned to the amino acid for which the tRNA is specific (Fig. 6-11).

We have explained how each amino acid has several (2 to 6) tRNA molecules to which it can become attached by a specific *activating enzyme* and how, at the other end of the tRNA mole-

---

Correspondence between observed and expected "anti-codons" of five tRNA molecules

Amino Acid	Codons	Anticodons in tRNAs	
		Expected	Observed
Alanine	GCU GCC----- GCA GCG	CGA CGG----- CGU CGC	CGI
Tyrosine	UAU UAC-----	AUA AUG-----	AUG
Serine	AGU AGC UCU UCC----- UCA UCG	UCA UCG AGA AGG----- AGU AGC	AGI
Phenylalanine	UUU UUC-----	AAA AAG-----	AAG
Valine	GUU GUC----- GUA GUG	CAA CAG----- CAU CAC	CAI

Figure 6-11 We can predict anticodons from known codons and show that the former, in fact, occur in the structure of the tRNA molecules so far studied. In the above table are listed five tRNA anticodons which have been obtained from those tRNA molecules which have been sequenced. In every case, it is possible to find a correspondence between the observed anticodon and *one* of the anticodons predicted from the codons.

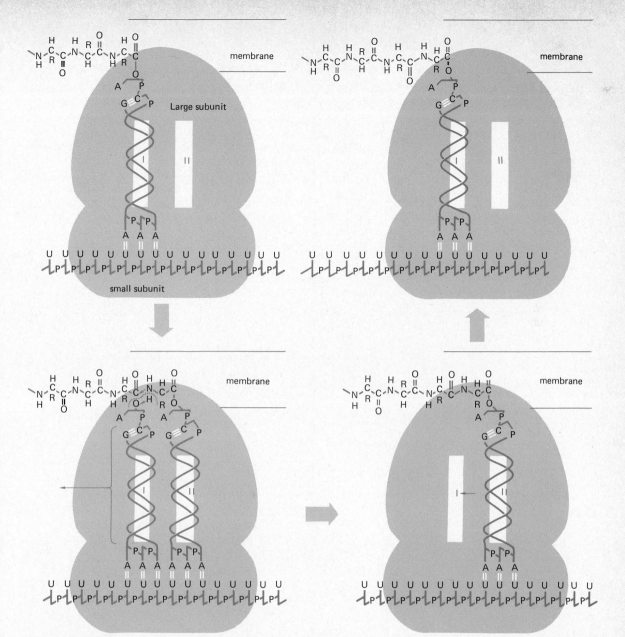

Figure 6-12 Our ideas of how the ribosome functions. Notice that the ribosome is made of a large and a small subunit. The mRNA, in this case the artificial polynucleotide polyuridylic acid (poly-U), is attached to the small subunit. We believe that the large subunit is attached to ER membranes and that it carries two sites for the binding of tRNA. Notice the complementarity between the codon (in this case, UUU) with the anticodon (in this case, AAA) and notice that the growing polypeptide chain (in this case, polyphenylalanine) is attached at the other end of the tRNA. Each new amino acid is attached to the growing polypeptide chain in a cycle of three steps:

Step 1 involves the binding of a tRNA molecule on site I of the ribosome.

Step 2 involves the binding on site II of another amino acid-tRNA molecule (in this case, phenylalanyl tRNA). At this point, the new amino acid is joined by its amino group to the polypeptide chain by a step which is still not well understood. The tRNA molecule on site I is now released from the ribosome.

Step 3 involves the movement of the tRNA molecule to site I and allows for another cycle of growth of the polypeptide chain.

cule, there is an anticodon which is complementary to a codon on the mRNA. The remainder is simply a matter of bringing together the amino acids and detaching them from the ribosome. Figure 6-12 is a schematic drawing of our present ideas regarding this exciting subject. Notice that the ribosome is made up of two subunits, a large one and a small one (Fig. 6-13). The small one interacts with the mRNA, which in Figure 6-12 happens to be poly U, a synthetic polymer of uridylic acid. It is thought that there are two sites for the attachment of tRNA on the ribosome. The amino acids are situated on the other end of the tRNA, and special sites of interaction probably exist between them and the ribosome. It is suggested that the synthesis of one peptide bond is the result of a cycle of reactions in which the tRNA shifts from site II to site I, thus allowing

for a second tRNA to occupy site II. Because more than one ribosome is attached to a given messenger at any one time, more than one polypeptide chain can be synthesized at the same time (Fig. 6-14). Using ^{35}S as a radioactive tag for protein molecules, one can, in fact, detect the synthesis of protein chains which are still attached to the polysome fractions.

Indeed, it has been possible in recent years to synthesize proteins *in vitro* if one puts the following complex mixture together:

Mixture of the 20 amino acids
Mixture of tRNA molecules
Mixture of tRNA activating enzymes
Some mRNA
Some ribosomes
Mg^{++}, GTP, ATP

If we then put in the synthetic messenger, poly U, we obtain a poly-phenylalanine polypeptide chain, which is what the code table shown in Figure 6-9 would predict.

We have come a long way in our discussion of the flow of information, yet still have to explain how the polypeptide forms the active enzyme.

THE LINEAR LANGUAGE FOLDS ITSELF INTO A THREE-DIMENSIONAL STRUCTURE

As we have pointed out, the enzyme owes its specificity for a particular substrate to its three-dimensional structure. We must therefore explain how a linear language can generate a three-dimensional structure. We still do not know many details of this process, but we know that it does happen spontaneously in a number of proteins we have studied. Thus a molecule like *ribonuclease* can be unfolded by dissolving it in a special reagent, and when the reagent is then removed, it can be shown that the ribonuclease can fold itself up to form a molecule with the same three-dimensional structure and the same enzyme specificity as before. It would appear that a given sequence of amino acids, because of the interaction of the amino acid side chains, produces a unique structure, that is, a polypeptide chain folded in a unique manner.

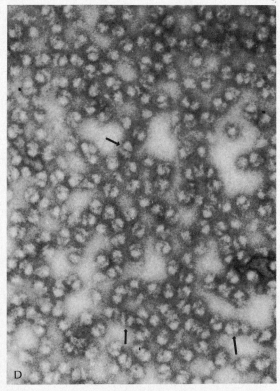

Fig. 6-13 Ribosomes as we have seen are composed of a large and a small subunit. These are ribosomes obtained from the bacterium *E. coli*. The dark grooves (arrows) between the large and the small subunit can be seen clearly (x 215,000).

In other words, our present view of the flow of information in the cell, which admittedly may be an oversimplification, is that the final three-dimensional shape of the enzyme molecule is determined by the sequence of the amino acids in the peptide chain, which is in turn determined by the sequence of nitrogen bases in the RNA, which is in turn determined by the base sequence in the DNA. In recent years a great deal of genetic data has accumulated that shows that a single mutation bringing about a change in a single base, thus altering a particular codon, will in fact bring about an amino acid substitution in the protein for which the DNA is coded. The classical observation in this area was made not with a microorganism but with hemoglobin obtained from human red cells. It was demonstrated that a genetic mutation causing the disease *sickle-cell anemia* brings about a substitution of a *valine* residue for a *glutamic acid* residue in one of the polypeptide chains of the hemoglobin molecule. This change in only one amino acid among 574 in the protein molecule brings about

a change in the properties of the hemoglobin molecule that makes it less reactive to oxygen, thus causing the humans carrying this mutation to be anemic. And thus for the first time the molecular basis for a disease was discovered.

HOW LARGE STRUCTURES ARE FORMED FROM SMALLER ONES

We have explained how a single polypeptide chain may fold itself into a molecule, but some protein molecules are made up of more than one chain. Hemoglobin, for instance, is made up of two α-chains and two β-chains which fit snugly together to form an almost spherical molecule (Fig. 6-15). Hemoglobin can easily be broken down into its separate chains and then be re-associated again in the test tube. Clearly, the four chains are so structured that they come together spontaneously. We can conclude that some of the information built into the polypeptide sequence of certain protein subunits brings about

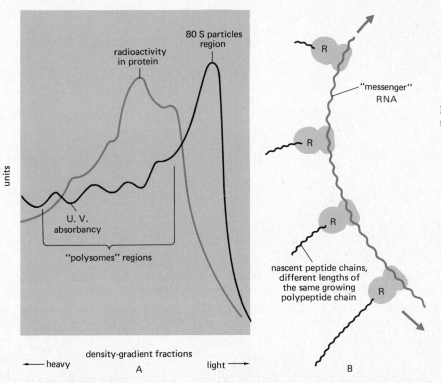

radioactivity in protein

80 S particles region

U. V. absorbancy

"polysomes" regions

units

density-gradient fractions

← heavy   light →

A

"messenger" RNA

nascent peptide chains, different lengths of the same growing polypeptide chain

B

Figure 6-14 Proteins are synthesized on polysomes.

A. Density gradient centrifugation experiment of a preparation from a pulse-labeled liver cell in which a fraction rich in radioactivity can be isolated.

B. We believe at present that the polysome fraction resembles this scheme: The messenger runs along a number of polysomes attached to it. A **nascent** polypeptide chain grows from each ribosome. The length of the chain depends on how far the messenger has moved.

properties that cause these subunits to associate with one another in a particular way.

Although we still do not know the details, we believe that the foregoing principles are true also for those molecular properties of proteins which cause them to associate with nonprotein substances such as lipids in the membranes, nucleic acids in ribosomes and chromosomes, and a host of other compounds such as cofactors, hormones, and pigments.

Thus, we have, at present, an admittedly vague and possibly over-simplified picture of the flow of information in the cell: the linear message of the DNA is translated through RNA into the linear language of the polypeptide chain, carrying within it all the information necessary to form the simpler structures of the cell such as membranes and ribosomes, and possibly even the larger structures, such as the organelles.

However, some suspect that other levels of heredity exist in the cell beyond that of DNA, and that some structures cannot be built only from the information residing in the DNA. It is argued that these structures might require patterns of organization which are present in the cell and without which they would not be elaborated. This

Figure 6-15 Hemoglobin, a globular protein molecule composed of four chains (2α and 2β) fitting snugly together. The folding of each chain as well as the interaction of the chains with each other is believed to be a function of the amino acid sequence which, as we have seen, finds its origin in the information stored in DNA and transferred to the protein by the RNA. We have here, therefore, an explanation of how information in the DNA can be used to build a complex biological structure.

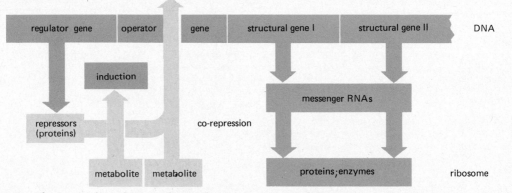

Figure 6-16 The operon hypothesis explaining how a gene can regulate protein synthesis by being turned off and on. The regulator gene synthesizes a protein (**repressor**) which acts on the **operator gene** and turns it off. When the operator gene is turned off then one or more **structural genes** are prevented from making messenger RNAs which in turn prevents the synthesis of the corresponding proteins. Protein synthesis is induced by the introduction of a given metabolite which reacts with the repressor protein, thus preventing it from turning off the operator gene. In some instances, a metabolite increases, rather than decreases the effect of the repressor (**co-repressor**).

is certainly an interesting problem, and no doubt we shall acquire much more knowledge about it in the near future.

HOW THE CELL SWITCHES THE PRODUCTION OF ENZYMES ON AND OFF

Now that we know something about how genes (DNA) control the synthesis of protein, it becomes important to ask how protein synthesis is *regulated* in the cell. That it *is* regulated there is no doubt. Here is an example.

The bacterium *E. coli* is known to synthesize the enzyme β-galactosidase only when galactose is present in the medium. Thus the presence of substrate (galactose) *induces* the synthesis of an enzyme (β-galactosidase) involved in the metabolism of the substrate. In fact we know that other enzymes involved in the metabolism of galactose, such as *galactose permease* and *transacetylase*, are also *induced* by introducing galactose into the growth medium.

To explain this phenomenon of the induction of enzyme synthesis by a substrate, Jacob and Monod proposed an ingenious model (Fig. 6-16). According to this model there are three categories of genes on the DNA: **regulator** genes, **operator** genes, and **structural** genes.

We already know about the structural genes which, through mRNA, produce the proteins of the cell. The operator gene, the model suggests, produces a substance (the **repressor substance**) which combines with the operator gene and prevents it from allowing the structural genes to operate. This is, then, the condition of *no protein synthesis* for the particular collection of regulator, operator, and structural genes called an **operon.** Now let us introduce galactose into the medium. After the galactose has entered the cell, it combines with the repressor substance, which prevents the repressor substance from combining with the operator gene. The operator gene therefore *turns on* and thus permits the structural genes to produce the messenger.

The Jacob-Monod model of regulation of gene activity may seem farfetched, but, to the de-

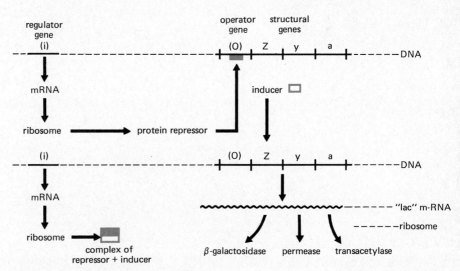

Figure 6-17   The lac operon, a well documented case for the operon hypothesis. In this case, we have three structural genes controlling the synthesis of three different enzymes. A great deal of genetic evidence has accumulated for the existence of the regulator, operator, and three structural genes. Recently, biochemical evidence has become available for the existence of protein repressor and its binding to the inducer (a galactoside) on the one hand and the operator gene on the other.

light of everyone in the field, considerable genetic evidence has accumulated in recent years to support it; also very recently the addition of direct biochemical evidence makes it virtually certain that this method of regulation of protein synthesis is utilized by bacteria and probably by all cells. Let us briefly look at the evidence supporting the Jacob-Monod operon model.

The best documented case we know is that of the so-called lac operon, which is responsible for the synthesis of the three proteins we mentioned before (Fig. 6-17). Genetic mapping work has shown that there is indeed a gene (the regu-

lator) which controls the activity of another gene (the operator), the latter being contiguous on the bacterial DNA with the three structural genes. Recently Gilbert and others have shown that the regulator gene indeed produces a protein (the repressor) which combines very tightly with an **inducer.** In this case the inducer is an analog of galactose that has the advantage of not being metabolized by the cell. Gilbert has also shown that in the absence of inducer the repressor protein combines with the operator gene on the DNA. Figure 6-18 illustrates this combination.

Hence we can conclude that the basic fea-

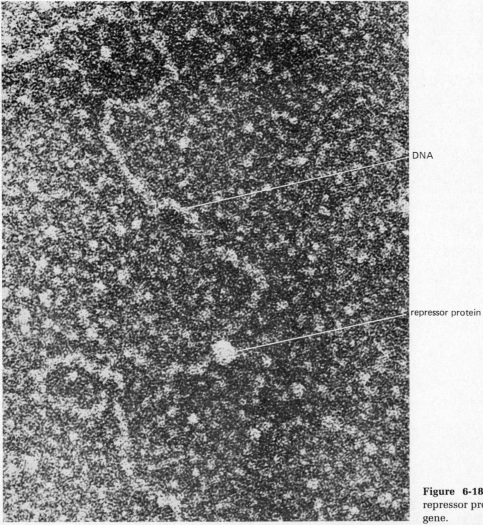

DNA

repressor protein

Figure 6-18 Combination of repressor protein with operator gene.

tures of the Jacob-Monod model are correct, and it only remains to be seen if and how widely it is used in eucaryotic cells for the regulation of protein synthesis.

There is, of course, no doubt that the cells of multicellular organisms do in fact regulate protein synthesis extensively. We know that during the development of an organism certain cells will begin to produce new proteins while turning off the production of others, and we might in fact view differentiation as being, among other things, correlated with the differential synthesis of proteins. More of this is discussed in Chapters 7 and 8.

The question remains whether protein synthesis is regulated at other levels. We have shown that it can be regulated at the transcription level. Can it be regulated at the translation level, and if so how? Unfortunately, we do not know very much as yet about regulation at this level. We can imagine that protein synthesis can be regulated (1) by controlling at the tRNA level, (2) by controlling ribosome synthesis, (3) by regulating amino acid production and (4) by a combination of these. Furthermore it is conceivable that an excess of a certain protein might indeed shut off its own production by a gigantic feedback loop acting on the DNA directly.

HOW ENZYME ACTIVITY IS REGULATED

So far we have discussed how the cell regulates the concentration of enzymes. However, metabolism can also be controlled by changing the *activity* of enzymes which are already present in the cell. How does this occur? Here again we can imagine a number of possibilities, but there is one we know something about — the mechanism of **feedback inhibition.**

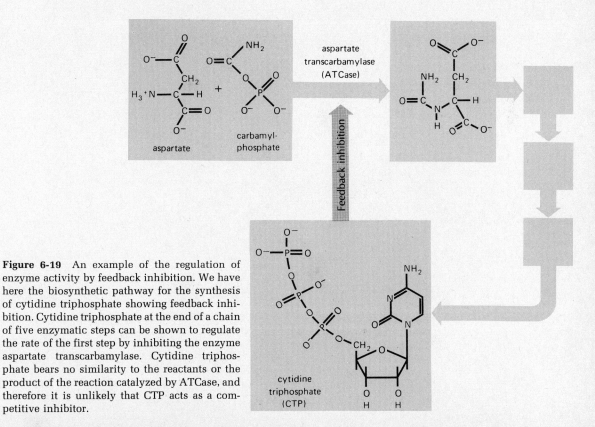

Figure 6-19 An example of the regulation of enzyme activity by feedback inhibition. We have here the biosynthetic pathway for the synthesis of cytidine triphosphate showing feedback inhibition. Cytidine triphosphate at the end of a chain of five enzymatic steps can be shown to regulate the rate of the first step by inhibiting the enzyme aspartate transcarbamylase. Cytidine triphosphate bears no similarity to the reactants or the product of the reaction catalyzed by ATCase, and therefore it is unlikely that CTP acts as a competitive inhibitor.

It has been observed, in a number of cases, that some enzymes are inhibited by metabolic intermediates which are several steps below them in the metabolic sequence in which they are involved. Take, for instance, the example shown in Figure 6-19. It involves a reaction sequence of 5 steps beginning with aspartate and carbamyl phosphate, and ending up with cytidine triphosphate (CTP). It was found that the enzyme involved in the first step, aspartate transcarbamylase or ATCase, is inhibited by CTP, the product of this particular reaction sequence. If the inhibitor had borne any similarity to the immediate product of this enzyme, its inhibitory property would not have been too surprising as it would have simply been a case of **competitive inhibition** in which a compound similar to the product inhibits an enzyme from acting on its substrate.

Many such cases of competitive inhibition have in fact been documented (Fig. 6-20). But in the example given in Figure 6-19 the inhibitor CTP bears no resemblance to the product of the reaction that ATCase is catalyzing. Because of this structural dissimilarity between the inhibitor and the product, this type of inhibition is called **allosteric.** A careful study of the ATCase enzyme has yielded a number of interesting results which are probably also applicable to other allosteric inhibitions.

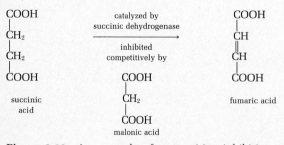

Figure 6-20 An example of competitive inhibition. Malonic acid is sufficiently similar to the substrate, succinic acid, so as to bind with the enzyme presumably on the active site, thus preventing the enzyme from acting catalytically. Since malonic acid cannot react, it will remain bound to the enzyme unless it is displaced by large amounts of substrate. One feature of competitive inhibition is, in fact, that it is reversed by increasing the concentration of substrate.

1. An allosteric enzyme is made of more than one polypeptide chain, with the molecule composed of more than one *subunit*.
2. There are two separate sites of action on the enzyme: (a) an **active site** which has the *catalytic* function and (b) a **regulatory** site which interacts with the *allosteric effector*.
3. The catalytic site and the regulatory site are on different subunits.
4. When the *allosteric effector* binds to the regulatory site, the catalytic site is affected in such a way as to reduce its catalytic activity.

We do not yet know the details of how the allosteric effector binding to one subunit of the macromolecule changes the reactivity of the catalytic site on the other subunit. We believe that the effect of small effector molecules on the structure of large protein macromolecules is likely to be an important and widespread phenomenon in the biology of the cell; a great deal of attention and study will be devoted to this in the near future.

HOW CELLULAR STRUCTURES MIGHT REGULATE METABOLIC ACTIVITIES

Although some enzymes appear to float as single molecules in the cell sap, others are linked together as **multienzyme complexes** (Fig. 6-21) which perform their role in a given reaction sequence better when they are spatially related to each other than when they are separate. Furthermore, a reaction sequence involving a multienzyme complex might be regulated more readily, although we still know very little about this.

Another example of the role of structure is, of course, the ubiquitous phenomenon of **compartmentalization** in eucaryotic cells. As we know from Chapter 2, the eucaryotic cell is full of membrane-bound compartments such as vacuoles, lysosomes, Golgi complexes, mitochondria, and chloroplasts. Numerous studies have shown that enzymes are unevenly distributed in these compartments. Certain enzymes, for instance, are found only in mitochondria while others are present mostly in the cytoplasm,

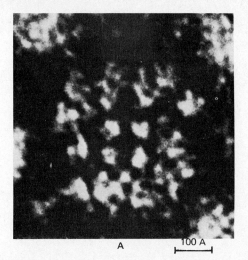

A   100 A

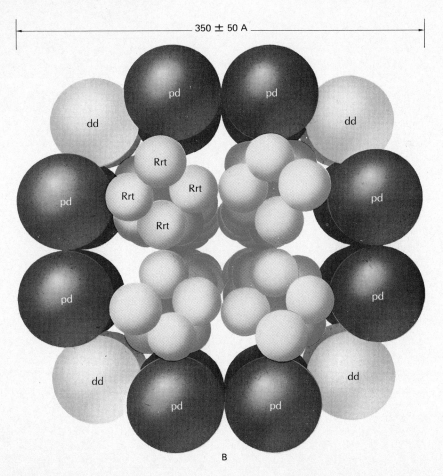

Figure 6-21 Enzymes often associate to form large complexes. In this case, we have the pyruvate dehydrogenase complex composed of 16 molecules of pyruvate decarboxylase (pd), 64 molecules of lipoic reductase transacetylase (Rrt) and 8 molecules of dihydrolipoate dehydrogenase.

350 ± 50 A

and so on. There is even the case of malic dehydrogenase in which the catalytic activity is found in the cytoplasm and in the mitochondria, although different enzymes, coded by different genes, are responsible for it.

Although we do not understand the details as yet, we conclude that the *structural relations* of enzymes in the cell must also have very important regulatory functions in the metabolism of the cell.

We have come to the end of our description of the flow of information in the cell. By now the reader must be keenly aware that we are beginning to unravel the puzzle of how the cell takes the energy and the materials supplied by the respiratory metabolism and ties the processes together to produce the delicate yet remarkably precise piece of machinery which, in a matter of minutes, can synthesize thousands of pieces, put them together to form a large variety of structures, and end up producing an exact copy of itself.

The problem we are going to consider in the following two chapters is how cells can cooperate with one another to produce an integrated, functioning organism. What we have learned here about the cell will make this task easier.

SUGGESTED READING LIST

Allen, J. (ed.), *Molecular Organization and Biological Function.* Harper and Row, New York, 1967.

Changeux, J. P., "The control of biochemical reactions," *Scientific American,* **212,** 36 (1965).

Cohen, G. N., *The Regulation of Cell Metabolism.* Holt, Rinehart and Winston, New York, 1968.

Cold Spring Harbor Symposia on Quantitative Biology, *Cellular Regulatory Mechanisms,* Vol. 26. Long Island Biological Association, Cold Spring Harbor, L. I., New York, 1961.

Crick, F. H. C., "The genetic code," *Scientific American,* **207,** 66 (1962).

Ebert, J. and I. Sussex, *Interacting Systems in Development,* 2nd Ed. Holt, Rinehart and Winston, New York, 1970.

Fraenkel-Conrat, H., *Design and Function at the Threshold of Life: The Viruses.* Academic Press, New York, 1962.

Goldsby, R. A., *Cells and Energy.* Macmillan, New York, 1967.

Hartman, P. E. and S. R. Suskind, *Gene Action.* Prentice-Hall, Englewood Cliffs, New Jersey, 1964.

Holley, R. W. et al., "Structure of a ribonucleic acid," *Science,* **147,** 1462 (1965).

Holley, R. W., "The nucleotide sequence of a nucleic acid," *Scientific American,* **214,** 30 (1966).

Hurwitz, J. and J. J. Furth, "Messenger RNA," *Scientific American,* **206,** 41 (1962).

Ingram, V. M., *The Biosynthesis of Macromolecules.* W. A. Benjamin, New York, 1965.

Lehninger, A. L., *Biochemistry.* Worth, New York, 1970.

Levine, R. P., *Genetics.* Holt, Rinehart and Winston, New York, 1962.

Loewy, A. G. and P. Siekevitz, *Cell Structure and Function,* second edition. Holt, Rinehart and Winston, New York, 1969.

Nirenberg, M. W., "The genetic code II," *Scientific American,* **208,** 80 (1963).

Rich, A., "Polyribosomes," *Scientific American,* **209,** 44 (1963).

Watson, J. D., *The Molecular Biology of the Gene,* second edition. W. A. Benjamin, New York, 1969.

Wolstenholme, G. E. W. and M. O'Connor (eds.), *Principles of Biomolecular Organization.* Little, Brown, Boston, 1966.

Zubay, G. L., (ed.), *Papers in Biochemical Genetics.* Holt, Rinehart and Winston, New York, 1968.

Part Two

The Emergence of Form

Here we consider how an
organism composed of billions
of cells develops from a
fertilized egg.

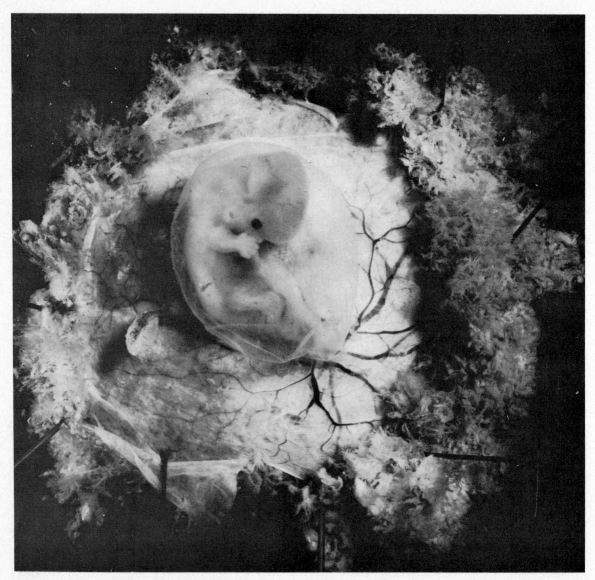

Figure 7-1   A human embryo at 39 days of development, shown with its protective membranes. (× 2.6).

The Shape of Things to Come

To understand the series of events through which the component parts of each animal or plant cell are fashioned, their interactions delicately balanced, and these parts duplicated in daughter cells generation after generation is not that simple. However, the complexities of cellular activity are multiplied many times when we observe the way a fertilized egg develops into a multicellular organism.

Higher animals and plants ordinarily begin as fertilized eggs. Eggs of fish and frogs (usually in masses) are discharged into the water and fertilized by sperm shortly afterward. The eggs of birds are fertilized in the upper reaches of the mother's oviduct; after the egg is laid, development stops until incubation in the nest or in the hatchery starts it up again. The fertilized human egg grows for about 280 days inside the mother's body. The eggs of many plants, fertilized by pollen, grow first into seeds, which then germinate and grow into new plants. Thus from the fertilized egg, or zygote, develops an entire organism—fish or frog, bird or man, radish or rutabaga.

The purpose of this chapter and the following one is not only to convey the fundamental facts and concepts of development, but an awareness of how they were obtained in the past and are being sought today. At the outset we wish to bring into sharp focus two concepts that underlie the study of development and that will permeate our discussion.

First, all development rests ultimately on the genes. The fabrication of a macromolecule and the final form of an organ must involve a series of interactions beyond the gene, beyond the individual cell, even with the external environment. Eventually, however, it will be necessary to trace the origin of these interactions to the structure of DNA and the control of its function.

Second, we shall stress the properties of embryos as distinct from the properties of adults. Too often in studying development we think of an embryo as an adult in miniature, failing to realize that the characteristic properties of an embryo or an embryonic cell change with time. Although the requirements of a nerve cell in the adult brain may differ from those of a cell in the adult peripheral nervous system, the requirements for the formation of either one during embryonic life may differ in still other ways. The requirements for *making* a nerve cell may differ drastically from those for *maintaining* it. The importance of recognizing the special properties of embryos is significant when we recall that agents like the German measles virus and the tranquilizing drug thalidomide, which have only minor effects later in life, produce drastic malformations in early human embryos.

There are developing systems other than the emergence of an individual from the fertilized egg, some of which we shall consider later. We begin, however, with the assumption that progressive change characterizes all development—whether of animal or plant, whether embryonic development proper or the regeneration of a salamander limb, or the growth of a potato tuber or a strawberry runner, or the formation of spores in a bread mold—and that the properties of each of these systems result from a sequence of molecular and cellular interactions dependent ultimately upon the organism's genetic endowment.

THE COMPONENT PROCESSES OF DEVELOPMENT

How may we characterize development? The cardinal criterion of development is progressive and cumulative change evident at all levels of biological organization (molecules, cells, tissues, and organs) during the life history of an organism. Beyond this an unfailing definition cannot be given. Certainly no one would doubt that the birth of an infant or the emergence of a flower is the culmination of developmental change, the unmistakable consequence of step by step events. But what of the individual steps? Here our task is more difficult. A given event, if taken out of context and considered without reference to whether its impact on the organism has been to produce a recognizable change in state, cannot be clearly defined as developmental or physiological or pathological. Such a distinction is not meaningful. In the shaping of a bird's wing a key step is the death of certain cells, a step that is highly ordered in space and time. Knowing the final product, several steps removed, we may speak of cell death as a developmental event—but only in that context. In examining a complex process there is some advantage in breaking it down into its component parts, if we realize that their separation is only an aid to concept formation. We consider four such component processes: growth, differentiation, determination, and morphogenesis.

The embryo grows

One obvious process in development is **growth.** The organism increases in size, in total mass. The fertilized egg divides, and then its daughter cells divide repeatedly until the organism contains millions or billions of cells. The human brain and spinal cord alone consist of billions of cells, arranged in an orderly manner with precise interconnections, all derived ultimately from the zygote. In the earliest stages of development of most animals, cell division occurs without growth, within the confines of the original egg, but eventually it must involve a great increase in size. A newborn baby weighs about 1 billion times more than the egg, barely visible to the naked eye, from which it grew.

There is another pattern of organ growth. After reaching a finite number, cell division may stop, and the cells themselves may become larger. At birth a human infant's heart has just as many cells as when he is grown. The infant's heart, however, is roughly one-sixteenth the size of an adult's. As the child grows the heart cells become larger until they too reach adult dimensions.

We may raise a number of fascinating questions related to the regulation of growth. What initiates growth? Does the newly fertilized egg grow, or do other steps have to be completed before growth begins? What factors determine the rate of growth? We know that different parts of the animal and plant grow at different rates and that their proportions gradually change. And not least in importance, why does growth stop? Consider this intriguing observation. A given number of segments (5, 10, 15) are removed from the tail end of a segmented worm. The worm will then regenerate a new tail of 5, 10, or 15 segments, usually no more or no less. Regeneration stops when the total number of segments, old plus new, approximates the number characteristic of the species (Fig. 7-2). Some clues as to possible mechanisms of growth regulation have come from studies of the regenerating kidney in rats and other animals. If part of a kidney or even one and one-half kidneys are removed from a rat, the cells in the remaining stump begin to proliferate. If we couple two rats surgically so that their blood circulations are connected (a technique known technically as *parabiosis*) and remove three kidneys from the two animals, the growth of the remaining kidney proceeds even more vigorously. Does this suggest that the kidney is constantly putting some specific factor or factors into the circulation that regulate kidney growth? Or does the organ "monitor" the amount of body

to be served and adjust its size accordingly? These ideas have occurred to many biologists, but proof has been elusive.

Its cells specialize

The second major process of development is **differentiation.** This means that parts of the organism increase in complexity, take on specialized roles, and in short become different from what they were to begin with. A child's cells are certainly different from the egg. The child has retinal cells that are sensitive to light so that he can see, muscle cells that contract, cells that secrete minerals which make up the hard structure of his bones and help support his body. A seed does not even suggest the complexity of the tree that will develop from it, with cells that carry out photosynthesis, others that provide channels for water and minerals, still others that transport food made in the leaves to other parts of the plant, and so on.

A tree depends on undifferentiated cells for much of its growth throughout its lifetime. During the growing season these cells divide rapidly and the new cells so produced differentiate into an array of specialized cells. The yearly rings of tree trunks are produced by the vascular tissue and its cells that enlarge and differentiate into vessels. In the spring these cells are larger and thinner walled than at other times of year. They contrast sharply with the dense wood of the previous summer. They thus leave a record of the annual growth pattern of the tree. The presence

Figure 7-2 Regeneration in the segmented marine worm, *Clymenella torquata.* Adults of this species have exactly 22 segments. When anterior and posterior ends are severed at various levels, leaving pieces 13 segments long, simultaneous anterior and posterior regeneration restores the normal number of segments.

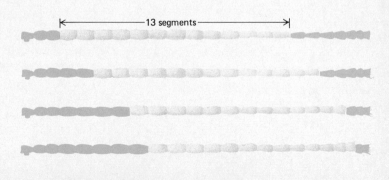

of large numbers of undifferentiated cells is typical of plants.

We cannot, however, make this generalization about animals. A small part of a coelenterate, like *Hydra*, or a flatworm, say *Planaria*, can regenerate the entire animal. An adult salamander can regenerate a lens, a limb, or a tail. But in the birds and mammals, including man, the capacity for regeneration is limited to a few organs, such as the liver and kidney, and those tissues which undergo continuing renewal, such as the skin. Thus in a newborn baby many cells have become specialized and are no longer able to revert back to more generalized roles. Most of the changes occurring after birth affect the child's shape and size. In this respect the typical course of differentiation has been carried out. Once a given line of cells has passed a certain point, its specialization cannot be reversed. However, the time at which they become *determined*, or restricted to their definitive roles, will not be the same for all types of cells. Some remain capable of differentiation throughout life; for example, the liver regenerates even in aged animals. However, the overall pattern of progressive restriction explains in part why drugs or diseases which interfere with development at crucial stages (thalidomide and German measles, for example) leave permanent defects. They have interfered with the developmental time-table, and the body can no longer correct the errors.

Earlier we used the word "differentiation" in the general sense to mean the full sequence of changes involved in the progressive diversification of cell structure and function that is the hallmark of development. However, it is increasingly useful in analyzing the mechanisms of development to recognize that differentiation—the actual appearance of new properties, whether defined in biochemical or structural terms (for example, the appearance of the contractile protein, myosin, in the muscle cell)—is preceded by **determination;** this is the process whereby a cell of an embryo becomes restricted to one particular pathway. Thus a cell may be "determined" to become a muscle cell long before it can be recognized as such.

The organism takes shape

The fourth major process of development is **morphogenesis** (from the Greek words for "form" and "to be born"). This involves the organization of various body parts into the proper form or shape. Conceptually the distinction between morphogenesis and differentiation is important, but in practice it is difficult to separate them. A functional eye obviously requires more than just the presence of the light-sensitive cells of the retina; these cells must also lie in proper position in relation to other parts of the eye. Thus it is important to examine the developmental events which

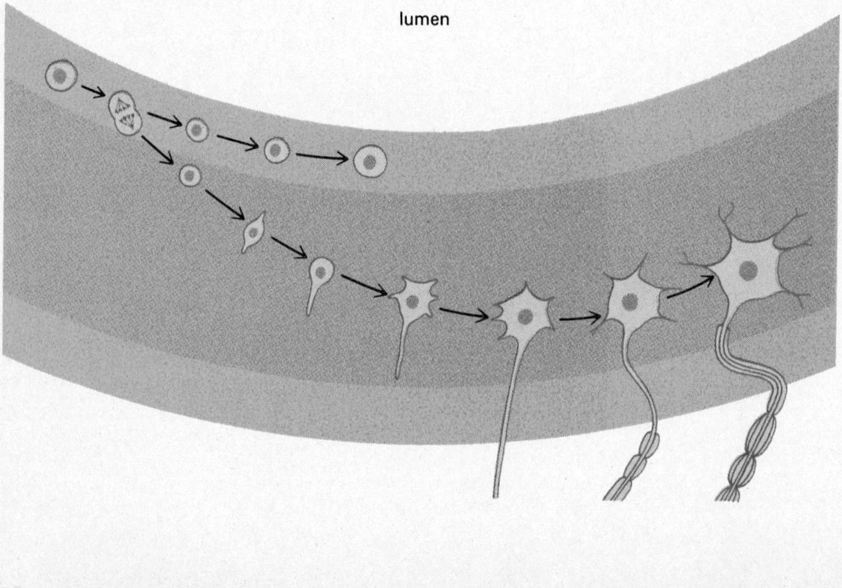

lumen

Figure 7-3 Progressive diversification of cell structure in animals. Stages in the development of a multipolar neuron.

are clearly part of the process by which new form is generated.

The developmental patterns of animals and plants differ most sharply in their form-building mechanisms. In animal development, apart from the "sculpturing" of structures by selective cell death, the principal morphogenetic mechanism is the regular redistribution and rearrangement of individual cells and organized groups of cells to bring them to new locations where they can interact with new environments. Several types of cells have their origin in one part of the embryo, but only after they have migrated to their definitive position in another part of the embryo do they differentiate to their functional state. Thus the cells of the medulla of the adrenal gland that will produce adrenalin, the melanin pigment-producing cells of skin and hair, cells of the sympathetic nervous system, and enamel-producing cells of the teeth all arise in the *neural crest*, which lies on each side of the developing spinal cord. However, they appear to undergo the major steps of their differentiation only after they have been distributed throughout the body. This movement of cells is typical of morphogenesis throughout the development of the animal embryo. Moreover, as we shall see, there are also oriented movements of sheets of cells. When such movements are experimentally prevented, or when they occur abnormally in nature, development is abnormal. How, then, does morphogenesis occur in plants where cell movements do not occur? Plant cells are bounded by rigid walls, and the cells of the tissue are firmly bonded together by extracellular cementing substances. Cell separation is thus prevented, and although cell shape may change during growth, these changes are not reversible and do not result in movement. Cells develop in the place in which they are first formed.

As we trace the course of development of a representative animal and plant it will be seen that differentiation and morphogenesis go hand in hand. At the same time we will begin to explore some of the molecular events which underlie development, considering, at the outset, nu-

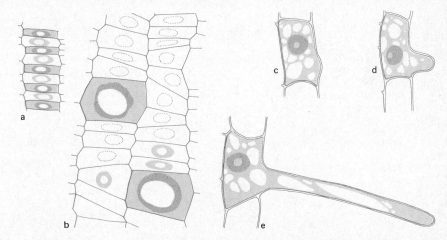

Figure 7-4 Progressive diversification of cell structure in plants. Differentiation of root hairs. (a) Epidermal cells that will develop root hairs are first identifiable by the increased density of their cytoplasm and nucleus. (b) As the root cells enlarge those that will form root hairs become much larger than adjacent nonroot-hair-forming cells. They retain their characteristic dense cytoplasm and the nucleus undergoes considerable enlargement. (c) Outgrowth of part of the cell to form the root hair is a local event and does not involve the entire outer part of the wall. The position of the outgrowth is a site of cytoplasmic accumulation. (d and e) Later stages of root hair growth showing the polarized nature of the cytoplasm in the growing tip and the nongrowing basal part of the hair.

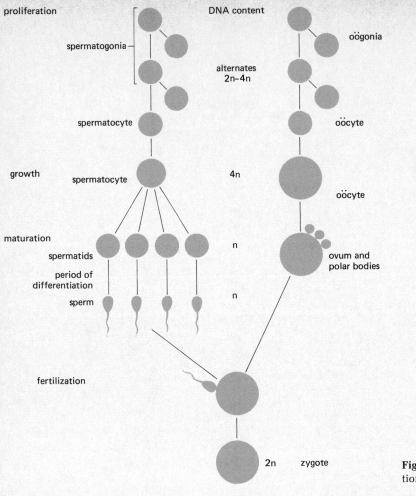

proliferation

spermatogonia

DNA content

oögonia

alternates
2n–4n

spermatocyte

oöcyte

growth

spermatocyte

4n

oöcyte

maturation

spermatids

n

ovum and
polar bodies

period of
differentiation

sperm

n

fertilization

2n   zygote

Figure 7-5 Diagram of the formation of eggs and sperm.

cleic acid and protein synthesis. Thus we may say that a muscle cell has differentiated when we can first detect the presence of proteins with the ability to contract. Differentiation, then, involves protein synthesis. Accordingly, when information about molecular processes is available, we shall consider them as well as the visible and more obvious signs of change.

However, if development were nothing more than nucleic acid and protein synthesis, our discussion would have been completed in the preceding chapters. The key to understanding development lies in unraveling the ways these processes are *organized* in both time and space. We must consider the *regulation* of molecular processes in orderly sequence, the *fabrication* of

their products (for example, proteins into macromolecular arrays), and the orderly arrangement of cells and tissues.

HOW ANIMALS DEVELOP

Origin and maturation of germ cells

At the beginning of this chapter we said that development begins with the activation of the egg by the sperm. It is a convenient starting point. But if we bear in mind that the development of an individual is but one cycle in an endless chain of generations, we see that the development of an individual actually begins with the production and maturation of the egg and sperm. The formation of the egg (*oögenesis*) in particular involves the laying down in an organized way of the machinery and, in some forms, of all the raw materials for development.

The primordial germ cells that mature in the ovaries and testes (which, upon maturing, become eggs and sperm, respectively) are like the other cells of the body in a very important respect—they have a full complement of chromosomes. Human cells, for instance, have 46 chromosomes. But if a sperm with 46 chromosomes were to fertilize an egg with the same number, the offspring would have twice the normal amount of hereditary material. Nature avoids this difficulty by reducing the chromosome complement during *meiosis.* In this process two successive cell divisions are accompanied by only one chromosomal duplication, the final result being a reduction in DNA content to half that in the primordial germ cells. Thus, as described in Chapter 3, the *gametes* produced have exactly

half the number of chromosomes as the somatic cells. The primitive germ and body cells are then *diploid.* The gametes are *haploid;* they have half as many chromosomes. When the haploid sperm fertilizes a haploid egg, the fertilized egg or *zygote* becomes diploid (Fig. 7-5).

In *spermatogenesis,* the divisions are equal. The four sperm resulting from one germ cell are the same size. They take on the spermlike shape after the second division (Fig. 7-6). In oögenesis, the yolk which will supply food for the embryo is not divided. Eventually one large haploid egg is formed, along with three smaller *polar bodies* which are also haploid. In animals the polar bodies soon disintegrate and disappear.

The genetic consequences of meiosis are the same for both sperm and eggs. The other consequences, however, are of course very different. Sperm are highly specialized for motility and for activating an egg once a collision has occurred. The egg is differentiated to provide the machinery for development.

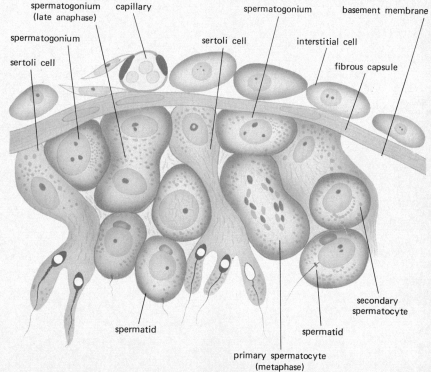

Figure 7-6 Section of part of human seminiferous tubule, showing sperm in all stages of maturation.

The sperm—small and mobile

A sperm has two functions. It starts the process of development by activating the egg and it supplies a haploid nucleus. These two functions are carried out by the sperm *head* which contains the *nucleus* and bears the *acrosome* which interacts with the egg to initiate development. To carry out these functions, the sperm must first reach the egg. The rest of the sperm provides its motility.

Proceeding "tailward" from the head, we next encounter the area about which we know least, the short *neck*, perhaps only 0.5 micron long, in which the principal elements that make up the midpiece and tail, or flagellum, arise. The sperm *tail*, like most cilia and flagella except for the bacterial flagellum, is composed of two central microtubules surrounded by a ring of nine double tubules (Fig. 7-7). These microtubules are attached to, and probably originate in, a *basal body* or granule in the neck. In the middle piece one additional major feature, the mitochondria, are located. They are arranged spirally around the microtubules.

Figure 7-7 Diagram of sperm, with representative transverse sections through the flagellum. Two central and nine double peripheral microtubules are typical of all such motile organelles. Mitochondria are present in the midpiece. An additional array of nine outermost filaments in the midpiece of mammalian sperm extends into the proximal part of the flagellum. The fibrous sheath of the tail is frequently ribbed.

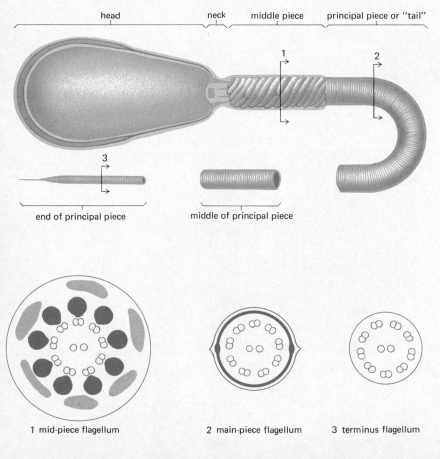

head   neck   middle piece   principal piece or "tail"

1

2

3

end of principal piece   middle of principal piece

1 mid-piece flagellum   2 main-piece flagellum   3 terminus flagellum

The sperm tail moves with an undulating snakelike motion. Despite intensive studies of the structure of microtubules and of the chemistry of proteins isolated from sperm tails, no proof has emerged that clearly defines the mechanism of sperm movements. There is no compelling evidence for contraction of the microtubules. There are numerous speculations: microtubules may bend, or they may slide up and down with respect to one another, or there may be a contraction or shortening of other elements in the flagellum running alongside the microtubules. Sperm movements are probably powered by energy generated in the mitochondria, but the manner in which this energy is translated into work is not clear nor is the mechanism by which these flagellar movements are coordinated.

Animal sperm come in many variations of this standard "model" (Fig. 7-8). They also cover a range of sizes. The sperm of sea urchins are only 40 microns long; those of some insects are 300 times as large, or 12 millimeters long, and fully visible to the naked eye. Human sperm are about 60 micrometers long.

Although sperm undergo maturation in the testes, the *primordial male germ cells*, in fact, germ cells of both sexes, originate in clearly demarcated areas outside the primitive gonad to which they must migrate. In some amphibians and mammals the migratory route appears to be through the tissues themselves, probably by amoeboid-like movement. In at least some birds the primordial germ cells appear to be dispersed in part through the circulation. The fact that the primordial germ cells are set aside in specific regions outside the gonad early in development makes it possible to produce sterile adults by removing the germ cells by radiation or surgery. If all the germ cells are destroyed, the rudimentary gonads develop but they will remain sterile throughout life.

When sperm are released, they are mixed with a thick fluid secreted by other glands. The combined secretions are called *semen*. The amount released at one time varies from species to species. In humans it averages 3 milliliters, but only about 10 percent of this volume is sperm. Each milliliter contains about 100 million cells. As in most animals the sperm outnumber the eggs tremendously, the large number being necessary to ensure fertilization. Studies of human fertility have shown that a sperm count of 40 to 50 million cells per milliliter means conception

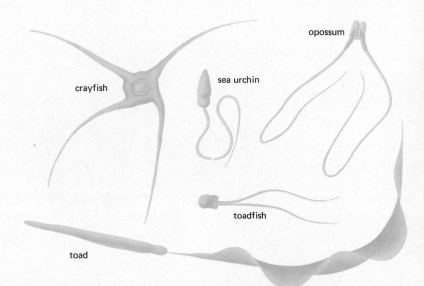

Figure 7-8 Variations in sperm morphology, showing sperm of crayfish, sea urchin, toadfish, toad, and opossum.

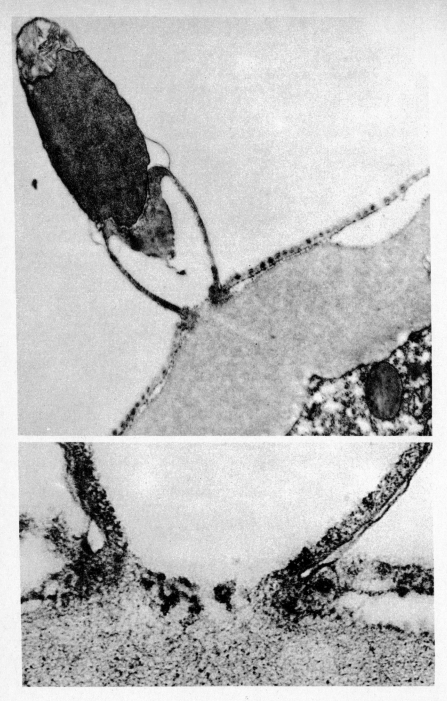

Figure 7-9 Electron micrographs of initial contact of sperm and egg membrane. (Top) Sperm plasma membrane meets egg envelope. Egg plasma membrane is still separated from sperm by egg envelope. × 37,000. (Bottom) Higher magnification of membrane interaction similar to one shown above. Acrosomal membrane is now inserted into sperm plasma and membrane. × 100,000. For details, see Fig. 7-10.

is likely; when the count falls below 20 million, fertilization is much less likely to occur. In rabbits no more than 50 to 250 sperm actually reach the site of fertilization in the female reproductive organs. Accordingly a low sperm count greatly reduces the number that reaches the eggs. The probability of fertilization is also reduced by the fact that, generally speaking, sperm have a short lifespan. In the human female reproductive tract they survive for about 36 hours.

The indispensable egg

In the normal course of development of most animals eggs are activated by sperm. However, in some species, as in certain rotifers, males are unknown, and eggs develop naturally without male intervention. This process is known as *natural parthenogenesis.* In the U. S. Department of Agriculture laboratories some strains of naturally parthenogenetic turkeys have been bred. In addition, over the years the eggs of a number of animals have been artificially activated. Thus both experimentally and in nature, sperm are dispensable. But the egg is indispensable. It not only supplies a haploid nucleus, it contains the machinery for development, and in many species carries with it all of the raw material and food resources on which the embryo grows until it can begin securing food for itself. How long this food reserve must last largely determines the size of the egg. In marine animals, which can feed themselves early in larval life, the egg may be small. In birds, on the other hand, the embryo must live off the stored food, or yolk, until it hatches.

Eggs of mammals are much like those of marine animals in terms of the amount of yolk they contain. The egg is small because it begins receiving nourishment from the mother's body within a few days of conception. Mammalian eggs are between 0.05 and 0.25 millimeter in diameter. The human egg is in the middle of this range, at 0.14 millimeter.

The female germ cells, like those of the male, are first found outside the gonad. Once in the gonad they proliferate. In the mouse the proliferative phase is confined to the period of intra-uterine life. All of the mature eggs to be ovulated by a mouse during her lifetime are derived from oöcytes laid down in embryonic life. They are present at birth and persist throughout adulthood. This pattern of oögenesis is probably not unique for the mouse; it seems to be a universal feature in mammals, including the human species. Before she is born, a girl probably has her life's supply of eggs. They will grow, mature, and be released at the rate of one a month from puberty until the end of the woman's reproductive period. A newly released human egg can be fertilized for about 24 hours. Eggs fertilized within

the body of an animal generally have a longer lifespan than those shed into water. Those in the latter category, such as eggs of most invertebrates, fish, and amphibians, must be fertilized immediately or within a few minutes at the most.

Egg and sperm unite

Egg and sperm get together largely by random collision. There are so many sperm that chances are very good such a collision will occur. Animals

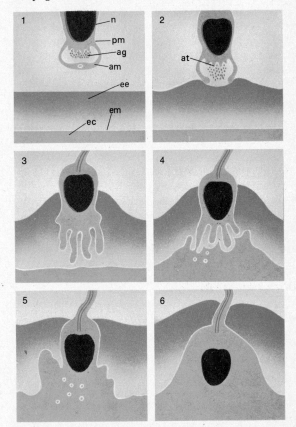

Figure 7-10 Principal steps in sperm-egg interaction. (1) Soon after the sperm contacts the egg envelope, the sperm tip opens, and the acrosomal and sperm plasma membranes become continuous. (2) and (3) The acrosomal wall is everted. (4) Sperm and egg plasma membranes meet; fertilization cone rises. (5) Sperm parts still protrude into egg envelope. (6) In zygote, internal sperm parts mingle with egg cytoplasm. *ag,* acrosomal granule; *am,* acrosomal membrane; *at,* acrosomal tubule; *ec,* egg cytoplasm; *ee,* egg envelope; *em,* egg membrane; *n,* sperm nucleus; *pm,* sperm plasma membrane

that shed eggs in water also shed sperm nearby; in mammals the sperm reach the egg to some extent by propelling themselves, but they also get help from the female reproductive tract, where cilia and contracting muscles move them toward the egg. In a few species chemical substances released by the egg attract sperm to its surface. However, such "chemotactic" mechanisms do not appear to be the general rule. Once the egg and sperm are in contact, however, there is nothing random about what follows.

The moment the sperm touches the egg, the *acrosome*, a structure at the tip of the head, produces a threadlike filament which connects the sperm to the egg surface. The acrosome then releases an enzyme which dissolves part of the egg envelope. Enzymes capable of doing this have been isolated from the sperm of several species, but the exact mechanism has not yet been detailed. Some of these enzymes break down the envelopes themselves, whereas others break down the cement which holds the envelopes together. Many disease-causing bacteria use similar enzymes to penetrate body tissues.

In most species once one sperm has entered the egg, no others are able to penetrate. There are exceptions among some birds and reptiles, but only one of the entering sperm finally fuses its nucleus with that of the egg.

Shortly after penetration the membranes of the acrosome and the egg become continuous, and soon the interiors of the two gametes are like one. Ultimately the head and middle piece of the sperm enter the egg; the tail is usually left outside. The two nuclei now move toward each other. Their nuclear membranes dissolve and their chromosomes take up positions on a spindle. The chromosomes are then duplicated and two new cells develop.

The sperm has thus activated the egg to begin division. As we have seen, however, the egg may divide, or cleave, without the help of sperm, in the process known as parthenogenesis. Frog eggs will begin dividing after they are pricked with a needle dipped in frog blood; sea urchin eggs start division after immersion in seawater to which butyric acid has been added. This suggests that the sperm triggers some initial step necessary for development of the egg. Artificial

methods of activation may also trigger this first step, but not surprisingly they succeed far less frequently. Nevertheless normal adults may result from parthenogenesis. The haploid egg may become diploid by duplicating its chromosomes once without cell division or by incorporating a polar body that would normally disintegrate.

Cleavage — Partitioning the zygote

To the casual observer cell division is the first overt sign of life in a zygote. The number of cells increases in geometric progression: 2, 4, 8, 16, 32. But unlike division in single-celled organisms, or even in later development of many-celled organisms, in many animals there is no growth of the zygote initially. The overall mass of the egg remains the same. This pattern of division without growth is called *cleavage*. During this time the zygote, which starts out larger than normal body cells, is partitioned into smaller units that are much closer to the size of the organism's body cells.

Although little else happens that we can readily observe, we know that a great deal is occurring at the molecular level. Each new cell contains a full complement of genes, so that nuclear materials are being synthesized during cleavage at the expense of the cytoplasm. DNA, RNA, and new cell membranes are among the major products of cellular activity at this time.

Patterns of cleavage vary widely in the animal kingdom, but within a species the timing of divisions and the physical orientation of new cells are strikingly similar. By the time a frog's egg has eight cells, half of them (those in the upper or *animal* hemisphere) are smaller in size than the others. All species show a typical rhythm of division, too, affecting all cells at once. Gradually, however, the process loses its synchronous nature and becomes more irregular as time passes. In both monkey and human embryos the *blastomeres* (as the daughter cells are called) show size differences from the start. Moreover the number of blastomeres does not increase by a regular doubling sequence but progresses arithmetically. Although there is variety throughout nature in these various processes, the overall pattern is nevertheless similar. The more closely the animals are related, the closer the resemblances. Often a fundamental similarity among a group of animals is taken as evidence of their close relationship.

It is difficult and possibly not important to say *just when differentiation begins.* Remembering that differentiation involves a restriction in capacity for change, we might phrase the following question: A zygote is capable of forming a whole new individual. How soon during cleavage do the daughter cells—if they can be isolated—lose their capacity to form a whole individual? Very soon—at the first or second cleavage. The daughter cells of several species can be separated at the two- and four-cell stages, surgically in frog and salamander embryos, or by growing the embryos in calcium-free seawater, as in the sea urchin. Whole frogs or salamanders can arise from individual daughter cells isolated at the two-cell stage. Cells of sea urchin embryos separated in the four-cell stage develop into normal embryos.

Experiments like these are more difficult to perform on mammalian embryos which must develop inside the mother. However, a few ingenious experimenters have separated the blastomeres of two-cell rabbit embryos, implanted them in foster mothers, and achieved normal development. Presumably human identical twins result when a fertilized egg divides and each daughter cell then for some unknown reason separates and produces an embryo; if the cells separate after two divisions, identical quadruplets might be born.

In many species, however, the first cleavage divides the zygote into two cells with unequal capacities for development. The eggs of many worms and mollusks fall into this category. Obviously the first cleavage has not equally partitioned some vital component or components.

An egg in action

To help understand cleavage and the other crucially important processes that follow it, we shall trace the development of a typical amphibian, the salamander. Salamander eggs, like frog eggs, are easily studied, and they can be observed at home or in the laboratory by using a magnifier or dissecting microscope. The two commonly used but more technical methods of study involve the use of time-lapse movies (in which photographs are taken at regular intervals a few seconds apart) or the tagging of cells with carbon particles, dyes, or radioactive materials.

The major steps in cleavage are illustrated in Figure 7-11. The changes shown occur over a period of slightly more than a day. Thirty hours after the egg is laid it contains hundreds of cells and a cavity has begun to develop inside it. The first two divisions are longitudinal and take place

Figure 7-11 Cleavage in the salamander. (a) Shortly after entrance of the sperm. (b) 2-cell stage. (c) 4-cell stage. Blastomeres still equal-sized. (d) 8-cell stage, showing initial inequality of blastomeres. (e) 12- to 16-cell stage. (f) Late cleavage.

more or less from top to bottom. The resulting four cells are equal in size and all contain nearly equal amounts of material from the upper and lower hemispheres of the newly fertilized egg. The upper hemisphere is known as the *animal* hemisphere, the lower, as the *vegetal* hemisphere. The lighter-colored vegetal hemisphere is rich in yolk and is heavier.

The third division is more or less horizontal. The top four cells now contain most of the material from the animal hemisphere, and the bottom four are made up mostly from the vegetal half. These lower cells are slightly larger. From this time on division becomes more irregular, the cells of the animal hemisphere cleaving more rapidly and being somewhat smaller.

The cavity inside the 30-hour-old embryo is called the **blastocoel,** and the embryo itself a **blastula.** Cleavage partitions the single-celled zygote into the multicellular blastula; as the number of cells increases, so does the opportunity for interactions following the cellular rearrangement that results from the next major step, **gastrulation.**

The cells are rearranged during gastrulation

About 20 hours later, when the embryo is approximately 50 hours old, the next major step begins. A small crescent-shaped opening appears near the union of animal and vegetal hemispheres. The cells around the opening begin moving toward the opening and then through it. The appearance of the opening or **blastopore** marks the beginning of gastrulation (Fig. 7-12).

During gastrulation the blastula is transformed into a three-layered gastrula (Fig. 7-13). Most of the cells that move through the blastopore make up the middle layer, or **mesoderm,** which lies between the outer layer or **ectoderm** and the yolk-laden **endoderm** cells. Shortly after the ninetieth hour of life the blastopore is closed and gastrulation is complete. The original blastocoel has disappeared and a new cavity, the **gastrocoel,** has formed. The gastrocoel represents the lumen (cavity) of the **archenteron,** or primitive digestive tract. The archenteron will ultimately be lined with endoderm.

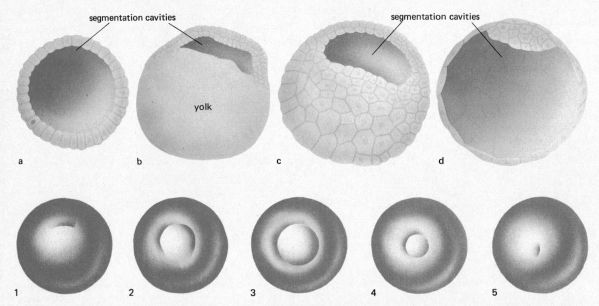

Figure 7-12 Upper: A diagrammatic comparison of chordate blastulae, demonstrating the basic anatomical pattern of the blastula stage. These drawings are not made to scale. Each blastula is composed of many cells surrounding (or partially surrounding) a central or eccentric segmentation cavity, sometimes termed a blastocoel. (a) *Amphioxus*, representing primitive chordates; (b) teleost fish; (c) amphibian; (d) mammal. Lower: Formation of the blastopore in the salamander blastula.

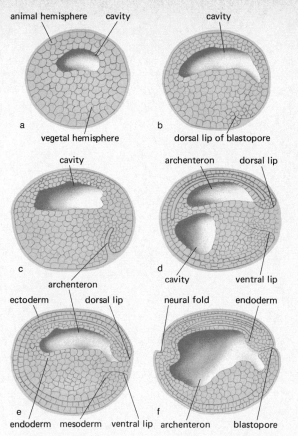

Figure 7-13 Germ layer formation in the salamander as seen in sectioned embryos.

The three layers formed during gastrulation provide the basis for all the parts of the organism. For example, the ectoderm will give rise to the nervous system and the outer layer of skin. The mesoderm will become muscle (including heart muscle), bone, kidneys, blood vessels, and gonads. The innermost layer, the endoderm, will develop into the lining of the stomach and the intestine, lungs, liver, and pancreas.

However, any listing of derivatives of these germ layers tells only a part of the story; of greater significance is their interaction during organ formation. We will see shortly the importance of such interactions for the formation of the nervous system. In fact they characterize the development of most organs. In addition, many organs are made up of tissues originating in more than one layer. While the outer epidermis of the skin origi-

nates from ectoderm, the underlying dermis is mesodermal. The lining of the stomach and intestine is formed from endodermal cells, but the muscle that makes up their walls is mesodermal.

The three-layered embryonic stage is typical of all vertebrates, whether they are amphibians, fish, birds, reptiles, or mammals. This similar pattern of development among animals destined to grow into very diverse forms as adults suggests their common evolutionary origin.

Origins of the nervous system

Just as the blastopore is closing, two folds or ridges of tissue are elevated on the opposite, or future *dorsal*, side of the embryo. They are called

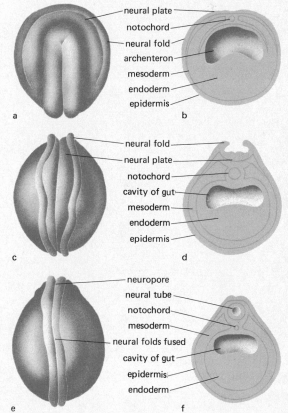

Figure 7-14 Neurulation in an amphibian embryo. Left, whole embryos, dorsal view. Right, anterior halves of embryos, cut transversely. (a, b) Early neurula; (c, d) middle neurula; (e, f) late neurula.

the *neural folds,* because they mark the beginning of the nervous system, the brain and the spinal cord. Gradually the groove deepens and the folds meet over it (Fig. 7-14). As they close the *neural tube* is formed, which is now beneath the surface of the covering epidermis. One end of the neural tube is noticeably wider than the other. Within it are three bulges that will become three parts of the brain: forebrain, midbrain, and hindbrain. This stage of development, which sets the stage for the entire nervous system including the sense organs, takes less than 20 hours. The embryo is now about 110 hours old. From now on the embryo begins to assume its larval form (Fig. 7-15). Gradually the head takes shape. The eye starts to form as an outgrowth of the forebrain. Gills begin to appear, and the tail is growing. At the same time, now about nine days since fertilization, most, if not all, of the animal's organ systems are developing. We see all the characteristics of a salamander larva that will begin to eat and otherwise fend for itself in another two weeks.

Embryonic induction

Let us continue to explore the progressive restriction of developmental capacity in the embryo. Earlier we pointed out that by the four-cell stage the capacities of the cleavage cells in the salamander embryo are restricted, so that none of them can form a complete embryo when isolated. We may now ask when the cells of the ectoderm are determined, some of them to form brain and spinal cord and others epidermis.

By direct observation and by following the fates of cells in different parts of the embryo which have been marked with vital dyes (which stain living cells without damaging them), we can learn just which regions of the blastula normally form epidermis and neural tube, respectively. Suppose we now exchange them, cutting out small pieces using fine glass or steel needles and transplanting them carefully into their new positions. In short we perform a reciprocal transplantation experiment. The grafts will heal in, and development continue normally. The cells which would normally have formed neural tube (*prospective* neural tube) now form epidermis; the cells which would normally have formed epidermis (*prospective* epidermis) now become neural tube. In other words, the cells in these regions of the blastula have not yet become determined. They are able to perform more than their prospective normal fate. However, if we do the same experiment just two days later, when gastrulation has been completed, the results are quite different. Prospective epidermis transplanted to the site of the neural tube now produces only a patch of epidermis. Prospective neural tube now produces only neural tube in the epidermal site.

Thus determination of ectodermal cells accompanies gastrulation. But is it a *result* of gastrulation? If we look at a section through a late gastrula, we see that the cells of the animal pole are now underlain by mesoderm, whereas those situated more laterally are not. We might speculate, then, that some interaction between meso-

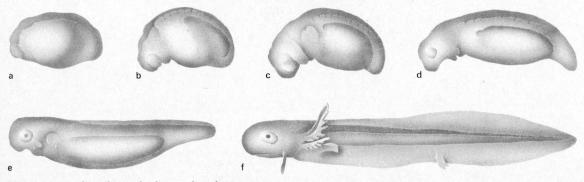

Figure 7-15   The salamander larva takes shape.

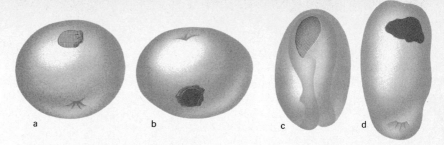

Figure 7-16 Exchange of prospective epidermis and prospective neural plate in the early gastrula stage of newt embryos differing slightly in density of pigmentation. (a and b) Embryos immediately after the operation. (c and d) The same embryos in the neural plate stage. (c) The graft forms part of the neural plate. (d) The graft forms part of the anteroventral epidermis.

derm and ectoderm results in the development of the ectoderm into neural tube. How might this hypothesis be tested?

First, according to the hypothesis, if the interactions were prevented, the brain would not develop. This problem has been posed for many species, and the answer is clear. For example, when salamander embryos are grown in lithium chloride, one consequence is *exogastrulation*; the mesodermal cells turn outward instead of inward; they do not take up their customary positions and the brain fails to develop.

Second, if the mesoderm has the capacity, by interacting with ectoderm at the animal pole, of *inducing* it to become brain, might it also induce the ectoderm of other regions to form brain? The answer to this important question was provided

in a brilliant series of experiments by the German embryologist Hans Spemann and his younger colleague Hilde Mangold. They knew that in salamander embryos the cells that will move inside the blastopore and become mesoderm are located just above the lip of the blastopore (Fig. 7-17). Spemann took this mesoderm-to-be from one gastrula and transplanted it to the opposite side of a second embryo. The results proved the inductive capacity of the blastopore lip, for not only did the host embryo develop its own neural tube in the location where it usually occurred, it also formed a neural tube at the site of the graft. The graft formed a second blastopore, cells moved inside the gastrula, and the nervous system was formed over it. In some hosts almost an entire *secondary embryo* was produced (Fig. 7-17).

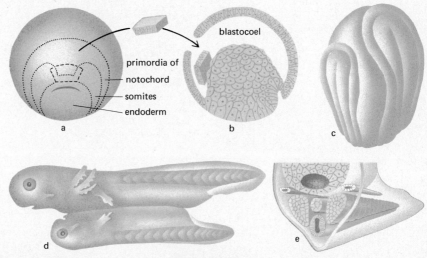

Figure 7-17 Diagram of the transplantation of a piece of the upper blastoporal lip into another gastrula (a and b) and the self-differentiation plus inductions of the graft (c-e).

From these experiments Spemann concluded that the area of the gastrula above the blastopore, called the *dorsal lip*, serves as an "organizer" of normal development of the nervous system. This process and others similar to it are called *embryonic induction*. This is a basic mechanism underlying the orderly development of many other body parts. Part of the forebrain, for instance, forms the eye cups which induce the ectoderm of the head to form the lens of the eye. If cells from the eye cup are transplanted to another embryo at the proper stage of development, the second embryo will develop an extra lens.

The words "at the proper stage of development" are significant. Timing in development is just as important as location. Development involves an orderly sequence of changes: mesoderm inducing neural tube, neural tube inducing lens, and lens in turn inducing cornea.

We may properly inquire into the mechanism of embryonic induction. How does mesoderm influence ectoderm to form neural tube? Although the "organizer" phenomenon was the first major example of embryonic induction to be described, and although it is still among the more widely discussed reactions of its kind, it has not proved to be a favorable system for biochemical analysis. In this, as in embryonic inductions generally, we postulate that a molecular signal passes from the inducing to the responding cells. We can give no unequivocal statement about the nature of the signal.

An eye is formed

Both morphogenetic movements of cells to new locations and embryonic induction are well illustrated in the development of the eye. The processes by which an eye is formed have been studied in many vertebrates, particularly in amphibians and the chicken. From these studies it is possible to outline some of the general characteristics of organ formation. One pattern is typified by formative movements through which the organ is "roughed in" in bold outline. Only after this are the specific cell types needed for the organ to function then differentiated. This means that in general, organ rudiments are formed before the specific tissues of the functional organ.

Eye formation involves a whole array of interrelated inductive steps. We shall not trace them all, but describe enough of them to suggest just how complex organ formation really is. The eye starts in a sense with the induction of the ectodermal neural plate by the underlying mesoderm. As the forebrain takes shape from the neural tube, it bulges on each side. These bulges, known as *optic vesicles* (Fig. 7-18), expand and contact the inner surface of the overlying ectoderm. Another contact is made as the wall of each vesicle pushes inward to form a double-walled *optic cup*. This cup will eventually form the retina and the optic nerve. The layer nearest the skin will differentiate into the light-sensitive layer of the retina. The cells immediately under it (the inner layer of the cup) will form the pigmented lining of the retina. The muscles that will focus the eye and the part of the iris that will control the amount of light admitted through the pupil are mesodermal in origin.

The optic cup also stimulates the ectoderm to start the processes that will lead to formation of the lens. The ectoderm over each optic cup thickens, then separates from the rest of the ectoderm to form the forerunner of the *lens vesicle*. This vesicle sits in the opening of the optic cup. The cornea, the transparent portion of the protective tunic of the eye, develops as a result of interaction of lens and retina with the overlying skin.

In the case of the eye, as in the development of other organs of the body, induction is not always a one-way street. It goes in one direction when the optic vesicle is inducing the lens to originate from ectoderm. But a piece of cellophane placed between the vesicle and the overlying tissue will prevent even the first step in lens formation. Another experiment shows, however, that the ectoderm also influences the rest of the eye. The optic cup of a large, rapidly growing species of salamander is covered by ectoderm transplanted from a smaller, slower growing species. The resulting eye is not misshapen because it has a lens too small for the eyeball. Instead, the lens is larger than in the species from which the tissue was transplanted, and the eyeball in the animal receiving the transplant is smaller than usual. The result is a working eye, intermediate in size, and well proportioned. The optic cup and

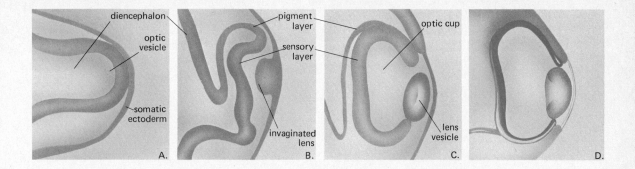

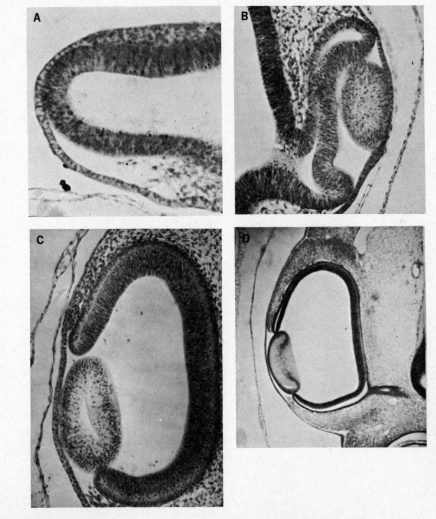

Figure 7-18 The optic vesicle of the embryonic brain induces the head ectoderm to form a lens, which is then incorporated into the structure of the eye. In the chick embryo, A: The optic vesicle at 33 hours extends laterally from the brain and touches the epidermis. B: At 48 hours the epidermis is induced to form a lens, and the optic vesicle becomes converted into an optic cup, which remains connected to the brain by means of a thick optic stalk. C: A 3-day-old eye, showing an optic cup, with a thick inner sensory layer and a thin outer pigment layer. The lens cells adjacent to the optic cup are beginning to elongate, a process which is continued in D.

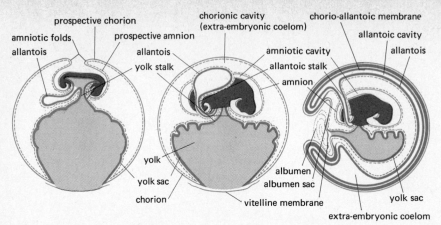

Figure 7-19 Stages in development of extraembryonic membranes of the chick. (Left) Early: (middle) later; (right) fully mature. Note the four different membranes. The *amniotic membrane* encloses a fluid-filled amniotic cavity which houses the embryo. The *allantoic membrane* forms a sac—the allantois—which collects the urinary waste of the embryo. The allantoic membrane fuses with the outer membrane called the *chorion* which surrounds the embryo and the other membranes. This combined *chorio-allantoic membrane* becomes the site of gas exchange for the developing embryo. The *yolk sac membrane* encloses the yolk, which is the food material of the developing embryo.

the ectoderm have influenced each other. Such reciprocal inductions have been observed in many animals and in the development of many organ systems: in wings, feathers, spurs, and beaks of birds, tail fins in salamanders, and hair in mammals.

"Space capsules" for development

The newly emerging organism, whether it is an amphibian, a bird, or a mammal, lives in a very special environment. The animal is very much like an astronaut hurtling high above the earth in a space capsule. It must be provided with everything it needs to live until it completes its journey.

A salamander's egg is laid in the water. Like a chicken's egg it contains enough yolk to nourish the larva until it breaks out of its protective membranes and can find food on its own. Oxygen diffuses through the outer membrane to supply the embryo, and wastes leave the same way. But how does the chick embryo, which also requires a watery world, develop when its egg is laid on dry land? In examining the inside of a hen's egg when the embryo is growing (Fig. 7-19), it is seen that there are really four parts. There is the embryo itself. Attached to it is the large *yolk sac,* surrounded by its own membrane, and another sac covered by the *chorioallantois.* The embryo is covered by the *amnion,* a membrane that keeps in fluids so that the young embryo is maintained in a watery environment.

But how does it breathe, and how does it dispose of wastes? The chorioallantois takes care of these needs. Many blood vessels course through it as it lines the inside of the shell. Thus it serves as a "lung." Oxygen which diffuses through the porous shell is absorbed by the blood vessels, and carbon dioxide is released. But the chorioallantois serves another important function; it is a sort of "septic tank" where liquid and solid wastes can be stored until the chick escapes from the capsule when it is old enough to break out of the shell. We may think of a bird's egg as being completely self-contained, and it is almost that. But it also breathes, and the chick would die if the porous shell were covered so air could not get in and out.

An inner "space capsule"

The embryo of a mammal, such as a baby growing in its mother's body, needs all the things a chick embryo needs inside the shell: food, oxygen, and a way of getting rid of wastes. The human egg is obviously too small to serve as a self-contained capsule for very long. What special environment does the body provide so the young can grow until the time of birth?

In many ways the membranes of the human or mammalian embryo resemble those of the chick. As Fig. 7-20 shows, the outermost membrane is a chorion that completely surrounds the fetus. Immediately inside the chorion is the amnion, also a fluid-filled sac, in which the fetus literally floats. There is a small yolk sac, but it does not contain yolk and has no nutritional role. The fetus also develops an allantois, but it does not function in respiration or waste disposal in the same way as a chick's allantois.

Notice in the illustration that on one side of the chorion there are fingerlike projections. These are the *villi*, which early in embryonic life push forward and protrude into the lining of the mother's uterus. Blood from the mother bathes these villi. Small blood vessels absorb food, oxygen, and other vital substances from the mother's blood and make it possible for the baby to develop. Waste products are also passed to the mother's circulation in this way.

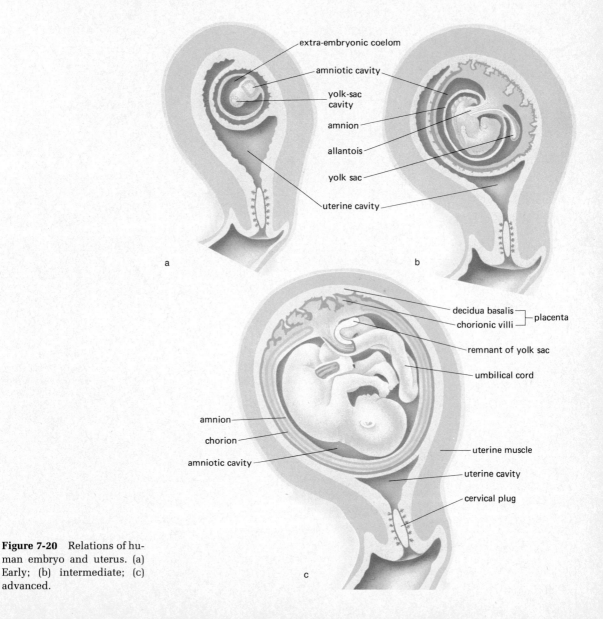

Figure 7-20   Relations of human embryo and uterus. (a) Early; (b) intermediate; (c) advanced.

Substances absorbed by the villi are circulated to the embryo through the *umbilical cord*. The cord includes two arteries and a vein that were once part of the allantois. The umbilical cord connects to the *placenta*, an organ composed of the villi and the adjacent parts of the uterus which provide the villi with blood circulation. Although the placenta brings the mother's blood in close contact with that of the baby, there is practically no interchange of blood between the two. The blood vessels of the villi separate the two blood supplies so that the baby's own circulation remains separate from that of its mother.

Thus we see that despite the close physical relationship between the mother's body and her offspring, the embryo of a mammal also lives in a sort of inner space capsule. Membranes like those of the chick surround the embryo, and the placenta nourishes the baby from, but not with, the mother's blood.

The course of human development

The pattern of cleavage, gastrulation, and neurogenesis that we observed in the salamander egg earlier in this chapter serves as a good basis for understanding the pattern of early human development. Egg and sperm meet in the upper end of the oviduct. The new zygote cleaves as it passes down the oviduct toward the uterus. The eventual result of cleavage, however, is not a single-layered blastula but a balloonlike **blastocyst**, at one pole of which is a group of cells that will become the embryo (Fig. 7-21). The cells on the surface of the balloon are destined to become the chorion, contributing to the fetal part of the placenta, which as we have just seen is a composite organ having fetal and maternal parts.

The cells inside the blastocyst form a plate called the **embryonic disc.** It is this disc which eventually undergoes gastrulation between eight and twenty days following fertilization, so that the three embryonic layers are formed: ectoderm, endoderm, and mesoderm. During this period blood vessels begin to appear within the villi.

The embryo attaches itself to the uterus six or seven days after fertilization. When the dividing egg reaches the uterus, the walls of this organ are well prepared to receive it. Thickened and

Figure 7-21 Photomicrograph of section of 107-cell (about five-day) human embryo, showing inner cell mass (top) and trophoblast. × 600.

spongy, with a large supply of blood in their tissues, the uterine walls have been readied for implantation of the egg during the normal course of the menstrual cycle (Chapter 15).

As is the case with the salamander the nervous system takes shape soon after gastrulation is completed. The neural plate forms first, followed by the neural ridges which fold in to form the neural tube.

By the time a human embryo is one month old, it is not even half an inch long, but the rudiments of its brain, eyes, heart, stomach, and liver can already be distinguished. Its heart beats regularly, and it weighs about 10,000 times what it did at the moment of conception in the oviduct. At the end of two months the embryo, now called a fetus, is a recognizable human being. It is still so small that it does not measure as long as the bone in the tip of one's finger, that is, only three-quarters of an inch long. But it is very much alive. The muscles are able to contract and the developing arms and legs can move. The heart circulates blood throughout the miniature body and the umbilical cord that connects it to the placenta. A month later the fetus has nail-beds on its finger tips and external sex organs. By the end of four months it has hair and may be large and vigorous enough to be felt by its mother when it moves. Within the next two months the baby becomes quite active. It even "sleeps" and is "awake" like a newborn.

The last three months are a time of growth. The baby may gain as much as five pounds of the seven or eight it will weigh at birth. It will be born about 280 days after conception. What was once only a single cell has become a living boy or girl, another member of the human species.

And so, with birth, one stage in the story of development ends. The baby, of course, will go on to develop into an adult, just as all living things do, if accident or disease does not intervene.

However, change continues, resulting, in the normal course of events, in senescence. What are the mechanisms of aging? Is the aging process "programmed" to follow development? Do all of the body's cells "age"? We have already remarked that cells in a few organs, like liver, have the capacity for proliferation throughout life. But for others, brain and heart, for example, there is a fixed number of cells. As these cells are lost or rendered inactive through the accumulation of the by-products of metabolism, they are not replaced, and the organism declines in vigor (Chapter 8). And so, as Seneca observed, "Old age is an incurable disease."

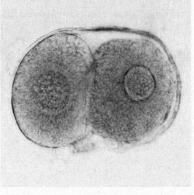
a

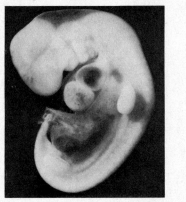
b

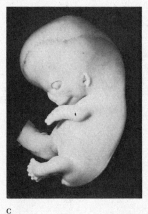
c

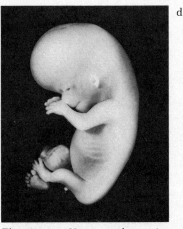
d

e

Figure 7-22   Human embryos: (a) 2-cell stage, × 500; (b) 30 days, × 12 (actual length, 7.3 mm.); (c) 44 days, × 3 (actual length, 23 mm.); (d) 56 days, × 2 (actual length, 37 mm.); (e) at the beginning of the fifth month, × 0.75.

HOW A FLOWERING PLANT DEVELOPS

Origin of gametes and fertilization

There is no evidence of the continuity of germ plasm in flowering plants. The reproductive organs and gametes are not formed until late in development, and any somatic cell of the plant appears to retain the capacity to form tissues that will give rise to reproductive structures. These structures include the *carpel,* which will produce ovules in which the female gametes are formed, and the *stamen,* which will produce the anthers in which the male pollen arises. We shall bypass the development of the reproductive organs themselves and focus first on maturation of the egg and the pollen. Figure 7-23 summarizes normal reproduction in flowering plants. Pollen grains are formed in the terminal anther that is attached to the filament of the stamen. Cells in the center of the anther undergo several mitotic divisions and then go into meiosis, each diploid microspore mother cell producing four haploid microspores. The latter, in turn, produce heavy cell walls and become pollen grains. Just as in animal gametogenesis, DNA doubling to the "4c" level occurs in premeiotic interphase and into early first pro-

phase, with no further DNA synthesis occurring during meiosis. Once the microspore is formed the process is completed by a final nuclear division so that the definitive pollen grain contains two haploid nuclei.

Somewhat similarly, an embryo sac, containing an egg, develops from a megaspore mother cell. The mother cell undergoes meiosis producing four haploid cells, three of which die. The fourth divides, forming the complex embryo sac containing eight haploid cells or nuclei, an egg cell, two synergid cells, three antipodal cells, and two "polar" nuclei which fuse together to form the "fusion" nucleus.

Once the gametes are formed they are brought together in two discrete stages, *pollination* and *fertilization.* In the first stage a pollen grain is deposited on the sticky stigma at the tip of the carpel. From this grain a *pollen tube* arises that grows through the style tissues, digesting its way toward the egg. The two male nuclei from the grain now move down the pollen tube. One fuses with the egg nucleus to form the zygote; the other joins the fusion nucleus to make the nutritive endosperm. The seed contains the diploid embryo and the triploid endosperm.

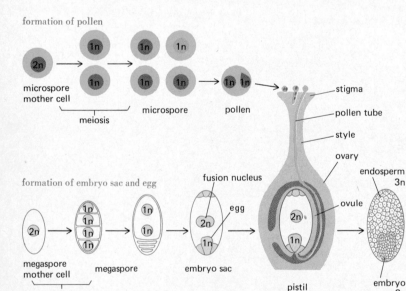

formation of pollen

microspore mother cell

meiosis

microspore

pollen

stigma

pollen tube

style

ovary

endosperm 3n

ovule

egg

embryo sac

pistil

embryo 2n

formation of embryo sac and egg

fusion nucleus

megaspore mother cell

megaspore

meiosis

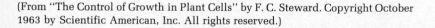
(From "The Control of Growth in Plant Cells" by F. C. Steward. Copyright October 1963 by Scientific American, Inc. All rights reserved.)

Figure 7-23 Normal reproduction in flowering plants. (Top left) Development of pollen after formation in anther, beginning with a 2n microspore mother cell and progressing through several divisions until one pollen grain lodges on a stigma and gives rise to a pollen tube. (Bottom left) An embryo sac containing an egg similarly develops from a megaspore mother cell. On the pistil, two of the nuclei that have developed in pollen grain start down pollen tube. One nucleus joins the egg to accomplish fertilization; one joins the fusion nucleus to create endosperm, a nutritive stimulant to embryo growth. The end product is a seed containing an embryo.

A common pattern of embryo development

One of the more intensively studied plant embryos is that of a common, weedy flowering plant named Shepherd's Purse (*Capsella bursa-pastoris*). This embryo, like the embryo of all flowering plants, develops within an ovule deeply embedded in the female parts of the flower. Because the embryo is relatively inaccessible, at least when it is very small, it is difficult to make direct visual observations of it. Therefore to follow the development of the embryo it is necessary to examine serial sections cut in various planes through embryos which have been killed and stained at successive stages of development. From light and electron microscopic studies it is possible to reconstruct three-dimensional interpretations of what the living embryo must have been like.

At fertilization the gametes of *Capsella*, in common with those of all flowering plants, are structurally unspecialized. The sperm consists simply of a haploid nucleus enclosed in a cytoplasmic sheath, which is transported in the pollen tube from the stigma of the flower to the vicinity of the egg.

The egg is a small, pear-shaped cell contained within the embryo sac in the ovule (Fig. 7-24). It consists of a haploid nucleus situated at the broad end of the cell and surrounded by cytoplasm which, because of the scarcity of aggregated ribosomes, Golgi apparatus, and endoplasmic reticulum, appears to be metabolically inactive. It does not contain reserve nutrients.

Soon after fertilization extensive changes take place in the cytoplasmic organization of the zygote. New ribosomes begin to be synthesized. These and the preexisting maternal ribosomes become aggregated into polysomes. The endoplasmic reticulum and the Golgi system become more abundant. The wall, which up to this time was very thin, becomes increasingly thick, and many of the plasmodesmata become blocked off, increasing the physiological isolation of the protoplast from the surrounding cells. Following the first nuclear division the embryo is separated into two unequal cells: a small terminal cell at the rounded end of the embryo and a larger vacu-

olated cell at the tapered basal end (Fig. 7-25). These two cells are structurally differentiated from one another and have different developmental fates. The basal cell divides transversely and forms a linear *suspensor* consisting of five

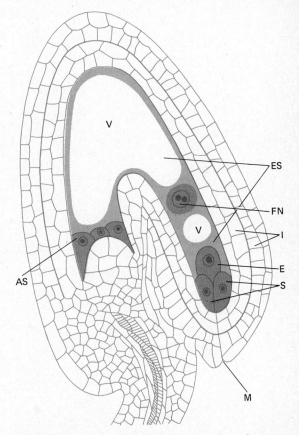

Figure 7-24   Ovule of *Capsella* showing the embryo sac just before fertilization. Two integuments (I) form the outer cell layers of the ovule. The tips of the integuments come together to form a micropyle (M) through which the pollen tube will grow as it enters the ovule. The embryo sac (ES) consists of eight cells or nuclei. At the end nearest the micropyle are two synergid cells (S) and the larger egg cell, the female gamete (E). In the center of the embryo sac is the fusion nucleus (FN) which is formed by fusion of two haploid nuclei. This nucleus will fuse with one of the male nuclei during fertilization to initiate the endosperm development. At the basal end of the curved embryo sac are three vegetative antipodal cells (AS). The cytoplasm of the embryo sac is vacuolated and several vacuoles (V) are visible in this section.

to seven large vacuolated cells. Meanwhile derivatives of the terminal cell divide in various planes to form a globular mass of small cells.

The globular part of the embryo is at first smaller than the suspensor, but it becomes much larger and forms all the embryonic organs. Early divisions in it are synchronous, but later the rate of cell division varies in different parts. When

about 50 cells have been formed in the *globular stage embryo*, three tissues begin to differentiate: (1) a distinctive surface layer of cells which will produce the *epidermal tissue* of the plant; (2) a central core of elongated *procambial cells* which are precursors of the vascular system; and (3) between these two a cylinder of cells which will form the *cortex*. Very soon, the terminal part of

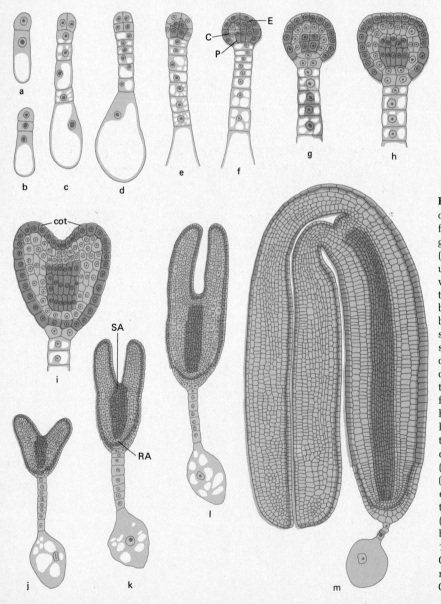

Figure 7-25 Embryo development in *Capsella*. The first division divides the zygote into two unequal cells (a) The lower cell forms the uniseriate suspensor (b-d) while the terminal cell forms the globular part of the embryo (c-f). In the globular embryo the three primary tissue systems are differentiated: a surface epidermis (E), procambium (P), and intervening cortex (C) (f-h). Subsequently the cotyledons (cot) are differentiated producing the heart-stage embryo (i). In later stages of development the terminal meristems are delineated: the shoot apex (SA), and the root apex (RA) (j-l). The mature, dormant embryo is curved to fit into the curvature of the ovule (m). (From *Plant Morphology* by A. W. Haupt. Copyright 1953 by McGraw-Hill Book Company, Inc. Used with permission of McGraw-Hill Book Company.)

the embryo loses its radial symmetry and becomes bilaterally symmetrical as two hemispherical mounds grow out from the end opposite the suspensor. These are the *initials* of the cotyledons, the embryo leaves. The shape they impart to the embryo has led to its being called the *heart stage.*

At the pole of the embryonic axis between the cotyledons a group of cells remains in a relatively undifferentiated state, and these form the *shoot apical meristem.* At the opposite pole of the axis, near the suspensor, another group of cells forms the *root apical meristem.* These two meristems, which do not begin to function until germination, are situated at the two ends of the procambial system and are progressively separated from one another by continued growth of the intervening embryonic axis, the *hypocotyl.* Because of its shape the elongated embryo has been called the *torpedo stage.* The ovule cavity in which the embryo is growing is curved in *Capsella,* and as the embryo continues to enlarge the cotyledons become curved to fit into this space. By this time the seed is maturing, and the rate of embryo growth gradually declines and is terminated as the seed becomes dormant. The entire process of embryo development in *Capsella* takes only about ten days.

Let us compare the embryo of *Capsella* with the animal embryos described previously. Since there is nothing comparable to a fertilization membrane formed in the plant zygote cytoplasm, multiple fertilization must be prevented in other ways. How this is done is not known. The changing shape of the plant embryo results entirely from the formation and growth of cells and there is no cell migration. In the absence of a stage of development comparable to gastrulation, the cells develop in the place where they are formed. This severely limits the kinds of interactions in which plant cells can participate. The fully developed embryo of *Capsella* does not possess all the organs that will be present in the adult plant. The shoot system and the root system are represented only by terminal meristems and there are no initials of vegetative leaves, lateral shoots or roots, or any of the reproductive structures. These will be initiated only after germination. Finally, there is no stage truly comparable to cleavage,

because the plant embryo begins to enlarge and differentiate soon after fertilization. However, the zygote is the largest cell of the plant embryo, and all later formed cells are smaller because the rate of division is greater than the rate of cell growth. The effect is to partition the embryo into a relatively large number of small cells, a situation not unlike the blastula of animal embryos.

How general is the pattern of plant embryo development we have just investigated? The sequence of early cell divisions is very regular in some species and variable in others. The size of the globular embryo, the number of cells formed before organs and tissues begin to differentiate and the extent of embryo development before the seed becomes dormant is quite variable. However, there are features in the development of all plant embryos which suggest that there are similar developmental processes involved.

In the conifers, where the embryo also develops enclosed within an ovule, the first nuclear divisions are not accompanied by wall formation. Free nuclei ranging from a few to several thousand may be formed in a common cytoplasm. In pine this stage is relatively brief, and wall formation begins in the embryo when there are four nuclei. The first four cells are located at one end of the zygote in which two further synchronous divisions occur, producing 16 cells in four equal tiers. At this time the embryo begins to enlarge — the four cells of the second or suspensor tier are expanding and push the terminal cells into the surrounding tissues of the ovule. The terminal cells form a globular cell mass within which distinctive tissue systems and organs similar to those of *Capsella* differentiate. Thus here also there is early differentiation of the embryo into a terminal small-celled organogenetic region and a region of larger suspensor cells.

In mosses and ferns the embryo develops in a free-living form and not in a nutrient-rich ovule. Perhaps the patterns of embryo development we have studied in the seed plants are adaptations to the abundant supply of nutrients surrounding the embryo. However, the development of embryos in mosses and ferns is remarkably like that in seed plants; the embryos soon become differentiated into two regions, one of enlarged vacuolated cells called the *foot,* which resembles the suspensor and is thought to function in the transfer of nutrients into the embryo, and a small-celled region within which the embryonic organs are differentiated. In the moss embryo the only organ to be formed is the capsule. The fern embryo, however, contains the terminal meristems of shoot and root, the first leaf, and the same three tissue systems as are differentiated in seed plant embryos. Moss and fern embryos are distinctive in that no period of dormancy interrupts their growth. Development is continuous into postembryonic stages.

SUMMARY

We have described embryogenesis in animals and plants, emphasizing the contributions which each of these two groups of organisms can make to understanding development. Fundamental biochemical pathways are the same in cells of plants, worms, and man. The basic laws of heredity operate in lower forms of life just as they do in man. Some aspects of the behavior of cells — their mode of division and death — are also similar in the many forms studied. Thus in considering some questions we have attempted a synthesis of the information on animals and plants. On the other hand, some questions must be considered in one kingdom, with only passing reference to the other. For example, changing form in animal embryos embraces the organized movements of cells and groups of cells, rearrangements involving the shifting of cells from one place to another, and bringing a given cell into a new environment in which it can be used in the ensuing steps of development. In this sense it is a process of redistribution. In plant embryos, because each cell is surrounded by a rigid wall secreted by the cell itself, cells cannot move. Therefore shape changes result not from cellular migration but from cell growth.

This is only one of the differences between animal development and plant development. However, rather than emphasize the contrast between animals and plants or the exclusivity of processes, we wish to stress instead the fact that different processes are used to a different extent by different organisms. For example, most animal embryos contain essentially all the organs and tissues that the adults will have; but insect larvae and plant embryos contain only rudimentary primordia of some of the organs, with most of development taking place in the postlarval and postembryonic periods.

It is important to emphasize that not all developmental phenomena are readily observable and analyzable in all animal or plant forms. Certain organisms lend themselves particularly well to the investigation of cell division. In others cell migration may be most easily studied; still others are particularly suited for studying gene action. The investigator therefore must choose the organism most likely to yield information on a particular phenomenon. Thus it is of strategic advantage not to restrict ourselves to the study of mice and men. We must exploit such diverse forms as sea urchins and green plants, frogs and fungi. This is our approach in the following chapter as we probe more deeply into the mechanisms of development.

SUGGESTED READING LIST

Austin, C. R., *Ultrastructure of Fertilization.* Holt, Rinehart and Winston, New York, 1968.

Balinsky, B. I., *An Introduction to Embryology,* third edition. Saunders, Philadelphia, 1970.

Barth, L. J., *Development: Selected Topics.* Addison-Wesley, Reading, Mass., 1964.

Blackler, A. W., "Transfer of primordial germ cells between two subspecies of *Xenopus laevis,*" *Journal of Embryology and Experimental Morphology,* **10,** 641 (1962).

Bonner, J. T., *The Evolution of Development.* University Press, Cambridge, England, 1958.

Corner, G. W., *Ourselves Unborn.* Yale University Press, New Haven, 1944.

Ebert, J. D., and I. M. Sussex, *Interacting Systems in Development.* Holt, Rinehart and Winston, New York, 1970.

Goss, R. J., *Principles of Regeneration.* Academic Press, New York, 1969.

Hay, E. D., *Regeneration.* Holt, Rinehart and Winston, New York, 1966.

Monroy, A., *Chemistry and Physiology of Fertilization.* Holt, Rinehart and Winston, New York, 1965.

Saunders, J. W., Jr., *Patterns and Principles of Animal Development.* Macmillan, New York, 1970.

Saxen, L., and J. Rapola, *Congenital Defects*. Holt, Rinehart and Winston, New York, 1969.

Spemann, H., and H. Mangold, "Induction of embryonic primordia by implantation of organizers from a different species" (1924). Reprinted in B. H. Willier and J. M. Oppenheimer (eds.), *Foundations of Experimental Embryology*, Prentice-Hall, Englewood Cliffs, New Jersey, 1964, p. 144.

Spratt, N. T., Jr., *Developmental Biology*. Wadsworth, Belmont, California, 1971.

Steward, F. C., *Growth and Organization in Plants*. Addison-Wesley, Reading, Mass., 1968.

Taussig, H. B., "The thalidomide syndrome," *Scientific American*, p. 29 (August 1962).

Torrey, J., *Development in Flowering Plants*. Macmillan, New York, 1967.

Torrey, T. W., *Morphogenesis of the Vertebrates*, second edition. Wiley, New York, 1967.

Twitty, V. C., *Of Scientists and Salamanders*. W. H. Freeman, San Francisco, 1966.

Whittaker, J. R., *Cellular Differentiation*. Dickenson Publishing, Belmont, California, 1968. (A collection of research articles)

CHAPTER 8

Developmental Processes and Their Controls

THE CELL'S INNER CONTROLS

During early development the nuclear DNA of the zygote is duplicated repeatedly, so that all the cells of an organism, embryo or adult, contain the same kind of DNA. Yet the cells differentiate; new proteins are synthesized; new cellular components emerge. If our earlier statements are correct, and there is no reason to doubt them, these syntheses are directed by DNA. We must conclude therefore that different segments of DNA—different genes—are functioning in different cells. According to this view cell differentiation is a result of *differential gene expression*. But what factors influence the expression of genes?

The nuclear control center

The nucleus is indispensable for development. This has been demonstrated in many organisms, but let us look at it first in one-celled amoebae and algae. When the nucleus is removed from an amoeba, the cell gradually loses the ability to move about. Unable to feed, it gradually declines, although it may live for as long as two weeks. For a while the changes are reversible. If the nucleus is replaced within two or three days, the organism resumes normal activity, including cell division at the appropriate time. But by the end of six days the organism no longer responds.

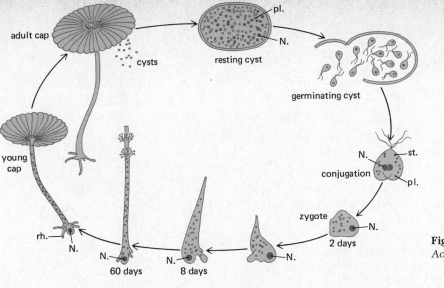

adult cap

cysts

resting cyst

pl.

N.

germinating cyst

young cap

N. st.

conjugation

pl.

rh.

N.

N.

60 days

N.

8 days

N.

zygote

2 days

N.

Figure 8-1 Life cycle of *Acetabularia mediterranea*.

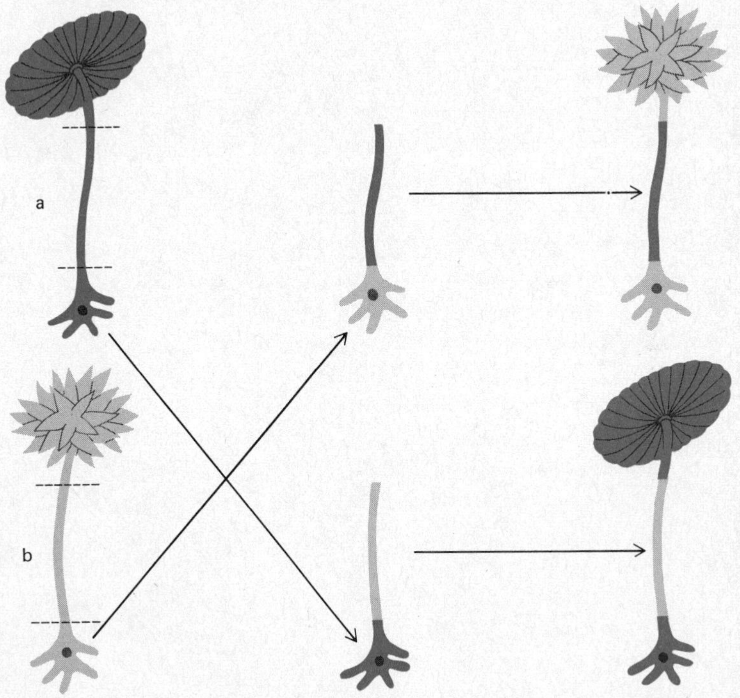

Figure 8-2 Transplantation experiments using two species of *Acetabularia* which differ in shape of the cap. The character of the cap is determined by the species of the nucleus. Arrows indicate successive decapitations. (a) *Med* is shaded to facilitate tracing it through the grafting experiments. In (a), stalk from *med* is grafted to a nucleus from *cren*. In (b) the reciprocal graft is diagrammed.

a

b

The contribution of the nucleus is even more strikingly demonstrated in the alga *Acetabularia*. Although one-celled, this marine alga is large and complex, reaching a height of 6 to 9 centimeters. One of the roots at its base contains the nucleus (Fig. 8-1). In one species, *A. mediterranea*, the cap, which tops the stalk, is flat and disclike. In another, *A. crenulata*, the cap is branched. Thus the species are easily told apart. In both species the alga is able to grow a new cap if one has been cut off.

In laboratory experiments, the base containing the nucleus of one species has been grafted to the stalk of the other species. This is illustrated in Figure 8-2 in which a *crenulata* (or *cren*) stalk is attached to a *mediterranea* (or *med*) base. The first time this is done the stalk regenerates a cap which is intermediate between the two species. However, if that cap is removed, all subsequent caps are characteristic of the *med* base (which contains the *med* nucleus). Products from the nucleus obviously direct cap structure. It is believed that the delay in complete change of characteristics comes about because some of the nuclear products are stored in the stalk. If the experiment is reversed, and the *cren* base is grafted to a *med* stalk, *cren* caps result when the alga forms new caps.

One of the first experiments which demonstrated the essential role of the nucleus during development of a higher organism was done by Hans Spemann, using a fertilized egg from a newt (or salamander). The zygote was pinched in the middle by tying a fine strand of hair around it (Fig. 8-3), forcing its nucleus into one side, but leaving a small bridge of cytoplasm connecting the two halves.

The part of the egg containing the nucleus will cleave normally, but the other half will not. If the constriction is relaxed—reopening the bridge between the halves—one of the nuclei resulting from cleavage will eventually slip across it. When this occurs, an identical twin of the first newt will begin developing. As Figure 8-3 shows the second embryo grows more slowly, but the animals are identical in all other respects.

Are the nuclei of older embryos equivalent? A normal embryo results when the constriction is relaxed even after four divisions have occurred in the cleaving half. By this time the embryo has 16 cells. Thus the nuclei of the 16-cell stage are like the original one in their ability to support development. But what happens to the nuclei later? Are they still equivalent after gastrulation has begun, for instance?

In trying to answer such questions embryologists have turned to nuclear transplantation. Briggs and King perfected a method for transferring nuclei of embryonic cells into enucleated frog eggs (*Rana*). The transplantation is carried out in two main steps (Fig. 8-4). First, the recipient eggs are activated parthenogenetically with a glass needle, and subsequently enucleated. Second, donor cells are isolated by disaggregating an older embryo. A given cell is drawn up into the tip of a micropipette with an inner diameter smaller than the cell. The cell surface is broken but not dispersed; thus the nucleus is protected by its own cytoplasm until the pipette is inserted into the recipient egg. One possible objection to the technique is that a small volume of donor cytoplasm has been included in the transfer . This objection can at least partly be met by performing a control experiment in which cytoplasm injected alone does not support development.

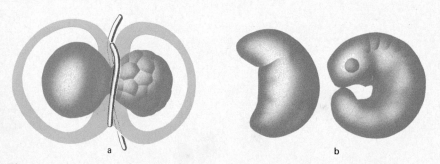

a   b

Figure 8-3 Twinning of newt embryos combined with delayed nucleation of one half. (a) Constriction of the egg along the advancing first furrow leading to the 2-cell stage has left the copulation nucleus in the right half. This half has cleaved. At the stage illustrated the nucleus of the blastomere nearest to the "bottleneck" escapes into the unsegmented left half, which presently begins to cleave. (b) Twin embryos developed from (a). The lag of the left partner owing to the initial delay still is appreciable.

It was found that late blastula nuclei can be transplanted in undamaged condition; they are equivalent to the nucleus at the beginning of development, for the majority of eggs that receive nuclei from late blastulae develop normally. Similar results were obtained by Fischberg and his colleagues in experiments with the South African clawed toad, *Xenopus*. Thus far our conclusion is clear: stable nuclear changes affecting the capacity to support development have not been observed through the blastula stage.

However, when nuclei of cells from older *Rana* embryos are transferred, the outcome of the experiment is different. For example, when nuclei taken from late gastrula endoderm cells are transplanted, they frequently, *but not invariably,* are unable to support normal development. Defective embryos are produced; although the endodermal structures develop normally (and to a remarkable extent also the mesodermal structures associated with the endoderm), the ectodermal structures are usually deficient. Thus some change has occurred in these nuclei. More-

over this change persists under conditions of serial nuclear transplantation.

What are these nuclear changes? Are these results meaningful in relation to our theory of differentiation? What other facts must be taken into account? It is possible that the defects result from deleterious effects of the operative methods on the nuclei. As cells get older they may become more "sensitive," more subject to damage in the course of the experiments. There is some evidence to support this contention. On close examination many of the abnormal embryos produced by transplantation of late gastrula nuclei are found to have chromosomal abnormalities.

We must also consider another line of evidence. As Gurdon and his colleagues have shown in experiments carried out on *Xenopus*, nuclei from the majority of differentiating endoderm cells and from fully differentiated cells of the intestine can support the formation of nerve and muscle cells after transplantation to enucleated eggs. Moreover a number of *fertile adult frogs have now been obtained from eggs whose*

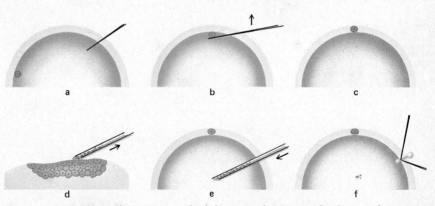

Figure 8-4 Diagram illustrating method for transplanting nuclei from embryonic cells into enucleated eggs (*Rana pipiens*). (Upper) Steps in preparing enucleated eggs. The egg is activated (a), and enucleated (b) with a clean glass needle. The enucleation operation results in the formation of an exovate containing the egg nucleus, shown trapped in the jelly surrounding the egg (c). (Lower) Procedure for transplanting nuclei. The donor cell is isolated on the surface of an intact blastula (which serves as an operating platform) and is then drawn up into a micropipette (d). Since the diameter of the pipette is somewhat smaller than that of the cell, the cell surface is broken. When the broken cell is injected, as shown in (e) the nucleus is liberated undamaged in the egg cytoplasm. As the pipette is withdrawn it tends to draw the surface coat of the egg with it, forming a small canal to the exterior which must be severed, as shown in (f), to prevent leakage.

only active nuclei were transplanted intestine nuclei. The evidence is by now convincing; in *Xenopus,* at least, nuclei of some differentiated cells can support the whole range of normal development. Do the differences in findings in *Rana* and *Xenopus* simply reflect the fact that *Rana* cells are more sensitive to experimental manipulation than *Xenopus* cells? Are *Xenopus* cells simply hardier than those of the frog? Or should it be argued that the time of onset of observable stable change is later in *Xenopus?*

Our questions must remain unanswered for the present. We are confronted with two major possibilities. First, it is possible that stable nuclear changes affecting the capacity to support development do occur. They may be cell-type specific, although fully convincing evidence is lacking thus far. For reasons that are not entirely clear, such changes are better demonstrated in *Rana* than in *Xenopus.* Second, it is possible that the changes so far observed in *Rana* are not meaningful to our theory of differentiation in their present context. We emphasize the finding that *some* nuclei of swimming tadpoles support normal development. Therefore it can be argued that the observed deficiencies are the consequence of the experiment and that they do not reflect fundamental restrictions in the capacity of the nucleus to support development. It is possible, for example, that the "deficiency" of the older nucleus lies in its inability to enter the mitotic cycle characteristic of the zygote.

It would be misleading to select one or the other of these points of view from the sketchy data just presented. However, there is considerable information which shows that many kinds of cells retain some ability to transform into other cell types throughout life. Both the invertebrates and vertebrates possess remarkable capacities for regeneration of lost parts; as we see in the next section, some cells from specialized tissues of mature higher plants are capable, when isolated, of giving rise to whole plants.

Equivalence of somatic cell nuclei in plants. There is direct evidence that the tissues surrounding the plant embryo influence its development. In flowering plants most of the nutrients used by the embryo for its growth accumulate in the *endosperm,* a tissue which completely surrounds the embryo. As the embryo grows it digests the endosperm and enlarges to fill the cavity it creates. The endosperm is initiated after fertilization when a second sperm nucleus carried in the pollen tube enters the embryo sac and fuses with one or more centrally placed nuclei. The early development of this tissue is much more rapid than that of the embryo, and the endosperm usually consists of many cells before the zygote divides. In some species endosperm is cellular from the start, but in others nuclear divisions are not followed by wall formation and a liquid endosperm is formed. This happens in the coconut in which the "milk" is actually endosperm.

Many experiments have shown not only that natural endosperm provides an optimal environment for development of embryos, but that the addition of endosperms like coconut milk to plant cells cultivated *in vitro* enhances their development.

Will a single differentiated cell produce both shoot and root meristems and thus give rise to an entire plant? If differentiation does not involve irreversible changes in the DNA, this should be possible. The studies of Steward and his colleagues have shown that mature cells can be made to repeat stages in embryo development, often with surprising regularity. They grew tissue from cultivated strains of carrot or from the wild carrot in rotating flasks in a liquid culture medium containing coconut milk. As the cells divided and grew some became detached and continued to develop while they were freely suspended in the medium. When the suspension, which consisted of some single cells and some cell clusters, was spread on the surface of an agar medium, almost all of the cellular units gave rise to embryolike structures. Because of the similarity between these and normal embryos, they have been called *embryoids.* They were formed most abundantly from suspensions made from wild carrot embryos, but mature cells of the petiole and root have also formed them (Fig. 8-5). Several other species of flowering plants and gymnosperms have now been found to produce embryoids, and haploid embryoids have even been produced from tobacco pollen.

Is the physical isolation of cells in carrot suspension cultures a necessary condition for embryoid initiation? Some carrot embryoids seem to arise from groups of cells, but in suspension cultures the development of individual cells cannot be followed because they are grown under conditions in which microscopic observation is not possible. In tobacco, however, there is conclusive proof that *entire plants arise from single somatic cells* (Fig. 8-6). Single tobacco cells from a suspension were tranferred to a drop of nutrient

Figure 8-5 Production of embryoids and whole plants from tissue cultures of carrot. On the left side of the illustration a carrot root has been used to obtain explants of phloem tissue which are then cultured in a liquid medium in slowly rotating flasks. Here the tissues proliferate and single cells and aggregates of cells separate and are released into the medium. Some of these continue to develop producing roots, shoots, and embryoids which can be transferred individually into test-tubes, and then into soil where they grow as typical carrot plants producing enlarged tap-roots and flowers. On the right side of the illustration such a plant obtained from tissue culture is shown as the source of an embryo which was placed in sterile culture to proliferate and to repeat the cycle of embryoid production to whole flowering plant.

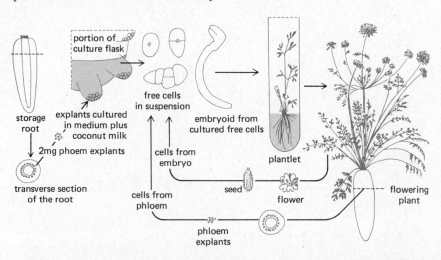

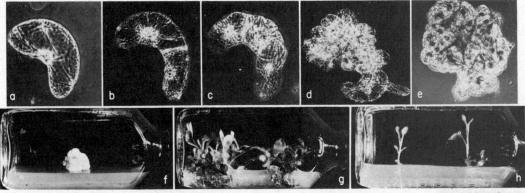

Figure 8-6 Development of tobacco plants from single tissue culture cells. In this technique single cells are removed aseptically from liquid suspension cultures and grown in microcultures where they can be continuously observed and photographed. In microculture the cells undergo division (a-e) forming small callus masses which can be transferred to the surface of an agar culture medium (f). After further growth on a medium containing both auxin and kinetin, shoots and roots are differentiated (g). Young plants are then transferred to fresh culture medium (h) and when large enough are transplanted into soil where they continue to grow vegetatively and later flower (right).

medium on a microscope slide and covered to exclude microorganism contaminants. The slide was then maintained on the stage of a microscope and photographed periodically. Under these conditions many of the cells were seen to divide, producing small tissue masses which were subsequently transferred to larger culture vessels containing media with plant hormones and growth factors. Roots and shoots were initiated on the tissues and developed into whole plants, which have since been transferred to soil where they completed their life cycle by flowering and setting viable seed.

Not all of the evidence is this dramatic, but many specialized cells are capable, under appropriate conditions, of extending their repertoire, of making new products and assuming new roles. Thus it is difficult to accept the idea of irreversible, stable changes in the capacity of all cell nuclei to support development.

Therefore while keeping an open mind pending the outcome of further experiments, we shall emphasize the regulation of genetic and other nuclear functions which depend on continuing interactions with the cytoplasm, rather than their irreversible restriction.

Genes in action

Is it possible to detect signs of gene activity along the length of a chromosome? Specifically, if the ideas we are developing are correct, we should be able to demonstrate that different gene loci show activity in different tissues and at different developmental stages within a given tissue.

We have already learned that no single material is suited for analyzing every problem; we must constantly be on the lookout for forms that are uniquely fitted for the study of specific questions. Certain organisms lend themselves particularly to investigating cell division or cell migration. In others, specific types of metabolic processes are most readily revealed; still others are particularly suited for studying heredity. No-

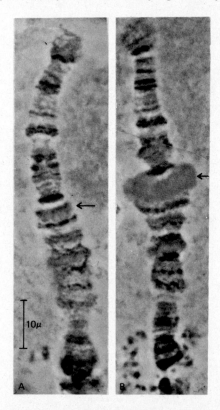

Figure 8-7 (Left) A segment of *Chironomus* salivary gland chromosome showing a moderately developed puff. (Right) Short arm of chromosome "C" at two stages of development of the fly, *Sarcophaga bullata*. (A) Late day 8 showing minimum puffing activity (beginning of puff at arrow); (B) At day 10 this same puff is much larger.

where, however, is this generalization more applicable than in the analysis of chromosome structure. The giant, banded chromosomes in the salivary glands and other tissues of several insects, including the fruit fly *Drosophila*, the gnat *Rhynchosciara*, and the midge *Chironomus*, provide exceptionally favorable material. During the development of these tissues, the chromosomes are replicated but not the cells. The newly formed chromosomes do not separate but remain together with the old ones, forming a functional unit or *polytene* chromosome. These giant chromosomes have banding patterns so distinct that an experienced observer can use them to identify a specific chromosome or even part of a chromosome. Moreover in *Drosophila* the banding pattern has been correlated with the genetic map as revealed by breeding experiments.

When examining these giant chromosomes carefully, we observe characteristic features along their length. At points along the chromosome, instead of a sharp band there may be diffuse, "puffed" regions (Fig. 8-7). These puffs reflect a "loosening" of the chromosome structure, the many strands of DNA in these giant chromosomes having separated to form loops.

Studies of the pattern of puffing activity during development have been especially revealing. First, the locations of puffs are different in different tissues at any one time. Second, they are different in the same tissues at different stages of development (Fig. 8-8). Finally, at any given time, all the cells of any one type in any one tissue show the same pattern.

These observations led to the proposal, first in the early 1950s, that the puffs indicate regions in which genes are especially active. It is now clear that puffs are *visible signs of gene activity*. Earlier we learned that, in molecular terms, gene activity means the DNA-directed synthesis of RNA. Thus it should be possible to demonstrate the active synthesis of RNA at the puffs. Such evidence is available. Experiments with labeled uridine, a derivative of uracil (that is incorporated into RNA but not into DNA), show that RNA is accumulated rapidly at the puffs. Biochemical studies of this RNA suggest further that it is messenger or DNA-like in its composition. Moreover puffing can be prevented by *actinomycin D*, an antibiotic known to inhibit the synthesis of RNA.

Isolation and identification of genes. Technical advances in molecular biology and electron microscopy have made possible the isolation and identification of specific genes (Fig. 8-9). To fully appreciate the thrust of this achievement it is necessary to consider the background of the work and the thinking behind it. A gene is best appreciated in a proper setting.

The time when any gene begins to act can be most accurately defined by determining the time when DNA-dependent RNA synthesis begins. In all the species that have been studied some of the genes of the new individual come into play at the very beginning of embryonic life, or very early. But do the genes for the several forms of RNA—messenger RNAs, ribosomal RNAs, and soluble (transfer) RNAs—begin to function as a coordinated unit or independently? Because the different stages of oögenesis and embryogenesis are so obviously different from one another with respect to rates of growth, cell division, and mor-

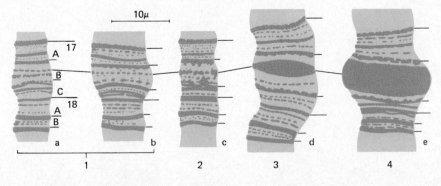

Figure 8-8 Various degrees of puffing at band I-17-B in *Chironomus*; (a-c) From untreated control animals; (d) and (e) from larvae that had been injected with the hormone, ecdysone.

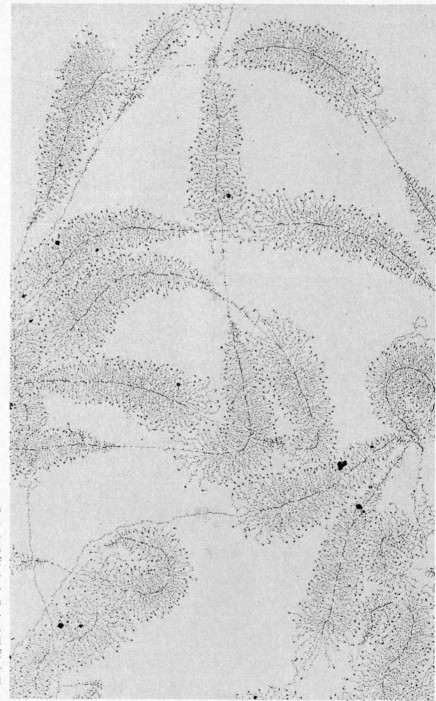

Figure 8-9 Nucleolar genes from an amphibian oöcyte. The presence of extrachromosomal nucleoli in amphibian oöcytes permits isolation of the genes coding for ribosomal RNA precursor molecules, which appear as a gradient of fibrils in progressive stages of completion. Active genes are separated by "spacers," stretches of DNA that are apparently inactive at the time of ribosomal RNA synthesis.

phogenesis, variations in the relative synthesis of the different classes of RNA might be expected. It is now clear that the three different functional classes of RNA are synthesized independently of one another at rates determined by the stage of the embryo.

It cannot be said that we have a comprehensive knowledge of RNA synthesis in any species; however, more is known of the synthesis of the ribosomal RNAs (rRNAs) than of any of the other classes; and more is known of the synthesis of ribosomal RNA in the amphibians than in any other forms.

Ribosomal RNA constitutes more than 80 percent of the RNA in all tissues of higher organisms studied thus far. At least three classes of rRNA are associated with animal and plant ribosomes. They are commonly designated by their sedimentation constants, which are measures of the velocity of the particles in a field of force in the ultracentrifuge. The values vary slightly from species to species; in the amphibians the three classes are designated as 28S, 18S, and 5S rRNAs. The corresponding molecular weights are estimated to be 1,600,000, 600,000, and 30,000 daltons, respectively.

In the clawed toad, *Xenopus*, and in other amphibians the pattern of synthesis of rRNAs is particularly striking. During oögenesis their synthesis is intense, greatly predominating over that of messenger RNA (mRNA) and soluble RNA (sRNA). This synthesis occurs in immature oöcytes. Once the growth phase of oögenesis is terminated, little or no further synthesis of the rRNAs occurs until gastrulation. In short most if not all of the rRNA synthesized during oögenesis is conserved not only during the entire period of oögenesis (a period of many months and in *Xenopus* perhaps years) but during a large part of the period of embryogenesis as well.

During oögenesis, then, the "maternal genes" are highly active in the synthesis of the rRNAs. However, during fertilization, cleavage, and blastula stages, the embryonic genes for the rRNAs appear to be "silent." The synthesis of rRNA, inactive since oögenesis, starts anew at the beginning of gastrulation. This synthesis is correlated with the first appearance of nucleoli in the embryonic cells.

When rRNA synthesis begins at gastrulation, its pace is relatively slow in *Xenopus*. It is not until later, when the tail bud is forming (at about the time of onset of yolk utilization), that there is a significant increase in the *net* amount of the rRNAs and of ribosomes. Thus the activity of genes for rRNAs appears to be controlled in a sensitive way during oögenesis and embryogenesis. The control of this activity involves both the availability of genes for transcription and frequency with which available genes are transcribed.

Evidence from many laboratories shows that the nucleolus is the site at which the 28S and 18S rRNAs are made. The most convincing demonstration of this was obtained in the *anucleolate* mutant of *Xenopus* in which the nucleolus was completely absent.

Embryos that are homozygous for the mutation do not synthesize rRNAs. The embryo contains enough rRNA made previously during oögenesis to support early development, but when in later stages the embryo fails to make its own new rRNA, it dies. The nucleolus originates from a specific chromosomal site, the "nucleolar organizer," which is visible in most animal cells as a secondary constriction on one or more metaphase chromosomes. In *Xenopus* each set of haploid chromosomes contains one autosome with a single nucleolar organizer. In the homozygous lethal anucleolate mutant both nucleolar organizers are absent. Moreover molecular hybridization studies show that at least 99 percent of the DNA coding for the 28S and 18S rRNAs is also missing. In other words, the anucleolate mutant is in fact a *deletion* of that part of the chromosome coding for these two rRNAs.

The occurrence of this deletion and the development of refined fractionation and molecular hybridization techniques have made possible the "molecular mapping" of these genes. We now know that each haploid set in normal *Xenopus* cells contains about 450 genes for 28S and about 450 genes for 18S rRNA. These genes are clustered on one part of one of the 16 chromosomes in the set, probably along with some "spacer" DNA that differs from that controlling the two rRNAs. Moreover the evidence suggests that the 28S and 18S genes are alternating. It is not known whether spacer DNA separates the 28S and 18S genes, but it seems unlikely.

There are over 20,000 genes for 5S rRNA in the normal set, which appear to be partly clustered, but they are not intermingled with the cluster of genes for 28S and 18S rRNA.

Thus in the normal somatic cell the genes for the rRNAs are *redundant*, that is there are repeated sequences. These presumably occur in all the cells of an organism and at all times in its life history. For convenience we shall refer to the genes coding for rRNA as the "ribosomal DNA" or rDNA.

Gene amplification. We are concerned, however, with an even more striking phenomenon, *specific gene amplification*. Oöcytes of several amphibians, an echiuroid worm, and the surf clam increase their synthesis of ribosomes by replicating specifically the genes for 28S and 18S early in oögenesis, subsequently using these extra genes as templates for massive rRNA synthesis.

An oöcyte of *Xenopus* has about 1000 times as many genes for 28S and 18S rRNA as does the nucleus of the somatic cell. These extra copies are made during a relatively short interval of oögenesis, that is, between two to four weeks after metamorphosis. Moreover, once made, these extra copies are isolated in multiple nucleoli. Since the growing oöcyte persists in the first meiotic prophase for an extended period, it is

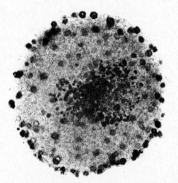

Figure 8-10 Photomicrograph of an isolated germinal vesicle of *Xenopus laevis*. The germinal vesicle was dissected from a mature oöcyte in 0.01 M MgCl₂ and 0.02 M tris buffer, pH 7.4, and flooded with cresyl violet stain. Its diameter is about 400 μ. The deeply stained spots are some of the hundreds of nucleoli.

tetraploid. Thus one would expect to find four nucleoli. Instead, however, in *Xenopus* oöcytes, there are about 1000 (Fig. 8-10). Several techniques prove not only that these nucleoli contain DNA but that the DNA contains sequences homologous to the 28S and 18S rRNAs; in other words, it is (at least in part) rDNA. These multiple copies function only during oögenesis. They are not replicated during cleavage, and are ultimately somehow rendered nonfunctional and discarded.

Why do oöcytes need multiple copies of these genes? The most obvious explanation is that they are needed to support the extraordinarily high rate of rRNA synthesis. An immature oöcyte of *Xenopus*—one cell—can synthesize rRNA at a rate comparable to an equal weight of liver tissue containing about 200,000 cells. It is interesting that the genes for 5S rRNA are not amplified. However, as we have noted, they are already highly redundant (20,000 copies per haploid complement, compared to 450 copies each for 28S and 18S). Clearly the 28S-18S gene cluster is somehow coordinated with the 5S genes, for the latter, although present, do not function in the anucleolate mutant.

The presence of extrachromosomal nucleoli in amphibian oöcytes stimulated Miller and Beatty to isolate the genes coding for rRNA precursor molecules and to observe them in the electron microscope (Fig. 8-9). The isolated nucleolus contains a compact fibrous core, containing DNA, RNA, and proteins, and a granular cortex that lacks DNA. When the fibrous core is isolated and "unwound," it is seen to contain a thin axial fiber, 100 to 300Å in diameter, that is periodically coated with matrix material. From studies using enzymatic digestion Miller and Beatty conclude that the core fiber is a double-helix DNA molecule coated with protein. Further studies with enzymes, coupled with electron microscope autoradiography, indicate that the matrix material is RNA, most likely a precursor of rRNA. Therefore it appears that each matrix-covered DNA segment codes for the pre-rRNA. The redundant structural arrangement of genes and intergene segments visually confirms the biochemical evidence for the organization of the genes coding for the rRNAs.

The "silk message" and its genes. The techniques that proved so successful in isolating the rRNAs and their genes are now being applied to the characterization of messenger RNAs and their genes. Until recently no animal cell message had been isolated and characterized, and although rRNA genes have been isolated from animal DNA, no purification of a gene which codes for a specific cellular protein had been successful. Now a "message" has been isolated and the first steps have been taken toward describing its genes. The secret of success lay in identifying an "optimal organism," often a crucial step in the development of a problem, even an entire field.

The "silk message" was chosen for analysis. The specific protein, *fibroin,* is made in the posterior silk gland of the silkworm, *Bombyx mori,* during a specific time late in larval life. The protein is highly unusual, containing alternating glycine residues in the crystalline region and about 73 percent of its total amino acids as glycine and alanine. This unique composition made it possible to predict the base composition of the corresponding mRNA and ultimately to isolate and identify it. Specific molecular hybridization of fibroin mRNA and its gene was demonstrated by direct characterization of RNA isolated from the hybrid, and then this exceedingly sensitive assay was used to determine the level of fibroin genes in DNA isolated from posterior and middle silk gland cells and from carcass. This measurement *ruled out specific fibroin gene amplification* in silk gland cells. A calculation limiting fibroin genes to three or fewer per genome indicates that massive fibroin synthesis is not facilitated by reiteration of this gene, but rather by the enormous translational yield of stable fibroin mRNA.

The cytoplasmic environment

It is clear by now that the nucleus does not operate alone. It influences and is influenced by the rest of the cell. In examining the influence of the nucleus, let us look again at the marine alga *Acetabularia.* The alga can regenerate a cap under control of the nucleus in the base. Even if the nucleus is removed a new cap can be produced and the cell will survive for several months, although the cell is incapable of further regeneration after one cap has been regenerated under these circumstances. Products synthesized earlier by the nucleus or under its control and in sufficient amounts to rebuild one cap are apparently stored in the stalk. This confirms the findings of the transplant experiment. The first cap regenerated by the stalk and the base from the other species has mixed characteristics, as will be recalled.

Other experiments will further illustrate how nuclei may be influenced by the cytoplasm. In the eggs of some marine snails and mussels part of the egg protrudes as a "polar lobe" immediately before the first cleavage; at division it remains attached to one blastomere into which it is withdrawn, only to reappear during the next division. This lobe makes an attractive "target" (Fig. 8-11). When the lobe is amputated, the embryo develops, but imperfectly, the resulting larvae lacking mesodermal structures. Thus this localized region contains cytoplasmic materials essential to formation of mesodermal structures.

Experiments in which sea urchin eggs are divided make the same point. As Driesch showed in 1892, at the two- or four-cell stage each isolated blastomere can give rise to a complete larva, which is normal and proportionate, although

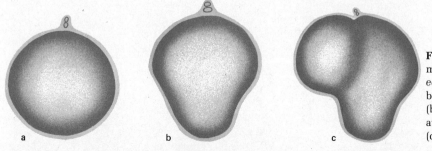

a   b   c

Figure 8-11 Early development of the mussel, *Mytilus edulis.* (a) Formation of polar bodies at the top of the egg. (b) Formation of polar lobe at the bottom of the egg. (c) First cleavage.

small. Thus at the outset of development it would *appear* that there were no regional cytoplasmic differences. In another experiment performed two decades later, two whole sea urchin eggs were fused. A single larva — giant but otherwise normal — developed.

We have, however, omitted an important point. When the embryos are fused, their axes must coincide; thus some organization is implied in the fertilized egg, to which we now return. Instead of allowing the egg to cleave normally and then separating the blastomeres, let us cut the unfertilized egg in half and fertilize each half (Fig. 8-12). If we make the cut vertically (as the first cleavage does), two normal larvae result; but if the egg is cut along the equator between the animal and vegetal hemispheres, the animal half forms a ball of ciliated cells and the vegetal half forms an incomplete embryo. In the normal development of the sea urchin embryo, the first two cleavages are vertical, but the third is horizontal.

What would one expect to find if he separated the animal and vegetal halves at the eight-cell stage? If you said the results would be similar to those obtained by cutting the fertilized egg horizontally, you would be correct.

From many investigations of this sort and from the evidence cited earlier in the chapter, it is clear that although the nucleus is serving as a control center for development, it does not act independently but in concert with the cytoplasm.

One of the most striking examples of the cytoplasmic regulation of nuclear activity is found in the following experiment. Recall that the synthesis of rRNA that predominates during oögenesis in *Xenopus* stops at maturation, not to be resumed in significant amounts until gastrulation. Nuclei from late embryos or young tadpoles were transplanted into eggs. Following transplantation, the nucleoli which were prominent in the embryonic nuclei disappeared within 40 minutes, and rRNA synthesis was not detected until gastru-

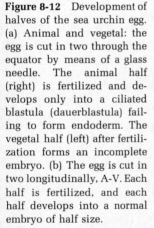

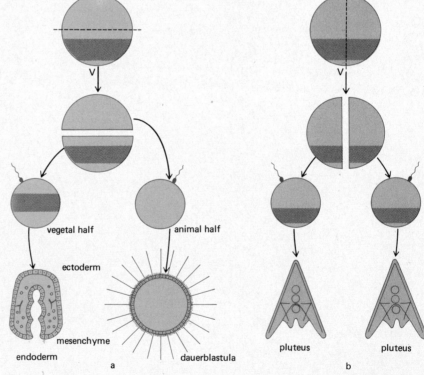

Figure 8-12 Development of halves of the sea urchin egg. (a) Animal and vegetal: the egg is cut in two through the equator by means of a glass needle. The animal half (right) is fertilized and develops only into a ciliated blastula (dauerblastula) failing to form endoderm. The vegetal half (left) after fertilization forms an incomplete embryo. (b) The egg is cut in two longitudinally, A-V. Each half is fertilized, and each half develops into a normal embryo of half size.

lation. In other words, the genes controlling rRNA synthesis are active in cytoplasm of cells at the tadpole stage, but become inactive when placed in cytoplasm of cleaving eggs.

It may be helpful next to consider some of the kinds of questions with which we shall be dealing. We raised one question in considering the rRNAs. Within a given genome, how are the activities of genes at different loci, even on different chromosomes, coordinated? If only a fraction of the genes in a given cell are acting at any moment, how are the neighboring genes held in check? **Levels of control.**   Gene action may be regulated at four levels. First, the number of copies of a given gene may differ from cell to cell or from time to time in the history of a given cell as a consequence of a *differential replication* or *amplification* of part of the genome. We have already presented conclusive evidence for the control of rRNA synthesis by *differential gene content.*

Second, the frequency with which a given gene is transcribed into RNA copies may vary. Such controls at the *level of transcription* appear to be commonplace. We refer to the third level of control as *intranuclear processing.* By this we mean that the immediate products of transcription, for example, the rRNAs, may not be identical with the final functional product we identify in the cytoplasm. The immediate product may have to be "processed" before it becomes functional.

Finally, transcribed copies of genes in the form of mRNAs may or may not be translated into products; that is, regulation may occur at the *level of translation.*

We know very little about how these control mechanisms function. To illustrate the "state of the art," let us consider control at transcription.

It is commonly assumed that since chromosomal DNA exists as a highly ordered complex with proteins, and perhaps with RNA, that one or more of the proteins "masks" the DNA and inhibits transcription. It is further assumed that the positioning, quality, or quantity of the "mask" is somehow regulated by molecular input from the cytoplasm. This hypothesis is based, in part, on the well-known regulator-operator-operon hypothesis discussed in Chapter 6.

Many of the predictions of the model have now been verified for bacteria and viruses. Two repressors have been isolated and characterized as proteins (in the *lac* operon in *E. coli,* which controls the enzymes β-galactosidase, permease, and transacetylase, and in cells containing bacteriophage λ). These proteins have a high affinity for DNA, suggesting that the repressor blocks transcription by binding directly to DNA. *It is not at all certain, however, that the regulatory mechanisms in the cells of animals and plants function in the same way as those in bacteria.*

To what extent may this scheme be applied to embryos? We begin this discussion by briefly reviewing the organization of DNA in chromosomes. The complex of DNA with protein and RNA is termed **chromatin.** It has two states, "condensed" and "extended," with the terms meaning just what they say. A typical example of condensed chromatin is a chromosome at metaphase when it is discrete and small. An example of extended chromatin is a "puff." Condensed chromatin is not active, that is, the DNA is not transcribed. The DNA in extended chromatin, however, is transcribed. There is evidence that most of the DNA of somatic cells is not transcribed, and that the fraction that is transcribed varies from one cell type to another. Different cell types use different amounts and parts of the genome. In maturing red blood cells, which have been "determined" for hemoglobin synthesis, the amount of condensed chromatin is high.

There is also biochemical evidence for organ-specificity of chromatin; that is, chromatin extracted from bone marrow and thymus, for example, contains readily transcribable DNA sequences specific for those organs. This kind of evidence is taken to mean that most of the DNA is masked. What is the nature of the mask?

It appears reasonably certain that in mammalian chromatin, ionically linked proteins mask DNA. Some of the masking may be due to *histones,* at least histones are logical candidates. These basic proteins (which are made in the cytoplasm) are found closely associated with DNA in chromosomes. Moreover there is impressive evidence that the removal of histones from chromosomes *in vitro* may derepress genes, leading to mRNA synthesis, and conversely that the addition of histones to DNA preparations will inhibit synthesis. There is also evidence that nonhistone, acidic proteins may also mask DNA. It is interesting that the repressor in the bacteriophage λ is an acidic protein. But how does one account for the specificity in the regulation of gene action? According to one of several ideas advanced, the specificity is embodied in a special class of RNA molecules linked to histones. The base composition of the DNA to be masked would be recognized not by histone but by RNA. A second proposal suggests that DNA-histone interactions are modified according to the acetylation of histones. Acetyl groups are added to histones at times of extensive gene activation.

A third scheme attempts to bring together ideas from bacterial (specific repressors) and animal (nonspecific masks or repressors) systems. Specific repressors would interact by reversible binding with one or a limited number of operators, and nonspecific histones or other nonspecific masks could interact with the DNA at any point. As in the usual bacterial scheme the regulatory action of a specific repressor would be reversible. Histone would turn a gene off only if it became complexed with the operator; only under these conditions would the initiation of RNA synthesis become sensitive to histone binding. If there are no inducers for histones, there would be only one time during the cell life cycle when a gene whose operator is complexed with histone could become free of histones. This would be during the period when both DNA and histone synthesis occur in the nucleus and cytoplasm, respectively. In other words, newly duplicated genes have the option of complexing with either specific repressors or histones. A gene previously masked nonspecifically with histones may simply reassociate again with histones. However, if other factors have changed since the previous replication (change in the cytoplasm, etc.), it may now associate with a specific repressor, thus opening up a new range of possibilities.

We must emphasize that all of these schemes, especially the third, are *highly speculative*. There is very little direct evidence of the existence of cytoplasmic inhibitors and stimulators in animal and plant cells.

The search for cytoplasmic inhibitors and stimulators. A common assumption remains that there must exist in the cytoplasm substances that interact with the genome or its products, thus regulating cellular activities. The entire history of the study of nucleocytoplasmic interactions argues for their existence. We have discussed one example, the cytoplasmic control of the rRNA cistrons in *Xenopus*. This experiment argues for an inhibitor (or repressor), although it is not known whether it might act directly or indirectly on the genome.

One of the more novel ways of approaching the interactions of nucleus and cytoplasm and, hopefully, ultimately identifying the factors involved, is the *fusion of animal cells* to form **heterokaryons** and **somatic cell hybrids**. A heterokaryon results from the fusion of two or more cells and consists of a single cytoplasmic mass containing two or more different nuclei. There are no further implications: the mass does not have to propagate itself to be a heterokaryon. A true *hybrid cell strain*, in contrast, is a propagating cell line containing the chromosomes from different parent cells within a single nucleus. In other words, it involves the *mating* of somatic cells. These recent innovations have already begun to prove their usefulness.

Heterokaryons are commonly formed by mixing two cell types with an inactivated virus, the so-called Sendai virus (named for one of the cities in Japan where it was studied). The exact manner in which the noninfective virus operates in promoting cell fusion is unknown; in some way, by modifying the cell membranes, cells are caused to fuse. The following example is especially instructive. Chicken red blood cells were fused with human cells. The latter were either of two types, cultured cell lines known as HA and HeLa. The *HA cells* have the capacity to produce *interferon*, a specific product whose role in the animal is to inactivate viruses. We are interested in it here only as a "marker," indicative of the onset of gene function. *HeLa cells* do not produce interferon nor do chicken erythrocytes. In fact in the latter the nuclei are considered to be inactive (although they are not extruded as are the nuclei of mammalian erythrocytes). The heterokaryon of HeLa cells and erythrocytes produces no interferon. However, interferon is produced in the HA-erythrocyte combination. But what kind? Interferons are known to be species-specific. Both chicken and human interferons are made. At some level the erythrocyte has been reactivated to make a product it probably never made in the course of its life history. *A long silent part of the genome appears to have been derepressed.*

Propagating hybrid strains have usually originated "spontaneously" when the cells of two established lines are mixed in culture. The underlying cause of such fusions, mating and subsequent establishment of propagating lines is not known. Evidence of interaction between the genomes in hybrids is given in the following experiment. Hybrids were made between cells of a pigmented Syrian hamster melanoma line (a *melanoma* is a highly malignant black tumor) and an unpigmented mouse cell line. The hybrid cells were isolated and maintained in culture up to 100 cell generations. The hybrid cells remain *unpigmented* and lack one of the key enzymes involved in pigment production. Somehow the functions of the hamster melanoma cells are suppressed by factors in the mouse cells.

The cytoplasm and heredity. Although the amphibian zygote and the unicellular alga, *Acetabularia*, will not continue to develop or regenerate repeatedly in the absence of a nucleus, some of their cytoplasmic elements or organelles determine part of their own inheritance and are self-duplicating. The chloroplasts of *Acetabularia* and several higher plants contain DNA. Significantly, chloroplast DNA and nuclear DNA are not identical, suggesting, along with genetic evidence, that it is manufactured in the chloroplasts and not the nucleus. Thus chloroplasts initiate their own reproduction in the cell with their own hereditary instructions. This does not mean, however, that the chloroplasts are completely independent of the nucleus.

Following suggestions from genetic studies that in yeasts mitochondria may be inherited by a similar mechanism, evidence has been obtained that mitochondria of several types of cells from organisms as diverse as the bread mold, *Neurospora*, and the amphibian embryo contain DNA. In the amphibians, in which it has been studied extensively, we know that the mitochondria con-

tain a circular DNA molecule (Fig. 8-13), containing about 16,000 base pairs. The function of this DNA is not fully understood. However, it is known that it codes for rRNAs that are characteristic for mitochondria. The mitochondrial rRNAs are smaller than the 28S and 18S rRNAs in the rest of the cytoplasm. Moreover these mitochondrial rRNAs are contained in a special mitochondrial ribosome. Thus a large part of the mitochondrial genome must function in the production of a specialized mitochondrial protein synthesizing system.

CELLULAR INTERACTIONS

In the embryo, as we have observed, kidney cells, muscle cells, and neurons do not develop or function in isolation. They develop as part of the whole. What properties of the many types of cells of which it is composed determine the shape and size of a worm, a man, or an oak tree? Clearly we must examine not only the cell's inner controls but the manner in which a cell interacts with its neighbors and its environment. How do cells impinge upon and influence each other?

We know that during the course of development few, if any, structures in the body of vertebrates—and of many invertebrates as far as the evidence goes—are elaborated without an initial interaction of their constituent tissues. We also

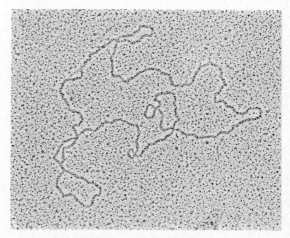

Figure 8-13 Electron micrograph of DNA from oöcyte mitochondria of *X. laevis.* × 85,000.

know that one effective way of illustrating such an interaction is by indicating the consequences of its failure to occur. In Chapter 7 we saw that when frog or salamander embryos are caused to exogastrulate, so that the chordamesoderm does not come in contact with the overlying ectoderm, the ectoderm fails to develop into nervous system. We emphasized that in normal development the process must be highly ordered in both space and time. Both the ability of one tissue to affect the other and the capacity of the second to respond must be considered. The flowchart in Figure 8-14 suggests the interactions that occur during the organization of just a single organ, the eye. When considering that similar flow sheets could be worked out for other organs, we acquire some conception of the complexity of development that we are just beginning to unravel.

Often such interactions are spoken of as "contact mediated"; this expression may have two implications. The first possibility is that there is no exchange of material between the cells—the surface of one cell reacts with the surface of an adjacent cell, the resulting surface changes being reflected in changes within the cell.

The second implication of contact mediated is that the reacting tissues remain in intimate association in such a manner, and for periods long enough, that an exchange or interaction of products is permitted. Even our limited understanding of the membrane systems of cells should permit us to realize that we are using the simple word "contact" to refer to what must be not one, but rather an array of complex interactions: the juxtaposition of pores in apposed membranes with a closely timed transfer of particulates, macromolecules, or small molecules; the elaboration of processes or products from apposed cells which interact; or the uptake by one cell of small molecules from another. There may be as many mechanisms as there are combinations; there may be a few large classes of mechanisms; or there may be one mechanism common to all.

The life history of muscle

During the earliest stages of development changes may occur synchronously in all of the cells of the embryo, irrespective of their eventual differentiation. At later stages, however, cells become progressively more divergent in their properties. During these phases we focus our attention upon specific embryonic cell types. In what Abercrombie has called "an astonishing stride forward in the history of biology," over 60 years ago Ross G. Harrison advanced a new technique of tremendous power, that of tissue culture, establishing that cells could be grown outside the body. Only recently have culture techniques been sufficiently refined to permit the derivation of clones, populations of cells derived by division from a single isolated cell. However, most "established" strains

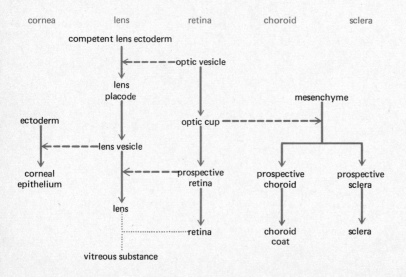

Figure 8-14 Tissue interactions in the development of the eye.

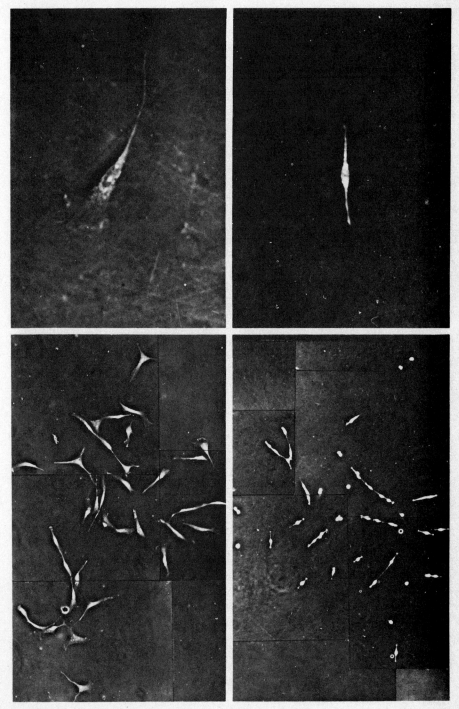

Figure 8-15 (Top) Single myoblast (right) and fibroblastic cell (left). As soon as single cells attach and flatten on the bottom of the petri plate, the myoblast can be distinguished by its spindle shape. (Bottom) Two living colonies photographed on fourth day of culture. (Right) embryonic muscle; (left) fibroblastic cells. The latter are more extensively attached and appear larger.

of cells isolated originally from animal tissues and maintained continuously in culture are unsuitable for studies of cellular differentiation, since it is exceedingly rare for such cell strains to bear the remotest resemblance to the major cell type of the tissue of origin. Such cell populations, during the course of their cultivation, lose the cell-specific properties that characterize the cells of the original tissue. In order to study cell-specific properties and the manner in which they are acquired, it is necessary to apply cloning techniques to newly isolated embryonic cells. Konigsberg first achieved this goal in studies of the growth and differentiation of embryonic skeletal muscle cells.

Early in the development of tissue culture as a research tool, Margaret and Warren Lewis demonstrated that fragments of embryonic chick skeletal muscle, embedded in clotted plasma, not only grow but form striated muscle. It is now clear that such cells develop equally well when grown by newer methods of cell culture in which the tissue is first enzymatically dissociated into its component cells. Cell suspensions prepared from the leg muscle of 12-day-old chick embryos attach to the bottom of the culture chamber, grow, and form a continuous sheet of cells. As the cell layer approaches confluency, large numbers of long fibers appear that within a few days begin to contract spontaneously. The contractions tell us that functional muscle has differentiated, and morphologic and biochemical evidence is provided by the presence in the elongated cells of the cross-striated pattern and the contractile proteins typical of muscle. Thus we can tell that embryonic muscle cells grown in cultures of randomly distributed individual cells are still capable of differentiating into structurally and functionally recognizable units.

Can a single cell give rise to a colony? By physically isolating a single primitive muscle cell or *myoblast* (Fig. 8-15) and culturing it in a small glass cylinder, under conditions which exclude contact with any other cell, it was proved that the single cell can produce a colony of differentiated muscle.

During the first four days of culture, such cells divide every 12 or 18 hours, producing small colonies of roughly 50 cells (Fig. 8-15). The first indication of further differentiation is observed on the fifth or sixth day of culture when cells fuse to form multinuclear "myotubes." At successively later stages these multinuclear myotubes increase in length and number until by the end of the second week they form a colony of interlaced fiberlike cells (Fig. 8-16). These colonies, which measure several millimeters in diameter, are apparent to the naked eye when appropriately stained. Moreover many of the myofibrils contain cross-striations; the level of differentiation attained in such cultures is illustrated in Figure 8-16. Not all of the single cells fuse, however; some remain and continue to divide even at the end of the second week. However, the abrupt appearance of multinucleated myotubes is paralleled by an equally abrupt drop in the overall rate of proliferation. The nuclei in the developing myotubes no longer divide. The fusion of myoblasts to form a multinucleated myotube exemplifies, albeit in a specialized form, the *surface interactions* of cells. It is now well established, by several techniques, that skeletal muscle is a *syncytium*, many nuclei lying in a common cytoplasm, and that it is formed by the fusion of myoblasts. Until recently all of the evidence was derived from studying cultured cells. Now, however, crucial evidence has been presented that this process does occur in the intact animal.

This evidence has resulted from two advances, one technical and one conceptual. Technically, ways were found to do in mammalian embryos what had been accomplished earlier in the sea urchin, that is, to *fuse* early embryos. Conceptually, the developmental questions were asked with insights provided from the field of genetics. Two or more cleaving mouse eggs may be fused in culture to form a normal blastocyst, which when transferred to the uterus of a foster mother produces a normal mouse. Embryos fused as late as the eight-cell stage routinely develop normally. This technique makes it possible to create new kinds of adult mice in which cells with two or more *different* (rather than identical) genotypes would be included. Cleaving embryos with distinct genotypes, bearing a wide range of genetic markers, may be fused and reared to adult-

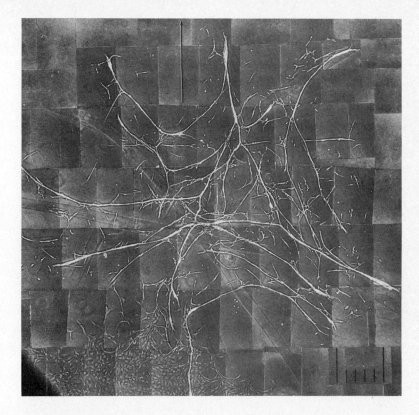

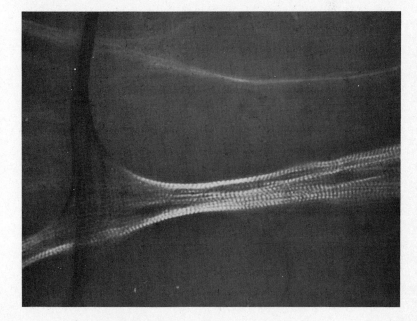

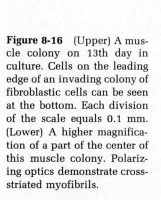

Figure 8-16 (Upper) A muscle colony on 13th day in culture. Cells on the leading edge of an invading colony of fibroblastic cells can be seen at the bottom. Each division of the scale equals 0.1 mm. (Lower) A higher magnification of a part of the center of this muscle colony. Polarizing optics demonstrate cross-striated myofibrils.

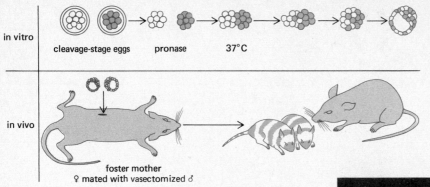

Figure 8-17 Diagram of the experimental procedures for producing allophenic mice from aggregated eggs.

in vitro   cleavage-stage eggs   pronase   37°C

in vivo

foster mother
♀ mated with vasectomized ♂

Figure 8-18 (Below) Adult allophenic mouse containing both homozygous black (CC) and albino (cc) melanocytes from C57BL6, and ICR genotypes, respectively.

hood (Figs. 8-17 and 8-18). These mice are called *allophenic* mice—individuals with a simultaneous, orderly expression of two or more allelic cellular phenotypes or allophenes, each with a distinctive genetic basis (*alleles* being alternative genes at the same locus).

Here then is an ideal situation in which to test the hypothesis that the normal development of muscle involves fusion just as it does in culture. We need only one additional bit of information, namely that many enzymes are made up of subunits. The two strains of mice used in the experiment clearly contain different forms of a specific enzyme, *isocitrate dehydrogenase*, which is known to be made up of subunits. If muscle fibers are made by fusion, then in the mosaic animal, myoblasts of both "parental" strains may contribute to the muscle fiber. Thus a myotube may contain two kinds of nuclei in a common cytoplasm. Since it is in the cytoplasm that the polymerization of enzyme subunits occurs, we should expect to find three kinds of enzyme molecules in a myotube formed by fusion of myoblasts of different origin: enzymes of each original type and "hybrid" enzyme. That is exactly what is found. Skeletal muscle from allophenic mice contained appreciable quantities of hybrid enzyme (Fig. 8-19). The hybrid enzyme was not found in tissues which are not syncytial, for example, *cardiac* muscle, liver, kidney, lung, and spleen.

In concluding this section, we return to one of the questions that lie at the very heart of developmental biology, the nature of cellular interactions. Our example, muscle formation, illustrates two of the major problems which confront

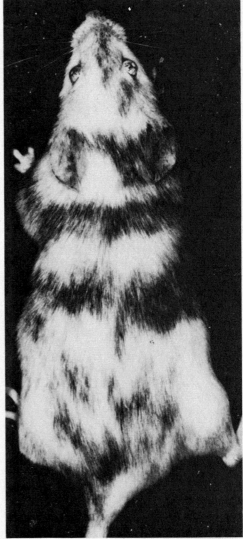

us. We have just considered one, the interactions of myoblasts among themselves. The second is illustrated by the interactions of myoblasts with other cells, notably fibroblasts.

Muscle tissue contains two major cell types, the muscle cell itself and the fibroblast which lays down the connective tissue framework of the organ. Mass cultures prepared by disaggregating embryonic muscle contain both cell types. Each type can be grown in the absence of the other. It has long been known that fibroblastic cells can be grown clonally, and we have just noted that clones of muscle cells have been developed, suggesting they do not require an interaction with fibroblasts. However, to clone muscle cells successfully, a "conditioned" medium had to be used initially; that is, medium recovered from mass

cultures containing muscle fibroblasts was reutilized to grow clones from myoblasts. The medium appears to be altered by the activities of the cells themselves, making it more suitable for supporting muscle differentiation.

What is the nature of the conditioning process? It now appears that the fibroblastic cells secrete the fibrous protein, collagen, which coats the surface of the culture dish, thereby enhancing muscle formation. At least the conditioned medium may be replaced by spreading a thin layer of pure collagen on the surface of the vessel in which myoblasts are cloned. Whether collagen plays this role in the normal development of muscle is yet to be determined. Nor do we know its role in the morphogenesis of any organ. Consider, for example, the development of the mouse

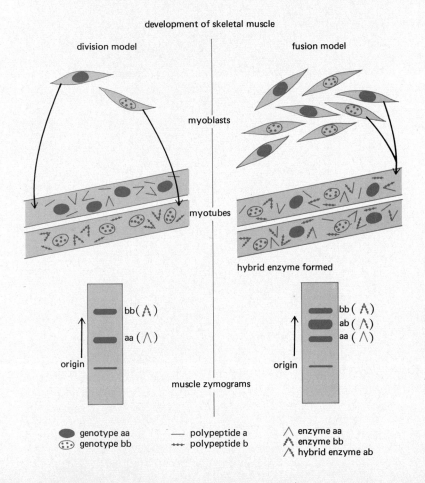

development of skeletal muscle

division model   fusion model

myoblasts

myotubes

hybrid enzyme formed

muscle zymograms

bb (⋀)
aa (⋀)

origin

bb (⋀)
ab (⋀)
aa (⋀)

origin

⬤ genotype aa   — polypeptide a   ⋀ enzyme aa
⊚ genotype bb   ⋯ polypeptide b   ⋀ enzyme bb
                                               ⋀ hybrid enzyme ab

Figure 8-19 Diagram of expected isozyme results in allophenic mice, on the "division" (left) versus "fusion" (right) models of skeletal muscle development. When homozygous cells of different NADP-isocitrate dehydrogenase genotypes coexist, heterokaryons would result in the event of myoblast fusion, and hybrid enzyme could be formed. Enzyme molecules are represented as dimers formed in the cytoplasm from polypeptide subunits.

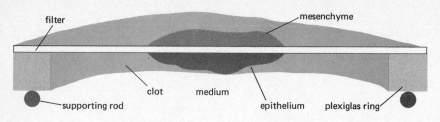

Figure 8-20 Filter assembly for preparing standard trans-filter cultures of epithelium and mesenchyme. Intact rudiments may be cultured either "in the clot" in the position of the epithelium, or "on the platform" in the position of the mesenchyme.

salivary gland. We know that the formation of the gland depends on the interaction between two layers of cells, the outer **epithelium** and an underlying **mesenchyme.** Using *tissue culture* techniques, these tissues may be isolated from the animal and cultivated in the laboratory. When the epithelium and mesenchyme are cultivated *separately*, they do not differentiate. When they are grown together, typical salivary gland tissue is formed. When the two tissues are grown together in culture, separated only by a fine filter which allows molecules to pass but prevents transport of cells (Fig. 8-20), the epithelium differentiates under the influence of mesenchyme.

The branched pattern of morphogenesis characteristic of the cultured salivary gland under these circumstances is shown in Figure 8-21. But if the tissue culture is treated with collagenase, an enzyme which breaks down collagen, normal morphogenesis does not occur; the epithelium grows as a simple disc.

Further studies have shown that during the interaction between epithelium and mesenchyme, the latter produces collagen in a soluble form. Collagen fibers are first found, however, as the result of a migration of molecules into the epithelium. The fibers may be polymerized, or built up from simpler units into the more complex fibers, under the influence of factors provided by the epithelium. It has been suggested that collagen plays a role in maintaining and possibly in initiating epithelial contours. Possibly collagen acts by putting a "jacket" around certain areas which no longer show morphogenetic changes.

Even these findings do not prove that collagen is the only factor involved in induction of the salivary gland, however. We can say only that collagen appears to be involved and that other components are also probably implicated. At

least in this system, there does not appear to be a simple one-step reaction but a series of several reactions. We have no reason to believe that other inductive tissue interactions will turn out to be very much simpler.

Although we have offered only a sample, abundant and compelling evidence exists to show how the changing chemical environment results in the differential expression of genes, with telling effect on the course of differentiation. Enough

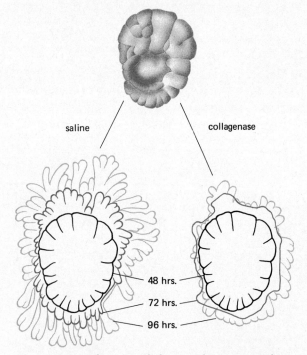

Figure 8-21 Salivary epithelium undergoing morphogenesis in vitro. (Upper) At 48 hours, untreated; (Lower left) at 72 hours, treated with Tyrode's solution at 48 hours; (Lower right) at 72 hours, treated with collagenase at 48 hours.

has been said to emphasize that the conditions for differentiation, achieved through the interactions we have referred to as inductive, may differ widely, and thus a search for a single, universal "inductive agent" is illusory. Molecules exchanged between cells probably affect mitosis, cell participation in morphogenesis, and synthesis and assembly of specific products. All three kinds of processes are essential for organogenesis.

Thus far we have focused on associations in which a large molecule—collagen—appears to be involved. Lest this emphasis be misleading, we should examine at least one example in which a cell with at least two alternative capacities is "directed" by a small molecule to take one course to the exclusion of another. When epidermis of the chick embryo is explanted to a medium of blood plasma, it produces stratified epithelium that is characteristic of skin, the outer layers of which become keratinized. When vitamin A is added to the medium, the cells form a columnar epithelium which produces cilia and secretes mucus. Only a short treatment with vitamin A is necessary; cells exposed briefly and returned to normal medium proceed to make mucus.

Cells that recognize their neighbors

It is becoming apparent from what we have said that the surfaces of cells must play critical roles in development. We have considered their importance in the transmission and reception of inductive stimuli. It almost goes without saying that cell surfaces must be involved in what happens when cells move about during processes such as gastrulation, the formation of the nervous system, and the formation of many other organs.

The experiments which give us the best clues about surface interactions had their origin in studies of marine sponges. A living sponge can be pressed through a finely meshed cloth into salt water so that it is separated into individual cells or small cell clusters. If these cells are allowed to stand, they will move together and ultimately reconstruct a sponge that is complete with all the internal structures characteristic of the species. A sponge whose cells have been dissociated chemically by being placed in seawater free of calcium and magnesium will also reassociate

in seawater. The reassociation of these cells is strikingly species-specific. If the cells of two different species of sponge are mixed, they reaggregate, not into one chimeric sponge but into two sponges, each composed of cells characteristic for the species.

Subsequently, methods of chemical dissociation have been used on embryos of amphibians, chicks, and mice, among others. Amphibian embryo cells are separated by placing them in alkaline solutions. Tissues from chick and mammalian embryos are treated with the enzyme, trypsin.

We must take immediate notice of one striking difference between the reassociation of marine sponge cells (which are, after all, adult cells) and the reassociation of embryonic cells. The former reassociate species specifically; the latter reassociate tissue specifically. A brief description of two experiments will make the distinction clear. Suppose we dissociate a mass of prospective cartilage and an embryonic kidney from a chick embryo, and mix the two populations of cells. In reassociating, a mass of cartilage will reform in the center of a mass of kidney. Reassociation has been cell-type or tissue specific. Suppose we perform the same experiment on a more ambitious scale, dissociating and mixing chick embryo precartilage and kidney and mouse embryo precartilage and kidney, stirring all "ingredients" in the same dish. The results: a chimeric mass of cartilage, with chick and mouse cartilage cells interspersed, surrounded by a chimeric kidney.

There is now evidence that animal embryonic cells sort themselves out on the basis of their cell or tissue specificity. Studies of six different types of chick embryo cells, mixed two at a time and allowed to reassociate, showed a regular relationship (Fig. 8-22). They could be listed in the form of a hierarchy in which each tissue always took the inside position in relation to those below it, and always surrounded those above it on the list.

These cell mixtures are able to carry out a surprising amount of morphogenesis. Suspensions of skin, liver, and kidney cells from chick embryos, placed on the vascular chorioallantois of an older embryo, will produce structures rivaling normal organs in complexity.

Some interesting observations have resulted from these experiments. At first the cells form aggregates in which the cells are distributed randomly. Then the cell types shift about and gradually group together. There is no cellular transformation, no change of type as the regrouping occurs; if one type is radioactively labeled, the cells with that label all turn up in the same place. Perhaps most significant, the abilities of the cells to move and to adhere to their own type neighbors change with time. Early embryonic cells have greater capacity for moving and adhering than older embryonic ones; adult cells show few if any signs of the same tendencies.

The mechanisms involved in these processes are not at all clear. Possibly embryonic cells produce some extracellular substance which binds them to like cells, promoting the proper orientation. If such substances do exist, is their presence sufficient to account for the formation of characteristically organized multicellular patterns both *in vitro* and *in vivo*? There is evidence that embryonic cells produce and release extracellular products which serve as binding agents and promote orientation during tissue formation. In this approach emphasis is placed on the isolation of molecules that have the capacity to promote tissue-specific aggregation.

A substance prepared from the culture medium in which embryonic chick neural retina cells had been grown enhances specifically the aggregation of freshly dissociated retinal cells. The specificity of the preparation is shown by the fact that cells from other tissues do not respond to it. The

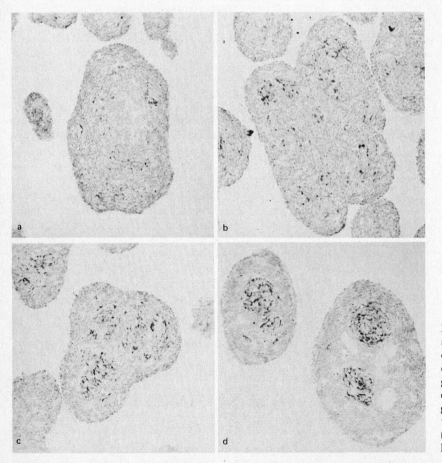

Figure 8-22 Time course of segregation of chick embryonic heart cells from chick embryonic retinal cells in gyratory shaker culture. Heart cells contain darkly stained glycogen granules. (a) After 17 hours; (b) after 24 hours; (c) after 31 hours; (d) after 66 hours. × 106.

binding activity is not removed from the preparation by liver cells, although it is by retinal cells. Moreover antibodies made in rabbits against the substance do not cross-react with liver cells. These and other lines of evidence suggest that such materials may play an important part in the formation of tissues. Again we need to know the nature of such factors and their origin and location within or on the surface of the cells. Only after these questions are answered can the larger question of the manner in which they act be examined.

ENVIRONMENTAL CONTROLS

We have now enumerated two levels of developmental control, the first at the level of the cell and the second at the intercellular level. However, there is yet a third; external environmental factors may exert a profound influence on the developmental response. A specific illustration of environmental control is the photoperiodic control of flowering—the fact that, in many plants, the length of the daily exposure to light determines whether flowers can develop and thus, in effect, controls sexual reproduction. Nor is the importance of environmental factors limited to plant development. Sexuality in birds is subject to photoperiodic control, and changes in photoperiod and temperature may exert profound effects on the development of insects.

We consider only two examples of environmental controls, drawing them from the plant kingdom which is an especially rich source.

Embryonic polarity

How does the fertilized egg develop into the bipolar embryo possessing a shoot pole and a root pole in plants, or in animals a head end and a tail end? This question has occupied the time of many developmental biologists, who are still unsure of the answer. But it is becoming increasingly possible to provide partial answers to this question, and each piece of evidence sharpens the questions which remain.

A purely descriptive approach to polarity has limitations, and answers are most likely to come from well-planned experiments. However, it is not easy to conduct experiments on the zygotes of higher plants which are so deeply embedded in other tissues. What is needed is a zygote which develops outside the parent organism. Zygotes of this type occur in brown seaweeds such as *Fucus,* and these have been used in many studies of polarity. Gametes of *Fucus* can be collected in large numbers and fertilization and embryo development can easily be followed in laboratory cultures. Prior to fertilization the egg is spherical and radially symmetrical, with a centrally placed nucleus and uniformly distributed cytoplasmic organelles. About 15 hours after fertilization a tubular *rhizoidal outgrowth* is formed on one side of the zygote, and this is followed in another eight or nine hours by nuclear division and formation of a wall perpendicular to the rhizoidal outgrowth (Fig. 8-23). The zygote is thus cut into two unequal cells and these have different developmental fates, a situation remarkably like that in *Capsella* (Chapter 7). Both cells continue to divide. Derivatives of the rhizoidal cell produce a holdfast which attaches the plant to the substrate, and those of the terminal cell form the frondlike body. The developmental axis is formed before the first nuclear division occurs and is independent of division. The rhizoid forms normally in zygotes treated with colchicine which blocks mitosis by preventing the assembly of spindle fiber microtubules. Polarity cannot therefore depend on nuclear differences in different cells, but must result from localized changes within the cytoplasm of individual cells. How are such changes induced in the cytoplasm of a *Fucus* zygote?

When *Fucus* embryos develop in darkness, the rhizoids arise at random positions, but if the embryos are very close together the rhizoids all develop on the side toward the center of the group. Apparently the position of the rhizoid, and therefore the axis of polarity, can be influenced by factors external to the embryo. A number of environmental factors are known to interact with the zygote affecting the position at which the rhizoid emerges, and what is important is that these factors are effective only when they are present as a gradient across the zygote. Thus the rhizoid is formed on the warm side in a temperature gradient, on the more acid side in a pH gradient, and on the shaded side in a gradient of white light. It is not necessary to continue the gradients indefinitely and once the cell has become visibly asymmetric in a light gradient, for example, its original polarity will be preserved if it is then placed in darkness or is illuminated by light coming from a new direction.

The environmental determinants seem to act as developmental triggers activating some intracellular mechanism which is insensitive to further environmental modification. What is the nature of the changes which occur in the cytoplasm of a polarized *Fucus* cell? It is here that the experimental evidence becomes less conclusive. There is some evidence that cytoplasmic vesicles accumulate in the region which will grow out as the rhizoid and that at this time an intracellular electrical gradient which coincides with the axis of polarity is established across the zygote. It is tempting to speculate that this gradient results in movement of vesicles through the cytoplasm and that the accumulation of vesicles at a particular site is involved in the out-

growth of the rhizoid, but more evidence will be required to confirm this. However, the conclusion that polarity in *Fucus* zygotes is environmentally determined is well established.

Meristems and the control of post-embryonic plant development

The way in which plants respond to environmental change represents one of the major differences between them and animals. Because of their mobility animals can respond by a change in their behavior pattern. Thus, they tend to move from an unfavorable to a more favorable environment as in seasonal migrations, and when disturbed with respect to illumination or gravity, for example, they may simply move so as to resume optimal orientation.

For plants which are fixed in position this mode of response is not possible and they respond to an environmental stimulus by a change in their developmental pattern. Some of these plant growth responses are rapid, occurring in

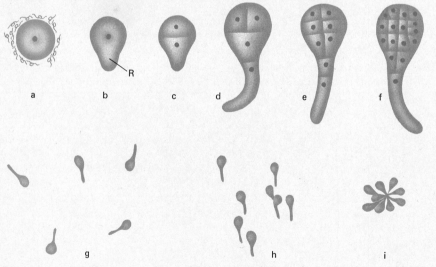

Figure 8-23 Development and growth responses of the embryo of *Fucus*. (a) The nonmotile radially symmetrical egg is shown surrounded by numerous small, motile sperm at the time of fertilization. Early developmental change in the zygote is seen in the outgrowth of the rhizoid (R) before the nucleus has divided (b). After division the two cells undergo numerous divisions so that newly formed cells are each smaller in size than the zygote (c-f). When embryos develop in darkness the rhizoids in a population are randomly oriented (g), but if illuminated from one side (here from the direction of the top of the page) the rhizoids develop predominantly from the shaded side (h). When embryos develop in closely spaced groups the rhizoids tend to develop toward the interior of the group (the so-called "group effect") regardless of other conditions (i).

minutes or hours. When a plant is laid horizontally so that its orientation in the gravity field is altered, or when it is illuminated from one side only, the plant alters its pattern of growth in such a way as to resume its normal orientation in respect to the stimulus.

In contrast, the responses of plants to changes in day length, as in flower induction (which we consider subsequently) take days or weeks for completion. These processes involve the formation of new lateral organs and considerable changes in growth.

You will have already observed that we have been speaking of effects "on plants"—not "on plant embryos." The reason is simple: much of plant development is post-embryonic. It is not unusual for differentiated cells in animals to divide. In fact, some cells continue division throughout the entire life-span in most organisms. The continued production and growth of new cells should result in continued enlargement of the organism, but in vertebrate animals, with the exception of fishes and a few others, the adult reaches a genetically determined size which is then maintained. How is adult size stability achieved in the face of continued cell production? The answer lies in a balance between cell formation and cell elimination. While new cells are being added to the organism, senescent cells are being removed. Cell replacement takes place within the framework of existing tissues so that while the cells are gradually replaced, the tissue maintains an unchanging size and appearance. This mode of selective cell replacement is possible in animals because the cells are mobile, and are capable of changing their shape and contacts with adjacent cells.

In plants, however, cell walls are rigid and firmly cemented together, and selective cell replacement cannot occur. Senescent and dead plant cells are either removed totally, as in shedding of leaves and bark, or are retained in the body of the organism throughout its life-span, as are the dead cells in wood. They are added in an accretionary manner in localized regions called **meristems.** This results in the continued enlargement of the plant throughout its life-span. The most important of the meristems are the apical meristems situated at the tip of each shoot and root. They consist of cells which retain the capacity for mitosis and produce a succession of differentiated structures (Fig. 8-24).

The apical meristems are delimited in the embryo but do not contribute significantly to its development. They become active in post-embryonic development, starting with germination, and most of the vegetative organs and all of the reproductive organs are formed by their activity during this period. The continued initiation of organs, which has no counterpart in animal development, has led to the meristems being called regions of "permanent embryogeny."

Apical meristems are continuously active in annual plants, and are still functional when the plant dies. Perhaps the meristems are capable of growing for unlimited periods of time under suitable conditions. This idea received striking confirmation when root tips were excised from a tomato plant and grown in sterile nutrient medium. The root tips were isolated in 1933 and have since been subcultured at weekly intervals by transferring tips to fresh culture medium. They were still growing actively in 1969, some 36 years later, far longer than the normal life-span of a tomato plant.

In perennial plants periods of dormancy intervene between periods of active meristem growth. The kinds of organs produced by the meristems may vary. The leaves of juvenile plants are often different from those of the adult; bud scales may replace vegetative leaves in advance of periods of dormancy; and reproductive organs ultimately replace vegetative ones. All of these changes suggest that the meristems are subject to developmental regulation in their growth. How is their activity regulated? Does each meristem function as a self-regulating cell-producing region, or is this activity regulated by stimuli transmitted to the meristem from other parts of the plant? Does a meristem organize the tissues and appendages it produces, or is the differentiation of tissues and organs controlled by interactions with older parts of the plant or the external environment?

To understand how the activity of a meristem is controlled by environmental stimuli, it is first necessary to understand how the stimuli are received by the plant. Consider for example the

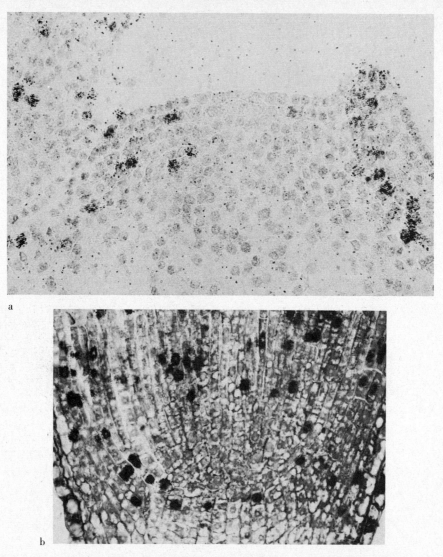

a

b

Figure 8-24 Longitudinal sections of terminal meristems after labeling with tritiated thymidine. (a) Shoot apex of sunflower *(Helianthus annuus)*. The meristem had previously been excised from the plant and grown in sterile culture on nutrient medium. After 48 hours in sterile culture, tritiated thymidine was added to the medium, and the meristem was allowed to incorporate thymidine into DNA for 24 hours. The shoot tip was then fixed, sectioned, and attached to microscope slides. The slides were then dipped into liquid photographic emulsion, and the emulsion-coated slides were stored in the dark. They were subsequently developed in photographic developer and stained to show the cells. In the figure, accumulations of silver grains in the emulsion show as dark spots and indicate nuclei that were synthesizing DNA during the experimental period. It can be seen that although several of the meristem cells did synthesize DNA, none of these was in the central part of the meristem. Thus this part is called the quiescent center. (b) Root apex of leafy spurge *(Euphorbia esula)*. The meristem was immersed in a solution of tritiated thymidine while still attached to the plant and then processed as the sunflower shoot tip described above. Here also it can be seen that none of the cells in the central part of the meristem has incorporated thymidine into DNA and this region is, therefore, the quiescent center.

effect of gravity on plants, a phenomenon called **geotropism** (from *geo,* meaning "earth" and *tropism* meaning "moving toward"). The fact that roots grow downward and shoots upward is the result of the response of the shoots and the roots to gravitational stimuli. In the roots gravitational stimuli are received by special receptor particles known as *statoliths* which are found in the root cap outside the root meristem. A shift in the position of the statoliths sets off a chain of chemical responses which culminate in a change in growth of the root, altering its position with respect to gravitational stimulation.

Similar to the gravity-induced orientation of a plant is orientation induced by light. This phenomenon is known as **phototropism.** As in the case of gravitational stimulation, the receptor particles, in this case pigment molecules, are located in or very near the growing tip of the shoot, the region responsible for the change in growth. Another similarity between gravity and light stimulated movement is that they both involve asymmetric growth of the root or shoot resulting in a curvature towards or away from the stimulation.

Photoperiodic effects

There is yet another way in which plants respond to the environment: *photoperiodic responses.* In photoperiodism the plant (or animal) responds to the relative lengths of the light and dark cycle in a 24 hour period. In plants, photoperiodic stimulation results in the initiation of new growth as in seed germination and the budding of shoots in the spring or in the formation of new structures such as over-wintering buds and flowers. The control of flowering is particularly important in the plant kingdom, so we will examine it more carefully.

In many plants flowering follows exposure to photoperiods of appropriate length. Plants fall into three general classes in their flowering response to photoperiod. Some are *short day plants* which initiate flowers only when the light period of the alternating light-dark cycle is less than a certain critical length. Short day plants include *Chrysanthemum, Poinsettia,* and soybean. In the second class are the *long day plants* which will flower if the critical light period is exceeded. Periods of daylight up to 24 hours of light will cause these plants to flower. Examples of long day plants are lettuce, radish, spinach, and many varieties of tobacco. Other plants such as dandelion, sunflower, and tomato flower regardless of the day length and are called *day neutral plants.* Some plants require only a single inductive photoperiod to cause flowering whereas others require a series of inductive photocycles.

Since the phenomenon of photoperiod was described, research has indicated that the length of the light period is not the critical factor determining the response but rather the length of the dark period. Therefore plants should more accurately be classified as long night, short night and neutral.

In contrast to the gravity and phototropic responses, the photoperiodic stimulus is received, not in the meristem but in mature leaves, the meristem itself being insensitive to direct stimulation by photoperiod. In the leaves the photoreceptor is probably *phytochrome,* a chromoprotein, which exists in interconvertible active and inactive forms. The particular balance between the two is determined by the spectral quality of the light. In a plant that has been kept in the dark for a day the phytochrome is generally in the inactive form. Phytochrome in this form will not trigger the flowering response or seed germination. The inactive phytochrome is designated as P_R because it absorbs red light (660 nm). If red light is shined on the plant the phytochrome changes to its active form, P_{FR} which absorbs light in the far red region of the spectrum (730 nm). Phytochrome in this form triggers growth responses. If red light is followed by far red light the phytochrome shifts to the inactive form and growth responses triggered by the red light are turned off. Under natural conditions, when a plant receives both red and far red light at the same time, the state of the phytochrome is determined by the relative amounts of red and far red light.

Although it is known that the activation of phytochrome is the first step in the induction of flowering, the processes between the initial light responses and the shift in growth of the apical meristem are not known.

The change in the pattern of meristem growth is extreme. Cells in the central mitotically quiescent zone resume division and synthesize increased amounts of RNA and protein. The floral meristem first enlarges, then becomes progressively smaller as lateral organs are produced. Finally all of its cells differentiate as floral parts and meristem function ceases. The floral organs, the sepals, petals, stamens, and carpels, are initiated sequentially, and the meristem appears to pass through a succession of developmental stages in each of which a new kind of lateral organ is initiated.

SENESCENCE

We have already observed (p. 203) that the study of *aging* properly lies within the scope of developmental biology. Not only is it difficult to state just when "development" ends and "aging" begins—the ill-defined term maturity usually being the border between the two—but it is actually difficult to distinguish development and aging in the most fundamental sense. Evidence is slowly accumulating that aging is more than just an accumulation of physiological "insults," the

random deterioration of parts. We often read the charge that manufacturers are programming obsolescence in their products; it seems entirely possible that senescence is "built into" the genome. We have already seen that many cell populations undergo programmed senescence in normal development.

Some animal cells apparently are capable of infinite growth in culture. These cells are usually of "established lines," that is, those that have been maintained in culture for many cell generations. Not infrequently they are grossly abnormal in **karyotypes,** the karyotype being the characteristic chromosome complex of an animal or plant. It is often suggested that the stability of these cell lines is somehow the result of their abnormal chromosomes.

How do normal diploid cells in primary cell strains behave in culture? The evidence is conflicting. It is frequently said that diploid cells have a limited lifetime in culture, that there is a period of active proliferation, followed by a decline to a state from which cells cannot be subcultured. Human cells are usually cited as examples. Various strains of diploid human cells generally survive for 40 to 60 generations, the only "immortal" cell strains being those that have

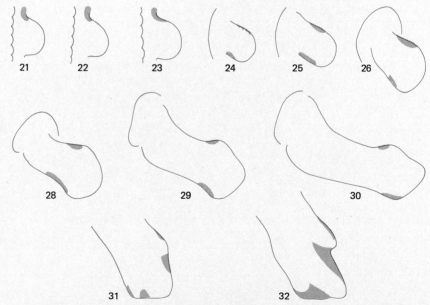

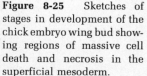

Figure 8-25 Sketches of stages in development of the chick embryo wing bud showing regions of massive cell death and necrosis in the superficial mesoderm.

undergone a change in chromosome number or composition. However, when one examines the literature carefully, he finds that the number of generations undergone by diploid human cells in culture varies widely with the medium used. Moreover several strains of diploid animal cells (derived from animals with shorter lifespans than man) are capable of subculturing for far longer periods.

At present, then, cell culture techniques have not provided compelling evidence for a built-in, finite lifespan for normal cell strains. However, in the intact animal, the deterioration of some functions with age clearly results from a decline in the number of cells that are capable of division. Especially noteworthy is the observation that old animals have a diminished capacity to make antibodies, and that their failure may be attributed to a decline in the *number* of cells capable of proliferating in response to stimuli of bacteria or other antigens.

Cell death in morphogenesis and senescence

One of the ways in which tissues and organs take shape is through differential cell death. Cell death is in fact a commonplace mechanism, as illustrated in the sculpturing of the limb. In an elegant morphologic and experimental study Saunders showed that the shaping of the chick wing was accomplished in part by the occurrence of localized zones of cell death. In particular, the shaping of the upper arm and forearm, especially the prospective elbow, and the elimination of tissues between the digits are effected in this manner (Figs. 8-25 and 8-26).

What is the utility of cell death? Is it mechanical? What is the fate of the breakdown products? We do not know. What factors control the onset of morphogenetic death? Its location in space? What "sets the death clock"? If a group of cells destined to die in the normal course of events at stage 24 in the wing of a chick are removed at stage 17 and grafted to the region of the somites, they die on schedule. If they are grafted to the dorsal side of the limb bud, they survive. However, if they are grafted later, at stage 22, they die in either site. These experiments suggest that although the "death clock" is set by stage 17, it can be turned off, by the imposition of external controls in an appropriate environment, up to stage 22. After stage 22, however, it proceeds inexorably.

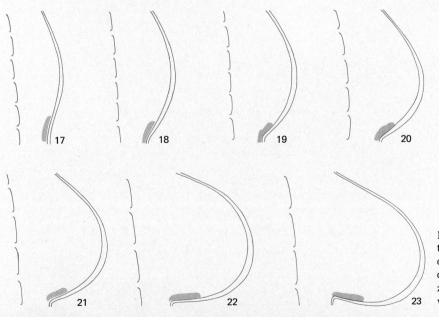

Figure 8-26 Gray area shows the approximate distribution of prospective degenerating cells of the posterior necrotic zone in the developing chick wing bud.

The necrotic zone of the wing bud ("determined" to die at stage 24) dies on schedule when cultured *in vitro*. However, death may be prevented by combining the necrotic zone with wing or leg mesoderm. In fact limb mesoderm prevents death even when the necrotic zone and the mesoderm are cultured by transfilter techniques. The filters preclude cell contact, but allow some "factors" to pass. Death does not occur in explants from embryos up to stage 20 cultivated with limb mesoderm of three- to four-day embryos. After stage 22 death is not prevented. Results are variable at stages 20 to 21.

How is the death sentence executed? In a number of systems, cells prepare for their own demise by producing or activating a battery of hydrolytic enzymes, to be used in breaking down their "remains." These enzymes are usually contained in distinct organelles, the *lysosomes*. However, in the limb, lysosomal enzyme activity does not seem to play an important role. The initial changes have not been recognized with certainty. Eventually, once general decay is advanced, cells and cell debris are engulfed by phagocytes.

CELLS OUT OF CONTROL

In his poem *Two Lives*, William Ellery Leonard wrote

> Of the cell, the wondrous seed
> Becoming plant and animal and mind
> Unerringly forever after its kind,
> In its omnipotence, in flower and weed
> And beast and bird and fish, and many a breed
> Of man and woman, from all years behind,
> Building its future.

These lines reveal the beauty and excitement of development, but they also express two of the principal themes of developmental biology which have coursed through the foregoing pages. They speak of the impact of genetics on development, "unerringly forever after its kind"; and of the scope of development, "in flower and weed and beast and bird and fish and many a breed of man and woman." It is the tremendous scope of development, the opportunity to explore its mechanisms over a wide range of animals and plants that is in large part responsible for the emergence of developmental biology as a focal field of research.

Indeed the scope is even wider than we have suggested, for we have barely touched upon abnormal development. The phrase *"Unerringly forever . . ."* is poetic license, for cells out of control and errors of development loom large in our consideration of the social impact of biology.

Congenital defects

Congenital defects are one of our more important unmet medical problems. They are exceeded only by childhood accidents as a cause of crippling and disability in children. They cause the death of 35,000 to 40,000 infants—one-fourth of all infants in the U.S. who are stillborn or die within the first month of life during an average year.

Until recently this field has been almost completely neglected. It is now at a stage in development where organized, concentrated, total efforts are bringing about improvement in care, substantial rehabilitation, and eventual prevention of a good many of the conditions that come under this general heading.

Each year about 250,000 children are born in the United States with significant birth defects. In about half of these children the condition leads to serious illness, to crippling, or to chronic disease. Combinations of defects in one child are not unusual.

The incidence of major anatomic defects is about 1.5 percent; and the total incidence of defects is estimated to be 5 to 6 percent. In short, one out of every 16 to 20 infants presents some defect. Some birth defects are widely known and recognized. Among them are cleft lip and palate, clubfoot, missing extremities, congenital cataracts, and feeblemindedness. There are over 600 different birth defects which, while not so familiar, are nonetheless serious medical problems.

Congenital defects are conditions due to factors acting *before* birth, not injuries occurring during the act of birth. The causes of defects are broadly speaking genetic or environmental, and we consider them briefly under these headings, bearing in mind that genes do not act in the absence of their environment.

The need of genic materials for development.
Earlier we learned that the nucleus is essential for development. Is it essential that there be a full haploid complement of chromosomes? Again there is ample evidence for an affirmative answer. Boveri showed that sea urchin eggs in which certain chromosomes were lacking failed to develop properly. In that classic object of study, *Drosophila melanogaster*, the necessity of having each of the four kinds of chromosomes represented in the fertilized egg has been proved. For example, one X chromosome must be present; without it (although there are a few cleavages), development stops; differentiation and morphogenesis fail to occur.

Muller's hypothesis of *dosage compensation*, formulated in 1932, held that the expression of genes on the X chromosome is the same in females and males although there are twice as many loci in the female. This hypothesis required essentially that one X chromosome in the *female* be inactivated or suppressed.

In 1949 Barr observed that in female mammals, including the human species, the two X chromosomes differ markedly in appearance. One is a "typical" chromosome; the other is *heteropycnotic*, that is, condensed and deeply staining. In the interphase nucleus, when the typical, fully extended chromosome cannot be seen by light microscopy, the heteropycnotic chromosome appears as a clearly defined chromocenter, which is now called the *sex chromatin* body. Thus cytologic evidence first suggested that one X chromosome in the female may be inactive, a conclusion later confirmed by autoradiographic methods which reveal it to be inactive in nucleic acid synthesis. At the outset of development, however, both chromosomes are active. In rabbits no X-derived chromatin body can be seen in two-day embryos, but it is observed at four and a half to five days. In the hamster embryo both X chromosomes are still active at the eight-cell stage. Hence one X chromosome is modified during early development, becoming inactive before implantation. Is the same X chromosome destined to be inactivated in each cell? Or is it a matter of chance?

Evidence from genetic studies on mammals, first brought together effectively by Lyon, favors the latter possibility. We know that certain traits are sex-linked. For example, in the mouse there are many sex-linked coat color mutants. Females that are heterozygous for these genes show *variegation*, their coats exhibiting patches of both mutant and wild-type coloration. In interpreting these observations, Lyon proposed that the genetic inactivation of one X chromosome must occur early in development paralleling the cytologic changes and that the inactivation must be random. After inactivation occurs in a cell, all its progeny would bear the same inactive loci. Thus if it arises from several cells present at the time of the first appearance of sex chromatin, a tissue should show variegation. In fact, analysis of X-linked markers should make it possible to estimate the number of progenitor cells from which a given cell type is derived. For example, use has been made of X inactivation in cell-lineage studies on populations of red blood cells. A number of human females heterozygous for the X-linked gene for the enzyme glucose-6-phosphate dehydrogenase have been found to contain hemizygous red cells; that is, their pooled red cells have enzyme levels no higher than those of deficient males (having only one X chromosome that contains the mutant gene for this enzyme). It has been estimated that more than 1 percent of all heterozygous females have red cells that are hemizygous. Assuming a random inactivation of X chromosomes it has been estimated that eight or fewer cells were *determined* to become red cells when X inactivation took place. Experiments were carried out to decide whether the inactivation occurred at a time when there were only eight cells in the embryo or eight red-cell progenitors. The latter alternative was shown to be more likely by demonstrating that other tissues such as skin and subcutaneous fat had enzyme levels characteristic of heterozygotes in the same individuals who had low (hemizygous) levels of enzyme in their red cells.

Studies of this kind complement those using allophenic mice. Thus the combination of a pigmented and an albino embryo produced the coat-color pattern shown in Figures 8-17 and 8-18. Mintz has interpreted each band on each side as a clone descended from one cell determined to be a melanoblast (prospective pigment cell). Thus the cells for coat color in the mouse are derived from 17 cells on each side (34 in all), which are determined as melanoblasts at an early stage.

In allophenic mice 50 percent of the fused animals are expected to contain a mosaic population of XX and XY cells and therefore could be expected to be hermaphrodites (individuals having both male and female organs). However, hermaphrodites have been found in only a very low percentage. The reason for the low incidence of hermaphrodites is unknown, but it may be related to selection of one kind of gamete during the proliferative phase. However, the presence of mosaic populations of germ cells in allophenic mice can easily be demonstrated by including autosomal markers in the embryos used in the original fusion. Allophenic mice were produced by fusing an embryo carrying a dominant coat-color gene with one carrying a recessive gene. The allophenic mouse was then crossed to a normal mouse with the recessive coat-color gene. Both color types were found in the progeny, showing the allophenic mouse to be mosaic in its germ cells. This finding shows that there must be two or more cells that are independently determined to be germ cells in the mouse. When germ cells can first be cytologically identified in the mouse embryo, they are ten in number. These two facts show the initial number of primordial germ cells to be between two and ten.

Chromosome imbalance.   It is a curious fact that although one X chromosome appears to be inactive, abnormalities are found when one chromosome is missing or when there is an imbalance in the diploid condition. One of the better known examples is found in the human species. When there is a deletion of an entire sex chromosome —the X chromosome being unaccompanied by either another X or Y chromosome (XO), so that the individual has only 45 chromosomes (Fig. 8-27)—the gonads fail to develop. The individual is female, but the rudimentary gonads consist only of connective tissue. The theoretically possible YO condition has never been observed in man, since this combination is probably incompatible with life.

What effects are produced by the presence of one or more *extra* chromosomes? Again several good examples have been described, one being the combination XXY in man. In these individuals, prenatal and postnatal development is normal until puberty, except for a reduction in spermatogonia. Then the seminiferous tubules atrophy, the testes remain small, and the individual becomes eunuchoid.

Another classic example is found in **Langdon Down's syndrome** or *trisomy 21* (formerly inappropriately called *mongolism*). It is now known that most individuals who are in this condition of mental retardation have one supplementary auto-

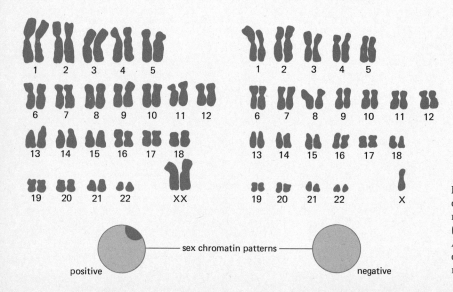

Figure 8-27 Chromosome complements and sex chromatin patterns in humans. (Left) Normal female. (Right) A patient with an XO sex chromosome complex (gonads remained rudimentary).

some, their total chromosome number thus being 47. This discovery was a milestone in the search for the origin of the defect. The earlier evidence had suggested a hereditary basis; if one of identical twins is affected, the other is also; in contrast, only very rarely would both dizygotic (nonidentical) twins be affected.

Defects produced by specific mutations. Large aberrations of genetic material are not required to alter the course of development. Discrete point mutations and deficiencies of single genes, or translocations of small blocks of genes, may produce marked changes in development. Rarely do we know the initial molecular "lesion," that is, the synthetic reaction first affected by the altered or absent gene. However, we have obtained some insight into the mechanisms by which defects are expressed morphologically. Let us begin with a familiar example. How might specific genetic defects be expressed in tissue interactions?

1. *Distortion of the inductive pattern.* **Polydactyly** (more than the normal number of digits) may result from an atypical distribution of the factor produced by limb mesoderm and required for maintaining the limb ectoderm, it being well known that tissue interactions are required for normal development of the limb (Fig. 8-28).

2. *Failure of the interactants to make proper contact.* An example of this type of failure is *anophthalmia* (eyelessness). The developmental events that precede eyelessness in certain strains of mice are the following. In the extreme expression the optic vesicle forms but does not make contact with the overlying ectoderm in which the lens normally is induced (page 198). As a consequence a lens does not form and the vesicle remains quite rudimentary. In some cases a reduced optic vesicle does make contact with the ectoderm and induces a small lens. It is presumed that eye development fails because the interacting tissues fail to make contact; however, the inducing and responding capacities of the two tissues are unimpaired.

3. *Loss of inductive capacity.* The wingless mutant in the chick is an example of loss of inductive capacity. Wing buds are formed, and develop for a time; then development stops.

If the tissue layers of normal and *wingless* limb buds are dissociated by the techniques we have already described, and the normal ectoderm combined with *wingless* mesoderm, the mesoderm begins to respond to the influence of the ectoderm. Shortly, however, the normal ectoderm begins to degenerate and outgrowth stops. Thus the normal ectoderm appears to depend on the mesoderm for its maintenance. In fact the existence of a *maintenance factor* has been postulated. It has not been identified.

4. *Loss of ability to respond.* In certain **brachyuric** (short-tailed) mice, part of the offspring have gross defects in the posterior regions of the body. Let us consider those in which the somites are grossly abnormal. Normally the longitudinal, paraxial mesoderm becomes segmented into paired cubical masses, or somites. In their further development, including the formation of cartilage, which is one of their principal roles, cells in the somites must interact with the adjacent neural tube. It has been shown that somites from the mutant strain cannot form cartilage when placed in contact with spinal cord; spinal cord from the mutant, however, can induce cartilage formation from genetically normal somites.

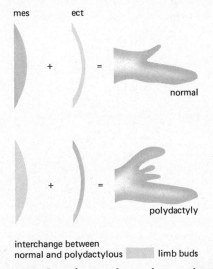

Figure 8-28 Interchange of mesoderm and ectoderm between genetically normal and polydactylous limb buds. The polydactylous condition develops from the combination of mutant mesoderm and normal ectoderm.

Environmental effects. In the course of evolution man has evolved a remarkable set of mechanisms for protecting the embryo. It is a subject about which we know far too little. It is clear, however, that the barriers may be breached, at times unexpectedly. How might one have predicted that rubella (or German measles) virus would be highly teratogenic (that is, it produces a large number of congenital malformations of eye and heart, to name only two "targets") when women contracted the disease during the first two to three months of pregnancy? Rubella is a "mild" virus for the child or adult, but highly damaging to the embryo. In contrast, more "severe" viruses, for examples, measles or mumps, with more serious consequences for children and adults, are not teratogenic.

We still do not understand the mechanism of action of thalidomide in producing major malformations of limbs and other organs in human embryos. This relatively mild sedative and tranquilizer had grave consequences when it was administered to mothers in the early months of pregnancy. The effects can be reproduced in some experimental animals, but only with great difficulty and by using extraordinarily high doses.

This fact points up the great difficulty we face in evaluating the vast stores of new products marketed as foods or drugs or poured into our environment almost daily. How do we evaluate potential mutagenic, teratogenic, or tumorigenic compounds? Thalidomide might not have been "caught" even in rigorous "screening" using experimental animals. Will we have to draw upon newer techniques of developmental biology, using clones of human cells as test systems?

Cancer

When we speak of "cells out of control," we most often think of one of mankind's greatest scourges —cancer—a general term for a variety of disorders characterized by the uncontrolled growth of relatively undifferentiated cells. These uncontrolled cells may exist as a confined mass, or **tumor;** they may circulate in the bloodstream (**leukemias**); or they may diffusely invade many tissues and organs (cancer in the strict sense).

There is no reason to assume that these disorders arise from a common cause. For many forms of cancer the factors that remove the controls on cell growth are unknown. For some, however, the causes are clear.

Many forms of cancer in animals represent the response to invasion by **viruses.** The viral origin of some cancers was identified a generation ago: leukosis in chickens, mammary tumors in mice caused by a milk-borne virus, and papillomas (skin tumors or warts) in rabbits. In mice a wide variety of tumors result from infection by the polyoma virus. Both DNA and RNA viruses may be tumorigenic. As of late 1972 viruses have not been indisputably identified as causes of human cancers, but they are the object of an intensive search in many laboratories around the world.

Other forms of cancer result from the action of *chemical carcinogens,* notably the polycyclic hydrocarbons. Observations in laboratory animals abound; and the extension of these observations to man is equally compelling. Cigarette smoke does contain chemical carcinogens. Heavy cigarette smokers do incur a high risk of cancer. Polycyclic hydrocarbons are found in the air, in emissions from the burning of fossil fuels, especially coal, but oil as well. The risk of cancer in highly industrialized areas with high levels of air pollution is higher than it is in the cleaner countryside.

Cancer research encompasses a range of activities spanning nearly the whole of biology and much of chemistry as well. Our understanding of normal differentiation and growth, as well as of abnormal growth, has been enhanced through cancer research. We limit our discussion to three examples.

Cell transformation is heritable. Cells infected with oncogenic viruses *in vitro* undergo heritable changes in form and growth properties. In the absence of the normal growth controls usually exhibited in monolayer cultures, cells pile up and form foci. The cells of these foci have the ability to form tumors when they are injected into appropriate hosts. Cells which have been modified in this fashion and can transmit these new properties to their progeny are called *transformed* cells.

Cells in clonal cultures, being suitable for genetic study and for biochemical research on a large scale, are used extensively for studies of neoplastic change. Both chemical and viral carcinogenesis may be studied *in vitro*. There is general agreement that such transformations are *heritable*. In regard to the mechanism of inheritance two principal ideas have emerged. According to the first, cell transformation results from a *qualitative* change in genetic information. Nucleic acids of infecting viruses insert new genetic information into cells. Chemical carcinogens, being mutagens, would, according to this view, also act by mutating the cellular genome, thus imparting *new* information.

Proponents of the second view hold that all of the information required for transformation is contained in the cell, and that tumor viruses, or other carcinogens, modify its expression. Thus transformation would result from variations in cellular control mechanisms. Accordingly the specific properties observed in transformed cells would reflect the expression of genes previously repressed or of products previously masked.

It is clear that oncogenic (tumor-causing) viral genes ("oncogenes") exist within "normal" cells. It has been said, only half in jest, that these "oncogenes" might well be called "archeo-genes" —genes introduced into the cell by viral infection eons ago and transmitted from generation to generation. In theory, then, viral and chemical carcinogenesis are reduced to a common denominator, the unmasking of an "oncogene." Unfortunately the oncogene hypothesis is unsatisfactory, for it cannot be disproved.

Biochemical "markers" of transformation.   Our definitions of transformation depend almost entirely on morphologic changes. There are few "exclusive" biochemical signs of transformation. One remarkable discovery, made independently by Baltimore and Temin, was the demonstration of a "reverse transcriptase," or "RNA-dependent DNA polymerase" in RNA-containing tumor viruses. However, the enzyme has been shown to exist in normal cells as well. Despite this disappointment the discovery remains highly significant. We have grown accustomed to the idea that DNA is transcribed to RNA; now it is clear that an RNA virus may reproduce itself by first forming a DNA intermediate provirus; to do so RNA is transcribed to DNA!

Another significant set of discoveries is centered about the observation that some cancer cells at least contain specific "marker" substances or antigens (so-called because they elicit specific antibodies and are recognized immunologically). Such antigens may be uncommonly useful in early diagnosis.

For example, naturally occurring tumors of the human colon contain a specific antigen also found in embryonic digestive tissues but lacking in the normal adult colon. This "carcinoembryonic" antigen can also be detected in the blood sera of patients with cancer of the colon and rectum. Thus it should be possible to screen patients suspected of having such cancers by blood tests.

Cancer chemotherapy.   The goal of cancer therapy is to reduce the rate of growth in such cells and, because the lesion is heritable within the affected cell line, to obliterate all such offending cells.

For solid tumors early surgical removal is the procedure of choice whenever possible. But distant metastases necessarily escape this procedure. Most other procedures now current rest on the fact that such a cell line engages more frequently in DNA synthesis than do normal cells. It is this process that makes them highly sensitive to X-rays. Frequently, however, irradiation cannot be used successfully. Such therapy necessarily damages the surrounding tissue; the beam can be narrow or broad but necessarily goes through healthy tissue, does not affect unsuspected metastases, and can be used only with difficulty in leukemia involving much or all of the bone marrow.

Accordingly research in cancer chemotherapy has been diligently pursued for two decades. To date only a few triumphs have been recorded. They stem from a knowledge of cellular metabolism. Folic acid is a vitamin required in the human diet. It serves as a coenzyme required for synthesis of the pyrimidine, thymine, which is required for synthesis of DNA. It was suggested that an antimetabolite of this vitamin should impair thymine synthesis, hence that of DNA, hence limit cell growth. A structural analogue of folic acid, methotrexate (amethopterin), has proved to be clinically useful. In early trials partial remissions were found in childhood leukemias. When tested in choriocarcinoma, an all too frequent invariably fatal cancer of young women of childbearing age, this compound effected cures in 50 percent of all patients.

Leukemias are also held in check for months or years by fluorinated uracil derivatives, particularly 5-fluorouracil and its riboside, 5-fluorouridine. Both are converted to 5-fluorouridylic acid, which then specifically inhibits the enzyme responsible for the final step in the normal biosynthesis of the thymine required to make new DNA. These compounds have shown some striking successes in some leukemias. In addition, they cause regression of lesions in perhaps 20 percent of all instances of carcinoma of the colon, rectum, breast, or liver.

The treatment of viral diseases by drugs has been singularly unsuccessful; the only useful measure has been immunization prior to infection. Recently it was discovered that the compound 5-iododeoxyuridine (iodoxuridine), an analogue of the thymine of DNA, inhibits the reproduction of DNA-containing viruses. When tested as an anticancer agent, it was discarded because of its undesirable effects on normal tissues. However, it has become the definitive treatment for a viral infection of the eye (herpes keratitis), previously a major cause of blindness in the United States, and can be administered with impunity in eye drops. This is but one of many examples of a valuable "spinoff" from cancer research.

SUMMARY

In animal development each differentiating tissue cell has its own inner controls. Yet in its development the cell is part of a larger whole; during its differentiation the cell must respond to control factors extrinsic to it. Thus one of the large tasks of developmental biology is to identify those external controls and understand the ways in which they impinge on the cell's inner controls.

In examining the external controls in animal tissue interactions, we have found first that they involve intimate association among cells. This association does not require "contact" in the mechanical sense of immediate juxtaposition and touching of surfaces; it does require that the cells communicate in a common microenvironment. This communication may take many forms. We have concluded that it would be unwise to exclude small molecules as agents in animal development, for under the proper conditions they could operate as substrates in differentiation triggered by the induction of an enzyme, or in feedback inhibition, or in repressing an operator. Large molecules contain more "possibilities" for accounting for changing specificities. But large or small, the agents act over short distances and are probably inactivated rapidly. Development requires mechanisms not only of synthesis but of degradation. Throughout this chapter we have stressed the sequential nature of development. All of the schemes proposed rely heavily on that premise. Once a given reaction in the sequence is initiated experimentally, the normal train of events follows unbroken.

In plants the meristems possess a high level of developmental independence. Given a supply of basal nutrients the apical meristems of both shoot and root are capable of normal development when grown in isolation from the rest of the plant, and are essentially autonomous.

The way in which the meristem regulates its own mitotic activity and the development of tissues and organs to which it gives rise is still not thoroughly understood. Because of the small sizes of the interacting parts, it has not been possible in most cases to determine whether the interaction is chemical or has some other cause. However, the evidence so far suggests that induction as it occurs in animal cell differentiation does not play a major role in plant development. The chemicals which control plant development are more likely to be hormones with molecular weights in the hundreds than to be macromolecules. Whether this is another consequence of the cell wall which increases the space between plant protoplasts and which may act as a molecular sieve to restrict the movement of large molecules is not known, but it provides an interesting basis on which to speculate about the different control mechanisms in the two groups of organisms.

Aging appears to be more than the random deterioration of parts. Senescence may be in large part "programmed." Cell death is a morphogenetic mechanism even during animal embryogenesis.

Congenital defects result from both internal and external causes—chromosomal aberrations and gene mutations on the one hand, and environmental factors, for example, viruses and drugs, on the other.

Cancer-cell transformation by viruses and chemical carcinogens is heritable. Current research is focused on understanding the mode of action of tumor viruses and carcinogens in order to make prevention possible. The existence of specific tumor antigens makes it increasingly likely that methods of early diagnosis can be developed for a number of cancers.

Successes in cancer chemotherapy have been limited thus far, but are sufficiently promising to warrant intensifying the search.

SUGGESTED READING LIST

Birnstiel, M., J. Speirs, I. Purdom, and K. Jones, "Properties and composition of the isolated ribosomal DNA satellite of *Xenopus laevis*," *Nature*, **219**, 454 (1968).

Bogorad, L., "Control Mechanisms in Plastid Development," in *Control Mechanisms in Developmental Processes*, M. Locke (ed.). Academic Press, New York, 1967, p. 1.

Bonner, J., M. E. Dahmus, D. Fambrough, R. C. Huang, K. Marushige, and D. Y. H. Tuan, "The biology of isolated chromatin," *Science*, **159**, 47 (1968).

Britten, R. J., and E. H. Davidson, "Gene regulation for higher cells: a theory," *Science*, **165**, 349 (1969).

Brown, D. D., and I. B. Dawid, "Specific gene amplification in oöcytes," *Science*, **160**, 272 (1968).

Coleman, J. R., and A. W. Coleman, "Muscle differentiation and macromolecular synthesis," *Journal of Cell Physiology*, **72**, suppl. 1, 19 (1968).

Dawid, I. B., and D. R. Wolstenholme, "Ultracentrifuge and electron microscope studies on the structure of mitochondrial DNA," *Journal of Molecular Biology*, **28**, 233 (1967).

Ephrussi, B., and M. C. Weiss, "Regulation of the cell cycle in mammalian cells: Inferences and speculations based on observations of interspecific somatic hybrids," in *Control Mechanisms in Developmental Processes*, M. Locke, (ed.). Academic Press, New York, 1968, p. 136.

Grobstein, C., "Cytodifferentiation and its controls," *Science*, **143**, 643 (1964).

Gurdon, J. B., "Nuclear transplantation in Amphibia and the importance of stable nuclear changes in promoting cellular differentiation," *Quarterly Review of Biology*, **38**, 54 (1963).

Holtfreter, J., "Tissue affinity, a means of embryonic morphogenesis," 1939. Reprinted in B. H. Willier and J. M. Oppenheimer (eds.), *Foundations of Experimental Embryology*. Prentice-Hall, Englewood Cliffs, New Jersey, 1964, p. 186.

Humphreys, T., "Chemical dissolution and *in vitro* reconstruction of sponge cell adhesions. I. Isolation and functional demonstration of the components involved," *Developmental Biology*, **8**, 27 (1963).

Konigsberg, I. R., "Clonal analysis of myogenesis," *Science*, **140**, 1273 (1963).

Mintz, B., and W. W. Baker, "Normal mammalian muscle differentiation and gene control of isocitrate dehydrogenase synthesis," *Proceedings of the National Academy of Sciences*, **58**, 592 (1967).

Moscona, A. A., "Aggregation of Sponge Cells: Cell-Linking Macromolecules and Their Role in the Formation of Multicellular Systems," *In Vitro*, **3**, 13 (1968).

Novikoff, A. B., and E. Holtzman, *Cells and Organelles*. Holt, Rinehart and Winston, New York, 1970.

Rawles, M. E., "Tissue interactions in scale and feather development as studied in dermal-epidermal recombinations," *Journal of Embryology and Experimental Morphology*, **11**, 765 (1963).

Rutter, W. J., J. D. Kemp, W. C. Bradshaw, W. R. Clark, R. A. Ronzio, and T. G. Sanders, "Regulation of specific protein synthesis in cytodifferentiation," *Journal of Cell Physiology*, **72**, suppl. 1, 1 (1968).

Steinberg, M. S., "The problem of adhesive selectivity in cellular interactions," in *Cellular Membranes in Development*, M. Locke (ed.). Academic Press, New York, 1964, p. 321.

Steward, F. C., M. O. Mapes, A. E. Kent, and R. D. Holsten, "Growth and development of cultured plant cells," *Science*, **143**, 1 (1964).

Temin, H., "RNA directed DNA synthesis," *Scientific American*, **226**, 24 (1972).

Vasil, V., and A. C. Hildebrandt, "Differentiation of tobacco plants from single isolated cells in microcultures," *Science*, **150**, 889 (1965).

Wilt, F. H., "The control of embryonic hemoglobin synthesis," in *Advances in Morphogenesis*, Vol. 6, M. Abercrombie and J. Brachet (eds.). Academic Press, New York, 1967, p. 89.

Yoshikawa-Fukada, M., and J. D. Ebert, "Interactions of oncogenic viruses and animal cells," *Bioscience*, **21**, 357 (1971).

Part Three

Why Organisms Behave As They Do

What factors determine the "life styles" of the earth's creatures? Here we explore the form, function and behavior of different groups in relation to their ecological roles.

The Versatile World of Microbes

With the exceptions of cultivated plants and domesticated animals, the microbes, especially bacteria, have had a more immediate effect on man's welfare than any other group of organisms. For thousands of years we have usefully employed the activities of some of these creatures and for even longer have suffered the ravages of diseases that some others cause. Yet it was little more than 300 years ago that microbes were first seen, and only about 150 years ago that biologists first began to study them seriously. This research has yielded many practical results: better wine and pickles and fewer diseases, for examples. More importantly we have learned of the great diversity of these organisms that we cannot even see with our eyes and of their profound significance to the economy of nature. But perhaps the most important result of our understanding of how microbes live has been the realization of the fundamental unity of living things which underlies their obvious diversity. This realization is not only of supreme importance to biology but has also profoundly affected man's view of himself and of his relationship to the rest of the universe.

EUCARYOTES, PROCARYOTES, AND VIRUSES

As we observed earlier, living creatures can be divided into three major groups: eucaryotes, procaryotes, and viruses. Eucaryotes include the organisms we are familiar with from everyday observations: ourselves, animals, and plants, as well as many other creatures too small to be seen with the naked eye, such as protozoa and many algae and fungi. Much of this book is devoted to eucaryotes. Procaryotes include bacteria and blue-green algae. The viruses include only viruses. This chapter deals with the biology of procaryotes and viruses.

Both eucaryotes and procaryotes are cellular organisms. This condition is signaled by the occurrence of both DNA and RNA in an organism. We have already learned that DNA is genetic material; in other words, the information for constructing an organism is encoded in its DNA. We have also learned that RNA is an essential part of the machinery which translates the genetic code into protein and hence into an organism. In contrast a virus contains *either* DNA *or* RNA (depending on the virus) *but not both*. Regardless of which nucleic acid is present, its sole function is genetic. Viruses have no machinery for translating the information in their genetic material into protein; they make use of the machinery of cellular organisms.

THE ROLES OF BACTERIA IN NATURE

Before looking at procaryotic organisms in detail it is well to consider broadly their roles in nature, because this accounts for many of their features. To do this we must anticipate some of the material in Chapter 16 (The Web of Life).

All life on earth depends on the ability of green plants to use the energy of sunlight to convert carbon dioxide and water into plant material and oxygen. The procaryotic blue-green algae share this ability with green plants. With very few exceptions all other creatures live by eating plants and degrading part of this material to obtain energy (ATP) which they use to convert the rest of the plant material into their own bodies. The mechanisms of these conversions have already been discussed in Chapters 4 through 6. A major share of the energy required by creatures that eat plants is obtained by the respiration of organic compounds (for instance, sugars) with oxygen to carbon dioxide and water. These organisms thus

use part of the energy of sunlight that plants originally stored in the form of organic compounds and oxygen. Everyday observation tells us that only some organisms eat plants themselves; many organisms live by eating other organisms that have eaten plants. Nevertheless the overall result is the same: the products of the activities of plants, namely plant material and oxygen, are converted in part to other organisms and in part to the substances plants use, that is, carbon dioxide and water. Clearly in the long run all the CO_2 fixed into plant material must be converted back to CO_2 if life is to continue, since the amount of carbon on earth is fixed.

As will be seen in Chapter 16 similar cycles exist for the other chemical elements in living things such as nitrogen, sulfur, and phosphorus. The conversion of organic compounds to inorganic ones (such as CO_2, NH_3, and SO_4^-) is termed *mineralization*.

Carbon and the other elements occur in a vast array of different compounds in living systems, as we have already seen in a small sampling in Chapters 4 and 5. A simple argument will be convincing that all the carbon in the organic compounds formed by organisms must eventually be converted back to carbon dioxide. Imagine what would happen if some one compound could not be mineralized. The compound would be made continuously and accumulate in the environment; year by year (or century by century) a larger and larger fraction of the available carbon (and nitrogen, sulfur, and so on, if the compound contained these elements) would become unavailable to plants and thus to other organisms. It follows that organisms must exist which can, either singly or in concert with other organisms, mineralize every naturally occurring compound.

Bacteria are the principal agents of complete mineralization of organic compounds. This implies that the bacteria must possess a repertory of metabolic abilities as wide as the chemical diversity of biologically produced substances. We shall see some examples later in this chapter. Some bacteria are able to mineralize a very restricted range of compounds; for example, certain bacteria can respire and grow on methane (CH_4) but can utilize almost no other compound. On the other hand, other bacteria can grow on any one of some 100 different compounds.

The metabolic diversity of such bacteria implies a highly plastic phenotype, which can be seen from the following considerations. From studying metabolism it can be appreciated that a large number of specific enzymes is necessary to metabolize 100 different organic compounds. Many of the compounds will be present in the environment only sporadically and it would be wasteful of the bacterium to make all the specific enzymes all the time. We expect, therefore, that many enzymes will be made only when their substrates are present in the environment. We examine several such examples later in the chapter.

Isolation of pure cultures

An obvious prerequisite to the study of any organism is to be able to study it in isolation from other organisms or at least from other kinds of organisms. This is easy in the case of macroscopic creatures; if one wants to investigate the anatomy or physiology of a daisy one simply picks a daisy. To isolate organisms that can be seen only under a microscope is a little more difficult. What is wanted in fact is a population of cells that is known to have come from a single cell. Such a population is called a *pure culture*. The technique almost universally used is based on the fact that when a single cell of, for example, a bacterium is deposited in or on a medium that has been solidified, the cell can grow and divide but cannot move, even though it may be motile in liquid. The progeny of the cell will therefore remain in the same area as the original cell. Continued growth of the cells leads in a reasonably short time to a mass of cells, called a *colony*, which is visible to the eye. In practice if a liquid containing several kinds of bacteria is spread on the surface of a solid medium, isolated colonies will appear and in general each colony will be a pure culture.

THE PROCARYOTIC CELL

Historically the study of procaryotes has been part of the field of microbiology, which is the study of organisms invisible to the naked eye. Although all procaryotes are small, not all small organisms are procaryotes; hence microbiology includes both eucaryotes and procaryotes. As we shall see, the distinction between these two groups is fundamental and far-reaching; it is certainly more fundamental than the distinction between organisms that are visible and those that are not visible to the unaided human eye. It makes more sense to treat procaryotes and eucaryotes separately, which is what we shall do here. The microscopic eucaryotes are covered in other portions of this book in conjunction with larger and more complex ones.

Procaryotes are separated from eucaryotes on the basis of a half dozen or so differences in the basic architecture of their cells. Therefore carefully review the material in Chapter 2 and compare it with the following description of a procaryotic cell. We shall describe an imaginary procaryotic cell to illustrate the features which are unique to them. A sketch of this cell is shown in Figure 9-1; electron micrographs of real procaryotes appear in Figures 9-2 and 9-3.

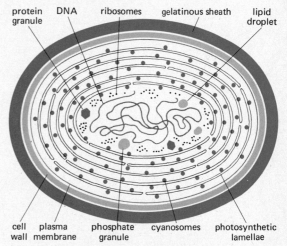

Figure 9-1 Diagram of a blue-green alga cell. As in all procaryotes the DNA is not segregated into a nucleus; no nuclear envelope is observed; nor have an endoplasmic reticulum, Golgi apparatus or mitochondria been found. As in bacteria, ribosomes and polysomes are "free." Cyanosomes are particles of a pigment called phycobilin.

Genetic material

The DNA of a procaryote is not separated from the cytoplasm by a nuclear membrane as it is in eucaryotes. It appears that the genetic information necessary to specify a procaryotic cell is contained in a single molecule of DNA with a molecular weight of the order of 10^9, which amounts to about 1.5×10^{-15} gram. The complete genetic information of a creature is called its *genome;* hence the genome of a procaryote is a single molecule. The sizes (molecular weights) of procaryotic genomes range from about 10^8 to 10^9. From what we have already learned about DNA and the genetic code, we can calculate that a genome with a molecular weight of 10^9 contains approximately

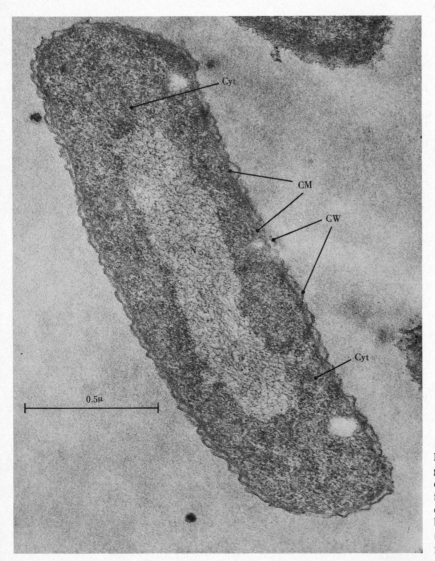

Figure 9-2 Electron micrograph of the bacterium *Escherichia coli*. The easily visible features are the DNA (N), the cytoplasm (Cyt), the cell membrane (CM), and the cell wall (CW). Note the absence of a membrane around the DNA.

$10^9/600 = 1.7 \times 10^6$ base pairs and can code for a maximum of $1.7 \times 10^6/3 = 6 \times 10^5$ amino acids in protein. A "run-of-the-mill protein" contains from 100 to 400 amino acids. Hence the DNA of a typical procaryote can specify between 2000 and 6000 different proteins. This is *not* a very large number.

The fact that the entire genome is a single molecule makes it unnecessary for a procaryote to have anything resembling mitosis. Recall that the purpose of mitosis is to ensure that at cell division each daughter cell receives a full set of chromosomes, that is, that it receives a complete genome. The machinery necessary to distribute a complete

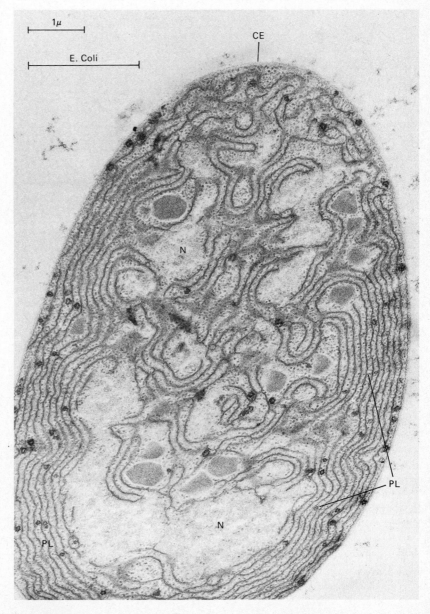

Figure 9-3 Electron micrograph of the blue-green alga, *Anabaena azollae*. The most striking feature is the photosynthetic membranes (PL) seen throughout the cell. Note the absence of a membrane separating these membranes from the cytoplasm. The DNA-containing region (N) is also not surrounded by a membrane. The cell envelope is clearly depicted in some regions (CE).

genome to each daughter cell can be considerably simpler if the genome is a single molecule. We return to the replication and distribution of DNA in procaryotes later.

Although the DNA is not separated by a membrane from the cytoplasm of procaryotes, it is not simply diffused throughout the cytoplasm but is sharply separated from it. This is clearly illustrated in Figure 9-2. The forces that maintain DNA as a distinct phase are unknown.

The cytoplasm and ribosomes

Many features of the behavior of eucaryotic cells indicate that the cytoplasm is at least partly liquid. Such diverse phenomena as amoeboid movement, pinocytosis, and cytoplasmic streaming all lead to this conclusion. Procaryotic cells do not show these features and it can be concluded that the cytoplasm is permanently in the form of a gel. This means that the only way of moving chemicals from one part of a cell to another is by passive diffusion rather than active mixing as can occur in eucaryotic cells. The various control systems encountered in Chapters 6 and 8 are all based on chemical interactions. For example, the repressor of a particular enzyme must diffuse from where it is made (on a ribosome in the cytoplasm) to where it works (a specific site on the DNA). Diffusion is a relatively slow process. It can be calculated, for example, that a protein molecule of molecular weight 20,000 will move about 1 micrometer through water in about 10 milliseconds (0.01 second). In jellylike cytoplasm, diffusion is slower than in water and it will take several tens of milliseconds for such a protein to move 1 micrometer, which is about the size of many procaryotic cells. This time may seem short to us, but it is not short compared to the rates of metabolic reactions. For example, in a rapidly growing cell only 25 seconds are needed to synthesize a protein molecule of molecular weight 40,000; in other words, an amino acid is added to the growing protein every 10 milliseconds or so. It seems possible, therefore, that the lack of active mixing of cytoplasm in procaryotes sets an upper limit to the size of the cell beyond which chemical communication between parts of

the cell would become too slow for the various control processes to function properly.

The ribosomes of procaryotes differ in several respects from those found in the cytoplasm of eucaryotes. First, they are never found associated with an endoplasmic reticulum because procaryotes do not have an endoplasmic reticulum. Second, procaryotic ribosomes are somewhat smaller, having molecular weights of around 2.7×10^6, whereas eucaryotic ribosomes have molecular weights of about 4×10^6. Third, procaryotic ribosomes are sensitive to a variety of antibiotics which do not affect eucaryotic ribosomes.

Mitochondria and chloroplasts (page 41) contain ribosomes and, interestingly, these are similar to procaryotic ribosomes. Thus we say that eucaryotes have two kinds of ribosomes, and procaryotes have only one kind.

The cell membrane

The cytoplasm of procaryotes, like that of all cells, is delimited by a cell (or plasma) membrane. The chemistry and structure of procaryotic and eucaryotic cell membranes are, so far as is known, very similar. As in eucaryotes the fundamental function of the cell membrane in procaryotes is to control the passage of chemicals between the cell and its environment.

In addition to the cell membrane, eucaryotic cells have a variety of other membranes which divide the cell into distinct spaces. The nucleus is surrounded by the nuclear membrane, the lumen of the endoplasmic reticulum is separated from the cytoplasm by a membrane, respiration occurs within mitochondria that are surrounded by a membrane, and photosynthesis occurs within similarly bounded chloroplasts. Procaryotes possess only the cell membrane and are therefore not compartmented as are eucaryotes.

The cell membrane of procaryotes carries out the functions of mitochondrial membranes and, in photosynthetic organisms, of the chloroplast membranes. We have already learned that in both organelles certain enzymes are integral parts of the membranes, whereas other enzymes are in a soluble form within the organelle. In procaryotes the membrane-bound enzymes are on the cell membrane, while the soluble enzymes are in the cytoplasm. In a functional sense, therefore, the procaryotic analogue of a mitochondrion or chloroplast is the entire cell. In many procaryotes, especially photosynthetic organisms, the cell membrane has developed rather elaborate invaginations (see Figs. 9-1 and 9-3). However, these invaginations are never surrounded by a distinct, outer bounding membrane separating them from the cytoplasm.

The cell wall

Almost without exception procaryotic cells are surrounded by a rigid cell wall. The primary function of the wall is to protect the organism from osmotic lysis. In the cell wall of all procaryotes is a structural polysaccharide called *murein*, which does not occur in any eucaryote. The unique constituent of murein is muramic acid (Fig. 9-4). The structure of murein is shown in Figure 9-5. Although the walls of a few procaryotes are composed largely of murein, most procaryotes have a variety of other polysaccharides and lipids in their walls.

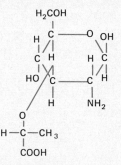

Figure 9-4 The structure of muramic acid.

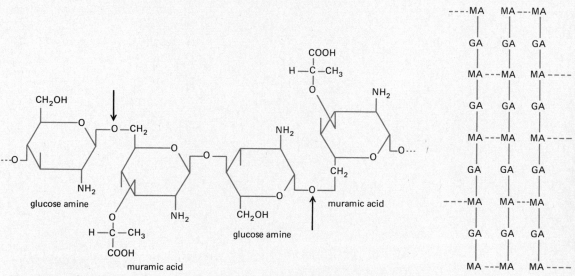

Figure 9-5   The structure of murein, the rigid component of cell walls of procaryotes. The upper portion of the figure shows the structure of the repeating unit of the backbone of murein, a polysaccharide composed of glucose amine and muramic acid (Fig. 9-4). The lower portion shows the interconnections between polysaccharide chains. Each interconnection consists of a small number of amino acids linked to each other and to muramic acid by peptide bonds.

Flagella and fimbriae

Many procaryotes are motile by means of organelles called *flagella*. The same word is used for eucaryotic organelles of motility, sperm tails, for example; the structures of the two kinds of flagella are, however, basically different. The sperm flagellum has been described in Chapter 7; it is clearly a very complex structure. The procaryotic flagellum is much simpler. It originates inside the cell membrane in a structure called a **basal granule.** These granules are much smaller than the basal bodies of eucaryotic cells. The flagellum penetrates the cell wall and is usually several micrometers long but only 12 to 15 nanometers in diameter, less than $1/10$ the diameter of a eucaryotic flagellum. A bacterial flagellum is composed of subunits of a single kind of protein

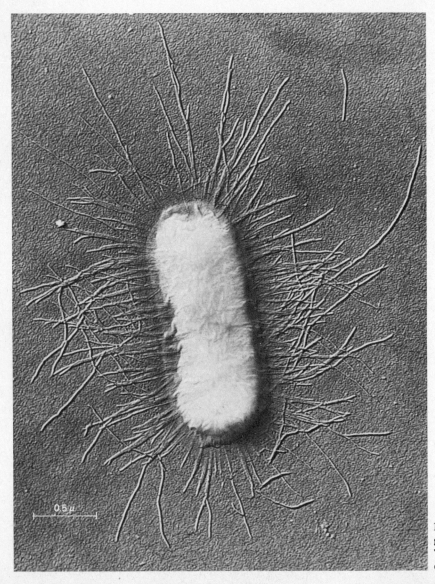

Figure 9-6 Electron micrograph of a bacterium with pili. The marker indicates 0.5 micrometer.

called *flagellin*. The flagellins of different species differ in amino acid composition, but all have molecular weights in the range 30,000 to 40,000. The flagellum is intrinsically helical in shape, and the pitch of the helix is of the order of 1 micrometer but varies from organism to organism.

Some bacteria bear a single flagellum at one or both ends of the cell; others have tufts of flagella at the ends, while others bear flagella over most of the cell surface. In those organisms with many flagella, the actual organelle of motion is a bundle of flagella. It is believed that flagella propel the cell by helical waves which pass down the flagellum from the cell. The flagellum itself does not rotate, but the effect of the moving waves is the same as if the organelle were a rigid corkscrew rotating around its axis. The mechanism of the generation of the waves and the way energy is coupled to flagellar motion are not understood.

Many bacteria bear small, proteinaceous appendages on their walls. These are called **pili** or **fimbriae** (Fig. 9-6). Pili are roughly 20 to 50 nanometers in diameter but may be a micrometer or so long. There are often several hundred pili per cell. Pili appear to be agents of specific cell-to-cell adhesion. As we shall see later sexual conjugation in bacteria is mediated by a special kind of pilus called the sex pilus. Some bacteria apparently use pili to form a thin film or pellicle of cells at the surface of a liquid which assures a good supply of oxygen. Mutants of these bacteria that lack pili grow much less rapidly in unstirred media than does the wild-type.

Resting stages

Certain groups of bacteria can form resting stages which serve to tide the organism over hard times. There are two sorts of these specialized cells, **microcysts** and **endospores.**

A *microcyst* is formed from a vegetative cell by the production of a very thick, multilayered wall. The microcyst is somewhat resistant to drying but not noticeably resistant to heat. Apart from the thick wall, microcysts do not differ greatly from the vegetative cells that give rise to them.

The formation of an *endospore* is a complex developmental process which we cannot explore in detail here. The endospore is formed within a vegetative cell. The fully formed spore is resistant to drying, to a variety of cell poisons and even to several hours of boiling. This resistance to boiling is why high pressure steam is used in canning and in sterilizing. A sketch of an endospore is shown in Figure 9-7. The resistance to cell poisons is probably due to the impenetrability of the spore wall. The extreme heat resistance most probably stems from the fact that the spore contains very little water; a typical endospore is only about 15 percent water. In spores the metabolism of the bacterial cell is turned off. The cell ceases to use energy and to synthesize proteins. The mechanisms that control endospore formation are not fully known but they involve the activation of a number of genes that are inactive in a normal bacterial cell and culminate in a resistant cell in suspended animation.

DIVERSITY OF PROCARYOTES

It is estimated that there are ¼ million species of flowering plants and conifers and somewhat more than 1 million species of animals (Chapters 23 and 24). On the other hand, less than 4000 species of procaryotes have been described. The word "species" certainly does not have the same meaning when applied to animals or plants as when applied to procaryotes; nevertheless the contrast in the numbers of species of these two groups is so great that it must be concluded that there is much less diversity among procaryotes.

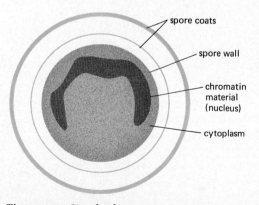

Figure 9-7 Simple diagrammatic representation of a cross section of a bacterial endospore.

The cause of this is probably the restricted size of the genome in procaryotes. The DNA content of a typical bacterium is from 10 to several hundred times less than that of any eucaryote. It has been pointed out that the genome of a bacterium is sufficient to code for a few thousand kinds of proteins. It is likely that the essential functions of metabolism which occur in all living creatures require a thousand or so enzymes; there is not very much genetic information left over in a procaryote for the elaboration of diversity.

The major groups of procaryotes

The ability to carry out oxygen-evolving photosynthesis separates the **blue-green algae** from the rest of the procaryotes. The blue-green algae have a photosynthetic pigment system which includes chlorophyll a and β-carotene. These two pigments are universally associated with oxygen-evolving photosynthesis. They also possess another type of pigment, the phycobilins, similar to those found in the red algae. Although the biochemistry of photosynthesis in the blue-green algae is in all respects similar to that of other algae and higher plants, it does not take place in chloroplasts, but rather in invaginations of the cell membrane.

Some blue-green algae are motile, but the motion is not due to flagella. The cells do not swim in liquid but rather glide over solid surfaces. The mechanism of gliding movement is not known. The cell forms include unicellular rods or spheres and multicellular filaments.

The blue-green algae are universally distributed in soil, fresh water, and the oceans. Some share with a few bacteria the ability to utilize atmospheric nitrogen as a source of cell nitrogen. Since they can also carry out photosynthesis, these blue-green algae can grow in an environment that contains only a few minerals in addition to carbon dioxide, nitrogen, and water.

The other procaryotes are a much more heterogenous assemblage. Three major groups can be distinguished: myxobacteria, spirochetes, and "true bacteria."

The **myxobacteria** are exclusively unicellular organisms which share with blue-green algae the capacity for gliding movement. Unlike blue-green algae and true bacteria the myxobacteria have thin, flexible cell walls. Most myxobacteria form resting forms called microcysts. In some representatives the microcysts are borne on structures called *fruiting bodies* which are often large enough to be visible to the naked eye. A fruiting body is the culmination of a remarkable life cycle. Vegetative cells grow by binary fission and because of their gliding movement form a thin, spreading colony on a solid surface (such as a piece of decaying wood). After a period of active growth, cells begin to move toward one or more centers within the colony and aggregate at the centers. These cell aggregates then differentiate into the fruiting bodies which are often branched and highly pigmented structures.

The **spirochetes** are a small group of unicellular organisms whose shape is always helical. They are characterized primarily by their unique form of motion. At either end of the cell is a tuft of flagella that are structurally rather similar to bacterial flagella. The flagella of spirochetes do not, however, penetrate the cell wall, but rather twist back over the cell between the cell membrane and the thin, flexible cell wall. The bundles of flagella from both ends overlap in the center of the cell. Contractions of the bundles send helical waves down the cell, which in turn propel it through a liquid.

The **true bacteria** are an exceedingly diverse group. Most of the metabolic diversity of procaryotes is found among these organisms. All have rigid cell walls and when motile possess flagella. There are three cell shapes: spheres (cocci), rods, and helices (spirilla). The vast majority are unicellular. One group (the actinomycetes) are *coenocytic*, that is, there are many nuclei with a single cytoplasm.

METABOLISM OF BACTERIA

It is remarkable that all the metabolic diversity of the living world reflects that of the procaryotes and especially of bacteria. Unless we recall some general features of metabolism, procaryotic metabolism can be bewildering rather than instructive. We begin then by briefly reviewing some general principles.

In bacteria the primary purpose of metabolism is growth, that is, the synthesis of new cell material. As we have already learned, this means the formation of macromolecules by polymerization of monomers. For example, the synthesis of protein is the polymerization of amino acids, and the synthesis of polysaccharides is the polymerization of sugars. As already seen the mechanisms of the reactions by which macromolecules are formed are remarkably similar in all organisms. To grow, a cell must supply itself with monomers for the various polymerization reactions and also with energy in the form of ATP to drive the polymerization reactions as well as those which supply the monomers.

In general the only two ways of obtaining monomers are (1) the organism can make them itself from simpler compounds or (2) it can acquire them ready-made from the environment. The mechanisms by which organisms synthesize monomers are also remarkably uniform. Some organisms can form all their organic compounds from carbon dioxide; these are called **autotrophs.** Others require at least one organic compound; these are called **heterotrophs.** As might be expected the reactions by which autotrophs convert carbon dioxide to cell material are quite uniform.

Energy metabolism has the function of generating ATP; every ATP generating metabolism involves the transfer of electrons from a compound in which they have a high potential energy (the electron donor) to one in which they have a lower potential energy (the electron acceptor). The difference in potential energies is used to form ATP from ADP and inorganic phosphate. Only three kinds of energy metabolism are known: fermentation, respiration, and photosynthesis. In fermentation both the electron donor and acceptor are organic compounds; in respiration the electron acceptor is an inorganic compound; in photosynthesis both the electron donor and acceptor are formed within the cell by the action of light, rather than being obtained from the environment as they are in fermentation or respiration. *It is in their energy metabolisms that bacteria display their diversity.*

These ideas lead to the following nutritional classification of organisms. First, those that utilize either fermentation or respiration are **chemotrophs,** meaning that they feed on chemicals; those that use light are **phototrophs.** Second, we have already seen that on the basis of their source of carbon, organisms are either *autotrophs* or *heterotrophs*. We can construct a table combining the two kinds of energy metabolism and the two kinds of carbon metabolism (Table 9-1). This table summarizes the nutrition of all living creatures.

Among eucaryotes only photoautotrophs (green plants) and chemoheterotrophs (animals, fungi, etc) are known. Among procaryotes all four nutritional categories are represented.

Phototrophs

The blue-green algae are procaryotes and carry out a typical green plant photosynthesis, that is, they reduce carbon dioxide to cell material using water as the reductant and releasing oxygen as a by-product. The details of green plant photosynthesis were treated in Chapter 5 and need not concern us here.

There is another group of phototrophic procaryotes, the photosynthetic bacteria. These include both photoautotrophs and photoheterotrophs. Let us first see how the photoautotrophic bacteria differ from blue-green algae and green plants. Any autotroph must reduce carbon dioxide to cell material; the empirical formula of cell material is approximately CH_2O; accordingly an autotroph needs a source of hydrogen to carry out the reduction. We already know that green plants use water as the source of hydrogen. In green plant photosynthesis water is the reductant and oxygen is the oxidized product. The photoautotrophic bacteria do not use water as the reductant, but use instead some other inorganic compound, most frequently a reduced sulfur compound such as hydrogen sulfide.

Table 9.1 Nutritional classification of organisms

Energy metabolism	Carbon metabolism	
	Autotrophic	*Heterotrophic*
Chemotrophic	Chemoautotrophs	Chemoheterotrophs
Phototrophic	Photoautotrophs	Photoheterotrophs

The reduction of carbon dioxide by any photoautotroph can be expressed by this equation:

$$CO_2 + 4H \longrightarrow CH_2O + H_2O$$

The symbol CH_2O stands for "cell material" and 4H are the four hydrogen atoms which must be supplied by the reductant. In green plants this is water, so the reducing reaction can be written

$$2H_2O \xrightarrow{\text{light}} 4H + O_2$$

The sum of these two reactions is

$$CO_2 + 2H_2O \xrightarrow{\text{light}} CH_2O + H_2O + O_2$$

which is the classical equation of photosynthesis.

In photoautotrophic bacteria the equation for reduction of carbon dioxide is, of course, the same. The reaction which supplies the hydrogen atoms, the reducing reaction, is different. When hydrogen sulfide is the reductant, the reaction is

$$2H_2S \xrightarrow{\text{light}} 4H + 2S \text{ (elemental sulfur)}$$

The sum of the two reactions is

$$CO_2 + 2H_2S \xrightarrow{\text{light}} CH_2O + H_2O + 2S$$

Thus bacterial photosynthesis requires a reductant other than water and does not evolve oxygen.

The photosynthetic pigments in these bacteria are chlorophylls, but are not identical to any of the plant chlorophylls.

Chemotrophs

The pathway of the fermentation of glucose by muscle cells was outlined in Chapter 5 (Fig. 5-17); this scheme illustrates the basic features of fermentation very clearly. Glucose is first converted to phosphoglyceraldehyde; this is oxidized with concomitant incorporation of inorganic phosphate to diphosphoglyceric acid. During the oxidation the cofactor NAD^+ is reduced to NADH. The phosphate groups in 2, 3-diphosphoglyceric acid are transferred to two ADPs forming two ATPs and pyruvic acid. Finally, pyruvic acid is reduced to lactic acid by the NADH formed earlier. Many bacteria use this fermentation, which is called the lactic acid fermentation, as

their energy metabolism. Other bacteria do not use pyruvic acid itself as the oxidant, but rather other compounds derived from it. Still other bacteria metabolize glucose by a different pathway which leads to carbon dioxide, ethanol, and lactic acid as the end products of the fermentation (Fig. 9-8).

Glucose is not the only substance which can be fermented. Amino acids are fermented by a variety of bacteria; the fermentation of alanine may serve as an example. One molecule of alanine is oxidized to ammonia and acetic acid:

$$CH_3{-}\underset{\underset{NH_2}{|}}{CH}{-}COOH \longrightarrow 4H + NH_3 + CH_3COOH + CO_2$$

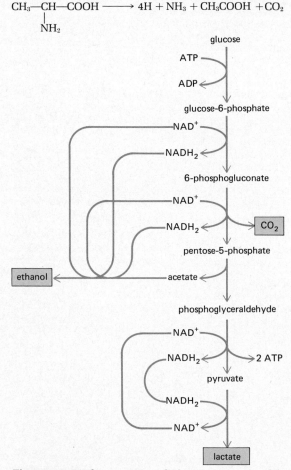

Figure 9-8 A fermentation of glucose. Note the differences between this fermentation and the one illustrated in Figure 5-17. The final products are enclosed by rectangles. Note that the net yield of ATP is one mole per mole glucose fermented.

The hydrogen atoms from this oxidation reduce two molecules of alanine, producing propionic acid and ammonia:

$$2CH_3\text{---}CH\text{---}COOH + 4H \longrightarrow 2CH_3CH_2COOH + 2NH_3$$
$$\quad\quad\quad |$$
$$\quad\quad\quad NH_2$$

In Chapter 5 we discussed the central role of the Krebs cycle in the respiration of glucose. Glucose is first converted to pyruvic acid, which is oxidized further to acetyl coenzyme A; this compound enters the Krebs cycle by condensing with oxaloacetic acid and is completely oxidized to two carbon dioxide, three NADH, and one reduced flavoprotein. The NADH and flavoprotein are oxidized via an ATP generating electron transport system by oxygen. The respiration of glucose is but one example of respiration of organic compounds by oxygen. Respiration of all these compounds follows the same general pattern. The substrate of respiration is converted by a series of reactions which may or may not involve production of reduced NAD and ATP to one or more intermediates of the Krebs cycle; the intermediates are oxidized completely by the cycle.

Most bacteria that respire organic compounds use oxygen to oxidize NADH. Some bacteria can utilize other inorganic oxidants for this reaction; these bacteria carry out anaerobic (without oxygen) respiration. The most common of these alternative oxidants is nitrate, which is reduced to nitrogen and in some cases to ammonia. This process is called *denitrification*.

We come finally to the metabolism of the chemoautotrophs. These bacteria can derive all cellular carbon from carbon dioxide. They obtain energy from the respiration of a number of reduced *inorganic* compounds; these compounds also provide the hydrogen needed for the reduction of carbon dioxide. The inorganic reductants that can be used include hydrogen gas, sulfide, ammonia, nitrite, and even ferrous iron. Respiration of each of these compounds is characteristic of a very restricted group of bacteria (see Table 9-2). The oxidation of ammonia to nitrite and of nitrite to nitrate by the nitrifying bacteria are important steps in the nitrogen cycle in nature.

THE GROWTH OF BACTERIA

In the foregoing section we said that the primary result of bacterial metabolism is growth. Having considered some aspects of metabolism we can now turn to an analysis of the growth of bacteria.

Growth of a biological system can be defined as an increase in the mass of the system such that the new mass contributes to further growth of the system. Normally growth results in an increase in the number of cells. In the case of unicellular organisms, such as bacteria, this is equivalent to an increase in the number of organisms. In the case of multicellular organisms, increase in cell number results only in an increase in the size of the organism. In order to study experimentally growth of a bacterium we must have a medium in which it can grow and methods for measuring increases in both mass and number of cells. We shall consider a bacterium which can grow in a medium with a single source of organic carbon (say glucose) plus inorganic sources of the other elements that it needs (nitrogen, sulfur, magnesium, and others).

Table 9-2   Bacterial respiration

Reductant	Oxidized product	Oxidant	Reduced product	Organism
Organic compounds	CO_2	O_2	H_2O	Many bacteria, plants, animals
Organic compounds	CO_2	NO_3	N_2, NH_3	Denitrifying bacteria
H_2	H_2O	O_2	H_2O	Hydrogen bacteria
NH_3	NO_2^-	O_2	H_2O	Nitrifying bacteria
NO_2^-	NO_3^-	O_2	H_2O	Nitrifying bacteria
S^{2-}	SO_3^-	O_2	H_2O	Sulfur bacteria
Fe^{2+}	Fe^{+++}	O_2	H_2O	Iron bacteria

The most commonly used method for estimating the number of cells per unit volume of a culture is to spread a small, known volume of a dilution of the culture over the surface of a solid medium. We have already seen (page 158) that each single cell will give rise, after suitable incubation, to a visible colony of bacteria. The number of colonies is thus a measure of the number of cells in the culture. Cell mass is usually measured by determining the amount of light scattered by a sample of the culture. Experience has shown that the amount of light scattered is proportional to the dry mass of cells per unit volume of culture.

Let us conduct an imaginary experiment. We inoculate an appropriate medium with some cells so that initially there are 2×10^7 cells per milliliter with a dry mass of 10 micrograms per milliliter. The culture is incubated and from time to time samples are removed to determine the number of cells and the dry mass per milliliter. Figure 9-9 shows typical results of such an experiment.

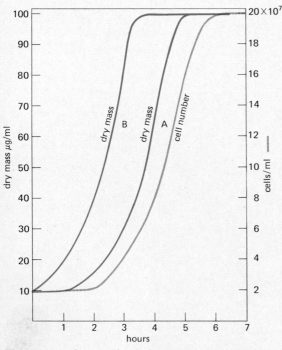

Figure 9-9 The growth of a bacterial culture. See text for details.

At the outset the average dry mass per cell was

$$(10 \text{ micrograms/ml})/(2 \times 10^7 \text{ cells/ml})= 5 \times 10^{-7} \text{ microgram per cell}$$

For the first hour or so nothing much happens; neither the mass nor the number of cells increases. After about 1 hour the mass of the culture begins to increase and increases faster and faster. Cell number, however, does not begin to increase until around 2 hours but soon is increasing in parallel with the mass. From the curves labeled A in Figure 9-9 you can calculate that from 1 hour to 2 hours the average mass per cell increases to about 7.5×10^{-7} microgram per cell and thereafter remains constant. After 5 hours cell mass stops increasing, but the cell number continues to increase (the cells continue to divide) for another hour or so. After a little more than 6 hours the size of the cells is again about 5×10^{-7} microgram per cell and all growth has ceased.

The first question we can ask about this result is: Does a delay always occur before growth begins or only sometimes depending on the previous history of the cells used to inoculate the culture? This question is easily answered by removing a sample of the culture while it is still growing and inoculating a second culture with it. The results of such an experiment are shown by curve B in Figure 9-9. It is obvious that the cells begin to grow immediately. We conclude that the delay observed earlier has something to do with the history of the inoculum and is not a necessary prelude to growth.

A second question we can ask is: Does growth cease because of some intrinsic cellular aging process or because of a change in the medium? The result of the second experiment already suggests that the latter alternative is correct. We might suspect that since all the constituents of the cells must come from the medium, growth ceases when some component of the medium has been depleted by growth. The limiting component might, for example, be the carbon source (glucose). If this is correct, the total amount of growth in a culture should depend on the initial concentration of glucose in the medium. The results of countless experiments of this sort all agree that bacteria will continue to grow indefinitely unless the medium becomes depleted of a component essential for growth or in some other way becomes unsuitable for growth. For example, a metabolic end product, such as an acid, may be inhibitory.

It is clear, therefore, that the interesting part of these growth curves is the central portion where the cells are growing rapidly. To analyze this let us return to our definition of growth and look at it more carefully: growth of a biological system is an increase in the mass of the system such that the new mass contributes to the further growth of the system. This means that the growing system (in our case a population of bacteria) is the cause of its own growth; in other words growth is *autocatalytic*. If this is true, then the rate of growth at any time must be proportional to the mass of the growing system at that time. Rate of growth means the increase in mass over a certain interval of time divided by that time. Inspection of curve B in Figure 9-9 shows that the rate of growth is constantly increasing. This can be realized (if it is not obvious by inspection) by noting that the mass increases in the first 15 minutes from 10 micrograms per milliliter to about 11.5 micrograms per milliliter; this is an increase of 6.0 micrograms per milliliter per hour; in the 15-minute interval from 2.75 hours to 3.0 hours the mass increases from 66 to 80 micrograms per milliliter, or an increase of about 56 micrograms per milliliter per hour. Thus the rate of growth will depend on the length of the time interval used to calculate it.

We must therefore make use of the instantaneous rate of growth; this is simply the tangent to the growth curve at any instant. In symbols the instantaneous growth rate can be written as dB/dt. Here dB is the increase in mass (or number) of cells in a very small interval of time, dt. Thus the statement which tells us that the rate of growth is proportional to the mass present can be expressed as

$$dB/dt = kB \qquad (1)$$

This is the fundamental equation which describes bacterial growth. The term k is constant for a particular bacterium growing under a particular set of conditions; in general its value is different for different bacteria or for the same bacterium under different conditions. From equation (1) we see that

$$k = (dB/dt)/B \qquad (2)$$

That is, k is the growth rate per unit mass of bacteria; for this reason it is called the *specific growth rate constant*.

In order to fix these ideas let us look at growth from a somewhat different point of view. Imagine observing under a microscope a single cell as it grows and divides. Remove a cell from a growing population just after it has been formed by division of its mother cell and watch it. The cell will increase in mass and after a time will become twice as large as it was initially. This time is called, naturally enough, the doubling time. The cell will divide to produce two new cells, each as large as the original cell was at the start of the observations. These two cells will grow and after another doubling time they will each have doubled in mass and will divide to form a total of four new cells. The process will continue so long as the medium remains adequate. The doubling times of all the cells will be nearly (but not exactly) the same; the average doubling time will not change with time. In other words, one of the properties a cell inherits from its mother cell (indeed the most important property) is the ability to repeat exactly the process that produced it. This statement is equivalent to the growth equation (1).

It is obvious that the number of cells increases geometrically; that is, after the first doubling there are 2 cells, after the second doubling, 4, after the third, 8, and after n doubling times there are 2^n cells. If the number initially was B_0, rather than one, the number after n doublings will be $B_0 \cdot 2^n$. Clearly, the same arguments apply to increase in cell mass.

A culture such as the one we have been considering is said to be dividing synchronously since every cell in it divides at the same time. Normally cultures do not divide synchronously even if derived from a single cell. The reason for this is that the doubling times of individual cells are not all exactly the same but rather are scattered about an average doubling time. The scatter of individual doubling times results in the divisions of individual cells getting out of step with one another. After a period the culture will be growing asynchronously, which means that no matter when it is observed the fraction of cells undergoing division will be the same. The important point is that even in an asynchronous culture it remains true that both the mass and number of cells per milliliter double during each interval of time equal to the average doubling time. If the average doubling time is t_D, then in an interval of time t there are t/t_D doublings. Thus we can write the equation

$$B = B_0 \cdot 2^{t/t_D} \tag{3}$$

In this equation, B_0 is the initial mass (or number) of cells and B is the mass (or number) after the time t.

Because time appears as an exponent in this equation, a culture whose growth follows the equation is said to be growing exponentially. By taking logarithms to the base 2 of both sides of the equation, we obtain

$$\log_2 B = \log_2 B_0 + t/t_D \tag{4}$$

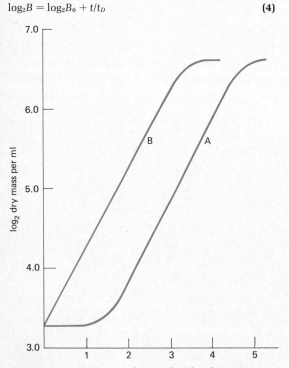

Figure 9-10 Exponential growth. The data of Figure 9-9 for cell mass per milliliter have been replotted on a logarithmic scale.

Thus it is easy to tell if a culture (such as in Fig. 9-9) is growing exponentially simply by plotting the logarithms to base 2 of the mass or number of cells per milliliter against time. If the graph is a straight line, growth is exponential. In Figure 9-10 the results from Figure 9-9 have been plotted logarithmically.

An important property of populations growing in a constant environment is that their chemical composition does not change with time. We have already seen an example of this in Figure 9-9. Recall that the dry mass per cell was constant during the period of exponential growth. The same is true of the amount of protein, or RNA, or any other constituent per cell. This is not surprising when it is remembered that the time needed for a cell to double itself remains constant in exponential growth and that this time must reflect the metabolic activities of the cell. The fact that the chemical composition of an exponentially growing population is constant implies that every component grows at the rate given by equation (1). Thus if C is the amount per milliliter of any component, then its rate of increase is

$$dC/dt = kC \tag{5}$$

REPLICATION AND DISTRIBUTION OF DNA

The genome of a bacterial cell is a single, probably circular molecule of DNA. How can such a molecule replicate itself and how can the two copies be distributed into the daughter cells at cell division? Experimentally it has been shown that replication occurs as shown in Figure 9-11. The "replication point" moves in one direction around the circle copying both strands of the DNA molecule simultaneously.

Recall that the two strands of DNA have opposite chemical polarities defined by the orientations of the 3'- and 5'-phosphate ester bonds. The fact that the replication point moves in one direction around the molecule implies that one strand is copied in the 3' to 5' direction and the other in the 5' to 3' direction. This rules out any simple role in DNA replication for the DNA polymerase discovered by Kornberg.

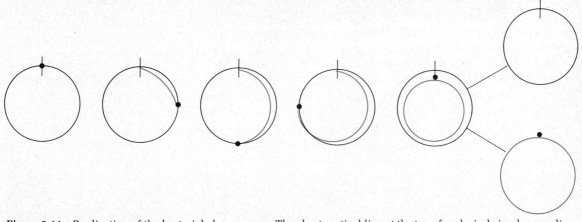

Figure 9-11 Replication of the bacterial chromosome. The short vertical line at the top of each circle is where replication begins; it proceeds clockwise. The position of the growing point is shown by the dot. The first four intervals each represent one quarter of a division time; the last stage takes a negligible time. For simplicity, the two strands of DNA have not been shown.

Many electron micrographs show a close association of bacterial DNA with the cell membrane. Experiments employing radioactive precursors of DNA indicate that newly synthesized DNA is bound to the cell membrane. On the basis of these observations it is plausible to suggest that the replication point is attached to the cell membrane and that the molecule moves past it during replication. At the completion of replication, the replication point, which is probably a protein, may itself be duplicated as shown in Figure 9-11; thus each new DNA molecule would be separately attached to the cell membrane. Growth of the membrane between the two attachment sites would push the two DNA molecules apart. If cell division occurred between the two sites of attachment, the newly replicated DNA would be neatly and almost effortlessly distributed to the daughter cells.

A given kind of bacterium can exhibit a wide range of doubling times; experimentally it is not difficult to obtain a tenfold range, from perhaps 20 to 200 minutes. Because all components of a cell must be precisely duplicated in a doubling time, in bacteria the rates of macromolecular synthesis are accurately controlled relative to the doubling time. We shall see later how the rate of protein synthesis is controlled; for the moment we focus our attention on how the rate of DNA replication is varied. In general two possibilities exist. First, the actual rate of replication, that is, the rate at which nucleotides are being added to the growing strands at each replication point, could be controlled. In terms of Figure 9-11 this means that the rate of movement of the replication point around the circle would be different in cells with different doubling times. The second possibility is that the rate of replication at a single replication point is constant, but the number of replication points per molecule varies. This possibility seems to be the correct one. The way it works is diagrammed in Figure 9-12. In this figure, for the sake of clarity, DNA has been pictured as an open linear molecule rather than as a closed circle.

The model in Figure 9-12 implies that the amount of DNA per cell should increase as the doubling time decreases. This is found to be true. The model implies a fixed time for the replication of one genome; but bacteria can have doubling times considerably in excess of this time. In this situation apparently DNA replication is completed some time before cell division. The interval between completion of DNA replication and cell division is longer the longer the doubling time.

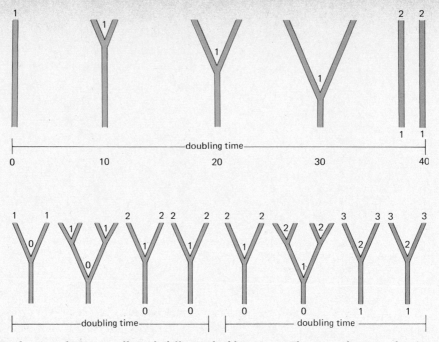

Figure 9-12 Replication of DNA in cells with different doubling times. The times shown in the center of the figure are minutes of the cell division cycle. (Above) Duplication of DNA in a cell with a doubling time of 40 minutes. We assume that at this doubling time a newly formed cell (at time zero) has an amount of DNA just equal to one genome. Thus it takes 40 minutes to duplicate one genome in this organism. (Below) Duplication of DNA when the doubling time is only 20 minutes. The rate of movement of a replication point in this case is the same as in the upper drawing; the point moves one-fourth of a genome in 10 minutes. The replication point marked 0 began its movement 20 minutes before zero time. At time zero replication at point 1 has just begun. At this doubling time a newly formed cell has 1.5 genome of DNA.

This method of varying the rate of DNA replication is an example of what seems to be a general principle of control of macromolecular synthesis: it is the frequency of *initiation* of polymerization reactions that is controlled and not the rates of the reactions themselves. The data shown in Table 9-3 indicate that this is how the rate of protein synthesis is varied in cultures with different growth rates. The last two lines of the table show that although the overall rate of protein synthesis in one culture may be nearly ten times that in another, the rate of protein synthesis per ribosome is almost the same.

METABOLIC CONTROL IN BACTERIA

It has already been pointed out that many, but by no means all, bacteria are highly versatile metabolically. Thus many bacteria can utilize any one of a large number of compounds as the sole source of carbon. This versatility appears to be based on

the operation of two kinds of control mechanisms introduced in Chapter 6. The first is control of the kinds of enzymes produced by a cell; the second is control of the activities of enzymes by allosteric interactions.

Let us examine the roles these two kinds of control play in the biology of bacteria. The mechanism suggested by Jacob and Monod for the regu-

Table 9-3 Rates of protein synthesis in cells with different doubling times

Doubling time (minutes)	30	120
Specific growth rate constant (sec^{-1})	3.8×10^{-4}	0.97×10^{-4}
Protein per cell (μg)	3.1×10^{-7}	1.3×10^{-7}
Peptide bonds per cell	18.5×10^{8}	7.7×10^{8}
Ribosomes per cell	41×10^{3}	5.4×10^{3}
Peptide bonds per cell per second	71×10^{4}	7.5×10^{4}
Peptide bonds per ribosome per second	17	14

lation of protein synthesis is outlined in Figure 9-13; this is merely a slight expansion of Figure 6-19. Figure 9-13A shows how an inducer (a small molecule present in the environment) turns *on* the synthesis of three enzymes simultaneously. The inducer does this by inhibiting the binding of the repressor, R_1, to the operator gene. When the inducer is removed, the repressor binds to the operator gene and the mRNA corresponding to structural genes S_1, S_2, and S_3 is no longer synthesized. Since mRNA is very rapidly degraded in bacteria, synthesis of the enzymes ceases within 1 or 2 minutes after removal of the inducer.

We have already mentioned that many bacteria can utilize a very large number of compounds as carbon sources. Metabolism of most of these requires at least one and often several specific enzymes. In the majority of instances these enzymes are inducible, that is, they are made by the cells only when their substrates are present in the medium. In assessing the advantage to the organism of control by repressor, it must be

borne in mind that the repressor is itself a protein. This mode of control will be advantageous only if the amount of enzyme protein not synthesized in the presence of repressor is considerably greater than the amount of repressor. This implies that the repressor must bind very tightly to the operator gene. As was pointed out in Chapter 6, it has been possible to isolate repressor for at least one control system; it turns out that there are only about ten molecules of repressor per cell. The number of enzyme molecules per cell in the absence of repressor (or in the presence of inducer) is several thousand in a typical case. Thus clearly this mechanism does effect a very considerable saving of protein. The saving is even greater when a single repressor can stop the synthesis of several enzymes, as is often the case.

Figure 9-13b shows how a small molecule from the environment can turn *off* synthesis of enzymes. In this case the product of the regulator gene (R_2) does not bind to the operator gene unless the small molecule (called a corepressor) is

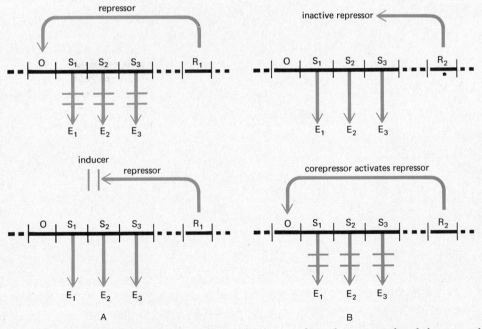

A   B

Figure 9-13 Control of protein synthesis by repressor. The figures show the repressed and derepressed states of two sets of enzymes each regulated by a single repressor. (A) is *induced* by an external inducer; (B) is repressed by an external corepressor. In each case there is an operator gene (O) and three structural genes (S_1, S_2, and S_3) which specify the structures of three enzymes (E_1, E_2, and E_3). In the top figures the external inducer or corepressor is absent; in the bottom figures it is present.

present and bound to R_2. Control by corepressors is used when enzymes synthesize compounds necessary for growth. For example, an amino acid is usually the corepressor of the enzymes necessary for the biosynthesis of that amino acid. Hence if a bacterium finds itself in an environment which contains an amino acid, it will stop making the enzymes required to make that amino acid. This clearly represents a saving in protein synthesis.

Compounds which act as corepressors are usually found to *inhibit* the activity of the same enzymes whose synthesis they repress. This is a further economy to the cell since inhibition of activity allows material that would otherwise be used in the biosynthesis of the amino acid to be put to other uses.

VIRUSES

Viruses are self-replicating units that, unlike other organisms, contain only one kind of nucleic acid, either DNA or RNA. They are obligate intracellular parasites that infect bacteria and cells of plants and animals. As a rule any one virus can infect only one kind of host; that is, a virus is specific for its host. A virus has a life cycle with two stages; one stage occurs within the host cell and the other is extracellular. In its extracellular stage a virus is an inert particle called a *virion*. The virion contains nucleic acid which is surrounded and protected by a protein coat or *capsid*. The capsid of certain large animal virions is itself surrounded by a membranous envelope.

All virions are extremely small, but there is a considerable range of sizes among them. One of the largest virions is that of cowpox, which is used in vaccination against smallpox. The virion is a cube about 20 nanometers on a side. On the other hand, a small virion may be only 2 nanometers in diameter. The sizes and some other properties of several plant and animal viruses are given in Table 9-4. Virions have very simple morphologies, since most of them are composed not of millions of molecules of thousands of different kinds, as are cells, but of only a few hundred molecules of several different kinds. As we shall see these few molecules are able to infect a cell and cause it to produce hundreds of new virion particles.

We shall emphasize viruses that infect bacteria—**bacteriophages,** or more briefly, **phages.** Most of what is known about the biology of viruses has come from the study of phages. Bacteriophages are not rare; probably there is a phage capable of infecting any bacterium. Recently viruses have been isolated which infect certain blue-green algae. Only a few phages have been intensively studied, but all have the same general characteristics. The most intensively studied phages attack the bacterium *Escherichia coli.* These "coli phages" are easy to isolate from nature. A culture of *E. coli* is inoculated with a small amount of sewage which has been filtered to remove bacteria; the phage virions are so small they pass through the filter. The culture will continue

Table 9-4   Properties of some plant and animal viruses

	Particle size[a]	Dimensions (nm)	Shape	Composition		
				Nucleic acid	Protein	Other
Tobacco mosaic	40×10^6	18×300	Thin rod	RNA	+	None
Tomato bushy stunt	9×10^6	30	Sphere	RNA	+	None
Turnip yellow mosaic	5×10^6	22	Sphere	RNA	+	None
Poliomyelitis	10×10^6	27	Sphere	RNA	+	None
Rabbit papilloma	50×10^6	45	Sphere	DNA	+	None
Influenza	280×10^6	80	Sphere	RNA	+	Carbohydrate, lipid
Cowpox (smallpox vaccine)	3×10^9	$28 \times 22 \times 22$	Brick-shaped	DNA	+	Lipid

[a]The weight of a single virion relative to the weight of a hydrogen atom.

to grow, but if "coli phages" were present in the filtrate, the culture will eventually lyse, that is, it will become clear. A small amount of the lysed culture can be filtered to remove any remaining bacteria; it will cause lysis of a second culture of *E. coli*. This process can be repeated an indefinite number of times. Thus the ability to lyse is not lost even after many transfers involving a very great dilution of the original filtrate; it can be concluded that the lytic agent multiplies. Phage particles cannot be seen under the ordinary light microscope; but their ability to lyse bacteria provides a means of counting virion particles. This is done by mixing a suspension of bacterial cells and phage particles and spreading the mixture over the surface of a solid medium. Growth of the bacteria will produce a turbid film peppered with clear areas or *plaques*. Each plaque is due to lysis brought about by the progeny of a single virion. The number of plaques is thus equal to the number of virions originally present.

With the advent of the electron microscope it became possible to examine the morphology of bacteriophages. The "coli phages" are com-posed of two parts: a head and a short tail. The head is a hexagonal prism with pyramidal ends. The tail, which is as long as the head but considerably narrower, is hexagonal in cross section (Fig. 9-14). Chemically the phage is composed of DNA and several kinds of proteins. The DNA is within the head; the wall of the head and the tail are protein. The DNA within the head is a single molecule. In one of the most studied phages the DNA in a single virion weighs about 6×10^{-13} microgram, which is only a few percent of the weight of a bacterial DNA molecule. Other bacteriophages contain RNA rather than DNA.

Life history of a phage

Let us examine more closely the development of phages in an infected bacterium. The process can be divided into three stages: (1) infection, (2) growth, and (3) lysis. Infection of the host cell occurs in two steps: attachment of the virion to the cell wall and injection of the phage DNA into the cell. The tip of the tail (which is really a proboscis) attaches to the cell wall of the host (Fig.

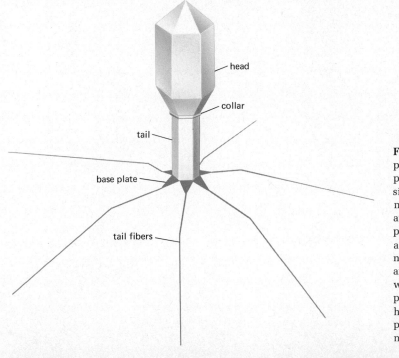

head

collar

tail

base plate

tail fibers

Figure 9-14 A diagrammatic picture of a bacteriophage particle. The head is a hollow, six-sided prism constructed of numerous protein subunits and containing the DNA of the phage. The head is attached to a six-sided tail. The tail terminates in an end plate to which are attached six long tail fibers which assist in binding the particle to the cell wall of its host. This beautiful and complex structure is only 200 nanometers long.

9-15). The tail binds to specific sites on the cell wall. An excellent example of the specificity of attachment is provided by certain phages that can infect male cells of *E. coli* but not female cells. The receptor sites for these phages are on the sex pili of the male cells (see page 278). Bacteria can become resistant to phage infection by mutations which alter the chemical configuration of the receptor sites.

Once attached to the cell wall, the virion injects its DNA. First a small area of the cell wall is dissolved by a phage enzyme; the mechanism of injection of the DNA is still obscure. Only the DNA enters the cell; none of the protein of the virion capsid enters. Since the infected cell will go on to form 100 or more complete virion particles, it is evident that DNA alone is sufficient to cause not only its own replication but also the formation of the protein parts of the virion. The importance of this observation in identifying DNA as genetic material has been considered in Chapter 3. Immediately after injection of the DNA, the metabolism of the host cell is profoundly altered. The synthesis of host-specific macromolecules ceases; that is, the bacterium stops forming those molecules that made it a particular kind of cell. Within a few minutes phage-specific macromolecules are formed—DNA and proteins. The proteins formed include not only the structural proteins of the complete virion but

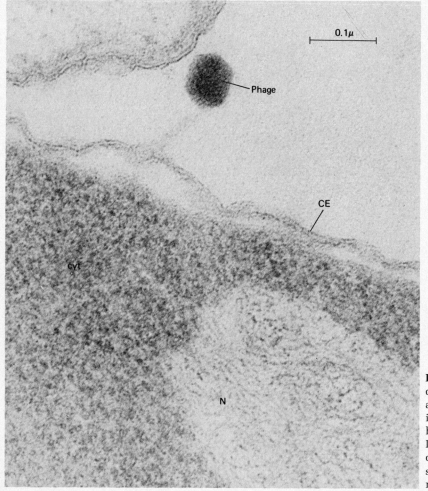

Figure 9-15 An electron micrograph of a phage particle attached to the cell wall (CE) of its host. The DNA of the phage has not yet been injected. The DNA (N) and cytoplasm (cyt) of the host are very clearly shown. The line represents 0.1 micrometer.

also certain enzymes that catalyze new biosynthetic reactions. For example, the DNA of certain phages does not contain cytosine but a closely related compound, 5-hydroxymethyl cytosine. The host cell does not possess the enzymes needed to make this compound; the phage DNA contains the information which specifies these enzymes. The mechanism of synthesis of phage-specific proteins is not unusual and involves the same mRNA-ribosome complex already discussed. The difference is that in an infected cell the mRNA molecules are transcribed from the phage DNA rather than from the host DNA.

Synthesis of phage DNA and phage protein go on independently. After a few minutes, complete phage particles begin to be assembled from the already formed DNA and protein; when the number of virion particles reaches 100 to 200, the cell bursts and the particles are released. This is lysis. The entire process—infection to lysis—takes about 20 minutes in E. coli.

The phage particles, it should be noted, do not divide or multiply; one phage particle does not give rise to two and these in turn to four. Rather, the complex of "phage + host cell" produces hundreds of phage particles. *The self-replicating unit is neither the host cell nor the virion; it is the complex of phage genetic material and host cell metabolism.*

The DNA of phage has the same genetic properties as the DNA of other creatures. It can mutate as does other genetic material. Also, two phage particles can infect the same host cell; the particles which emerge will include recombinants of the genetic information of the two infecting particles.

Temperate phage and lysogeny

The kind of phage just described, in which infection invariably leads to lysis, is called a *lytic phage*. There is another kind of phage that does not invariably cause lysis of its host, which is called *temperate phage*.

When a suspension of host cells is mixed with virions of a temperate phage, only a fraction of the cells will lyse; the remainder continue to grow and are apparently normal. The development of phage in the cells that do lyse is exactly as described previously. The cells that do not lyse have two peculiar properties: they are resistant to infection by the same phage and they can induce lysis of cells that are sensitive to the phage. Such cells are called *lysogenic* (lysis inducing).

Careful examination of a culture of a lysogenic bacterium reveals that occasionally a cell will spontaneously lyse and release phage virions. Any cell of the culture can do this; how-

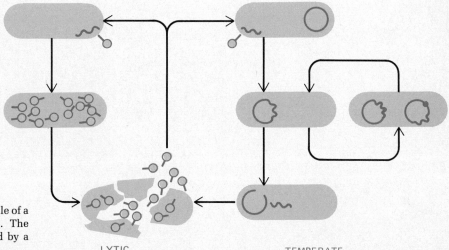

Figure 9-16 The life cycle of a temperate bacteriophage. The viral DNA is represented by a wavy line.

LYTIC

TEMPERATE

ever, when lysogenic bacteria are artificially broken open, no phage particles are found. Thus lysogenic bacteria are not full of phage virions; instead they possess the heritable property of being able under appropriate conditions to form the virions.

A temperate phage can thus exist in either of two alternative states: (1) as free phage virions which can infect and lyse a bacterium or (2) as part of the genetic information of the host cell. The latter state is called the "prophage" state. This term is the name given to the genetic information for producing phage particles when it is not expressed but transmitted from host cell generation to generation. Usually, as we have seen, a lysogenic bacterium containing prophage only rarely produces phage particles. In certain circumstances, for example, after irradiation with ultraviolet light, all the cells of a culture of a lysogenic bacterium form free phage. This is called "induction of the prophage." The development of phage virions in an induced cell closely resembles that of lytic phage, except that, of course, infection is unnecessary.

As stated earlier lysogenic bacteria are immune to infection by the temperate phage they release when induced. This immunity is specific: a bacterium lysogenic for one phage can be infected with a second temperate phage. The second phage may either lyse the cell or become a prophage. In this way a single cell can be lysogenic for more than one temperate phage. Immunity is not the same as resistance to phage infection. As we have seen resistance is due to the inability of the virion to attach to the host cell wall. On the other hand, in the case of an immune cell the virion does attach and indeed injects its DNA into the cell, but the immune cell does not produce phage particles. The development of the phage is repressed in an immune cell. Repression of phage development is a property of the cytoplasm of the host. The cytoplasm contains a protein repressor. The repressor is the product of a gene of the temperate phage. When a temperate phage is in the prophage state, the repressor gene is the only phage gene that is expressed; hence phage development does not occur.

In some temperate phages the *prophage* (that is, the phage genome) is physically incorporated into the bacterial chromosome. The prophage of any given temperate phage is always found at the same place on the chromosome. Genetically such a prophage is indistinguishable from any other region of the chromosome. When such a prophage is induced, it leaves the chromosome and initiates the formation of virions in the cytoplasm. The prophage of other temperate phage is not incorporated into the host cell chromosome but exists as a separate piece of DNA which replicates autonomously but in synchrony with replication of the host's chromosome.

Plant and animal viruses

The biology of plant and animal viruses, including the tumor viruses (p. 246), is basically the same as that of a typical bacteriophage. Plant and animal viruses differ from phage and from one another in chemical composition, shape, and the kind of host cell they infect. The properties of a few viruses are shown in Table 9-4. Notice that some animal viruses contain lipid and carbohydrates. These components of the virion are derived from the cell membrane of the host. The virion is released from the cell in such a way that the virus particle is enveloped by a portion of the cell membrane.

SEXUALITY OF BACTERIA

In all fundamental aspects the genetics of bacteria is the same as that of other organisms. Each heritable character of a bacterium is determined by a specific portion of its genetic material. As far as is known bacteria are normally haploid. The genetic material of a bacterium can mutate, and since the cell is haploid, mutations are immediately expressed. As we shall see sexual reproduction occurs in bacteria, but so far has been found in only a few species. It is not known therefore how important this process is in the natural history of bacteria. It is worth exploring this question further before taking up some details of sexuality in bacteria.

Like all organisms bacteria have been faced during their evolution with a changing and limited environment. In adapting to the environment bacteria, like all organisms, have had to find a compromise between genetic variability and stability. Variability is necessary to permit adaptation to changing conditions, but stability is also required to maintain characteristics which have proven advantageous. Although bacteria are haploid and mutations are immediately expressed, a population of bacteria in nature does not show the expected variability. One reason for this is that because of the high growth rates of bacteria, deleterious mutations will be rapidly eliminated by selection. By the same token advantageous mutations will be immediately selected for. These considerations seem to suggest that sexual reproduction is not significant in the natural history and evolution of bacteria. However, this analysis overlooks a significant fact: most phenotypic characters, which are what selection operates on, do not reflect the properties of a single protein determined by a single gene, but rather the properties of several or many proteins. For example, the biosynthesis of a single amino acid may easily require the sequential action of a half dozen or so specific enzymes, specified by an equal number of genes. One can imagine that in a population of bacteria random mutation and selection could result in an improved set of enzymes for the biosynthesis of one amino acid and in another population for the biosynthesis of a second amino acid. A mechanism which would allow these two sets of enzymes to be produced together in a single cell would clearly be advantageous. By definition sexual reproduction is such a mechanism. Therefore it is likely that sexual reproduction is quite widely distributed among bacteria and is important to their natural history even though it has so far been demonstrated in relatively few kinds.

Methodology of bacterial genetics

An obvious requirement for studying the genetics of any organism is to be able to separate the progeny of a mating from the parents. In most organisms this presents no difficulty; the progeny can usually be physically separated from their parents. Primarily because of their small size this is impossible, however, with bacteria. Hence use is made of genetic recombination that is the result of mating. The principle is most easily explained by an example. We wish to see if sexual reproduction can occur in some bacterium. The standard, or wild-type, of the bacterium can make all its own amino acids and is sensitive to streptomycin. By mutation we derive two strains. Strain A cannot make leucine (leu) and strain B cannot make phenylalanine (phe) or methionine (met) but is resistant to streptomycin (Sm^r). We can symbolize the genotypes of these strains thus: A, $leu^-phe^+met^+Sm^s$, and B, $leu^+phe^-met^-Sm^r$.

On a medium containing streptomycin but no amino acids the only cells that can grow will have the genotype $leu^+phe^+met^+Sm^r$. This genotype can arise from either strain A or B separately only by a double mutation. Any single mutation will be infrequent and the joint occurrence of two in one cell will be very infrequent. For example, if in strain A the mutation leu^- to leu^+ occurs with a frequency of 1 in 10^5 cells and the mutation Sm^s to Sm^r with a frequency of 1 in 10^6 cells, then we expect both to occur together in a single cell with the frequency of 1 in 10^{11} cells. Hence if we place 10^8 cells of strain A on the selective medium, we do not expect any colonies to appear. Similar arguments apply to strain B.

If we mix populations of the two strains beforehand and place say 10^8 cells on the selective medium, any colonies which appear must be the result of genetic recombination: the phe^+ and met^+ genes of strain A have combined with the Sm^r gene of strain B to produce cells with the genotype $leu^+phe^+met^+Sm^r$ which can grow on the selective medium.

Conjugation

The three ways by which bacterial DNA can be transferred from cell to cell are conjugation, transduction, and transformation. Transformation was discussed adequately in Chapter 3.

Conjugation has been shown to occur in several kinds of bacteria, but is best known in *Escherichia coli*. There are two mating types in *E. coli*. Mixtures of cells of the same mating type do not give recombinants, but mixtures of cells of opposite mating types do. The two mating types do not have equivalent roles in the production of recombinants and may therefore be designated as male and female. The male cells contribute only their DNA to the zygote; they then can be removed or killed without reducing the number of recombinants formed. The zygote is formed within the female cell; the female parent must therefore remain viable after conjugation for recombinants to be formed.

Conjugation is the pairing of a male cell with a female cell. The male cell is equipped with a special kind of pilus—the sex pilus (see page 260). In conjugation these attach to the cell wall of the female cell. DNA is then transferred from the male to the female. It is possible that the DNA travels through the sex pilus, but this is not certain at the present. Since female cells do not have these special appendages, *E. coli* is sexually dimorphic, which perhaps is surprising for so small a creature. The male and the female DNA within the zygote can produce recombinant DNA in a manner precisely similar to that discussed in Chapter 3.

The ability of male cells to make sex pili and therefore to be able to conjugate is determined by a genetic element called the F (for fertility) factor. Like other genetic elements, the F factor is composed of DNA. The amount of DNA in an F factor is only a small fraction of the DNA of the bacterial cell. The F factor can exist in one or the other of two alternative states in a male cell. It can occur as a cytoplasmic particle (the F particle) which is not integrated into the chromosome of the cell. When it is in this state, the F factor replicates independently of the chromosome. The F factor can also be integrated into the bacterial chromosome, in which case it behaves like any other chromosomal element and in particular is replicated in concert with the chromosome. Its integration is reversible; that is, the F factor can leave the chromosome and take up an independent existence in the cytoplasm again. Incorporation into the chromosome can occur at a number of its sites. A population of cells derived from one cell in which the F factor was cytoplasmic contains cells in which the F factor is integrated at various places on the chromosome.

When a male cell in which the F factor is in the cytoplasmic state conjugates with a female cell, an F particle is transferred to the female which becomes male. The F particle is therefore infectious. The chromosome of the male is not transferred, however, and recombinants are not formed. For this reason males in which the F factor is cytoplasmic are called infertile males.

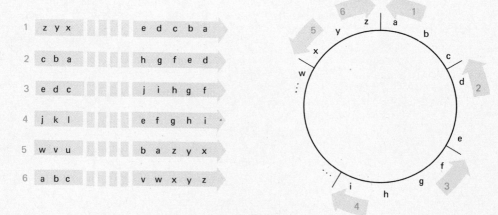

Figure 9-17   Patterns of gene transfer in fertile males of *E. coli*. For simplicity we assume that the genetic map of *E. coli* can be adequately represented by the 26 letters of the alphabet. At the left are shown six possible patterns of gene transfer. The origins are indicated by the arrowheads. Because of spontaneous breakage of the chromosome, only those genes to the right of the vertical lines on the arrows will be transferred ordinarily. To the right is a circular map. The six patterns can be generated by cutting the circle at the indicated points and reading the alphabet in the directions shown by the little arrows.

When a male in which the F factor is incorporated in the chromosome conjugates with a female, the chromosome of the male is transferred to the female; such males are said to be fertile. The passage of the chromosome into the female is slow; it may take up to 2 hours for the entire chromosome to be transferred. Under ordinary circumstances transfer is interrupted before the entire male chromosome has entered the female. Only those male characters that are transferred can appear among the progeny of the mating. The recombinants are almost invariably female. It can be inferred, therefore, that the F factor itself is seldom transferred when it is incorporated into the chromosome.

Chromosome transfer can be artificially interrupted by vigorously shaking the mating mixture. If the transfer is interrupted very early, only a few male genes appear among the progeny of the cross. If transfer is allowed to go on for longer and longer times, more and more male genes appear in the progeny. This suggests that transfer of the male chromosome is oriented; that is, that a particular gene enters the female first and is followed by the rest in order according to their positions of the chromosome. The first gene to enter defines the origin of the chromosome during transfer. For any one strain of fertile males the origin and the order are constant. Various strains of fertile males can be obtained in which either the origin or the order or both are different. As has already been mentioned, chromosome transfer is ordinarily spontaneously interrupted during mating. The various male strains will therefore differ with respect to what genes appear among the progeny of their matings. If a sufficient number of different strains is used, transfer of the entire genome of E. coli can be demonstrated. These points are illustrated in Figure 9-17.

It is possible to derive all the patterns of gene transfer among fertile males in the following way. Imagine that the complete genetic map of E. coli is circular (see Fig. 9-18). Now cut the circle between the letters a and z (the letters correspond to genes) and read in a clockwise direction. The letters will be read in the order a, b, c, . . . z. This gives the pattern of transfer labeled 1 in the figure. If the same cut is made but the letters read in the counterclockwise direction (z, y, w, . . . a), the pattern labeled 5 is obtained.

It is believed that chromosome transfer is linked to DNA replication. The site of integration of the F factor is thought to determine the point at which DNA replication begins. Conjugation

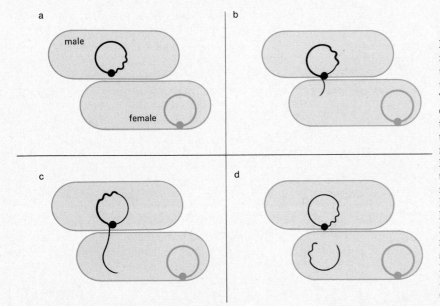

Figure 9-18 Coupling of DNA replication and chromosome transfer. The chromosome is the large circle within each cell. The wavy portion of the male chromosome represents the integrated F factor. (a) conjugation has occurred, but DNA replication has not begun. (b) DNA replication has begun in the male; the daughter DNA molecules are shown by thin lines. (c) The DNA molecule is half-replicated. (d) Replication is complete and the female has received a complete chromosome including the integrated F factor. Complete transfer happens only rarely under ordinary circumstances.

triggers the onset of replication, which proceeds around the molecule as shown earlier (Fig. 9-11). Replication during conjugation differs from normal replication in that only one of the new DNA molecules remains attached to the cell membrane of the male cell; the other daughter DNA molecule enters the female cell as it replicates (Fig. 9-18). Transfer of the F factor in its cytoplasmic state occurs by the same mechanism; in this case only replication of the F particle is triggered by conjugation and this is the only DNA transferred.

An F factor can exchange genetic material with the bacterial chromosome. Such an exchange gives rise to an F factor carrying one to several bacterial genes; this is called an F' factor. An F' factor can still exist as a cytoplasmic, autonomously replicating particle and the bacterial genes on the particle are still functional. If such a particle is transferred to a female cell, the female becomes not only male but also diploid for the genes carried by the F' factor.

Genetic elements similar to F factors in many respects have recently become considerably significant medically. These elements are called resistance transfer factors, or RTF. Any one RTF can carry genes determining resistance to one or several antibiotics. Like F factors, RTF is easily transferred from cell to cell. An RTF carrying genes for resistance to, for example, streptomycin, chloromycetin, and sulfanilamide will cause a cell to which it is transferred to become resistant to these antibiotics simultaneously. It is easily seen how resistance to a variety of antibiotics could very rapidly spread through a natural population of bacteria. The following story illustrates this dramatically.

Bacterial gastroenteritis (often called food poisoning) as anyone who has suffered it knows is not a trivial disease. It is caused by the bacterium *Salmonella typhimurium*. In England during 1962 only 3 percent of the strains of *S. typhimurium* isolated from patients with gastroenteritis were resistant to several antibiotics. By 1963 the frequency of resistant strains had risen to 21 percent and by 1964 more than half (61 percent) of the strains were resistant. A particularly insidious feature of this phenomenon is that *E. coli*, a normal inhabitant of the intestinal tract,

can acquire RTF, probably by selection through indiscriminate use of antibiotics. The RTF of *E. coli* can be transferred to *S. typhimurium*. Hence if someone whose intestinal flora contains *E. coli* carrying RTF is infected with *S. typhimurium*, the RTF from *E. coli* can be transferred very rapidly to the pathogenic *Salmonella* which in turn makes the disease impossible to treat with antibiotics.

Transduction

In transduction genetic material is transferred from one cell to another by means of a temperate bacteriophage. Recall that there are two sorts of temperate phage distinguished by the physical state of the prophage. The prophage may either be physically incorporated into the bacterial chromosome at a specific site or exist as an autonomously replicating piece of DNA. The first kind of temperate phage mediates *specialized* transduction; the second kind of temperate phage mediates *generalized* transduction.

Specialized transduction comes about in the following way. Very occasionally when the prophage leaves the bacterial chromosome after induction, it carries with it a small piece of the bacterial DNA amounting to only one or two genes. Since the bacterial genes are on the same DNA molecule as the phage genes, they are replicated along with the phage genes and become incorporated in the phage particles. When a phage particle produced in this way infects a sensitive cell, it will inject a molecule of DNA containing both viral and bacterial genes. The DNA may become incorporated into the chromosome of the host as prophage. The newly formed lysogenic cell will thus contain two copies of one or two genes: one copy brought in by the phage DNA and the other on the bacterial chromosome. The bacterial genes transduced will be limited to those lying close to either side of the site of prophage incorporation in the chromosome. Since only specific genes are involved, the process is called specialized transduction.

In generalized transduction any bacterial gene can be carried by the phage particle. In this case it is believed that after induction of the prophage the bacterial chromosome breaks down and any part of it can be put into a phage head. It does not seem to matter to the phage head whether it is full of phage DNA or bacterial DNA. Since the ability to infect cells and inject DNA is determined by the proteins of a virion and not by the DNA in that virion, bacterial DNA in a phage can be injected into a cell. Once inside the cell the transduced DNA can recombine with the host chromosome normally.

The amount of bacterial DNA that can be transduced by a single virion can be no larger than the amount of phage DNA in a normal virion. As we have seen this is a few percent of a bacterial genome.

MICROBIOLOGY AND MEDICINE

Our discussion has emphasized today's interplay of genetics, microbiology and molecular biology. We have only touched upon the key importance of microbes to man. Today genetics is a cornerstone of medicine. It was not always so. Genetics and microbiology, for example, evolved separately. In fact the pioneering work of Koch and Pasteur, in demonstrating that bacteria were causes of disease, focused attention on their relation to the health and welfare of man and on practical problems of pressing importance. Bacteriologists were "microbe hunters," identifying organisms and attempting to develop both specific antibodies for *passive*, and vaccines for *active, immunization*.

It was not until 1935 that an article appeared that ultimately drastically changed the attitude of microbiologists toward genetics and forced them to probe more deeply the mechanisms underlying bacterial variation.

In 1935 Domagk found that the dye Prontosil inhibited the growth of streptococci in man. However, although it was effective in treating infections, it was without effect on bacteria *in vitro*. It was soon discovered that Prontosil was degraded in the body and that its active antibacterial component, both in man and in the test tube, was *sulfanilamide* (Fig. 9-19). This discovery gave rise to a class of chemotherapeutic agents, the sulfonamides. The sulfonamides inhibit the growth of bacteria by preventing their synthesis of the vitamin folic acid. In the normal course of synthesis, p-aminobenzoic acid (Fig. 9-19) is fitted into the folic acid molecule. Sulfanilamide, being structurally similar to p-aminobenzoic acid, usurps its place; this opened the way to a major principle of drug design, the concept of *antimetabolites*.

Domagk's findings and their extension in turn sprung from the work of the German chemist Paul Ehrlich, who coined the term *chemotherapy* at the turn of the century. It means simply the treatment of disease with chemicals. Ehrlich had some success using dyes to treat African sleeping sickness and arsenicals against syphilis. For a time (between Ehrlich and Domagk) interest in chemotherapy waned because of the rising interest in the possibility that bacteriophages might be used as therapeutic agents. Bacteriophage therapy was never effective; but the idea captured the imagination, as Sinclair Lewis's novel *Arrowsmith* reveals.

As the concept of antimetabolites caught hold, microbiologists realized that antagonism among microbes exists in nature and reasoned that this phenomenon might provide the basis for more antimicrobial agents, or *antibiotics*. The antibiotic era was ushered in by Fleming's chance observation of the inhibitory properties of the fungus *Penicillium notatum* which secretes the antibacterial substance *penicillin* (Fig. 9-20). The drama of the discovery and development of penicillin in the early 1940s has been recounted too often to warrant a full-scale description here. It was the calamity of World War II that forced a frontal attack on the purification and mass production of penicillin. The first clinical trial in the United States was in 1942. By September 1943 enough of the drug was being produced to satisfy the demands of all the allied forces.

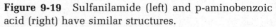

Figure 9-19 Sulfanilamide (left) and p-aminobenzoic acid (right) have similar structures.

Penicillin proved to be ineffective against one class of bacteria (the gram-negative type, so-called simply because of their failure to stain with a specific dye). Thus a search was begun leading to the isolation of *streptomycin*, which has been shown to be effective in treating many diseases, including tuberculosis.

Since the 1940s over 1000 antibiotics have been described and over 50 produced on a commercial scale. The search continues, with the roll of antibacterial compounds now including tyrothricin (which proved to contain two inhibitors, gramicidin and tyrocidin), bacitracin, chloromycetin, aureomycin, and terramycin. There are two inhibitors of fungal infections, amphotericin B and nystatin.

Today the availability of antibiotic therapy is taken for granted, and we easily lose sight of the fact that despite our advances there is a long way to go. First, the task is far from complete simply in fulfilling the classical mission of microbiology, the identification of agents responsible for disease. As we saw in Chapter 8, the viral etiology of cancer is a case in point. Second, the actual mechanism of action of many of the antibiotics,

even penicillin, is not completely understood. Penicillin is thought to block the final stage in the assembly of murein, the structural polysaccharide of the cell wall, leading to the ultimate lysis and death of the bacterium.

Poliomyelitis—Conquest of a viral disease by mass immunization

The general principle of immunization was established by Jenner's work on smallpox in the last century; but before the principle could be applied to poliomyelitis much fundamental knowledge had to be acquired. The modern era of research on viruses began in 1935 with the isolation of tobacco mosaic virus. There followed two decades of fundamental research on viruses. The keys to success in the campaign against polio which was spearheaded by the National Foundation for Infantile Paralysis (now the National Foundation) were the development of techniques for growing the virus in cultured mammalian cells and the recognition of three strains of polio virus. Each strain was grown in cultured cells and it was virus from this source that was used for preparing killed vaccines. Later, mutant forms of each strain were developed which had lost their disease-producing capacity but not their antigenicity (immunizing capacity), making possible the development of live virus vaccines.

In 1950 to 1954 there were more than 20,000 cases of polio yearly in the United States, 14.6 per 100,000 population. In 1961 the rate had dropped to 1.8 per 100,000. In 1965 there were only 61 cases reported in the whole country.

MICROBIOLOGY AND INDUSTRIAL TECHNOLOGY

In stressing the antibiotics we may have narrowed our focus too sharply, creating the impression that the only import of microbes to man lies in their relation to disease. However, another role looms large: "microbes as manufacturers." Microbiologists, chemists, and engineers have pooled their talents to develop industrial processes in which microorganisms (and their enzymes) produce a variety of useful compounds.

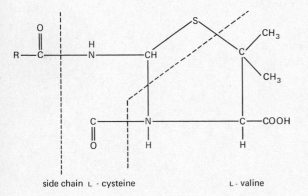

side chain ∟ - cysteine   ∟ - valine

R = benzyl, *n*-heptyl and
2-pentenyl in penicillins G,
K, and I, respectively

Figure 9-20   Structure of penicillin.

These techniques center about the **fermentation** process, whereby chemical changes are produced in an organic substance through the action of microbial enzymes.

All of us know about the role of yeasts in the production of beer, ale, and other malt beverages, wines and the fermented mashes from which whiskey and other distilled spirits are produced. However, we overlook the comparable role of bacteria in producing lactic acid, vinegar, dextrans (which are widely used as substitutes for blood plasma, as "stabilizers" in ice cream, as protective coatings on seeds, and numerous other roles), sorbose (employed in the synthesis of ascorbic acid), the flavor enhancer, monosodium glutamate, the amino acid lysine, used to supplement the food value of cereal grains, and vitamin B_{12}. Some bacteria like *Bacillus thuringiensis* produce *bioinsecticides* which kill insect larvae which consume them. They can be used to control insects that destroy alfalfa, cabbage, tobacco, and tomato.

The fermentation industry is an important resource. Looking toward the future, one of its major roles may be to exploit the synthetic prowess of microorganisms not only to manufacture complex substances but to destroy waste products of our complex civilization.

SUGGESTED READING LIST

Brock, T. D., *Biology of Microorganisms*. Prentice-Hall, Englewood Cliffs, New Jersey, 1970.

――――*Principles of Microbial Ecology*. Prentice-Hall, Englewood Cliffs, New Jersey, 1966.

――――*Milestones in Microbiology*, Prentice-Hall, Englewood Cliffs, New Jersey, 1965.

Edgar, R. S., and W. B. Wood, "Morphogenesis of bacteriophage T_4 in extracts of mutant-infected cells," *Proceedings of the National Academy of Sciences*, **55**, 498 (1968).

Handler, P. (ed.), *Biology and The Future of Man*. Oxford University Press, New York, 1970. Chapter 16, "Science and medical practice," Chapter 18, "Biology and industrial technology," and Chapter 19, "Environmental health."

Levine, R. P., *Genetics*, second edition. Holt, Rinehart and Winston, New York, 1968.

Luria, S., and J. E. Darnell, *General Virology*, second edition. Wiley, New York, 1967.

Nester, E., C. E. Roberts, Jr., B. McCarthy, and N. Pearsall, *Microbiology*. Holt, Rinehart and Winston, New York, 1973.

Sistrom, W. R., *Microbial Life*. Holt, Rinehart and Winston, New York, 1969.

Stanier, R. Y., M. Doudoroff, and E. A. Adelberg, *The Microbial World*, third edition. Prentice-Hall, Englewood Cliffs, New Jersey, 1970.

Zubay, G., *Papers in Biochemical Genetics*. Holt, Rinehart and Winston, New York, 1968.

Plants—The Food Providers

The procaryotic organisms discussed in the previous chapter are all basically unicellular. To understand how they function involves learning how a single cell works. In contrast, most plants and most animals are multicellular and are composed of many different kinds of cells. The biggest problem in analyzing how even one small part of a plant or an animal works is that so many cells are involved and so many different processes are occurring at the same time. Even a plant as small as a blade of grass or an animal as small as a housefly is composed of a million cells. Fortunately we can simplify this task by concentrating on the important functions, one at a time. This is not as artificial as it seems, because different parts of both plants and animals have become more or less specialized to perform particular functions. The specializations are evident at the cellular level where the millions of cells have specialized into dozens or hundreds of distinct types. As a result of these specializations only a few of the cells are able to live independently outside the organized body of which they form a part. Yet working together they form a plant or an animal that does things that a population of independent cells could not do.

Groups of similar cells such as those making up muscle in an animal or epidermis in a plant are called a **tissue.** Tissues in turn are usually combined in an organized fashion to make up an **organ** (such as a bone, kidney, or heart in an animal or a leaf in a plant). Groups of organs usually function together to make up an **organ system,** such as the skeletal system or the excretory system of an animal or the shoot of a plant. The individual activities of tissues and organs are regulated and integrated by chemical controls in both animals and plants and also by nervous controls in animals. These integrated activities enable plants to manufacture and transport food, to transport water and minerals, to grow and to reproduce; they also enable animals to move, feed, digest, circulate their body fluids, excrete, breathe, and reproduce.

This chapter describes how plants are put together and how they work. It is followed by a discussion of how animals are put together and how they work.

STRUCTURE OF A GREEN LAND PLANT

There are many different sorts of plants, but the best studied, the most familiar, and the most important are the great group of green land plants known as the *angiosperms* or flowering plants. This chapter focuses largely on these plants. There are about 250,000 different kinds of flowering plants and they are divided into two large groups, the **monocots** and the **dicots.** They differ from each other in a number of anatomical and other features that will be discussed in Chapter 23. The most obvious difference is that in monocots the veins in the leaves tend to run parallel to one another, whereas in dicots the leaves form a netlike pattern. Common monocots are grasses, irises, palms, and orchids. Dicots include most woody plants—bushes and trees—and herbaceous plants such as beans, mustards and buttercups. "Herbaceous" plants have soft, succulent stems in contrast to "woody" plants.

Another major group of land plants are *gymnosperms*, which include the conifers. Gymnosperms will be mentioned from time to time in this chapter, but are treated in detail in Chapter 23.

Green plants are first and foremost living machines which trap the energy of sunlight and use it through photosynthesis to convert carbon dioxide into organic compounds. Much of their structure and behavior reflects adaptations that have evolved over several hundred million years which enable them to be successful in diverse environments. To understand how a higher green plant functions we must learn how it is put together.

Plant cells and cell walls

The key to understanding plant structure is the plant cell and its **cell wall.** This wall surrounds the cell membrane of a plant cell and is secreted by the cell. It is often less than 1 micron in thickness. It is composed of a network of tiny threadlike fibrils of *cellulose,* a large molecular weight polymer formed by hooking large numbers of glucose molecules together. The cell wall also contains substances called *glycoproteins,* which are composed of sugars and proteins hooked together. When they are first secreted, the cellulose fibrils are laid down in a regular pattern, but as the cell grows they become less regularly arranged and finally form a meshwork. Unlike the cell membrane which allows only certain molecules in and out of plant cells, the cell wall is

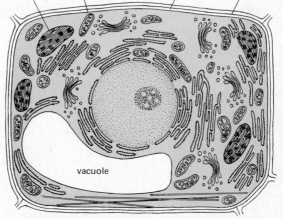

chloroplast   plasma membrane   cell wall   middle lamella

vacuole

Figure 10-1 A highly diagrammatic view of the structure of a plant cell. In addition to cytoplasmic organelles such as the nucleus, nucleolus, endoplasmic reticulum, mitochondria, and microtubules, which would be found in both plant and animal cells, a typical plant cell might contain chloroplasts, a large water-filled vacuole, a cell wall, and a middle lamella that glues the cells together.

permeable to most substances. It provides rigidity
to the plant cell and enables it to withstand considerable internal pressure. Animal cells do not
possess a cell wall and consequently function and
interact differently from plant cells.

Between the walls of two plant cells is a layer
called the **middle lamella** which glues the cells
together. The lack of motility in plant cells is
due to the fact that adjacent cells are glued together. The main component of this glue is a
polysaccharide called **pectin.** When a fruit ripens,
the pectin in the middle lamella may be dissolved and the cells become less closely glued
together so that the fruit becomes softer. The same
thing happens when a fruit is bruised and gets
infected by fungi or bacteria: enzymes produced
by these organisms digest the pectin and produce
large soft spots in the fruit.

The primary wall of a plant cell and the
middle lamella are usually quite thin. In addition, some plant cells also secrete a tough thick
secondary wall on the inner surface of the primary cell wall that may be many times thicker
than the primary cell wall. This secondary wall
is formed by adding large amounts of additional
cellulose and other polysaccharides to the inner
surface of the primary wall. Frequently this

Figure 10-2   Cellulose molecules may each contain as
many as 2500 glucose units. Hundreds of these molecules are bound together into microfibrils which comprise the cell wall of a plant. This electron micrograph
of a plant cell wall shows the meshwork of cellulose
microfibrils, 10–20 nanometers in diameter, which
remain after other cell wall materials and all cytoplasm
has been removed from the cell ($\times$ 24000).

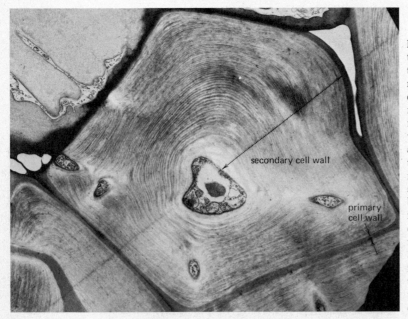

secondary cell wall

primary
cell wall

Figure 10-3   You may have noticed
that pear fruits have a "gritty" texture. The "grit" in the pear fruit
represents aggregates of *stone cells*,
cells which have deposited a remarkable amount of secondary wall
material inside the primary cell
wall. Here we see a cross section of
a stone cell as seen with the electron microscope. The stone cell remains alive inside the very thick
secondary wall it has built. A
branching network of cytoplasm-
filled channels penetrates the secondary wall and leads to the plasmodesmata in the primary cell wall.
Stone cells are able to obtain the
food they need to remain living and
to synthesize their thick secondary
walls by means of these cytoplasmic
channels ($\times$ 4000)

fibrous network is impregnated with *lignin,* a complex polymer built up of benzene ring derivatives that cements the cellulose fibers together to form a highly rigid structure. Rigid secondary walls are particularly obvious in the cells that form the woody parts of trees and the fibers of stems, such as that of the flax plant. Many cells die when their secondary wall is finished. They leave behind the tough tube formed by their walls which functions to transport fluids in the plant and to maintain support.

Although plant cells are enveloped by their cell walls, they are able to communicate with one another, for the cell wall has tiny holes in it through which cells make contact with thin threads of protoplasm. These *plasmodesmata* provide intercellular communication and are thought to enable plant cells to exchange chemicals with one another. Similar sites of intercellular communication also occur in animal cells and were discussed in Chapter 2.

Most mature plant cells have a fluid-filled space bounded by a membrane called a **vacuole** which occupies most of the cell. The fluid in this vacuole exerts a tremendous pressure, called **turgor pressure,** on the cell wall, so that the mature plant cell is quite rigid.

Plant tissues and organs

Mature plants are made up of two basic kinds of tissues, **meristematic tissues** and **mature tissues.** These tissues are usually composed of cells of one type but sometimes consist of several different kinds. Meristematic tissues are composed of embryonic, undifferentiated cells which can divide indefinitely. Mature tissues are composed of cells which may undergo only a limited number of divisions during normal development and usually cease dividing entirely. Such mature tissues often have highly specialized functions. Among the mature tissues of higher plants are the *surface tissue* or skin of the plant, the *vascular* or *conducting tissue* through which water or various solutions move from one part of the plant to another, and the *ground tissue* which is between the skin and the vascular tissue.

Meristematic tissues have already been mentioned in Chapter 8. In plant embryos cells divide throughout the embryo, but as the plant matures most cells cease to divide and cell division becomes limited largely to the meristems. **Apical meristems** are found at growing tips of stems or roots and enable plants to grow in length. An-

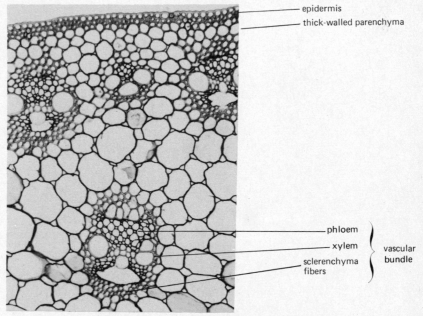

epidermis
thick-walled parenchyma

phloem
xylem
sclerenchyma fibers
} vascular bundle

Figure 10-4 Photograph of a cross section of part of the stem of a corn plant showing several types of tissue. In addition to thin-walled parenchyma cells, the outer part of the stem contains thick-walled parenchyma cells which help give the cornstalk its stiffness. The stem also contains tough sclerenchyma fibers which surround many of the vascular bundles. These fibers are the dead remains of elongated sclerenchyma cells and also provide stiffness to the cornstalk.

other meristem, the **vascular cambium,** occurs as a thin layer between the bark and the wood of woody plants. It gives rise to the vascular tissues and enables stems and roots to increase in girth. The vascular cambium is an example of a *secondary* or *lateral meristem.*

Surface tissue. Surface tissue forms the protective outer covering or skin of the plant body, and living epidermal cells are the main surface tissue of roots, stems, and leaves. Sometimes the **epidermis** secretes a waxy, water-resistant **cuticle** which protects the plant from water loss and injury. Frequently the epidermis produces hairlike structures, some of which secrete sticky substances that protect the plant against attack by insects. Figure 10-4 shows the epidermis and several other tissues of a corn stalk.

Ground tissues. Ground tissues comprise all the tissues except vascular tissues and surface tissues. Most ground tissues consist of unspecialized cells; the most common kind is called **parenchyma.** Parenchyma consists of cells that are mainly of one kind and usually have thin walls. Frequently stems contain two kinds of parenchyma. One, the **cortex,** occupies a peripheral position within the stem just under the epidermis, whereas the other, the **pith,** is found in the center of many stems. In leaves parenchyma cells are filled with chloroplasts and do most of the photosynthesizing. Some parenchyma cells retain the capacity to divide and may become meristems. Cortical parenchyma of stems may contain cells modified for support, whereas other parenchymal cells may function in food storage, such as the parenchyma of the potato tuber. In fact most of the edible parts of plants consist of parenchyma. An important feature of many parenchyma tissues is that the cells are loosely packed together and have interconnecting air spaces in between. Thus the intercellular spaces in most tissues of a land plant are filled with air. This characteristic of plant parenchyma contrasts with the intercellular spaces in animal tissues which are filled with fluid. The importance of this difference will become clear when we consider how essential it is for gases to move through plants for both photosynthesis and respiration to occur.

In addition to parenchyma tissue, other ground tissue (Fig. 10-4) provides for both storage and support. For example, **sclerenchyma** tissues secrete thick secondary walls and become extremely tough and strong. They often die when they are mature. Commercial hemp and flax as well as shells of nuts are composed of strands of tough sclerenchyma tissue.

Vascular tissue. Vascular tissue is the distinguishing feature of higher plants. It is an intricate network of internal plumbing that runs continuously from the tips of the leaves and buds through the stem and main root down to the smallest root. It is visible in leaves as the familiar "veins." The two major kinds of vascular tissue that occur side by side are **xylem** and **phloem.** Xylem transports water and dissolved minerals upward in the plant body and also supports the plant body. It is a complex tissue composed of several kinds of cells which form a series of minute continuous tubes from the roots, through the stems, to the leaves. Several kinds of cells that make up the xylem secrete a thick secondary wall and do not perform their conducting and supporting functions until they die. Phloem transports organic materials manufactured by the plant, such as sugars and amino acids, down from the leaves where the food was produced toward

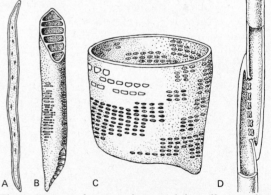

Figure 10-5 A drawing of some of the cell types that might be found in the xylem of woody plants. (A) A fiber from yellow poplar showing the pits through the secondary wall. (B) A vessel element from the spring wood of yellow poplar. (C) A vessel element from the spring wood of oak. (D) A tracheid from the summer wood of a white pine. Note the pits on the lateral surfaces through which water travels from cell to cell.

the roots and other food-consuming tissues. It also transports food materials up the plant body. Like xylem it provides a continuous transport system from leaves to roots. It too is composed of several kinds of cells, but unlike xylem cells, phloem cells, transport functions depend upon their being alive.

Structure of xylem and phloem

Let us take a brief look at the cellular structure of xylem and phloem. Xylem contains a number of different types of cells including two types called **tracheids** and **vessel elements** which play a major role in water and mineral transport (Fig. 10-5). Both of these cells are unique in that they can function in conducting materials only after they have died. During their brief life they produce a thick secondary cell wall composed largely of cellulose, other polysaccharides, and the complex polymer, lignin. Tracheids are elongated, tapering cells with thick secondary walls that are interrupted by numerous *pits* that have thin walls. Water and dissolved minerals can pass from tracheid to tracheid through the pits (Fig. 10-5).

Vessel elements are shorter and wider than

tracheids and are found only in flowering plants. They have some pits along their sides, but the ends of the cells are really well perforated and may be missing altogether. A series of such vessel elements, stacked one above the other, forms a continuous multicellular tube, the vessel. In trees vessels range in diameter from about 20 to more than 400 microns. In addition to vessels and tracheids which are dead, xylem also contains various types of living cells. These cells, however, can be killed without stopping water transport up the xylem. Apparently living cells of the xylem are not necessary for water transport.

Phloem like xylem is composed of a number of different types of cells. The phloem cells responsible for conducting the products of photosynthesis in flowering plants are known as **sieve tubes.** A sieve tube is analogous to a vessel in that it is composed of a large number of individual cells linked end to end to form a single conducting unit. The individual cells making up the sieve tubes are known as **sieve cells** or **sieve elements.** A sieve element is a long thin cell that resembles a tall tea strainer. Its end walls are called *sieve plates*, because they contain an un-

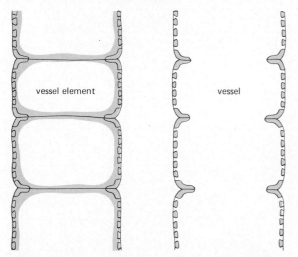

Figure 10-6A A diagram showing two stages in the formation of a vessel. Secondary wall material has been laid down to form spiral thickenings on the lateral walls of a vertical file of cells. When secondary wall deposition is complete the end walls, which did not receive any secondary thickenings, disintegrate as does the cytoplasm.

vessel element

vessel

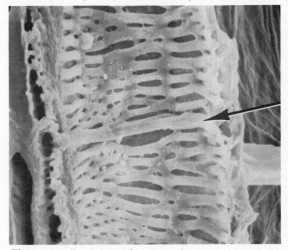

Figure 10-6B An inside view of two mature vessel elements stacked one on top of the other. This remarkable scanning electron micrograph was made by cutting the cylindrical vessels lengthwise and scanning the inside of the half cylinder. The thickened junction of two vessel elements is indicated by the arrow. The wall between them has largely disappeared as in 10-6A so that they form a continuous vessel ($\times$ 450).

usually large number of pores. These pores tend to be clustered into many sieve areas on each sieve plate. In addition, there may be sieve areas scattered over the lateral walls of these unique cells. The sieve plate at each end of the sieve element joins the sieve plates of sieve elements immediately above and below it. The pores through the sieve plates represent passages which lead from one sieve element into the next.

Unlike the vessel element of the xylem the functioning sieve element of the phloem is very much alive. The cytoplasm of the sieve element contains a fine network of a special protein, designated "P" protein. This fibrous protein extends through the pores of the sieve plates, and it is thought to function in the movement of food materials from sieve element to sieve element. When the sieve element is injured, the P protein aggregates as a slime plug over the sieve plate, blocking the further movement of materials through the injured sieve tube. Another unique

aspect of the sieve element is that its nucleus gradually disintegrates as the sieve element differentiates. Despite the loss of its nucleus it will remain alive for a year or more, before dying. Small nucleated cells called **companion cells** associated with the sieve element apparently provide nuclear functions. When the sieve element finally dies it also stops functioning in food conduction. Thus new sieve tubes must be formed each year to take over the duties of those which have ceased to function.

The transport functions of xylem and phloem can be demonstrated by many simple experiments. One of the most familiar is called *girdling*, in which a ring of bark is stripped off a tree. In stems of trees and other woody plants the xylem forms a solid central cylinder called the

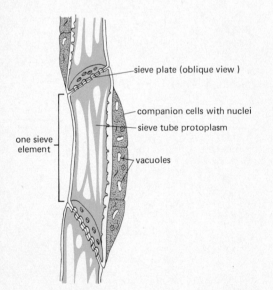

Figure 10-7   A diagram showing longitudinal view of a mature sieve tube in relation to some of the other types of cells found in the phloem of angiosperms. The mature sieve tube consists of a number of sieve elements without nuclei stacked end to end. They are always associated with smaller companion cells which have a nucleus. The end walls separating functional sieve elements are highly perforate, forming a structure known as the sieve plate.

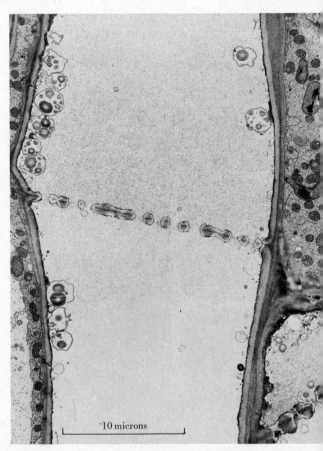

Figure 10-8   An electron micrograph of a mature sieve tube showing one sieve plate (× 3300).

wood which is surrounded by an external layer of phloem comprising part of the *bark* of a tree. If the bark is peeled off all the way around the trunk of a tree leaving the woody core of the stem, the roots get cut off from their food supply, cease growing, and finally die. This well-known observation indicates that the bark has vascular tissue which conducts food and that the roots need the food from the leaves. The leaves continue to be provided with water for a long period, because the water comes to them through the xylem. Finally, however, the leaves also die as the roots die because they cease to absorb enough water.

Although the vascular tissues of a plant look superficially like the circulatory system of an animal, they are basically different in design and function. Both animal circulatory systems and plant vascular tissues provide for the distribution of nutrients through the organism to places they are needed. However, in the plant vascular sys-tem, the xylem and the phloem are not directly interconnected, and the water that moves up in the xylem evaporates into the air instead of returning downward in the phloem. In contrast, in the vertebrate vascular system, the arteries and the veins are interconnected by the capillaries (see Chapter 12). Also plants have no "hearts" which actively pump their fluids through the xylem and the phloem. We shall examine how they move fluid around through their vascular tissues later in this chapter. Another difference is that in most cases the vascular tissues of plants do not transport oxygen from the air to the cell as they do in most animals. A final difference is that the vascular tissues of plants also serve as a skeleton and support the plant body mechanically.

Well-developed vascular tissue is limited to the higher green plants and is not found in simpler plants, such as algae and mosses. In many tall forest trees the distance between the

Figure 10-9 This stem of a black cherry tree was girdled two years before the photograph was made. Growth in diameter, brought about by the vascular cambium, continued above the girdle, utilizing food produced by the leaves of the branch, and water and minerals which were still transported through the xylem. However, no growth occurred below the girdle because no food could be transported to it. When food reserves present in the lower stem and roots became exhausted, the roots died and so did the entire plant.

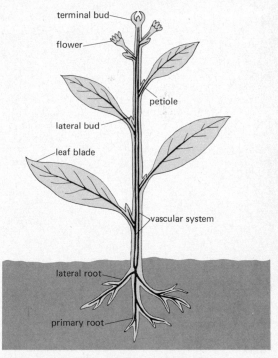

Figure 10-10 A blueprint for a higher plant. This diagram shows the principal parts of a higher plant. It also shows how these parts are connected by the plant's plumbing, the vascular system.

roots and leaves may be enormous, but the xylem and the phloem each form continuous pathways between them, and roots and leaves exchange materials with ease. The vascular system has clearly been a key to success for land plants and has enabled them to become large and upright and to live in places which are not constantly moist.

Plant organ systems

These several kinds of tissues make up the organs and the two basic and familiar organ systems of higher plants, the **root system** which grows downward into the soil and the **shoot system** which grows upward into the air. The root and the shoot display numerous differences in structure and function. The main function of roots is to procure and transport water and inorganic nutrients, to store nutrients, and to anchor the plant to its spot of earth. Roots are fed by sugar produced in the leaves. The shoot, which consists largely of the stem, the green leaves, and the reproductive organs, serves many functions. Leaves function in photosynthesis and produce sugar. The stem serves to transport minerals from the roots to the shoot and the leaves, to transport sugar from the leaves to the roots, to support the plant body, and to bear reproductive organs for continuing the species.

Structure and growth of a root— Making it underground

Figure 10-11 shows the cross section of a root of a young dicot plant. It is surrounded by a layer of cells, which is one cell thick, called the *epidermis*. Unlike the epidermis on the stem and the leaves, that of the root is not waterproofed so that the root is able to absorb water. Immediately below the epidermis is a tissue known as the *cortex*. It consists of a large number of thin-walled parenchyma cells which are sometimes arranged in definite layers. The innermost layer consists of a single layer of cells which are quite different from those of the rest of the cortex. This layer is known as the **endodermis,** and as we shall see later in the chapter, it controls the direction of water movement. The central core of the

root is called the **vascular cylinder** and contains the xylem and the phloem. The xylem cells are commonly arranged as a star-shaped figure with bundles of phloem cells between the xylem arms. As we mentioned earlier the primary function of the xylem is the movement of water and minerals from the roots, where they are absorbed, into the shoot. The main function of phloem is to transport food materials from the leaves toward the roots and other food-consuming tissues.

To understand how a root performs its principal duties of water and mineral absorption, it is necessary to know something about how a root grows. A very simple experiment, which can be performed at home, tells a great deal about root growth. Some pea seeds are germinated in moist paper towels. When the roots are about 2 centimeters long, a series of parallel lines is drawn across the root with india ink. The lines should be

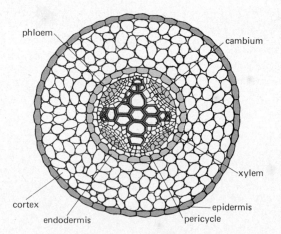

Figure 10-11 Diagram of a young dicot root as seen in cross section. Note the alternate arrangement of xylem and phloem. In a stem these tissues lie on the same radius with the phloem outside the xylem. When a root branches, the lateral root originates from the outermost tissue of the vascular cylinder called the *pericycle*. The lateral root pushes its way through the endodermis, cortex and epidermis.

exactly 1 millimeter apart and extend from the tip of the root to its base. The seeds are then returned to moist paper towels inside a glass jar. When the marked roots are examined several hours later, it is found that all the lines are still 1 millimeter apart, except for those near the root tip. The lines near the tip of the root, however, may now be several millimeters apart. Careful observation for several hours shows that the most rapidly growing part of the root is the region about 2 millimeters from the tip. Regardless of how long the root might become, the only part of the root that would increase in length would be the terminal few millimeters.

We can obtain a more detailed picture of how a root grows by examining a longitudinal section of the growing portion of the root (Fig. 10-13). Dividing cells are largely confined to the terminal 1 to 2 millimeters of the root. The dividing cells near the root tip are comparatively small. In the region between 1 millimeter to about 4 millimeters from the tip the individual cells of the root become progressively larger. Beyond about 4 millimeters, cell size does not increase and the cells do not divide—growth is complete.

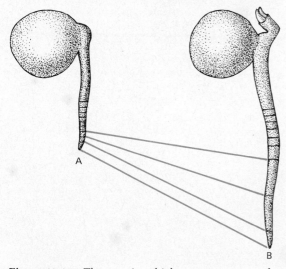

Figure 10-12 The way in which a root grows can be illustrated by placing india ink marks 1mm apart on the root of a pea seedling (A). A few hours later the distance between some of the marks will have increased, showing that the root was growing in those regions (B).

This means that the root grows by two different processes—*cell division* and *cell enlargement*. Both processes take place throughout the growing tip region. However, the small cells within 1 millimeter of the tip are growing primarily by cell division, whereas the cells between 2 and 4 millimeters from the tip are growing primarily by cell enlargement. The region of the root in which the cells become progressively longer is known as the *zone of elongation*. The small, dividing cells at the tip of the root make up the *root apical meristem*. To understand root growth it must be realized that elongation is not only a zone near the tip of the root but a developmental stage through which all of the mature cells of the root have passed. A new cell is born when an existing cell divides. If the new cell is formed within the apical meristem, it will probably divide several more times. If it comes into existence in the zone of elongation, it might never divide again. In either case it will eventually enter into the elongative phase of its growth cycle. During elongative growth it may become as much as ten times longer than the small cells of the apical meristem.

Besides producing cells in the direction of the main part of the root, the apical meristem also produces cells in the opposite direction which form a cap over the extreme tip of the root. As the apical meristem divides, the new **root cap** cells which are produced push the older root cap cells farther away from the main part of the root. The cells of the root cap are expendable. The growing root pushes the root downward or out-

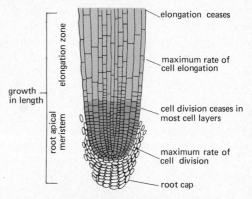

Figure 10-13 Longitudinal section of a root.

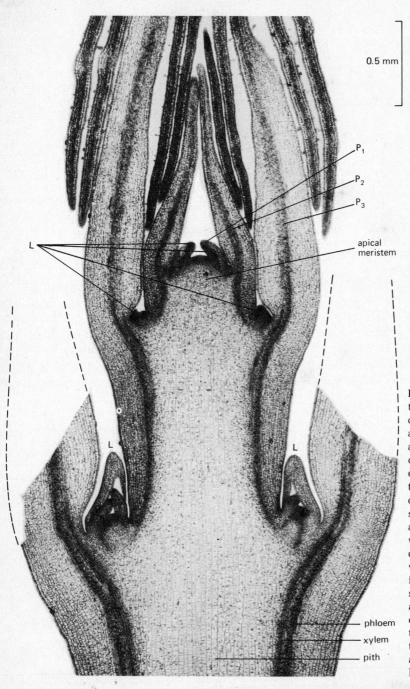

0.5 mm

P₁

P₂

P₃

apical
meristem

L

L

L

phloem

xylem

pith

Figure 10-14 A longitudinal section through a growing shoot tip of a lilac (*Syringa vulgaris*). The apical meristem (A) has produced all the structures shown. Not only does the growth of the shoot tip extend over a much larger area than that of the root tip, but various structures originate along the sides of the shoot apex. The shoot apex forms lateral appendages as well. These lateral structures include the leaf primordia from which new leaves will form (designated P₁, P₂, P₃, and P₄ to represent successively older primordia) and the lateral buds (L). Active cell division occurs not only in the apical meristem, but also in the developing stem below the apical meristem and in the young leaves.

ward through the soil. As the root cap brushes past rock particles and other obstacles, the outermost cells of the cap are scraped away. Thus the root cap protects the apical meristem as the root grows through the soil. Before they are lost, the root cap cells produce and excrete large amounts of a mucilagenous carbohydrate material. This mucilagenous material may help lubricate the movement of the root through the soil.

Structure and growth of a stem— Connecting roots and leaves

Stems are the middlemen between the roots and the leaves. They support the leaves and are the transportation route along which water and dissolved minerals move up to the leaves, and down which food travels from them. It is usually easy to distinguish stems from the roots and leaves. Some stems, however, are short and concealed under a crown of leaves, as in carrots or beets. Others grow underground, as in potatoes. Vines and thorns are highly modified stems.

The xylem tissue in the stem of higher plants is very similar to that of the root with respect to the kinds of water-conducting elements it contains and the way in which these elements are formed. However, the arrangement of the xylem tissue within the stem is quite different from that found in the root. Anatomically the shoot is considerably more complicated than the root. The shoot apical meristem not only produces cells which will become the primary tissues of the stem but cells which will become leaves, flowers, and buds as well. Furthermore growth is not confined to the terminal few millimeters as it is in the root. The growing region of the shoot may extend over several centimeters (Fig. 10-14).

Let us examine a cross section of the stem of an herbaceous dicot such as a soybean (Fig. 10-15). The first formed vascular tissues of a soybean consist of a number of discrete bundles which run vertically through the stem. Each *vascular bundle* contains both xylem and phloem. The xylem and the phloem of a particular vascular bundle are in the same radial plane, with the phloem toward the outside of the stem and the xylem toward the inside. Usually the stem has somewhere between four and ten discrete vascular bundles arranged in a circle. In the center of this circle is the *pith. Cortical cells* lie between the vascular bundles and the epidermis.

Anyone familiar with wood knows that this description does not apply to a tree (Fig. 10-17). The woody part of a tree is solid xylem, except for a very small amount of pith at the center. The bark of the tree is, in part, phloem. Obviously neither the woody part of the tree nor the bark is arranged in discrete bundles. It may start out that way, but very early in the growth of the seedling tree, a *vascular cambium* appears. The vascular cambium is a secondary meristem, as opposed to the root and shoot apical meristems which are primary meristems. It gives rise to so-called

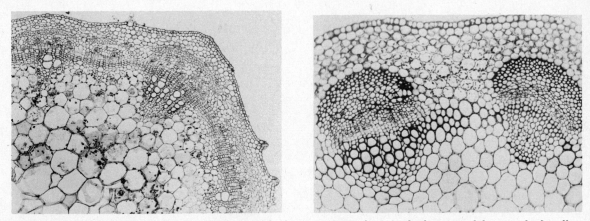

Figure 10-15 Photographs of the cross sections of herbaceous stems showing the location of the vascular bundles. The photograph on the left is of a soybean stem while the one on the right is of a clover stem. (Reprinted by permission from *Botanical Microtechnique*, 3rd ed., by John E. Sass, © 1958 by The Iowa University Press, Ames, Iowa.)

"*secondary tissues*" which are not derived from apical meristems. Growth of these secondary tissues, called *secondary growth,* enables trees to increase in girth. As Figure 10-17 shows, the vascular cambium is a layer of dividing elongated cells which forms between the xylem and the phloem of each vascular bundle and also extends across the tissues between the vascular bundles to form a continuous cylinder. As this cylinder of cells divides, the daughter cells produced toward the outside of the stem differentiate as phloem elements, whereas those produced toward the inside of the stem differentiate as xylem elements. As new cells are formed by the cambium, the diameter of the stem increases. The first phloem cells formed, the primary phloem, tend to be crushed by this outward expansion. After a few years' growth it cannot be seen at all in cross section of the stem. The primary xylem is not crushed, but it is often difficult to find since it represents such a small portion of the xylem tissue. When we cut the trunk of a tree, we see a solid central cylinder of xylem, the wood, which is surrounded by an external layer of phloem which comprises part of the bark.

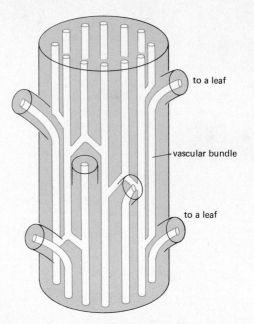

Figure 10-16 A diagram showing the vascular system of a flax plant. Note how the vascular bundles in the stem send extensions into leaves.

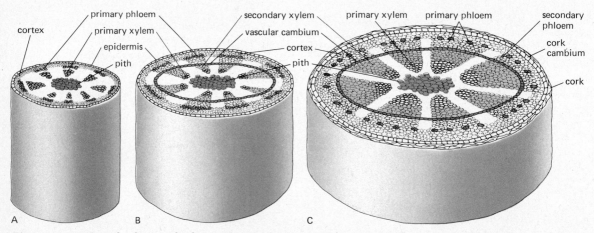

Figure 10-17 Growth of a woody dicot. (A) Diagrammatic cross section of the stem of a dicot before the start of secondary growth. Note the discrete bundles of xylem and phloem. (B) Secondary growth results from the activity of the vascular cambium which becomes a continuous thin cylinder between the first formed or primary xylem and primary phloem. The vascular cambium produces secondary xylem and phloem. The closer cells are to the vascular cambium, the more recently they were formed. (C) After further growth the secondary xylem and secondary phloem produced by the vascular cambium become continuous cylinders around the trunk. At about this time a new layer of cambial cells—the cork cambium—often appears just under the epidermis. The cork cambium gives rise to wax-containing cork cells which finally die and form a tough protective layer, the cork, around the stem. In this process the epidermal cells also die. Thus the bark of a tree usually consists of the phloem and the cork.

The diameter of the xylem elements formed early in the spring in trees growing in temperate climates is much larger than the xylem elements formed during the summer. This gives a cross section of the tree a banded appearance. Each year the tree first produces a light region of large diameter elements followed by a darker region of xylem elements with narrower diameters. This constitutes an annual ring of xylem. It represents the wood produced by a tree in a given year. The age of the tree can be determined with considerable accuracy by counting the number of annual rings.

The width of an annual ring indicates the vigor of the tree. It also tends to reflect the climatic and environmental conditions the tree was subjected to while it produced the annual ring. For example, a pine tree might struggle for many years in the shade of a large old oak. The pine would produce very narrow annual rings which would reflect the generally poor growth of the tree. However, if the oak were cut down, the pine might begin to grow very rapidly and produce much larger annual rings. In the South-

west the width of annual rings is correlated very closely with the amount of rainfall the trees receive during the growing season. The pattern of annual rings tends to be a good indication of when a particular piece of wood was made. Annual rings have been used to determine the age of artifacts in Southwestern archeological sites.

Cellulose and lignin— Paper and pollution

We have mentioned cellulose on several occasions and some special comments about this remarkable polymer are appropriate at this point. Cellulose not only is important to the plant but to man as well. Cotton fibers, which are produced by the fruit of the cotton plant, are almost pure cellulose. Wood, which is the xylem tissue of trees, owes its great strength and utility to the fact that it is about 50 percent cellulose by weight. From man's standpoint lignin, which makes up about 20 percent of the weight of the wood, is mainly a nuisance. Although it helps preserve the wood from attack by wood-rotting organisms,

Figure 10-18 A cross section of the trunk of a pine tree showing the annual rings in the wood.

it must be removed from the cellulose fiber before paper can be made from the wood. A high quality paper might be made entirely from cotton fibers, but newsprint, cardboard, and many other papers are made from cellulose fibers which are prepared from wood. Although vanilla has been synthesized from it, chemists have found few other uses for the lignin removed from wood pulp in the process of making paper. As a result the lignin extracted during the manufacture of paper represents a waste product with little or no value. All too often the waste lignin is simply discharged into our streams and rivers where it destroys fish and renders the water biologically useless.

PHOTOSYNTHESIS— HOW LEAVES WORK

Higher plants are immobile. They stand rooted to the same patch of soil throughout their lives, whether they live for just a few days or for thousands of years. This means that a plant must obtain everything it needs to grow and reproduce from the air, soil, and water of its immediate surroundings. Fortunately its needs are very simple. It must obtain carbon dioxide and oxygen from the air as well as sunlight, water, and a few minerals from the soil. From these simple raw materials it is able to manufacture all the proteins, carbohydrates, lipids, nucleic acids, and vitamins it needs to carry out its life processes.

At one time it was thought that plants obtained their food from the soil. In fact even today many organic gardeners seem to think that plants obtain nourishment from organic compounds added to the soil. For the scientist this view was discounted over 300 years ago by an experiment performed by Jan Baptista van Helmont (1577-1644). He planted a willow branch which weighed 5 pounds in a large pot containing 200 pounds of soil. Five years later when the sapling had grown into a small tree, van Helmont carefully removed the soil from the roots and again weighed the tree and the soil. He found that the tree had gained 164 pounds during those five years, whereas the soil had lost only a few ounces. Van Helmont concluded, correctly, that this meant that the tree could not have obtained from

the soil the food it needed to make 164 pounds of roots, bark, wood, twigs, and leaves. However, he also concluded, incorrectly, that the tree obtained its food from the water he had given it.

As we learned in Chapter 5 green plants produce their own food by photosynthesis. The raw materials needed for photosynthesis are carbon dioxide, water, and sunlight. The energy of light is trapped by the chloroplasts and some of this energy is used to split water into hydrogen ions and oxygen gas. The oxygen gas is given off as a waste product of photosynthesis, and the hydrogen ions are used to reduce carbon dioxide. Since most green plants produce all their food by photosynthesis, this means that the plant synthesizes itself literally out of air, with a little help from water. For simplicity we can consider glucose to be the principal product of photosynthesis. Glucose has the empirical formula $C_6H_{12}O_6$. The six carbon atoms and the six oxygen atoms of this molecule all come from carbon dioxide. Only the hydrogen atoms were obtained from water. The molecular weight of carbon is 12. Oxygen has a molecular weight of 16, and hydrogen has a molecular weight of 1, disregarding isotopes. Using these figures we can quickly calculate that about 7 percent of the weight of glucose is due to the hydrogen it contains. This means not only that 93 percent of the weight of glucose was derived from carbon dioxide, but that over 90 percent of the organic matter of a plant was originally a minor component of the air, for carbon dioxide composes only 0.03 percent of the air.

Importance of leaves

All green parts of plants can photosynthesize. Algae, mosses, and some simple vascular plants are almost entirely green, and any part of their body that is exposed to light photosynthesizes. Most higher plants, however, have well-developed green leaves which are the principal organs of photosynthesis. Leaves are usually thin flat structures and contain veins of vascular tissue which connect up with the conducting tissues of the stem. Xylem occupies the upper part of the veins and phloem, the lower part. The veins reach close to every photosynthetic cell to which they conduct water and minerals and from which they conduct the products of photosynthesis. They also provide the leaf with a rigid, supporting skeleton and enable it to stay flat. The large flat surfaces enable the leaf to expose large numbers of chloroplasts to light. The thinness ensures that carbon dioxide can diffuse rapidly to reach the chloroplasts.

If a leaf of a typical higher plant is cut through and examined in the microscope (Fig. 10-19), it looks something like a sandwich. The photosynthetic tissue, which is known as the **mesophyll,** is sandwiched between two layers of non-photosynthetic cells, the upper and lower *epidermis*. The mesophyll consists of several layers of parenchyma cells, even in very thin leaves. Frequently one or two layers of rectangular cells are found just below the upper epidermis. Because they are fairly closely packed together with their long axes at right angles to the plane of the leaf, they have been called the *palisade mesophyll*. The *spongy mesophyll* lies beneath this palisade mesophyll. It is called spongy because the cells are highly irregular in shape with large air spaces between them, something like a sponge. These mesophyll cells are so packed with chloroplasts that there may be as many as half a million chloroplasts per square millimeter of leaf surface in an average leaf.

Unlike the mesophyll cells typical epidermal cells have no chloroplasts. Their function seems to be to protect the mesophyll from drying out. The cells of the epidermis secrete a waxy *cuticle* which permits passage of light but protects the internal tissues of the leaf from losing water and also protects against injury. This is of great advantage to the leaf, but at the same time the cuticle blocks most of the inward diffusion of carbon dioxide needed for photosynthesis and the oxygen needed for respiration. The plant gets around this problem by providing leaves with special pores in the epidermis called **stomates** which connect to the air spaces in the mesophyll, through which carbon dioxide, water vapor, oxygen, and other gases can move in and out of leaves. A stomate is a lens-shaped pore through the epidermis. It is surrounded by two highly specialized cells known as **guard cells** (see Fig. 10-20). Most plants have somewhere around 200 stomates per square millimeter of leaf surface. A single oak leaf may contain 350,000 stomates. The majority of these stomates are found in the lower epidermis. The guard cells surrounding the stomates control the opening and closing of the stomates in the following way.

Unlike the other epidermal cells the guard cells contain chloroplasts. Furthermore they commonly have very thick cell walls next to the stomatal opening, whereas the cell walls furthest from the stomate are thin. When leaves are exposed to light, water enters the guard cells and they swell, which causes the cells to be distorted. Typically the swollen guard cell assumes the shape of a crescent. As the guard cell on each side of a stomate swells, the thick cell wall of each guard cell becomes bowed, opening the stomate so that carbon dioxide can diffuse into the leaf while water vapor and oxygen diffuse out. When leaves are placed in darkness, water rushes out of the guard cells. They lose their crescent shape and their thick cell walls become

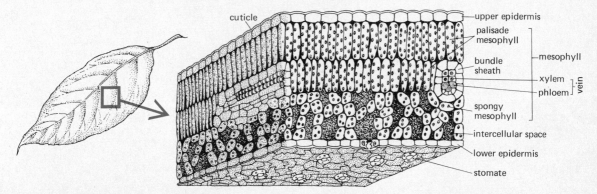

cuticle
upper epidermis
palisade mesophyll
mesophyll
bundle sheath
xylem
phloem
vein
spongy mesophyll
intercellular space
lower epidermis
stomate

Figure 10-19   Section of leaf showing both its internal and external structure.

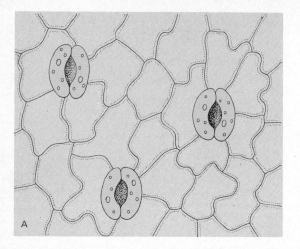

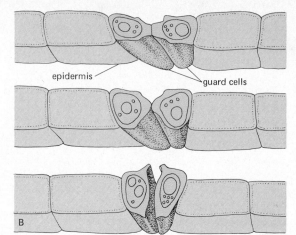

epidermis

guard cells

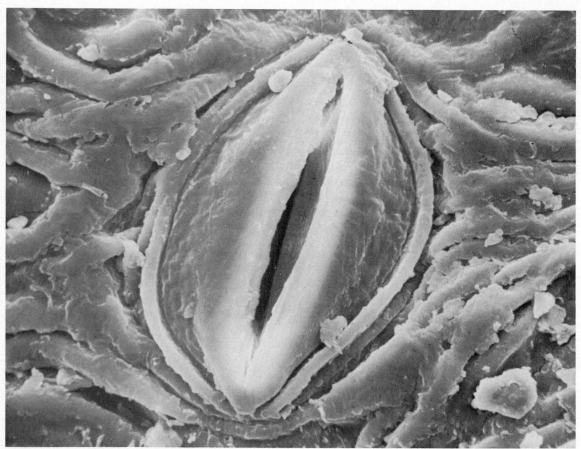

Figure 10-20   How stomates work. (A) Epidermal cells of a leaf showing the position of stomates among the epidermal cells. (B) Cross section through a stomate. Changes in turgor pressure alter the shape of the guard cells, opening and closing the stomate. (C) Scanning electron micrograph of an open stomate on the lower surface of a pear leaf (×3700).

straight. This closes the stomate. When the stomates are closed, gas exchange between the leaf and the air is often reduced to near zero.

In most plants stomates are open during the day to permit carbon dioxide to enter for photosynthesis. At the same time the open stomates permit large amounts of water vapor to escape, a process called **transpiration.** This process is important in transporting water up the stems of plants as we shall soon see. At night the stomates usually close, reducing water loss.

Although many studies have been made to determine how light controls stomatal opening and closing, we do not know how this is accomplished. The fact that the guard cells have chloroplasts suggests that photosynthesis might play some part in stomatal opening. This suggestion is supported by the fact that the wavelengths of light most effective in bringing about photosynthesis are also the most effective in causing the stomates to open. In any case by regulating the opening of the stomates, the plant is able to maintain a finely adjusted compromise between the loss of water and the diffusion of carbon dioxide and oxygen needed for photosynthesis and respiration. All terrestrial plants including mosses have stomates, a fact which suggests that stomates evolved early in the adaptation of plants to land.

Carbon dioxide and the rate of photosynthesis

Carbon dioxide accounts for a very small percentage of the mixture of gases we call air. Air is nearly 80 percent nitrogen and 20 percent oxygen, whereas CO_2 makes up only 0.03 percent of the volume of air near the earth's surface. As CO_2 is fixed by the dark reaction of photosynthesis in the mesophyll cells of the leaf, the air spaces within the mesophyll become depleted of CO_2. This establishes a CO_2 concentration gradient between the air inside the leaf and the air outside the leaf. As a result CO_2 rapidly diffuses from the region of relatively high concentration outside the leaf through the open stomate and into the air spaces around the mesophyll cells in which the CO_2 concentration was reduced. The CO_2 then dissolves in the water layer that surrounds each mesophyll cell and diffuses into the cell. Thus even though CO_2 makes up only a small fraction of the air, the normal physical force of diffusion ensures that carbon dioxide constantly moves into the leaf, as long as it is being used in photosynthesis and the stomates are open.

In addition to needing carbon dioxide for photosynthesis, green leaves also engage in cellular respiration and use oxygen. Day and night the mitochondria in the cells of green leaves and in all other cells of the plant oxidize sugars and provide the metabolic energy needed for growth and other processes. During the day this oxygen may be provided by photosynthesis, but at night oxygen must diffuse in from the air.

An important feature of this diffusion of gases is that the mesophyll cells inside a leaf are surrounded by air and not by a fluid. This is especially important because diffusion of a gas like oxygen takes place 300,000 times as fast in air as in a fluid. These air-filled spaces enable gases to diffuse rapidly through the stomates into the air spaces of the leaf where they can reach each individual cell. Most other plant tissues besides leaves also have empty intercellular spaces filled with air or gas. These intercellular gas spaces are connected together and open to the external air either through stomates or through other openings. Even thick tissues like a potato tuber have continuous intercellular gas spaces from the center of the potato to the skin through which gases rapidly diffuse. Because most of the intercellular spaces in animal tissues are filled with fluids, gas exchange in animals usually involves transporting fluids loaded with dissolved gases to and from the tissues.

Plants are very efficient in removing carbon dioxide from the air. It has been estimated that leaves are about half as efficient as concentrated potassium hydroxide of equivalent surface area when it comes to absorbing CO_2 from the air. In cold climates plants are commonly grown in greenhouses which must be kept closed during the winter. Under these conditions the CO_2 concentration in the air frequently falls as low as 0.01 percent, even though greenhouses are not airtight. Photosynthesis takes place at a faster rate than the supply of CO_2 can be replenished by diffusion under these conditions. The same phenomenon can also be observed outdoors. On a calm, warm, still summer's day the CO_2 concentration in the air over a field of corn has been observed to drop to 0.02 percent. Under these conditions CO_2 limits the rate at which photosynthesis can take place. When the CO_2 concentration falls below a certain level, the plant can trap more light energy than it can use for CO_2 reduction. Adding more CO_2 to the air would allow the plant to make more food with the same amount of light.

The rate at which plants grow is influenced by many factors, including the photosynthetic rate. Thus low levels of CO_2 not only limit the amount of food that can be produced through photosynthesis but they also result in slower growth. Commercial growers of carnations, chrysanthemums, and roses have found that they can stimulate the growth of these plants during the winter by raising the CO_2 levels in their greenhouses. This is done either by releasing CO_2 from tanks into the closed greenhouses or by burning methane, which produces not only water vapor and CO_2 but heat as well. Of course, supplementing the CO_2 in the air in order to stimulate photosynthesis would be practical only when plants are growing in an enclosed space such as a greenhouse.

Polluted air

The efficiency of gas exchange by leaves makes them very susceptible to industrial pollutants. In many urban areas, sulfur dioxide produced in the combustion of fuel oil, and peroxides and ozone generated photochemically from automobile exhausts, cause serious damage to plants. These agents appear to poison the chloroplasts and weaken the whole plant. Vegetable and orchard crops as well as natural vegetation are affected. Many long-lived trees seem particularly susceptible because they are liable to the cumulative effects of airborne pollutants.

Recently, active programs have been developed to breed pollution resistant strains of crop plants and street trees. Hopefully these strains will work for both agriculture and horticulture. However air pollutants also affect man and as one distinguished biologist put it, "unless the breeding and propagation of smog-tolerant strains of *Homo sapiens* are also contemplated, one would think that the air pollution menace might best be combatted at its source."

Light and photosynthesis— Shade plants versus sun plants

Different plants vary considerably in the amount of light they need if they are to grow and reproduce. Most gardeners know that some plants grow best in full sunlight, whereas others require partial or even complete shade if they are to thrive. Many pines and most grasses are **sun plants.** They require full sunlight if they are to survive in competition with other plants. However, the mosses growing in the shade of a pine would die if they were continually exposed to direct sunlight. Mosses, many ferns, and even some trees such as dogwood are shade-requiring or **shade-tolerant plants.**

As might be expected a plant's light requirements are related to the effect of increasing light intensities on the rate of photosynthesis. At high noon near sea level the summer sun has an intensity of about 8000 foot candles. The individual leaves of a plant cannot trap and utilize all of the energy of full sunlight. As dawn breaks each morning the sky gradually becomes brighter. Even after the sun comes up, the intensity of light striking any particular point gradually increases until the sun reaches its zenith. Photosynthesis begins at dawn, even before the sun is directly on the plant. As the light intensity increases during the morning, the amount of carbon dioxide fixed by photosynthesis also increases. However, long before the sun reaches its zenith, the photosynthetic rate stops increasing. The chloroplasts of the leaf become light-saturated. Further increases in light intensity cannot bring about any additional increase in the rate of carbon fixation once the light saturation point is reached. The individual leaves of sun plants become light-saturated when the light intensity is somewhere between 2000 and 4000 foot candles. With the leaves of shade plants light that is only about 10 percent as bright as full sunlight is sufficient to saturate photosynthesis.

These values are somewhat misleading, however. Even though the individual leaves of a sun plant might be light-saturated at 3000 foot candles, for example, the rate of carbon dioxide fixation by the whole plant often will reach a maximum only in full sunlight; even then the system might not be completely saturated. The reason for this is that although the leaves of a plant are arranged in a mosaic pattern to expose the maximum amount of leaf surface to the sun, many leaves overlap. Some leaves are partially or completely shaded by other leaves of the same plant. The shaded leaves receive only a fraction of the light received by leaves directly exposed to the sun. As a result the shaded leaves may not be light-saturated even though leaves exposed to direct sunlight are saturated.

Although shade plants and sun plants differ in the amount of light required to saturate their photosynthetic systems, this does not tell us why shade plants can grow at very low light intensities while sun plants cannot. Part of the answer is that shade plants simply have many more chloroplasts per cell than sun plants and therefore are much more efficient in trapping light. The light intensity at ground level in a dense tropical jungle or in an old redwood forest may be only 1 percent of the light intensity at the tops of the trees. Under these conditions a pine seedling is unable to fix enough carbon in

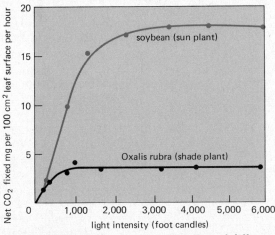

Figure 10-21 Graph comparing the effect of different light intensities upon the rates of photosynthesis in a sun plant (soybean) and a shade plant (*Oxalis rubra*). Note that very low light intensities saturate photosynthesis in the shade plant but that considerably more light is required to saturate photosynthesis in the sun plant. (A 100-watt incandescent bulb produces about 19 foot candles on a surface one meter distant.)

photosynthesis to keep itself alive. It starves to death. A fern or a moss, however, is more efficient in trapping the little light that filters down to the forest floor and in using this light energy to fix carbon dioxide. These shade-tolerant plants also consume food in respiration much more slowly than a pine seedling does. As a result a fern or a moss is able to make enough food not only to sustain itself but even to grow, slowly.

Moving chloroplasts around

Many plants have a mechanism for changing the distribution of chloroplasts within the mesophyll cells according to the intensity of light falling on the leaf. In darkness the chloroplasts are distributed at random around the cells. In dim light the chloroplasts move from the lateral walls into the cytoplasm along the upper and lower radial walls. Here they are in the best position to trap as much of the low intensity light as possible as it passes through these cells. As the light intensity increases during the morning, the chloroplasts may leave the upper and lower parts of the cell to move into the cytoplasm along the lateral cell walls. Here they may line up in rows parallel to the path light would take as it passed through the cells. This reduces the efficiency of these cells in harvesting light, because the chloroplasts will tend to shade each other when they are so arranged. Possibly these chloroplast movements represent an evolutionary adaptation, not only to increase the efficiency with which weak light is absorbed but also to prevent the harmful effects of very bright light. High intensity light may actually bleach chloroplasts by a process called *photooxidation* and make them useless in photosynthesis.

OBTAINING WATER AND MINERALS —HOW ROOTS WORK

Importance of soil—The good earth

Water is essential for plant life. Most land plants obtain it from the soil. A good agricultural soil is a mixture of small particles of rock, organic matter, air, and water containing just the right amount of dissolved minerals. Although we think of the soil as inert, it is literally teeming with life. It not only contains the living roots of plants and the familiar earthworm, but an abundance of insects, nematodes, fungi, and bacteria as well. A cubic foot of a good soil contains more bacteria than there are people on earth. These organisms are very important in breaking down the non-living organic matter of the soil, thereby returning important minerals to the soil.

From the plant's standpoint the most important characteristics of a soil are (1) its texture, (2) the availability of minerals, (3) the availability of water, and (4) the availability of oxygen. The texture of the soil is important because the plant's roots must grow through the soil. Also the soil texture has an important effect on the availability of water and oxygen to the plant. Coarse, sandy soils have relatively large spaces between the particles of rock we call sand. Oxygen readily diffuses into such soils, as does water. However, water also drains out of these soils very rapidly.

A clay soil is the opposite of a sandy soil. The inorganic particles of a clay soil are very fine and can become tightly packed together. This means that the spaces between the particles also will be very small. Water penetrates into clay very slowly, but it is equally slow in draining away. Thus clay soils easily can become "waterlogged"; that is, water displaces all the air between the clay particles. Without oxygen for their vital respiratory processes, roots will die.

The best soils for most crops are loams. A mixture of sand and clay, a good loam combines the water retention of clay with the ease of root penetration and good aeration of sand.

Hydroponics

After learning that the soil is very important to land plants, it may be surprising to discover that these plants can be grown perfectly well without any soil through a technique known as hydro-p'onics. This is a method for growing plants in water. The shoot of the plant is suspended just above the water and its roots grow down into it. Only two conditions must be met in order to obtain vigorously growing plants. First, air must be bubbled through the water so that the roots receive sufficient oxygen. Second, the water must contain an adequate supply of certain elements essential for plant growth. These elements usually are supplied in the form of water-soluble mineral salts. Under most circumstances it is simply not economically feasible to raise fruits and vegetables by hydroponics for commercial purposes. Nevertheless during World War II fruits and vegetables were grown by hydroponics on South Pacific islands to feed troops stationed there. The fact that it is possible to grow plants hydroponically is significant in that it demonstrates that the dead organic matter of the soil is of no direct nutritive value to higher plants.

Essential elements — The salt of the earth

The essential elements are those needed for plant growth and reproduction. Carbon, hydrogen, and oxygen are the most significant elements for plant growth. These elements represent the basic building blocks from which all organic molecules are constructed. In addition to these three elements, plants need relatively large amounts of nitrogen for the synthesis of proteins and nucleic acids. The synthesis of both RNA and DNA requires a substantial amount of phosphorus. Since three amino acids contain sulfur, plants require relatively large amounts of this element. Other elements required in fairly substantial amounts are calcium, magnesium, and potassium. Magnesium is an essential part of each chlorophyll molecule, whereas calcium is an important constituent of the cell wall. Plants need relatively large amounts of potassium even though this element is not an integral part of any important organic molecule the way these other elements are. Presumably potassium is important in maintaining the necessary concentration of ions in the cytoplasm for cellular reactions.

As has been seen, plants get the carbon and oxygen they need from the air, and their requirements for hydrogen are satisfied by splitting the water molecule. Plants must obtain the other six elements — nitrogen, phosphorus, sulfur, calcium, magnesium, and potassium — from the soil. Fairly substantial amounts are needed by growing plants. An acre of corn plants removes and uses somewhere between 100 and 200 pounds of each of these elements from the soil during a growing season. Because of the large quantities needed, these elements are known as macronutrient.

Table 10-1   Mineral elements essential for most green land plants

Element	Form in which usable by plants	Approximate concentration needed per liter of nutrient solution	Relative number of atoms in whole plant compared to molybdenum	Some functions in plants
Macronutrients				
Nitrogen (N)	N	15	1,000,000	In proteins, nucleic acids, chlorophyll, coenzymes
Potassium (K)	K^+	5	250,000	Increases permeability of cell membranes, activates certain enzymes in glycolysis, involved in protein synthesis
Calcium (Ca)	Ca^{++}	3	125,000	Structure and permeability of cell membranes. Combines with pectin in middle lamella
Phosphorus (P)	$H_2PO_4^-$, $HPO_4^=$	2	60,000	In nucleic acids, coenzymes, ATP, phospholipids
Sulfur (S)		1	30,000	In proteins, coenzymes
Magnesium (Mg)	Mg^{++}	1	80,000	In chlorophyll, required cofactor for many enzymes
Micronutrients				
Iron (Fe)	Fe^{+++}, Fe^{++}	0.1	2,000	In enzymes of electron transport chain (cytochromes, ferredoxin). Necessary for chlorophyll synthesis
Chlorine (Cl)	Cl^-	0.05	3,000	Activates photosynthetic reactions in which oxygen is produced
Boron (B)	$BO_3^=$, $B_4O_7^=$	0.05	2,000	Uncertain. Affects uptake and utilization of calcium in cell wall formation in meristems. Affects sugar translocation
Manganese (Mn)	Mn^{++}	0.01	1,000	Cofactor of various enzymes. Formation of oxygen in photosynthesis
Zinc (Zn)	Zn^{++}	0.001	300	In several dehydrogenases of respiration and nitrogen metabolism. In carbonic anhydrase
Copper (Cu)	Cu^+, Cu^{++}	0.0003	100	In cytochrome oxidase and other enzymes
Molybdenum (Mo)	$MoO_4^=$	0.0001	1	In nitrogen metabolism, especially nitrate reductase

The elements are arranged in order of the amounts required by plants. But all are essential and complete deprivation of any of these elements leads to death. The form in which they are usually absorbed and usable by the plant is also indicated. Thus, nitrogen in the form of nitrates (NO_3^-) or ammonia (NH_4^+) is absorbed and usable by plants. However, nitrogen in the form of nitrogen gas (N_2) is not usable by plants. The ions listed, as well as ions not listed such as sodium (Na^+), can affect the differently charged regions of protein molecules and by this means affect the configuration or folding pattern of many proteins including enzymes. Also, all of the principal ions have osmotic effects and also influence the distribution of negative and positive charges on the cell membrane.

A few other elements are required only by certain groups of higher green plants and others are required by certain algae. *Sodium* is required by salt-loving plants like the saltbush (*Atriplex*), by sugar beets and by blue-green algae. Its role is unknown. *Silicon* is essential for horsetails (*Esquisetum*) and probably for rice and some other grasses; diatoms need it for cell wall function. *Iodine* is needed by certain brown and red algae but its role is unknown.

Another group of essential elements that plants need but only in very small amounts are the *micronutrient elements*. They include iron, copper, manganese, zinc, molybdenum, boron, and chlorine. Although plants do not need large quantities of the micronutrient elements, they are just as essential for plant growth as the macronutrient elements. Most of these elements, including iron, copper, manganese, and zinc, are involved in vital enzymatic reactions. Both iron and copper are an essential part of key respiratory enzymes. The roles of boron and chlorine are not as clearly defined, however. In fact it is very difficult to demonstrate that these two elements are even essential for plant growth since the tiniest amount (sometimes called a *trace* amount) present as an impurity in other chemicals is enough to satisfy the plant's requirements.

It is interesting to note that, for the most part, animals also require these elements for their growth and reproduction. In fact animals usually depend on the plants they eat to supply them with the minerals they need. In many cases animals not only require the same elements that plants do but they also use them for the same purpose. There are some differences, however, in the elements required for plant and animal nutrition. For example, animals need relatively large amounts of sodium; for an animal it is a macronutrient. In contrast plants do not appear to need any sodium. If this element is required at all for plant growth, such small amounts are necessary that this requirement is met by the sodium present as an impurity in even the most highly purified chemicals. Higher animals, including man, require iodine, which plants do not need.

Under natural conditions the elements needed for plant growth come from the particles of rock in the soil. Acids produced by plants and bacteria dissolve compounds containing these elements from the rock, making them water soluble and thus available to the plant. Most of the essential elements absorbed by the roots of plants become part of the molecular structure of the plant. These absorbed and utilized elements are unavailable to other plants as well as to more recently formed cells in the same plant. Once a calcium molecule has become part of a cell wall inside the third leaf of a particular plant, the calcium will be unavailable when it is needed to form the walls of cells which will make up a new leaf. This means that a growing plant must continue to absorb essential minerals from the soil throughout its life. This process can deplete the soil if they cannot somehow be returned to it.

Under natural conditions decay organisms in the soil break down the organic molecules of plants and animals after they die, returning the elements they contained to the soil where they may be reused. When dead plants are not returned to the soil, as when crops are harvested, the elements those plants removed from the soil cannot be recycled and the soil may become depleted of one or more of the essential elements. Nitrogen, potassium, and phosphorus are the three elements that most commonly become deficient in agricultural soils. However, in certain soils on which crops have been raised for a few years other macronutrient or even micronutrient elements may become deficient. For instance it was discovered in 1942 that large areas of potentially useful land in Australia were deficient in molybdenum. When this deficiency was corrected by scattering *a few ounces of molybdenum compounds per square mile*, the land became productive. Similarly, most mineral deficiencies can be corrected easily by adding the right chemical fertilizers to the soil. Often the results are startling and dramatic. In Australia alone some 300 million acres of mineral-deficient soil may be reclaimed. In many parts of the Near East and Asia the soil has become so depleted of minerals through several millennia of farming that the land is now a desert, even though it may receive adequate rainfall. At the beginning of this century Israel was a desert, but its land has been returned to its former productivity through chemical fertilizers, irrigation, and proper agricultural practices. In other parts of Asia the use of chemical fertilizers alone would be sufficient to greatly increase the productivity of the land. It has been estimated that Pakistan could achieve up to a tenfold increase in food production through the proper use of chemical fertilizers.

The root system and water absorption

Higher plants are a little like icebergs. Although part of them, the shoot, extends above the surface into the air, there is another part, the root system, which we rarely see. Furthermore the root system frequently is larger than the shoot it supports both in mass and in surface area. This has been dramatically demonstrated by H. J. Dittmer, who carefully removed the soil from the root system of a rye plant and determined the number and length of all its roots. The rye plant is a bunch grass. Like other bunch grasses a single plant has an extensive fibrous root system with many branches which supports a clump of shoots. The particular rye plant studied by Dittmer had 80 shoots with a total of 480 leaves. These shoots had 143 main roots which had a combined length of 214 feet. There were 35,600 secondary roots coming from these 143 main roots. The combined secondary root length was 17,000 feet. Coming from the secondary roots were over 2 million tertiary roots. Even these tertiary roots were branched. Dittmer estimated that this one rye plant had over 380 miles of roots with a total surface area of 2500 square feet.

Figure 10-22 A photograph of a radish seedling root showing the extensive development of root hairs. The seed was germinated in a moist atmosphere. If the plant had been growing in soil, the older root hairs would have been broken off.

The roots of many plants grow to tremendous depths. Alfalfa roots may extend 25 feet down in a two-year-old field and have been found as deep as 130 feet in mine shafts. Mesquite trees have roots which may penetrate 175 feet or more. Obviously roots are well suited for their job of anchoring the plant and absorbing water and minerals from the soil. In some plants roots may have additional functions. For example, the familiar carrot has a large, fleshy tap root which acts as a reservoir for food produced by photosynthesis in the leaves. The Irish potato serves a similar function, although a potato tuber is not a root—it is a modified, underground stem.

Water absorption—Root hairs and "fungus roots"

Water absorption by the root occurs just behind the growing tip of each rootlet. In this region the epidermal cells produce elongated tubular extensions called **root hairs** which enormously increase the surface area of the root exposed to the soil. Root hairs are very fine and have extremely thin cell walls. They may be several centimeters long. Because they are so delicate, they are also short-lived. A root growing through soil usually will have root hairs only over a short stretch of the tip, starting just behind the growing tip and extending back a few centimeters or so. A given root hair is extended out from the root where it works its way into the spaces between the soil particles and absorbs the water and minerals surrounding these particles. A few hours later

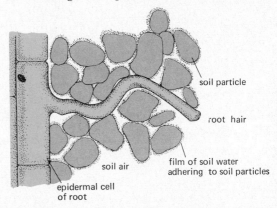

soil particle

root hair

film of soil water adhering to soil particles

soil air

epidermal cell of root

Figure 10-23 A root hair penetrating the soil illustrating its relationship to soil air and water.

the root hair will die, having completed its task. Meanwhile new root hairs are formed nearer the growing tip to begin the process all over again.

Not all plants have root hairs. Forest trees in particular tend not to form root hairs. In these species the function of the root hairs is taken over by soil fungi which invade the growing region of the lateral roots. These lateral roots with their associated thin filaments of fungi are known as **mycorrhiza,** a name which literally means "fungus roots." This is an association in which the fungus feeds on materials it finds in the roots. In turn the fungal filaments greatly increase the surface area and thus the water-absorbing capacity of the root. In a sense the fungus provides the plant with water in exchange for food. Such biological associations in which two species benefit each other are called **mutualistic.**

Whether plants have mycorrhiza or produce root hairs, these structures are the principal points of entry of both water and minerals. There are two different pathways water and dissolved minerals might follow as they move from the root hairs into the xylem. Cell walls have a great affinity for water. Thus the walls of the cortical cells and the spaces between these cells actually represent channels through which water can

move as it crosses the cortex. However, all of these channels end at the endodermis. The radial walls of the endodermal cells have a special waterproof thickening known as the *casparian strip.* At this point water must pass from the space around the cells into the living cytoplasm of the endodermal cells (Fig. 10-25).

The second pathway for water movement across the cortex is through the cytoplasm of the cortical cells. The cytoplasm of each cortical cell is connected to the cytoplasm of its neighbors through cytoplasmic strands, the *plasmodesmata,* which penetrate the cell walls separating them. Thus the cytoplasm of the cortical cells forms a continuous network from the epidermis to the endodermis, which provides a pathway for the movement of water and minerals. Regardless of how water and minerals reach the endodermis, at

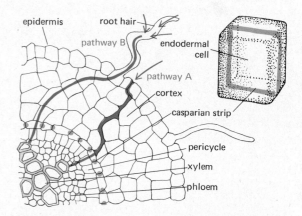

Figure 10-25 The two pathways which may be taken by water and minerals as they move into a root from the soil. The structure of the root has been simplified. Note especially the casparian strip which extends around the circumference of each endodermal cell. This strip is waxy and forms a waterproof seal between adjacent cells. Pathway A represents the diffusion of water and solutes into the root through the cell walls and the intercellular spaces but *not* past cell membranes. When water and solutes reach the endodermal cells they are blocked by the casparian strip from diffusing further *between* cells. To get into the vascular cylinder of the root, water and solutes must pass through the cytoplasm of the endodermis. Along pathway B water moves by osmosis and solutes by active transport through living cell membranes and finally through the endodermal cells. Thus the endodermal cells control the uptake of water and dissolved substances by the root.

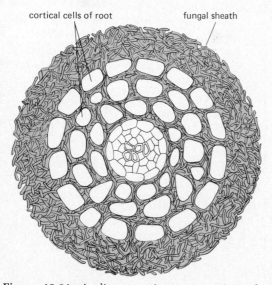

Figure 10-24 A diagrammatic representation of a mycorrhizal root in cross section. A sheath of fungal hyphae surrounds the root and penetrates into the spaces between the cortical cells of the root.

this point they must pass through the living cytoplasm of the endodermal cells. This means that the endodermis is in an extremely important position. It is the only living cytoplasm through which water must pass on its journey through the root and into the shoot. The endodermis is in a position to act as a molecular sieve to determine which mineral ions are to be permitted to enter the plant and which are to be excluded.

Diffusion and osmosis

Water enters roots by diffusion and osmosis. As we have learned diffusion simply is the movement of a substance from a region where its concentration is high to another region in which the concentration of that substance is lower. For example, the odor of a gardenia very quickly diffuses from the flower where the volatile odor-causing substances are manufactured into the room in which it is placed. Similarly, water diffuses from the soil, where it is more abundant, into the cell walls and intercellular spaces of the root cortex, where it is less abundant. The reason that there is less water in the cell walls and the space around cells than in the soil is because there is a strong tendency for water to move from the wall of the cell into the living cells of the root by osmosis and for water to be removed from the root as it passes through the xylem into the shoot.

Osmosis is a special kind of diffusion. It is the diffusion of water across a membrane which is permeable to water but not completely permeable to other substances. As we learned previously the plasma membrane is such a semipermeable membrane. Some molecules can pass readily through the plasma membrane, whereas others cannot. The size of a molecule is one of the factors that determines whether or not it can pass through the cell membrane. The plasma membrane is freely permeable to water molecules and to many other molecules of a similar size or smaller. However, the plasma membrane of plant cells and most other cells acts as a barrier to molecules as large as sucrose. Since sucrose has a molecular weight of 342, it is obvious that larger molecules such as proteins, nucleic acids, and polysaccharides cannot readily penetrate the plasma membrane.

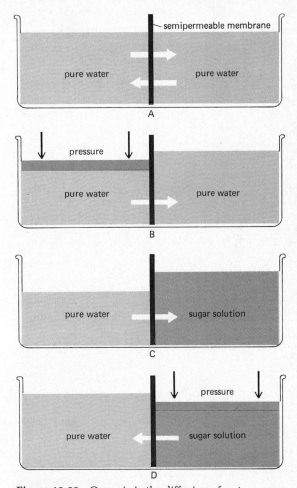

Figure 10-26 Osmosis is the diffusion of water across a membrane which is permeable to water but not completely permeable to other substances. Here the arrows show the net movement of water through a semipermeable membrane which separates 2 solutions. This membrane is permeable to water but impermeable to sucrose. (A) Water will diffuse across the membrane to the same extent in either direction. (B) Pressure applied to the solution on the left will force water to move through the membrane to the solution on the right. (C) Placing a solution of sugar on the right will result in the net diffusion of water from left to right, i.e., the water diffuses from the left side where the concentration of water is high to the right side where the concentration of water is lower. (D) Applying sufficient pressure to the sugar solution will force the movement of water out of the sugar solution, not only opposing but reversing the situation found in (C).

Consider what happens when a cell is placed in water. Water is free to pass in and out of the cell through the plasma membrane. However, most of the substances that compose the living cytoplasm cannot pass through it. To repeat, all substances tend to diffuse from a region of high concentration into regions where they are less abundant. Thus organic molecules in solution in the cytoplasm would diffuse into the water outside the cell if it were not for the fact that they cannot penetrate the plasma membrane. The concentration of *water* outside the cell is much greater than the concentration of water inside the cell. Since water can readily penetrate the plasma membrane, water diffuses into the cells. This movement of water across a semipermeable membrane from a higher concentration of water to a lower concentration of water is osmosis.

As a result of osmosis, when living cells are placed in fresh water, they tend to accumulate water. This presents all freshwater or soil-dwelling organisms with a problem, namely what they are to do with the excess water. When red blood cells, or other mammalian cells, are placed in fresh water, they take up water by osmosis. As water enters the cells, the cells swell. Very quickly the plasma membrane reaches the point where it can stretch no further and the cells explode. This can be prevented by adding salt to the water. Adding salt to the water outside the cells *decreases the water concentration* of the solution surrounding the cells. By trial and error it is possible to add just the right amount of salt to the water so that the concentration of dissolved substances outside the cell is the same as the concentration of dissolved substances in the cytoplasm. Such a solution would be called *isosmotic*. That is, it would have the same osmotic concentration as that of the cell. When placed in an isosmotic solution, a cell would neither swell nor shrink. The cells of most land animals are continually bathed with such an isosmotic solution—the blood.

Turgor pressure

Plant cells do not explode when immersed in fresh water because they are surrounded by a partially elastic cell wall. As water enters the cytoplasm of a plant cell by osmosis it swells and be-

gins to exert pressure against the cell wall. This pressure is known as **turgor pressure.** The wall of a plant cell surrounds the living cytoplasm the way stretch pants surround a leg. It will expand a bit as additional material is packed into it, but ultimately a point is reached when it can expand no further. At this point the plant cell must stop taking in water. Once the cell wall has reached the limits of its ability to expand, the turgor pressure that has built up will prevent more water from entering the cell. The cell cannot get any bigger and there is simply no more room for any additional growth.

A plant cell that has taken up all the water it can by osmosis is said to be turgid. Many leaves, flowers, and even stems are rigid because of the turgor pressure of their individual cells. Potted plants wilt if they are not watered. Water has evaporated from the leaves of such plants to the point that there is not enough water in the plant to maintain the normal cell turgor pressure. When wilted plants are watered they may become turgid again. However, a prolonged water shortage as in a drought will cause plants to become permanently wilted.

The reader may have performed a significant experiment in plant—water relation without knowing it. If sliced carrot sticks are placed in pure cold water before they are served, they will be crisp, because the cells will have developed

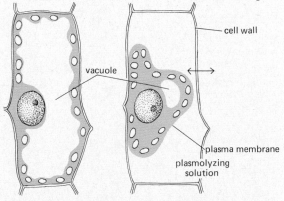

Figure 10-27 An illustration of what happens when a cell becomes plasmolyzed. The cell shown in A has been placed in a solution which contained more dissolved substances than would be found in the internal cell sap. As is shown in B, water moved out of the vacuole and into the external solution, causing the cytoplasm to pull away from the cell wall.

considerable turgor pressure as a result of the movement of water into the cells by osmosis. However, if the carrot sticks are salted and then placed in water they will become limp and soft. This is because the salt dissolves in the water, creating a solution of greater osmotic concentration than that in the cells of the carrot. Hence water leaves the cells by osmosis and the carrot sticks lose their turgidity. If we could examine the cells of the turgid and the flaccid carrot with a microscope, we would see that the cytoplasm of the flaccid cells had pulled away from the cell walls. Such cells are said to have been *plasmolyzed*—they have lost water by osmosis to the external solution. When a plasmolyzed cell is again placed in pure water, it will take up water by osmosis until the cytoplasm is again pressed against the cell wall. The cells of a wilted plant are plasmolyzed.

In addition to determining whether carrot sticks give a healthy crunch when eaten, plasmolysis by salts has many important practical consequences. Many desert areas have soils which contain a very high concentration of salt. Most crop plants cannot grow there even if they are watered because their cells will become plasmolyzed by the high levels of salt. Furthermore this also partially explains why seawater cannot be used to irrigate arid lands which lie near seacoasts. If irrigation water contained a high concentration of dissolved salts, as seawater does, the roots of most crop plants simply could not absorb the water. Osmosis no longer results in the movement of water into the root cells from the soil under such conditions. It is as though the plants were not watered at all.

A few plants are able to tolerate saline soils and brackish water, however. A plant known as

Figure 10-28   The effect of salt water on coleus plants. The plant on the right was placed in water containing 15% salt. It was unable to take up water and wilted. Instead of taking up water, its cells were plasmolyzed—they lost water by osmosis and the plant wilted. The unwilted plant on the left was placed in fresh water.

salt sage of our Western deserts is a good example. It is commonly found in salt flats where the surface of the soil may be white with an encrustation of salt. The salt sage is able to grow under such conditions because its cells accumulate large quantities of salt. This plant becomes even more salty than the soil it grows in so that water will move into its root cells by osmosis from the soil when it rains. This plant is so salty that its leaves taste like oversalted, roasted peanuts.

TRANSPORT OF WATER AND FOOD—HOW STEMS WORK

Root pressure and transpiration— Up the stem with water

How does water move from the roots into the stem and the leaves of a plant? This might not seem to be much of a problem, particularly if we confine our attention to petunias, wheat, and similar small herbaceous plants. But consider a redwood tree. It may be as much as 300 feet tall and most of its leaves are near the top. Somehow the redwood must pump water over 300 feet straight up into the air in order to meet the needs of its leaves. And the demands of the redwood tree, or any other large plant, for water are not inconsiderable. Even modest-sized plants will use an enormous amount of water. For each pound of plant material produced, potato plants need about 80 gallons of water, wheat about 60, and corn about 40. It has been estimated that over 400,000 gallons of water will pass through the corn planted on an average acre of ground during the growing season. Most of this water is lost from the leaves of the plant through *transpiration*.

Transpiration is the evaporation of water from the surfaces of the mesophyll cells of the leaf and the diffusion of this water vapor out of the leaf into the atmosphere. This diffusion occurs largely through the stomates, but in some plants it' also occurs through the waxy covering of leaves and stems. The water-conducting xylem tissue forms a continuous network extending from the roots to the leaves. The vessels of flowering plants are essentially continuous from near the tip of a root up through the stem and out into

a leaf. The vessels are filled with water. Water is pumped into the lower ends of these fine tubes by the osmotic forces in the root and it is removed from the upper ends, ultimately by the process of transpiration. This is an oversimplification of the forces involved in water movement in plants, but in essence it is correct.

The osmotic forces tending to push water up the xylem are known as **root pressure.** Its existence can be demonstrated dramatically by cutting the top from a plant near the ground. Water will exude from the cut stump for some time after the plant is decapitated. This is because water is being forced up the xylem by root pressure, even though it no longer has any place to go. However, since not all plants exhibit a positive root pressure, it seems improbable that root pressure is a major force in the upward movement of water in the xylem. Also, even in plants which have positive root pressure, the pressure that is developed is only enough to push the water a few feet up the stem.

Instead of being pushed up the xylem from below, it looks as though water is pulled up the stem from above. Water has many remarkable properties. Among these is the great cohesiveness of water molecules, that is, their ability to stick to one another. This cohesiveness results from the fact that the water molecule is electrically polar with positive and negative ends. The positive part of one water molecule tends to attract the negative part of another water molecule. Because water molecules tend to adhere to one another, a pull on one end of a thin column of water, such as is found in a xylem vessel, will lift the whole column as a unit. In the plant this pull is provided by transpiration. As water evaporates from the surface of the leaf mesophyll cells, it will be replaced by water from inside these cells. This creates a water deficit inside the mesophyll cells and water moves into them by osmosis from the vessel elements which end near them. Thus transpiration is tugging at the columns of water in the vessels, pulling water up into the leaves. Since water is being pulled up the stem from above, the columns of water inside the vessels must be under tension. Actual values of 25 to 32 atmospheres of tension have been measured in the xylem of trees. This is more than enough force to pull water up to the top of the tallest trees.

Transport of food—Down and up the stem with sugar

Earlier in this chapter we stated that the primary function of the phloem is food conduction. Food moves from the primary producers, the leaves, into various consumer and storage tissues such as the roots, the vascular cambium, and even the shoot apex. The pattern of movement of food within a plant is more complex than the movement of water. Water movement in the xylem is almost entirely in one direction—up. In contrast, food may move in almost any direction, as long as the conducting tissue is present.

The food produced by the leaves is chiefly in the form of phosphorylated sugars such as fructose-6-phosphate and glucose-6-phosphate. This, however, is not the form in which sugars are transported by the sieve tubes. The mobile form of food is sucrose, a disaccharide in which one molecule of glucose is linked to one molecule of fructose. Sucrose is common table sugar. Sucrose is obtained from either a grass known as sugar cane or a close relative of the table beet known as the sugar beet. These happen to be somewhat unusual plants in which sucrose is not only the form in which food is transported but it represents the storage form of food as well. Most plants store food as starch. Starch is a polysaccharide which is produced by linking a large number of glucose molecules together. The most common situation is for leaves to manufacture phosphorylated sugars in photosynthesis. These phosphorylated sugars are then modified enzymatically and used to synthesize sucrose, which is then moved by the phloem to the parts of the plant that need food. If it reaches an actively growing region of the plant, sucrose is broken down into glucose and fructose. These sugars are then phosphorylated again and "burned" by the reactions of respiratory metabolism to provide the energy necessary for growth. If the sucrose reaches a storage organ, such as a potato tuber, the sucrose is converted to glucose phosphate and this phosphorylated sugar is used to make starch.

Using aphids to measure phloem transport

In recent years scientists have used aphids to study the movement of materials in the phloem. Aphids are insects which have long hypodermic needlelike stylets which they can insert into a plant to obtain a free meal. The aphid's stylet probes around in the tissues of a leaf or stem

Figure 10-29 (A) An aphid (*Longistigma caryae*) feeding on a basswood stem (*Tilia americana*). The aphid drives its sharp mouth part or stylet like a hypodermic needle through the epidermis of the plant into a single sieve tube element. The high turgor pressure of the sieve tube forces the sieve tube material not only into the aphid's stomach but out its anus as well. Note a drop of this "honeydew" forming at the aphid's anus.

until it penetrates a sieve tube. Somehow this does not trigger slime-plug formation as other forms of injury would. The penetrated sieve tube continues to transport sugar. However, because of the high turgor pressure of the sieve tubes, some of the material being transported is forced up the aphid's stylet and into his stomach. Actually the turgor pressure is so great that considerable quantities of sieve tube exudates are not only forced into the aphid's stomach but out its anus as well. This very sticky material is known as aphid honeydew. It is sticky because the aphid can utilize only a small fraction of the sugar present in the phloem exudates that pass through his digestive tract. In the spring the infestation of aphids on the new growth of maples and other deciduous trees often is so heavy that a gentle, sticky rain of honeydew falls from the trees.

For scientific purposes the aphid can be anesthetized with carbon dioxide after his stylet has penetrated a sieve tube. Then the aphid can be severed from his stylet, leaving the stylet implanted in a sieve tube. Sieve tube exudates will continue to flow from the severed stylet, this time unchanged by passage through the insect's digestive tract. This aphid stylet technique has been used to analyze the composition of the material moving through the sieve tubes as well as the rate at which such movement occurs. The sieve tube exudates commonly have been found to contain from 15 to 25 percent sucrose, with only small amounts of sugars other than sucrose. They also contain amino acids and minerals. The aphid stylet technique has been used on willows to obtain an estimate of 100 centimeters per hour for the rate at which sucrose moves through a sieve tube. This is not a rapid rate of movement when compared to the rate at which blood is pumped through the vascular system of an animal. However, considering the very high concentration of food materials in the sieve tubes, it appears to be a reasonably efficient method for transporting food.

Mechanism of phloem transport

What is the actual mechanism by which sucrose and other food move through the sieve tubes? It cannot be diffusion because sucrose diffusing from a 15 percent solution would take many months to diffuse 100 centimeters. The mechanism is not completely known, but it appears to involve in part the mass transport or *mass flow* of sieve tube sap. Evidence for mass flow comes from the fact that foreign substances that enter the sieve tubes, such as dyes and virus particles, are carried along in the sieve tube stream even though they are not useful to the plant. Clearly such foreign materials are not being actively transported.

The mechanism of mass flow is easy to visualize. At the leaf end of a sieve tube, sugar must be transported into the sieve tube. This tends to increase the osmotic concentration of the sieve tube sap and leads to osmotic water absorption and a high turgor pressure. At the delivery end of a sieve tube such as a root or a fruit, sugar must be released. This tends to lower the osmotic concentration in the sieve tube and leads to the loss of water and a lower turgor pressure. The end result is a pressure difference between the two ends of the sieve tube and the sieve tube sap simply flows from one end of the sieve tube to the other.

According to this mechanism, the transport of sieve tube sap is a passive physical process. Why then are living sieve tube cells needed? The answer is that the living cells develop and maintain the turgor pressure that drives the process. In addition the living sieve tube cells are necessary to actively transport sugar from the leaf cells to the sieve tubes because the concentration of sugar in the sieve tubes may be ten times as high as the concentration in leaf cells.

From this analysis we conclude that by a combination of mass flow and active transport sieve tubes are able to move sucrose and other materials rapidly down and up the stem.

INTEGRATING PLANT ACTIVITIES BY REGULATING GROWTH— PLANT HORMONES

Plants are able to orient themselves in their environment so that there is maximum opportunity for their shoots to photosynthesize, their roots to absorb minerals and water, and their flowers to be fertilized. This orientation of body parts is accomplished largely by regulating growth: shoots grow toward light and roots grow downwards. Regulation of growth also permits changes in the size and structure of various organ systems which ensure improved function. Thus, when a shoot grows, roots usually respond by sending out additional branches, ensuring a balance between mineral nutrition, water absorption and photosynthesis. The key agents of this growth regulation are the plant hormones.

Hormones are chemical messengers synthesized by one tissue and transported to a target tissue, where they produce an effect. They are ordinarily produced in response to some environmental stimulus or as part of an unfolding developmental program. Most hormones are active in extremely small amounts. The effects produced depend upon both the hormone and the target tissue. The same hormone may provoke different responses at different concentrations. A noteworthy feature of hormones is that organisms have evolved effective mechanisms to inactivate or eliminate their hormones. This is especially important, because if a hormone accumulated, its "message" would be present all of the time and the hormone would cease to be useful as a "messenger."

The hormones of animals are commonly produced by special hormone-synthesizing tissues in organs called *endocrine glands*. Animal hormones regulate many processes including carbohydrate metabolism, excretion, digestion, and growth. The situation in plants differs in several ways. In the first place, plant hormones are commonly produced by meristematic tissues —the growing tips of shoots and roots—as well as by enlarging fruits and the vascular cambium. Unlike the endocrine glands of animals, these meristematic tissues obviously have many other functions in addition to hormone secretion. Secondly, plant hormones are involved primarily in growth and development. However, they affect indirectly most of the activities of the plant, because, as we noted above, integrating the activities of a plant is accomplished largely by regulating growth.

We shall examine several hormones, the auxins, gibberellins, cytokinins, ethylene, and abscisic acid. The life of a plant is integrated by the timely release and transport of these hormones. Another important growth regulator, phytochrome, was discussed earlier in Chapter 8. Because it is not transported from one location to another, it is not usually considered to be a hormone.

Auxins and phototropism

In Chapter 8 we described phototropism, the bending of shoots toward light. This process has great adaptive value for plants and ensures that they grow toward light. It is controlled by a plant hormone. Charles Darwin, the originator of the theory of evolution by natural selection, and his son Francis made the first key discovery in phototropism in about 1880. They used seedlings of canary grass and of oats and showed that when the growing tip of the shoot was illuminated from the side it bent toward the light. If the entire shoot except for the tip was covered with a blackened tube which excluded light, the shoot still bent toward the light. However if the tip was covered with the blackened tube, the shoot failed to bend toward the light. This result led them to conclude that the shoot tip perceived light and played a key role in the phototropic response. They confirmed this by cutting off the tip and showing that the shoot failed to bend. Apparently the tip transmitted some influence to the lower part of the shoot causing it to bend.

The mechanism of this phototropic effect was uncovered nearly 50 years later by Fritz Went when he was a graduate student working in his father's laboratory at Utrecht in the Netherlands (Fig. 10-30). Went cut the tips off oat seedlings and placed their cut surface on a small block of agar, a gelatinlike material. After about an hour he removed the tips from the agar blocks and placed one of these agar blocks against the cut surface of the remaining stump of the oat seedling. The stumps which had stopped growing began to elongate again as if their tips had been replaced. Plain agar blocks used in control experiments had no effect.

In the next experiment Went deliberately placed the agar blocks off-center on the cut surface and placed the shoots in the dark. Within an hour the shoot began to bend *away* from the side on which the block was placed just as in the phototropic response. A plain agar block which had not been exposed to the oat seedling tip failed to cause this bending. Thus Went proved that the tip of the oat seedling produces a chemical which can be absorbed by the agar and causes the shoot to elongate. He called this substance **auxin** from the Greek word "auxein" which means "to increase."

By using Went's method of assay for auxin, it became possible to purify, isolate, and identify several different natural auxins from plants, the most common of which is **indoleacetic acid,** abbreviated IAA.

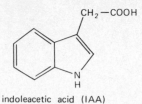

indoleacetic acid (IAA)

In a living plant auxin is produced mainly by the rapidly dividing cells of the meristem in the shoot tip, by young leaves, embryos, and fruits. It is normally transported from the tip down the shoot so that it reaches and stimulates the elongation of cells only directly under the point at which it is released. The experiments in which agar blocks with auxin were placed off-center

Figure 10-30 Went's experiments demonstrating the existence of auxin.

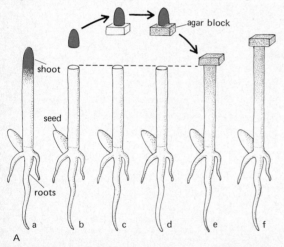

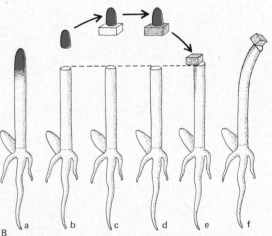

(A) Went cut the tip off oat seedlings and placed them on agar blocks. With the tip removed, the stump of the seedling stopped growing. When he put one of the exposed agar blocks against the cut surface of the stump, the stump resumed growing. He concluded that a hormone, auxin, diffused from the tips into the agar blocks and this hormone then diffused from the blocks into the stump of the seedling. In this diagram the auxin is denoted by color.

(B) When he placed an exposed agar block off center on the cut surface of the stump and placed the stump in the dark, the stump grew and bent away from the side on which the block was placed. Apparently the hormone from the block is readily transported downward in the shoot but not laterally across the shoot. As a result only the cells directly below the block receive auxin and grow faster than the cells on the opposite side.

When plant tissues are exposed to plain agar blocks, the blocks can be tested for auxin activity by placing them off center on the cut surface of a seedling stump and measuring the amount of bending. This provides a biological assay (or bioassay) for auxin and other chemicals with auxin activity.

show that there is *normally* little *lateral* movement of auxin after it has been released. However, in phototropism, there is some lateral movement of auxin from one side of the shoot tip to the other. Light stimulates the migration of auxin from the light side to the dark side of the shoot tip (Fig. 10-31). The cells on the dark side end up with more auxin and elongate. As a result, the plant bends toward the light. Thus, phototropism results from the lateral movement of auxin across the shoot tip. This causes an unequal distribution of auxin which in turn leads to unequal growth rates on the two sides of the shoot tip. In uniform light, the auxin is distributed in equal amounts on all sides and the shoot grows upright. Unequal distribution of auxins also underlies the geotropic responses of shoots and may cause the geotropic responses of roots which we mentioned in Chapter 8 (Fig. 10-32).

How auxins stimulate cell enlargement

Auxins cause cells to enlarge by changing the mechanical properties of the cell wall. As we noted earlier, plant cell walls consist of several different kinds of polysaccharides including cellulose and pectins, as well as proteins. These substances form a fairly rigid jacket around the

cell. When a cell is treated with auxin, its wall becomes pliable. This allows water to enter the cell and the cell to enlarge in response to the turgor pressure of the water within its vacuole. Auxin makes the cell wall pliable by causing chemical bonds within the wall to be broken. It does not do this directly, but rather it causes the cell to secrete substances, possibly enzymes, into the cell wall which break chemical bonds in some polysaccharides, softening the wall. Cell enlargement continues until it is stopped by the resistance of the wall. Finally the cell wall becomes stiffer again.

Inhibitory effects of auxins

Besides these growth stimulating effects, auxins also have growth inhibiting effects in many plants. Gardeners have known since ancient times that if you snip off the *terminal bud* of many plants, the *lateral buds* closest to them begin to grow extensively and produce a bushier

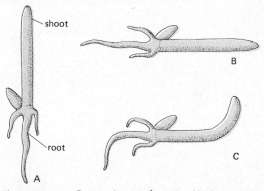

Figure 10-32 Geotropism and auxin. (A) In a normal seedling shoots grow upward and roots grow downward. (B) Even when a seedling is placed on its side its shoots grow upward and its roots grow downward. In such a seedling, auxin (denoted by color) accumulates in the lower side of the shoot. (C) This promotes cell enlargement on the lower side and the shoot turns upward.

The cause of roots growing downward is less clear. One possible explanation is that auxin accumulates on the lower side of the root where it blocks cell enlargement. Auxins will certainly block cell enlargement in roots but no one has shown yet that there are differences in the auxin concentration in the underside and upper side of a root placed on its side.

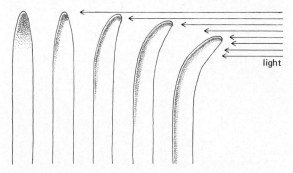

Figure 10-31 Phototropism and auxins. This diagram illustrates the way in which light is thought to alter the distribution of auxin within an oat seedling. Auxin is indicated by color. Under the influence of light auxins migrate from the light side to the dark side of the shoot tip. The auxins then move down one side of the shoot and cause the cells on that side to elongate. The events shown here would occur in about 2 hr in a seedling about 2 cm high.

plant (Fig. 10-33). Apparently the terminal bud normally keeps the lateral buds in check. Auxin appears to be a key factor, because if the terminal bud is cut off and auxins applied, the lateral buds are inhibited from growing. However, the mechanism by which auxin inhibits lateral bud formation is not yet known.

Although terminal buds prevent the growth of the lateral buds just below them, one lateral bud does not prevent the growth of another lateral bud. The reason for this is that auxins only move downward in stems, and never upward. Why this is true is not fully known, but the cells of shoots are only able to transport auxin in one direction. Most of this auxin transport takes place through the general parenchyma of the stem but some may be transported through the vascular system. The ability of auxins to inhibit certain buds from growing has been used agriculturally to prevent potatoes from sprouting during storage. In a potato plant, the edible part is a modified stem and the eyes are buds. When auxin is applied, bud growth is blocked and potatoes can be stored for as long as three years.

From the foregoing, it appears that some tissues are *stimulated* by auxin while others are *inhibited*. Whether a given concentration of auxin stimulates or inhibits growth depends upon what kind of tissue it is applied to. Roots, shoots, and buds respond very differently to auxin, as is shown by Figure 10-34. Root growth is only promoted by very low concentrations of auxin, whereas shoot growth is not affected at all until the auxin concentration reaches levels that strongly inhibit root growth. As the concentration of auxin is increased still further, even shoot growth is inhibited. For a long time, plant physiologists did not know why high concentrations of auxin inhibited growth. While the exact mechanism has not yet been uncovered, it is now known that many of these inhibitory responses occur because high levels of auxin cause plant tissues to produce a gaseous hormone, ethylene. Ethylene, which will be discussed later in the chapter, is a powerful inhibitor of cell division and cell elongation.

Auxin effects on fruit growth and abscission

Auxins are important for fruit growth. Fruits usually develop from ovary walls or the flower

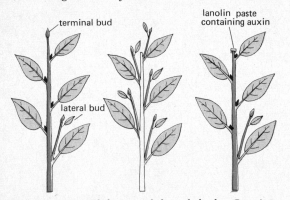

Figure 10-33 Inhibition of lateral buds. Growing, auxin-producing buds are shown in brown, whereas non-growing lateral buds are black. Auxin transport is indicated by brown shading. (Left) A terminal bud inhibits the lateral buds near it. The lowest lateral bud in this drawing is so far away from the terminal bud that it is not inhibited from growing, presumably because it is far away from the source of auxin. (Center) When the terminal bud is removed, the lateral buds near it begin to grow. (Right) When the terminal bud is removed and is replaced with a lanolin paste containing auxin, the lateral buds nearby are inhibited. Note that the auxin does not inhibit the lateral branch that was already growing.

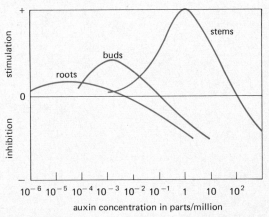

Figure 10-34 Stimulation and inhibition of roots, buds, and stems by auxin. Auxins may either stimulate or inhibit the growth of plant tissues depending on the concentration. The amount that stimulates shoot growth inhibits both buds and roots.

receptacles. Ordinarily ovary walls become fleshy and fruits grow only if the flowers are pollinated and eggs fertilized. Embryos appear to be the source of most of the auxin for fruit development. If the seeds containing the embryos are removed from the developing fruit, the fruits cease their growth. If only some seeds are removed, then fruits may grow only partially. However, if auxins are applied to a fruit from which seeds have been removed, it will continue its growth. In a few plants such as navel oranges the ovaries produce so much auxin that fertilization is unnecessary for fruit development and seedless fruits are produced. In other plants such as cucumbers and tomatoes, seedless fruits can be produced by applying auxin to the stigmas of unfertilized flowers. Farmers sometimes treat fruit crops with auxins to supplement normal pollination and ensure that a larger amount of fruit develops.

Auxins also affect the dropping of fruits, leaves, flowers, and stems, a process called **abscission.** Abscission results from the formation of a weak layer of thin-walled cells, called an *abscission layer.* As a leaf grows older, the cells in the abscission layer become drier and more fragile. By the end of the growing season the layer breaks under the slightest strain and the leaf falls. The same process occurs when unfertilized flowers fall off a stem. It also takes place in the stems of fruit, allowing fruit to drop when it is ripe. Auxins normally block abscission. When auxin production in leaves and fruits decreases, abscission is likely to occur. Fruit growers often spray their groves with synthetic auxins in order to delay fruit drop until harvesting is convenient.

Auxin effects on organ formation

In addition to their effects on cell enlargement and fruit development, auxins are responsible for the formation of new organs. This is especially evident in the case of *adventitious root* formation. Adventitious roots are roots which arise from an unusual place such as a stem or a leaf. For example, if a *Coleus* stem is cut off from its roots and kept in a moist atmosphere, adventitious roots will form at the base of the stem, near the cut. This has been shown to be the result of auxin which is transported to the base of the shoot and accumulates near the cut. The leaves of some species such as *Begonia* also will form roots if detached from the plant and kept in a moist atmosphere. This stimulation of root formation by auxin is of adaptive value to the plant. If the plant is injured, auxin will accumulate toward the base of the plant. This increase in auxin concentration promotes the regeneration of new roots from the tissue just above the wound and increases the chances of survival of the shoot.

One synthetic auxin, naphthalene acetic acid, NAA, has become quite important commercially in stimulating root formation. Although the stems of some plants such as *Coleus*, mint and poplar will readily form roots if placed in water or soil, many plants such as holly will not root readily. In many cases auxins will stimulate root formation, making it possible to propagate numerous important plants from cuttings.

Auxins in weed control

The discovery that high doses of auxins are toxic to plants suggested that auxins could be used as weed killers and herbicides. This encouraged efforts to synthesize artificial auxins in the laboratory. The most well known of these artificial auxins is 2,4-D which is used as an herbicide. Synthetic auxins have proved more useful in agriculture than the natural auxins because they are not broken down readily by enzymes in the plants. 2,4-D is particularly useful because dicots such as most weeds are much more sensitive to it than are monocots such as grasses. As a result 2,4-D can be used to eliminate weeds from crops of corn and other cereals, from pastures and from lawns and virtually eliminate the necessity for weeding. The reasons for this selectivity are not yet understood but it is extremely valuable agriculturally. Efforts are underway to develop selective herbicides which will kill grasses but not dicots, as well as highly specific herbicides which will affect a particular species of weed but will not affect other plants growing around it. The use of herbicides has increased agricultural production immensely, but it also has created a number of problems which are discussed at the end of this chapter.

Gibberellins

At about the time Went performed his experiments with auxins, a Japanese scientist, E. Kurosawa, discovered another group of plant hormones called the **gibberellins.** Like the auxins, the gibberellins stimulate cell elongation and cell division. Kurosawa discovered that rice seedlings infected by a fungus, *Gibberella fujikuroi*, grew unusually tall and had long spindly stems. The strange appearance of these infected plants led to the name "foolish-seedling disease." Kurosawa showed that extracts of the fungus would produce the same effect and cause rapid stem elongation. The chemical responsible was soon identified and called gibberellic acid. Subsequently more than 30 different substances with similar structures that produce the same effect have been isolated not only from fungi but also from higher plants. Thus, gibberellins are important natural growth regulators in higher plants.

gibberellic acid

(gibberellin A$_3$)

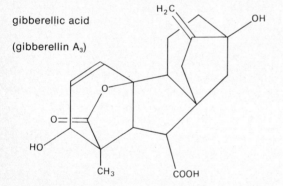

Dramatic effects of gibberellins are seen in dwarf varieties of certain plants, corn for example (Fig. 10-35). Such dwarf plants result from a genetic mutation and their stems fail to elongate. However, if the dwarf plants are treated with gibberellins, both the rates of cell elongation and cell division are accelerated and the dwarf plants grow essentially like normal plants. These effects cannot be mimicked by auxins or any other known plant hormone and dwarf mutants have been used in specific bioassays for gibberellins.

Gibberellins also have striking effects on certain biennial plants such as cabbage which normally take two years to flower. Normally, during its first year, a cabbage stem undergoes little elongation. As a result, the developing leaves have virtually no stem separating them and they form a compact head or rosette. In its second year, after exposure to an appropriate photoperiod and low temperatures, the stem elongates between the leaves and flowering occurs. Exposure to gibberellins causes the stem of cabbage to elongate between the developing leaves so that instead of a compact head or rosette, the cabbage develops a long stem (Fig. 10-36). Gibberellins also stimulate certain seeds to germinate and break the dormancy of buds of various plants. They also stimulate flowering in a number of plants.

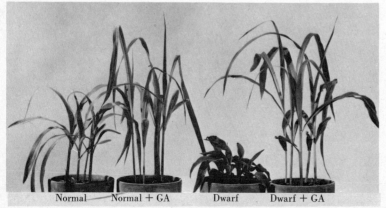

Figure 10-35 Effects of a gibberellin on normal and dwarf variety of corn. When the seedlings were a week old they were treated with 0.01 milligrams of gibberellin A$_3$ every other day until the seedlings were three weeks old, at which time this photograph was made. Normal plants were stimulated to some extent, but the dwarf plants responded much more and grew as rapidly as the treated normal plants.

The mode of action of gibberellins is not fully known but some important clues have come from studies on the germination of barley seeds. The seeds of barley and other grasses have a layer of special cells—called aleurone cells—rich in protein, just inside the seed coat. When the seeds take up water and start to germinate, the embryo synthesizes and releases gibberellin. This gibberellin stimulates the aleurone cells to produce hydrolytic enzymes which digest the starchy endosperm tissue of the seed and release sugars needed for further embryonic development. At the same time, other hormones, *auxins* and perhaps *cytokinins*, are released from the endo-sperm as a result of its enzymatic digestion. Thus, in germination of seeds, gibberellin seems to control both the mobilization of food and the release of other hormones which may function in embryonic development.

Cytokinins

Both auxins and gibberellins control cell enlargement and sometimes also affect cell division. However the principal regulators of cell division in plants are a group of hormones called the **cytokinins** (from *cytokinesis*, which means cell division). Cytokinins have been isolated from numerous plants and are present in high amounts in actively dividing tissues including seeds and fruits. They promote cell division in plant cells grown in tissue culture and in plant embryos. The best characterized natural cytokinin is *zeatin*, a derivative of the purine base adenine of DNA and RNA. More than 100 synthetic cytokinins have been produced, of which the most well known is *kinetin*.

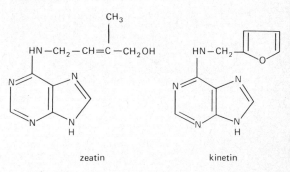

zeatin   kinetin

Figure 10-36 Effect on gibberellin on growth of cabbage. The plant on the left is normal and untreated. The plant on the right was sprayed several times with gibberellin A₃.

The cytokinins are necessary for cell division but they usually require the simultaneous presence of auxins. Figure 10-37 illustrates a typical experiment in which a tissue culture of undifferentiated plant cells is exposed to various concentrations of kinetin and auxin. If auxin alone is present, the cells enlarge but do not divide. However, if an appropriate amount of kinetin is added to the auxin, rapid cell division occurs and large numbers of tiny undifferentiated cells are formed. Kinetin alone has no effect, a result which emphasizes that the two hormones act together. If the tissue culture is exposed to a high concentration of auxin and a low concentration of kinetin, it develops roots. If it is exposed to a lower concentration of auxin and a higher concentration of kinetin, it forms buds. Evidently, the balance between cytokinins and auxin regulates the development of these plant tissues.

In a normal growing plant, the two hormones often enhance each other's effects, but in some cases they oppose each other. Thus, whereas auxins inhibit the growth of lateral buds, cytokinins stimulate their growth. Cytokinins also have curious effects on leaves and appear to prevent them from aging. When leaves get old, they commonly turn yellow. If a drop of cytokinin is applied to part of a yellowing leaf it will often prevent the cells in that part of the leaf from turning yellow and it may turn green again.

Cytokinins are found in xylem sap and are thought to be conducted from the root to the shoot. If the shoot is broken off, the cytokinins accumulate in the sap which bleeds out at the cut surface and may stimulate bud formation. This probably accounts for the growth of new shoots from the top end of a dandelion root when the shoot has been broken off.

The mode of action of cytokinins is unknown but some cytokinins are present as bases in transfer RNAs. What role, if any, they play in tRNA function remains to be discovered.

Ethylene

Another hormone important in plant development is the gas, **ethylene** ($H_2C = CH_2$), which is produced by many growing plant tissues, particularly ripening fruit. The production of this gas by ripe apples has given rise to the old saying that "one rotten apple spoils the barrel" because the ethylene produced by a single rotten apple will cause all of the other apples to ripen badly. Ethylene promotes fruit ripening in many different plants. It also promotes processes such as the withering of flowers and the dropping of leaves which we associate with aging. It is effective at concentrations well below 1 part in 10 million of air. Never put fresh flowers in the same refrigerator with a ripe apple or place blooming flowers in the same room with a ripe apple, for many kinds of flowers will rapidly wither because of the ethylene.

Ethylene also promotes abscission in a number of plants, such as oranges. This has proven of tremendous value to farmers who spray their trees with a chemical which releases ethylene when it breaks down. The ethylene causes the rapid formation of an abscission layer at the base of virtually every fruit on a tree. As a result, the entire crop can be harvested at the same time with

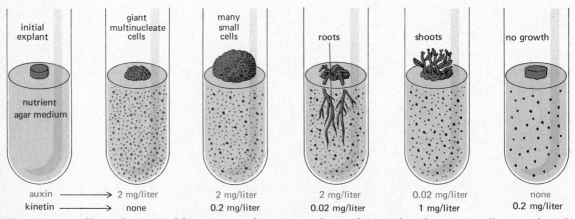

Figure 10-37 Effects of auxin and kinetin on a plant tissue culture. The initial explant is a small piece of sterile tissue from the pith of a tobacco stem. The tissue is placed on a medium containing various nutrients solidified with agar. Specific concentrations of auxin and kinetin were added to the medium and after several weeks, different growth patterns were observed. For details see text.

a mechanical orange picker that merely shakes the tree, causing the fruit to fall off.

Treatment with auxin causes a burst of ethylene production by some plant tissues. This result has led to the belief that some of the effects usually attributed to auxin are actually due to ethylene. For instance the inhibition of lateral bud formation and of root cell enlargement by auxin may actually be due to auxin-stimulated ethylene production.

Abscisic acid

In many plants buds form in one growing season and develop in the next. The newly-formed buds usually remain dormant until they are exposed to low temperatures or to an appropriate photoperiod. **Abscisic acid** is a hormone that was first isolated from dormant buds and fruits and occurs in maximum concentration just before the fruit

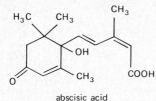

abscisic acid

drops. It was shown to accelerate the dropping or abscission of leaves, flowers, fruits, and stems.

Abscisic acid has proven especially interesting, not because it causes abscission, but because it often acts as a direct antagonist of the other growth regulators. For instance, abscisic acid can block the elongation of the stems of plants and this effect can be reversed by gibberellin. Similarly abscisic acid promotes abscission in leaves and fruits and this can be blocked by auxins. Likewise the stimulating effects of auxin on stumps of seedlings can be blocked by abscisic acid.

Hormonal interactions

From these examples the picture emerges of plants producing a powerful battery of growth regulators—auxins, gibberellins, cytokinins, ethylene and abscisic acid. These hormones affect many target tissues, interact with each other and sometimes antagonize each other. Their

production, transport, and action are often influenced by environmental factors such as light, gravity, and temperature. The balance of these hormones regulates the rate of growth and development and enables the plant to adapt to environmental changes and function effectively.

Plant growth regulators and human affairs

Many of the growth regulators we have discussed are being used in agriculture to improve food production. However, like all powerful discoveries, man must use them prudently or he may find himself the victim of their effects rather than the beneficiary. Perhaps by using growth regulators properly we can have more effective and efficient agricultural production and more and better food for mankind. However even weed-killing by a biodegradable material like 2,4-D presents serious problems. 2,4-D is so easily blown around that crop dusting of a wheat field by an airplane may kill bean fields nearby and trees in a nearby woodland. However if we restrict the use of herbicides like 2,4-D, then a sizeable percentage of our population will have to return to the land to do weeding and other field work. The result would be an increase in the cost of food and a decrease in its quality. Unless our present population size decreases, it is likely that herbicides will continue to be necessary to ensure adequate food production. However they must be used with caution.

It also appears that herbicides should not be used in any area for a prolonged period. Vast quantities of herbicides and defoliants were used in Vietnam to reduce chances for concealment. Many scientists belive that these activities may have done long-term damage to the ecology of huge areas of that country and made the land unfit for agriculture and forestry. We shall have to wait a number of years to answer that question. There are also reports that the large scale use of some herbicides has had deleterious effects on human beings and other animals. These facts taken together indicate that we must exercise great caution in applying these growth regulators on a large scale and that their use should not be continuous.

SUGGESTED READING LIST

Biddulph, O., and S. Biddulph, "The Circulatory System of Plants," *Scientific American,* February, 1959 (Offprint 53).

Corner, E. J. H., *The Life of Plants.* The New American Library, Inc. (Mentor Books), New York, 1964.

Crafts, A. S., *Translocation in Plants.* Holt, Rinehart and Winston, New York, 1961.

———, and C. E. Crisp, *Phloem Transport in Plants.* Freeman, San Francisco, California, 1971.

Esau, K., *Plant Anatomy* (second edition). Wiley, New York, 1965.

Galston, A. W., *The Green Plant.* Prentice-Hall, Inc., Englewood Cliffs, New Jersey, 1968.

———, and P. J. Davies, *Control Mechanisms in Plant Growth,* Prentice-Hall, Englewood Cliffs, New Jersey, 1970.

Hillman, W. S., (Ed.), *Papers in Plant Physiology.* Holt, Rinehart and Winston, New York, 1970.

Ledbetter, M. C., and K. R. Porter, *Introduction to the Fine Structure of Plant Cells.* Springer-Verlag, New York, 1970.

Meyer, B. S., D. B. Anderson, and R. H. Bohning, *Introduction to Plant Physiology.* Van Nostrand, Princeton, New Jersey, 1960.

O'Brien, T. P., and M. E. McCully, *Plant Structure and Development: A Pictorial and Physiological Approach.* Macmillan, New York, 1969.

Raven, P. H., and H. Curtis, *Biology of Plants.* Worth Publishers, Inc., New York, 1970.

Ray, P. M., *The Living Plant* (second edition). Holt, Rinehart and Winston, New York, 1972.

Richardson, M., *Translocation in Plants.* St. Martin's Press, Inc., New York, 1968.

Salisbury, F. B., and R. V. Parke, *Vascular Plants; Form and Function,* Wadsworth Publishing Co., Inc., Belmont, California, 1970.

———, and C. Ross, *Plant Physiology.* Wadsworth Publishing Co., Inc., Belmont, California, 1969.

Steward, F. C., *About Plants.* Addison-Wesley Publishing Company, Inc., Reading, Mass., 1968.

———, "The Control of Growth in Plant Cells," *Scientific American,* October, 1968 (Offprint 167).

———, *Growth and Organization in Plants.* Addison-Wesley Publishing Company, Inc., Reading, Mass., 1968.

———, *Plants at Work.* Addison-Wesley Publishing Company, Inc., Reading, Mass., 1964.

Street, H. E., and H. Opik, *The Physiology of Flowering Plants: Their Growth and Development.* American Elsevier Publishing Co., Inc., New York, 1970.

Sutcliffe, J., *Plants and Water.* St. Martin's Press, Inc., New York, 1968.

Troughton, J., and L. A. Donaldson, *Probing Plant Structure.* McGraw-Hill Book Company, New York, 1972.

van Overbeek, J., "The Control of Plant Growth," *Scientific American,* July, 1968 (Offprint 1111).

Wilson, C. L., W. E. Loomis, and T. A. Steeves, *Botany* (fifth edition). Holt, Rinehart and Winston, New York, 1971.

Zimmermann, M. H., "How Sap Moves in Trees," *Scientific American,* 208(3), 132-142, 1963.

CHAPTER 11

The Mobile Animals

Chapter 10 described how tens of thousands of cells are put together to form a plant and how plants function. This chapter and the three that follow describe how animals are constructed and how they function. At the outset it is useful to identify the significant features that enable us to distinguish an animal from a plant. For example, what are the basic differences between a Sequoia tree and a rabbit? At first glance there are so many differences between them that it is difficult to know where to begin. Yet one difference stands out above all the rest: animals *do* things, that is, animals are responsive. When a Sequoia tree is stimulated with mechanical stimuli such as heat, sound, light or chemicals, it shows no immediate response. In contrast a rabbit responds quickly to each of these stimuli by moving.

A moment's reflection will convince you that most plants do not move after being stimulated. It is true that *some* plants can move. Touch the leaves of the sensitive *Mimosa* and they curl up. In this case movement results from changes in the water pressure (turgor) of special cells located at the base of each leaf. A similar mechanism may play a role in the closing of a Venus fly trap around its insect prey. Most plant movements, however, are *growth* movements which are very slow. For example, every plant enthusiast carefully rotates his potted plants so they will "grow evenly," knowing that plants grow toward the sunlight. In contrast almost all animals move rather quickly. Even animals like barnacles and corals which are attached in one spot for most of their lives move parts of their bodies.

The presence or absence of movements characterizes the entire lives of animals and plants. In most plants, except for those that have motile sperm cells, there simply are no cells that move. During the development of a Sequoia tree from a fertilized egg, none of the cells move except when they are pushed from one place to another by other cells which are dividing or enlarging. In the mature tree not a single cell moves. Each one of its 2000 trillion cells lives out its life, motionless in the cellulose embrace of its own rigid cell walls. In contrast, during the development of an animal like a rabbit, many cells move actively from place to place and one group of cells, muscle cells, has become specialized for movement. When large numbers of muscle cells contract in the same direction, part of the organism moves.

What were some of the factors that may have led to the ability to move? Was there early in the history of life a fundamental difference between plants and animals which favored the evolution of animal movements? Or, more simply, why do animals move? Movement is tied up with the way animals obtain food. Most plants manufacture their own food from sunlight, carbon dioxide, water, and minerals which bathe them. Animals cannot do this; hence most animals move to obtain food. Animals either move through the environment and graze—lions graze on antelope, snails graze on grass, earthworms graze on decaying matter in the soil—or they move the environment through themselves, such as a clam that sets up a current of water which brings food to itself.

Effective food-getting requires controlled movements. Consider the earthworm. Earthworms move through the soil and eat dirt; they extract decaying organic matter from the soil and eliminate the rest. To do this they must be able to move. In addition, their movements must be controlled. Moving straight up or down does not accomplish very much nor does wriggling in one place. Earthworms have muscles which contract and cause movement. They also have nerve cells which insure that the muscles contract in a coordinated way to produce effective movements. Almost all other animals use muscles controlled

by nerves for food-getting activities. Muscle and nerve cells are peculiar to animals and are not found in plants.

The most basic difference between plants and animals is the method of obtaining food. Most other functions of plants and animals are distinct as a result of this difference. What plants or animals do depends largely on how each is put together. The cells of plants are arranged to make the plant efficient for synthesizing its food while staying in one place. The cells of animals are arranged to make the animal efficient at gathering food from its environment. Let us examine how animals are put together and how they function.

AN APPROACH TO THE STUDY OF ANIMAL FUNCTION

How does the machinery of the body work? There are several different kinds of "hows." As an example let us consider how a rabbit hears. First, how did the structure associated with hearing—the ear—evolve? Mammals evolved over a period of several hundred million years; during that time the upper and lower jaws and certain gill arches of fishes slowly changed in structure into the various parts of the mammalian ear. Second, how is hearing important—what is in it for the rabbit? It enables the rabbit to sense sounds, including those made by its enemies such as the coyote. Third, how does the ear as an organ function? The eardrum vibrates in response to sound waves, and these vibrations are transmitted by small bones in the middle ear to a fluid in the inner ear; this fluid in turn sends messages over nerves to the brain, which interprets the messages as sound.

Thus in one sense hearing by the rabbit is explained as the end result of a series of evolutionary changes. In another sense hearing is explained as the effect of evolutionary processes which have given rabbits the ability to detect their predators. In still another sense hearing is explained as the activity of a specific structural mechanism.

Our discussion of animal physiology emphasizes how the various parts of the body function, but we also consider how these body structures and their functions evolved.

AN EMPHASIS ON MAN AND THE VERTEBRATES

Most of the questions about physiology that we face in our lives are about our own physiology— human physiology. Fortunately most of the answers that we have to questions about physiology also involve vertebrates. The functioning of the mammalian body is more elaborate and complex than that of any other living system on our planet. The human brain is probably the most complex single structure in the known universe. Understanding the processes that go on within the sun is uncomplicated compared to understanding the processes that go on inside a man's brain.

Man is the most complex vertebrate, and indeed the most complex animal. This idea could easily be ridiculed as a product of human vanity, but this is not so. Man and the other mammals do stand apart from other animals, because they can survive under the most varied and extreme conditions. Man's survival and success, for example, are due partly to adaptations as widely different as the lavish way he cares for his young, the high arch of his foot which is unique even among mammals, and most important, his brain. Adaptations such as these have given man—and only man—the capacity to dominate and control the world around him. Only man is capable of turning an unfavorable environment into a livable one; unfortunately he is also capable of ruining the entire biosphere.

However, the focus on man alone fails to give us some important answers. For one thing human beings simply are not suitable for many sorts of experiments. For every problem in physiology there seems to be an animal ideally suited for its solution. Many of the animals which hold the solutions to major problems of organ function live in environments where great stress is put on the body. Studying these animals tells us "what it takes" to survive in an extreme environment. For example, desert rats have remarkable kidneys which enable them to live in the desert without ever drinking water. Arctic terns have circulatory systems which prevent their toes from getting frostbitten on ice floes. Seals have respiratory adaptations which enable them to stay sub-

merged for 30 minutes. The study of such seemingly specialized cases provides basic clues to how organ systems work.

We shall also consider animals other than vertebrates, especially insects. There are several reasons for considering the physiology of insects. One is because they are so economically important to man. Just how important they are can be illustrated by an example. If we could control just one insect pest, the rice stem borer, we could, without any improved agricultural techniques, harvest enough additional rice each year to feed 120 million people. In addition, insect structures are scaled down, and their functions are reduced to essentials. Yet they solve the same problems that man must solve, and quite successfully too, for most of the animals living today are insects. Usually an insect's solution to a problem is quite different from the way man handles the same problem, and understanding the solution provides insight into insects' success specifically and the way different animals solve problems in general.

Finally, to illustrate how one of the simplest animals functions, we consider how hydra, a small, freshwater animal, solves the same kind of basic problems that confront man and insects. Studying various animals along with man not only illuminates the physiology of man but also helps us to understand the lives of the many animals upon which our human lives depend.

BASIC BODY PLANS

It is impossible to understand how an animal functions without some knowledge of its structure. Therefore let us look briefly at the basic body plan of a hydra, an insect, and man—the three animals we shall emphasize. These three basic plans are shown in Figure 11-1.

The hydra's body is a sac with body parts radiating from the open end of the sac like spokes from a wheel. The hydra is **radially symmetrical.** The body wall, which is composed of three layers of cells, surrounds a cavity which serves in digestion, circulation, and elimination. The single opening in the sac functions as both a mouth and an anus. The tentacles are extensions of the body wall. There is some division of labor among the cells and certain functions are concentrated in particular regions. For instance, digestion is principally delegated to cells lining the cavity, the gut cells, whereas the cells which surround the animal, the epidermal cells, are specialized for protection and defense. This division of labor, however, is not complete and certain cells carry out several important functions simultaneously. Many epidermal cells and gut cells also serve as muscle cells and enable the hydra to move. In insects and man (and most other animals) these two functions would be performed by separate cell types. In hydra there are about ten different types of cells, whereas in insects there are prob-

ably a hundred, and in man, several hundred. Also in hydra there are no organs. Although there is a mouth end and a foot end, the regions in between are not sharply defined; there is no **segmentation.**

The middle layer between the epidermis and the gut is a simple sort of **connective tissue** that contains some cells along with a sticky jellylike material secreted by the cells. This connective tissue looks like a simplified version of the kind widespread among other animals where it holds organs together.

Insects and man are **bilaterally symmetrical,** that is, each side of the body is a mirror image of the other (some organs are not paired, however).

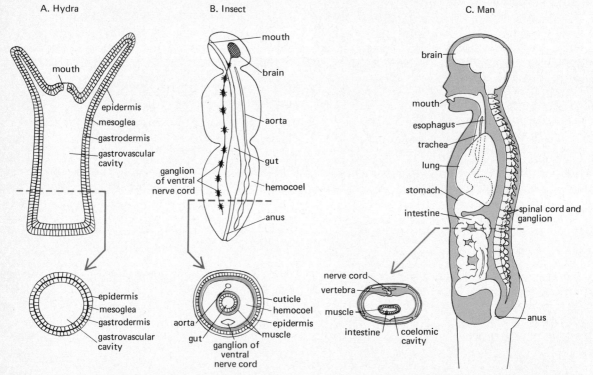

Figure 11-1 Longitudinal and cross-sectional views of the basic body plan of A. Hydra, B. An adult insect, and C. Man. All three animals have three body layers. Man and insect have a distinct middle layer or mesoderm which gives rise to various organs, including muscles. In contrast, the middle layer or mesoglea of hydra is very thin. In man, some organs (kidneys and gonads, for example) lie within the body cavity or coelom, which is lined by mesoderm. The body cavity is moist but not filled with fluid, and organs are suspended in it. The body cavity of insects is called a hemocoel, and is filled with blood or hemolymph. The organs are suspended in the hemocoel and are bathed with this fluid. Hydra has no coelom. The body cavity is called a gastrovascular cavity because it serves the animal in digestion, elimination, and circulation. (The coelom is discussed in Chapter 24.)

There is a head end with many sense organs that probe the environment and a tail end. Both have a tubular digestive tract in which food moves from mouth to anus, undergoing digestion along the way. In addition both man and insects are segmented. True they are not divided into a series of identical segments like an earthworm, but their bodies are divided into various regions, and some body structures—the vertebrae and ribs in man and the nerve centers or ganglia in the nerve cord of an insect—are repeated.

Both insects and man have specialized organs to perform major functions. But in insects most of these organs consist of only one or two tissues, whereas in man the organs usually consist of many different tissues (Fig. 11-2).

Major differences between the insect body plan and that of the human are seen in the skeleton and the circulatory system. Insects have a stiff body covering, or **exoskeleton,** which not only supports the animal but also acts as a site for attaching the muscles. These firm attachments are necessary for the rapid muscular movements that make both running and flight possible. The

exoskeleton is also the skin and provides a formidable defensive armor. Furthermore its surface is waterproof, and this feature has enabled insects and some other arthropods to join the vertebrates as the major animal inhabitants of a dry land environment. Insect blood circulates in the body cavity and is not confined to vessels. This fact prevents insects from developing any real blood pressure and means that insects cannot use hydraulic pressure to perform many important functions (like excretion) as we shall soon see.

Man's skeleton, like that of insects, protects, supports, and aids in locomotion. Like the insect skeleton, it is jointed, thus permitting appendages like legs to perform various functions. However, because man's skeleton is internal, it can support more weight than an exoskeleton; hence man and most vertebrates can grow ·to a considerably larger size than insects. An internal skeleton itself can grow and in general provide for greater flexibility in movement than an exoskeleton. Man's circulatory system, which is highly specialized, is enclosed in a continuous system of

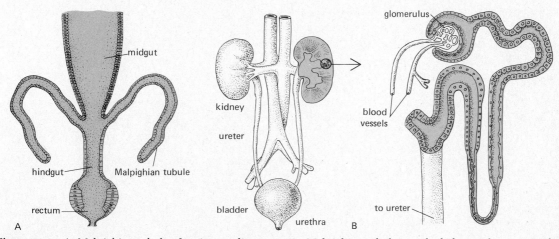

Figure 11-2 A. Malpighian tubule of an insect, diagrammatic. Malpighian tubules are the kidneys of an insect. They connect up with the gut. Each tubule (there may be anywhere from 2 to 150, depending on the kind of insect) is usually composed of only one kind of tubule cell. B. A mammalian kidney (left) is a complex structure composed of tens of thousands of individual kidney units called nephrons (right). Each nephron is composed of many tissues, including epithelial tissue of the tubules, tissues forming blood vessels, connective tissue, muscle, and epithelial glandular tissue. Moreover, each of these tissues may be composed of several kinds of cells. For instance, there are at least five different kinds of tubule cells.

vessels; his digestive, respiratory, excretory, and reproductive organs, which have also reached a high degree of complexity and specialization, are located within the body cavity.

In the remainder of this chapter we consider skin, skeleton, and muscle. All of these structures protect, support, aid in the locomotion of, and give form to, the body. In addition they play important roles in some of the other body functions that we discuss in the following two chapters.

SKIN

Animal cells are extremely delicate. If they are to function properly, their cell membranes must be protected from the external environment. At the same time those same cells must get rid of the carbon dioxide they produce, and they must take in oxygen and other materials necessary for life. The problem of protecting cells from the environment and at the same time allowing them to exchange substances with the environment could be an enormous one for an animal to solve if it were not for the remarkable properties of **skin** or **integument,** which acts as a boundary between the tissues and the external environment.

Human skin

Man's skin has been studied more than any other animal's, and what we have learned shows in a dramatic way its importance.

The skin is man's largest and, with the possible exception of the liver, his most versatile organ. Its average thickness is about 1 millimeter. Its area increases about seven times between birth and maturity so that it is about 20 square feet in the adult. If we include the underlying tissue (fat, connective tissue), we would find that its weight is about 20 percent of the total body weight.

What are the skin's principal functions?

1. It is nearly *waterproof* and prevents drying out (man is 60 percent water). It also *protects* internal structures from radiation, mechanical and chemical injury, and invasion by many bacteria and parasites.

2. It serves as one of the principal channels of *communication* between man and his environment because of its many nerve endings and sensory cells.
3. It plays an essential role in *regulating body temperature.*
4. It plays roles in *fat metabolism, control of blood volume,* and *nutrition.*

The breakdown of these functions is protected by the skin's amazing ability to heal itself.

Skin is composed of two principal layers, the outer **epidermis** and the inner **dermis** (Fig. 11-3). It is a turbulent tissue in which life and death go on side by side at a rapid rate. Cells in the innermost layer of the epidermis divide continuously and the new cells are slowly pushed to the surface. As the cells move to the surface they produce tremendous amounts of a single fibrous pro-

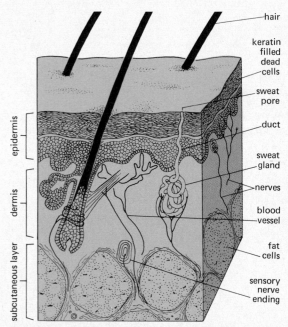

Figure 11-3 Section of human skin. Embedded in the dermis, which makes up most of the bulk of skin, are blood and lymph vessels, nerve endings, and gland cells. When the smooth muscle shown near the hair contracts, the hair rises or "stands on end." Supporting the dermis (but not considered as part of skin) is the subcutaneous layer, containing many fat cells. Epidermal cells lying near the dermis are living but the keratin-filled outer epidermal cells are dead.

tein called **keratin.** At the surface the epidermal cells become separated from the vital blood supply in the dermis and the cells die, leaving a legacy of keratin. Thus skin behaves like a giant gland whose principal secretory product, keratin, is released when the cells die. It is this horny layer of dead cells, packed with keratin, that helps waterproof and toughen the skin. If vertebrates had not evolved keratin, they would have found survival in the dry, inhospitable, terrestrial environment difficult. Man sheds these dead skin cells continuously every time he washes himself or scratches his head. In contrast insects shed their cuticle, the noncellular product that is secreted by their epidermal cells, only at molts.

More keratin is found on certain body parts than on others. For example, there is more keratin on the calloused soles of the feet than on the delicate eyelids. The structure of the epidermis itself varies. Over movable joints like elbows, knees, and knuckles the skin is pleated and is more flexible. Over the fingertips the skin is ridged and thus has greater gripping power.

Also located in the epidermis are the *melanocytes,* cells which produce *melanin,* the most important skin pigment. All human beings have about the same number of melanocytes but their activity varies in different people; albinos are white because melanin cannot be formed and blacks are black because much melanin is formed.

The dermis is composed of loose connective tissue in which glands, lymphatic and blood vessels, muscle fibers, and sensory structures are embedded. It also contains elastic fibers that decrease in number as a person ages; hence cosmetic-defying wrinkles appear. The skin also contains several different kinds of glands (Fig. 11-5) of which the sweat glands are the most in-

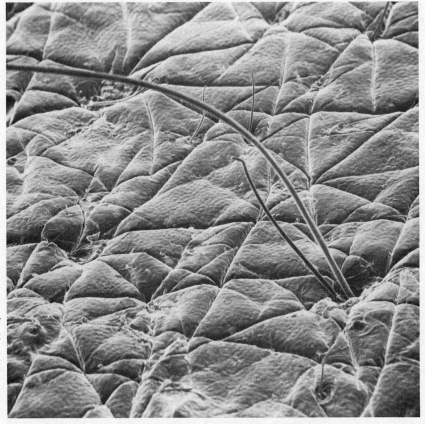

Figure 11-4 A scanning electron micrograph of the skin of a young girl (×85). It is evident from this picture that "smooth as a baby's skin" is not very smooth at all. Even a child's skin is covered with dead keratin which continuously flakes off.

teresting. As we shall see in Chapter 12 sweating is of tremendous importance to man in regulating his body temperature, and the ability to sweat has enabled man to inhabit the whole earth.

For unknown reasons man is becoming increasingly naked, and his hair has become ornamental rather than protective. For insulation man does not depend on his fur but on the layer of fat beneath the dermis. Women have more fat than men, and thus are insulated better than men. A few other mammals, notably the whale and seal,

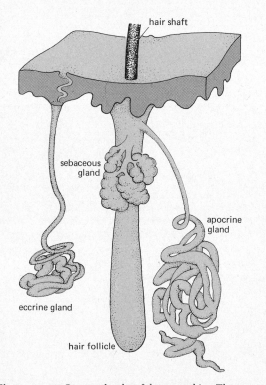

Figure 11-5 Sweat glands of human skin. There are two kinds of sweat glands, eccrine and apocrine. Some eccrine glands secrete water which evaporates, thus cooling the body. Others are active only when one is excited or under stress. The apocrine glands, along with the sebaceous glands, produce substances which give each individual his own particular odor—a function which may have helped our ancestors recognize each other. Sebaceous glands are not sweat glands. Their adaptive significance is not known although they may help to lubricate the skin. Some of their products are actually toxic to living tissues, and may cause blemishes. Sebaceous glands are at least partially responsible for acne.

are protected more by the thick layer of fat under the skin than by hair or fur. The whale is hairless for all practical purposes, and the seal's fur, when wet, is a poor insulator.

This layer of fat beneath the skin also plays other roles. It is a splendid shock absorber, and we are sometimes painfully aware of those spots where it does not cover our bones. (How many pro quarterbacks are sidelined each autumn because of knee injuries?) Fat is also a *food reserve*. In early May of each year, bull fur seals weighing perhaps half a ton arrive at their rookery in the Pribilof Islands. Each establishes his own territory, defends it, and mates with each cow in his harem (there may be 50 to 100). For this entire period which lasts about three months, the bull eats nothing. The Atlantic salmon migrates into the fresh waters of the rivers flowing to the sea. They, too, cease feeding, yet they can lead an active life for as long as a year. Both the bull seals and the salmon live on stored fat. Flying animals like birds and insects can use fat as an immediate energy store because fat has more calories per gram than any other food.

Skin plays a role in the body's water balance. Several quarts of water may be lost as sweat during strenuous exercise on a warm day. This is why major league "fastball" pitchers sometimes lose 10 or more pounds of weight during a nine-inning game on a hot day. This lost water must be replaced from water in tissue fluids and blood. On the other hand, should the water content of the blood drop, water from the skin (which normally holds about 40 percent of the body's water) enters the blood. This is just one mechanism whereby the water content of the blood and the blood volume are held constant. Just how important skin is in maintaining the blood volume is illustrated by persons who have been badly burned. Water leaves other cells of the body and enters the blood to compensate for the water that normally would be supplied to the blood by the skin. As a result a severely burned person may dehydrate to the point where he dies.

Other skins

Clearly the skin plays several crucial roles and is a vital organ for man. In other animals it may play additional roles. Amphibians partly respire

Figure 11-6 Outgrowths of vertebrate skin. The versatile skin gives rise to numerous structures which include the horn of a rhinoceros, the scales of an armadillo, the quills of a porcupine, the nail of a human, the scales of a lizard, of a snake, and of a fish, the antlers of a deer, the feathers of a bird, and various claws and hooves.

through their thin, moist skin. In fact some salamanders have no lungs and rely completely on their skin for respiration. About 40 percent of the respiratory exchange of frogs takes place across the skin. The horns of a rhinoceros and the talons of an eagle are all outgrowths of skin. Figure 11-6 shows some other outgrowths of skin of vertebrates. In addition to these structures some animals have special glands which are derived from the skin; an example is the gland that produces the skunk's unmistakable odor.

In insects the skin also serves as a skeleton, which we discuss when we examine the skeleton. The most important of its conventional "skin" functions is its elegant, waxy waterproofing which is especially important to a land animal that must conserve water. Indeed, as we shall see, many of the insect structures and functions are geared to conserving water. An insect's integument, like that of man's, is also the location of important sensory structures and is covered with sensory hairs projecting through the cuticles. Many of these are touch receptors; others like those in the forelegs of houseflies and butterflies are taste receptors.

The epidermis of the hydra is also waterproof. This waterproofing is important for freshwater animals. Without it hydra would be flooded by fresh water. A special feature of the hydra epidermis and especially of the tentacles are special cells called *cnidoblasts* which contain miniature barbed harpoons called *nematocysts*. These bizarre intracellular weapons are loaded with poison and are discharged when the surface of the hydra is brushed by another organism. Nematocysts are crucial to a hydra's food-gathering activities.

THE SKELETON

A few aquatic animals such as hydra and jellyfish depend on their skin and on the buoyancy of the water in which they live to support their body. A number of other animals such as most of the segmented worms—clam worms, earthworms, leeches—and the roundworms have a "hydrostatic skeleton." They entrap water or body fluid between their body walls and use this fluid for support. Imagine a balloon filled with water and tied at one end; it will hold its shape on land or under water. Similarly, an earthworm is also filled with fluid and will hold its shape in or out of water. When the muscles in the body wall contract, the force of the contraction is directed against the "skeleton" of the incompressible body fluids. This enables the earthworm to move, as we shall see.

Most animals depend upon a hard internal or external skeleton of some sort to support their bodies. Sponges and corals use skeletons mainly for support and protection. Other animals, such as clams, lobsters, fish, insects, and vertebrates, attach muscles to their skeletons and use them as an aid in locomotion. However, nature is opportunistic, and we find skeletons used for all sorts of additional purposes besides these conventional functions of support, protection, and locomotion. In mammals, for example, the skeleton is the major site of blood cell production and is the principal storehouse and exchange depot for calcium and phosphorus.

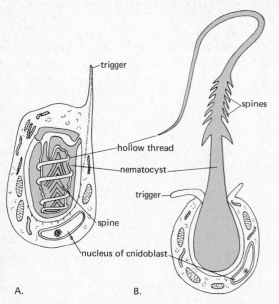

A.

B.

Figure 11-7 Nematocyst of hydra before (A) and after (B) it is fired. Nematocysts are enclosed in special cells called stinging cells or cnidoblasts, which are especially numerous on the tentacles. A combination of mechanical and chemical stimuli causes the cnidoblast to eject the nematocyst. The freed nematocyst then injects poison into the prey or intruder.

Figure 11-8 Some examples of endoskeletons and exoskeletons. (A) Skeleton of glass sponge, often called Venus' flower basket. (B) Living paper nautilus in its shell. (C) Skeleton of purple sea urchin. (D) A living scorpion. (E) Skeleton of a Gaboon viper.

The human skeleton

Man's skeleton has been shaped by two key factors. One was the adoption of life in the trees by some ancestral primate more than 70 million years ago. The skeletal and muscular systems gradually changed and allowed jumping, swinging, and grasping. The second was his erect stance which began about a million years ago. This led to further changes in his skeletal and muscular systems. No other animal has such long legs relative to the arms, a foot with such a high arch, or such a remarkable hand with an opposing thumb —hands which can make and use tools.

Figure 11-10 shows a human skeleton. We shall concern ourselves not with the anatomists' names for the 206 bones of the adult skeleton but with the general arrangement and function of these bones. The skeleton is divided into two parts, **axial** and **appendicular.** The axial skeleton includes the *skull,* which protects the brain and sense organs of the head; the *vertebrae,* which surround and protect the spinal cord; and the *rib cage,* which encloses the heart and lungs. The appendicular skeleton includes the *pectoral* and *pelvic girdles,* which serve as the base for movements of the arms and legs, respectively. The pelvic girdle also forms a stable platform for the flexible vertebrae.

Watch a baby put its toes into its mouth, or a yoga in the lotus position, or a ballet dancer

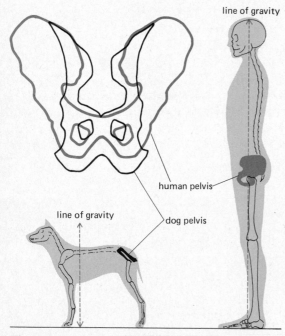

Figure 11-9 Man's upright posture has caused tremendous changes in his pelvis and the curvature of his spine. Notice the large pelvis of man with the huge flared arches which support his entire torso. The much smaller pelvis of a dog serves mainly as an articulation point and muscle attachment for its hind legs and its spine, and does not carry nearly as much weight.

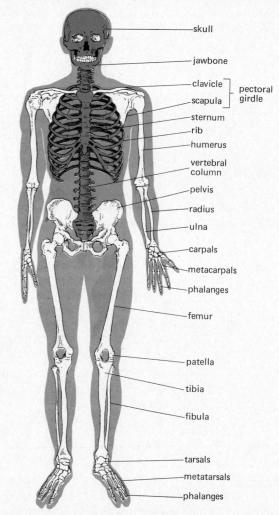

Figure 11-10   The human skeleton. The axial skeleton is shaded, and the appendicular skeleton is unshaded.

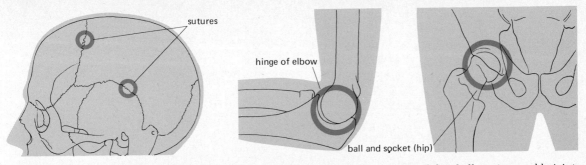

Figure 11-11 Some different types of joints in the human skeleton. The sutures of the skull are immovable joints. Movable joints include the hinge (elbow, knee) and ball-and-socket (femur and hip, humerus and shoulder). Hinge joints permit back-and-forth movements. Ball-and-socket joints permit movement in all directions.

and you realize how flexible a human skeleton can be. This flexibility is provided by **joints** (Fig. 11-11), which also absorb weight and shock. Many joints are *movable* (shoulder, knees, fingers); joints where bone movements would be dangerous are *immovable* (skull). Movable joints are protected by **cartilage,** a type of connective tissue; cartilage is "oiled" by fluid produced by cells of the joint. Bones at a joint are tied together by **ligaments** (tough connective tissue). Joints and ligaments provide an incredible degree of mobility. True we cannot turn our head completely around or scratch our right elbow

with our right index finger, but such limitations are few.

Human childbirth appears to be more painful than in many other mammals. Perhaps this is part of the price we pay for standing erect. However, the skeleton of the human female is adapted for childbirth. The pelvic joints are tightly bound by cartilage but they are not immovable. During the last days of pregnancy these joints loosen and separate slightly, widening the opening for the baby. In addition the baby's skull joints, unlike those of the adult, are not rigid; they "give," thus making birth easier than it would be otherwise.

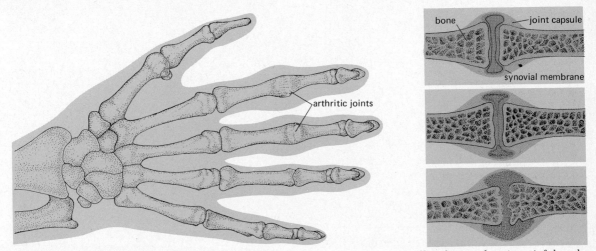

Figure 11-12 The figure at the left shows an arthritic hand. Arthritic joints are swollen, frequently quite painful, and, in severe cases, frozen in place. The figure at the right shows progressive stages in an arthritic attack in a joint. Synovial fluid, which oils and lubricates the joints, is secreted by cells of the synovial membrane lining the joint. In one form of arthritis, these cells start multiplying at a rapid rate, causing swelling. In later stages of the disease, the cells may even attack the bone itself. The causes of arthritis were unknown at the time of writing.

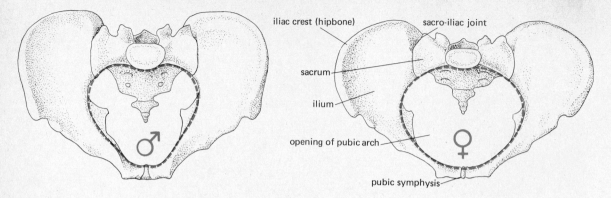

Figure 11-13   Comparison of the size and shape of the pelvic opening in the male and the female. In the female this opening becomes the birth canal.

The shape and size of bones vary from the tiniest bone (the stirrup in the ear) to the largest (femur or thigh bone), but almost every bone is adapted for a particular role. An exception is man's tail bone (coccyx) which has no obviously useful function. Bone's strength is similar to that of cast iron but it is much lighter. Its strength is partly the result of the calcium phosphate and other minerals and collagen (protein) of which it is made and the way these substances are cemented together. Its lightness results from its tubular construction and its porosity. So light and porous is bone that only 27 pounds of the weight of a 150-pound man is due to bone.

The thigh bone is an extraordinary example of strength. It must withstand tremendous pressures. When a pole vaulter lands from his vault the pressures on certain parts of the femur may be more than 12,000 pounds per square inch! It is shaped like a hollow cylinder—a superb design from the engineering viewpoint for maximum strength with minimum weight. In addition the hollow part of the cylinder has numerous thin ridges of bone built up in those planes where stresses occur, and these ridges provide added strength. Should the stresses change during life—as a result of injury, for example—the bone would grow new ridges at the new stress points.

The backbone is another exquisite example of tubular construction; the tube opening houses the spinal nerve cord. Tubular construction is often found where light weight is a crucial factor.

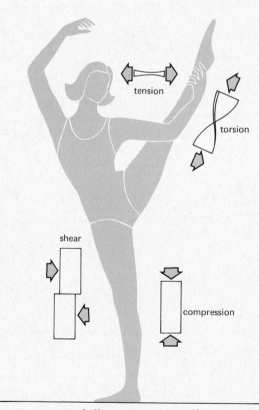

Figure 11-14   A ballerina in action illustrates the stresses that bones must withstand.

For example, many of the same bones that are relatively solid in man, such as the humerus, are hollow, spongy and essentially tubular in birds (Fig. 11-15).

However, not all of man's bones are as well adapted as those mentioned. Our backbone, for example, leaves much to be desired. In four-legged animals the backbone is a horizontal girder; in man the backbone is vertical, and thus is subjected to weight, bending, and compression. The straight spine we were born with became S-shaped when we started walking. As a result we have aching backs and other ills — again, the price of standing erect.

Bone is an ever-changing tissue. It is continuously being reformed and reshaped by living bone cells which lie in small cavities within the bone and are perpetually reabsorbing and secreting new bone in response to changes in our activity from birth to old age. The embryonic skeleton is largely cartilage but as the individual grows, much of the cartilage is gradually replaced by bone. Some cartilage remains in the outer ear and in the nose. The skeletons of some vertebrates (sharks, skates and rays) are entirely cartilaginous throughout life. In man's old age calcium and other minerals are reabsorbed leaving bones soft and brittle, and thus easily broken or compressed

("dowager's hump"). But even the bones of elderly persons are filled with living cells and continue to respond to the stresses of daily life, capable of knitting together when they are broken.

Man's skeleton plays other important roles in addition to support and protection. For example, bone is one of the sites of blood cell production. In the adult the bones of the ribs and the skull and the long bones, like those of the upper arm and thigh, are filled with fatty *marrow*. Marrow is active in producing both red and white blood cells as well as platelets — structures that aid in blood clotting. Also bone stores calcium and phosphorus, and releases those elements to the blood when they are needed. The release is triggered by a blood-borne hormone secreted by the parathyroid gland; the effects of the parathyroid hormone, in turn, are checked and balanced by a hormone from the thyroid gland. Un-

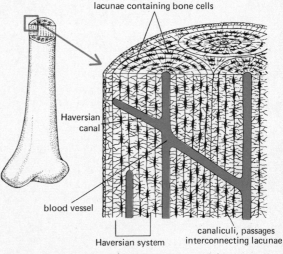

Figure 11-16 Microscopic structure of bone. Bone cells lie within spaces called lacunae, which are connected by a network of tiny channels or canaliculi. Nutrients and other materials diffuse from blood vessels to bone cells by way of these canaliculi. Blood vessels and nerves lie within the Haversian canal. The Haversian canal and its associated lacunae and canaliculi form an elongated cylinder called a Haversian system. Bone is many Haversian systems in a "cement" of protein fibers and calcium salts.

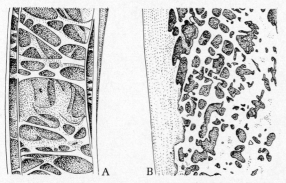

Figure 11-15 Longitudinal section through part of a long bone of a bird (A) and a man (B). Although long bones of birds are hollow, they are strong; their strength is due to thin horizontal bars of bone called struts. So light are bones of birds that their feathers may weigh more than their entire skeleton. The skeleton of a four-pound frigate bird with a seven foot wingspan weighs only about four ounces.

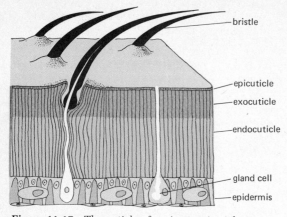

Figure 11-17   The cuticle of an insect. At right is a scanning electron micrograph of cuticle.

fortunately bone will absorb radioactive strontium which is chemically similar to calcium. It can also absorb beneficial substances such as fluoride which helps prevent tooth decay.

The insect skeleton

Man's skeleton is an **endoskeleton;** it is located *within* the body with muscles attached outside the framework. In contrast an insect or a clam has an **exoskeleton** which is located *outside* the body with the muscles attached to the inside of the skeleton. In an insect the muscles move the body parts the way our muscles move our legs and hands so that a beetle can walk and a grasshopper can open and close its jaws. The exoskeleton of insects is secreted by a single layer of living cells which envelops the insect. These epidermal cells secrete layer after layer of a tough fibrous material called **chitin.** This is chemically related to polysaccharides like the cellulose of plants, and is a polymer of a substance related to glucose. It is stiffened with a horny material called *sclerotin*. Sclerotin consists of proteins which the insect tans with chemical secretions very much the way hides are tanned to produce leather. Just as the wood of trees is stiffened by the incorporation of lignin with cellulose, so the hard parts of insects are stiffened by the incorporation of sclerotin with chitin. The combination is extremely tough and resilient, just the properties needed for a skeleton. The outermost layer is waxy and waterproof. The entire structure is called the **cuticle** (Fig. 11-17).

The insect's exoskeleton suffers from one serious drawback. Once it is laid down it cannot grow or expand. Before insects and other arthropods can grow they must secrete a new cuticle and shed the old one. This process is called **molting** and occurs periodically (Fig. 11-18). During molting the epidermis comes loose from the hard old cuticle and secretes a soft new cuticle with pleats and folds. The animal then sheds its old cuticle and expands to stretch the new larger cuticle before it hardens. The insect now has a new enlarged rigid skeleton (often somewhat compressed and telescoped) into which it gradually grows until the next molt. This process occurs in all arthropods. So-called softshelled crabs are captured shortly after molt and before the new cuticle has hardened. At other times their exoskeleton is quite hard.

Figure 11-18   A larva of the Cecropia silk moth molting from the fourth larval stage to the fifth larval stage. Just prior to molting, the larva spins a tiny pad of silk to which it clings. Then it secretes a new cuticle and crawls forward out of the old cuticle which remains attached to the silk pad. In this illustration the old cuticle can be seen clinging to the hind half of the caterpillar like a loose stocking.

We see that an insect is boxed in by its skeleton that cannot grow, and thus must make a new skeleton periodically. In this regard vertebrates have a clear advantage, for bone is capable of continuous growth for many years.

MUSCLE

When an animal is stimulated it responds. The organs that respond are called **effectors.** The most important of these effector organs are muscles which cause movement. Other effectors include the sweat glands which help cool man's body, the pigment cells of chameleons and fish which change in size or shape and alter the animal's color, the light-producing organs of fireflies which signal mating behavior, the nematocysts of hydra which immobilize prey, and the "batteries" of electric eels, to name just a few. These are the "executive" organs of the animal body—the parts by means of which an animal responds to changes in both the external environment and within its own body. In this section we consider effector organs concerned with movement. In Chapters 13 and 15, we discuss another type of effector, the endocrine glands.

Movement without muscles

All protoplasm is capable of some movement. The movement of the chromosomes during cell division is a well-known example. Look at a plant cell through a microscope and you will see the

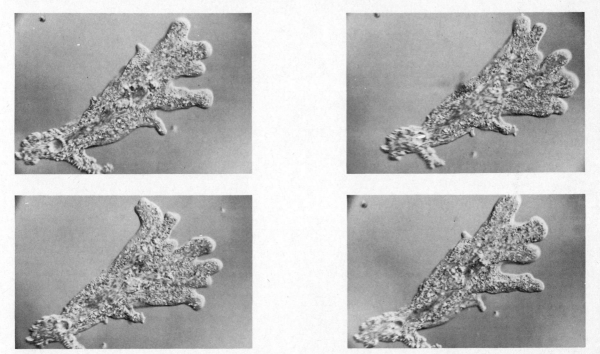

Figure 11-19 Amoeboid movement is a well-studied example of nonmuscular movement. This sequence of four time lapse photos of an *Amoeba proteus* shows the changes in cell form as blunt projections of the cell called pseudopodia (or pseudopods) extend and retract (sequence is clockwise, starting at upper left). The exposure time for each photograph was about 0.5 seconds and the blurring is caused by the movement of the inner part of the cytoplasm, the endoplasm. In this amoeba, the inner part of the cytoplasm (endoplasm) is more fluid than the outer part (ectoplasm), but both parts change their consistency readily. The endoplasm flows forward into the pseudopods, and then becomes ectoplasm. Meanwhile, ectoplasm at the opposite end of the organism becomes endoplasm, flows forward, and the process is repeated.

cytoplasm streaming around the cell. Sometimes this streaming results in the movement of an entire cell, as in a moving amoeba (Fig. 11-19). As the amoeba wanders about it sends out projections, or **pseudopodia,** and some protoplasm flows into the projections. Our white blood cells behave in much the same way when they engulf bacteria. Many other animal cells such as skin cells also send out long projections and may move slowly when they heal wounds.

Some small animals—many protozoans, flatworms—and sperms of most animals and many plants move by lashings of hairlike **cilia** or **flagella.** These tiny outgrowths of cells move rapidly like flexible oars and can propel small animals. Even large animals use them for many purposes. For example, in man ciliated epithelial cells move small particles along the trachea or windpipe. In clams ciliated cells sweep currents of water through parts of the body, bringing in oxygen and food and getting rid of wastes.

The molecular basis of these *nonmuscular movements* is not fully known, but in many cases it is associated with the presence within the cell of rigid thin rods called microtubules (Chapter 2). Cilia, flagella, and the spindle fibers that move chromosomes are all made up of these microtubules. We now look at *muscular movement.*

Structure of muscle cells—
Fibrous cells and fibrous molecules

Many cells move and contract, but muscle is a tissue specialized for motion. Although there are several different kinds of muscles in man and in other animals, all of them have the same basic structure and seem to work in the same way. Muscles, wherever they are found, consist of numerous elongated fibers that are usually supplied with nerves. When nerve impulses reach the muscle fibers, the fibers contract or shorten and the muscle becomes quite hard. When the nerve impulses cease, the muscle relaxes and becomes limp. Muscles can be made to contract and relax even after they are removed from an animal.

Man and other vertebrates contain two main types of muscle, **smooth** and **skeletal.** A third type of muscle, *cardiac* or *heart muscle,* resembles skeletal muscle but has some special features

discussed at the end of this chapter. A few other kinds of muscles are found in some invertebrates, but the basic structure and function of muscles, wherever they occur, are similar. Smooth muscles are composed of thin tapering cells with a single nucleus. Bundles or sheets of these cells line walls of many internal organs including uterus, bladder, gut, and blood cells. Skeletal muscles make up most of what we call "meat" in man and other animals and are usually found attached to bones—hence the name. They are composed of long cylindrical fibers containing many nuclei. As we observed in Chapter 8 they are formed by the fusion of individual *myoblasts* (embryonic muscle cells). Tens of thousands of these skeletal muscle fibers are bound together by connective tissue to form muscles. They usually contract more rapidly and vigorously than do smooth muscle cells.

A striking feature of skeletal muscle (Fig. (11-21) is the presence of large numbers of long *myofibrils* embedded in its cytoplasm. Skeletal muscle also shows alternating light and dark bands or *striations* that are caused by the regular arrays of myofibrils in the cytoplasm. Each myofibril is made up of a series of cylinders called *sarcomeres,* which shorten during muscle con-

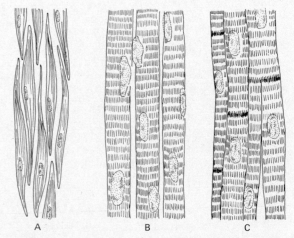

A   B   C

Figure 11-20 (A) Smooth muscle consists of separate cells with single nuclei; (B) Skeletal muscle cells contain many nuclei; (C) Cardiac muscle cells branch and interconnect. The dark bands represent thin transverse walls between individual cardiac muscle cells.

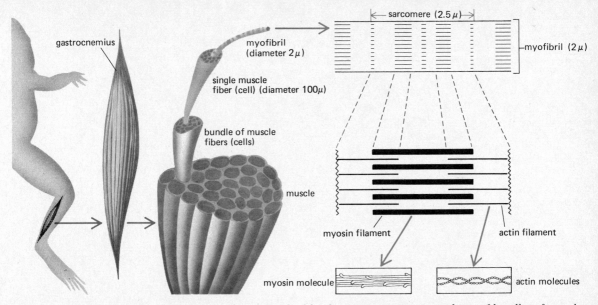

Figure 11-21 Organization of skeletal muscle. A muscle, like the gastrocnemius, is made up of bundles of muscle fibers encased in a connective tissue sheath. A single fiber is a muscle cell. The fiber, in turn, is made up of many myofibrils. Myofibrils are composed of two kinds of myofilaments, thin (actin) and thick (myosin).

traction. Each sarcomere consists of a number of fine myofilaments. The thick myofilaments are concentrated in the dark bands and the thin ones in the light bands. When muscles are treated with chemicals that dissolve away particular kinds of proteins, it is possible to cause the thick filaments and the dark bands to disappear, leaving the thin filaments intact. Such experiments show that the thick filaments are composed of a protein called **myosin** and the thin filaments of another protein, **actin.** It is possible to extract these two proteins from muscle and to combine them in solution to form *actomyosin* which will shorten when treated with ATP.

Do these proteins combine to cause contraction in muscle? If so, how? The answer to the first question apparently is "yes." Unfortunately we still do not know the complete answer to the second question. There is good evidence, however, that the thin actin filaments slide into the region of thick myosin filaments. This can be seen in electron micrographs that actually show filaments in various phases of the contraction process. It has been suggested that there are hooks or cross bridges on the myosin filaments that mo-

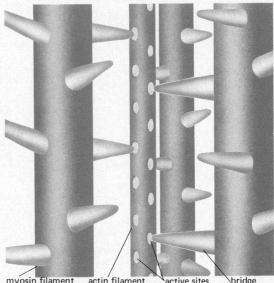

Figure 11-22 The interaction between actin and myosin causes muscle contraction. Projections (cross bridges or hooks) on the thick myosin filaments react with active sites on the thin actin filaments, causing them to slide along among the myosin filaments. Exactly how this process occurs is not known.

mentarily catch and hold the actin filaments (Fig. 11-22). Contraction, then, is really the sliding of the thin filaments between the thick ones. This causes each sarcomere to shorten, although the filaments themselves do not shorten, and the muscle as a whole to contract.

A contracting muscle does work, and work requires energy. Where does the energy for muscle contraction come from? The immediate energy source is ATP, which is formed from the breakdown of glycogen, a polysaccharide that is the main fuel for muscle contraction. In some way one of the contractile proteins of the muscle, myosin, splits the ATP molecule to liberate energy, which is then used for contraction.

Interestingly ATP is also important in the process of relaxation. When ATP disappears from a muscle after an animal dies, the muscle becomes stiff and rigid in *rigor mortis.*

How muscles move things

How do muscles do work? They pull — they cannot push. Even pushups are possible only because the muscles are pulling against the skeleton. Muscles pull in two ways: (1) they change their length and move objects (*isotonic* contrac-

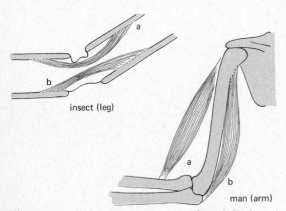

insect (leg)

man (arm)

Figure 11-23 How contracting muscles of the leg of an insect and the arm of a man cause movement. In each case a pair of antagonistic muscles is shown. Insect: When *a* contracts, the leg extends; thus *a* is an extensor muscle. When *b* contracts, the leg is flexed (bent); *b* is a flexor muscle. Man: When *a* (biceps) contracts, the arm is flexed; thus the biceps is a flexor muscle. When *b* (triceps) contracts, the arm is extended. The triceps is an extensor muscle.

tion) or (2) they develop tension (tightening) and exert a force without any change in their length (*isometric* contraction). It is easy to see how muscles do work when we compare the movements of man's arm and an insect's leg (Fig. 11-23). In man the muscles are attached to bones by *tendons;* in insects the muscles are attached to the exoskeleton. In both man and insects, in this particular case, one end of a muscle is attached to one part of the skeleton and the other end to a different part; between the two skeletal parts is a movable joint. Whenever man's biceps muscle contracts, the lower arm is pulled up; whenever the triceps muscle contracts, the arm is pulled down. The situation, however, is reversed in the insect. Contraction of muscle *a* pulls the leg down and contraction of muscle *b* pulls the leg up. This difference results from muscles being attached to the inside of the skeleton in insects rather than to the outside as in man. In both cases the muscles act in *antagonistic* pairs, that is, when one muscle shortens, the other lengthens. Most of man's 600 skeletal muscles are in antagonistic pairs, and their action is under the control of the nervous system. The same is true of insects.

Even in animals without real skeletons, muscles are arranged in antagonistic pairs. In earthworms the body wall is composed of a set of longitudinal muscles that run lengthwise along the body and a set of circular muscles that run around the body (Fig. 11-24). When the circular muscles contract, the longitudinal muscles relax and the incompressible body fluids act as a hydrostatic skeleton which transmits the pressure so that the worm becomes longer and thinner. Fine bristles along the body prevent the worm from sliding backwards and the front end of the body is pushed forward. The longitudinal muscles then contract and, because the bristles prevent the front end from moving backward, the hind end of the worm is pulled forward.

In hydra there are no true muscle cells. Instead some epithelial cells in the outer layer of the body wall and some digestive cells in the inner layer of the digestive tract contain myofibrils. In the outer epithelial cells they are arranged vertically, and when the outer cells contract, the hydra gets shorter and fatter. Myofibrils in the digestive cells run at right angles to those in the epitheliomuscular cells, and their contrac-

Figure 11-24 A. Movement of an earthworm. The body of an earthworm is made up of fluid-filled segments separated by sheets of tissue called septa. Each segment has its own set of circular and longitudinal muscles that act as antagonists. When any muscle in the body wall contracts, it presses the body wall against the fluid inside the body cavity. This in turn causes the stretching of any relaxed muscles. Movements occur as the animal alternately contracts the circular and longitudinal muscle layers in certain parts of the body. In this diagram, every fourth segment is shaded, thus indicating how the shapes of these segments change as the worm crawls. Bristles called setae give the worm traction and prevent it from slipping. When the longitudinal muscles contract, the setae stick out, whereas when the circular muscles contract, the setae are withdrawn.

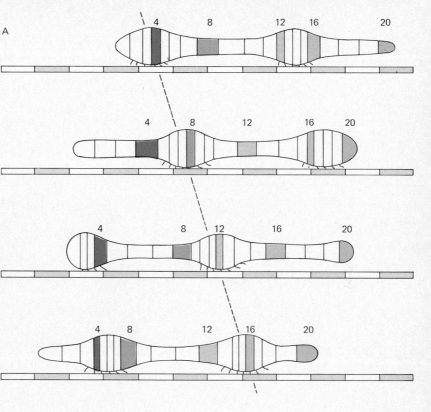

tion lengthens the hydra's body (Fig. 11-25). These movements are perfectly good muscular movements even though hydra has no true muscles. Apparently in hydra, muscle contraction is only a sideline for cells which are specialized to do something else. In most animals muscle contraction is performed by specialized muscle cells.

Activation of muscle cells

Normally muscle cells are stimulated to contract by nerve impulses. To understand the way in which all muscle cells respond to a nerve, let us consider the ways cells in general respond to stimulation. All cells are *excitable*, that is, they react to changes in the environment by changing their activities. At the molecular level cellular excitability appears to depend on changes in the cell membrane. If we insert a fine electrode with a microscopic tip through the cell membrane and into a cell, and another electrode on the surface of the cell, we can measure a voltage difference

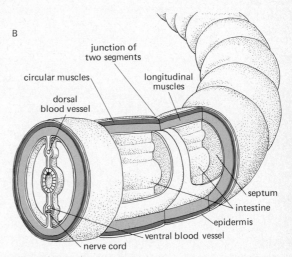

B. Segments of an earthworm showing circular and longitudinal muscle layers. The intestines of man have circular and longitudinal muscle layers like those of an earthworm, and contractions of these muscles move food along the intestine. The diagram shows how the septa separate the body into fluid-filled segments.

of about 0.1 volt. Moreover the inside of the cell is more negative than the outstide (Fig. 11-26). This is caused by the presence of more negative ions on the inside of the cell membrane than on the outside. The cell keeps its interior more negative than its exterior by transporting out certain positive ions—usually sodium. This causes a redistribution of potassium and chloride and results in an excess of negative ions inside the cell. We have not worked out the exact mechanism of this transport of sodium ions, but we do know that it requires energy. It is usually referred to as the **sodium-potassium exchange pump,** and the voltage difference across the cell membrane is commonly called the **membrane potential.** When the cell dies, the sodium-potassium exchange pump stops, positive sodium ions pour in, potassium ions pour out, and the voltage difference disappears.

Sodium-potassium exchange pumps and membrane potentials are found in most animal cells and are especially important in the functioning of muscle cells and nerve cells. Whenever the cell membrane is stimulated by a nerve or experimentally by adding certain chemicals, the delicate balance of ions is disturbed, and the voltage difference or membrane potential

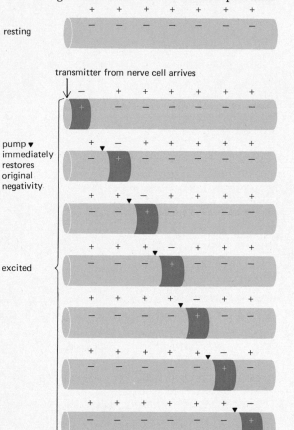

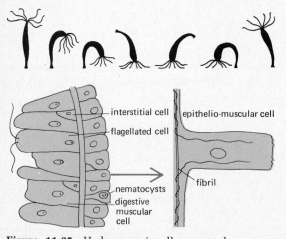

Figure 11-25 Hydra occasionally moves by somersaulting from place to place (as shown at top). Movements of hydra are caused by the contraction of fibrils in certain cells, especially epithelio-muscular cells in the epidermis and digestive-muscular cells in the gastrodermis. Fibrils in the epithelio-muscular cells run longitudinally; when they contract, the hydra becomes shorter and fatter. Fibrils in the digestive-muscular cells are oriented transversely (at right angles to the long axis of the body); when they contract, the hydra becomes longer and thinner. Longitudinal and transverse fibrils are antagonists. Fibrils are not visible in the illustration at the left.

Figure 11-26 Depolarization of a muscle cell membrane. The inside of a resting muscle cell membrane is negative with respect to the outside, that is, it is polarized. When the membrane is excited by the transmitter substance released by nerve endings, the membrane becomes more permeable to certain ions, especially Na^+. The inside of the cell becomes less negative or depolarizes. A wave of depolarization sweeps down the fiber. An active transport mechanism—"pump"—restores the original negativity of depolarized regions, that is, they are repolarized.

changes. This means that when a cell is stimulated there is an electrical change in the cell membrane. Once we understand this property of cells we can make progress in explaining how muscle cells, nerve cells, and receptor cells respond to stimuli.

As we have said, muscles are supplied with nerves. Nerve cells send out long fibers called **axons** which end very close to, but do not actually touch, the muscle cell membrane at a region called the **neuromuscular junction** (Fig. 11-27). An excited nerve sends waves of excitation—*nerve impulses*—down the axon. When a nerve impulse arrives at the neuromuscular junction, the axon releases from tiny sacs near its end a chemical called a **transmitter substance.** The transmitter, in a way which is not yet understood, momentarily makes the membrane of the muscle cell at that point permeable to positive ions (Na^+ and Ca^{2+}). As the positive ions enter the muscle cell, the cell becomes less negative

than before and the membrane potential falls. Almost immediately, however, the sodium pump starts pumping out Na^+, restoring the membrane potential, and the inside of the cell once again becomes negative compared to the outside. The fall in membrane potential of the muscle membrane is always followed by contraction.

How is excitation of the muscle cell membrane coupled to contraction? Efforts to answer this question on individual skeletal muscle cells of vertebrates have failed, largely because vertebrate skeletal muscle cells, although fairly large (0.1 mm thick and 10 mm long in frogs), are quite small for chemical study. Fortunately some crabs and barnacles have gigantic muscle cells 2 mm thick and 50 mm long which are large enough to inject chemicals into and to take out samples. Studies on these giant muscle cells indicate that the excited membrane permits calcium ions, as well as sodium ions, to enter the cell, and that calcium ions actually trigger the contraction. Apparently calcium ions in some way permit the interaction of actin and myosin. This results in contraction of the muscle cell. In this process of contraction myosin breaks down ATP within the muscle cell to provide the energy needed for contraction. When the excitation of the muscle membrane ceases, the calcium ions are removed from the immediate environment of the actin and myosin, these contractile proteins no longer interact, and the muscle relaxes.

Regulation of degree of contraction

Muscles can produce an astonishing variety of movements. The same hand that can deliver a karate chop can also caress. The muscles in a crab's claw can snap the claw shut rapidly to seize prey or can close it slowly and powerfully to crush a clam. What determines whether contraction is slow or fast, light or strong? The answers are the kinds of muscle and the nervous control of that muscle. Skeletal muscle contracts much more rapidly than other muscles, but smooth muscle remains contracted longer than skeletal muscle. Contraction rates vary from one long slow contraction every 5 minutes in the smooth column muscles of some sea anemones (relatives of hydra) to 30 per second in man's eye to 1000 per second in the striated flight muscles

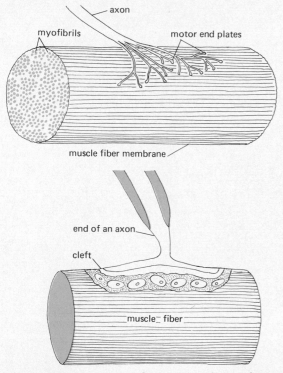

Figure 11-27 Neuromuscular junction of a vertebrate. The ending of an axon on a skeletal muscle fiber in a vertebrate is called a motor end plate. There is a cleft between the two cells.

of some insects. Mammalian skeletal muscles appear to have several different kinds of fibers. Thick skeletal muscle fibers contract rapidly; thin skeletal muscle fibers contract slowly. These "fast" and "slow" muscle fibers permit either extremely rapid or slow contractions in the same muscle. Some invertebrates have only smooth muscle whereas others, such as insects, have only striated muscle. Still others have both. For example, scallops use striated muscles in swimming rapidly away from danger, but smooth muscles in keeping the shell closed for long periods. In general, as we move from the simpler to the more complex invertebrates and then to the vertebrates, the proportion of striated to smooth muscle increases. This reflects increasing activity of the animals.

There are two basic patterns of nervous control of muscle contraction. One is characteristic of vertebrate skeletal muscles and the other of vertebrate smooth muscles and of the skeletal muscles of arthropods.

A single vertebrate skeletal muscle is usually supplied by branches from many nerve cells, perhaps thousands. But each muscle cell is innervated by only one branch of one axon. When nerve impulses excite the muscle cell, the cell contracts. When nerve impulses cease to stimulate the muscle cell, the muscle cell relaxes. If a vertebrate striated muscle cell receives several nerve impulses following rapidly on each other, the cell will not relax completely between impulses, but it contracts more strongly than it normally would. If the impulses arrive extremely rapidly, the muscle will not relax at all but goes into a state of sustained contraction, or *tetanus*. (This is not to be confused with the disease "tetanus" or lockjaw caused by certain bacteria.) In fact most of our ordinary muscle activity involves tetanus rather than simple muscle twitches. Body muscles are seldom completely relaxed. Rather, they are in a state of constant partial contraction, which is referred to as *muscle tone*. We do not feel fatigued by these partial contractions, because not all muscle cells contract at the same time; first one and then another group of muscle cells contracts. Muscle tone disappears when muscles become inactive as a result of disease or injury. For example, if the nerve supplying a muscle is cut or destroyed by a virus like polio

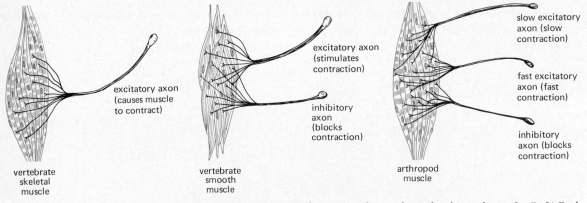

Figure 11-28 Innervation of vertebrate skeletal muscle, vertebrate smooth muscle, and arthropod muscle. (Left) Each fiber in a vertebrate skeletal muscle in innervated by a single branch of an excitatory axon. When this axon is excited, nerve impulses cause the muscle fibers to contract. When this axon ceases to send impulses, the muscle fibers relax. There are no inhibitory nerves to vertebrate skeletal muscles. (Middle) Most smooth muscle fibers are innervated by both an excitatory axon and an inhibitory axon. The muscle fibers contract in response to excitatory impulses and relax in response to inhibitory impulses. If both nerves stimulate the muscle cells, they respond in some intermediate fashion. (Right) Arthropod muscle cells, particularly those of crustaceans, may receive three kinds of axons. Fast excitatory axons, when stimulated, cause the muscle cells to contract rapidly. When slow excitatory axons are excited, they cause the muscle cells to contract slowly. When inhibitory axons are excited, they block the activity of either slow or fast excitatory fibers, and prevent the muscle cells from contracting.

virus, the muscle is inactivated, and it gradually becomes smaller or *atrophies.*

It is worth emphasizing that skeletal muscles in man and other vertebrates are innervated only by excitatory nerves. To cause a skeletal muscle in a vertebrate to relax it is necessary to stop sending nerve impulses to it. The situation is quite different in vertebrate smooth muscle and in the skeletal muscles of some invertebrates. Most smooth muscle cells in vertebrates are innervated by two nerve fibers; one is a parasympathetic and the other is a sympathetic nerve fiber. Nerve impulses in the parasympathetic nerve fibers cause them to release a transmitter substance called acetylcholine which usually, but not always, excites a smooth muscle cell to contract. Nerve impulses in sympathetic nerve fibers cause them to release a transmitter substance called noradrenalin which usually inhibits the smooth muscle from contracting. Thus to inhibit a vertebrate smooth muscle cell from contracting requires inhibitory nerve impulses.

Many arthropod muscles, particularly in crustaceans, are also supplied with both excitatory and inhibitory nerves. In crabs, for example, the skeletal muscle cells of the claw receive both excitatory and inhibitory nerve fibers. Frequently there are two kinds of excitatory nerves. Impulses from the *slow excitatory nerve* cause a slow powerful contraction of the muscle cells and permit the claw to crush slowly. Impulses from the *fast excitatory nerve* cause a rapid contraction of the muscle cells and enable the claw to snap shut swiftly. When the *inhibitory nerve* is stimulated, the muscle relaxes and impulses from the excitatory nerves are no longer effective.

Cardiac muscle and some other special muscles

Not all muscles require nerve impulses to contract. Vertebrate heart muscle and some types of smooth muscle are intrinsically rhythmic; even an isolated piece of heart muscle will contract and relax spontaneously at intervals. Another special feature of vertebrate heart muscle is that the muscle fibers are not insulated from one another. As a result, when a few heart muscle fibers are excited, the other fibers also become excited and the entire heart contracts. This is very different from most skeletal muscles in which the individual fibers are insulated from each other. In skeletal muscles electrical changes and contraction occur only within excited fibers and do not spread to other fibers.

Another kind of muscle which contracts without nerve impulses are the *flight muscles* of many insects. These special striated muscles contract and relax more rapidly than any other muscles known. In some insects the wings beat as frequently as several hundred or even 1000 times per second and the flight muscles contract and relax that rapidly. Surprisingly nerve impulses stimulate these muscles much less often than this. Apparently the muscles, skeleton, and wings of the insect start vibrating at a rapid rate and only an occasional nerve impulse to the flight muscles is needed to keep the vibrations going.

Other fascinating muscles include the muscles which hold clam shells shut; these muscles seem to have some kind of a catch mechanism that keeps them contracted for very long periods. The electric organs of electric eels and most other electric fish are made up of highly modified muscle cells which can generate strong electric currents. The electric eel uses its high voltage discharge in stunning prey and for protection. Other electric fish use electric organs as a kind of electric sonar to detect the presence of other fish and objects near them. Why electric fish are not electrocuted by their own electric discharges is still a puzzle.

SUGGESTED READING LIST

Allen, R. D., "Amoeboid Movement," *Scientific American*, February, 1962 (Offprint 182).

Beck, W. S., *Human Design.* Harcourt Brace Jovanovich, New York, 1971.

Bloom, W., and D. W. Fawcett, *A Textbook of Histology* (ninth edition). W. B. Saunders Company, Philadelphia, Pennsylvania, 1968.

Gans, C., "How Snakes Move," *Scientific American*, June, 1970 (Offprint 1180).

Gordon, M. S. (in collaboration with G. A. Bartholomew, A. D. Grinnell, C. B. Jorgensen, and F. N.

White), *Animal Physiology: Principles and Adaptations* (second edition). The Macmillan Company, New York, 1972.

Griffin, D. R., and A. Novick, *Animal Structure and Function* (second edition). Holt, Rinehart and Winston, New York, 1970.

Guyton, A. C., *Function of the Human Body* (third edition). W. B. Saunders Company, Philadelphia, Pennsylvania, 1969.

Hoyle, G., "How is Muscle Turned On and Off?" *Scientific American*, April, 1970 (Offprint 1175).

Huxley, H. E., "The Contraction of Muscle," *Scientific American*, November, 1958 (Offprint 19).

Huxley, H. E., "The Mechanism of Muscular Contraction," *Scientific American*, December, 1965 (Offprint 1026).

Huxley, H. E., "The Mechanism of Muscular Contraction," *Science*, 164, 1356-1366, 1969.

Marshall, P. T., and G. M. Hughes, *The Physiology of Mammals and Other Vertebrates*. Cambridge University Press, New York, 1965.

Montagna, W., "The Skin," *Scientific American*, February, 1965 (Offprint 1003).

Novikoff, A. B., and E. Holtzman, *Cells and Organelles*. Holt, Rinehart and Winston, New York, 1970.

Smith, D. S., "The Flight Muscles of Insects," *Scientific American*, June, 1965 (Offprint 1014).

Winton and Bayliss, *Human Physiology* (sixth edition), rev. and ed. by O. C. J. Lippold and F. R. Winton. Williams and Wilkins, Baltimore, Maryland, 1969.

CHAPTER 12

Food and Energy for the Animal Way of Life

Both plants and animals must eat to live. For green plants the dietary requirements are simple: minerals and water from the soil and carbon dioxide. As noted in Chapters 5 and 10 green plants utilize the energy of sunlight to synthesize sugars and all of the important biological molecules from these simple materials. They use some of the molecules for structural purposes. They break down others to release chemical energy for the work of the cells. Accordingly some glucose molecules are used to form the cellulose of plant cell walls, whereas others are used to provide energy. Animals, however, cannot synthesize all of the molecules they need; they must use substances prefabricated by plants. This inability to synthesize all of the necessary molecules presents problems. First, unless it wallows in its food like a tapeworm, an animal must have some kind of feeding mechanism. Second, it needs digestive systems to break down or convert food into a usable form.

In this chapter we discuss first the nutritional needs of man and some other animals, how they obtain these nutrients, and how foods are broken down during the process of digestion into molecules that the animal's cells can use. We then examine some of the processes, such as breathing, circulation, excretion, and temperature control, for which energy derived from food is needed.

NUTRITION

A complete diet—
Wool, wood, milk, or fish

It is only about 100 years since biologists recognized that in man and other animals the need for food actually involves two interlocking requirements. No matter how much food an animal eats, it will die unless the food contains certain chemical ingredients. Nor will a diet rich in these chemicals keep an animal alive if there is not enough of it. Although there are some differences in the kinds of chemicals needed by different groups of animals, for the most part their nutritional requirements are remarkably similar. Both men and insects, for example, need the same ten amino acids in their diet. Vision in both man and insects requires vitamin A. Thus although clothes moths feed on wool and termites on wood, Masai herdsmen on milk and blood, and Eskimos on raw fish and seal meat, they all reduce the substances they eat to the same few basic molecules that all animal cells need to do their job.

Nutritional requirements of man

The three main classes of foods are *carbohydrates*, *fats*, and *proteins*. The structure of these molecules was discussed in Chapter 4. These foods are needed in bulk and provide energy and building materials for man's cells. Man's diet must also include *vitamins* and *minerals* which play essential roles in certain cellular reactions.

Carbohydrates.   *Sugars* and *starches* are typical carbohydrates. Some African tribes (for example, the Masai) and Eskimos rarely, if ever, eat carbohydrates, yet they obviously get along without them. Glucose and other products of molecules derived from carbohydrates are essential to the cell; for example, a sugar forms part of the structure of the DNA molecule. Still carbohydrates are not necessary in man's diet because cells form glucose and other carbohydrate derivatives from protein. However, the diets of most of the people of the world consist largely of carbohydrates, especially in underdeveloped countries.

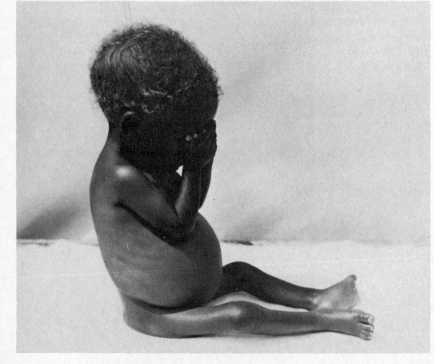

Figure 12-1A   Child with kwashiorkor. Usually a baby or young child does not develop this disease as long as it is breast-fed. But in many parts of the world, when another baby is born to the family, the older child must be weaned on a protein-poor diet, the cause of kwashiorkor.

Fats. As we have already seen, more energy can be obtained from fat than from an equal weight of carbohydrate or protein. Fats are found in many meats and dairy products and in some fruits (nuts, avocados). There is no doubt that fats are important to the cell; for example, they form part of the structure of all cell membranes. However, the amount of fat eaten by people in the highly developed Western countries is far beyond what the average person needs, and may even contribute to diseases such as circulatory ailments.

Proteins. No organism can live without protein. Its breakdown products, the amino acids, are essential for synthesizing cell structures and for forming enzymes. In addition proteins can be used as an energy source, but they seldom are except during starvation. Fortunately proteins are found in many types of foods, such as meats, milk and dairy products, cereals and grains, nuts, and some beans. Most fruits and vegetables, however, are relatively low in protein. Unfortunately the diets of most of the people in the world consist of vegetables, and malnutrition is widespread in parts of South America, Africa, and Asia. *Kwashiorkor* is a severe nutritional disease that is caused by a lack of protein. It particularly affects babies and children (Fig. 12-1A) because their need for protein is greater than an adult's. For example, a newborn baby needs five times as much and a young child needs about two and a half times as much protein as an adult, pound for pound. However, the *amount* of protein is not as important as the *amino acid content* of the protein. Cells use about twenty different kinds of amino acids; some of these amino acids can be synthesized by the animal body, and others, the so-called **essential amino acids,** must be supplied in the diet. The term "essential amino acid" is unfortunate, for

Figure 12-1B Child with rickets. The term rickets is derived from *wrikken,* an Anglo-Saxon word meaning "to twist." The disease is the result of a deficiency of vitamin D, which is necessary for proper metabolism of calcium and phosphorus, two important elements in bone. In severe cases, not only the legs but also the spine and the pelvis may be twisted. Rickets was first described in the 17th century; at that time London was becoming an industrial city, and heavy smoke often blanketed the city (air pollution is not a 20th century phenomenon.) Doctors noted that children of poor families living in London developed the disease more often and with greater severity than country children or children of wealthy families who escaped to the country. Today rickets seldom occurs except in young puppies and in children in cultures which practice purdah — the seclusion of women and children from the outside world.

it implies that some amino acids are not necessary to the cells. This is not true; the term applies only to amino acids which cells need *and* cannot synthesize.

Vitamins.   Man also needs in his diet very small quantities of certain substances called *vitamins.* Some of these vitamins, their natural sources, and symptoms of their deficiency are shown in Table 12-1. Vitamins participate in a wide variety of cellular reactions; for example, many vitamins function as *coenzymes* without which the protein portions of enzymes are ineffective. What is a vitamin for man is not necessarily a vitamin for another animal. For example, man's diet must include vitamin C because he cannot synthesize it; but this substance is not a vitamin for the rat because it can be synthesized in its liver.

A well-balanced diet supplies the vitamins needed by most normal healthy individuals; the daily vitamin pill is usually unnecesary. In fact the body cannot store excess water-soluble vitamins; it excretes them. Apparently some fat-soluble vitamins can be stored for a few weeks in the liver, and if the daily vitamin pill leads to an excess of some fat-soluble vitamins they can become toxic rather than beneficial. Incidentally some vitamin D can be synthesized by the skin in the presence of the ultraviolet rays of sunlight, but it is not known whether all of man's vitamin D requirement can be obtained this way (Fig. 12-1B).

Of course, not everyone is healthy, and a vitamin lack can cause disease or even death. Up until the 1930s persons with pernicious anemia—a disease in which the number of red blood cells steadily declines—slowly died because they could not transport about a millionth of a gram of vitamin B_{12} across the intestinal walls. A substance (the nature of which we are still ignorant) which is produced in the stomach aids in the absorption of vitamin B_{12} across the gut wall; if this substance is missing, the vitamin B_{12} synthesized by intestinal bacteria cannot be absorbed, and it is useless. There is still no cure for

Table 12-1.   Vitamins Needed by Man

Vitamin	Major food sources	Results of deficiency
A* (Retinol)	Milk and dairy products; liver and liver oils; vegetables	Night-blindness; dry, flaky skin
B_1 (Thiamin)	Whole-grain or enriched cereals; nuts; wheat germ	Beriberi
B_2 (Riboflavin)	Milk and dairy products; leafy vegetables; whole-grain or enriched cereals	Cracking of corners of mouth; sores on skin; blood-shot eyes
B_6 (Pyridoxine)	Whole-grain or enriched cereals; liver; kidney; fish; fresh vegetables	Dermatitis; convulsions; lowered resistance to infections
B_{12} (Cyanocobalamin)	Liver; milk and dairy products	Pernicious anemia
Biotin	Liver; milk; eggs; vegetables; whole-grain cereals	Scaly, itchy skin
Folic acid (Folacin)	Liver; leafy vegetables	Anemia; diarrhea
Nicotinic acid (Niacin)	Liver and other meats; eggs; whole-grain cereals	Pellagra
Pantothenic acid	Present in most foods	Abnormal functioning of adrenal gland
C (Ascorbic acid)	Citrus fruits; tomatoes	Scurvy
D† (Calciferol)	Liver oil; milk and dairy products	Rickets
E (Tocopherol)	Present in most foods	Destruction of red blood cells
K‡ (Phylloquinone)	Liver; green vegetables	Slow clotting of blood or hemorrhage

* Some vitamin A may be formed in small intestine from carotene, the pigment that makes carrots yellow. † Enough may be synthesized by action of sunlight on skin. ‡ Enough may be produced by bacteria in the large intestine.

pernicious anemia, but it is now possible for those who have it to take regular injections of vitamin B_{12} and then lead normal lives.

Minerals. A well-balanced diet also supplies the kinds and amounts of minerals that man needs. Minerals are part of the structure of many important biological molecules. For example, iron is found not only in the hemoglobin molecule which carries oxygen throughout the body but also in the cytochromes which are key oxidative enzymes in cells. Some other necessary minerals are sodium, potassium, calcium, magnesium, and phosphorus; traces of iodine, fluorine, copper, cobalt, zinc, manganese, selenium and chromium are also essential. For example, cobalt is part of the structure of vitamin B_{12} which we discussed, and calcium makes bones and teeth hard. Recent studies indicate that traces of silicon, tin, vanadium and nickel may also be necessary.

FEEDING MECHANISMS

Some internal parasites, such as tapeworms, passively wait for their food to come to them and then absorb the molecules of food that has already been digested by the host. Most animals, however, must seek and ingest their food. They have evolved many specialized mechanisms which enable them to feed. In fact, most of the organs we find in simpler animals have something to do with capturing, eating, and digesting food.

Feeding mechanisms are related more to the *kind* of food an animal eats than to the taxonomic group to which the animal belongs. For example, some of the whales, which are mammals, ingest microscopic plants and animals as do many other unrelated animals such as clams. These whales and clams are **filter feeders.** Filter feeders eat almost continuously. Many are sluggish like the clam; a few others, like whales, are quite active. Some filter feeders, such as clams, gather their food by sweeping water through parts of their body with cilia and by secreting mucus which traps food particles in this water.

Many animals, including man, most insects, and hydra, are **bulk feeders,** that is, they eat a relatively large amount of food at one time. As a result bulk feeders usually do not feed continuously. Most bulk feeders break up the large mass of food into small chunks, which can then be attacked easily by their digestive enzymes. Animals use all sorts of mechanical and chemical techniques to break up food (Fig. 12-3). Man and most other mammals *chew* their food before it is swallowed. Birds have a muscular chamber, the *gizzard*, containing pebbles or grit to grind food. A chicken or a turkey will die of starvation even though it has plenty to eat unless

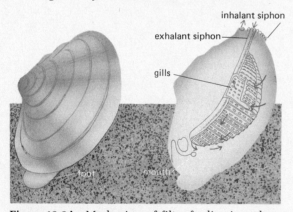

Figure 12-2A Mechanism of filter feeding in a clam. The gills are covered with beating cilia which continuously sweep water through the inhalant siphon and over the gills. Food particles in the water are first trapped in the mucus on the gills. The mucus is moved by the cilia to the mouth, like a "continuously moving flypaper." The mucus with the entrapped food is digested. The water, freed of food particles, passes behind the gills and out the exhalant siphon. Arrows show the directions of water movement.

Figure 12-2B A bulk feeder such as the large mouth bass takes in large amounts of food only at intervals.

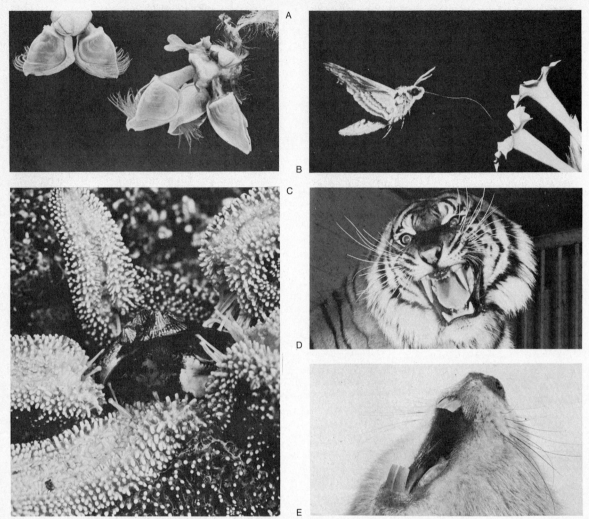

Figure 12-3   Some adaptations for feeding. A. Adult barnacles like these buoy-making barnacles are unlikely looking, sessile crustaceans. The feathery limbs or cirri projecting from the shell sweep through the water to capture small animals. As the famous biologist Louis Agassiz once noted, the barnacle is ". . . nothing more than a little shrimplike animal standing on its head in a limestone house and kicking food into its mouth." B. Sphinx moth feeding on tobacco blossoms. The tongue is much longer than the moth. C. Starfish feeding on abalone. Using the combined suction of thousands of tiny tube feet, the starfish pulls the valves of the abalone apart, everts its stomach through its mouth, and wraps it around the inside of the abalone. Powerful enzymes then start to digest the abalone. D. The basic diet of carnivores and herbivores reflects the kind and shape of their teeth. Carnivores, like this Bengal tiger, have sharp incisors, long pointed canines to seize and tear flesh, and premolars and molars (when present) that are ideal for tearing and cutting. Teeth of carnivores are not well adapted for chewing, and these animals chew little if at all. E. The teeth of herbivores such as beavers are admirably adapted for gnawing and chewing. The incisors, which are much longer than those of carvivores, are constantly worn down and regrown. Their inside surface, unlike the front surface, has no enamel covering; consequently, it wears down more rapidly than the front, producing a beveled edge ideal for gnawing. Herbivores have no canine teeth. Their premolars and molars have surfaces suited for grinding. Man's teeth, like those of other omnivores, resemble some of the teeth of both carnivores and herbivores.

it also has pebbles in its gizzard. Some birds also drop their prey against stones to expose soft body parts, and parrots use their beaks to disintegrate their food. The mouth parts of "chewing" insects like grasshoppers tear the food into smaller pieces. Mollusks often have rasping mouth parts with which they drill through hard objects, such as shells, in order to get to the soft parts of their prey. Earthworms are toothless like birds and also use a gizzard.

Some animals break up their food by chemical means before they eat it. A spider which has just captured a fly injects digestive enzymes into its victim's body, and then the spider sucks the partially digested fluid body parts of the

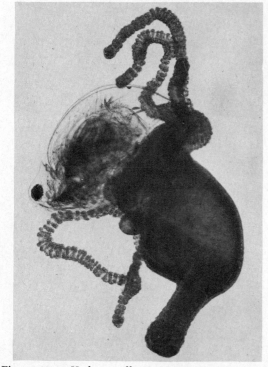

Figure 12-4 Hydra swallowing a minute crustacean called a water flea. Nematocysts capture and poison the prey; the tentacles draw it to the mouth and stuff it into the gastrovascular cavity. There it is partially digested and broken up into small pieces; cells lining the cavity engulf these pieces. Since the breakdown of food to molecules the cells can use occurs largely within the digestive cells, digestion is mostly **intracellular.** Waste materials go out the mouth.

fly. Similarly, starfish actually protrude their stomach between the opened halves of a clam's shell and secrete digestive enzymes which partially digest the clam.

Some animals, such as bees, mosquitoes, and leeches, have mouth parts which are adapted for sucking fluids from the plant or animal on which they feed.

DIGESTION

The spectacular concentration of animals on our earth depends in no small measure on their ability to digest almost any kind of food. Hummingbirds digest nectar, clothes moths digest wool, termites digest wood.

Digestion is the breaking down of food into molecules which the animal can use. It has three purposes: (1) to provide chemical energy for the work of the cells, (2) to supply "building blocks" for protein and other structural molecules for growth and tissue repair, and (3) to supply vitamins and minerals needed by the body.

A simple organism like a hydra, after capturing its tiny prey such as a water flea, swallows it (Fig. 12-4). Within the body cavity digestive enzymes are relased from cells lining the cavity. These enzymes break down the food into particles small enough to be engulfed by these cells, and the rest of digestion is completed inside these cells. Most of the waste matter is voided through the mouth. More complex animals such as man and insects break down most of their food into individual molecules before the cells of the digestive tract absorb them. Their digestive tracts are spectacular disassembly lines that take in big chunks of food and split them into individual molecules. Different parts of the digestive tract are specialized for different steps in this disassembly.

Human digestive system

The human digestive system is shown in Figure 12-5. The digestive tract from the top of the esophagus to the anus is about 30 feet long in the adult. And yet the lining of the entire digestive tract is renewed by cell division every three days.

To understand how digestion takes place let us follow the fate of a hamburger on a bun you have just eaten. The bread, of course, is starch (carbohydrate) and the hamburger is mostly protein with some fat. Some digestion occurs in the mouth. Before you even took a bite of food, the brain—perhaps remembering the last cookout and how good *those* hamburgers were—signaled the **salivary glands** to start secreting saliva; the flow of *saliva* continues when food is in the mouth. Saliva has two functions: (1) it moistens food, making it easier to swallow and (2) it contains a digestive enzyme, *amylase* (ptyalin), which breaks down starch into molecules of a complex sugar. Actually very little starch is digested in the mouth because food remains there only briefly, and when food reaches the stomach, the amylase is inactivated.

After food is broken into smaller bits by chewing, it is swallowed. We swallow more or less at will, but the events that accompany swallowing are automatic and beyond our control. The openings to the nose and the windpipe are closed. Therefore food has only one place to go—from the mouth, through the **pharynx,** and down the **esophagus.** In the esophagus food is squeezed toward the stomach by waves of muscular contraction or **peristalsis.** Normally the esophagus end of the stomach is closed by a valve, but the valve opens when a peristaltic wave bearing food arrives from the esophagus.

The **stomach** is a large muscular pouch which actually lies on the left side of the abdomen just below the ribs. It stores and helps digest food. Man's stomach will hold about half a gallon of food; this is one reason why he can go for rela-

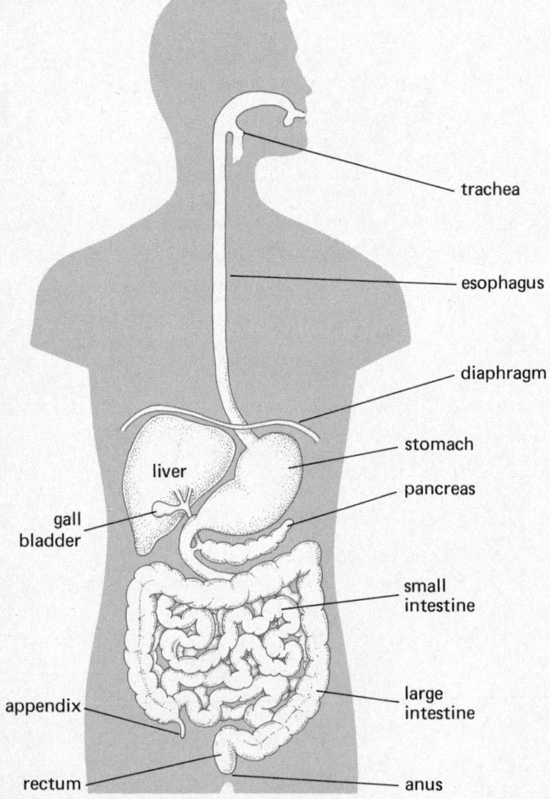

liver

gall bladder

trachea

esophagus

diaphragm

stomach

pancreas

small intestine

appendix

large intestine

rectum

anus

Figure 12-5 Human digestive system. Food is broken down to individual molecules in the gut; thus digestion in man is entirely **extracellular.**

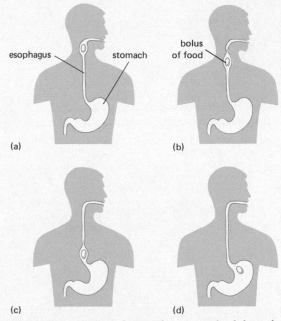

esophagus   stomach

bolus of food

(a)   (b)

(c)   (d)

Figure 12-6 A wave of peristalsis moves food through the esophagus to the stomach (and through the intestines). Peristalsis occurs when circular muscles in the wall of the esophagus (or gut) contract behind a lump of food, while muscles in the wall in front of the lump relax. Food travels from the upper esophagus to the stomach in just a few seconds. Fluids flow through the esophagus by gravity. (See Fig. 11-24B of earthworm where the circular and longitudinal muscles are arranged like those of the intestine.)

tively long periods without eating. Food remains in the stomach until it can be admitted to the first part of the small intestine which can handle only small amounts of food at a time.

The stomach wall has several layers. The middle layer is composed of smooth muscle. Powerful contractions of these muscles churn, mix, and disintegrate the food. The inner wall contains many gland cells. Some of them secrete the digestive juice and others secrete mucus which lines the stomach and helps prevent its being digested by its own juice. The digestive juice contains *hydrochloric acid* and a powerful enzyme called *pepsin* which splits proteins. The gland cells do not store the strong pepsin—they would be destroyed; rather, they store and release an inactive form of the enzyme called *pepsinogen*. In the presence of the acid the pepsinogin is activated and becomes pepsin. It is pepsin which attacks the hamburger protein and breaks it into smaller units. If you had a bottle of beer with your hamburger, some of the alcohol now would pass directly through the stomach walls into the bloodstream.

In the 1800s an Army surgeon, William Beaumont, observed the activity of the stomach in a series of spectacular experiments. Beaumont treated a French-Canadian fur trader, Alexis St. Martin, who had been shot in the stomach. St. Martin recovered from what then was usually a fatal wound, but a 2½-inch opening to the outside remained in the stomach. Beaumont was excited because he could look "directly into the cavity of the stomach, and almost see the process of digestion . . ." He persuaded St. Martin to cooperate in experiments in which Beaumont tied pieces of food to a string, and dropped the food into the stomach at intervals to see what would happen. Much of what we know about the stomach's role in digestion came from Beaumont's observations. It appears that St. Martin was hardly an ideal patient. He reportedly drank too freely in the middle of some of Beaumont's most important experiments. It is said that he poured the alcohol directly into his stomach in order, he claimed, "to save time." Even so he outlived Beaumont by several years.

Most digestion takes place in the **small intestine.** From the stomach the partially digested food passes into the first part of the small intestine, the **duodenum.** This is probably the most important site of digestion. Man can live without a stomach (by simply eating smaller meals more often), but losing the duodenum can be critical. Food coming from the stomach is acidic. This acid food stimulates the intestine to secrete *intestinal hormones* which enter the blood and stimulate the **pancreas** to secrete *pancreatic juices* and the **liver** to secrete *bile*. The pancreas lies just below the stomach, and is connected to the duodenum by a passageway, or *duct*. Pancreatic juices contain enzymes that digest carbohydrates, proteins, and fats. The fat-digesting enzyme, *lipase*, is aided by bile. Bile is stored in the **gall bladder.** It is not a digestive enzyme, but it acts on fats in the same way that soap acts on grease. Bile breaks up fat particles so they can be broken down by lipase to *fatty acids* and *glycerin*.

At this point, then, your hamburger and bun have been reduced to long chains of amino acids, complex sugars, fatty acids, and glycerin. These substances pass from the duodenum into the other parts of the small intestine, the **jejunum** and the **ileum.** The long chains of amino acid are broken down further into individual *amino acids*, and the complex sugars become *glucose*. After the hamburger and bun have been broken down to their simplest molecules, *absorption* of the useful molecules begins, and the unused materials pass on to the large intestine or *colon*.

If you were to run your finger over the inner lining of the small intestine, you would find that it feels soft and plush, much like velvet. If you look at a bit of this "velvet" through a microscope, you will see thousands of tiny fingerlike projections called **villi.** Food is absorbed by the villi. Electron micrographs of villi show that each *villus* is in turn composed of many smaller fingers or **microvilli** (Fig. 12-7).

Man's small intestine is beautifully designed to absorb food molecules. In the adult the small intestine, including the foot-long duodenum, is more than 20 feet long and about 1 inch in diameter. The inner lining is draped into numerous folds, each square inch of which is covered with thousands of villi and thousands more microvilli. The total area for absorption is more than 1000 square feet! In addition food remains in the small intestine from 4 to 8 hours, thus permitting sufficient time for absorption.

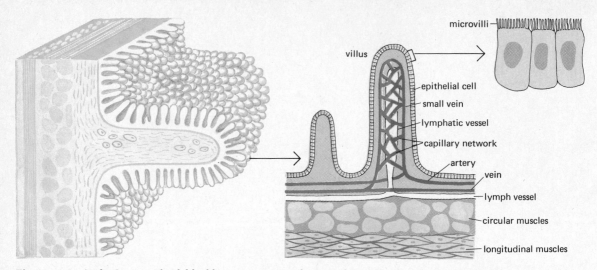

Figure 12-7 (Left) Section of a fold of human intestine showing the tremendous increase in surface area produced by villi. (Center) The microscopic structure of a villus. (Right) Enlargement of several epithelial cells of a villus, whose surface is increased further by the presence of microvilli. As a result of these modifications, one square inch of intestinal wall has 600 square inches of absorptive surface.

Each villus is supplied with fine blood vessels called **capillaries** and other fine tubes called **lymph vessels.** Amino acids and sugars are picked up by the blood supplying the villi and transported to the liver. The liver also rearranges and stores most of the incoming sugars, amino acids, and fatty acids. Sugar from the bun is stored in the liver as *glycogen;* if there is too much sugar, the liver converts the excess to fat. Later on as the body needs fuel, the liver breaks down its stored glycogen into glucose, releasing it bit by bit into the blood. Animal cells cannot function without glucose. Yet surprisingly no other tissues except muscle store appreciable amounts of glycogen that can be converted to glucose. Most tissues and organs, including the brain, depend instead on the liver to release glucose to the bloodstream, which then transports the glucose throughout the body. The cells needing glucose then pick it up from the blood. Indeed, if the glucose level of the blood supplying the brain drops by as little as one part in a thousand, one feels dizzy and sick; a further drop could cause unconsciousness and even death. The liver, then, is vitally important because it regulates the fuel supply which keeps the metabolic fires burning in the cells (Fig. 12-8).

The liver also stores some of the amino acids from the hamburger that are later sent gradually to the cells of the body; there the amino acids are used for growth and tissue repair. The liver converts the nitrogen of any excess amino acids into *urea,* the form in which nitrogen is excreted from the body by the kidneys. The fatty acids from the hamburger are carried from the villi by lymph vessels which pass them to the bloodstream. From the bloodstream the fatty acids move into the liver which synthesizes cholesterol and other fatlike substances from them. These substances form part of the structure of cell membranes and the sheath which insulates nerve cells. In addition to these functions the liver detoxifies many substances absorbed by the body. For example, the alcohol in beer is broken down by the liver to acetic acid, a relatively harmless substance.

The liver performs many other functions in addition. The importance of this vital "chemical factory" to the body is underscored by the fact that not only has nature provided us with an excess of liver tissue—we could get by on about one-fourth of what we have—but also the liver can regenerate itself quickly when damaged.

Unused food material in the large intestine is attacked by bacteria that may supply several of

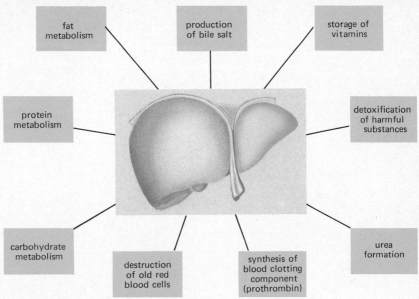

fat
metabolism

production
of bile salt

storage of
vitamins

protein
metabolism

detoxification
of harmful
substances

carbohydrate
metabolism

destruction
of old red
blood cells

synthesis of
blood clotting
component
(prothrombin)

urea
formation

Figure 12-8 Some of the major roles played by the mammalian liver.

the vitamins that man needs. These bacteria make up about half the dry weight of feces. One of the functions of the large intestine is to absorb certain ions and also release excess minerals that the body cannot use. In addition it reabsorbs water from the unused material passing through it. Man, like all terrestrial animals, has a water conservation problem. If the water in his intestine were lost he would be severely desiccated. Sometimes material is moved too quickly through the large intestine for water to be reabsorbed, and *diarrhea* results. On the other hand, if materials move too slowly, too much water is reabsorbed and *constipation* results. The large intestine ends in the **rectum** which stores feces until they are voided through the **anus.**

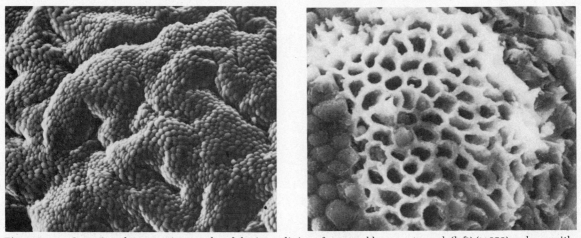

Figure 12-9 Scanning electron micrographs of the inner lining of a normal human stomach (left) ($\times$ 350) and one with ulcers (right). ($\times$ 1150). In the ulcerated stomach the tops of some epithelial cells have been destroyed, leaving holes in the protective mucous coating.

What controls digestion? We decide what and when we eat, but the process of digestion itself is controlled by the nervous system. It is a well-known fact, however, that our emotions can and do affect digestion. Executives get ulcers, athletes have trouble holding down food before games, and you get "butterflies" before an examination. Emotions can inhibit or accelerate the flow of digestive enzymes. For example, anxiety in many persons causes the stomach to secrete too much acid. Should some of the excess acid come in contact with the duodenum or even penetrate the protective coat of mucus in the stomach, an ulcer can develop (Fig. 12-9).

What is *hunger*? How does it differ from *appetite*? No one knows all the answers to these questions, but we now believe that in part of the brain known as the *hypothalamus*, there are two groups of nerve cells. One group forms a *satiety center* and the other a *hunger center*. If the hunger center of a rat that has just gorged itself is stimulated, it will continue to eat. What normally regulates these centers? The answer appears to be the *glucose level* of the blood (Fig. 12-10). After a meal the glucose level is high, then it begins to drop off gradually. At some point the hunger center in the hypothalamus responds to the falling glucose level and "tells" the animal to eat. High glucose levels affect the satiety center which tells the animal not to eat. Again emotions affect these centers and no doubt complicate man's responses.

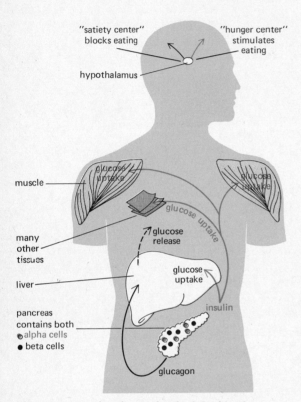

Figure 12-10   The glucose level of the blood is controlled by a number of interlocking systems. Two of these systems are illustrated.

The region of the brain called the **hypothalamus** has at least two groups of nerve cells sensitive to the glucose level of the blood. When blood sugar falls, one group of cells, the "hunger center," stimulates the animal to eat. High blood sugar stimulates other cells in a "satiety center" and decreases the desire to eat.

The **pancreas** has two groups of cells which respond to changes in blood sugar. Low blood sugar causes alpha cells to release a hormone called *glucagon* which acts on the liver, and causes the liver to release glucose to the blood. High blood sugar stimulates another group of cells, beta cells, to release a hormone called *insulin*, which enables cells in many tissues to take up glucose.

Digestion in other vertebrates— Some specializations

Man, like a few other animals, is **omnivorous,** that is, he eats both plants and animals. Most animals, however, are either **carnivorous** (meat-eating) or **herbivorous** (plant-eating), and their entire digestive systems have evolved adaptations which enable them to break down particular kinds of food. For instance, the digestive system of a lion cannot break down the cellulose in hay. When we compare the digestive tract of man with that of cows, lions, and other vertebrates, we discover that the basic plan of all vertebrate digestive systems is similar, but different animals have evolved strategic specializations for disassembling particular foods. These specializations are almost as varied as the animals themselves, and we can give only a few examples here.

As an example consider man's **appendix.** This is a small sac not much larger than a finger which projects from the **caecum,** an area of the large intestine close to where it joins the small intestine. In man the appendix has only nuisance value when it becomes infected and has to be removed; the caecum is also small and not very important functionally. In many herbivores, however, the caecum is large and contains many bacteria which digest the cellulose of plant material. Man, incidentally, cannot digest cellulose. The horse's caecum is large, but since it is located beyond the area of greatest absorption, the

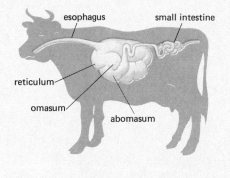

small intestine, the horse is not benefited very much by its caecum. Cattle have evolved still another solution; they have an enlarged area of the esophagus, the **rumen,** which also contains bacteria (Fig. 12-11). These bacteria degrade plant food, including cellulose; some of this material, the cud, is regurgutated and chewed before being passed on to the stomach and intestine. Microbial action precedes intestinal absorption; thus the cow gets more benefit from her food than does the horse.

It is more difficult to digest plant material than animal material. As a result, herbivores have longer intestines than carnivores. This specialization is seen even in tadpoles, which are herbivores; adult frogs are carnivores and have correspondingly shorter intestines (Fig. 12-12).

We have seen that man's stomach, like that of other mammals, acts like a storage organ. One reason that mammals have been able to evolve complex behavioral patterns is because they do not feed continuously. Thus they have time for other activities.

Digestion in insects

The digestive system of insects is shown in Figure 12-13. Like man's digestive system that of the insect is *complete,* that is, it has both a mouth and an anus.

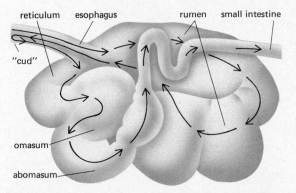

Figure 12-11 Digestion in the cow, a ruminant. There are four chambers in the stomach: rumen, reticulum, omasum, and abomasum. Of these the rumen is the largest. When food is first swallowed, it enters the rumen and reticulum. There, large numbers of bacteria and protozoa break down cellulose and other substances. Some of the breakdown products move on to the omasum, where some water is reabsorbed, and to the abomasum, the true stomach. Other substances are regurgitated — the "cud" — and rechewed before being reswallowed and passing in turn to the omasum and abomasum.

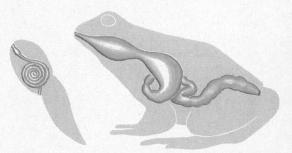

Figure 12-12 Intestine of an herbivore (tadpole) and a carnivore (adult frog). Herbivores have relatively longer intestines than carnivores. The longer the intestine, the greater the area for absorption of nutrients. This is of particular importance for herbivores; because of the large amount of cellulose in plants, herbivores' diet is less nutritious and more difficult to digest than that of carnivores.

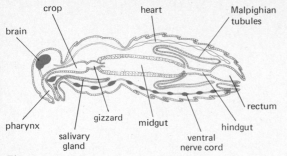

Figure 12-13 Insect digestive system. The exact structure depends not only on the species of insect but also on the stage of the life cycle. Insects with a liquid diet, such as adult mosquitoes, have a somewhat different digestive system from caterpillars which eat tough plant material. However the structures shown here occur in almost all insect digestive systems.

Some digestion occurs in the **crop** which functions mainly as a storage organ. Some insects, like the female mosquito, have a large sac called a *diverticulum* that branches off the anterior part of the digestive tract and extends into the abdomen. In fact a female mosquito can store enough of her blood meal in this diverticulum to enable her to go for several days without feeding —time enough to find a place to deposit her eggs.

In many insects the muscular *gizzard* is lined with chitin which helps grind the food. Most of digestion and absorption take place in the **midgut.** Since the **foregut** is lined with cuticle, little absorption can occur there; the **hindgut,** as we shall see when we discuss excretion, func-

tions primarily in reabsorbing water from the waste products of digestion and metabolism.

Once food has been digested by an insect, a man, or any other animal, it must be delivered to the cells. Within the cells food either is used as building blocks or as a source of energy. During the process of cellular respiration, which uses oxygen, food molecules release energy for the work of the cell. Thus the processes of the intake of oxygen through breathing, which we now consider, and circulation go hand in hand with digestion.

THE BREATH OF LIFE

Both a candle flame and a cell require oxygen— the one to burn wax and produce light energy and the other to burn food molecules and release chemical energy for its other tasks. Of course, a cell burns its food molecules at a much lower temperature than a candle. Also it uses the oxygen to extract chemical energy from molecules such as glucose by means of the cellular reactions described in Chapters 4 and 5. But as a result of burning organic molecules, both a flame and a cell produce carbon dioxide that must be removed from them. Carbon dioxide can be very useful, indeed essential, to the cell, but if its concentration gets too high, it is toxic.

Animals, then, must obtain oxygen and get rid of carbon dioxide. The respiratory mechanisms they have evolved fulfill both needs.

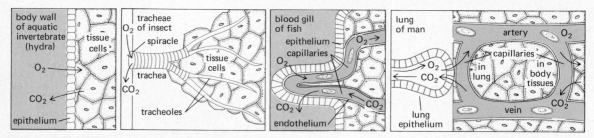

Figure 12-14 Whatever the nature of specialized respiratory tissues and organs, oxygen and carbon dioxide move across cell membranes by diffusion. In hydra (A) oxygen and carbon dioxide diffuse between the water in which it lives and the cells of the animal. In insects (B) gases diffuse between the air and the cells through air-filled tubes. Large tubes, called tracheae, connect with small branches called tracheoles which may even penetrate cells. In fish (C) and man (D) gases diffuse between the environment (water, air) and special respiratory surfaces (gills, lungs) supplied with blood, and between the circulating blood and the cells of the rest of the body.

At this point we should say exactly what we mean by "respiration." In Chapters 4 and 5 the chemical reactions by which energy is made available to the cell were discussed. These processes frequently are referred to as **cellular respiration.** When we speak of "respiration" in this chapter, we mean the exchange of oxygen and carbon dioxide between the cells of an animal and its environment, or **breathing.**

For small animals such as hydra breathing poses no problem. They simply exchange oxygen and carbon dioxide with the water around them by the process of **diffusion.** As long as their skin is permeable to oxygen and there is a higher concentration of oxygen in the water around them than inside their cells, oxygen will diffuse into the animal. However, diffusion alone simply will not work for larger animals such as a goldfish. The reason is simply that diffusion of oxygen through body fluids or tissues is a very slow process. Once an animal is more than a few millimeters thick it will die of suffocation unless it has a specialized respiratory system to aid diffusion. Getting rid of carbon dioxide poses less of a problem than getting oxygen, because carbon dioxide diffuses faster than oxygen through tissues. The most common breathing devices we find in animals are **lungs,** which are *inward* foldings of tissue through which gases diffuse. These are most common in land animals. In aquatic animals we usually find **gills,** which are *outward* foldings of tissue through which gases diffuse. As we shall see, large animals also use a circulatory system for breathing. They pump blood through capillaries in their lungs or their gills, thus bringing oxygen to the tissues inside their bodies and transporting carbon dioxide from the tissues to the outside.

Human respiratory system

Man's respiratory system is shown in Figure 12-15. Air entering the *nose* is warmed, and dust particles are filtered out by hairs or trapped by mucus before the air passes to the **pharynx.** From the pharynx air moves through the open **glottis** into the **larynx.** During swallowing the glottis is closed by the **epiglottis,** a flap of tissue. The larynx is the "voice box" of man and many other animals (but not birds). From the larynx air flows into the **trachea** or windpipe, a hollow tube about 4 inches long and about 1 inch in diameter. The tracheal lining is ciliated, and the cilia sweep mucus and particles away from the lungs and toward the pharynx. The lower end of the trachea divides into two **bronchi,** which in turn divide and subdivide still further to form the **bronchioles.** The smallest bronchioles end in a cluster of tiny sacs or **alveoli.** The lung is honeycombed with these tiny sacs.

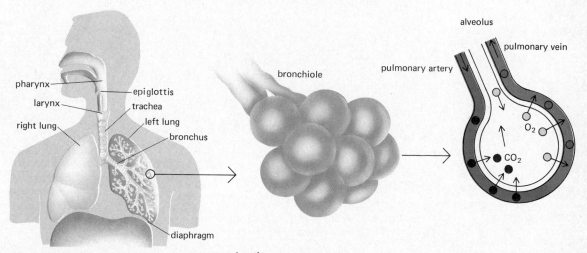

Figure 12-15   Human respiratory system. For details, see text.

Each alveolus is about 0.25 millimeter in diameter. Its membrane is very thin—about one cell layer thick—and permeable to oxygen and carbon dioxide. The alveolus is richly supplied with capillaries, and the actual exchange of oxygen and carbon dioxide takes place across the membrane. Air has more oxygen than the blood in the capillaries; therefore, oxygen in the alveolar air diffuses into the blood. There is more carbon dioxide in the blood than in the alveolar air, so carbon dioxide diffuses into the alveoli, and is exhaled. The total area for gas exchange across the 300 million alveoli is enormous—600 square feet is probably a conservative estimate. Yet a man's lungs weigh only about 2½ pounds.

Breathing movements during which air moves into and out of the lungs are obvious in man (Fig. 12-16). These movements are caused by contraction of the muscles of the rib cage and the **diaphragm,** a muscular sheet that separates the upper part of the body or thorax from the abdominal cavity. When the diaphragm muscles contract, the thoracic volume increases, and air rushes into the lungs; when the diaphragm relaxes, the thoracic volume decreases, and the elastic walls of the lungs squeeze air out.

All respiratory surfaces, whether in man or in other animals, share the following characteristics: (1) they are permeable to oxygen and carbon dioxide, (2) they are moist, and (3) they have a relatively large surface area for gas exchange. As they became increasingly more specialized during the process of evolution they were supplied more and more by the circulating body fluids. In man the circulatory system is so important in respiration that it is impossible to discuss respiration without considering the role of blood in gas transport.

Transport of oxygen and carbon dioxide

The blood of man and many other animals contains special substances called *oxygen-carrying pigments* which loosely combine with oxygen. In man this pigment is **hemoglobin** (Fig. 6-15), and it is packaged in red blood cells; each of man's red blood cells contains about 280 million molecules of hemoglobin. As each drop of blood moves through fine blood vessels, the capillaries in the lung, hemoglobin picks up oxygen molecules which diffuse into the blood from the alveolar air. The combination of hemoglobin with oxygen is called *oxyhemoglobin.* Because of the hemoglobin, the blood can hold 20 times as much oxygen as could be dissolved in the same amount of water. The blood moves from the lungs and

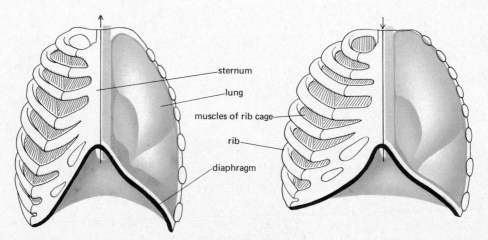

sternum

lung

muscles of rib cage

rib

diaphragm

Figure 12-16 How breathing occurs. When the diaphragm, a large sheet of muscle stretching across the bottom of the chest cavity, relaxes, it becomes somewhat dome-shaped. As a result the volume of the chest cavity decreases, and air is forced out of the lungs. This is *exhalation* or *expiration.* When the diaphragm contracts, it flattens, the volume of the chest cavity increases, lung pressure decreases, and air is sucked into the lungs. This is *inhalation* or *inspiration.* Although one can control inhalation and exhalation to a certain extent, they are normally under involuntary control.

then is pumped by the heart to capillaries which penetrate all of the tissues of the body. The oxyhemoglobin loses some of its oxygen, which diffuses from the blood into tissue cells that have less oxygen than the blood. The blood then returns to the lungs where it picks up another load of oxygen.

The blood also transports carbon dioxide from the tissues to the lungs. Carbon dioxide diffuses from the tissues where it is formed, into the blood arriving from the lungs because this blood is low in carbon dioxide. When the blood reaches the lungs, carbon dioxide diffuses from this returning blood into the alveoli, because the carbon dioxide concentration in the returning blood is higher than in the alveoli. The carbon dioxide then is breathed out of the lungs.

Control of respiration

Man is a fast-living creature, and his body requires considerable oxygen. Moreover he must be able to vary the rate of respiration to meet his changing needs. How is the rate of breathing regulated to meet the needs of a man sleeping on the beach and one minute later to meet a tenfold increase when he dashes into the water answering a cry for help? The principal regulators of our breathing are groups of cells in the medulla of the brain which are called collectively the respiratory center. Part of this center spontaneously discharges, and sends nerve impulses to the diaphragm and muscles of the rib cage. These muscles contract, causing inhalation. In normal quiet breathing, exhalation occurs passively as a result of the elastic recoil of the lungs. Under certain conditions such as coughing or speaking, however, cells in the "expiratory" part of the respiratory center actively cause exhalation. The respiratory center is exquisitely sensitive to the carbon dioxide concentration of the blood passing through the medulla. If the carbon dioxide concentration in the blood increases only slightly, breathing immediately becomes deeper and faster; as a result more carbon dioxide leaves the blood. Breathing slows as the carbon dioxide level returns to normal. The respiratory center is also influenced by other parts of the brain, especially a region of the hindbrain called the *pons*.

When you take a series of deep breaths before swimming under water, you can stay under longer, not because you have taken in more oxygen than usual but because you have expelled more than the usual amount of carbon dioxide from the blood. It takes a minute or two for carbon dioxide to build up to the level at which it stimulates the respiratory center to trigger breathing. It is customary to use a mixture of oxygen and carbon dioxide in resuscitation; the carbon dioxide stimulates the respiratory center. From the foregoing discussion we come to the unexpected realization that the supply of oxygen to the cells is controlled by a completely different gas, carbon dioxide.

The respiratory center in the brain is virtually insensitive to oxygen lack. However, if there is a drastic decrease in the oxygen concentration in the blood, a second control mechanism is activated. There are special receptors in the *aorta*, the large artery which leaves the left ventricle of the heart, and in the *carotid arteries*; these receptors are sensitive to oxygen lack, and they send impulses to the respiratory center when the oxygen level of the blood gets too low. The impulses from the respiratory center then stimulate breathing.

A great deal of what we know about carbon dioxide's roles in respiration has been learned during mountaineering expeditions to the Himalayas and other high ranges. Around 10,000 feet an unacclimatized man will experience difficulty breathing. Even moderate exertions will make him pant. As a result he blows off most of his carbon dioxide so that there is little carbon dioxide to stimulate the respiratory center. At this point the receptors in the aorta and carotid arteries help keep respiration going. As one becomes acclimatized, however, the respiratory and circulatory systems adjust themselves to the new environment.

Some problems in human respiration

Respiratory distress in the newborn. Before birth a human fetus breathes like most other aquatic animals; it obtains the oxygen it needs from an oxygen-rich fluid—in this case its mother's blood. As soon as he is born, however, man suddenly becomes terrestrial and must get his oxygen from air. The exact mechanism by which he makes this changeover is not known. Even today many babies have respiratory difficulties at birth which we cannot cure. This is especially true of infants delivered by Caesarean section. There is some evidence that normal passage through the birth canal "wrings out" the infant's lungs, but this has not been shown conclusively.

Asthma. The diameter of the fine ends of the bronchioles is controlled by smooth muscle. During an asthma attack these muscles contract and constrict the passages through which air moves into and out of the alveoli. Breathing *out* becomes especially difficult and forced, and the victim is exhausted by a severe attack. The muscles of the bronchioles can be relaxed by adrenalin or atropine.

Emphysema. If a rubber band is repeatedly stretched too far, it loses its elasticity. Often diseases such as asthma and other respiratory allergies overinflate and stretch the alveoli so that they lose their elasticity. This may be one of the causes of emphysema. This disease may also be caused by the breaking down of the membranes separating the alveoli. This decreases the surface area for gas exchange, and less oxygen can be delivered to the blood.

Air pollution

It is not known at this time just how seriously air pollution affects man's health. It has been estimated that each day each resident of many of our major cities breathes polluted air that has as much toxic material as nearly two packs of cigarettes. There are probably dozens of poisonous substances in our atmosphere, and no doubt hundreds more will be added by the highly industrialized nations of the world. From time to time air pollution combined with particular weather patterns such as fog and low wind speed has contributed to a number of deaths. It is true that most of the deaths associated with smog occurred in persons with a history of respiratory and circulatory ailments. However, symptoms of respiratory distress are showing up in many previously healthy persons living in cities which are becoming larger and more industrialized.

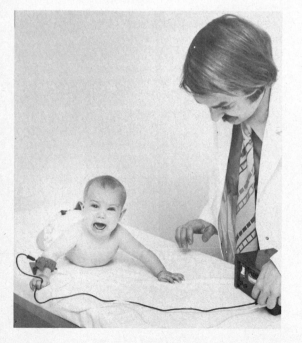

Figure 12-17 Child with cystic fibrosis. Respiratory distress is only one of the many symptoms of this disease. Mucus, which is produced by many of the body's tissues and organs, including the lungs, becomes as thick as glue. The lungs become clogged and breathing is difficult. Bacteria which cause pneumonia and lung infections have an opportunity to multiply. Presently there is no cure for C/F. Treatment includes having the child breathe special mists in a mask or tent. The mists soften the mucus and contain antibiotics to fight infection. In 90 percent of patients with cystic fibrosis, the secretion of sodium and chloride in sweat is increased two to five times normal levels. A diagnosis of cystic fibrosis is commonly confirmed by analyzing the concentrations of sodium and chloride in sweat. The photograph illustrates the collection of sweat, the production of which is stimulated by a mild electric current. The disease, which is inherited, is believed to be due to a single recessive gene. About 5 per cent of the adults in the United States carry the gene. Should two such persons marry, the chances are one in four that their children will have C/F.

Because weather conditions, the concentration of pollutants in the air, and the people breathing the air are all so variable, it is very difficult to say what an unsafe level of air pollution is. For example, the death rate from various respiratory diseases and lung cancer in middle-aged men living in the United Kingdom is greater than the death rate from the same ailments in a similar group in America, presumably because of the greater air pollution in the British Isles. However, other factors such as cigarette smoking, occupation, and living conditions must also be considered before we can determine *how much* of the increased death rate from respiratory ailments is due to air pollution. Although the facts are not all in, the evidence at hand indicates that in many areas of the world air pollution is becoming an important health hazard that is shortening human life.

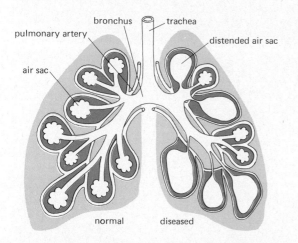

Figure 12-18 Effects of emphysema on the lungs. Normally there are many tiny alveoli or air sacs. But in emphysema the membranes of several alveoli break down, forming one large sac. Since the combined surface area of several small alveoli is greater than that of one large sac, the area over which oxygen and carbon dioxide can be exchanged between the alveoli and the blood is reduced. In addition to a decrease in the number and the total surface area of the alveoli, their walls become less elastic than before. Furthermore, the diameter of the bronchi decreases, and air movement is restricted. Emphysema is currently the leading respiratory disease. However, many of its victims die of heart disease. Why?

Warning: The Surgeon General Has Determined That Cigarette Smoking Is Dangerous to Your Health.

There seems to be little doubt that cigarette smoking contributes to respiratory and other diseases. In the 40 years between 1930 and 1970 the actual number of deaths from lung cancer increased nearly 17 times, while the number of deaths from other cancers declined. Can we blame these deaths on cigarette smoking? Medical scientists have presented persuasive evidence that cigarette smoking is "hazardous to your health."

Figure 12-19 Lung cancer.
A. Section of normal human lung.

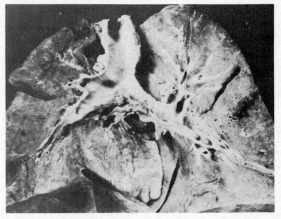

B. Section of lung of a chronic smoker who died of lung cancer. Arrow indicates an area of cancer.

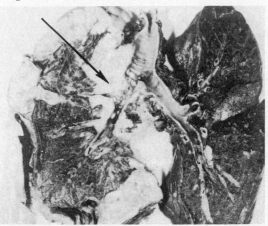

1. More smokers have lung and other types of cancer than nonsmokers. Moreover the death rate from ulcers is several times higher among smokers than for nonsmokers.

2. Substances in cigarette smoke may inhibit the action of mucus and cilia in the respiratory tract; hence the lungs cannot be "swept clean" of dust and other substances which might be harmful.

3. The inside diameters of a smoker's bronchioles are constricted more than they normally would be. This restricts the movements of oxygen and carbon dioxide.

4. Nicotine increases the demands of the heart for oxygen and, at the same time, carbon monoxide from tobacco smoke reduces the amount of oxygen available to the heart. As a result many smokers experience shortness of breath. The heart has to work harder than it normally does to pump enough oxygen to the tissues.

5. Smoking increases the tendency of the blood to clot. We shall examine the relationship between clotting and heart attacks in the following section on circulation.

At the moment we can say that smoking causes lung cancer, but we cannot say how. Research on the effects of smoking on man's health continues. One suggestion is that although smoking itself does not cause cancer, it may make tissues more susceptible to viruses or other substances which do cause cancer.

Respiration in vertebrates other than man

Diving mammals. How does a diving mammal like a seal manage to stay under for so long? A seal has little superiority in the oxygen-carrying capacity of its blood or its lung volume, but it is nonetheless well-adapted for diving. (1) As soon as its nostrils go under water, nerve impulses inhibit the respiratory center and make it much less sensitive to carbon dioxide; thus the animal can stay submerged much longer—sometimes from 20 to 30 minutes—than most mammals. (2) Much of the blood flow to the abdominal organs is shut off and redirected to the vital heart and brain. (3) The heart itself slows its rate of pumping so that the rate is about one-tenth the normal rate. (4) The muscles have large amounts of *myoglobin,* an oxygen-carrying pigment that delivers oxygen only when the oxygen concentration in muscles gets quite low.

Although man cannot remain submerged as long as other diving mammals, the diving women of Korea and Japan can remain under water longer than any other human beings. Their dives at depths from 15 to 80 feet last about 2 minutes. They have become physiologically adapted in several ways for their profession; they have a greater lung volume than nondivers; their heartbeat slows from about 100 beats per minute just before a dive to about 60 beats per minute as a dive nears its end.

Fish. Breathing in the water like a seal and actually breathing water like a fish are very dif-

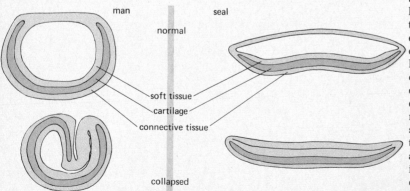

man   seal

normal

soft tissue
cartilage
connective tissue

collapsed

Figure 12-20 Unlike the round trachea of man which cannot collapse much without serious damage, that of the Weddell seal is flat and easily compressed. This is just one of several structural and physiological adaptations that enable it to dive routinely to depths of 400 feet or more, and to descend to and ascend from that depth in about a minute. The seal, instead of resisting pressure changes, yields to them. The rib cage and the lungs also collapse when the animal is in deep water. Thus there is little difference between the external and internal pressure.

ferent processes. Real water breathing has its problems. Even when water dissolves as much oxygen as it can, there is only about 1 part of oxygen per 100 parts of water. In contrast there are 21 parts of oxygen in 100 parts of air. Therefore a fish has to pump 21 times as much water to get as much oxygen as a seal or an insect breathing air. Water is a lot harder to pump than air; thus the fish has to spend a lot more energy than a seal or an insect to get the same amount of oxygen. However, even though water breathing has its problems, fish have solved the problem with great efficiency by means of their gills.

The arrangement of the gills of a bony fish is shown in Figure 12-21. Several openings, the *gill slits*, are on each side of the pharynx. Gas ex-

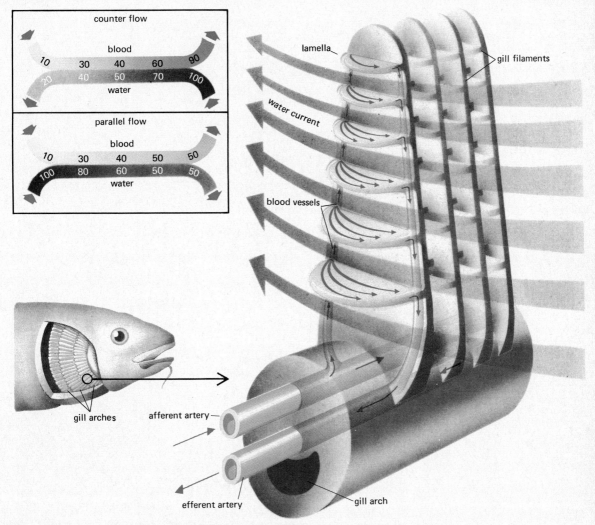

Figure 12-21 Countercurrent exchange of oxygen across the gills of a bony fish. The operculum has been removed from the side of the head. The gill slits are between the gill arches. The afferent artery brings oxygen-poor blood to the gills, and the efferent artery returns oxygen-rich blood to the body. As a result of countercurrent exchange, fish extract up to 90 percent of the oxygen in water. In contrast, mammals extract only about 25 percent of the oxygen in the air. The insert illustrates the advantage of countercurrent flow compared with parallel flow. The numbers are simply examples of the relative concentrations of oxygen which might occur in blood and water under the two arrangements.

change occurs across the gills, which are thin flaps of tissue with a very large surface area. Embedded in them is a huge number of capillaries through which blood flows. These flaps are located externally just off the pharynx. The capillaries lie close to the surface of the flaps and are bathed by water. They are protected from injury by a bony plate called the *operculum* that covers them. Water circulates from the mouth through the gill slits, over the gills, and out the opercula. Some fish move water over the gills by contracting muscles in the mouth and pharynx. Others, such as mackerel, swim rapidly with their mouths open to circulate water past the gills. If they stop swimming, they suffocate.

One of the really great engineering features of gills is that the blood flows through the gill flaps in a direction opposite or *counter* to the direction of water flow. This permits a tremendously efficient **countercurrent** exchange of oxygen to occur between blood and water. How it works is shown in Figure 12-21. Blood entering

Figure 12-22 Florida's walking catfish, *Clarias batrachus*, has gills for breathing in water and lung-like air sacs for breathing on land. A native of Asia, some specimens walked out of their retaining ponds and now infest canals in large areas of southeast Florida.

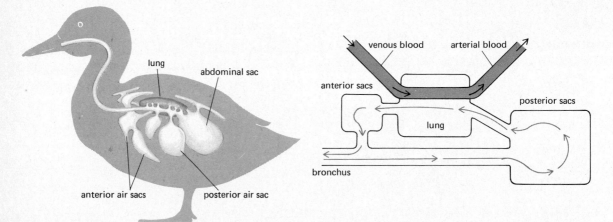

Figure 12-23 Respiratory system of a bird. Like mammals, birds have a trachea and two bronchi, each of which leads to a lung. Unlike the mammalian bronchus which divides to form bronchioles ending in many alveoli, that of the bird ends in many small tubes called **parabronchi,** which function like the alveoli of mammalian lungs. The parabronchi pass *through* the lungs and end in large, thin-walled **air sacs** which are outside of the lungs. These air sacs, in turn, are connected to air spaces in the bones. When the bird inhales, most of the air goes to the posterior air sacs. From the posterior sacs the air goes to the lungs and then to the anterior sacs before it is breathed out. Thus there is always a continuous flow of air through the lungs.

The direction of most of the air flow is opposite that of the blood flow through the lungs. Thus a countercurrent exchange of oxygen occurs in the lungs of a bird. It is this factor that enables birds to fly at 20,000 feet or more, altitudes at which mammals have difficulty breathing. The block diagram illustrating this countercurrent air flow should be compared with Fig. 12-21, which illustrates countercurrent water flow across gills.

the gill from the body is poor in oxygen; water flowing across the gill surface has a high concentration of oxygen. As they flow counter to each other, oxygen diffuses from the water into the blood. There is never quite as much oxygen in the blood as in water; therefore oxygen continues to diffuse into the blood. If blood and water flow in the same direction, a point is reached where the oxygen concentration in both is the same. The advantage of a countercurrent exchange system is that water can always give up some oxygen to blood.

Countercurrent exchange is a familiar engineering concept. A common example in engineering is the use of exhaust gases from a boiler to preheat incoming air before it goes into the combustion chamber. Heat is saved which would normally be lost. Countercurrent exchange is used to great advantage by animals in many ways besides the exchange of oxygen across the gills of fish. Birds use it to provide adequate oxygen during flight at high altitudes (Fig. 12-23). As we see in later sections, countercurrent exchange is also used to produce a concentrated urine in the mammalian kidney and to conserve body heat in many vertebrates.

A few fish (Figs. 12-22 and 24-73) are air breathers. Their "lungs" usually are saclike regions of the digestive tract. Many biologists believe that the primitive ancestors of modern fish had ventral lungs that evolved into the swim bladder of modern fish. The swim bladder functions primarily in "depth control"; there is some evidence that it may aid the fish during a respiratory emergency, but this is uncertain.

Respiration in insects

Insects and most other land arthropods transport oxygen and carbon dioxide by means of hollow tubes called **tracheae,** which open to the air by **spiracles** (Fig. 12-25). The amount of air flowing into the tracheae is regulated in many insects by spiracular valves. Tracheae branch and rebranch, becoming smaller and smaller, finally ending in **tracheoles;** in many cases the tracheoles end within the cells that they supply (Fig. 12-14). Diffusion accounts for most of the gas exchange. However, diffusion is a slow process and especially in large insects it must be supplemented, particularly during flight, by pumping movements of the abdomen or **ventilation.** Insects have remained small for several reasons. One is the inability of the tracheal system, which depends largely on diffusion, to supply oxygen to active cells which are more than a few millimeters away from the external air.

Some insects, like the pupae of wild silkmoths, open their spiracles only once every few hours to let carbon dioxide escape. Their oxygen

Figure 12-24 Gills have evolved in many groups of animals, both invertebrate and vertebrate, and vary greatly in their structure and complexity. Two types are shown here.

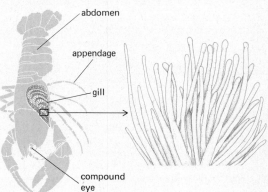

A. Gills of the fan worm, *Sabella pavonia,* a marine polychaete annelid. The gills, which are also food-gathering structures, can be withdrawn into the tube in which the animal lives.

B. Gills of a lobster. To the right is an enlarged view of a small portion of a gill, showing a few hollow respiratory filaments through which blood courses. Each gill has hundreds of such filaments.

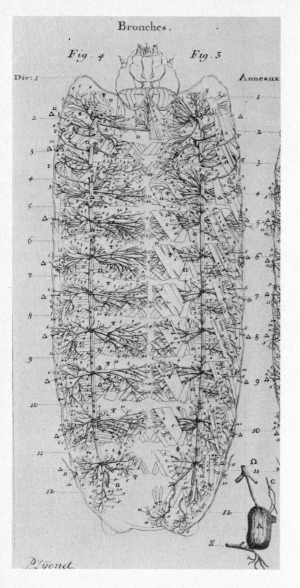

Figure 12-25 This remarkable drawing made in 1760 by Pierre Lyonet depicts the respiratory system of the caterpillar of the goat moth *Cossus*. Air enters and leaves the insect via tiny openings called spiracles. From the spiracles many air tubes or tracheae carry air throughout the insect. The tracheae end in fine tubes called tracheoles. No cell in the animal is more than a few microns away from a tracheole. The tracheoles are not shown in the illustration. Figure 12-14 illustrates the fine tracheoles branching among the cells of the body.

demands are low, and enough oxygen seeps in past the spiracles to meet their needs. Since spiracles are the primary site of their water loss, this kind of respiration enables them to conserve water. Thus respiration in this instance is tied up with water conservation—a problem for all terrestrial animals, and especially for these insects. They never feed or drink during the pupal stage which lasts for several months.

Another interesting respiratory mechanism is found in diving beetles. A diving beetle carries its oxygen supply with it when it goes beneath the surface of the water. The oxygen is present in an air bubble which the insect traps beneath its wings. As it uses oxygen from the bubble more oxygen diffuses in from the water, thus replenishing the oxygen supply of the bubble.

MOVING THE BLOOD

Every cell in an animal or plant lives its own life. It synthesizes its own enzymes, makes its own membranes, makes its own ATP, and carries out whatever else is necessary for growth and activity. It requires nutrients and oxygen, and must rid itself of metabolic waste products and carbon dioxide. In small and simple aquatic organisms like a hydra the water of the environment carries nutrients and waste products to and from the cells. Hydra has no special circulatory system to transport materials. The movements of its body move water in and out of its body cavity and across its body surface. But as animals and plants became larger, more specialized, and more complex, and as they migrated to the land, they were faced with the problem of transporting materials to and from their many cells. As already discussed in Chapter 10 higher plants evolved a vascular system which moves liquids from roots to leaves and from leaves to roots. Most animals have solved the problem by evolving a special transport or **circulatory system** which moves oxygen, carbon dioxide, products of digestion, waste products, and hormones from one part of the body to another. In some animals, such as vertebrates, the circulating fluid, or blood, moves from place to place through a network of tubular blood vessels. In other animals like insects there are only a few blood vessels and the blood bathes the tissues rather than flows through them. In all of these circulatory systems, fluids are usually pumped through the body by a heart of some sort. The heart is therefore a good place to begin our study of circulation.

The human circulatory system

In December of 1967 the first human heart transplant was made. Perhaps no other feat of modern medicine has so captured the public interest and imagination. This is understandable, for no other organ has ever been so identified with life itself as the heart.

It was not until the middle of the seventeenth century that William Harvey of England proved that blood circulates because of the pumping action of the heart. For nearly 1500 years before scientists had believed that the blood ebbed and flowed through the body like the tides of the ocean. Like most great scientific discoveries Harvey's approach was beautifully simple in design. He determined the volume of blood that the heart of a cadaver could hold, and then calculated the volume of blood pumped by the heart in half an hour. Although his figures were wrong, he made his point: the amount of blood the heart pumps in a half-hour weighs far more than all the

blood a man has. He concluded that the heart pumps the same blood over and over again, and therefore the blood must circulate.

It is the blood which supplies life-sustaining oxygen and food and other essential substances to the individual cells, and removes carbon dioxide, wastes such as urea, and cellular products from them. So vital is blood, in fact, that one becomes unconscious if the blood flow to the brain stops for 3 to 5 seconds; the brain is damaged if it goes without blood for more than about 5 minutes. But blood must circulate if it is to perform its tasks, and the heart circulates it to every part of the body through a closed system of tubes, the blood vessels. Before we consider the many functions of the blood, let us look first at the structure and function of the heart and blood vessels.

Heart. Of all living structures perhaps only man's brain is more awe-inspiring than his heart. This fist-sized organ, weighing slightly less than a pound in the adult, puts on an incredible performance. During each minute of normal activity it pumps about 5 quarts of blood—a volume equal to the total amount of blood in the body; in a lifetime of 70 years it pumps about 46 million gallons of blood! As a result of the efforts in the last few years to make artificial hearts we have learned even more to appreciate the natural one.

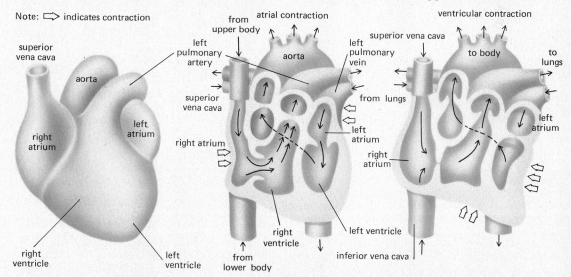

Figure 12-26 Structure of the human heart. Solid arrows indicate the direction of blood flow.

In contrast to any mechanical heart designed and made thus far, the heart can go on functioning even when half of its vitally important left ventricle is damaged or destroyed by disease. It uses the same simple molecules that other organs, tissues, and cells use for fuel. It beats spontaneously as long as it is supplied with oxygen, certain ions, and nutrients, and is kept at the proper temperature. It can pump more rapidly and with greater force when the body uses more oxygen to meet a sudden emergency.

The heart is really two separate pumps, which lie side by side, connected by a thick wall and bound together by circular and spiral layers of muscle and connective tissue (Fig. 12-26). The right pump sends blood to the lungs, and the left pump sends blood to the rest of the body. Each of these pumps consists of a thin muscular reservoir called an **atrium** and a much thicker muscular chamber called a **ventricle.** The muscles in both chambers are composed of cardiac muscle fibers which were discussed in Chapter 11. A valve separates each ventricle from its atrium, and controls the direction of blood flow. Blood flows in from the tissues of the body into the great veins, the *inferior vena cava* and the *superior vena cava* which empty into the right atrium. The right atrium then contracts and blood flows into the right ventricle. When the right ventricle contracts, it sends blood to the pulmonary artery and thence to the nearby lungs where the blood unloads carbon dioxide and picks up oxygen. From the lungs the blood goes back to the heart, this time to the left atrium. When the left atrium contracts, the blood passes into the left ventricle. When the left ventricle contracts, it sends blood surging out into a great artery, the **aorta,** and from there to all the tissues of the body where it unloads some oxygen and picks up carbon dioxide. Back the blood goes to the heart, this time to the right atrium and the cycle begins again. The two atria contract at about the same time and the two ventricles follow a few hundredths of a second later. When the atria contract, the force opens the *atrioventricular valves.* When the ventricles contract, the force shuts the atrioventricular valves and forces open the valves leading to the pulmonary artery and aorta. If your heart beats 72 times each minute, all of this will take place in about 0.8 second. Figure 12-27 summarizes these events and illustrates the path taken by blood leaving the right atrium and making a full circuit of the body. It takes a mere 20 seconds for a drop of blood to make the full trip and return.

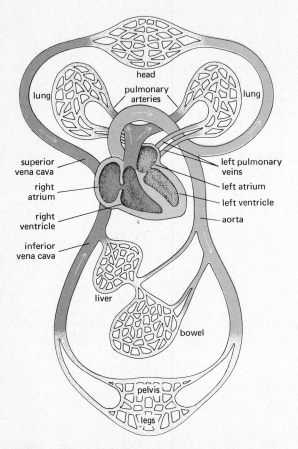

Figure 12-27 Major circulatory routes in man. A drop of blood going from the heart to the big toe, for example, is pumped from the left ventricle into the aorta. From the aorta it moves to the arteries of the leg, foot, and big toe. On its return trip, the blood leaves the capillaries of the toe, enters the venules, and from there it goes to the veins of the legs. Blood in these veins is collected, along with blood from other parts of the lower body, in the inferior vena cava vein which empties it into the right atrium. From the right atrium blood flows into the right ventricle, which pumps blood into the lungs via the pulmonary arteries. Oxygenated blood leaves the lungs and enters the left ventricle via the pulmonary veins. Blood flows from the left atrium into the left ventricle, thus completing one round trip.

Since the atria have to pump blood only to the ventricles a few centimeters away, they are not very muscular. The right ventricle has the much harder job of pumping blood through the lungs, and is thin and muscular. But the left ventricle has the hardest work of all the heart tissues: it pumps blood through the rest of the body. Not surprisingly the muscular walls of the left ventricle are thick and powerful. An animal like a giraffe has a blood pressure three times that of man which enables it to pump blood up its long neck to its brain. Not surprisingly it has a very thick left ventricle.

Once it was believed that atrial contraction forces blood into the ventricles. Now we know that the ventricles will fill with blood even if the atria are too damaged by disease to contract. This fact was welcomed by the physicians and the engineers who design and make artificial hearts, because it greatly simplified their task; in fact mechanical hearts without contracting atria have been made and used successfully in dogs.

As part of every routine physical examination the doctor listens to the heart sounds, and from the quality and rhythm of these sounds he judges whether or not the valves are functioning properly. The closing of the valves is responsible for the two sounds which are especially clear — a long "lubb" followed by a short, sharp "dupp." If the valves leak, perhaps as a result of rheumatic fever, the doctor will hear a "murmur." If their openings have been narrowed, he will hear a hiss. Fortunately it is now possible to replace faulty valves with artificial ones.

What controls the rate at which the heart beats? If a frog's heart is removed from its body and placed in a physiological salt solution along with some glucose, it will continue to beat — sometimes for several days. What makes it beat? What determines its rhythm? What initiates contraction? In the rear wall of a frog's right atrium as in man's there is a tiny mass of specialized muscle tissue that spontaneously contracts and sets the rhythm for the entire heart. In man this knot of tissue is called the **sino-atrial node** or the **S-A node,** or simply the **pacemaker.** We do not know exactly why cells in the S-A node spontaneously contract rhythmically. We do know that the pacemaker begins its duties in the em-

bryo and continues throughout life. Should the S-A node fail because of damage or disease, a second pacemaker region, the **A-V (atrio-ventricular) node,** located between the atria and the ventricles takes over. Sometimes both of these pacemakers fail; in this case artificial ones can be implanted directly in the heart. These maintain a steady rhythm, but they cannot alter that rhythm in response to the body's changing needs. The action of the pacemakers and the electrical changes that occur in the heart muscle with each heartbeat can be recorded electronically; the record is called an *electrocardiogram* or "ECG." A sample ECG is shown in Figure 12-29. A specialist can interpret the tracings and determine whether or not the heart is functioning normally; often he can diagnose the nature of malfunctions. Normally the heartbeat is triggered by

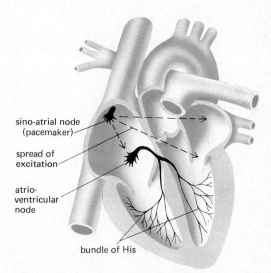

sino-atrial node (pacemaker)

spread of excitation

atrio-ventricular node

bundle of His

Figure 12-28 Spread of excitation through the heart of a mammal. Impulses start spontaneously in the specialized muscle cells of the S-A node. From there the impulses spread through the atria, which contract simultaneously. The impulses finally reach the A-V node. The fibers of the A-V node conduct slowly and the atria contract before impulses reach the ventricles. From the A-V node the impulses spread throughout the ventricle, through specialized muscle cells called the bundle of His. This rapid spread of impulses enables the ventricles to contract as a whole.

the heart's own pacemakers, but it can be influenced by the nervous system. Chemicals released by two sets of nerves will either speed up the heart rate or slow it down, depending upon the body's needs at any given time. We have more to say about this particular action of the nervous system in Chapter 13.

Blood vessels. Blood leaving the left ventricle is distributed throughout the body. On its journey from the heart and back again, a drop of blood flows through, in succession, **arteries, arterioles, capillaries, venules,** and **veins.** So vast is this network of tubes that if they were stretched end to end, their total length would be more than 60,000 miles.

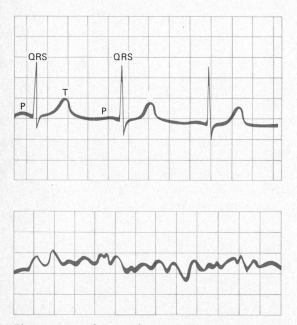

Figure 12-29 Electrocardiograms (ECG) are records of voltage changes occurring during each heartbeat. A single heartbeat is represented by three distinct peaks—P, QRS, and T. P is caused by atrial depolarization, QRS by ventricular depolarization, and T by ventricular repolarization. Portions of a normal (above) and an abnormal (below) ECG are shown. The abnormal record is of a patient whose heart muscle was beating irregularly and wildly during a heart attack caused by a blood clot in one of the blood vessels supplying the heart tissue. Carefully controlled electric shock is usually required to stop this type of irregular beating, which is called **fibrillation.**

Arteries and the smaller arterioles carry blood away from the heart; veins and the smaller venules carry blood toward the heart. Not all arteries carry the bright red oxygenated blood, however; the pulmonary artery transports nonoxygenated blood from the right ventricle to the lungs where the blood is oxygenated. Nor do all veins carry nonoxygenated blood, which is a deep, dull red (not blue); the pulmonary vein takes oxygenated blood from the lungs back to the left atrium of the heart.

The walls of arteries and veins are similar in structure; each has an outer elastic layer, a middle muscular layer, and an inner layer lined with simple endothelium (Fig. 12-30). However, the arterial walls, which are under more pressure from the pumping of the heart, are thicker and more elastic than are those of the veins. Only the endothelial lining remains in the capillaries, and it is across these extremely narrow (less than 3 microns) capillary walls that the exchange of substances between the blood and the individual cells takes place.

The capillaries link the arterial and venous systems. William Harvey never knew how blood leaves the tissues and enters the veins; after all, he had no microscope. Just four years after Harvey's death, the great Italian physiologist, Mar-

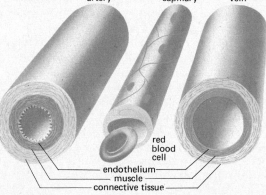

Figure 12-30 Structure of the walls of different types of blood vessels. The walls of both arteries and veins consist of three layers, an inner endothelial layer, a middle muscular layer, and an outer connective tissue layer. However, the muscular layer of arteries is thicker than that of veins. Capillary walls are simply a single layer of endothelium. Arterioles look very much like small arteries and venules like small veins.

cello Malpighi, discovered the capillaries while looking at a frog's lungs through a microscope.

No cell in the human body is ever more than about 130 microns from a capillary. So numerous are the capillaries that if they could be lumped together they would form an organ more than twice as large as the liver. Yet a capillary may be smaller in diameter than a red blood cell. In fact if you look at a frog's tongue through a microscope, you will see red blood cells twisting single file through the capillaries.

If all the capillaries were open at once, they would hold the body's entire supply of blood. This, of course, does not happen normally. Capillaries open in areas where an increased supply of blood is needed. For example, the capillaries in our skeletal muscles open when we exercise, and at the same time most of the capillaries supplying the digestive organs close. The flow of blood in the capillaries is controlled by a muscular ring in the arteriole near the beginning of the capillary, and this ring in turn responds to commands from the nervous system.

When blood leaves the left ventricle and enters the great artery, the aorta, the blood is under pressure from the pumping of the heart; this pressure is the *blood pressure*. As the aorta divides into other arteries, and these arteries divide and subdivide to form the arterioles, the blood pressure drops. At the same time the rate of flow also decreases. The blood pressure in the capillaries is relatively low, as is the rate of flow. The slow rate of flow through the capillaries permits more exchange of substances between the blood and cells to take place than would occur if the blood went rushing through. Although the rate of flow again increases once the blood leaves the capillaries, the blood pressure continues to drop in the venules, and finally drops to its lowest level in the great veins leading into the right atrium. Not only is blood in the veins under less pressure than in the rest of the circulatory system, but the blood below the level of the heart also must return to the heart against the pull of gravity. To help prevent the backflow of blood, veins have valves in them which keep blood moving in the direction of the heart (Fig. 12-31). Movements such as those of the legs also help to keep blood moving toward the heart.

Blood. Blood is the only fluid tissue. Its functions are many. (1) It transports oxygen and carbon dioxide, nutrients, hormones, and other molecules and ions needed by the cells. (2) It helps fight diseases and invading organisms such as bacteria. (3) It helps regulate the body temperature, the pH of the body fluids, and the water balance. (4) It forms clots and thus guards against its own loss from the body.

The liquid **plasma** of the blood, which accounts for slightly more than half of the blood volume, carries in it the **red blood cells** or **erythrocytes, white blood cells** or **leukocytes,** and **platelets.** The yellowish plasma itself is about 90 percent water, with the other 10 percent being composed of various substances such as ions (K^+, Na^+, Ca^{2+}), hormones (adrenalin), nutrients (glucose, fats, amino acids), and plasma proteins (fibrinogen, globulins, albumins).

Red blood cells. Earlier in this chapter we discussed the role of the red blood cells in gas transport. The red blood cells contain a reddish pigment, hemoglobin, which is a complex protein combined with iron. The iron atoms loosely combine with oxygen molecules, and certain amino acids of the protein combine with some of the carbon dioxide. Most vertebrates have hemoglobin in red blood cells; notable exceptions are larvae of eels and fish found in Antarctica called icefish. If man's hemoglobin were not packaged in the red blood cells, the blood would be quite thick, and the heart would have to work far harder than it does to pump the blood. The mature red blood cell of man is about 7 microns in diameter

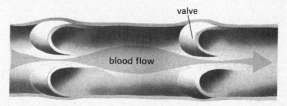

Figure 12-31 Valves in a vein regulate the direction of blood flow. Without these valves blood would pool in a person's feet every time he stood up. The valves permit blood to flow toward the heart but they prevent blood from going the wrong way. If the valves become defective, then blood tends to pool in the veins of the legs. The blood stretches the veins, thus causing varicose veins.

when it leaves the bone marrow where it was formed. It circulates in the blood for about 120 days and then it is destroyed by the liver or the **spleen,** an organ lying behind the stomach. In fact at least 2 million of your red blood cells are destroyed every second. However, equally as many new cells are released from the marrow at the same time. At sea level a normal man has close to 25 trillion red blood cells in his body (women have slightly fewer). At high altitudes the number of red blood cells in a cubic millimeter of a man's blood may increase from 5 million to 7 or 8 million. If one has too few red blood cells or too little hemoglobin in the red blood cells, he is anemic. **Anemia** can be caused by a number of factors, ranging from an iron-poor diet to hemorrhage to hookworm to a lack of the substance which transports the vitamin B_{12} needed for the formation of red blood cells.

The red blood cells of all vertebrates except mammals have a nucleus. The mammalian red blood cell loses its nucleus when it matures, and like all enucleated cells it cannot live long thereafter. Some biologists believe that this loss of the nucleus makes room for more oxygen-carrying hemoglobin. If you look at a drop of your blood magnified many times, you will see that red blood cells resemble discs rather than spheres (Fig. 12-32). This flattening of the cell may shorten the distance oxygen has to diffuse before attaching itself to the iron of the hemoglobin. If red blood cells were spherical, man would need about nine times as many cells as he has to supply oxygen to the tissues at the same rate.

White blood cells. The red blood cells outnumber the larger, nucleated white blood cells by about 700 to one. However, during certain diseases and infections the white blood cell count rises several times; they are part of the body's first line of defense against invading substances and organisms, such as bacteria.

There are at least five different kinds of leukocytes in man. Many are formed in the bone marrow as are the red blood cells and platelets. Still others are formed in tissues such as the lymph nodes and the spleen. Sometimes the tissues which form the white blood cells grow abnormally large, and in so doing they crowd out

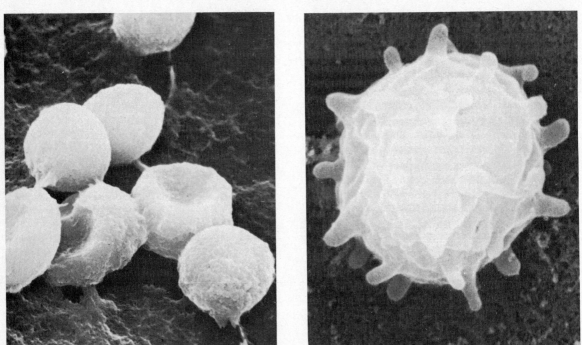

Figure 12-32 Scanning electron micrographs of human red blood cells (left) and a human white blood cell (right). Note the many pseudopods of the white blood cell.

the tissues which form the red blood cells. This condition is called *leukemia,* and one of its results is severe anemia. In some types of leukemia abnormal white blood cells which cannot fight diseases are formed. Many of these leukemia victims die from infections which the body can no longer fight.

The white blood cells move along with the bloodstream, or by amoeboid movement; they can squeeze through the walls of the blood vessels into the tissues. Some leukocytes ingest bacteria. Others destroy or react with foreign substances which can cause disease.

Platelets. Fragments of the cytoplasm of certain cells in the bone marrow are released to the bloodstream as *platelets* (thrombocytes). A normal adult will average more than a million platelets in each cubic millimeter of blood; yet so tiny are these disclike particles that all of his platelets will fill only about two teaspoons. Their lifespan is about eight to ten days. Platelets release *thromboplastin,* a substance that is essential for the clotting of blood.

Plasma. We have already discussed some of the roles of the nutrients, hormones, and ions transported by plasma. Let us now consider some of the **plasma proteins.** These proteins have many functions; two important roles are the clotting of blood and the development of immunity.

Blood clotting is an extremely complex process involving at least two plasma proteins. We do not understand all of the reactions which occur during clotting, but the main reactions appear to be as follows. The platelets release substances called thromboplastins whenever they encounter a rough surface such as torn tissue. In the presence of thromboplastin and Ca^{2+} the inactive plasma protein *prothrombin* is converted to the active form *thrombin.* Thrombin in turn activates another plasma protein, *fibrinogen,* which forms the network of tiny fibers of the *clot* (Fig. 12-33). In summary:

Ruptured platelets $\longrightarrow$ Thromboplastin

$$Prothrombin \xrightarrow[Ca^{++}]{Thromboplastin} Thrombin$$

$$Fibrinogen \xrightarrow{Thrombin} Fibrin \longrightarrow Blood\ clot$$

If any substance that participates in the clotting reaction is missing, clotting cannot occur. Hemophilia is a hereditary disease caused by the absence of platelets. Any wound, even a tiny scratch, can be fatal because the blood flow cannot be stopped. The disease is sex-linked and usually shows up only in males. Several male descendants of England's Queen Victoria were hemophiliacs, including the last Crown Prince of Russia, Alexis. The monk Rasputin gained enormous power because he claimed he could "cure" the little Tsarevitch; Rasputin's abuse of his power is believed to be one of the factors leading to the downfall of Imperial Russia. It is fascinating to speculate whether Russian history might have been different if the Tsarevitch had not had hemophilia.

Sometimes substances deposited on the inner walls of blood vessels stimulate clotting. Should these clots become large enough to obstruct the blood vessel or should they be swept along to block a smaller vessel they could be fatal (Fig. 12-34).

Globulins are plasma proteins which enable man and other vertebrates to develop immunity to foreign substances called **antigens.** Antigens are usually, but not always, proteins, such as the proteins in polio virus. They stimulate the production of **antibodies.** Antibodies are globulins; they are synthesized by special cells called **plasma cells,** which are in fact leukocytes that have become differentiated or specialized for secreting globulins.

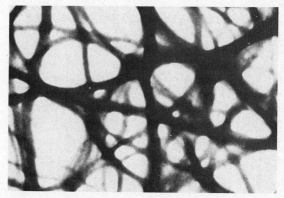

Figure 12-33 Blood clot, showing protein fibers of fibrin.

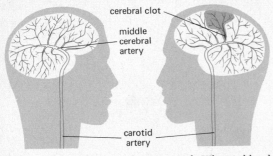

Figure 12-34 How a stroke is caused. When a blood clot interrupts the flow of blood in part of the brain, cells in that part do not get the oxygen they need, and die. The parts of the body that are normally served by those cells cease to function properly.

When an antigen enters the body for the first time, it stimulates the plasma cells to produce antibodies. These antibodies are specific, that is, each type of antibody fits only one type of antigen. After antibodies have been produced, they combine with the antigen in some way, exactly how is not known. For example, suppose we take a typhoid vaccination. The injection will contain dead typhoid bacteria. Substances in the flagella and bodies of these dead bacteria act as antigens to stimulate the formation of typhoid antibodies by the plasma cells. After several weeks the blood contains typhoid antibodies. These plasma cells and their descendants retain the capacity to produce these specific antibodies. Another exposure to typhoid bacteria, even years later, results in the rapid mobilization of anti-typhoid antibodies. Because of this "immunological memory," one is "immune" to typhoid. Of course, persons who have had the disease also have the antibodies and are immune to further attacks of typhoid.

Immune reactions are sometimes undesirable. One of the problems of organ transplants is the rejection of the organ by the recipient's body. The proteins of a transplanted heart, for example, act as antigens to stimulate antibody production by the recipient. These antibodies will reject the transplanted heart. Antibody production can be suppressed by massive irradiation, or by drugs that inhibit DNA synthesis (required for antibody formation), but then the patient becomes extremely susceptible to disease and infection.

Blood transfusions can cause serious trouble if the donor's blood does not have the same kinds of antigens as the recipient's blood. As noted in Chapter 3, blood is typed on the basis of several genetic groups, the best known of which is the **A-B-O group.** For example, you may have type A, type B, type AB, or type O blood.

Normally the plasma does not have antibodies against an antigen to which the body has not been exposed. However, if you have type A blood, your plasma already has antibodies against the red blood cells of a person who has type B blood. If you were given a transfusion of type B blood by mistake, the B-antibodies in your plasma would attack the type B red blood cells (that have B-antigen) and cause those cells to clump. Clumps, like clots, can be fatal if they block a major blood vessel. Table 12-2 lists the type of antigens and antibodies present in each major type of human blood.

Note that a person who has type O blood (and most of us do) has no A- or B-antigen. Persons with type O blood can donate blood to persons with type A, B, AB, or O blood, and for this reason they are called "universal donors." The plasma of type O blood does have A- and B-antibodies, but unless the recipient with type A, B, or AB receives large amounts of type O blood, the A- and B-antibodies are so diluted in the recipient's blood that no difficulties arise. Persons with type AB blood have no A- or B-antibodies in their plasma; therefore they can receive transfusions of A, B, AB, or O blood. Hence they are called "universal recipients." Table 12-2 also shows the relationship between the blood of a potential donor and recipient.

Table 12–2. Human Blood Groups

Blood type	Antigens on red blood cells	Antibodies in plasma	Can give blood to	Can receive blood from
A	A	anti-B	A, AB	A, O
B	B	anti-A	B, AB	B, O
AB	A and B	None	AB	A, B, AB, O
O	None	anti-A and anti-B	O, A, B, AB	O

Equally important as determining which A-B-O type blood a person has is determining whether his blood is *Rh positive* or *Rh negative*. (The **Rh** is a symbol for the antigen which was first discovered in the blood of the *Rhesus* monkey.) The red blood cells of Rh positive persons have the Rh antigen; the red blood cells of Rh negative persons do not have the antigen. In the United States about 85 percent of the white population have Rh positive blood and 15 percent have Rh negative blood. The percentage of persons who have Rh negative blood is much smaller among blacks.

The blood of an Rh negative person will normally contain no Rh antigens nor antibodies against the Rh antigen, unless he has been exposed to Rh positive blood. However, once an Rh negative person has been given a transfusion of Rh positive blood, exposure to Rh antigens sensitizes him and stimulates his plasma cells to produce anti-Rh antibodies. If he should receive a second transfusion of Rh positive blood, the anti-Rh antibodies in his plasma will cause clumping of the transfused Rh positive blood cells and he may die. Thus it is crucial for both donor and recipients in blood transfusions to be typed for Rh factor as well as for ABO blood group.

Rh factor also becomes important in pregnancy. For example, if an Rh negative woman has an Rh positive fetus, and if there is some defect in the placenta which allows some of the maternal and fetal bloods to mix (normally this mixing never occurs), some of the fetal red blood cells get into the mother's blood. The fetal red blood cells, which are Rh positive, stimulate Rh antibody production by the mother. During the first pregnancy of this sort few if any difficulties will arise because birth occurs before enough of the mother's Rh antibodies get into the fetal circulation. However, if the fetus in subsequent pregnancies is Rh positive and if there is again a placental defect which permits some mixing of the maternal and fetal blood, the mother's Rh antibodies may cause clumping of the fetal blood. When this condition, known as *erythroblastosis fetalis* or Rh disease occurs, the baby's blood is replaced by massive transfusions as soon after birth as possible.

Recently it has become possible to immunize Rh negative mothers so that they will not form Rh antibodies. If an Rh negative woman gives birth to an Rh positive child, she is injected promptly after birth with an antiserum against Rh antigen. This Rh antiserum kills any Rh positive fetal cells that may have entered her bloodstream before they can stimulate her to produce Rh antibodies. Soon after the Rh positive cells are destroyed, the injected Rh antibodies disappear and the woman is no longer sensitive to Rh antigens and can safely bear another Rh positive child.

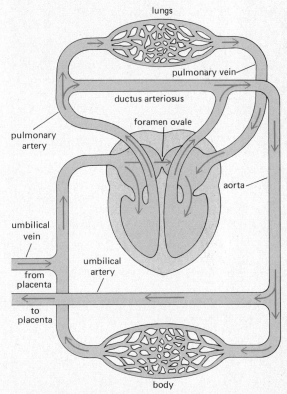

Figure 12-35 Diagram of fetal circulation. Oxygenated blood enters the fetus from the placenta via the umbilical vein (not the umbilical artery). Blood then flows into the right atrium. Some blood goes directly to the left atrium via a small opening, the *foramen ovale*. Most of the blood bypasses the lungs, which are not used for breathing until after birth; this bypass is the *ductus arteriosus*. Blood then flows into the aorta, and throughout the rest of the body of the fetus. It is then returned to the placenta via the umbilical artery.

Diseases of the human circulatory system

Heart diseases continue to be the leading cause of death in the United States and account for about 40 percent of all deaths. What are some of these diseases, and what are the prospects of alleviating or eliminating them?

Congenital heart defects. The fetal blood exchanges oxygen and carbon dioxide with the maternal blood. Then the fetal blood moves to the right atrium of the fetal heart, and from there most of the blood passes directly into the left atrium through an opening between the two atria. Some blood does flow into the right ventricle, but instead of going from the pulmonary artery to the lungs, the blood is shunted from the pulmonary artery to the aorta by a special blood vessel (Fig. 12-35). At birth the opening between the atria closes, and growing cells block the vessel opening. Sometimes, however, this opening does not close, and oxygenated and nonoxygenated blood are mixed. The result is a blue baby, so-called because the mixed blood gives the infant's skin a bluish color. The vessel can be tied off rather easily, and now that open heart surgery is possible the opening between the atria can be sewed up.

Heart diseases. Most cases of heart disease are not "heart" diseases as such. They develop as a result of changes within the blood vessels. Skeletal muscle can contract in the absence of oxygen; cardiac muscle cannot. Cardiac muscle extracts about 80 percent of the oxygen brought to it by its own system of blood vessels which includes the **coronary arteries** (Fig. 12-36). Sometimes these vessels become constricted or blocked. When they are constricted, the heart works harder than it does normally to push blood through them to supply its own needs. This constriction and the resulting efforts by the heart cause the intense pain of *angina pectoris* (literally "pain of the chest"). Angina pectoris can be relieved temporarily by drugs such as nitroglycerin which dilate the vessels. Curiously, adrenalin, which normally would dilate the arteries, does not dilate them in this disease.

When coronary vessels are blocked by a clot, the result is *coronary thrombosis.* There is evidence that fatty substances and cholesterol are culprits in this disease. These substances form deposits which harden on the inner walls of the vessels; platelets rupture against these deposits and trigger clotting. Sometimes anticlotting agents such as heparin help reduce the number of coronary attacks, as does restricting the fats in one's diet. When coronary arteries are blocked, the tissues they normally serve become fibrous, and may even die. New blood vessels will sometimes grow around the diseased area of the heart. Exercise helps stimulate the spreading of these new arteries; this is why many doctors often recommend mild to moderate exercise for patients whose hearts have not been badly damaged by a previous coronary attack.

Coronary thrombosis in young people is sometimes caused by *syphilis.* For some unknown reason the bacterium that causes syphilis attacks coronary arteries and pits them. The pits cause platelets to rupture, and clotting results.

One of the chief causes of heart trouble is *hypertension* or high blood pressure. The immediate cause of hypertension is constriction of the arteries and arterioles. However, many factors contribute to constriction; one is "hardening of the arteries," or *atherosclerosis,* in which cholesterol coats the arteries and then hardens

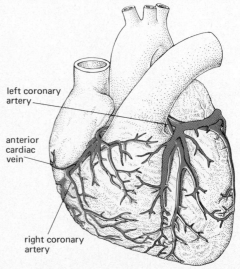

left coronary artery

anterior cardiac vein

right coronary artery

Figure 12-36 Coronary circulation, the routes which convey blood to the heart muscle. The arteries are colored lighter than the veins.

(Fig. 12-37). Whatever the cause of constriction, the end result is that the heart has to work harder and harder to push blood through the narrowed vessels which no longer stretch when the blood pressure rises at each heart beat.

The development of the heart-lung machine has made open heart surgery possible, and thus many circulatory defects can be corrected. But what about the estimated tens of thousands of persons whose hearts are so badly damaged that they need new ones? Even if the legal, moral, and physiological problems of heart transplants are solved, it seems unlikely that there will ever be enough donors to meet the demands. Consequently a great deal of research is being done on artificial hearts. As one might expect a number of conditions must be met before artificial hearts can be used routinely. Some of these criteria are (1) the atria must be collapsible so that a vacuum will not form as blood flows into the ventricles; (2) the two ventricles must pump equal amounts of blood; (3) the rate of pumping and the amount of blood pumped must meet the needs of the body at any given time; (4) a minimum amount of maintenance, if any, should be required; (5) the power supply must be small enough not to get in the patient's way. When we try to meet these demands and make a device which is compatible with the body fluids and tissues, and still fit in the space intended for it, we begin to really appreciate what nature has already accomplished.

A second circulatory system — The lymphatic system

Blood in the capillaries is under pressure. Because the capillaries are slightly leaky, some plasma and plasma proteins are forced by the pressure through the capillary walls and form a fluid around the tissues. This fluid is called **lymph,** and it transports substances from the blood to the cells. Lymph, which is really blood plasma without the blood cells and platelets, is the true environment of cells. Lymph diffuses from the tissues into thin-walled tubes called *lymphatic capillaries* that end blindly in most tissues; these capillaries join to form larger lymph vessels. The largest lymph vessels drain into the veins (Fig. 12-38). Thus plasma and proteins are transported by lymph and returned to the bloodstream by the lymphatic vessels which act as a drainage system. Lymph transports other substances, too; as we have already seen lymph transports fats from the villi of the small intestine to the bloodstream.

Lymph filters through small structures called **lymph nodes** before it is returned to the veins. These nodes, located at intervals along lymph vessels, are especially numerous in the armpit, groin, and neck. Not only do they remove foreign particles such as bacteria, dust, dead cells, and soot from lymph but they are also sites where certain white blood cells are formed. During an

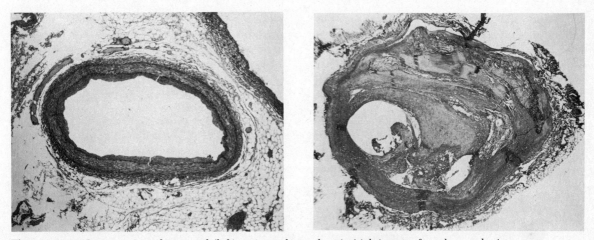

Figure 12-37   Cross section of a normal (left) and an atherosclerotic (right) artery from human brain.

infection, lymph nodes step up their activity, and may become sore and swollen.

Although lymphatic vessels can contract, lymph is moved toward the heart mostly by movements of the body. Like the veins the larger lymph vessels have valves which prevent a backflow. Sometimes lymph is not returned rapidly enough to the heart, and swelling results.

Some other circulatory systems

The circulatory systems of most vertebrates are basically similar to the system in man. Their blood flows through a *closed* system of tubes, it is distributed under pressure from a heart, and it eventually flows through capillaries into all the tissues. However, the design of the heart varies widely (Fig. 12-39). For example, the fish heart, unlike that of birds, mammals, and crocodiles, has only two chambers. Blood is pumped by the heart to capillaries in the gills. From the gills it goes straight to capillaries in the body tissues and back to the heart.

The circulatory system of insects is *open*, that is, blood does not flow through an interconnecting system of vessels. In fact blood flows through only one vessel in most insects (Fig. 12-40). The hind part of the vessel functions as a "heart"; one-way valves in the heart permit blood to flow from the body cavity into the heart. Contractions of the heart propel blood toward the head through an "aorta," and from there the blood moves again into the body spaces. One of the most significant differences between the insect heart and man's heart is that almost no blood pressure is developed by the insect heart. As we see in the next section the blood pressure of man's pumping heart is crucial in the functioning of his excretory organs. As we might expect, excretion in insects cannot occur in the same way as in man for this reason. Like man's heart that of many insects beats spontaneously, and the rate may be influenced by nerves supplying the heart. Rapid circulation of blood depends on the movements of the animal: the more active the insect, the more rapidly its blood circulates.

The blood of most insects is colorless, since few insects have an oxygen-carrying pigment. Gas transport in insects takes place by diffusion of the gases through the tracheae. However, many aquatic arthropods, including the lobster, have a blue oxygen-carrying pigment, hemocyanin, which functions much like the hemoglobin of man and the vertebrates.

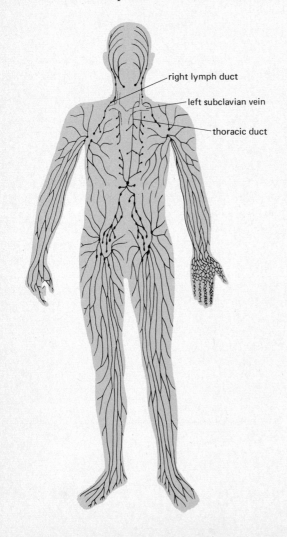

right lymph duct

left subclavian vein

thoracic duct

Figure 12-38 The lymphatic system of man. Lymph vessels penetrate every part of the body, and eventually empty into veins. The right lymph duct drains the upper right side of the body (including neck and head), the lungs, and the heart, and then empties into the right subclavian vein. The thoracic duct drains most of the rest of the body, and empties into the left subclavian vein. Major lymph nodes are shown at intervals along some of the lymph vessels.

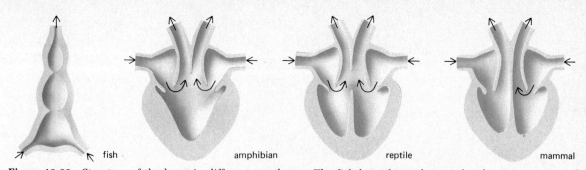

Figure 12-39 Structure of the heart in different vertebrates. The fish heart has only two chambers, an atrium and a ventricle. However, there is no mixing of oxygenated and non-oxygenated blood; oxygenated blood leaving the gills does not return to the heart but circulates throughout the body. The heart simply pumps non-oxygenated blood returning from the body back to the gills. Amphibians and many reptiles have three chambers, two atria and a single ventricle. This ventricle is partially divided but the two halves are not completely separated. As a result, some oxygenated and non-oxygenated blood are mixed. In crocodiles, birds, and mammals the ventricles are completely separated and the heart has four chambers. Thus no mixing of oxygenated and non-oxygenated blood occurs. This is one reason why circulation in birds and mammals—animals with a high metabolic rate and a fast-paced way of life—is more efficient than in amphibia and most reptiles.

REGULATING THE WATER SUPPLY

Most biologists believe that life began in ancient seas. Numerous exits from water to land have been made during the course of evolution. However, although many animals can survive on land, their cells cannot. Indeed most animal cells seem to function properly only when bathed in a fluid

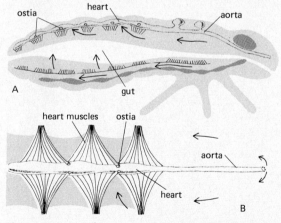

Figure 12-40 A. Insect circulatory system. The heart is a dorsal tube running most of the length of the body. Openings (ostia) permit blood to enter the heart, but not to leave it. When the heart contracts, blood is forced out of the anterior end of the heart or the aorta. Blood then percolates throughout the body cavity. B. Dorsal view of anterior end of an insect's heart.

which resembles seawater in the concentration and kinds of ions it contains. Most animals carry their own ocean to bathe their cells. They have evolved osmoregulatory mechanisms which adjust the ionic composition of their inner oceans so that the cells have just the right amounts of salts and water. They also have excretory mechanisms which dispose of various metabolic wastes, particularly the breakdown products of protein metabolism, such as ammonia and urea which are frequently quite toxic and must be eliminated.

In simple animals like a hydra the regulation of water, salts, and disposing of nitrogenous wastes is carried out by the individual cells of the body. Each cell fends for itself. More complex animals have evolved special excretory systems which are able to maintain ionic composition and get rid of wastes simultaneously.

Human excretory system

Man's principal excretory organs are the **kidneys,** two dark red, bean-shaped structures that lie just behind the stomach and liver (Fig. 12-41). Urine formed in the kidneys flows through the *ureters* into the *urinary bladder,* where the urine is stored until it is voided through the *urethra.*

Each kidney is composed of more than a million individually functioning units called **nephrons.** A nephron is a long, thin tube or tubule

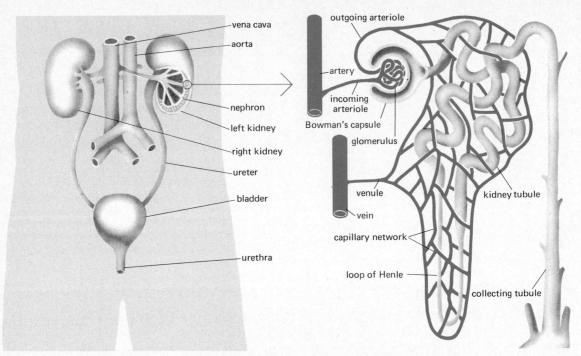

Figure 12-41 (Left) The human excretory system. (Right) A nephron. For details see text.

which is bent like a hairpin. The ends of the tubule are coiled; one end of the tubule expands to form a cup-shaped structure or **Bowman's capsule;** the other end leads into a **collecting tubule.** Fitting inside the Bowman's capsule like a ball held in the palm of a hand is the **glomerulus,** a tiny ball of capillaries which arise from a branch of the renal artery called an *incoming arteriole.* These capillaries connect with an *outgoing arteriole* — not a venule as they usually would — which passes out of Bowman's capsule; this outgoing arteriole then forms another capillary network around the rest of the tubule before forming the *venules and renal vein.*

The short renal artery that supplies the kidney branches directly off the aorta; hence little of the pressure of the blood in the aorta has been lost by the time blood enters the kidney. In fact the blood pressure in the glomerular capillaries is twice as high as it is in any other capillaries. This high pressure initiates urine formation.

Urine is formed in the following way. Each day the heart pumps about 8000 quarts of blood, of which about 1900 quarts are carried to the kid-

neys by the two renal arteries. About 180 quarts of this blood actually pass through the walls of the glomerular capillaries which act as filters. Under pressure developed by the heart everything in blood except the blood cells, platelets, and most plasma proteins passes across the glomerular capillary walls into Bowman's capsule and then into the tubules.

However, **filtration** of the blood through the glomeruli is only part of the process of urine formation; equally important is **reabsorption.** While the filtrate does contain metabolic wastes such as urea (which is formed in the liver, *not* in the kidney), most of the substances found in the filtrate — water, salts, glucose, amino acids — are extremely valuable to the body. Think what would happen if water were not reabsorbed. We would void about 180 quarts of urine a day, and this loss would have to be replaced by drinking. As it is, about 99 percent of the water which passes into the kidney is reabsorbed, and we lose only 1 or 2 quarts of water a day as urine. Similarly, about 2.5 pounds of salt pass into the tubules each day, yet no more than about one-third of an ounce is lost in the urine. The kidneys reabsorb glucose so efficiently that the appearance of significant amounts of glucose in the urine indicates that the person may have a disease called diabetes. How does the reabsorption of water and various substances take place in the nephron?

As an example let us consider the reabsorption of sodium. Look at the structure of the nephron again. The sharp bend of the hairpin is called the **loop of Henle,** and cells in the ascending part of the loop can actively transport Na^+ out of the filtrate and into the tissue fluids (Fig. 12-42). As the filtrate passes through the descending part of the loop, some Na^+ will diffuse out of the tissue fluids and into the filtrate. This process, which is repeated over and over, is our third example of countercurrent exchange. The net result of this exchange is that the filtrate that

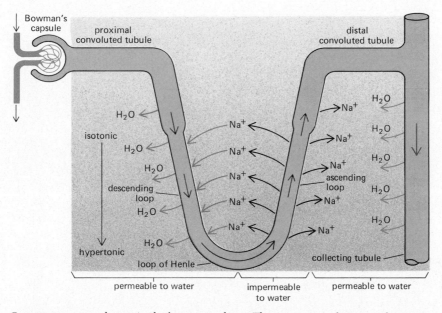

Figure 12-42 Countercurrent exchange in the human nephron. The movement of water and ions, especially sodium, across different parts of the nephron tubules makes it possible for man to form urine that is more concentrated than his blood and tissue fluids, thus conserving water. We speak of man forming *hypertonic* urine. (A hypertonic solution has a higher osmotic pressure than the blood and is therefore more concentrated. An *isotonic* solution has the same osmotic pressure as the blood, whereas a *hypotonic* solution has a lower osmotic pressure than the blood.)

Blood is filtered in Bowman's capsule. Blood proteins and blood cells do not pass into the kidney tubule but everything else does. The filtrate entering the proximal convoluted tubule is isotonic with the blood. As the filtrate flows toward the loop of Henle, it becomes hypertonic because Na^+ diffuses into the filtrate from the surrounding tissue fluids where the Na^+ concentration is high. When the filtrate passes through the ascending loop, Na^+ is actively pumped from the filtrate back into the surrounding tissue fluids. The filtrate becomes hypotonic and the surrounding fluids develop a high Na^+ concentration. The Na^+ pumped back into the surrounding fluids then can diffuse into the filtrate flowing through the descending loop. Thus Na^+ is recycled constantly and the loop of Henle and the lower section of the collecting tubule are continuously bathed in salty fluid. As hypotonic filtrate flows out through the collecting tubule, water diffuses from the filtrate into the surrounding fluids, which have a high Na^+ concentration. The final filtrate, urine, is hypertonic.

This mechanism depends on the recycling and the countercurrent exchange of Na^+ which causes the fluids surrounding the loop of Henle and the lower section of the collecting tubule to be very salty. It also depends on the fact that cells in the ascending loop are impermeable to water whereas those in the collecting tubule are very permeable to water.

flows into the collecting tubule is not as concentrated as the blood or original glomerular filtrate; yet the tissue fluids around the tubules are more concentrated than the original filtrate.

The collecting tubule runs parallel to its nephron. As the filtrate flows through the collecting tubule water diffuses from the filtrate into the tissue fluids, which are more concentrated than the filtrate as a result of the countercurrent exchange of Na^+. As a result the urine which enters the ureters from the collecting tubule is more concentrated than the original blood or glomerular filtrate.

The permeability of the collecting tubules to water is controlled by **antidiuretic hormone (ADH)** formed by the hypothalamus of the brain. The brain responds to an increase in the salts of the blood by releasing ADH. ADH in turn makes the collecting tubules more permeable to water; more water is reabsorbed by the kidneys, and the concentration of the body fluids drops. The brain responds to this drop by shutting off the production of ADH. This is a beautiful example of *feedback mechanisms* which we discuss in detail in the following chapter. Alcohol inhibits the production of ADH, with the result that the urine output increases. The increased urine flow that follows a beer-drinking spree is not due to alcohol inhibiting the release of ADH—there is not enough alcohol in beer; it is the large volume of fluid in beer that increases the volume of urine produced.

The final step in urine formation is called **secretion.** The tubules secrete certain chemicals, penicillin, for instance, from the blood in the capillaries surrounding the tubules into the filtrate in the tubules. Filtration of blood through the glomeruli depends on blood pressure. Secretion does not depend on blood pressure; it depends upon active processes like those involved in "sodium-potassium exchange pumps."

Like so many other organ systems in man the excretory system has many functions in addition to, in this case, reabsorption of valuable substances, water balance, and excretion of urea. The kidneys also play a vital role in holding the pH of the body fluids constant. The tubules secrete both acids and ammonia which would otherwise tend to alter the pH.

Kidney diseases

The kidney is a complex structure, and like all complex structures there are many ways that things can and do go wrong. For example, sometimes the glomeruli become clogged, and blood cannot flow through them. Or the renal arteries will harden, thus restricting or shutting off the blood flow to the kidneys. A number of diseases can lead to complete kidney failure; urea and other metabolic wastes accumulate, and unless the composition of the body fluids is somehow returned to normal, death will follow.

Kidney transplants between identical twins have been quite successful. Transplants between persons who are not identical twins have not been so successful for the same reason that heart transplants usually fail—rejection of the transplanted organ by the recipient's body. Even if transplants between persons who are not identical twins are made more successful than they are now, the demand for healthy kidneys probably will still be greater than the supply. Artificial kidneys are available, but at this time they are expensive, time consuming, and available only in certain clinics and hospitals. A great deal of research is being done on a portable kidney which the patient can operate at home.

Osmoregulation and excretion in some other vertebrates

Only the kidney tubules of birds and mammals have a loop of Henle. The relative length of the loop indicates whether or not an animal can produce a concentrated urine; the longer the loop, the more concentrated the urine. Each kidney tubule of the tiny kangaroo rat (Fig. 12-43), a desert mammal, has a very long loop of Henle. This rat never drinks; all of its water is "metabolic water"—the water formed as a result of the metabolism of the dry seeds it eats. It conserves body water in several ways. First, the animal conserves water by burrowing during the heat of the day; in addition it has very few sweat glands so little water is lost by this route. Furthermore it produces only a small amount of urine, and that is extremely concentrated; in fact its urine is twice as salty as seawater and contains four times as much urea as man's urine. Man cannot drink seawater; his body loses even more water in eliminating excess salt. In addition magnesium sulfate—one of the salts found in seawater—causes diarrhea, which adds to man's water loss. But the kangaroo rat can be forced experimentally to drink seawater with no ill effects; its long loop of Henle concentrates its urine and rids its body of excess salt.

Marine birds and reptiles, however, do drink seawater. Reptiles have no loop of Henle, and that of birds is not as highly developed as the mammalian loop of Henle. How do these animals cope with the excess salt? The excess salt is secreted from **salt glands** located in the head. In marine birds this gland is many times larger than a similar structure found in nonmarine forms. The fluid from this gland may be five times as salty as the bird's blood and body fluids. The "tears" a giant female sea turtle sheds when she comes ashore to lay her eggs have nothing to do with the egg-laying process itself; she is getting rid of excess salt.

The structure of the nephron varies in different vertebrates. For example, freshwater fish have body fluids which are more concentrated than the water in which they live; hence water moves into these animals across the permeable gills. As a result freshwater fish must "bail out"

the excess water by forming large volumes of a very dilute urine. They have evolved large glomeruli which are well suited for filtering out large volumes of water. Since there is little reabsorption of salts by their tubules, special cells in the gills of freshwater fish actively absorb salt. The glomeruli are predictably large in amphibia, too; amphibia also produce a dilute urine. In birds the glomeruli are small, and in reptiles they are smaller still. Mammals have relatively large glomeruli, but, as we have seen, conserve water and salts through efficient tubular reabsorption.

It may be surprising to learn that marine fish have a water conservation problem. Their body fluids are less concentrated than seawater; therefore they lose body water to their surroundings. To compensate for this loss they drink almost continuously and form as little urine as possible. Their glomeruli are greatly reduced; in fact some marine fishes have no glomeruli, only tubules. They excrete excess salt across the gills. Sharks and their relatives, however, have evolved another solution to the problem of conserving body water. Their tubules reabsorb much of the urea they form, and this urea is used to make the concentration of the body fluids similar to that of the seawater. Hence they stop the loss of their body water to the sea.

Figure 12-43 The kangaroo rat conserves body water so efficiently that it can thrive even in barren, dry regions like Death Valley. Its normal diet includes seeds with a high fat content. The oxidation of this fat produces much more metabolic water than would be produced by the oxidation of carbohydrates or proteins. This metabolic water provides the animal with all of the water it needs, and it may never drink.

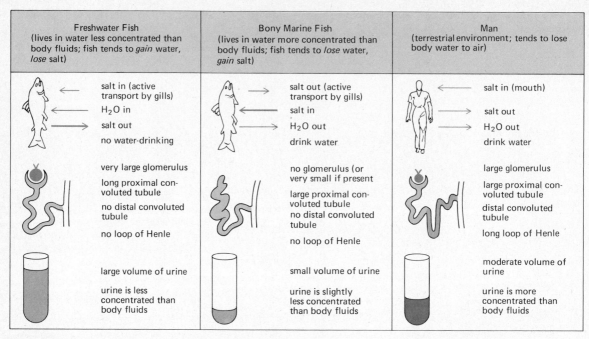

Freshwater Fish (lives in water less concentrated than body fluids; fish tends to *gain* water, *lose* salt)	Bony Marine Fish (lives in water more concentrated than body fluids; fish tends to *lose* water, *gain* salt)	Man (terrestrial environment; tends to lose body water to air)
salt in (active transport by gills)	salt out (active transport by gills)	salt in (mouth)
H₂O in	salt in	salt out
salt out	H₂O out	H₂O out
no water-drinking	drink water	drink water
very large glomerulus	no glomerulus (or very small if present	large glomerulus
long proximal convoluted tubule	large proximal convoluted tubule	large proximal convoluted tubule
no distal convoluted tubule	no distal convoluted tubule	distal convoluted tubule
no loop of Henle	no loop of Henle	long loop of Henle
large volume of urine	small volume of urine	moderate volume of urine
urine is less concentrated than body fluids	urine is slightly less concentrated than body fluids	urine is more concentrated than body fluids

Figure 12-44 Correlation between the environment and the structure of the nephron of the kidneys of various vertebrates.

Osmoregulation and excretion in insects

Insects have a well-developed excretory system (Fig. 12-45). It consists of the *hindgut* and a variable number of fingerlike projections called **Malpighian tubules.** One end of each tubule opens into the hindgut at its junction with the midgut; the other end of the tubule, the "fingertip," is closed. The blood bathes the tubules. But because the insect's circulatory system is an open one, there is little blood pressure to drive fluid from the blood into the Malpighian tubules. Instead the Malpighian tubules secrete waste products from the blood into the fluid in the tubules. Nitrogenous material is secreted in a semisolid form as *uric acid*; salts and some water are reabsorbed in the tubule. It is not known exactly how insects produce a urine which is more concentrated than the body fluids, but it is known that most of the water is reabsorbed by the rectum. The urine and the feces are released together as a rather hard dry pellet.

We have seen that several of the insect's body systems play some role in helping these organisms conserve water; the skin is largely imper-

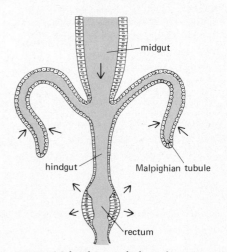

Figure 12-45 Malpighian tubules of an insect. For details, see text.

meable to water, the spiracles of many insects are closed most of the time, thus minimizing water loss, and the excretory product is practically dry rather than fluid as in man. Yet the same sort of situation exists in most animals; the prevention of water loss, for example, is not the province of any one system. Let us examine this in more detail.

Osmoregulation in vertebrates— A second look

One of the breakdown products of nitrogen metabolism is **ammonia.** However, ammonia is very toxic to tissues, and it must be removed as rapidly as it is formed. This presents no problem to many aquatic organisms which simply release the ammonia to the surrounding water as it is formed. Terrestrial animals cannot do this, obviously; they and the marine fish could flush out the ammonia with large volumes of urine but they must conserve water. How do they solve this dilemma? There are two principal ways of getting around the problem. Animals like man that can afford to expend some water, detoxify ammonia to **urea,** which can be excreted with a moderate amount of water loss. Some animals, however, cannot afford to lose even that small amount of water; these include snakes and lizards, most birds, and insects. They convert ammonia to nontoxic **uric acid** which can be stored in the body for relatively long periods and can be excreted as a paste with little water or in a dry form with almost no water.

An animal may excrete nitrogen in various forms at different times during its life. For example, the tadpole, which is aquatic, excretes ammonia, but the adult frog, faced as it is with water loss through its very permeable skin, cannot spare enough water to excrete ammonia, and so it excretes urea.

Interestingly the nitrogenous waste product which an animal excretes depends on where its embryo develops. Birds, insects, and some reptiles develop in eggs which have tough, waterproof coverings. They can manage this because the embryo secretes nontoxic uric acid which can be stored until the animal hatches, and no harm is done. The embryos of man and of many other animals that develop within the body of the mother excrete urea; the developing human fetus, for example, excretes urea which is carried away from the placenta by the mother's blood.

These few examples illustrate how several different physiological systems interact with each other and control the animal's water balance. Let us conclude this chapter by examining briefly another physiological function which requires the proper interlocking of several physiological systems—temperature regulation.

TEMPERATURE REGULATION

On warm summer evenings in many parts of the world the sounds of night are dominated by a chorus of crickets. The rate at which these insects chirp is so closely related to temperature that some people can tell within a degree or so what the air temperature is. The chirping of crickets, like all activities of most animals, depends on the temperature of the environment. Only a few animals—birds and mammals—are relatively independent of the environmental temperature. Animals like a cricket or a hydra are **poikilothermic,** or "cold-blooded" animals; mammals and birds are **homeothermic,** or "warm-blooded" animals. But cold-blooded and warm-blooded are misleading terms, as we shall see. What we really mean is that the homeotherms can maintain a constant body temperature that is fairly high and usually above that of the environment. Poikilotherms are not able to do this and their body temperature varies with the environmental temperature.

The rates of chemical reactions, including those in the cell, depend on temperature. Thus temperature regulation is quite important to many animals. The degree to which an animal regulates its body temperature determines how independent the organism is of the environmental temperature and how free it is to make a living in various habitats. By keeping their body temperature high, animals can keep their metabolism high and remain active in cold weather. This is particularly important for terrestrial animals, because the temperature of the air changes much more rapidly than that of water and may go to great extremes.

Animals have evolved some fascinating behavioral and physiological adaptations to temperature changes (Fig. 12-46). We shall consider some of the adaptations of both homeotherms and poikilotherms.

Temperature regulation in homeotherms

Man. As we mentioned in Chapter 11 man is becoming increasingly hairless, and what hair he has left seems more ornamental than protective. The "naked ape" has evolved a mechanism for

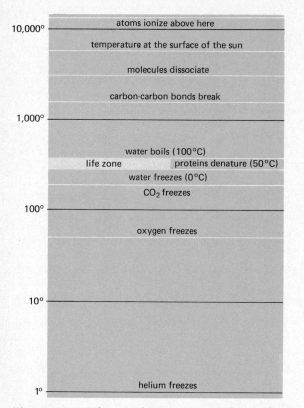

Figure 12-46 Life as we know it exists within a relatively narrow range of temperature. The temperature scale illustrated here is the absolute or Kelvin scale. At absolute zero (0° K) all molecular motion ceases. It is equivalent to −273.1° C or −459° F. Most organisms live somewhere between 273.1° and 313° K, that is, between 0° and 40° C. Above 40° C the three-dimensional structure of many proteins begins to alter, a process called *denaturation.* There are some fascinating exceptions. Certain blue-green algae survive at 88° C (190° F). A species of fish in Ceylon lives in water at 50° C (122° F). A brine fly inhabits salt pools at 46° C (115° F). The horned toad (a lizard) of southwestern American deserts is most active at about 41° C (105° F). On the other hand, a polar codfish swims in water at −2° C (29° F). The Alaskan stonefly mates at 0° C (32° F). Some crustaceans die of heat at 11° C (52° F).

holding his body temperature constant, in which the sweat glands play an important part.

Like weather observers reporting to a central weather bureau, sensory cells in the skin keep the cortex informed of their local temperature. More important in the regulation of deep body temperature, the hypothalamus responds to the temperature of the blood flowing through it, and sends instructions to the sweat glands and the muscles controlling the diameter of blood vessels and the erection of hairs of the skin. These in turn obey the commands from the hypothalamus to conserve or to give off body heat (Fig. 12-47).

Body temperature is a balance between heat lost to the environment and heat produced as a by-product of cellular activity. When man's surroundings get too cool, blood circulation in his skin decreases, and the heat of the blood is conserved. The muscles supplying the hairs in the skin contract, causing "gooseflesh"; this increases the depth of the insulating hair over the surface of the skin. This last process is not as important to man as it is to most mammals who endure cold environments, such as the Arctic husky.

It is more difficult for man and other homeotherms to cool themselves than to warm themselves. Few homeotherms in fact can survive environmental temperatures much higher than that of their body for any length of time. Even man, who can produce twice as much sweat per unit area of skin as any other mammal, cannot tolerate prolonged exposure to high temperatures. On the other hand, warming the body can be done in several ways. Man, other mammals, and birds shiver. Shivering involves uncoordinated muscle activity which liberates heat.

Homeotherms have the most highly evolved nervous systems and brains of all animals. Apparently one reason they have been able to evolve such complex brains is because they have extremely efficient temperature control systems. In effect nature has placed this intricate, sensitive computer — the brain — in a temperature-controlled room — the human body. But the body temperature of even homeotherms varies. In man, for example, there is a daily temperature cycle. Man's maximum body temperature of about 100°F occurs late in the afternoon, and his minimum temperature of about 97°F occurs early in the morning. However, recent experiments indicate that different persons have different cycles, which explains what many have known intuitively all along—that some of us are night owls, whereas others function best during the daytime. Our greatest activity and feelings of well-being coincide with our temperature maximum.

Of course, the body temperature sometimes gets too high, and we have a *fever*. What happens during a fever? Essentially the "thermostat" in the hypothalamus is "reset" to a higher temperature, and one acts as though he were cold. He feels chilly, he begins to shiver, and continues to shiver until his body reaches its new temperature. When the cause of the fever is removed, he acts as though he were too warm, and the hypothalamus triggers sweating and increased blood circulation in the skin. Fever has several causes, but the most common is the presence of substances called *pyrogens* in the blood. Recent evidence suggests that *prostaglandins*, discussed in Chapter 15, may act as pyrogens. We do not know how pyrogens affect the hypothalamus, but we do know that one of the best ways to combat them is to take aspirin. The greater the fever, the more effective is the aspirin. Interestingly, aspirin is now known to inhibit the production of prostaglandins.

Other mammals and birds.   Many homeotherms help regulate the body temperature by behavioral means. Kangaroo rats burrow during the heat of the desert day. Outside the temperature may be more than 100°F but within the burrow the temperature is about 80°F. How important is this burrowing behavior to this animal? If we place a kangaroo rat in a cage out in the heat of the open desert, the rat will die within an hour. The Arctic husky, although well protected from the cold by its heavy fur, curls up in the snow, thus reducing the area for heat loss from the body.

Since most mammals and birds sweat very little—wet fur could lead to chilling and wet feathers would create a drag in flight—they utilize evaporative cooling in other ways. For example, many mammals such as dogs pant. Panting moves air rapidly over moist respiratory surfaces and, as in sweating, the evaporation of water cools the animal. Other mammals, such as cats, lick the fur of the face and belly, and evaporation from the licked areas cools the animals.

Some homeotherms migrate or hibernate when the seasons change. They do so probably as much in response to a dwindling food supply as to a lowered environmental temperature. In hibernation the body temperature may fall to a few degrees above freezing and the animal goes

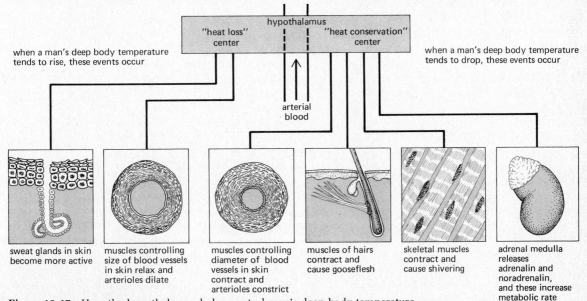

Figure 12-47 How the hypothalamus helps control man's deep body temperature.

into a prolonged sleep (Fig. 12-48). But the temperature control center remains alert and the animal will stir if the temperature gets too low.

Some animals seem to hibernate almost every night and come out of hibernation each morning. We see this clearly in both the smallest mammal, the shrew, and the smallest bird, the hummingbird. Like all small homeotherms they have extremely high metabolic rates. For example, a shrew may weigh less than a penny, and a hummingbird weighs less than a dime; yet gram for gram the metabolic rate of a hummingbird *at rest* is a hundred times that of an elephant! Since they have very high metabolic demands, these small homeotherms feed continuously during the day. At night they do not feed, and their body temperature drops. They actually hibernate overnight. In effect they are homeotherms by day and poikilotherms by night.

Camels do not hibernate at night, but their body temperature does drop to about 93°F. As a result it takes several hours during the following day for their body temperature to rise to 105°, the temperature at which they start sweating. This means that they lose less water than they would if their body temperature remained at 105°. (Contrary to popular belief camels do not

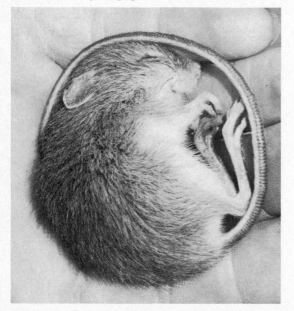

Figure 12-48 A hibernating meadow jumping mouse.

store water for unusually long periods of time.) Several years ago the camel's temperature fluctuations were interpreted as meaning that the camel is a poor temperature regulator. Now that we understand its physiology better it is clear that this mechanism is a beautiful adaptation of the camel to its desert environment.

Many birds cool themselves by wading or bathing in water. You probably have seen chickens cooling themselves on a hot day by spreading or dragging their wings in the dust. Some desert birds are nocturnal and avoid the heat of the day. (We have to be careful how we interpret nocturnal behavior; it is also possible that these birds' food supply is available only at night.)

Animals have evolved various structures which aid in temperature regulation. The whale and seal are more protected by the thick layer of fat under the skin than by hair or fur. The whale is hairless for all practical purposes, and the seal's fur, when wet, is a poor insulator. In contrast much of the camel's body fat is stored in one place, the hump; this is why the camel's body temperature fluctuates as widely as it does. The thickness of the body covering of birds and mammals changes seasonally, with heavier coats in winter than in summer. Arctic mammals have up to nine times as much insulation as tropical forms. Appendages frequently aid in heat conservation or loss. For example, the Arctic hare has much smaller ears than a desert rabbit. As a result the Arctic hare loses less heat from its ears than does its desert cousin (Fig. 12-49). The webbed feet of aquatic birds, which probably evolved originally for wading and paddling, also increase the surface area for heat loss.

Temperature regulation in poikilotherms

Vertebrates. Many reptiles control their body temperature by seeking sun when they are cold or shade when they are too warm. Figure 12-50 illustrates the way in which some lizards regulate their temperature by means of their behavior. In the morning when the sun first rises the animal is buried in the sand except for the head which is exposed to the sun. In the head is a large blood sinus or cavity. Blood in this cavity is warmed by the sun, then flows throughout the animal's body and warms the animal. Only then does the lizard emerge and run about. When it gets too warm the lizard seeks shade under a rock or places itself parallel to the sun's rays to expose the minimum of its body surface to the radiant heat. At sundown it buries itself in the sand again. Other lizards warm themselves by aligning the body so that its long axis is at right angles to the rays of the sun. A lizard in the Andes Mountains of Peru can regulate its body temperature so that it is 87° F while the air temperature is close to 32° F. In addition to this behavioral regulation of temperature, some lizards pant when the temperature gets too high.

Fish are poikilotherms, yet some great sport

Figure 12-49 The long ears of the desert jackrabbit (left) are efficient radiators of body heat. In contrast, the small ears of the Arctic varying hare (right) enable it to conserve body heat.

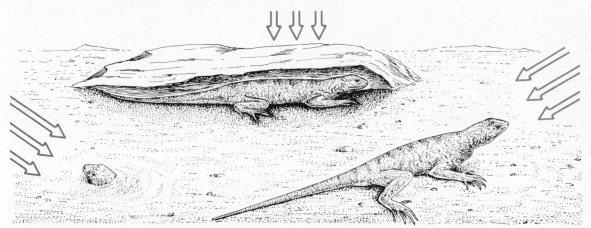

Figure 12-50 Many lizards regulate body temperature by their behavior. Arrows show direction of sun's rays. (From "How Reptiles Regulate Their Body Temperature" by Charles M. Bogert. Copyright © April 1959 by Scientific American, Inc. All rights reserved.)

fish—the mako and porbeagle sharks and bluefin tuna—are literally "warm-blooded." A mako's body temperature, for instance, may be about 70° F when the water temperature is about 50° F. The reason for this is countercurrent exchange of heat. The muscles involved in swimming liberate large amounts of heat which warms the blood flowing through the body. Arteries carry warm blood from the interior to the surface of the animal. As the arteries approach the surface they come to lie parallel to veins carrying cooler blood from the surface to the interior. Much of the heat of the warm arterial blood is passed to the cooler venous blood, which becomes warmer. This countercurrent exchange of heat is exactly the same as the countercurrent exchange of oxygen across the gills of fish, and is the fourth example of countercurrent exchange we have encountered. Because of their high body temperature, these fish can cruise at about 20 miles per hour and make spectacular leaps which fishermen report are 15 to 20 feet out of the water.

Insects.   Watch a butterfly or moth poised on a flower for takeoff. If the day happens to be rather cool, the insect appears to shiver. Actually it is warming up its flight muscles. The temperature of a moth's flight muscles can be as much as 95° F when the air temperature is 68° F.

Bees keep the temperature of the hive close to 95° F. Worker bees warm the hive by remaining inside and fanning their wings. They cool the hive in one of two ways. (1) If the outside air is cooler than the hive, they move to its entrance and fan their wings, thus moving cool air into the hive. (2) If the outside air is warmer than the hive, some workers carry water into the hive. Evaporation of this water cools the hive just as evaporation of sweat from your skin cools you.

With the onset of winter some insects form glycerin in their body fluids; glycerin acts as an "antifreeze," and helps the insect to withstand subzero temperatures.

The temperature tolerance of insects varies tremendously. For example, some insect larvae live in frozen mud and ponds in the arctic. Yet they can withstand repeated thawing and freezing. Exactly how they do this is not understood. Although most insects are most active at relatively high environmental temperatures, one unusual and primitive insect will die a "heat death" if held in the palm of one's hand.

In the following chapter we discuss some of the ways that temperature regulation and the other physiological processes that we have discussed are modified, controlled, and balanced with respect to one another.

SUGGESTED READING LIST

Adolph, E. F., "The Heart's Pacemaker," *Scientific American*, March, 1967 (Offprint 1067).

Beck, W. S., *Human Design: Molecular, Cellular, and Systematic Physiology*. Harcourt, Brace, Jovanovich, New York, 1971.

Bogert, C. M., "How Reptiles Regulate Their Body Temperature," *Scientific American*, April, 1959 (Offprint 1119).

Davenport, H. W., "Why the Stomach Does Not Digest Itself," *Scientific American*, January, 1972.

Frieden, E., "The Chemical Elements of Life," *Scientific American*, July, 1972.

Gordon, M. S. (in collaboration with G. A. Bartholomew, A. D. Grinnell, C. B. Jorgensen, and F. N. White), *Animal Physiology: Principles and Adaptations* (second edition). The Macmillan Company, New York, 1972.

Griffin, D. R., and A. Novick, *Animal Structure and Function* (second edition). Holt, Rinehart and Winston, New York, 1970.

Guyton, A. C., *Function of the Human Body* (third edition). W. B. Saunders Company, Philadelphia, Penn., 1969.

Heinrich, B., and G. A. Bartholomew, "Temperature Control in Flying Moths," *Scientific American*, June, 1972.

Hong, S. K., and H. Rahn, "The Diving Women of Korea and Japan," *Scientific American*, May, 1967 (Offprint 1072).

Irving, L., "Adaptations to Cold," *Scientific American*, January, 1966 (Offprint 1032).

Johansen, K., "Air-Breathing Fishes," *Scientific American*, October, 1968 (Offprint 1125).

Krogh, A., *The Comparative Physiology of Respiratory Mechanisms*. Dover Publications, Inc., New York, 1968.

Macey, R. I., *Human Physiology*. Prentice-Hall, Inc., Englewood Cliffs, New Jersey, 1968.

Marshall, P. T., and G. M. Hughes, *The Physiology of Mammals and Other Vertebrates*. Cambridge University Press, New York, 1965.

Mayerson, H. S., "The Lymphatic System," *Scientific American*, June, 1963 (Offprint 158).

Nossal, G. J. V., *Antibodies and Immunity*. Basic Books, Inc., Publishers, New York, 1969.

Ramsay, J. A., *Physiological Approach to the Lower Animals* (second edition). Cambridge University Press, New York, 1968.

Schmidt-Nielsen, K., "Salt Glands," *Scientific American*, January, 1959 (Offprint 1118).

———, *Desert Animals*. Oxford University Press, New York, 1964.

———, *Animal Physiology* (third edition). Prentice-Hall, Inc., Englewood Cliffs, New Jersey, 1970.

———, "The Physiology of the Camel," *Scientific American*, December, 1959 (Offprint 1096).

———, "How Birds Breathe," *Scientific American*, December, 1971 (Offprint 1238).

———, and B. Schmidt-Nielsen, "The Desert Rat," *Scientific American*, July, 1953.

———, *How Animals Work*. Cambridge University Press, New York, 1972.

Smith, H. W., *From Fish to Philosopher*. Doubleday Anchor Books, Garden City, New York, 1961.

U. S. Public Health Service, *The Health Consequences of Smoking. A Report of the Surgeon General: 1972*. DHEW Publication No. (HSM) 72–7516, 1972.

Vander, A. J., J. H. Sherman, and D. S. Luciano, *Human Physiology*. McGraw-Hill Book Co., New York, 1970.

Wigglesworth, V. B., *The Principles of Insect Physiology* (sixth edition). Methuen, London, 1965.

———, *The Life of Insects*, World Publishing, Cleveland, Ohio, 1964.

Williams, C. M., "Insect Breathing," *Scientific American*, February, 1953.

Wilson, J. A., *Principles of Animal Physiology*. Macmillan Company, New York, 1972.

Winton and Bayliss, *Human Physiology* (sixth edition), rev. and ed. by O. C. J. Lippold and F. R. Winton. Williams and Wilkins, Baltimore, Maryland, 1969.

Wood, D. W., *Principles of Animal Physiology*. Edward Arnold (Publishers) Ltd., London, 1968.

Wood, J. E., "The Venous System," *Scientific American*, January, 1968 (Offprint 1093).

Zweifach, B. W., "The Microcirculation of the Blood," *Scientific American*, January, 1959 (Offprint 64).

CHAPTER 13

Controlling the Animal Way of Life—The Nervous System

Multicellular organisms, such as a man, an insect, or a Sequoia tree, are composed of millions—often millions of millions—of cells. Each cell is a separate life, "separately born and destined separately to die." Yet we are not conscious of the separateness of these millions of cellular lives. What impresses us instead is the life of the animal or plant of which each cell is a part. The fact that animals and plants function so well even though they are communities of millions of separate cells shows that the individual lives of these vast populations of cells are coordinated. The activity of cells in one part of the organism is coordinated or *integrated* with the activity of cells in another part. Integration is brought about by two distinct mechanisms, *chemical control mechanisms* and *nervous control mechanisms*.

The principal chemical control mechanisms for integrating the activities of different cells are **hormones.** These are "chemical messengers" produced by certain specialized cells of the body that exert specific effects on certain other tissues of the body. Chapter 10 discussed plant hormones like auxins that are produced by actively growing parts of plants. In animals hormones are produced by special groups of cells called **endocrine**

glands. Animal hormones are transported to all parts of the body by the blood or other body fluids. For example, as we learned in Chapter 12, a hormone (*secretin*) produced by cells in the duodenum stimulates the pancreas to release pancreatic juices.

Since they are blood-borne, animal hormones may take several seconds or even minutes to get to the tissues which are sensitive to them, and to affect these tissues. Sometimes a hormone affects several different tissues at once. Often the effects last for minutes or even hours. Animals use hormones to integrate many activities where speed is *not* required—digestion, reproduction, and growth. For example, the promotion of milk secretion by human mammary glands does not require a split-second response. The release of the lactogenic hormone over a period of days and weeks stimulates the growth and milk-secreting capacities of the mammary glands.

When speed is essential, animals use their nervous systems, and not hormones. When you inadvertently put your hand on a hot stove, you do not wait a minute for a chemical message to circulate by way of the blood to muscles of the arm to stimulate them to contract. A speedy response is important here. Nerve impulses fly from pain receptors in your hand along nerves to the spinal cord. The spinal cord sends nerve impulses immediately to the muscles of the arm which withdraws your hand. All this occurs within a fraction of a second (actually *before* you feel anything). Animals have evolved both muscles and nervous systems to make such rapid movements possible. Since the life-style of a plant does not demand rapid movements, plants never evolved either muscles or nervous systems.

The nervous system can act so quickly for two reasons. First, all the muscles of the body are connected to the spinal cord and brain by nerves. Second, nerve impulses travel over nerves far more rapidly than the blood circulates hormones. As a result nerve cells excite tissues within thousandths of a second. In addition the nervous system can act selectively and can exert far more localized effects than any circulating hormone because the nerve endings release transmitter chemicals "locally," that is, only a few microns away from the cell to be stimulated.

All multicellular animals more complex than hydra have both a slow (endocrine) and a fast (nervous) control system. To guarantee that these two control systems do not conflict with one another, they are closely integrated, usually by the brain. This chapter is concerned largely with nervous control systems. Chapter 15 examines some endocrine control mechanisms.

Nervous control systems

The simplest nervous systems like that of hydra contain only a few hundred or thousand nerve cells. A bee's nervous system may contain half a million cells. The most complex nervous system—man's—contains more than 10^{10} nerve cells. Fortunately all nervous systems, whether simple or complex, are composed of the same basic kinds of cells—nerve cells. As far as we know nerve cells function in much the same way wherever they are found—in a hydra, a bee, or a man. The nerve cell is where we begin our examination of the nervous system.

STRUCTURE OF NERVE CELLS

True nerve cells

Nerve cells or **neurons** carry messages, and they do this in two ways.

1. Messages may be sent from one part of the nerve cell to another part of the nerve cell, that is, *intra*cellularly. Such intracellular messages, called **nerve impulses,** are *electrical* changes in the cell membrane which are very similar to those which occur in the membrane of a contracting muscle cell (Chapter 11). Nerve impulses are *not* electric currents in the sense of electrons moving down a wire, and therefore they do not move with the speed of light. They are movements of electric charges in and out of the cell membrane and are conducted very rapidly from one part of a nerve cell to another part.

2. Messages may be sent between cells, that is, *inter*cellularly. Such intercellular messages are transmitted usually in the form of a **transmitter chemical.** As might be expected,

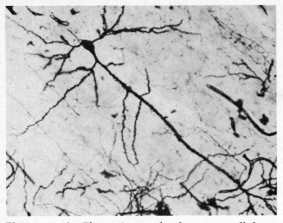

Figure 13-1A Photomicrograph of a nerve cell from the brain of a rat. Nerve cells may differ in shape and size of the cell body, in the length of the axon, and in the presence or absence of a fatty layer called the myelin sheath.

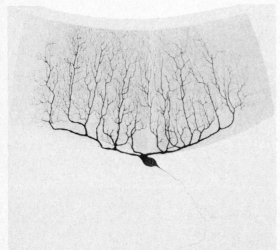

Figure 13-1B A neuron from a region of the human brain called the dentate gyrus. This drawing was prepared in 1886 by Camillo Golgi, a brilliant Italian neuroanatomist who developed a technique in which only a few of the nerve cells in a piece of tissue are stained. Moreover, these few cells are stained in their entirety. This method opened the way for a detailed analysis of the anatomy of the brain which has continued actively to this day. Note the faintly stained fine axon leaving the cell body and the hundreds of branching dendrites from this single cell.

neurons are specialized for sending and receiving these messages.

Look at a mammalian nerve cell under a microscope (Fig. 13-1A). The cell body is about 0.1 millimeter in diameter and it contains many ribosomes. This should not be surprising, for nerve cells are very active in synthesizing

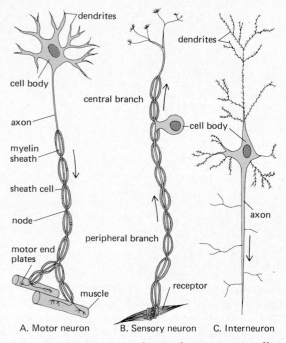

A. Motor neuron   B. Sensory neuron   C. Interneuron

Figure 13-1C Structure of some human nerve cells. (1) **Motor neurons** have long axons that run from the spinal cord or brain to an effector like a muscle. The axons of motor nerve cells usually have a myelin sheath. The dense granules in the cell body and dendrites are mainly ribosomes and endoplasmic reticulum. Stimuli are received by the dendrites which conduct them to the cell body and to the axon. The power for the nerve impulse is generated by the neuron. The nerve impulse itself is a minute voltage difference which is self-propelled down the neuron. (2) **Sensory neurons** usually have only one fiber which branches soon after it leaves the cell body. One branch goes to the receptor, usually a sense organ, whereas the other branch goes to the spinal cord or brain. These branches have the same structure as axons, even though the peripheral branch from the receptor conducts impulses toward the cell body. (3) **Interneurons** often have short axons. Notice the spiny appearance of the dendrites in contrast to the relatively smooth appearance of the axon.

proteins. There are also many structures in the cell that we usually associate with secretion. Some of these structures may play a role in making transmitter chemicals.

Unlike the cell body of most other cells, the cell body of a neuron has branches. Most of the branches are short, very thin, and have a spiny look. They are called **dendrites.** Dendrites act as "antennae," receiving incoming messages in the form of transmitter chemicals from other nerve cells. When enough of the dendrites of a receiving cell get excited by transmitter chemicals, the electrical charges spread to the cell body. Most nerve cells also have a single long branch, the **axon,** which conducts nerve impulses away from the cell body. Some axons, such as one running from the spinal cord to the big toe, may be extremely long. In a giraffe some axons that run down the neck are 10 feet long, and in a blue whale some may be many yards long. The axon of some neurons, however, may be only a few microns long. When the nerve impulse reaches the end of the axon, a transmitter is released onto a neighboring cell's dendrites, and the whole process starts again in the neighboring cell. In man the diameter of axons ranges from less than 1 to about 100 microns. Although axons and dendrites usually look different, the basic distinction between them is in how they work. Dendrites are excited by other cells, whereas axons usually are not. Also axons can excite other cells by releasing transmitters, whereas dendrites cannot. Although neurons have many, many dendrites, they usually have only one axon but the axon may be branched.

Like the rest of the nerve cell, the axon is covered by the nerve cell membrane. The axon of many neurons is also covered by layers of a fatty substance called **myelin** (Fig. 13-2), which is composed of the cell membranes of special cells, called *sheath cells,* which are wrapped tightly around the axon like a jellyroll. There are gaps in the myelin covering called nodes (Fig. 13-1C); at these gaps the axon is bathed by body fluids. As we shall see later these gaps in the myelin greatly speed up the conduction of impulses in *myelinated* axons.

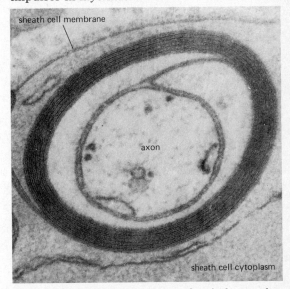

Figure 13-2A Electron micrograph of the myelin sheath which shows clearly that the myelin sheath is the cell membrane of a sheath cell (also called a Schwann cell) wrapped around the axon like a jelly roll. The nodes in Figure 13-1C are the points at which one sheath cell ends and another begins.

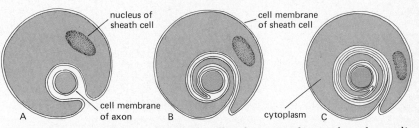

Figure 13-2B Cross section of a sheath cell and an axon showing how the myelin sheath forms around an axon. Initially the axon lies in an infolding of a sheath cell. The infolded area begins to wrap tightly around the axon, ultimately producing the final pattern seen in Figure 13-2A.

The axon itself branches, and the branches end in tiny buttonlike structures called **synaptic knobs.** Synaptic knobs almost, but do not quite touch other cells. The regions where the knobs come so near other cells are the *synapses.* If the knobs end near a skeletal muscle cell, the synapse is called the *neuromuscular junction.* Within the synaptic knobs are tiny sacs called *synaptic vesicles* containing the transmitter chemical (Fig. 13-4). It is here that the chemical transmitter is released.

Most biologists believe that neurons have become so specialized that they do not even divide. Parts of a damaged nerve cell may regenerate if the cell body is still healthy, but nerve cells do not divide to produce new ones. For example, polio virus destroys the nuclei of nerve cells. These nerve cells, like all enucleated cells, then die; the muscles they innervate become paralyzed because no new nerve cells replace the old ones. If, as many biologists believe, learning and memory are tied up with the structure of nerve cells, it may be essential that they *not* divide. If they did, our memory might change even more than it does! Most biologists believe that very shortly after we are born we have all the nerve cells we shall ever have. In fact several thousand nerve cells die each day of our life, especially as we grow older. This may be one reason why we become senile.

The supporting cast—Glial cells

Most cells in the nervous system are not even nerve cells. **Glial cells** outnumber nerve cells by at least ten to one (Fig. 13-5). They get their name from the Greek word for "glue." We are already familiar with one type of glial cell, the sheath cells which insulate many axons. We do not know very much about the role of other types of glial cells. Many of them cling to nerve cells and some apparently move. They may structurally support nerve cells or perhaps provide nutrients for them. A neuron with a long axon has more glial cells associated with its cell body than a neuron with a short axon. Some biologists interpret this fact as evidence that glial cells sustain axons in some way.

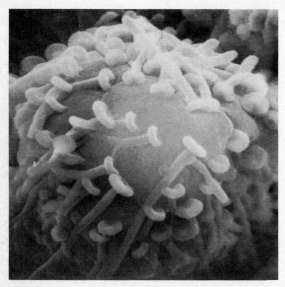

Figure 13-3   Scanning electron micrograph of synaptic knobs of nerve cell of a sea hare (*Aplysia californica*). (× 4500)

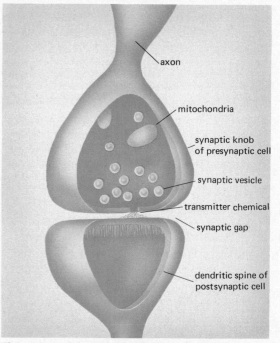

Figure 13-4   Structure of a synapse. The cell before the synapse (presynaptic cell) does not actually touch the cell after the synapse (postsynaptic cell). There is a gap of about 200 Å between the two cells into which the transmitter chemical is released by the presynaptic cell.

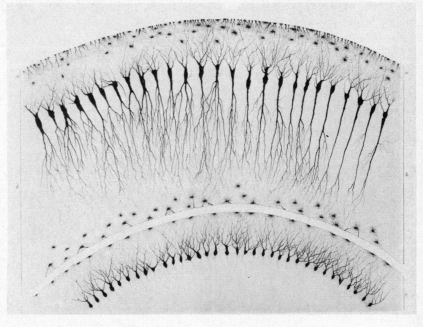

Figure 13-5 A part of the brain called the **hippocampus** showing both neurons and glial cells. This drawing by Camillo Golgi shows only a few of the cells in this region. The small cells which look like asterisks (**) are glial cells called **astrocytes.** The large neurons in the upper part of the drawing are hippocampal pyramidal cells. The small neurons in the bottom of the drawing are dentate granule cells. The challenge of unravelling the wiring of the brain becomes clear from such a drawing.

Some biologists believe that after injury glial cells form a tube which, in effect, "guides" a regenerating axon back to the cells which it originally innervated. Glial cells make up about 40 percent of the brain's bulk; what they do there is a mystery. No doubt future experiments will show that glial cells play vitally important roles in the overall functioning of the nervous system.

HOW NERVE CELLS FUNCTION

Conduction of nerve impulses

Several teams of biologists, most notably two British biologists, Allen Hodgkin and Andrew Huxley, and two Americans, Howard Curtis and Kenneth Cole, determined what happens during the passage of a nerve impulse, and the crucial role which the cell membrane plays in conducting impulses. Their work was made possible by the discovery at the Marine Biological Laboratory in Woods Hole, Massachusetts, of the *giant nerve cells* of squids (Fig. 13-6). The living squid uses these large nerve cells to send rapid nerve impulses to the muscles which operate its jet-

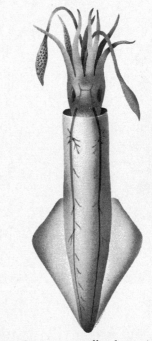

Figure 13-6 Giant nerve cells of a squid.

propulsion system. They permit the squid to change course rapidly. The axon of these neurons is about 1 millimeter in diameter—large enough to poke electrodes into them (in order to measure electrical changes) and large enough to be able to follow the movement of radioactively labeled substances through them.

Hodgkin and Huxley inserted tiny electrodes into these giant axons. They confirmed that the inside of a resting nerve cell—one which is not conducting impulses—has a negative charge compared to the outside of the nerve cell. This is true of the entire surface membrane of the cell, including the axon and the cell body. When they stimulated the axon, a nerve impulse was rapidly conducted down it. They found that when a nerve impulse reached the electrodes, the inside of the axon became positively charged; as the impulse moved on past the electrodes the inside of the axon recovered its original negative charge (Fig. 13-7). Exactly the same thing happens in a contracting muscle cell, as we have already learned (Chapter 11). Further work showed that the electrical changes across a nerve cell membrane, like those across a muscle cell membrane, are caused by changes in the permeability of the membrane to sodium, potassium, and calcium ions. The electrical and ionic changes that travel down an axon *are* the nerve impulse.

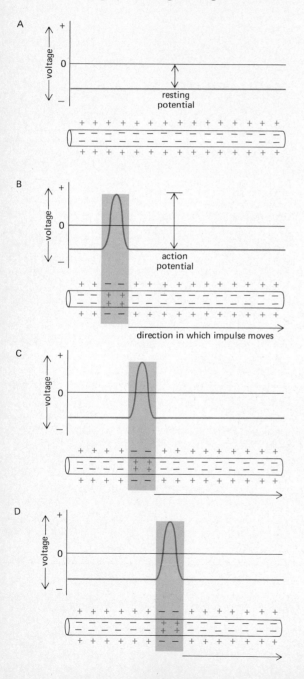

Figure 13-7 Conduction of a nerve impulse. The curves represent electrical changes, and the cylinders represent an axon and the electrical charges inside and outside its membrane. An impulse is conducted in the following way. A. In a resting nerve cell (one that is not conducting an impulse) the inside of the cell is negative to the outside. Thus a voltage difference exists which is referred to as the **resting potential** (electrical potential = voltage). B. When the axon is stimulated, the inside becomes positive or **depolarizes** at the point of stimulation. This increased voltage difference is called the **action potential.** Although the excited region quickly recovers its original negativity or **repolarizes** the impulse does not disappear. C. Instead it moves forward along the axon. At the same time that the original excited region repolarizes, the region in front of it depolarizes. D. These events are repeated down the length of the axon, which in this way conducts an impulse.

In a resting nerve cell, sodium ions are pumped out, and potassium ions are pumped in. But when the axon is stimulated, it becomes leaky to Na+ which then enters faster than it can be pumped out. As a result, an action potential is initiated. An instant later, K+ diffuses out of the axon. During repolarization the cell becomes less permeable to Na+, and the Na+-K+ exchange pump can once again pump excess Na+ out and K+ in.

Other biologists used the squid's giant axon to show that the nerve cell membrane, not the cytoplasm, is responsible for conducting impulses. They squeezed the cytoplasm out of the axon much as you squeeze toothpaste out of a tube and replaced the cytoplasm with salt solutions. The axon still conducted several hundred thousand impulses, even without its cytoplasm.

Many different stimuli—heat, touch, electricity, chemicals, for example—will excite a neuron. These stimuli change the permeability of the membrane in a way which we do not yet understand. In an intact animal, of course, it is usually the transmitter chemical arriving from other cells that excites a nerve cell and triggers impulses. One of the really great discoveries of physiology is that the cell membranes of both nerve cells and of muscle cells respond to stimuli in exactly the same way, namely, they conduct impulses. In nerve cells the impulses conducted by the nerve cell membrane stimulate the synaptic knobs at the tips of the axons to release transmitter chemicals. In muscle cells the impulses conducted by the muscle cell membrane stimulate the contractile machinery in the cytoplasm to contract.

Nerve cells can conduct several hundred or up to a thousand impulses each second. Yet the cell still has time for an incredibly brief rest after the passage of each impulse, and will not conduct any more impulses during most of its rest period.

Rate of impulse conduction

The rate at which nerve impulses travel along an axon was first determined in the mid-1800s by Hermann von Helmholtz, the same German scientist who later made very important contributions to the study of electricity and magnetism (Fig. 13-8). In one of his earliest experiments he determined the rate at which a frog nerve conducts impulses. Perhaps he was spurred by his teacher's claim that no one would ever be able to do this.

Again and again we find that the truly great scientific experiments are simple in design. Helmholtz's was no exception. In fact it can be easily repeated in the laboratory. He made a frog muscle contract by stimulating the nerve to the muscle in two places, near the muscle and at a point distant to the muscle. In each case he determined the time which elapsed between stimulating the nerve and muscle contraction. He found that the interval was longer when the nerve was stimulated farther from the muscle. Knowing the two time intervals and the length

Figure 13-8 Experimental setup to determine the speed at which a nerve conducts impulses. The nerve is stimulated at two places, near the muscle (*a*) and near the opposite end of the nerve (*b*). Stimulating the nerve causes the muscle to contract; the contractions are recorded on a moving drum. The velocity of conduction is calculated by dividing the length of the nerve between the points of stimulation by the time required for an impulse to travel that distance.

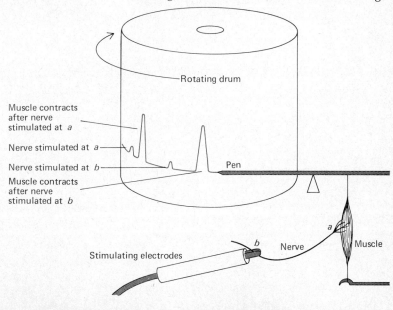

of the nerve, he calculated the rate and found it to be about 30 meters per second (65 mph).

We now know the rate at which many different axons conduct impulses and some of the factors that affect the rate. Usually axons with a large diameter conduct impulses more rapidly than those with a small diameter. Myelinated axons conduct impulses as much as ten times faster than nonmyelinated axons of the same diameter. Most axons of vertebrates are myelinated, whereas those of most other animals are nonmyelinated. Therefore vertebrate axons generally conduct impulses more rapidly than those of other animals. In mammals the rate may be more than 100 meters per second (220 mph). Recall that there are gaps in the myelin. Apparently impulses skip from gap to gap in myelinated axons, and the skipping speeds up the rate of conduction.

Animals like man and the squid have a fast-paced way of life that depends on the rapid conduction of nerve impulses. Although the squid's giant axons are nonmyelinated, they conduct impulses rapidly because they are large in diameter. Since the squid has relatively few axons, their large size presents no problem. However, man has millions of axons packed into his spinal cord alone. Obviously the diameter of each axon must be quite small if all of them are to occupy such a small space. The mammalian solution to the problem of how to make thin axons conduct impulses rapidly was to insulate the axons with myelin.

The neural language

Although the normal stimulus to a nerve cell is a transmitter chemical, biologists usually use electrical stimuli in their experiments with nerve cells. The reasons for this are that the size and duration of the electrical stimulus can be controlled easily, and gentle electrical stimuli do not damage neurons. If a neuron is stimulated with a weak electrical stimulus, nothing happens. If the strength of the stimulus is increased, a point will be reached at which the nerve cell conducts an impulse. The weakest current or other stimulus which excites a neuron is the *threshold stimulus* of that neuron. Each neuron has its own threshold.

When a neuron is excited, the size or voltage and the duration of the impulses it conducts can be measured by simple electronic recording devices. When this is done you discover that all the impulses generated by a given neuron have the same size and the same duration. Increasing the strength of a stimulus beyond the threshold does not increase the size or duration of the nerve impulse. The transmission of a nerve impulse is an **all-or-none reaction.** It either goes completely or not at all.

If all nerve impulses are just alike, how does a neuron send different messages? This question can be answered by a simple experiment. Suppose you stimulate a neuron for 2 seconds with a threshold stimulus. You find that the neuron conducts 3 impulses per second for 2 seconds. Now stimulate the neuron for 2 seconds with a much stronger stimulus. The neuron now conducts 6 impulses per second for 2 seconds. In both cases the neuron starts conducting nerve impulses when you stimulate it, and stops conducting when the stimulus stops (Fig. 13-9). Thus the neuron sends information about the **intensity** of the stimulus by varying the *frequency* of nerve impulses. Under normal conditions (not experimental electrical stimulation), a neuron sends information about the intensity of a stimulus, not only by varying the frequency of impulses but also by varying the pattern of grouping of impulses. The start and stop of these impulses tell the duration of the stimulus.

Synapses

Impulses are conducted away from the cell body along an axon toward the synapse. At the synapse

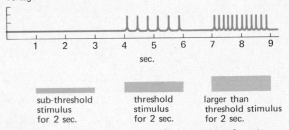

Figure 13-9 Effect of varying the intensity of a stimulus on the size and frequency of nerve impulses. Stimuli of varying strengths (but lasting two seconds) affect the frequency but not the size of impulses.

they trigger the release of a transmitter chemical from the sacs in the synaptic knobs of the **presynaptic cell** (cell before the synapse). The transmitter chemical diffuses across the tiny gap next to the presynaptic neuron and reaches the **postsynaptic cell** (cell after the synapse). It combines with receptor molecules on the membrane of the post-synaptic cell, changing the permeability of the membrane. In this way messages are sent from a nerve cell to the postsynaptic cell, which may be another neuron, a muscle cell, a gland cell, or some other effector.

A neuron can conduct impulses in both directions. But only impulses that move toward the synaptic knobs of the axon can excite the postsynaptic cell. Impulses that move from the cell body toward the dendrites die out when they reach them because the dendrites do not release transmitter chemicals. For this reason synapses can be thought of as *one-way valves* which control the direction of nerve impulses.

Transmitter chemicals are promptly destroyed by specific enzymes after they have been released and have transmitted their signal across the synapse. This destruction is necessary for the control function of the nervous system. If a transmitter persisted, then the postsynaptic cell could no longer be controlled.

A synapse may be either **excitatory** or **inhibitory**, depending on whether the transmitter chemical released by the presynaptic cell excites or inhibits the postsynaptic cell. An individual neuron apparently synthesizes only one type of transmitter chemical. Several transmitter chemicals have been identified in man and other animals. The excitatory transmitter chemical released at neuromuscular junctions between neurons and skeletal muscle in vertebrates and many invertebrates is **acetylcholine.**

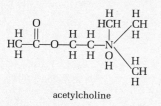

acetylcholine

It is also the excitatory transmitter chemical for synapses between certain neurons. The chemical nature of the inhibitory transmitters of

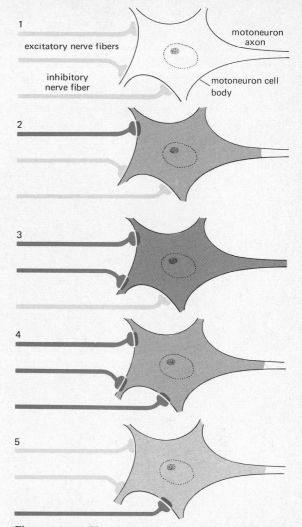

Figure 13-10 How a neuron is excited or inhibited. In this example, two excitatory fibers and one inhibitory fiber synapse with a motor neuron. The motor neuron will conduct an impulse (fire) when there is enough excitatory transmitter substance present to depolarize the neuron to its threshold level. A resting neuron does not fire (1). The amount of transmitter substance released by one excitatory fiber may not be enough to make the neuron fire (2), but both excitatory fibers may release enough transmitter substance to cause firing (3). If, at the same time, the inhibitory fiber also releases enough of its transmitter substance, and prevents depolarization, the neuron will not fire (4). Or the inhibitory fiber alone may release its transmitter substance, and prevent the neuron from firing (5). (From "The Synapse" by J. C. Eccles. Copyright January 1965 by Scientific American, Inc. All rights reserved.)

vertebrates is not yet known. All excitatory transmitter chemicals make the inside of the postsynaptic cell more positive than usual. On the other hand, inhibitory transmitter chemicals make the inside of the postsynaptic cell more negative than usual (Fig. 13-10).

Anyone who drives knows that he can regulate the movement of a car more accurately by using both an accelerator and a brake. In a similar way, having both excitatory and inhibitory synapses permits much more precise control of nervous activity than is possible with just one kind of synapse. Strychnine prevents inhibitory synapses from doing their usual job—blocking nerve impulses. If you inject strychnine into a frog, and then pinch the animal's toes, all of the frog's skeletal muscles—not just those which normally cause the animal to withdraw its foot—contract. In fact the animal becomes rigid. Because the entire nervous system is interconnected, the absence of inhibitory synapses permits a nerve impulse from the toe to travel throughout the frog's entire body. If we did not have inhibitory synapses, we would be just like the strychnine-injected frog. Whenever part of the body was stimulated our whole body would react convulsively. Coordinated movement would be impossible.

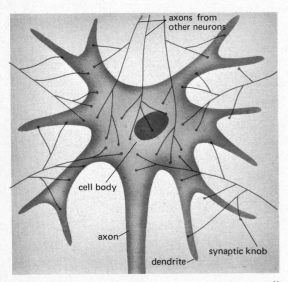

Figure 13-11 Synapses on a single motor nerve cell body. Only a few are indicated. There may be thousands.

To fire or not to fire.   Usually a postsynaptic neuron synapses with many presynaptic neurons. In fact some neurons in our spinal cord may be supplied with as many as 15,000 synaptic knobs, and nerve cells in the brain probably receive even more! Some of the presynaptic neurons release an excitatory transmitter, whereas others release an inhibitory transmitter. What a postsynaptic neuron does depends on the amounts of these excitatory and inhibitory transmitters that are released by the cells that synapse with it. A postsynaptic neuron will conduct impulses only when it receives enough of an excitatory transmitter chemical to overcome any inhibitory transmitter chemical released at the inhibitory synapses.

The firing of a nerve cell in response to excitatory transmitters released by one group of cells can be blocked by the inhibitory transmitters released by another group of cells. This interplay and balancing of excitation and inhibition are nervous integration. To understand the way nervous integration works in specific cases we must understand how nerve cells are arranged in the nervous system.

THE HUMAN NERVOUS SYSTEM

How nerve cells are arranged

The neurons of almost all animals including man occur in four characteristic arrangements. (1) Some neurons form an interconnecting **nerve net.** We find this in simple organisms such as hydra and also in parts of complex organisms, as in the neurons which innervate the smooth muscles of man's intestines. (2) Neurons may become concentrated into **nerve cords** containing cell bodies, synapses, and axons. Nerve cords are found in almost all organisms more complex than hydra and include the human spinal cord. (3) Neurons also become aggregated into clusters known as **ganglia** which, like nerve cords, consist of many interconnecting nerve cells. Ganglia are also found in all organisms more complex than hydra. The spinal cord of man has ganglia closely connected to it and the brain of man represents the fusion and development of the anterior part of the spinal cord and a number of ganglia. (4) Nerve cells also form **nerves** which are bundles of axons that run from nerve cords and ganglia to effectors and from sense organs back to nerve cords and ganglia.

We examine first the arrangement of the human nervous system. The brain and spinal cord of man make up the **central nervous system.** Information about the environment is fed into the central nervous system in the form of nerve impulses from receptors in all parts of the body (Fig. 13-12). The neurons that relay these sensory impulses to the central nervous system are called **sensory neurons.** They make up the **afferent** (from Latin words meaning "to carry to") **nervous system.** Sensory neurons usually synapse with one or many **interneurons** in the central nervous system. An interneuron is a nerve cell which acts as an intermediary between other nerve cells. Interneurons synapse with many other interneurons. Indeed, about 99 percent of human nerve cells are interneurons. They are "managerial" cells. Together they sort, integrate, balance and sharpen all of the sensory input, and finally send out nerve impulses to neurons which innervate muscles and other effectors. The neurons that relay these nerve impulses to effectors make up the **efferent** (from Latin words meaning "to carry away") **nervous system.** Some nerve impulses go by way of **motor neurons** to skeletal muscles. Other efferent neurons called **autonomic neurons** conduct nerve impulses to other effectors such as the heart, smooth muscles, and glands. We shall have more to say about autonomic neurons later.

The central nervous system—Spinal cord

Our spinal cord is like a rope no larger in diameter than a little finger. Figure 13-13 shows how the cord looks in cross section. The H-shaped *gray matter* is made up of the cell bodies of interneurons and motor neurons, as well as their dendrites and synapses, and glial cells. The cell bodies of sensory neurons are aggregated in clusters of nerve cell bodies known as the **dorsal root ganglia** which lie just outside the spinal cord. The surrounding *white matter* is composed of bundles of axons called **fiber tracts** which conduct impulses between different levels of the spinal cord and between the spinal cord and the brain. Many of the axons in the fiber tracts are myelinated; the myelin gives this part of the spinal cord a glossy white appearance. Sensory axons extend from the dorsal root ganglia and motor axons from the ventral root of the spinal cord. A short distance from the cord itself, the axons from both roots join to form a **spinal nerve.**

A nerve is composed of thousands of separate axons bound together, each able to transmit a separate message, just like a telephone cable composed of many telephone wires bound together. Each nerve contains both motor and

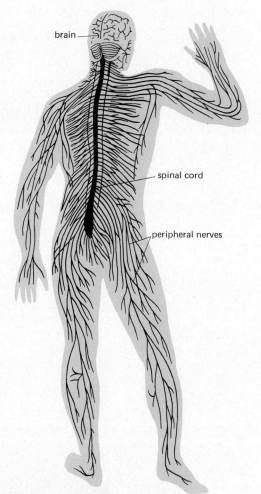

brain

spinal cord

peripheral nerves

Figure 13-12   Human nervous system.

sensory axons. There are no synapses in nerves, just axons. Synapses occur in the gray matter of the spinal cord, in dorsal root ganglia, and at the endings of motor nerves. There are 31 pairs of spinal nerves entering and emerging from the spinal cord through spaces between the vertebrae.

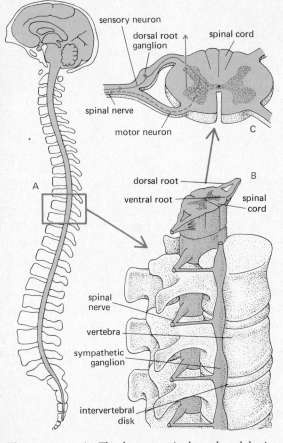

Figure 13-13 A. The human spinal cord and brain (central nervous system) are housed in the vertebrae and skull, respectively. Nerves running from the spinal cord in spaces between the vertebrae connect the central nervous system with all parts of the body. B. How the spinal cord is protected by vertebrae. Adjacent vertebrae are separated by an **intervertebral disk,** a cartilaginous structure that cushions the vertebrae and further protects the spinal cord. When one of these disks gets out of place (a "slipped disk") it exerts pressure on a spinal nerve, thus causing an extremely painful condition known as a "pinched nerve." C. Cross section of the spinal cord. For details, see text.

Each pair of nerves innervates a different area of the body. They branch and rebranch until at last they extend into almost every part of the body below the head. They are the major means of communication between the central nervous system and most of the body.

The role of the dorsal and ventral roots can be demonstrated by observing people, parts of whose spinal cord is damaged. For example, suppose that part of one's spinal cord controlling the right arm is injured. If the injury is in the dorsal root, the arm will feel "dead," but it can still be moved. If the ventral root is injured, your arm cannot be moved, but it will still have "feeling" in it. Hence afferent or sensory axons enter the dorsal root, and efferent or motor axons leave the ventral root.

Journey of a nerve impulse.   Where does an impulse go after it arrives in the spinal cord by way of a sensory neuron? It may follow millions of possible pathways. The sensory neuron may synapse directly with a motor neuron; in this case the impulse quickly leaves the spinal cord and the central nervous system. Usually, however, the sensory neuron synapses with one to many interneurons; interneurons link most afferent and efferent neurons. The more interneurons an incoming nerve impulse excites, the more possible pathways the nerve impulse may follow. Thus the nerve impulse may even spread to parts of the brain, including the parts involving consciousness. Such pathways as these are complex, involve many neurons, and therefore are difficult to trace. We can, however, get some idea of how the nervous system works by following an impulse along routes that involve only a few neurons. The simplest such pathways in man and other vertebrates are called **spinal reflexes,** and the simplest spinal reflex is the *knee-jerk reflex.*

Spinal reflexes.   One of the things a doctor does when he gives you a routine physical examination is to test your response to a tap on the knee. The tap stretches the tendon which attaches the **extensor** muscle above the knee to the knee and excites stretch receptors buried in the tendon. These stretch receptors in turn excite a sensory neuron that conducts nerve impulses to the spinal cord. Within the spinal cord itself the sensory neuron synapses with a motor neuron to the extensor muscle. This is an excitatory synapse; the motor neuron conducts impulses to the extensor muscle which contracts, causing the leg to jerk. This knee-jerk reflex is the simplest nervous pathway in our body because it involves only two neurons, sensory and motor, and there is only a single synapse in the central nervous system. Most reflexes involve hundreds or even thousands of neurons.

So far we have considered the reflex activity of only the muscle which was stretched. This is only half of the story. As we pointed out in Chapter 11 most skeletal muscles are arranged in antagonistic pairs. The knee-jerk reflex depends not only on the contraction of the extensor muscle but also on the relaxation of its antagonist, the **flexor** muscle in the back of the thigh. Its relaxation is accomplished in the following way. The same sensory neuron which excited the motor neuron to the extensor muscle also synapses with an interneuron. This particular interneuron makes an inhibitory synapse with the motor neuron that goes to the flexor muscle. Con-

sequently this motor neuron does not fire, and the flexor muscle relaxes. Thus the reflex contraction of the extensor muscle is accompanied by the reflex relaxation of the flexor muscle. Most antagonistic muscles are "wired up" this way. In daily life the knee-jerk reflex helps us to remain upright when we stand. Should the leg start to buckle, the reflex would again extend the leg.

Spinal reflexes are automatic and need not involve the brain, although the brain can modify them, as we shall see. When you touch a hot stove, you jerk your hand away—a reflex—*before* the message about pain gets to your brain. Even a mammal with its brain severed from the spinal cord still shows some reflexes. Similar reflexes occur in insects and most other animals. We discuss reflex behavior in the next chapter.

The emerging importance of the brain

The nerve pathway of a spinal reflex generally includes neurons which synapse and interconnect with other parts of the nervous system, including the brain. For this reason "simple" reflexes often become quite complex, especially in man. Man's brain can modify or even override a spinal reflex. For example, consider what may happen when a doctor tests a small boy's knee-jerk reflex. The child is probably frightened—he remembers the last time he visited the doctor and received an injection. He is wary and tense, and stiffens his knee. As a result there is no knee-jerk when the doctor taps the boy's knee.

The child can inhibit his knee-jerk reflex for the following reason: the sensory neuron conducting impulses from the stretch receptors in the extensor muscle also synapses with a number of interneurons in the spinal cord. Some of these interneurons eventually channel the impulses to the child's brain. His brain blocks the reflex by sending out impulses to his flexor muscles which cause him to stiffen his knee. (The doctor's way around this problem is to tell the child to twiddle his thumbs, count to ten, or spell his name. The boy then concentrates on the task, and he will usually forget to stiffen his knee.)

Our brain can do more than alter spinal reflexes; in man, as in other mammals, it takes over jobs which the spinal cord or nerve cord performs

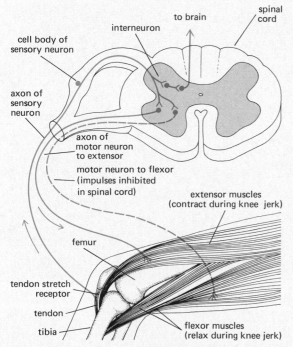

Figure 13-14 Pathway of the knee-jerk reflex. For details, see text.

in less complex animals. For example, if the brain of an insect or even a frog is destroyed, each of these animals can still make coordinated movements. A spinal frog (one without a brain but with the spinal cord intact) can jump about when stimulated. However, a spinal mammal cannot even stand. In the course of vertebrate evolution the brain became increasingly more important at the expense of the spinal cord; movements once controlled by the spinal cord came under the ever-increasing control of the brain.

Autonomic nerves

Most reflexes which involve skeletal muscles enable an animal to adjust to the outside world. Something happens in the external environment, and the animal responds in some way, usually by moving. But an animal finds itself at the mercy of two environments, an external and an internal. Let us now see how the nervous system helps to regulate the internal environment.

Internal organs are innervated by sensory nerve fibers which send input to the central nervous system. The central nervous system responds to this input by means of two sets of motor nerves which innervate and control the activities of internal organs. These motor nerves are called **autonomic nerves.** They are further divided into a **sympathetic** and a **parasympathetic** branch. The autonomic nerves and some of the tissues which they control are shown in Figure 13-15. What autonomic nerves do can be seen when we consider how the rate at which the heart beats is controlled.

If a frog's or even a man's heart is removed from the body, cutting off the heart's autonomic nerve supply, the heart continues to beat. As we discussed in Chapter 11, vertebrate heart muscle contracts spontaneously. What, then, do the autonomic nerves do? They affect the *rate* at which the heart beats. When a man asleep on the beach awakens in response to a cry for help, his heart immediately begins to pump blood faster than before. It does so because the heart-accelerating center in the medulla of the brain sends impulses over a *sympathetic* nerve to the pacemaker of the heart. The pacemaker then increases its

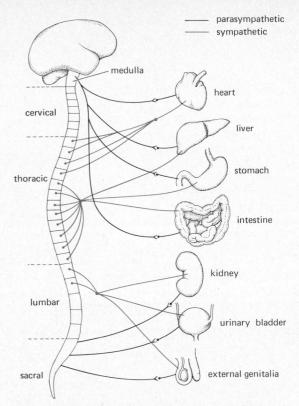

Figure 13-15 The autonomic nervous system, and some of the organs that it innervates. Most large organs are served by both sympathetic and parasympathetic nerves which function in opposition to each other. If, as is usually the case, the sympathetic nerve excites an organ, then the parasympathetic nerve produces the opposite effect and inhibits the activity. The balance between these two systems produces the normal state of activity of the various organs. A few tissues, such as small blood vessels, sweat glands, and hair muscles, are innervated only by sympathetic nerves. The diagram illustrates a basic anatomical difference between autonomic motor nerves and the motor nerves which innervate skeletal muscle, called somatic motor nerves. The cell bodies of somatic motor nerves are located within the central nervous system and send a long axon running without interruption all the way to the skeletal muscle. The autonomic nerves also have cell bodies in or close to the central nervous system. However, their axons do not travel directly to the target organ, but synapse instead with a second set of neurons which innervates the cells of the internal organ. These postganglionic neurons represent a distinct difference between autonomic nerves and somatic motor nerves.

rhythm, and the rate at which the heart beats increases. On the other hand, impulses over the vagus nerve, the *parasympathetic* nerve to the heart, originate in the heart-inhibiting center in the medulla, and slow down the rate at which the heart beats.

Thus we see that sympathetic and parasympathetic nerves have opposite effects on the tissues they innervate. These nerves offset and balance each other's effects. For example, if one autonomic nerve stimulates the smooth muscle of the gut to contract, the other autonomic nerve inhibits the muscle from contracting. As you might expect each type of autonomic nerve releases a different transmitter chemical. Acetylcholine is released at the synapses of parasympathetic neurons and the tissues which they innervate. **Noradrenalin** (norepinephrine) is the transmitter chemical released at the synapses between sympathetic neurons and the tissues which they innervate.

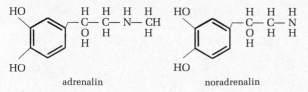

adrenalin   noradrenalin

Noradrenalin is very similar to the hormone released by the adrenal gland sitting atop each kidney. The sympathetic nerves to the adrenal glands stimulate them to release **adrenalin** (epinephrine) in greater amounts than usual whenever an animal is in danger or under stress. Together adrenalin and noradrenalin bring about several changes within the animal. Consider a mouse being chased by a cat. These chemicals cause the blood vessels of the mouse to constrict. As a result much of the blood is rerouted from most of the internal organs and is sent to the brain, heart, and skeletal muscles. Adrenalin and noradrenalin cause the heart to pump blood faster than before and increase the blood pressure. They relax the bronchioles, thus permitting a greater exchange of air in the lungs. They stimulate the liver to release glucose, which is used to refuel the mouse's skeletal muscles as it scampers away. They cause the pupils of its eyes to widen, en-

abling it to see in the dim light under the furniture. And if the mouse is lucky enough to escape with just a scratch, adrenalin and noradrenalin speed up the clotting time of the mouse's blood. In short these two chemicals prepare the mouse for "fight or flight." An animal without its adrenal glands can also get prepared for "fight or flight," but the massive release of adrenalin from the adrenal glands probably helps.

The biggest difference in the workings of autonomic neurons to smooth and cardiac muscles and motor neurons to the skeletal muscles is in how the activity of the muscles is inhibited. You learned in the previous section that the firing of a vertebrate motor neuron, and the subsequent contraction of the skeletal muscle innervated by the neuron, is inhibited *in* the central nervous system, not *out* at the neuromuscular junction. In contrast smooth and cardiac muscles that are innervated by autonomic neurons are inhibited out at the synapses of the neuron with the muscle it innervates.

It is extremely difficult to experiment directly on synapses within the central nervous system, especially in vertebrates where the spinal cord is housed within the bony vertebral column. However, synapses outside the central nervous system, such as those of autonomic neurons with their effectors, are relatively easy for physiologists to work with. In fact the first demonstration that a transmitter chemical is released at nerve endings was done by an Austrian physiologist, Otto Loewi, working with the vagus nerve to the amphibian heart (Fig. 13-16).

Loewi began his experiment by stimulating the vagus nerve to an isolated toad heart that he had placed in a salt solution. The heartbeat slowed. He then applied some of the fluid within this heart to a second isolated heart. The second heart then responded as though its vagus nerve had been stimulated, that is, its rate of beating slowed too. Thus Loewi demonstrated that it is not the stimulation of the vagus nerve itself that inhibits the heartbeat; it is the transmitter chemical released by the stimulated vagus nerve. Loewi called this chemical "Vagusstoff." Later it was identified as acetylcholine. Loewi's experiment is a classic and he was awarded a Nobel Prize.

The autonomic nervous system was so named because we usually have little control over its activities. In addition to regulating the rate at which the heart beats, it controls respiration, excretion, reproduction, the diameter of blood vessels, digestion, secretion by the sweat glands, and the release of hormones from many endocrine glands. These processes go on without our even being aware of them. In fact Claude Bernard, one of the greatest of all physiologists, wrote that "Nature thought it prudent to remove these important phenomena from the caprice of the ignorant will." This removal from the will is incomplete; for example, you can make yourself salivate by thinking of food. The daring physiologist Rudolph Weber stopped his heart in class; after he was revived, he resumed his lecture.

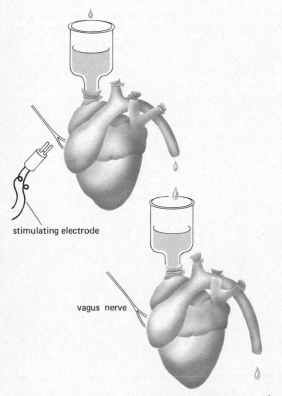

stimulating electrode

vagus nerve

Figure 13-16 Diagram of Loewi's experiment. The vagus nerve (parasympathetic) innervating the upper heart was stimulated, and the heart slowed its beating. When fluid from this heart was applied to the lower heart, its rate of beating also slowed.

It now seems likely that autonomic processes can be learned in much the same way that acts involving motor neurons and skeletal muscles can be learned. Experiments have been done with rats to investigate learning in the autonomic nervous system. Some rats were trained to vary the diameter of their blood vessels in such a way that more blood flows to one ear than to the other! Biologists do not know yet whether man can learn autonomic responses to the degree that rats apparently can. It seems likely that he can. We all know individuals who become genuinely ill when faced with an unpleasant task. It has been suggested that such a person learned to bring on such illness when as a child his mother allowed him to stay home from school when he "developed" a headache. No doubt as we learn more about how the autonomic nervous system functions, we will uncover the roots of many similar psychosomatic illnesses. It also appears likely that many of the "spiritual" exercises of Eastern yogas will turn out to involve autonomic learning.

RECEPTORS AND SENSE ORGANS

Changes constantly occur within our bodies and all around us. Survival depends on our nervous system receiving input about these changes, integrating this input, and acting on it. In man and other multicellular animals the monitors that inform the animal about what is happening in its environment and within itself are called **receptors.** Sometimes receptors are single cells, such as some pain and touch receptors in man. More often receptor cells are collected into groups called **sense organs,** such as the human eye. Receptors are the first step in all reflexes. To understand how the nervous system works and how an animal behaves, we must know what stimuli its receptors detect and how the receptors work. As we shall see, man and other animals have receptors sensitive to a wide variety of physical, chemical, and mechanical stimuli.

Although receptors differ in structure and function to enable them to respond to different kinds of stimuli, all receptors share one common feature: they change the environmental stimulus into a nervous signal. Some receptors are the ends of sensory nerve cells. This is the case with various pain receptors in our skin. Usually receptors are specialized cells, often not nerve cells, that make close contact with sensory neurons. Such receptors include taste receptors of various sorts, for example. Receptor cells are extremely sensitive to one type of stimulus, such as heat, light, sound, chemicals, touch, and pressure. A stimulus causes changes in the membrane of a receptor cell. These changes in turn trigger nerve impulses in the sensory neuron.

Some of our receptors have reached the limits of sensitivity. For example, if our ears were more sensitive than they are, we could hear the random movements of molecules. Some receptor cells in our eyes are excited by a single photon—the smallest "packets" of light—in the blue-green part of the visible spectrum. As insect collectors know, some male moths have such a sensitive sense of smell that they can detect a female moth of the same species several miles away. Pit vipers have temperature receptors so sensitive that these snakes can detect the temperature change caused by a passing shadow—a change of about 0.003°C. These receptors line a pit located in front of each eye. In addition to detecting warm-blooded animals, these pits can detect temperature changes of the air resulting from the evaporation of water from the skin of a frog nearby. Thus pit vipers can locate both warm-blooded and cold-blooded prey.

A receptor cell usually responds only to the stimulus for which it is specialized. However, another type of stimulus may, if it is very strong, excite the receptor cell. For example, light receptors usually respond only to certain wavelengths of light. But if you receive a sock in the eye, the blow may excite light receptors; in this case you will "see stars."

We usually speak of man's five senses: vision, hearing, touch, smell, and taste. Actually we have many more, such as warmth, cold, pain, and pressure. In addition we have all sorts of internal senses of which we are not conscious. Stretch receptors buried in our tendons and our muscles are important in reflexes like the knee-jerk and in maintaining posture. These receptors provide the central nervous system with information about the movements and position of arms, legs, and other parts of the body. We also have receptors which respond to osmotic pressure, and to the glucose, oxygen, and carbon dioxide concentration of the blood. Sometimes when these receptors are stimulated we may become conscious of changes in the body. For example, when the glucose level of the blood gets too low, we become hungry.

Many animals may not respond to the same stimuli which excite our receptors. For example, we respond to the color red; bees cannot see red, and may ignore red flowers. On the other hand, bees see ultraviolet patterns in flowers; we cannot. Bats hear sounds pitched up to 150,000 vibrations per second; the upper limit of our ears is about 20,000 vibrations per second. Many animals can detect smells far too faint for our smell receptors. We have no receptor cells sensitive to some types of environmental changes. For instance, we do not receive any information about radio waves or radioactivity. We can, however, build radios and geiger counters which detect radio waves and radioactivity, and then convert these stimuli into sounds we can hear. Radios and geiger counters are extensions of our senses; they are man-made receptors sensitive to environmental stimuli that we normally ignore.

Receptor cells have survival value, and animals generally have evolved receptors that respond to changes in the particular environment they inhabit. For example, animals living in ocean depths where there is no light rely on receptors sensitive to chemicals in the seawater. Such *chemoreceptors* enable them to find their food and their mates, and to avoid their enemies. As another example consider the species of moths which have evolved tiny "ears." So perfectly attuned are these ears to the ultrasonic cries of bats that prey upon the moths that a moth can escape when it "hears" a bat (Fig. 13-17). If there are other animals in the far reaches of the universe, they probably evolved and live under conditions vastly different from those on our planet. As a result the receptors that they have evolved probably respond to stimuli that these animals encounter in their own environment. Conceivably the ability to detect radioactivity might be an advantage to animals on certain kinds of planets, and such animals might have receptors sensitive to radioactivity.

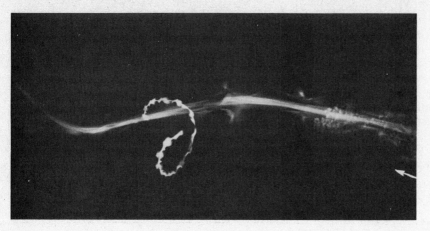

Figure 13-17 A moth (short streak in middle of photograph) eludes a bat (streak across photograph) by flying up just as it seemed about to be captured by the bat. The moth's "ears" enable it to sense the bat's sounds, thus giving the moth a better chance to escape.

Decoding the sensory message

These various receptors send input about environmental stimuli into the spinal cord and brain. Information about touch comes into the central nervous system along sensory neurons which contact touch receptors. Information about stretch in a particular muscle comes into the central nervous system along sensory neurons which contact the stretch receptors in that muscle (Fig. 13-18). Thus each receptor has a private telephone line— its sensory neuron—to the central nervous system. Nerve impulses along that particular sensory neuron inform the central nervous system that a certain kind of environmental stimulus, touch or stretch, for example, has stimulated a particular receptor. However, in order to respond appropriately to a stimulus, the central nervous system needs more information than the mere fact that the stimulus was present. It also needs information about how strong a stimulus is. In other words, is the light bright or dim? Is the touch a caress or a right hook? The strength of a stimulus received by a receptor is communicated to the spinal cord and to the brain in three ways. As the strength of a stimulus increases there is (1) an increase in the number of receptor cells which are excited, (2) an increase in the frequency at which nerve impulses are conducted along sensory neurons to the spinal cord, and (3) a change in the pattern of grouping of impulses.

A great deal of what we know about how sensory nerves communicate the intensity of a stimulus came from some experiments on muscle

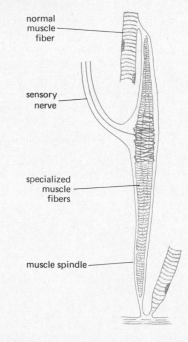

normal muscle fiber

sensory nerve

specialized muscle fibers

muscle spindle

Figure 13-18 Stretch receptor in a skeletal muscle. Muscles may possess hundreds of stretch receptors called muscle spindles, each of which is a bundle of specialized muscle fibers. Wrapped around the middle of the bundle are endings of a sensory nerve; in this region the muscle fibers have few striations, and contract little, if at all.

When the muscle is stretched, the sensory nerve endings in the spindle are stimulated and send impulses to the spinal cord. These impulses activate the motor neurons that serve the muscle and cause the muscle to contract.

spindles by Edgar Adrian and Yngve Zotterman. Within skeletal muscles there are muscle cells which are wrapped by the endings of sensory neurons; these modified muscle cells are stretch receptors called muscle spindles. When the muscle is stretched, the muscle spindles are stimulated, and impulses are conducted by the sensory neuron to the central nervous system. Adrian and Zotterman isolated a small piece of frog muscle which contained only a single muscle spindle. When they stretched the piece of muscle by a small amount, the sensory neuron conducted a barrage of impulses; when they stretched the muscle further, the frequency of impulse conduction increased. These impulses, like all nerve impulses conducted by a particular neuron, were of "one size," but the frequency at which they were conducted depended on the strength of the stimulus. This experiment with muscle spindles serves as a model of how receptors function generally. When a sense organ receives an intense stimulus, many of its receptor cells become excited, and they generate nerve impulses in sensory neurons at a high frequency. In addition groups of impulses are usually bunched more closely together.

Sensations

All the messages which receptors send to the central nervous system are in the form of nerve impulses. How, then, do we "see" a flower or "smell" its fragrance? How do we know where a stimulus, for example, a sound, is coming from? Seeing, smelling, hearing, tasting, feeling—these are *sensations,* and *they exist solely in our brain.* Sensations are not a part of the stimulus itself or even the receptor cells which the stimulus excites. Nerve impulses from the receptors must reach a particular part of the brain if we are to experience sensations. Visual receptors send sensory impulses to a specific part of the brain. When an impulse arrives at that part of the brain, you experience a visual sensation. Warmth receptors also send sensory impulses to a specific part of the brain, and you experience a sensation of warmth whenever impulses from those receptors arrive at that part of the brain. If you were to electrically stimulate that part of your brain,

you would experience warmth. A sensation is your brain's interpretation of certain incoming stimuli.

As already noted, not all sensory messages produce sensations of which we are conscious. For example, we normally are unaware of the minute-to-minute changes in our blood pressure, yet special receptors, like sentinels, keep the medulla of the brain informed of blood pressure changes. However, these events in the medulla do not reach consciousness. Nonetheless our bodies respond to these stimuli even though we are not conscious of them. Clearly, responding to a stimulus does not require consciousness. Now you see why we cannot say that an insect or an earthworm "feels" pain. If you poke such an animal with a sharp needle, it responds, and its actions may be those that we associate with pain. But the brains of these animals do not appear to have any areas that would give rise to conscious sensations such as pain. Let us now see how the brain interprets what the stimulus is and where it comes from.

Nature of the stimulus.   Nerve pathways linking the receptors with the brain end at a specific place in the brain. It is the brain that decides *what* the stimulus is. For example, the optic nerves from the eyes conduct impulses to regions of the brain involved in vision (often called the "visual center") in the back part of the cerebral cortex (Fig. 13-19). No matter how the visual center is stimulated—by light exciting the receptors in the eye or by that sock in the eye—the brain interprets

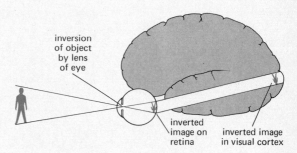

Figure 13-19 Visual center of the brain. The lens of the eye acts like a camera lens so that the retina receives an inverted image. The receptor cells in the retina transmit this inverted image to the visual center by the optic nerves. We use many other cues such as touch and gravity to interpret the image as right-side-up.

the messages as light. If nerves from the ear could be led to the visual center instead of to the hearing center (and vice versa), one could "hear lightning and see thunder."

Localization of the stimulus.   The brain also interprets the origin of a stimulus by where in the brain the nerve pathway from receptors end. A good example of this is the "phantom limb phenomenon." Persons who have lost an arm or a leg may actually complain of sensation in the missing limb. Apparently many of these "phantom limb sensations" are due to the irritation of some of the remaining stumps of nerves which originally innervated the missing limb. These nerves send messages to the brain which interprets them as real sensory events originating in the limb. In other cases the brain *learns* to locate a stimulus. An example is the way the brain learns to locate a sound. Sounds are not interpreted as occurring within the ear but rather outside the body.

Interpreting groups and patterns of stimuli.   At any moment we are on the receiving end of not just a single stimulus but of patterns composed of hundreds of stimuli. How we interpret these patterns of stimuli depends largely on our past experience and what we have learned to expect from similar patterns of stimuli. For example, if you were to see an unfamiliar object, such as a dinosaur, some distance away, you could easily estimate its size because you know the size of familiar objects around the dinosaur. Furthermore, when you see only part of an object, your mind "fills in" the rest of the pattern, again from past experiences. We are in a very real sense always guessing about what is happening in the external environment, and our guesses are based on our past experiences. We shall have more to say about this in Chapter 14 when we discuss learning.

How receptors work

Nerve cells either fire or they do not; that is, they respond in an "all-or-none" way. Receptor cells behave differently. As the stimulus gets stronger a receptor becomes more and more excited; that is, it responds in a "graded" way. Physiologists have a special name, the **generator potential,** for

this graded response. It is so called because once it gets strong enough, it generates nerve impulses in the sensory neuron which links the receptor with the central nervous system. As the generator potential gets larger and larger it generates more frequent impulses in the sensory neuron.

How the receptor generates a nerve impulse is a key question, and one to which we do not know the answer. Experiments indicate that the excited receptor somehow alters the permeability of the membrane of the sensory neuron to sodium ions. For example, light causes chemical changes in the visual pigments of light receptors in the eye. These chemical changes may trigger the release of an excitatory transmitter chemical which, like all excitatory transmitter chemicals, changes the permeability of the membrane to sodium ions.

Adaptation

We do not react to all of the stimuli which affect our receptors. A great deal of information never reaches the central nervous system because many receptor cells *adapt.* That is, they are not excited by a constant stimulus. For example, when you first put on your clothes you are very much aware of them, but you quickly become accustomed to them and soon forget them. One stimulus to which the receptors never adapt is pain. Pain, of course, is nature's way of telling us that something is wrong with our body. It is a means of helping an animal correct a situation, thus increasing the animal's chances of survival. A torn ligament causes pain and the animal favors the injured limb. If adaptation to pain occurred, pain's purpose would be defeated.

Recently there has been increasing interest in blocking out pain by means of **acupuncture.** In this procedure, needles inserted into one part of the body appear to produce analgesia (loss of sensation of pain) in another part of the body. How acupuncture works is not at all clear. Indeed, pain is one of the least understood of all sensations.

Major sense organs of man

Now that we have discussed the basic way in which receptors work, we are prepared to understand how some important human receptors and sense organs work. Figure 13-20 shows several receptor cells and sense organs and describes how they convert environmental stimuli into nerve impulses.

THE CENTRAL NERVOUS SYSTEM—THE BRAIN

The human brain is a dense jungle of 10 billion nerve cells and ten times that number of glial cells. A single cubic inch may contain 100 million cells, each of which may connect to as many as 50,000 others. It is the most complex single structure in the known universe. Analyzing its structure and function seems almost impossible. Yet the rewards are immense. For within the brain lie reason, memory, creative talent, and consciousness itself—all the things that can make human life worthwhile. What makes the human brain even more challenging is that each human brain is different from every other human brain. One human kidney works pretty much like every other human kidney, but the brain of a man is the man himself.

One of the great surprises of modern neurobiology is that notwithstanding its complexity, the human brain can be analyzed. This section describes briefly the basic way the human brain is put together and how we believe it works. The cells of the brain, like those of the spinal cord, are organized as gray matter—neurons and glial cells—and white matter—bundles of axons organized into fiber tracts. The brain as a whole can be divided into three major regions, which for convenience we consider separately: the **hindbrain,** the **midbrain,** and the **forebrain** (Fig. 13-21). Do not get the idea that specific functions are precisely localized in each of these regions. In fact we shall find that most processes require several different parts of the brain, all working together. In most activities the brain functions as a unit with many parts of the brain participating.

Another feature of the brain that we must keep in mind is its flexibility or plasticity in terms of where various functions are performed by the brain. Normally movement of the left foot is controlled by a specific region of the brain. However, if that region is damaged by injury or disease, another part of the brain takes over that function. This plasticity makes the brain immensely adaptable.

The brain communicates with the body by two main routes. The major route is the spinal cord which extends from the hindbrain to the last vertebra of the backbone. Its 31 pairs of spinal nerves provide inflow and outflow paths for the brain and part of the body below the neck. The

Figure 13-20 (shown on pp. 422-423) Some major sensory structures of man. A. **The eye,** a sphere about one inch in diameter, is fluid-filled. Its shape depends in part on a fluid called the vitreous humor. The lens and the cornea are bathed by another fluid, the aqueous humor. Light enters the eye first through the transparent cornea and then through the pupil. The amount of light entering the pupil is controlled by the smooth muscles of the iris, the colored ring surrounding the pupil. Behind the pupil is the lens, which focuses light on the light-sensitive retina. Ciliary muscles change the shape of the lens, thus making focusing possible. The retina acts like a film which is sensitive to both color and black-and-white.

There are at least three kinds of cells in the retina (right)—photoreceptor cells, bipolar neurons, and ganglion cells. The photoreceptors in the outer layer of the retina (nearest the back of the eyeball) synapse with the bipolar cells, which are interneurons. These synapse in turn with the ganglion cells which are in the innermost layer. Light striking the photoreceptors sets up nerve impulses in bipolar cells and ganglion cells. Axons of ganglion cells join to form the optic nerve, composed of a million nerve fibers carrying impulses to the brain. The visual center then "develops" the picture taken by the eyes. Because there are interneurons and several sets of synapses in the retina, the eye can modify the information during transmission from the receptor cells to the brain. It is thought that the interneurons in the retina sort out and sharpen raw sensory data before transmitting it to the brain.

In each retina there are about 125 million photoreceptors, called rods and cones because of their shapes. Rods are more light-sensitive than cones, and make vision in dim light possible. However, they do not permit sharp vision. For this, cones are necessary. Animals known for their visual acuity (eagles, for example) may have six times as many cones per square millimeter of retinal surface as man. Cones are also responsible for color vision. Exactly how we see color is not very well understood; there is some evidence, however, that there are three different kinds of cones, each of which is excited by certain wavelengths of light. Again it is the brain that "perceives," in this case, the color of an

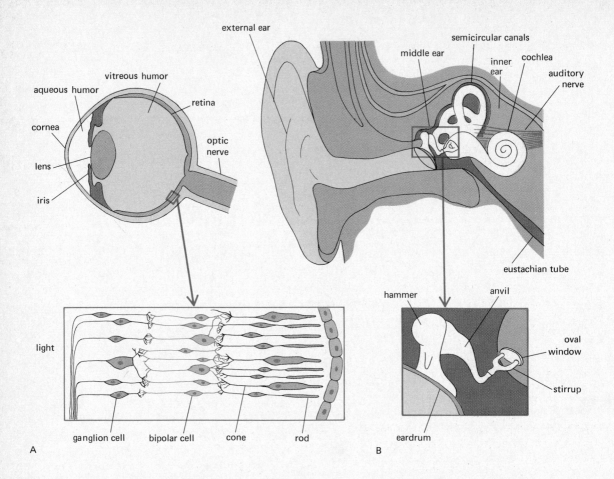

A

B

object. Some nocturnal animals such as rats, bats, and toads have only rods and can see in dim light, but without color. In contrast, some animals such as some reptiles that are active in the daylight hours have only cones.

B. **The ear.** Sound waves entering the external ear cause the eardrum (tympanic membrane) to vibrate. These vibrations, in turn, set up vibrations in the three tiny bones (the hammer, anvil, and stirrup) of the middle ear. Movement of one of these bones, the stirrup, exerts pressure on a membrane called the oval window. The oval window, in turn, transmits pressure to the fluid in a snail-shaped spiral called the cochlea. Receptor cells in the cochlea in the inner ear, excited by the pressure changes in the fluid, stimulate the auditory nerve which carries nerve impulses to the auditory centers of the brain.

The normal human ear responds to frequencies between 16 and 20,000 vibrations per second. Children hear even higher frequencies (up to 40,000 vibrations per second) but as we get older, the eardrum loses some of its elasticity and does not respond as readily to high frequencies. The ear is so sensitive that the eardrum will respond to some sounds by vibrating no more than one billionth of a centimeter—less than the diameter of a hydrogen atom. And a membrane in the inner ear is 100 times more sensitive than this!

The inner ear contains not only receptor cells sensitive to sound but also cells that respond to changes in the orientation of the head. These cells are located in the semicircular canals.

The eustachian tube connects the pharynx or throat to the middle ear. Connecting the middle ear to the outside makes it possible to have the same pressure on both sides of the eardrum. When you dive under water, the outside pressure increases. To equalize the pressure, you swallow to open up the eustachian tube and permit the passage of air to the middle ear.

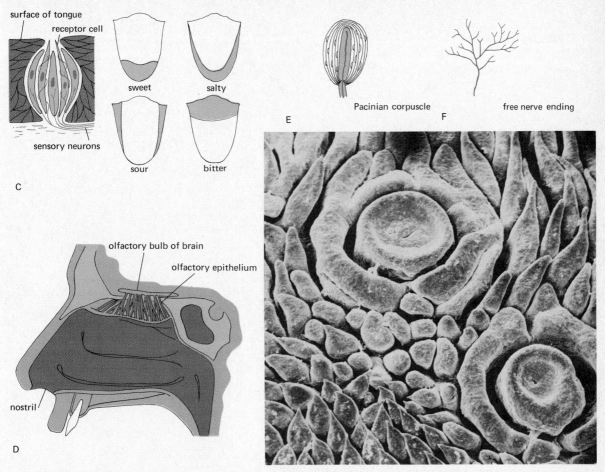

Figure 13-20   C. Taste receptors are chemoreceptors. They are not neurons but specialized cells with hairlike processes at their ends. Sensory neurons lie close to them. When the receptor cells are stimulated, they generate nerve impulses in the sensory neurons. These receptor cells, grouped into structures of the tongue called taste buds, respond to four different tastes. Those on the tip detect sweetness and saltiness; those on the sides detect sourness, and those in back detect bitterness. Most of what we "taste" is actually what we smell, as anyone with a cold knows. The scanning electron micrograph of part of the tongue of a dog shows a taste receptor called a circumvallate papilla.
D. **Olfactory receptors.** In the upper nasal passages is the olfactory epithelium, a small area with slightly more than half a million olfactory cells. Certain molecules in the air flowing through the nasal passages are dissolved in the mucus covering the olfactory epithelium, and in some unknown way excite the olfactory cells which transmit impulses to a region of the brain called the olfactory bulb. Although man's sense of smell is very weak compared to that of many animals, he is still able to recognize about 10,000 different kinds of molecules.
E. **Mechanoreceptors.** Many of the receptor cells in the skin respond to mechanical stimuli. For example, the onion-shaped Pacinian corpuscle is stimulated by pressure changes. When the Pacinian corpuscle is excited, it sets up nerve impulses in the axon of a sensory neuron (color). Other mechanoreceptors have different structures but they all help to make man's skin—especially his lips and fingertips—exquisitely sensitive.
F. **Pain receptors** in the skin are neurons with many free nerve endings.
Many animals have one or more senses which are far keener than man's. For example, dogs have a far better sense of smell than man, and birds of prey have better vision. But man's superior brain enables him to interpret information coming from his sense organs, limited though they may be, in a way that no other animal can.

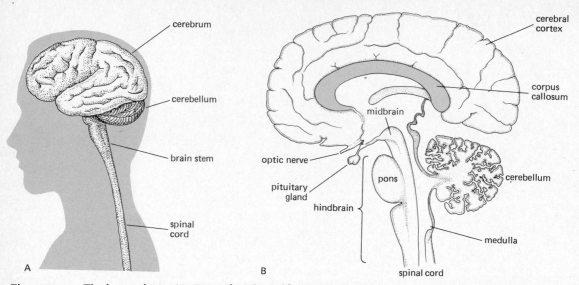

Figure 13-21   The human brain. (A) External surface of brain. (B) Longitudinal section through the midline of brain, showing the inside of the right hemisphere.

brain also communicates directly with the organs and tissues of the head and neck by 12 pairs of cranial nerves. One of these cranial nerves, the vagus nerve mentioned earlier, also innervates several organs below the neck including the heart and the lungs. Some cranial nerves, like spinal nerves, are made up of both sensory and motor neurons. Other cranial nerves are either sensory or motor.

Hindbrain—The medulla

The lower part of the hindbrain is the **medulla.** We have already learned that the medulla controls respiration and heartbeat. In addition it is the region through which nerve impulses must pass in both directions between the spinal cord and the rest of the brain.

Running through the medulla and on into the midbrain and part of the forebrain is a complex organization of nerve cells called the **reticular formation.** This alerts and arouses other regions of the brain, especially the **cerebral cortex** (the covering of the forebrain) and **thalamus,** which we shall discuss. For example, if someone knocks on your door, sensory nerve impulses set up by the knocking reach the cortex of your brain. How-

ever, you will not react to the knock unless the reticular formation is excited and unless it simultaneously excites the cortex. Fortunately for us the reticular formation is quite choosy and does not make the cortex aware of every stimulus. If the reticular formation made us conscious of every stimulus which is bombarding our body and we tried to respond to them, we would be overwhelmed. The reticular formation is very efficient; for example, all of us at some time have become so engrossed in a book or a television program that all movements and noises around us were completely blocked out. A mother often awakens from a sound sleep at the slightest whimper from her baby, yet the infant's father may sleep blissfully through a tantrum. Both of these examples of *selective attention* illustrate the reticular formation at work.

If a person's reticular formation is destroyed, his cerebral cortex can no longer be aroused, and he becomes permanently unconscious. General anesthetics, such as ether, may work by shutting off nerve impulses in the reticular formation. Exactly how sleep-inducing drugs such as barbiturates cause sleep is not known; they probably interrupt the flow of nerve impulses from the reticular formation to the cortex.

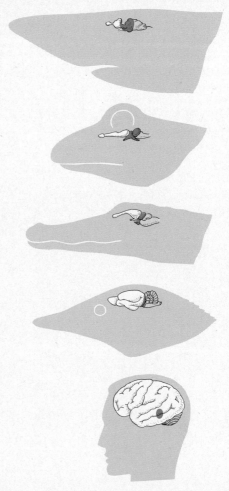

Figure 13-22 The reticular formation is a network of neurons that runs through the medulla and on into parts of the forebrain, including the thalamus and cortex. Since it is a network and not a discrete structure, it is represented by arrows in the diagram.

Hindbrain—The cerebellum

Lying above the medulla is the baseball-sized **cerebellum,** the covering of which is even more wrinkled than the cerebral cortex. The cerebellum seems to be larger in relationship to the rest of the brain in animals that move in a three-dimensional world (fish and birds) than in animals that live mostly on the surface of the earth. Surprisingly, perhaps, man's cerebellum is quite large relative to the rest of his brain even though he is biologically a surface creature (Fig. 13-23). His large cerebellum makes sense when we learn that the cerebellum coordinates skeletal muscle movement. Certainly there is no other animal which has as large a repertoire of voluntary movements as man.

However, the cerebellum does not initiate movements. We can demonstrate this by removing the cerebellum from a dog. The animal can still move, but it has no muscle tone and its movements are uncoordinated. A man whose cerebellum is damaged can move, but he may reel and stagger as though he were drunk. Persons with cerebellar damage have in fact been arrested for drunkenness. It is the cerebral cortex which initiates movements. We might say, then, that it is the cerebral cortex which makes the policy but the cerebellum which sees that the policy is carried out efficiently.

Figure 13-23 Relative size of major regions of the brain in fish, frog, alligator, shrew (a primitive mammal), and man. Notice that in the series from fish to man, the relative size of the midbrain (brown) decreases whereas the forebrain (white) expands enormously.

The events that go on within the cerebellum illustrate especially well how a group of neurons within the brain works (Fig. 13-24). Basically these groups of neurons do three things.

1. They receive *input* in the form of nerve impulses from receptors and from other parts of the nervous system.
2. They *sort out* all this input. This is the real decision-making process. It is the summing up

of all the things going on at the synapses within the cerebellum. Some input impulses from the environment are ignored because of other input impulses from the cerebral cortex. Other input impulses may be amplified for the same reason. We at last begin to understand why there are so many cells in the brain: it receives so many messages that one of its biggest jobs is "talking to itself"—sorting out these messages and deciding what is important.

3. On the basis of this sorting out and talking to itself the cerebellum sends *output* in the form of nerve impulses to other parts of the brain, and may send nerve impulses by way of the spinal cord and efferent neurons to effectors.

For example, suppose you wish to turn this page. Out go commands from the cerebral cortex to the skeletal muscles involved in the act; at the same time the cortex tells the cerebellum that it should expect certain sensory input from stretch receptors in the hand and arm, and from touch receptors in the fingers and eyes. In other words, the cerebral cortex tells the cerebellum what these muscles are supposed to be doing. Receptor cells in the fingertips, muscles of the hand and arm, and eyes send nerve impulses to the cerebellum and keep it informed of what is happening. Should your fingers miss the edge of the page the cerebellum would send impulses to the cortex to

the effect that the original commands must be changed to correct for the error. In other words, the cerebellum compares what *is* happening with what *should be* happening. In this way it coordinates the movement.

Midbrain

If you compare the brain of a fish and a man, you will see that the fish's midbrain is relatively much larger than man's. Furthermore, if you also compare the brain of a frog, an alligator, and a shrew, you will find that the midbrain becomes increasingly smaller with respect to the rest of the brain as the animal becomes more complex, and that the forebrain becomes correspondingly larger (Fig. 13-23). What you have observed reflects what happened during the long evolution of vertebrates. The midbrain became smaller and less important in the higher vertebrates than in fish or frogs; it was in effect sacrificed to permit the huge forebrain of mammals to develop. However, even in man the midbrain is still important. It is not only a connecting link between the hind- and forebrain but it also receives some impulses from light receptors in the eyes, and is primarily responsible for control over the movements of the eye muscles. But most of the functions (primarily sensory) which the midbrain performed in our vertebrate ancestors have been taken over by our forebrain.

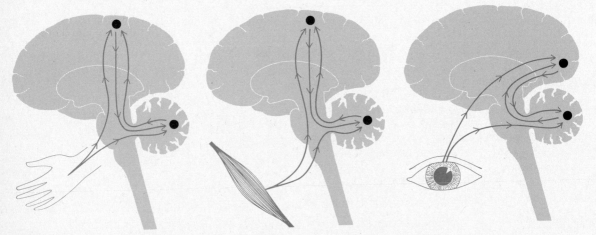

Figure 13-24 Some sources of input to the cerebellum. If one turns a page, for example, input enters the cerebellum from the cerebral cortex, the fingers, muscles of the hand and arm, and the eyes.

Forebrain

Man has the largest forebrain of any animal, almost exclusively caused by the enormous growth of the cerebral cortex. It is the cortex that really makes us what we are. Because the forebrain is so fascinating to us, many different experiments have been done on man and other vertebrates to determine what its different areas do. In addition we have learned a great deal about it simply by observing what happens when parts of it are affected by a tumor or are injured.

The gray matter covering the cerebral cortex is made up of the cell bodies of neurons. In fact most of the nerve cells in the brain are in the cortex. And yet this covering is only about 2 millimeters thick. It is, however, very wrinkled or convoluted. These convolutions provide for an increase in the surface area of the forebrain without requiring a larger skull. (Man has the most extensive cortex of all animals. Birds and reptiles have little more than a hint of a cortex, and fish and amphibians have none.) When looking at the cerebral cortex, it appears to be split down the middle into a right and a left hemisphere. The two halves are actually connected by a tract of myelinated axons from one hemisphere crossing to the other hemisphere. As we shall see when we discuss memory, this axon tract is a freeway over which impulses move between the two hemispheres. Lying at the base of the cerebrum and above the midbrain is the *thalamus*, and beneath it the **hypothalamus.** Wedged between these areas and the cerebral cortex is the **limbic system.**

Forebrain—Hypothalamus. We have already learned that the hypothalamus controls eating and body temperature and produces the antidiuretic hormone. By electrically stimulating various parts of the hypothalamus in experimental animals we can show that it is also involved in regulating thirst, pleasure, pain, and sexual and reproductive activity. In fact, all the nervous mechanisms involved in holding the internal environment constant and in determining the use that is made of the resources of the body are in some way related to the hypothalamus. Yet this critically important area constitutes less than 1 percent of the human brain.

The effects of certain chemicals and drugs on the hypothalamus have also been explored. Some fascinating results were obtained in an experiment which has become known as "the case of the mixed-up rat." A male sex hormone, testosterone, was injected into a certain area of a male rat's hypothalamus. Instead of becoming sexually aggressive as one might have expected, he acted like a new mother, and set about building a nest for some newborn rat pups which were placed in his cage. Normally he would eat them. However, later injections of the same hormone at a nearby site in the hypothalamus did result in the expected male sexual behavior. Clearly there are discrete areas in the hypothalamus, each of which regulates some aspects of the animal's behavior.

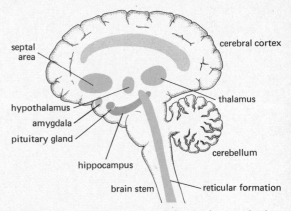

Figure 13-25 Various parts of the forebrain. The hypothalamus lies at the base of the cerebral hemispheres close to the pituitary gland. It is of special importance in regulating the internal environment and processes such as eating, drinking, and sleeping. It interconnects with many regions of the brain including the limbic system. The limbic system is an integrated network that includes the hippocampus, the septal area, the amygdala, parts of the reticular formation, and the hypothalamus itself. These structures seem to be important in aspects of behavior such as emotion and motivation. For instance, the amygdala become active whenever we encounter anything new or unexpected, whereas the septal area depresses emotional reactions. The thalamus is located just above the midbrain. It relays sensory information to the cerebral cortex from specific sensory pathways. It also plays a role in processes such as sleep and wakefulness, and closely interacts with the limbic system.

We examine this in more detail in discussing motivation.

Forebrain — Thalamus.   The thalamus, a pair of oval bodies, is the region of the brain through which most sensory nerve impulses finally travel on their way to the cortex. Impulses are integrated in the thalamus, and then relayed to other areas in the cortex. The thalamus is very much like a giant relay through which the body and the rest of the brain are able to reach the cerebral cortex. With the single exception of smell, all sensory systems funnel through it.

Forebrain — The limbic system.   It is only recently that biologists have begun to understand what functions the limbic system may serve (Fig. 13-25). At first it was thought to be the brain center for smell, but we now know it has a far more important role in mammalian behavior. It is here that emotion is organized. If, for example, an electric current stimulates a certain part of the limbic system of your affectionate house cat, she will suddenly go into a fit of rage — clawing, spitting, and yowling. She will have nothing to do with you. Stimulate nearby cells in the limbic system, and she becomes affectionate once more, purring and rubbing against you.

Forebrain — The hippocampus.   Another section of the limbic system has long puzzled brain scientists because it had no known functions. It is not tied to sensory inputs or to motor systems and it apparently plays no role in emotion. However, in 1958 a possible function for the hippocampus was discovered in a dramatic and unexpected fashion. The first case of a human being without a hippocampus was found that year and psychological studies revealed that this lack produced only one clear-cut effect: the man was unable to transfer information about his surroundings into long-term memory (see Chapter 14). He could, for example, read a magazine and then several hours later read the same magazine as though he had never seen it before.

Cerebral cortex.   What happens if you surgically remove a frog's tiny forebrain? You may be surprised to find that your frog acts almost like any normal frog: it can jump, swim, and see. If you repeat your experiment with a rat, it can do many things that a normal rat does — even walk about

and drink — but it moves sluggishly. However, removing the cerebrum from a monkey leaves the animal helpless, and it will soon die. The cortex of a newborn baby is not functional, but those tragic infants which are born without a cortex, *anencephalic babies,* never live very long.

We can get some idea of how a decorticated man (the equivalent of one without a cerebral cortex) would act simply by observing a man after he has had several drinks. An intoxicated person is in effect a partly decorticated animal. How does alcohol affect the brain? When more alcohol is consumed than the liver can handle, the excess goes to the brain, the direct target of alcohol. Alcohol *depresses — it does not stimulate —* the activity of the brain. The first part to be affected is the reticular formation; as a result a stimulus such as the appearance of a "stop sign" will take an unusually long time to be properly registered by the cortex. If still more alcohol is drunk, the function of the cortex itself is affected; one becomes emotional, giddy, and his tongue trips up. He is emotional because the limbic system centers that control emotion are no longer held in check by the cortex; his speech is slurred and his steps are unsteady, because the alcohol depresses the areas of the cortex that control these functions. Thus one role of the cortex is to regulate the rest of the brain. It stops us from eating until dinner, it allows us to suppress our temper, and organize behavior so that we can carry on social life. But even this monumental task is not the totality of its function as you see below.

Mapping the cortex.   Much has been learned about the cortex by electrically stimulating or recording brain activity in different parts of it

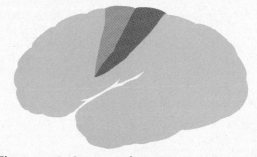

Figure 13-26   Sensory and motor areas of the cerebral cortex. Sensory cortex is darker.

during some types of brain surgery. The patient receives only a local anesthetic; he is conscious and can cooperate with his doctors. (The brain itself has no pain receptors.) As a result of such experiments it is possible to map the **sensory** and **motor areas** of the cortex. These studies reveal that there are parts of the cortex which will respond only to *one* kind of sensory input. For example, a cell in the so-called visual area of the cortex will respond only to input from the optic nerve but not to input from the auditory nerve. A cell in the auditory cortex will respond only to input from the auditory nerves. You can shine a light in the patient's eye, but cells in the auditory cortex will not respond.

A part of the cortex is also associated with bodily sensations such as touch. There is a spot in this somatic sensory cortex, as it is called, for each part of the body. Just as you can find Chicago at a certain place on a map of the United States, so you can find a spot on the somatic sensory cortex that receives sensory impulses from touch receptors in the left big toe. If we were to draw a picture of the different parts of the body as they are represented on the map of the sensory cortex, we would see a grotesque, dismembered little man (Fig. 13-27). He is so oddly shaped because the amount of cortical surface related to a specific body part such as the lips is proportional to the number of receptors the part of the body has, not to the actual size of the part. For example, a man's lips have many touch receptors which send messages to a large area of the sensory cortex, whereas a man's back has only a few touch receptors which send messages to a small area of the sensory cortex.

The motor cortex is also arranged in a similar way. Its specific parts control the movement of specific body parts such as the hand and the thumb. Man's hands take up a larger area of the motor cortex than his back, because he can move his hands so much more than his back.

Association areas of the cortex.   Three-fourths of man's cortex is neither motor nor does it respond

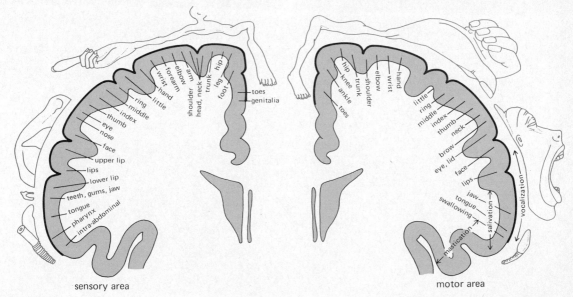

sensory area   motor area

Figure 13-27   Map of the surface of part of the human cerebral cortex indicating areas associated with movement (motor cortex) and bodily sensations (somatic sensory cortex). Stimulation of a specific area in the motor cortex causes a movement in the corresponding part of the body. Stimulation of various parts of the body evokes electrical activity in corresponding parts of the somatic sensory cortex. Notice that the area of cortex devoted to each part of the body is proportional to the extent of motor activities and the sensory capabilities of that part, not to its size. In man the mouth and the thumb occupy a huge area of the cortex. As you might expect when you map the brain of a pig, the snout occupies a large area of the cortex whereas in a spider monkey, the tail occupies a large area of the cortex.

to a specific sensory input as do the visual and somatic sensory cortex. These areas of the cortex have been called **association areas.** Presumably they are the places where different kinds of inputs are integrated. Evidence comes from the discovery that some individual cells in the association cortex will respond to three kinds of sensory input: touch, sound, and light. In this respect the cells in these association areas behave differently from individual cells in the sensory cortex which respond only to one kind of sensory input.

There are also individual cells in the association area of the cortex that appear to "count." If you present a series of stimuli to an anaesthetized cat (sounds to the ear, light flashes to the eye, mild electric shock to a paw), certain individual association cells respond by discharging after a specific number of stimuli. For instance, a "number 7" cell will respond after 7 stimuli, but not after 5 stimuli. The response of the "counting" cell has little to do with the intensity or rate of presentation of the stimuli. Nor does it matter whether the stimulus is sound, light or electric shock. These cells seem to be responding to the number of stimuli presented. About one percent of the cells in the association areas of a cat appear to be "counting" cells.

Other association cells respond to no sensory stimuli. What they are doing no one knows, but presumably they are involved in putting together and evaluating the complex stimuli which the animal receives.

The more complex the brain of a mammal, the larger is the association area. The association area of man is the most complex of all. Most psychobiologists believe that the association area of the cortex is the part of the brain which enables us to interpret, remember, learn, and reason. How these particular kinds of behavior take place is largely unknown. A few of the things we do know about them are considered when we discuss behavior.

DRUGS AND THE BRAIN — A TWO-EDGED SWORD

The more complex a structure is, the more opportunities there are for things to go wrong. The brain is no exception. We cannot discuss the hundreds of things that go wrong with the brain because, tragically, so little is known about its function. In many cases the symptoms of mental illness and brain disease are all too well known, but the physiological basis of these symptoms is poorly understood, if it is known at all.

There is hope, however, that certain drugs not only will buy us a little time until research shows us why and how things go wrong in the brain, but also they will give us clues as to how the brain works. Certain drugs frequently help relieve the distress and suffering which accompany mental illness and brain damage. Several years ago mental patients received electric shock or, even worse, part of the association area of their brain was removed (prefrontal lobotomy). These drastic treatments usually achieved only one positive result: quieting a difficult or violent patient, and making it easier for overworked staffs of mental institutions to care for him. Often the patient's personality changed and his mental ability decreased. Now drugs such as *reserpine* can quiet upset patients and help them to lead more productive lives. The symptoms of Parkinson's disease in which certain cells that work with the cerebellum in coordinating movements are damaged can now be relieved temporarily by drugs such as L-DOPA (dihydroxyphenylalanine). Scientists today are at work trying to develop drugs to help the slow learner.

However, other drugs which affect the brain have kicked up a storm of controversy. We have already seen how one of these drugs, alcohol, affects the brain. What do drugs such as **LSD** do to it? Some evidence suggests that LSD (lysergic acid diethylamide) blocks the activity of certain nerve cells which form serotonin, a transmitter substance in the brain. On the other hand, other evidence indicates that LSD can "fool" nerve cells into accepting it in the place of serotonin. Certainly the chemical structure of LSD and several other **"hallucinogens"** or "psychodelics" — mescaline, peyote, psilocybin, psilocin — is very similar to that of serotonin. However, LSD cannot transmit messages in the same way that serotonin can; as a result the message becomes "garbled" as it crosses the synapse between two nerve cells. This may explain the wildly exaggerated sensations that one experiences on a "trip." In addition LSD may overstimulate the reticular formation, which in turn alerts the cortex more than it usually does; this may account for the "heightened awareness" that LSD users often report.

We do not know exactly how **marihuana** affects the brain. So much seems to depend on the purity of the drug and the person using it. Many users report feelings of well-being and elation—a "high." Marihuana may, like alcohol, do this by depressing the activity of the cortex which nor-

mally holds a tight rein on the emotional centers of the brain. There is good evidence that it interferes with skills such as driving a car.

The question that concerns us here is what dangers do marihuana and the hallucinogens pose to the brain and the rest of the body? Unfortunately we do not know the answer to this question. Although there is evidence that LSD can lead to mental illness, there is not enough evi-

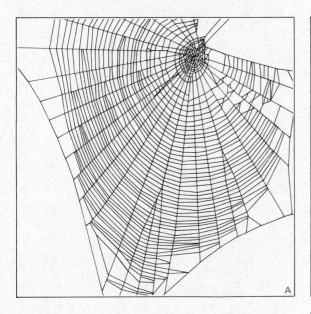

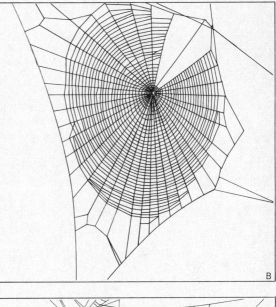

Figure 13-28 The nervous system of many animals besides mammals is affected by various stimulatory and hallucinogenic drugs. (Nicotine, for example, was used for many years as an insecticide!) These drawings show the webs spun by spiders which were drugged by pharmacologist Peter Witt. The *normal web* (A) has the familiar precision of spider architecture and is surely one of the most effective snares in nature. However it has numerous flaws such as branching radii and unequal spacing between the concentric circles. *LSD* (B) causes the spider to produce a nearly perfect web, as if it were "totally concentrating." *Caffeine* (C) gives the spider the equivalent of "coffee nerves" and the web is a disorganized tangle.

dence at present to show that marihuana is physiologically dangerous. Recent suggestions that it causes chromosomal damage have not been borne out. Indeed many scientists believe that marihuana may be a less dangerous drug to the whole body than alcohol. Others remind us that it took many years before all the effects of alcohol were known and that the impairment of protein synthesis in the brain by alcohol has been discovered only recently. They also remind us that it took 300 years for carcinogenic and other adverse effects of tobacco to be clearly demonstrated. Further research and careful followups on long-term marihuana users are clearly necessary.

It is said that neither marihuana nor any of the hallucinogens is *physiologically addictive,* that is, the cells do not change their metabolism so that their functions depend on the drug. These substances, however, may well be *psychologically addictive,* that is, a user may feel that he simply cannot face his world and its problems without his drug. We do not understand the physiological functions of the brain that lead to psychological addiction. An increasing number of biologists believe that there is no such thing as "psychological addiction." They reason that psychological problems have their root in the physiological activities of nerve cells. Thus psychological problems are really physiological problems.

There is no doubt that narcotics such as opium and its derivatives (morphine, heroin), alcohol in immoderate amounts, amphetamines, and barbiturates are physiologically addictive.

The body does increase its *tolerance* to LSD and the other hallucinogens; this simply means that with continued use of a drug, larger doses are needed to get the same effects. However, a user who suddenly stops taking a hallucinogen does not experience the agonizing "withdrawal symptoms" that a narcotic addict endures when he comes off his drug. There is no increased tolerance to marihuana, whereas an alcoholic does have increased tolerance to alcohol.

The amphetamines, generally known as speed or pep pills, have a chemical structure which resembles adrenalin. When amphetamine or adrenalin is released in the bloodstream, it gives a sudden burst of energy as would be needed for a flight, fight, or fright response. They appear to act partly by stimulating the sensory cortex and partly by causing the release of noradrenaline and other stimulatory chemicals within the body. When used in low doses, amphetamines produce a general increase in alertness, a sense of well-being, and decreased feelings of fatigue. They suppress appetite, increase heart rate and blood pressure, and change sleep patterns. When taken for prolonged periods, increased tolerance develops as in the case of LSD and increased doses are required to induce desired effects. These gradually increasing doses of amphetamine can lead to highly abnormal behavior manifested by unfounded suspiciousness, hostility, hallucinations, and persecutory delusions; in other words, what we commonly term paranoid behavior. Another pattern of amphetamine use is the "speed binge" which involves repeated intravenous injections. This use of amphetamines is associated with aggressive and assaultive behavior, exaggerated self-confidence, increased tendencies to take physical action quickly without consideration of future consequences, hostility, and distrust. The evidence indicates clearly that chronic amphetamine use does serious harm to the individual and to those around him.

It appears almost certain that man will continue to use various chemicals such as drugs, alcohol, and tobacco to change his behavior. It is hoped that further research may make it possible to avoid many of the dangers of drug and other chemical use and yet obtain particular behavioral effects that may be desired under certain circumstances.

EVOLUTION OF NERVOUS SYSTEMS

Nerve nets

What were the evolutionary steps involved in producing a structure as spectacular as the human brain? To answer this question we must consider the origin of the nervous system itself. Most zoologists believe that primitive nervous systems were probably very much like the *nerve net* of a hydra (Fig. 13-29). Nerve cells form a network throughout the animal's epidermis. Each nerve cell has several short extensions or fibers. However, the fibers are not specialized as dendrites or axons, and they transmit impulses to neighboring nerve cells in both directions. This is very different from the situation in more complex animals where synapses between nerve cells permit one-way traffic only. In hydra the synapses between nerve cells permit two-way traffic. There are many synapses which an impulse must cross as it moves from one part of a hydra's body to another. Since each synapse represents a delay in the transmission of an impulse, conduction is quite slow.

The hydra has sensory cells in its epidermis. Most of the sensory cells of hydra double as both receptors and sensory neurons, that is, they both receive sensory stimuli and transmit impulses. One end of the sensory cells may have a sensory bristle, whereas the other end of the sensory cell has nervelike branches that synapse with nerve cells. These sensory cells and some other cells respond to a few stimuli such as touch and to some chemicals. For example, prodding the hydra in one part of its body excites the sensory cells in that part, and nerve impulses are initiated. The impulses spread very much the way circular ripples do when a pebble is dropped into a pond. How far the impulse spreads to neighboring nerve cells depends on the strength of the stimulus. The stronger the stimulus, the

more nerve cells will be excited, and the more pronounced the hydra's response. The response is usually limited to a local contraction of the animal's body.

This simple nervous system has no obvious coordinating center, no brain or ganglia where several nervous inputs may be sorted out and nervous outputs initiated. As a result hydra is capable of only a limited amount of behavior. It can feed in an organized way and it may somersault slowly from place to place. Rapid movement and complex behavior are not possible for animals with only a nerve net.

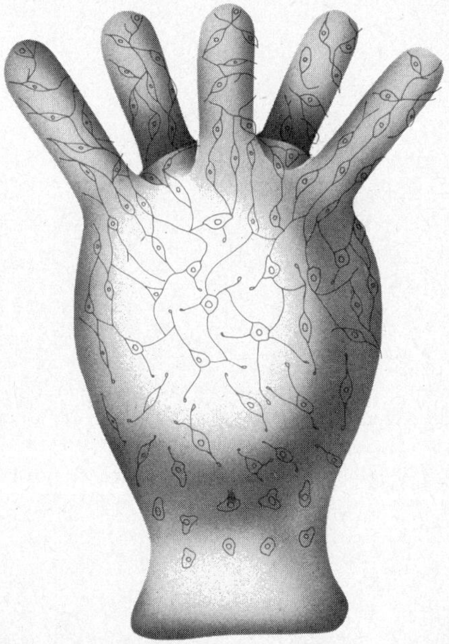

Flatworm nervous systems

The nervous system of a flatworm such as planaria is considerably more complex than a nerve net. Planarians retain a nerve net in some parts of their bodies (and so do we for that matter, in the wall of our intestine), but they also have the beginnings of a central nervous system. Most planarians have two main nerve trunks in which most of the nerve cell bodies lie, and numerous cross connections between the nerve trunks. The nerve cells of planarians have synapses which permit one-way conduction. Therefore nerve impulses can be channeled in specific directions and not merely spread in all directions as they do in hydra. Planarians have the beginnings of an afferent system which receives sensory input and an efferent system which sends nerve impulses out to muscle cells and other effectors.

At the front end of a planarian is a large concentration of nerve cells that forms a brain. This brain is a true nervous center; it receives input from various sense organs and sends output to the effectors. Near the brain are sense organs that send impulses to the brain. There is a pair of eyes that are sensitive to the direction and intensity of light. Most planarians move away from light. Near the eyes are chemoreceptors. These sense organs are important in locating food. Since the front end of the animal is the part which is first poked into a new environment as the animal moves along, there is adaptive value in having these receptors and the brain located in the head region. If a planarian's brain is removed, the animal can still move about from place to place and respond to stimuli, but it may take a

Figure 13-29   Nerve net of hydra.

little longer to do so than normally. The brain maintains a certain level of excitability in the animal and enables it to respond more readily.

The planarian's nervous system shows several evolutionary advances over the hydra's nerve net. The nervous system is *centralized* with nerve cords and an *anterior brain* which receives sensory input, sorts it out, and sends output to the nerve cords. There are afferent and efferent fibers, permitting *one-way*—hence more efficient—*conduction* of impulses. In addition a planarian has *more numerous sense organs* and *more specialized sense organs* than a hydra, and many of these sense organs are located near the head end. Presumably the brain is located near these sense organs at the head end because it is advantageous to get sensory input to the brain as rapidly as possible. The brain functions mainly to funnel nerve impulses from sense organs to the cords. Thanks to its more complex nervous system, a planarian leads a more active life than a hydra and has a larger repertoire of behavior. Some scientists believe that planaria are even capable of learning, but this question has not been settled. If they do learn, you can be certain that they learn very little.

All bilaterally symmetrical animals use the basic design of the planarian nervous system. It was apparently a good design. In addition the evolutionary trends which we saw when we compared hydra and planaria are continued in other groups of animals. In the annelids, arthropods, and the vertebrates, centralization of the nervous system into nerve cords and ganglia is much greater than in flatworms. There is also a tremendous concentration of major sense organs such as eyes in the front end, and along with it the concentration of ganglia to form large brains.

Figure 13-30 Nervous system of a flatworm. The beginnings of a central nervous system are evident in the nerve cords running lengthwise down the animal and an anterior brain.

Insect nervous systems

In insects the nervous system consists of a large brain and a nerve cord. The nerve cord, like that of almost all animals except the vertebrates, is

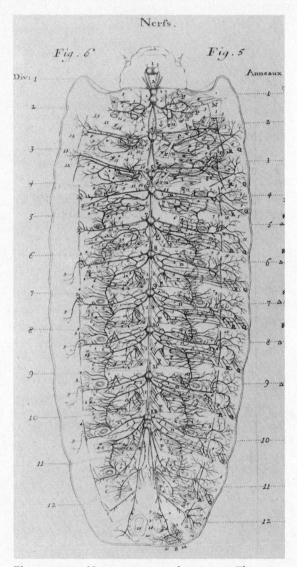

Figure 13-31 Nervous system of an insect. This 1760 drawing of the nervous system of a goat moth caterpillar by Lyonet illustrates the basic features of the insect nervous system. The anterior brain is connected to a nerve cord consisting of a chain of ganglia connected to each other by two large bundles of nerve fibers. There is usually one ganglion in each segment.

double. It consists of a chain of ganglia, usually one in each segment, which are connected to each other by two huge bundles of nerve fibers (Fig. 13-31). The ganglia contain most of the nerve cell bodies. The bundles of nerve fibers which connect the segments provide for a through-conducting system. Each ganglion coordinates many of the responses in its own segment.

There are some large and complex sense organs such as compound eyes in the head (Figs. 13-32 and -33). But most of the sense organs of insects are small structures with a tiny receptor cell and a single nerve cell. These tiny sense organs are scattered over the surface of the body. If you look closely at the skin of an insect, you will see numerous tiny hairs which project everywhere from the surface. These are all tiny sense organs called *sensory hairs* (Fig. 13-34). Sensory hairs of this kind are particularly plentiful on the feet of insects, where they enable the insect to sense its contact with the ground. The movement of the hair in its socket compresses or stretches the ending of the sensory nerve fiber. A sensory nerve cell body lies just below each sensory hair and sends a nerve fiber directly to the base of the hair. A nerve impulse is initiated and travels along the axon of the sensory cell body to a ganglion in the central nervous system. There the sensory nerve fiber synapses with an interneuron.

This picture of the insect's nervous system emphasizes a major difference between the nervous system of most invertebrates and that of vertebrates. In insects and most other invertebrates the cell bodies of sensory neurons are not in the central nervous system as they are in vertebrates. Instead the sensory nerve cell bodies of insects and most other invertebrates are located out in the skin near the sensory receptors. Another major difference is the location of the whole nervous system. In insects and most other invertebrates the nerve cord lies underneath the gut in a ventral position. In vertebrates the nerve cord lies in a dorsal position above the gut.

The biggest difference between the nervous system of an insect and a vertebrate has to do with the brain. An insect brain is little more than a large ganglion. In fact if the brain of an insect is removed, the animal still does many of the things it normally does. The brain of an insect does not dominate the other ganglia nearly as much as the brain of a mammal dominates the spinal ganglia. Also the brain of insects permits only a limited variety of behavior compared to the extensive repertoire of behavior made possible by the brain of a mammal.

The differences in brain function between insects and vertebrates are probably related in part to the completely different ways in which their

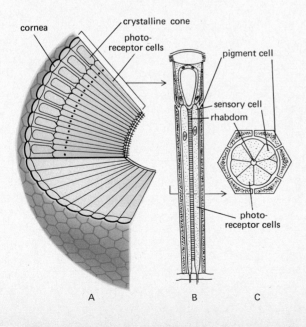

cornea
crystalline cone
photo-receptor cells
pigment cell
sensory cell
rhabdom
photo-receptor cells

A   B   C

Figure 13-32   A. Compound eyes of an insect. Each eye is made up of hundreds (often thousands) of ommatidia, each of which is a small complete eye which connects to the brain. B. Each ommatidium has a cornea of transparent cuticle and a crystalline cone which focuses light onto a clear rod called a rhabdom. The rhabdom transmits the light to eight photoreceptor cells which surround it. C. Cross section of an ommatidium. The sides of the ommatidium are coated with an opaque pigment which blocks transmission of light from one ommatidium to its neighbors. Each ommatidium points in a slightly different direction and responds to the light directly in front of it. The photoreceptor cells are modified neurons, and their axons form the optic nerves which carry impulses to the insect's brain where they are integrated. In addition, the photoreceptor cells of the eye itself may modify the visual input before it is transmitted to the brain. The precise nature of the message that the insect's brain receives is not known. Whether it is a mosaic picture composed of many dots like a television picture or some other representation of the world is unclear.

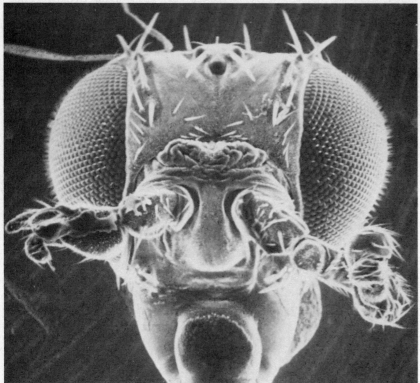

Figure 13-33 Scanning electron micrograph of the compound eyes of a fruit fly.

brains evolved. The insect brain, like that of lobsters, spiders, snails, and octopods, was formed by the gradual migration of preexisting groups of nerve cells to the animal's head. Such a brain is largely a rearrangement of groups of nerve cells that had evolved earlier. In contrast the vertebrate brain appears to have evolved by the addition of new nervous tissue to older, preexisting nervous tissue. The most spectacular of these new additions is the cerebral cortex of mammals. This remarkable organ arose as an entirely new structure. It first appeared as a tiny bump on the forebrain of the reptilian ancestors of the mammals. We still see this bump on the forebrain of some living reptiles such as alligators. In more specialized mammals such as cats and man, the cerebral cortex has increased in size until it covers the entire forebrain.

SUGGESTED READING LIST

DiCara, L. V., "Learning in the Autonomic Nervous System," *Scientific American,* January, 1970 (Offprint 525).

Eccles, J. C., "The Synapse," *Scientific American,* January, 1965 (Offprint 1001).

Fisher, A. E., "Chemical Stimulation of the Brain," *Scientific American,* June, 1964 (Offprint 485).

French, J. D., "The Reticular Formation," *Scientific American,* May, 1957 (Offprint 66).

Galambos, R., *Nerves and Muscles.* Doubleday, New York, 1962.

Griffin, D. R., "More About Bat 'Radar'," *Scientific American,* July, 1958 (Offprint 1121).

Grinspoon, L., "Marihuana," *Scientific American,* December, 1969 (Offprint 524).

Handler, P., *Biology and the Future of Man.* Oxford University Press, New York, 1970.

Heimer, L., "Pathways in the Brain," *Scientific American,* July, 1971 (Offprint 1227).

Held, R., "Plasticity in Sensory-Motor Systems," *Scientific American,* November, 1965 (Offprint 494).

Figure 13-34 Scanning electron micrograph of several different sorts of sensory hairs on the leg of a fruit fly.

Hodgkin, A. L., "The Ionic Basis of Nervous Conduction," *Science*, vol. 145, 1148-1154, 1964.

Hubel, D. H., "The Visual Cortex of the Brain," *Scientific American, November,* 1963 (Offprint 168).

Kagan, J., "Do Infants Think?" *Scientific American,* August, 1972.

Katz, B., "The Nerve Impulse," *Scientific American,* November, 1952 (Offprint 20).

————, "How Cells Communicate," *Scientific American,* September, 1961 (Offprint 98).

————, *Nerve, Muscle and Synapse.* McGraw-Hill, New York, 1966.

Keynes, R. D., "The Nerve Impulse and the Squid," *Scientific American,* December, 1958 (Offprint 58).

Mae, R., and C. Mueller, *Light and Vision* (Life Science Library). Time-Life, New York, 1966.

Merton, P. A., "How We Control the Contraction of Our Muscles," *Scientific American,* May, 1972.

Michael, C. R., "Retinal Processing of Visual Images," *Scientific American,* May, 1969 (Offprint 1143).

Noton, D., and L. Stark, "Eye Movement and Visual Perception," *Scientific American,* June, 1971 (Offprint 537).

Pettigrew, J. D., "The Neurophysiology of Binocular Vision," *Scientific American,* August, 1972.

Pribram, K. H., "The Neurophysiology of Remembering," *Scientific American,* January, 1969 (Offprint 520).

Snider, R. S., "The Cerebellum," *Scientific American,* August, 1958 (Offprint 38).

Sperry, R. W., "The Great Cerebral Commissure," *Scientific American,* January, 1964 (Offprint 174).

Stettner, L. J., and K. A. Matyniak, "The Brain of Birds," *Scientific American,* June, 1968 (Offprint 515).

Thompson, R. F., *Foundations of Physiological Psychology.* Harper & Row, Publishers, Inc., New York, 1967.

Vander, A. J., J. H. Sherman, and D. S. Luciano, *Human Physiology.* McGraw-Hill, New York, 1970.

Von Bekesy, G., "The Ear," *Scientific American,* August, 1957 (Offprint 46).

Warskofsky, F., and S. S. Stevens, *Sound and Hearing* (Life Science Library). Time-Life, New York, 1965.

Willows, A. O. D., "Giant Brain Cells in Mollusks," *Scientific American,* February, 1971.

Wilson, J. R., *The Mind* (Life Science Library). Time-Life, New York, 1964.

Witt, P., "Spider Webs and Drugs," *Scientific American,* December, 1954.

Wood, D. W., *Principles of Animal Physiology.* Edward Arnold (Publishers) Ltd., London, 1968.

Young, R. W., "Visual Cells," *Scientific American,* October, 1970.

CHAPTER 14

Animal Behavior

Behavior is what an animal does. It is the outcome of the activities of the animal's nervous system, its muscles, and other effectors. The study of behavior covers processes as simple as spinal reflexes and as complex as learning and thinking.

THE SURVIVAL VALUE OF BEHAVIOR

A snail pulls its body back into its shell when you touch it. A peacock spreads his splendid fan of tail feathers when a peahen is near. A man puts up with the frustrations and irritations of rush-hour traffic as he drives to and from work. The behavior of each of these animals has something in common: in each case behavior is directly or indirectly related to survival. In fact almost everything that animals other than man do is somehow related to their own survival or that of their species. Let us look at a specific example.

The egg-laying behavior of the female *Aedes* mosquito, the carrier of the yellow fever virus, is geared to the survival of her species. First, she lays her eggs just above the water line of a pond or some other body of water rather than in the water. Second, she does not "put all her eggs in one basket" but distributes them in clusters in several different locations. Her behavior has several results. The eggs will not hatch until the water level *rises*. Had she laid her eggs in a small body of water they might have hatched. And if the water had dried up, the young larvae would have died. But if the eggs hatch only when the water level rises, this usually means that drying-up will not occur at least before the larvae are safely on their way to adult development. Since she distributes her eggs, the chances are good that at least some of them will survive and hatch, even if others are destroyed. Furthermore, distribution of her eggs helps prevent overcrowding of newly hatched larvae that would compete for food.

This pattern of behavior seems quite "intelligent" and "purposeful." The *Aedes* mosquito "insures" that her offspring will have the best chance of surviving. However, this is not the case at all. Neither intelligence nor purpose is involved. Both intelligence and purpose require that an animal be consciously "aware" of what it is doing. The *Aedes* brain is so different from the brain of a man or a horse or a dog that it is unlikely that she is consciously aware the way humans and some mammals are. Nonetheless her behavior assists the survival of her species.

BUILT-IN BEHAVIOR—REFLEXES AND INSTINCTS

Let us analyze this small piece of the female *Aedes'* behavior a little further. All *Aedes* mosquitoes are orphaned by their parents as soon as the eggs are laid. Each mosquito is strictly on its own. It has no opportunity to learn this special kind of egg-laying behavior; yet all female *Aedes* mosquitoes behave in this way. We come to the inescapable conclusion that this egg-laying behavior must be inherited, and the control of it is built into the insect's nervous system. The nerve cells of an *Aedes* mosquito are "wired up" in such a way that females will automatically display this egg-laying behavior when they are filled with eggs and provided with the right environmental stimuli.

Many behavioral patterns that we see in animals are inherited. They are as much a part of an animal as the color of its eyes. They are in a sense the "color of its nervous system." We have already discussed the simplest kind of inherited behavior—reflexes. Reflexes are automatic responses, and many, such as spinal reflexes in vertebrates, require no brain although they may be modified by the brain. They are the result of the way an animal's behavioral machinery is "wired together." Thus an earthworm that has had its head chopped off can still wriggle from place to place because the reflex pathways that cause wriggling are still intact.

The egg-laying behavior of the *Aedes* mosquito is too complex to be called a reflex. But nonetheless it is programmed in the insect's nervous system. Biologists call such inherited patterns of behavior **instinctive** or **innate** (inborn) **behavior.** There are many familiar examples. Some of the easiest instincts to study are those which leave a permanent record for us to examine. The cocoon of a silkworm and the web of a spider are records of the instinctive behavior of the animals which spun them (Fig. 14-1). Worm tracks in Cambrian rocks 500 million years old are fossil records of the instinctive behavior of ancient annelid worms.

Instinctive behavior cannot be altered very easily. For example, if you plug up the *spinnerets*—the little spigots through which the silk flows out as a thread—a silkworm will still go through all the motions of spinning a cocoon! When the job is "completed," the insect will lie down and pupate within its invisible cocoon. Or take the case of a female hunter wasp, *Sphex flavipennis*. She captures a grasshopper and drags her prey back to her burrow. She leaves her victim outside the burrow while she inspects her nest; then she drags the grasshopper by its antennae into the burrow. Jean Henri Fabre, a great student of insect behavior, discovered that if he snipped off the antennae and other structures about the grasshopper's head which the wasp could use as a "rope," she would leave her victim outside the burrow. She simply could not grasp the fact—obvious to us—that she could grab one of the grasshopper's legs and use it to drag the animal into her burrow. Her nervous system, like that of all insects, is made up of a relatively small number of nerve cells arranged in such a way that most of the responses that she makes are rigidly determined. She is largely incapable of coping with situations for which her species has not evolved instinctive behavior.

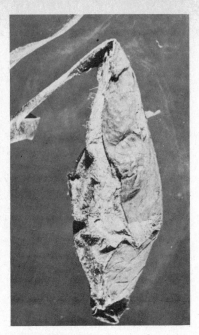

(a) *Rothschildia forbesi*   (b) *Callosamia promethea*   (c) *Samia cynthia*

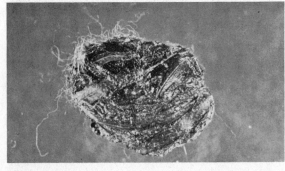

(d) *Automeris io*   (e) *Actias luna*

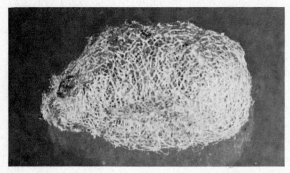
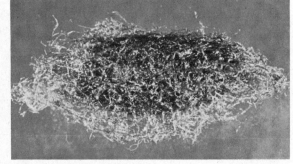

(f) *Agapema galbina*   (g) *Calosaturnia mendocino*

Figure 14-1 Cocoons of 7 species of wild silkmoths. Intricate as it is, a caterpillar's cocoon is nonetheless the result of instinctive behavior; learning is never involved. In its genes, each animal carries not only the design of the silken structure characteristic of its species but also how to make it.

This is not to say that wasps cannot learn anything. Figure 14-2 describes a convincing experiment with another species of wasp, a digger wasp, *Philanthus triangulum*, which clearly demonstrates that the wasp can perceive form and pattern and can memorize the location of its nest by means of landmarks. This ability enables it to find its way to its nest after a distant flight. Evidently, even an animal with a tiny nervous system can learn. However, only animals with large nervous systems can learn very much. The insect nervous system simply does not have enough nerve cells to permit extensive learning.

THE ROLE OF ENVIRONMENT IN THE DEVELOPMENT OF BEHAVIOR

All forms of behavior, even automatic behavior like reflexes and instincts which are inherited, are still profoundly affected by the animal's environment, both internal and external. For example, the way a bird builds its nest is instinctive, but hormonal changes within the animal trigger the nest-building activity. Even reflexes are "set" by the environment. For example, chimpanzees reared in the dark from birth until they are about three years old cannot blink their eyes when first exposed to light. Apparently light is needed before the eye-blink reflex can be made to work properly. An even more impressive ex-

ample comes from studies of kittens. Kittens were reared for the first few months of life in a room painted with vertical stripes. When the kittens were later placed in a normal environment, they ignored all horizontal images in their environment, and responded only to vertical images. Thus they would respond to upright rods placed in front of them by avoiding them, but would ignore horizontal rods and bump into them. This behavior persisted throughout their lives. Clearly, sensory experience in early life was shown to have produced more or less permanent changes in behavior.

The precise nature of the changes induced in the brain by exposure to certain kinds of sensory input early in development is not definitely known. It may, however, involve changes in the structure of the nerve cells themselves. Rats were raised in an environment with little sensory input—darkness, quiet, few objects. When these rats were returned to a normal environment, they learned poorly, were asocial, and generally behaved "quite stupidly" compared to rats raised in a normal environment. The cortex of a sensory-deprived rat actually weighed less than the cortex of a normal rat. Also when the cortex of sensory-deprived rats was examined microscopically, the brain cells had many fewer dendrites and synapses than the brain cells of normal rats (Fig. 14-3). *Sensory deprivation* had led not only to a permanent change in the rats' behavior but also

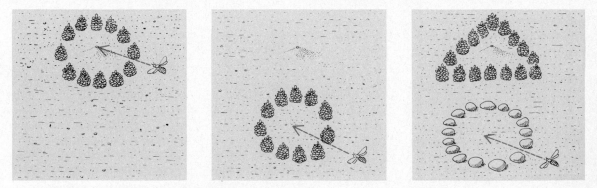

Figure 14-2 Wily wasp. Dr. Niko Tinbergen, a famous student of animal behavior, tested learning in female digger wasps. Initially a female wasp's burrow was surrounded by a ring of pine cones. But when the ring of cones was moved a foot or two away, she was unable to find her nest, which had not been moved. When the cones were rearranged in a triangle around the burrow, and a decoy ring of pebbles placed nearby, she flew to the ring of pebbles. These results indicate that the female wasp memorizes the location of her nest, using landmarks to guide her to it, and that it is the circular arrangement rather than the cones themselves to which the insect was responding.

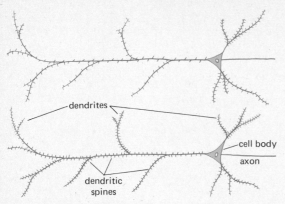

Figure 14-3 Dendritic spines are tiny projections from the dendrites of a neuron. This diagram shows the neurons of a "naive" rat (top) and an "experienced" rat (bottom). The neuron of the "experienced" rat has more dendritic spines.

to an actual structural change in their nervous systems. Apparently, sensory input to the nervous system not only provides information but actually stimulates the development of the brain.

These experimental results have implications for man that cannot be ignored. If the first two or three years of a child's life are spent in an environment which is poor in sensory stimulation—no colors, no toys, no music, no conversation, few experiences—the child's brain may develop abnormally. This effect may be more or less permanent, and an asocial adult who has difficulties in learning may result. Add to this nutritional deficiencies in early childhood which also affect brain development and the problem is compounded. What is especially important here is that the stage in life when the brain requires stimuli and nutrition for full development is very specific. The "windows of the nervous system" are open for a brief time in an animal's early life. If the brain does not receive enough stimuli during this period, the windows close.

BIOLOGICAL CLOCKS

The world is filled with cues that show the passage of time. There are the daily cycles of light and dark, the monthly phases of the moon, seasonal changes in length of day and temperature. All of us are familiar with the fact that many animals

behave differently at different times of day. Cockroaches and mice scurry about at night. Sparrows are active during the day. It seems reasonable to suppose that animals respond to the light-dark cycle. However, when such animals are placed under constant conditions (for example, constant darkness or constant light at a constant temperature) so that they have no indication of what is going on in the environment, they still show rhythmic cycles of activity. The cycles usually are about 24 hours long, just as occurs normally. Furthermore these cycles may persist for weeks.

What is especially interesting is that when animals are kept under such constant conditions, their cycles may get slightly longer or shorter than 24 hours. Figure 14-4 shows a record of the behavior of a cockroach. This insect normally becomes active at about 10 P.M. and quiets down at 4 A.M. That is, it is active for 6 hours and inactive for 18 hours. When such a cockroach is placed in constant light, its rhythmic activity continues. However, the cycle is no longer exactly 24 hours. It runs for 6 hours and 15 minutes and is inactive for 18 hours. In other words, its entire cycle is now 24 hours and 15 minutes. This cycle is innate and continues for weeks. Gradually the cockroach gets completely out of phase with the actual conditions in the environment. In this case after 28 days under constant conditions the cockroach is 7 hours out of phase and starts to run at 5 A.M. and stops at 11:15 A.M. The cockroach behaves as if it had an internal "clock" of some kind which tells it when to be active and when to rest. Here the clock has a cycle 24 hours and 15 minutes long, instead of the 24-hour cycle of the environment. The clock is apparently innate and does not depend on earthly indications.

Many animals show evidence of having biological clocks. They have 24-hour cycles of all sorts of activities, such as behavior, secretion, growth, and body temperature. There is also evidence that many plants have biological clocks. Some flowers open and close in a cycle of about 24 hours, even when the plants are kept under constant conditions. Other plants secrete nectar at certain specific times of day. Bees and hummingbirds which have their own biological clocks get in the habit of visiting these flowers at these times, thereby ensuring food for themselves and cross pollination for the flowers. These observations on plants provide convincing proof that a nervous system is not required for the generation of these 24-hour cycles of activity. Indeed it is becoming increasingly clear that even cells and organs may show 24 cycles of activity when kept under constant conditions. These rhythms of about a day's length are called **circadian rhythms** (from the Latin *circa*, meaning "about," and *dies*, meaning "days"). The rhythms can be reset by changes in the illumination cycle and are normally kept in phase with day and night.

In addition to these daily rhythms there are monthly and yearly rhythms. One of the most familiar yearly biological clocks is the one which triggers bird migration. If European warblers are hatched and raised in closed, soundproofed rooms at summerlike temperatures, they still become restless in the autumn when, if they were free, they would migrate. For many nights they remain awake, flitting over their perches (they are night travelers by nature). In fact they will act like this for as many nights as they normally would take to fly to their winter quarters in Africa. In the spring the whole thing is repeated.

What makes a biological clock tick? We do not know. External stimuli "set" the clock just as we set our watches to a certain hour, but the clock itself is located within cells or populations of cells. One such clock has been identified in giant nerve cells of a sea slug called *Aplysia*. A ganglion was isolated from an *Aplysia* and kept alive in a salt solution for days. Individual cells in this ganglion continued to fire with a 24 hour rhythm even when isolated from environmental variables. Efforts are now underway to discover the chemical changes in these giant nerve cells that cause this 24-hour discharge. There is some evidence that mammals and birds have a master clock, the *pineal gland*, a tiny bit of tissue deep within the brain; the pineal clock may be set by the illumination cycle (Fig. 14-5).

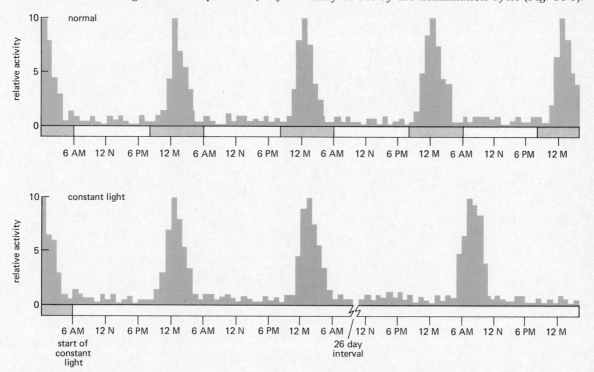

Figure 14-4 The activity of a cockroach occurs in distinct daily cycles of the kind diagrammatically shown here. When the insect is transferred to constant light, the cycles persist but are not exactly 24 hours long. After a month the cycles are out of phase by 7 hours.

Man has several biological clocks. Body temperature may be 2° F higher at 4:00 P.M. than at 4:00 A.M. Heart rate, blood pressure, hormone secretion, and the excretion in the urine of calcium, sodium, and potassium all have a circadian rhythm. The greatest tolerance for alcohol is at 5:00 P.M. Even birth and death have a circadian rhythm: a man is most likely to be born and most likely to die between 3:00 and 4:00 A.M One biological clock is familiar to anyone who has ever made a transcontinental or transoceanic flight. The symptoms are called "jet lag." You probably felt generally "out of sorts"; your eating and sleeping schedules did not jibe with the local time. You felt mixed up because the clocks which time many of your body's rhythms had not yet been "reset." Jet lag increases decision time and reaction time and decreases mental alertness and the ability to concentrate. Your body may take as long as a week to readjust after such a flight. Commercial airlines know this, and they schedule their pilots' flights to account for changing biological rhythms.

MOTIVATION

We know from experience that the same stimulus given to the same animal at different times does not always produce the same response. Sometimes a very small stimulus may be enough to cause a powerful response. At other times far stronger stimuli are ineffective. For example, your response to a hamburger when you are hungry and when you have just finished a T-bone steak is completely different!

When an animal changes its response to the same sort of stimulus at different times, we can explain its behavior in several ways. The animal may be *tired*. It may have *learned* to respond differently. Or it may have *matured*. (Young birds do not really learn to fly from habit or from practicing flying; their ability to fly improves as they mature.) If none of these explanations is satisfactory, we attribute a change in behavior to a change in **motivation.** Motivation can be described simply as *a change in the animal's internal environment that leads to a particular behavior pattern*. Motivated behavior often appears

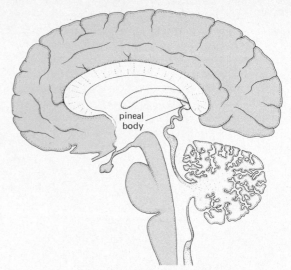

Figure 14-5 Location of the pineal gland.

to be directed to some specific goal and to meet some bodily need. Psychologists sometimes speak of drives, such as hunger drives, cold drives, and sex drives. Apparently when the internal environment becomes unbalanced in some way (low blood glucose, for example), the animal's behavior changes in specific ways. These changes in behavior favor reestablishing the balance in the internal environment. Consider the following example. If a dog is kept without food for a day, it will move about restlessly. It becomes highly responsive to the sight or smell of food. It will also respond to other stimuli which it has learned to associate with food, such as the sound of a refrigerator door opening or the sight of a food bowl. It will not play or be diverted in any way, and remains restless until it has eaten. The important point here is that motivation is highly specific and directed, and leads to a *particular* behavioral pattern.

We already discussed the mechanisms of motivation. In Chapter 12 we saw that certain areas in the hypothalamus are stimulated by a drop in the level of blood glucose. In the experiment just described, keeping the dog hungry lowers its blood glucose level. The drop in blood glucose stimulates the eating areas in its hypothalamus and leads to restless behavior until the animal is fed. If, after the animal has gorged itself, we electrically stimulate an eating area in its hypothalamus, motivation to eat reappears and the animal starts to eat once again. Interestingly there are other areas in the hypothalamus which affect eating. If these areas are stimulated in a hungry animal, the animal ignores food and behaves as if it had eaten enough. These areas have been called the "satiety center." They measure when the animal has eaten enough, and block the motivation to eat. If these areas are surgically destroyed, then the animal will eat so much that it becomes grossly obese.

These areas which regulate particular kinds of behavior must not be thought of as specific places in the hypothalamus. They actually involve many groups of cells in different parts of the hypothalamus, and areas controlling activities such as eating, drinking, and sexual behavior may overlap. The hypothalamus appears to regulate a great deal of motivated behavior in mammals. This is not surprising since we know that the hypothalamus has something to say about almost everything that goes on within the internal environment of mammals.

Consider the effects of hormones produced by the hypothalamus in motivating sexual behavior. The females of many species of mammals (cats, dogs, rats, for example) are receptive to males only when the females are "in heat." Their breeding season depends on the level of sex hormones in their blood, and this level is controlled by the hypothalamus.

A striking example of how internal changes affect behavior is "mother love." This phenomenon varies among different species, but most female mammals show a pattern of behavior which resembles human mother love to a degree. There is little question that in the human female mother love is an intensely emotional experience. But we cannot say that the care which other female mammals lavish upon their young is an emotional thing. We have to take a more scientific, if less inspiring and romantic, approach.

Experiments investigating the causes of mother love were performed with female goats which had just given birth to kids. If the mother is allowed a few minutes of contact with her kid, during which time she sniffs it, she will thereafter recognize and accept it as her own, and allow it to nurse. She will reject other females' kids. Is mother love in the goat triggered by the smell of her kid? Apparently not. A female ready to give birth had the inside of her nose anesthetized so she could not smell. After she had given birth, she was permitted contact with her kid even though she could not smell it. She later accepted it *and also* kids that were not hers. There is something about the birth process itself that evokes mother love. There is evidence that the stimulus is *oxytocin*, a hormone produced by the hypothalamus. The level of oxytocin in the blood increases as the neck of the uterus widens, and the young mammal enters the vagina or birth canal. At birth it stimulates contractions in the uterus. In addition it apparently stimulates mother love—at least in the female goat.

LEARNING

As animals became more complex with correspondingly complex nervous systems, the balance shifted slowly from behavior which is mostly instinctive, as in insects, to behavior which is mostly learned, as in man. **Learning** has a clear advantage over instinct; learning enables an animal to vary its behavior when it faces a change in situations.

What do we mean when we say that we have learned something? Learning can be described as *a more or less permanent change in behavior as a result of experience.* This change in behavior results from some change in the nervous system. The fact that an elderly person can vividly recall events which occurred many years before certainly suggests that the changes in the nervous system involved in learning may be more or less permanent changes in neurons and synapses of the brain. Electrical stimulation of the cortex in humans often causes them to recall experiences from childhood that were "long forgotten."

Learning is often subdivided into several kinds of learning, such as trial-and-error, conditioning, and reasoning. But there is no compelling evidence yet that there is really more than one kind of learning. All learning may have the same neural basis. However, because many different kinds of experiments on learning are performed by biologists and psychologists, we consider several different kinds of learning situations that are likely to be discussed.

1. *Imprinting.* Konrad Lorenz, one of the foremost students of behavior, first coined the term "imprinting" to describe the results of his experiments with newly hatched goslings. They can waddle about, following their mother, as soon as they hatch. Lorenz showed that the first moving object which they see upon hatching becomes their "mother." If the object happened to be something besides their mother—even Lorenz himself—they followed the object and ignored their real mother. In other words, the object was imprinted upon them. One of the most significant results of imprinting is that the young geese, when sexually mature, attempt to mate with a member of the species, or object, with which they were imprinted. In nature, of course, everything works out fine. The first thing a gosling sees upon hatching is its real mother. Then when it mates it does so with its own kind. Imprinting probably helps an animal recognize its own species.

There is some evidence that imprinting also occurs in mammals. Anyone who has ever reared a lamb rejected by its mother knows the truth of "Everywhere that Mary went the lamb was sure to go."

2. *Habituation.* Habituation is a simple and widespread learning phenomenon. If an animal is repeatedly given a stimulus which is not accompanied by any reward or punishment, it ceases to respond. A snail crawling across a pane of glass will withdraw into its shell when you tap on the glass. After a moment it will emerge and begin to move again. If you tap a second time, it will withdraw again, but this time it emerges more quickly than before. Finally the snail ceases to respond at all to the taps. Even an animal as simple as a hydra will show habituation to stimuli like shaking.

Habituation is important in the development of behavior in many young animals. A baby rabbit may begin by showing escape responses to anything which moves. But very soon it learns to ignore the waving of grass in the wind. One highly adaptive value of habituation is that it frees an animal's brain of the work involved in directing the animal's attention to insignificant stimuli. Thus an animal is freed from the "chores" of responding to situations it need not respond to, and instead can engage in behavior that will help it survive.

Figure 14-6 Konrad Lorenz and two of his geese. As a result of imprinting, the geese respond to him as their mother. (Nina Leen, Life Magazine, © 1972 Time, Inc.)

3. *Conditioning.* The great Russian physiologist Ivan Pavlov demonstrated another learning situation, conditioning. He rang a bell each time he fed his dogs. Soon they began to drool at the mere sound of a bell even when no food was in sight. There are many examples of conditioning. For instance, if you always go to the refrigerator to get milk for your cat, the cat will learn to associate refrigerator and milk. When you go to the refrigerator to get yourself a snack, your cat will probably run up and rub against your leg. The cat has learned the association between refrigerator and milk. Conditioning has been observed in many different animals, from earthworms to chimpanzees. They may involve punishment instead of reward. Once a bird has experienced the bad taste of a cinnabar moth, the bird avoids them. Inci-dentally, another species of insect, which probably would be quite tasty to birds, has evolved a cinnabarlike coloration which deceives the birds. In effect, this mimicry advertises to would-be bird predators that "I taste awful. Leave me alone!" This type of mimicry is a widespread phenomenon among animals, especially insects.

4. *Trial-and-error.* We are all too familiar with still another example of learning, trial-and-error. Many of the highly coordinated movements that we learn to make, such as driving a car with a standard shift, playing a musical instrument, making a figure 8, are all examples of this type of learning. In trial-and-error learning animals learn to eliminate behavior leading to no reward and increase the frequency of rewarded behavior.

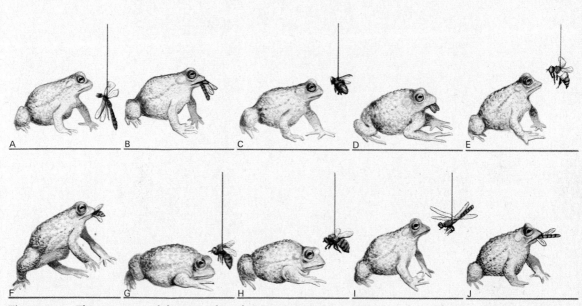

Figure 14-7 This sequence of drawings describes a convincing experiment conducted by Lincoln and Jane Brower which shows how learning takes place in a toad. Prior to this experiment this particular toad had seen and eaten many insects but had never seen either a bumblebee or a stingless insect which resembles it called a robber fly. A. The toad is shown a dragonfly which is wiggled on the end of a string. B. The toad eats the dragonfly. C. The toad is presented with a robber fly which looks like a bumblebee but has no sting. D. It eats the robber fly. E. The toad is presented with a bumblebee. F. The toad tries to eat it but gets stung on the tongue and spits out the bee. G. The toad refuses a second bumblebee. H. It also refuses a robber fly, taking the fly for a bee. I. To prove the toad was not simply suspicious of all food, another dragonfly is offered. J. This it eats readily. Evidently the toad had learned good from bad.

5. *Insight or reasoning.* This is perhaps the most important form of learning from our human viewpoint. We can all recall occasions when the solution to a problem has "come in a flash," perhaps after several minutes of hard thinking. If an animal correctly solves a new problem the first time by applying what it al-already knows, we say that the animal can reason. Reason does not belong to man alone. The great psychologist Wolfgang Köhler placed a chimpanzee in a cage along with two short bamboo poles. The end of one could fit inside the end of the other to make a long pole. Köhler then placed a bunch of bananas outside the cage, too far for the animal to reach with one pole. The chimp first tried to reach the bananas. When this failed he grabbed a pole and tried to reach them. He still had no luck. Finally he put two poles together, and got the bananas. Köhler's chimps had played with poles before, but they had never faced this kind of problem. They were using knowledge that they already had and applied it to a new problem. This is true reasoning.

Figure 14-8 This photograph is from a film which Wolfgang Köhler made between 1913 and 1917. It shows a chimpanzee named Sultan making a double stick with which he pulls bananas to him from outside his cage. Köhler concluded from experiments of this kind that chimpanzees are capable of reasoning.

MEMORY

These several examples describe some of the forms of learning. They do not tell us, however, how learning and memory take place. When we learn something, a more or less permanent record of the learning is made in the brain. In other words, the learning experience is stored as a "memory." How is a memory formed? This is a particularly fascinating question. Who among us has not been frustrated by the fact that he can remember vividly the details of his first date but cannot recall his own telephone number?

We know, of course, that memories sometimes persist for a lifetime of 70 years or more. Some biologists believe that memory involves permanent changes in the structure of neurons themselves, perhaps an increase in the number of synapses. Other biologists take a different view. There is no increase in the number of synapses or other structural change, but rather there are chemical changes within the neurons. These chemical changes somehow make it easier for transmission to occur across the synapses of certain nervous pathways. Thus they argue that learning and memory are the result of biochemical changes within neurons.

Neurons definitely synthesize a great deal of protein. Proteins are needed to synthesize the transmitter substance, and the enzyme which destroys the transmitter substance is itself a protein. However, there is evidence that the formation of memories is accompanied by the synthesis of more protein than is needed for making and destroying the transmitter substance. Is this "extra" protein somehow being used to make memories? In an effort to answer this question scientists have studied the effects of puromycin, an antibiotic which blocks protein synthesis. Experimental animals were injected with the drug and then used in various learning experiments.

In one such experiment mice were trained to run into the left arm of a Y-shaped maze. Several weeks later the same animals were retrained, this time to run into the right arm of the maze. Immediately after this retraining, half of the mice were injected with puromycin, the other half with a physiological saline (salt solution). They were put into the maze. The puromycin-injected mice ran into the *left* arm of the maze; the saline-injected mice ran into the *right* arm. Apparently the puromycin-injected mice did not remember their recent training; the drug had blocked protein synthesis and prevented a memory of the recent training from being formed. The saline-injected mice, however, remembered their training; saline does not affect protein synthesis.

This experiment also demonstrated the fact that there is both "long-term" and "short-term" memory. The long-term memory of the puromycin-injected mice, remembering to run into the left arm, was not affected by the antibiotic. Can a short-term memory become a long-term memory? If so, how? Biologists believe that this conversion occurs and that it depends on some kind of continuous activity — exactly what we do not know — of the neurons. If anything happens during this conversion process, for example, a concussion, the continuous activity is interrupted, and no memory is formed. This may be why one does not remember very much of the events occurring just prior to suffering a concussion.

As we know protein synthesis depends on various kinds of RNA molecules. We also know that there is a great deal of RNA in the cytoplasm of neurons. RNA can store genetic information. Can it store memories too? There have been many attempts to answer this question. Holger Hydén and his co-workers in Sweden trained rats to walk a thin tightrope to reach their food. The rats had to learn to balance themselves. Then Hydén examined the RNA of neurons isolated from a part of the brain which receives messages from receptors in the middle ear (the part concerned with balance, not hearing). The structure of the RNA of these neurons changed in trained rats. The nature of the change is still unknown. Hydén suggests that memory is stored in RNA. He suggests too that the changes in the RNA affect the kind of protein synthesized, and ultimately the synthesis of transmitter substances for new neural connections.

Where in the brain are memories stored? The memory of yesterday's lunch, for example, is probably not stored in one neat little package in the cortex, or any other place for that matter. We remember how the food looked, smelled, tasted, and the sound of our friends' voices and what they talked about. The "lunch" memory is actually many memories. The location of each of these may be different. That this is true was dramatically demonstrated with trained rats. A psychologist, Karl Lashley, trained rats to run a maze. He then removed various parts of the cerebral cortex to see what effects these areas have on a rat's ability to remember how to run the maze. The startling conclusion he reached was that it is not the region removed which affects the rat's "maze-memory," it is the total *amount* of cortex removed which is important.

Even more surprising was the discovery that memories are stored in duplicate and perhaps triplicate. It is possible to train a mammal such as a monkey and then split its brain into two parts by cutting the connections between the cerebral hemispheres. If one hemisphere is then damaged experimentally, the other still remembers, and the animal's behavior is not noticeably affected. This does not mean that the two hemispheres are equal in their control of many activities in all mammals. In man one hemisphere dominates the other. For example, damaging some portions of the *left* hemisphere in right-handed persons (and vice-versa) interferes more with memory involving a verbal component than damaging the other hemisphere. Similarly, damaging the right hemisphere in right-handed persons interferes more with nonverbal visual and auditory memories. Obviously the two hemispheres in man have evolved slightly different functions.

SOME PATTERNS OF ANIMAL BEHAVIOR

Social behavior

Most of the higher animals are social at some time during, if not throughout, their lives. For example, many species of fish move in schools, lions hunt in prides, prairie dogs build "towns." Some of the social insects — ants, bees, termites, wasps — also display social behavior (we discuss this again in Chapter 24). In many instances two or more species will live close together. For example, impalas and baboons travel together through the African savannas. The baboon's eye-

Figure 14-9. Many animals including insects, communicate by means of pheromones. This is often important in locating a mate. In some species the male releases the pheromone and "calls" the female. In others, the female calls the male. The male tassar silkmoth, *Antheraea mylitta,* above, has broad feathery antennae which enable him to detect the minute amounts of pheromone released by female tassar silkmoths. The antenna of the female tassar silkmoth are not as well developed as those of the male.

sight and the impala's sense of smell are very keen; these two groups, in effect, pool their senses, and both profit from the association.

Animals that live together must be able to communicate with other members of the group. Communication takes many forms. We have already learned that males of some species of moths can "smell" a virgin female of the species a mile or so away (Fig. 14-9). Many insects, especially the social ones, communicate with *pheromones.* As we shall see in Chapter 24 these are substances released by an animal which affect other members of the same species. For example, some ants mark a trail with pheromones which their comrades can follow.

Many animals, of course, communicate with sound. In fact, Konrad Lorenz, the modern-day Dr. Doolittle, has learned the language of his snow geese and can "talk" with them. The antennae of male *Aedes* mosquitoes are tuned to buzzing sounds produced by the wings of flying female *Aedes* mosquitoes. A male mosquito will even home in on a tuning fork if it vibrates at the right frequency. Interestingly the male's antennae "pick up" the female's signals only when he is sexually mature (Fig. 14-10).

Bees communicate with other bees by dancing (Fig. 14-11). In fact, Austrian and Italian honey bees even have different "dialects"— variations in their dances! "Body language" is not restricted to bees, of course. Who has not heard of the "pecking order" of barnyard chickens, the way these birds have of letting others know who stands where on the social scale. Man himself has a whole vocabulary of body language. If you doubt this, watch a man scratch his head when he is perplexed or embarrassed.

The way wolves in the same pack communicate with each other is fascinating. Wolves are wanderers. Moving wolves find more prey, and the pack is on the move almost constantly. Life within the pack follows certain ground rules. Each wolf knows whom he can dominate and to whom he must submit. One wolf, the "alpha-wolf," leads the pack (Fig. 14-12); he has his pick of the females and generally runs things until a younger, stronger wolf deposes him. They seldom fight among themselves, probably because they can communicate so well with each other. Communication is by sound (barks, growls, howls) and by body language.

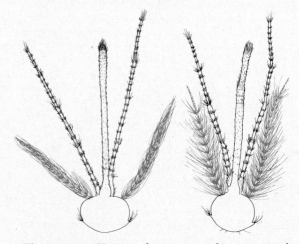

Figure 14-10 Hairs on the antennae of immature male mosquitoes (left) lie close to the antennae. But in a sexually mature male (right) the hairs stand out, and are very sensitive to the vibrations produced by the wings of a female.

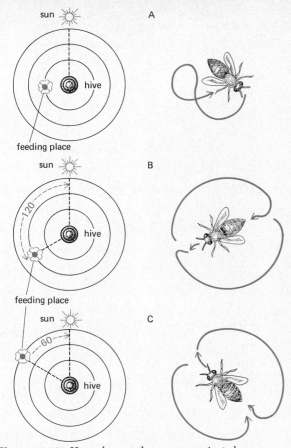

Figure 14-11 Honeybee workers communicate by performing variations of two basic dances, the round dance and the wagging dance. The round dance A consists of circling first in one direction and then in the other direction. When a worker does this dance, she is informing her fellow workers that there is a source of nectar nearby. In the wagging dance B, the worker runs forward in a straight line, all the time wagging her abdomen, circles, runs forward again, circles in the other direction, and runs forward again. The orientation of the run indicates the direction from the hive of the food source and the number of turns per unit time indicates the distance. In B the run was traced at an angle of about 120° with respect to the sun; the abdomen was wagged rapidly indicating a relatively close location. In C the run was traced at a 60° angle; the abdomen wagged slowly indicating a more distant location.

The way a wolf holds his tail indicates a threat, surrender, or his social standing in the pack. Biologists and naturalists who study wolves claim these animals have facial expressions which can be interpreted by their fellow wolves. And a wolf low on the social scale of the pack makes appeasing gestures to his superiors.

A wolf growl is clearly a warning. His barks may be staccato (an alarm signal) or rolling (a challenge or warning to trespassers). His unforgettable howl, however, makes up much of his vocabulary. Its range may be about two and a half octaves (no mean feat as any singer will agree). A howl may have up to a dozen harmonics or overtones. (Harmonics are what make a note played on a piano sound different from the same note played on a violin.) We can hear some harmonics, but wolves apparently are far more sensitive to them than we are. A wolf probably uses harmonics to distinguish the howls of all other wolves in his pack (Fig. 14-13). Wolf howls keep the pack together, for although they move as a unit, the members may be a mile apart. Wolf howls also proclaim territory to another pack.

Migration, homing, and orientation

The autumn skies in temperate regions of the world are filled with migrating birds. Salmon leave the oceans and fight their way upstream against rapids and other hazards to spawn. These are two of the most familiar examples of animal migration, but there are many more.

We do not know how birds find their way to their summer and winter quarters, especially birds that are making their first such trip. No doubt the older birds in a flock guide the young birds in some species, for example, the swallows. Yet other birds fly solo, and the older birds may precede the young birds by several days (red-winged blackbird) or follow them (starlings).

Birds that fly by day and birds that fly at night use the position of the sun and stars as a compass. No one knows how birds account for the changing position of the sun and stars, but they must have a clock of some kind synchronized to the natural day–night cycle. In at least one bird that has been studied, the biological clock is located in its brain. It is probably the same mechanism that causes circadian rhythms.

Both American and European eels spawn in the Atlantic Ocean between the West Indies and Bermuda. The young eel larvae make their way

Figure 14-12 Wolves, showing the dominant, or alpha, male in the center, receiving the "solicitous affection" of the pack during a "greeting ceremony."

back to the freshwater streams where they will live most of their adult lives. The larvae "know" which continent is to be their home. When their time comes several years later to spawn, they return to the place in the Atlantic where they themselves hatched. There is evidence that eels and salmon and other migrating fish depend on chemical cues to guide them back. Apparently there are chemical differences in those waters to which the fish are sensitive.

Everyone is familiar with the bats' sonar system for orienting themselves. Many other animals, whales, for example, depend on echo location too. In fact blind persons develop this ability to a certain extent. And some electric fish use weak electric fields which enable the fish to orient themselves in a somewhat similar manner.

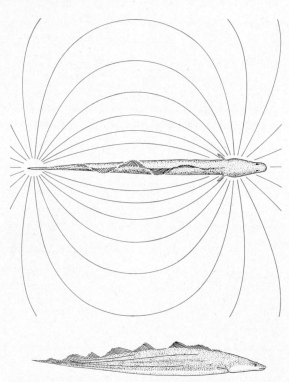

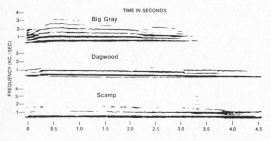

Figure 14-13 This sonogram of the howls of three different wolves shows that each wolf has a distinctive voice. The groups of lines represent different harmonics; thus Big Gray's howls always had at least five harmonics or overtones. Wolves hear these different harmonics, and recognize each other by them.

Figure 14-14 Some fish orient themselves by producing a weak electric field. The lines on the drawing represent such a field. Any disturbance of the field, such as the approach of other animals or objects, is sensed by the fish.

HUMAN BEHAVIOR—
SOME PROSPECTS

Recent discoveries have shown that some behavior is the result of events within the animal at the level of the cell and even the gene. Phenylketonuria, a congenital disease which often results in mental retardation, is caused by the absence of a single enzyme. In this case a retarded child's behavior stems from a specific genetic disorder. And this is just one of several known genetic defects that lead to abnormal behavior. A much publicized but less well authenticated example is that of violent "supermales"—males who have more than one Y-chromosome.

We do not know very much about the interaction of an animal's behavioral machinery and the environment. Most psychologists agree, for example, that young children must be exposed to learning situations if their mental abilities are to develop properly. But how much of intelligence is innate, and how much and how intelligence is shaped by the environment we do not know. As we understand more about the physiological basis of behavior, we may be able to change harmful or abnormal behavior. For example, can we develop chemicals that will safely depress the eating areas of the hypothalamus of a person who is obese because he grossly overeats? We know that many animals limit the size of their population when overcrowding threatens. How do they do this? What effects does overcrowding have on man's mental health and behavior? Answers to such questions have obvious importance. A few years ago the study of behavior was primarily the concern of psychologists. Now the field of psychobiology includes scientists from many different areas of study, all trying to contribute to our understanding of how animals, especially man, behave.

SUGGESTED READING LIST

Atkinson, R. C., and R. M. Shiffrin, "The Control of Short-Term Memory," *Scientific American*, August, 1971 (Offprint 538).

Brown, Jr., F. A., J. W. Hastings, and J. D. Palmer, *The Biological Clock*. Academic Press, New York, 1970.

Burnett, A., and T. Eisner, *Animal Adaptation*. Holt, Rinehart and Winston, New York, 1964.

Cooke, I., and M. Lipkin, Jr., *Cellular Neurophysiology, A Source Book*. Holt, Rinehart and Winston, Inc., New York, 1972.

Dethier, V. G., and E. Stellar, *Animal Behavior* (third edition). Prentice-Hall, Inc., Englewood Cliffs, 1970.

Eibl-Eibesfeldt, I., *Ethology: The Biology of Behavior*. Holt, Rinehart and Winston, New York, 1970.

Geschwind, N., "Language and the Brain," *Scientific American*, April, 1972.

Harlow, H. F., "Love in Infant Monkeys," *Scientific American*, June, 1959 (Offprint 429).

Harlow, H. F., J. L. McGaugh, and R. F. Thompson, *Psychology*. Albion Publishing Co., San Francisco, 1971.

Hasler, A. D., and J. A. Larsen, "The Homing Salmon," *Scientific American*, August, 1955 (Offprint 411).

Hess, E. H., "Imprinting in Animals," *Scientific American*, March, 1958 (Offprint 416).

————, "'Imprinting' in a Natural Laboratory," *Scientific American*, August, 1972.

Krogh, A., "The Language of the Bees," *Scientific American*, August, 1948 (Offprint 21).

Lorenz, K. Z., *King Solomon's Ring*. Thomas Y. Crowell Co., New York, 1952.

————, "The Evolution of Behavior," *Scientific American*, December, 1958 (Offprint 412).

Marler, P., and W. J. Hamilton, *Mechanisms of Animal Behavior*. John Wiley and Sons, Inc., 1966.

McGaugh, J. L., N. M. Weinberger, and R. E. Whalen, *Psychobiology*. W. H. Freeman, San Francisco, California, 1967.

McGill, T. E. (ed.), *Readings in Animal Behavior*. Holt, Rinehart and Winston, New York, 1965.

Mech, L. D., "Where the Wolves Are and How They Stand," *Natural History*, 80(4), 26, 1971.

Menaker, M., "Nonvisual Light Reception," *Scientific American*, March, 1972.

Rosenzweig, M. R., "Brain Changes in Response to Experience," *Scientific American*, February, 1972.

Thompson, R. F. (ed.), *Physiological Psychology*. Freeman, San Francisco, 1972.

Thorpe, W. H., *Learning and Instinct in Animals*. Harvard University Press, Cambridge, 1956.

Tinbergen, N., and the Editors of *Life*, *Animal Behavior* (Life Nature Library). Time-Life, New York, 1965.

Van der Kloot, W. G., *Behavior*. Holt, Rinehart and Winston, New York, 1968.

Van der Kloot, W. G., C. Walcott, and B. Dane, *Readings in the Biology of Behavior*. Holt, Rinehart and Winston, 1972.

Von Frisch, K., "Dialects in the Language of Bees," *Scientific American*, August, 1962 (Offprint 130).

————, *Bees: Their Vision, Chemical Senses, and Language* (rev. ed.). Cornell University Press, Ithaca, New York, 1971.

Wilson, E. O., *Insect Societies*. Belknap Press, Cambridge, Mass., 1971.

CHAPTER 15

The Hormones in Human Reproduction

REPRODUCTION—
THE GENERAL SCHEME

In *The Science of Life*, H. G. Wells, Julian Huxley and G. P. Wells wrote, "Cleared of the complication of sex, reproduction is seen to be simply the detachment of living bits of one generation, which grow up into the next." In primitive animals and in most of the world's plants, nonsexual reproduction by the "detachment of living bits" is a way of life, sometimes the only way, but often alternating with a sexual mechanism.

The flatworm, *Planaria*, and the ciliated protozoan, *Cothurnia*, reproduce by **vegetative fission**, that is, dividing in two. We did not say "*simply* dividing in two." We can hardly call a process we do not fully understand "simple." In the protozoa the lengthwise division of the single cell, with the formation of new organelles, is generally a long and complex process taking 7 to 9 hours for completion. The multicellular flatworm pinches in two transversely, just behind the pharynx, whereupon each half regenerates the missing portions. We can reproduce the latter process in the laboratory; *Planaria* cut in two also regenerate. Undifferentiated cells accumulate at each wound surface. These cells could be derived from one (or both) of two sources, the "dedifferentiation" of specialized cells or the "holding-over" of embryonic cells in the adult—the so-called *neoblasts*. Although conclusive proof is still lacking, experimental evidence favors the indispensable role of neoblasts in planarians (Fig. 15-1).

Both unicellular and multicellular animals have evolved variations of this theme. They may not divide into two equal cells, or two more or less equal halves, but extrude daughter cells as mere buds which break away while small. In the common freshwater polyp *Hydra* (Fig. 15-2) the buds develop from the side of the tubular parent. In some sponges the buds (or "gemmules") are internal, and are freed only when the parent dies.

When the bud is not a complete replica of the mature organism (an adult in miniature) but is incomplete—even a single cell—we call it a **spore.** Sporulation occurs in tissues especially earmarked for spore production. As we learn in Chapters 7 and 23, nearly all the world's plants employ this mechanism, usually alternating a *sporophyte* generation with a sexual (*gametophyte*) generation. The nonsexual and sexual forms may resemble each other, but more frequently, especially in the animals, they are strikingly different. In *Obelia,* a hydra-like form, the bud becomes a free-swimming jellyfish, which in turn produces a generation of polyps like the original hydroid (Fig. 15-3).

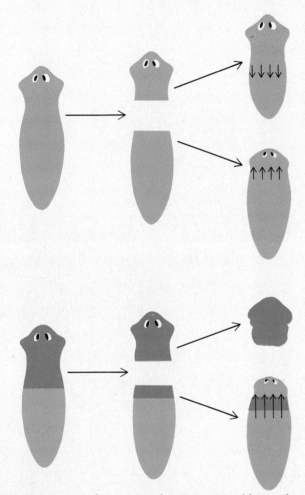

Figure 15-1 Above: Normal regeneration of bisected planarian illustrating the local source of blastema cells (short arrows). Below: Experiments by Dubois in which a partially X-rayed worm is cut through the irradiated (shaded) region. The irradiated head degenerates for lack of healthy cells from which to regenerate. The tail piece grows a new head, presumably from unirradiated neoblasts.

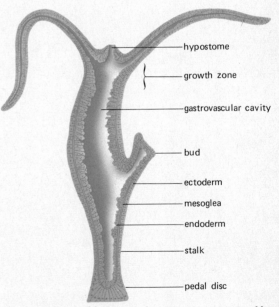

Figure 15-2 Budding in *Hydra.* Form is maintained by the turnover of cells originating in a growth zone just below the tentacles. Asexual budding occurs in a region intermediate between the head and foot.

WHY SEX?

As George Corner wrote, "No characteristic of man and the other animals is so fundamental, so completely taken for granted, as the existence of two sexes. It is the first fact the Bible mentions about the human race: '. . . male and female created He them.' In every nature myth the animals enter two by two. In primitive song and story every Jack that cracks his crown has a Jill that tumbles after. Man that is born of woman finds it impossible to think of a race with only one sex. . . ."

Yet sex is not necessary for the production of ample numbers of "replicas" of existing individuals. Why then have sexual mechanisms evolved to the point at which, in the higher animals, nonsexual mechanisms have been lost? Sexual reproduction provides the powerful advantage of *genetic recombination* (Chapter 3) whereby genetic change may be shared with other members of the population.

The gist of this brief preface to human reproduction is that the higher animals, ourselves included, reproduce their kind by the production and union of egg and sperm cells. It is, in outline, a simple task, yet to accomplish it requires the interaction of an elaborate set of organs. In Chapter 7 we laid the groundwork for this part of our story when we considered the formation of eggs and sperm, their union, the development of the fertilized egg, and (in mammals) the attachment of the embryo to its mother. Now we must consider the subtleties: the timing of events, the adjustment of the chemical and physical environment as each step leads to the next. As we know the body has two important ways of linking the action of its separate organs: the nervous system, discussed in Chapters 13 and 14, and the endocrine system, discussed in this chapter.

Figure 15-3   Life cycle of *Obelia*.

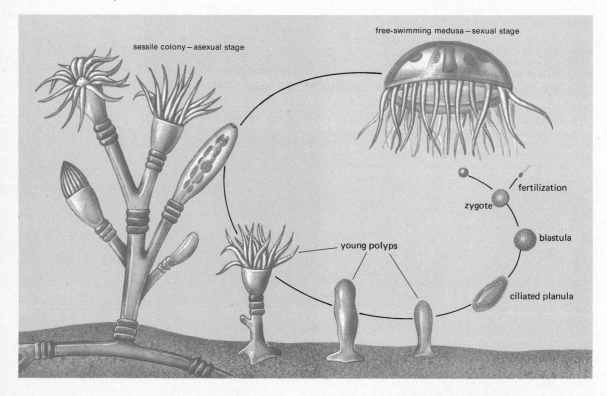

Figure 15-4 Reproductive organs of the human female.

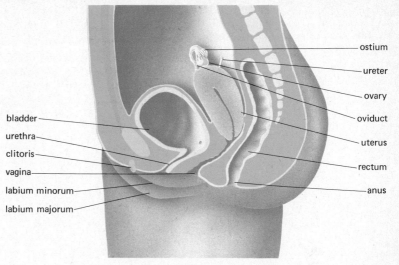

THE REPRODUCTIVE SYSTEM IN THE HUMAN FEMALE

"What can we say of the ovary, an organ so remarkable that it is able to produce the egg?" (Fig. 15-4). We should begin by relating that it has *two* prime functions: (1) producing eggs and (2) producing hormones that play a number of roles, such as development of secondary characters, affecting sex drive, and regulating the uterine environment.

The eggs of mammals, which average about 0.1 millimeter (0.004 inch) in diameter, are produced by ingrowth from the surface cells of the ovary, where they undergo maturation in *ovarian follicles,* which are round, fluid-filled egg chambers (Fig. 15-5). At birth the two human ovaries contain their lifetime supply of oöcytes (about 400,000 in all) which begin to mature, one at a time, at puberty (11 to 16 years). Whether a particular egg has waited 15 years or 40 years for an opportunity to mature, once its turn comes it

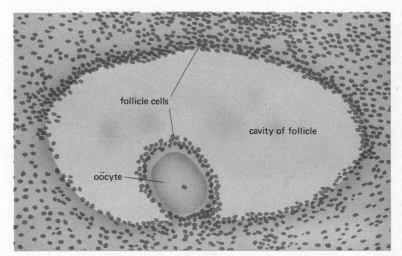

Figure 15-5 Diagram of a mammalian oöcyte surrounded by follicle cells.

sinks deeper into the ovary; the cells around it multiply to form a thick mass, which hollows out to form a follicle. The follicle continues to grow, and as it does it pushes its way to the surface. The wall of the follicle and the overlying capsule of the ovary stretch and become thinner, until at last they rupture and the contents of the follicle, including the egg, are expelled. In humans only one egg is usually expelled each month, although it occasionally happens that two or more eggs may be released, thereby resulting in twins and other multiple births.

Once released into the abdominal cavity, an egg is "picked up" by the oviduct (or fallopian tube). Cilia on the surface of the oviduct beat toward the uterus, transporting the egg out of the abdominal cavity into the tube and toward the uterus. If the egg encounters sperm in the upper reaches of the oviduct and is fertilized, it begins to divide and, upon reaching the uterus, implants and proceeds to develop as we have already learned (Chapter 7).

The ovary as timepiece

In introducing his chapter under this title, George Corner drew from a passage in Catlin's *The North*

American Indians: "Our Western plainsmen used to watch, in August and September, milling herds of bison, blackening the prairie for miles. It was no uncommon thing to see thousands of them, eddying and wheeling about under a dense cloud of dust raised by the bulls as they pawed in the dirt or engaged in desperate combat. In these herds the males were continually following the females and mating with them. The whole mass was in constant motion, all bellowing at once in deep and hollow sounds, which mingling together seemed at a distance of a mile or two like the noise of distant thunder. This was the yearly period of **estrus,** the mating time, when the females were ready to produce their eggs and the males to fertilize them."

The word estrus (or "heat"), for a recurrent period of sexual excitement, comes from the insect described by Virgil, ". . . the gadfly ('oestrus' is the Greek name for it)—a brute with a shrill buzz that drives whole herds crazy. . . ."

A mating season is almost universal in nature; robins nest in the spring, dogs mate two or three times a year, and if we could inspect their ovaries when they show signs of being ready, we would see the follicles enlarging, with a crop of ripening eggs.

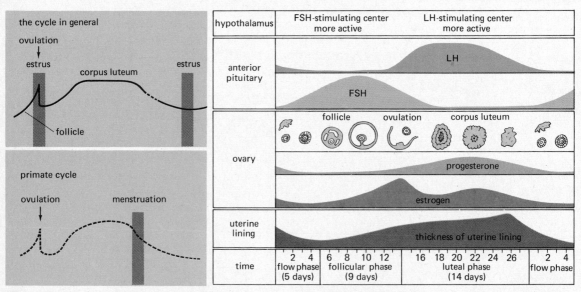

Figure 15-6 Left: Diagram comparing the estrus cycle with the menstrual cycle of higher primates. Right: The sequence of events in the human menstrual cycle.

"The ovary as timepiece runs at curiously different rates in different animals." Breeding may be timed with the seasons of the year, or with the tides, or at particular phases of the moon. Many animals, however, have cycles which are not related directly (or at least obviously) to natural phenomena: chickens lay an egg a day; in obedience to some obscure signal, the locusts swarm at intervals of 17 years.

Rats and mice ovulate every 4½ to 5 days, guinea pigs every 15; cows, mares, and swine have periods at 21-day intervals; dogs, cats, and lions have two or three estrus periods yearly; sheep are in heat several times in the late summer but are anestrus during the winter.

The most peculiar variation of all occurs in humans, apes, and higher monkeys. The length of the cycle is about four weeks, but there is no well-defined estrus period. Mating can occur at all times.

The follicle ripens and discharges its egg without a period of "heat." In some women ovulation is accompanied by a period of heightened sexual excitement, but it is by no means obvious or universal. About two weeks after ovulation the lining of the uterus breaks down, with bleeding. This periodic **menstruation** occurs only in humans and higher primates.

Let us examine the monthly cycle in the human female (Fig. 15-6), looking first at events within the ovary itself. As the follicle matures it produces the primary female sex hormones or *estrogens*, which maintain the uterus and vagina, play a role in the development of the breasts, indeed the feminine form, and regulate the distribution of hair. After shedding the egg, the follicle is transformed in appearance and function; the cells are laden with fat. In animals whose fat is yellow, the transformed follicles are bright yellow, which is why they were originally named *corpora lutea*, or yellow bodies. They have the appearance of secretory glands, as indeed they are, producing the second major female hormone, *progesterone*, whose role is to

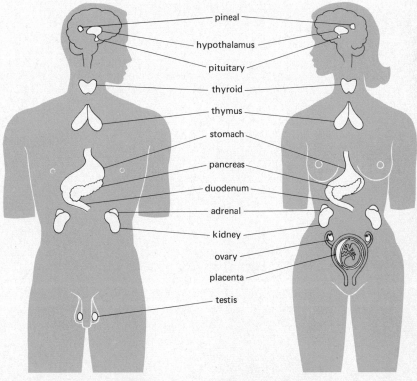

pineal
hypothalamus
pituitary
thyroid
thymus
stomach
pancreas
duodenum
adrenal
kidney
ovary
placenta
testis

Figure 15-7 The major endocrine glands in man.

prepare the uterus for implantation and nourishment of the embryo. Progesterone acts on the uterine lining, where it stimulates glandular activity, without which implantation cannot occur. Progesterone is, then, the hormone of pregnancy.

The ovary does not act alone; it functions in concert with the *anterior pituitary gland*. The monthly cycle is dependent on the release by the anterior pituitary of two hormones, follicle-stimulating hormone (FSH) and luteinizing hormone (LH). The release of these *gonadotropins* is in turn dependent on signals from the closely related part of the brain, the *hypothalamus*. This pattern involving the interaction of hypothalamus, anterior pituitary, and gonad is not an uncommon one. It is representative of the pattern exhibited by many of the endocrine glands (Figs. 15-7, 15-8; Table 15-1).

FSH stimulates some of the primary follicles of the ovary to begin growing each month, promoting proliferation of the cells surrounding the egg. These cells then begin to secrete estrogens. Thus the two functions of the follicle-stimulating hormone are to stimulate proliferation of the ovarian follicular cells and to stimulate secretory activity by these cells. As the follicles approach their maximum size, the anterior pituitary begins

Table 15-1 Principal vertebrate hormones

Source	Hormone	Functions
Adrenal cortex	Glucocorticoids (corticosterone, cortisone, hydrocortisone, etc.)	Conversion of noncarbohydrate to glycogen; help maintain normal blood-sugar level
	Mineralocorticoids (aldosterone, deoxycorticosterone, etc.)	Regulate sodium-potassium metabolism
	Cortical sex hormones (adrenosterone, etc.)	Stimulate secondary sexual characteristics, particularly those of the male
Adrenal medulla	Adrenalin	Increases blood sugar; augments sympathetic part of the autonomic nervous system
	Noradrenalin	Constricts blood vessels
Duodenum	Secretin	Stimulates secretion of pancreatic juice
	Cholecystokinin	Stimulates release of bile by gall bladder
	Enterogastrone	Inhibits gastric secretion
Hypothalamus to anterior pituitary	Releasing factors, RFs	Specific releasing factors are delivered by a vascular portal system and specifically stimulate the release of anterior pituitary hormones
	Prolactin-inhibiting factor	Inhibits prolactin secretion
Hypothalamus to posterior pituitary	Oxytocin	Stimulates uterine muscle contraction; releases milk
	Vasopressin (vasotocin in amphibia, reptiles, birds)	Stimulates water reabsorption by kidney tubules; stimulates smooth muscle contraction
Kidney	Erythropoietin	Stimulates production of red blood cells in bone marrow
Ovary	Estrogens	Initiate, maintain female secondary sexual characteristics; initiate periodic thickening of uterine mucosa; inhibit FSH
	Progesterone	Cooperates with estrogens in stimulating female secondary sexual characteristics; supports and glandularizes uterine mucosa; inhibits LH and FSH
	Relaxin	Stimulates relaxing of pelvic ligaments

to produce LH. It is LH that influences the mass of follicle cells to undergo their transformation into a corpus luteum and to begin producing progesterone.

How does the cyclic pattern of release of pituitary and ovarian hormones begin? The cycle is established by the ovary. The developing hypothalamus and pituitary are noncyclic, and become cyclic only after estrogen is produced by the ovary.

A description of menstruation

Having discussed the cellular events in which the sex hormones participate, we now return to the tissue level to examine the consequences of hormone deprivation. About the fourteenth or fifteenth day after ovulation in the human, the corpus luteum begins to degenerate. The uterus, thus deprived of progesterone, undergoes a violent reaction. The circulation of the blood in the inner lining is disrupted and tissues break down. Blood trickles toward the vaginal canal. The surface layer is sloughed away. In a few days the debris is cleared up and repair sets in.

What happens to cause this breakdown? To understand the phenomenon we must first look at the blood circulation of the *endometrium*, or uterine lining (Fig. 15-9). The lining is fed by arter-

Source	Hormone	Functions
Pancreas	Insulin	Causes reduction in amount of glucose in the blood; stimulates formation and storage of carbohydrates
	Glucagon	Causes increase in blood sugar by breakdown of glycogen to glucose in the liver
Parathyroid	Parathormone	Regulates calcium and phosphorus metabolism
Pineal	Melatonin	Possibly helps regulate pituitary function *via* hypothalamic releasing centers
Pituitary (anterior lobe)	Growth-stimulating hormone (GSH)	Stimulates growth
	Adrenocorticotrophic hormone (ACTH)	Stimulates adrenal cortex
	Thyroid stimulating hormone (TSH)	Stimulates thyroid
	Follicle-stimulating hormone (FSH)	Stimulates ovarian follicle development; tubule development in testes
	Luteinizing hormone (LH)	Stimulates conversion of ovarian follicular to corpus luteum; stimulates progesterone and testosterone production
	Lactogenic hormone (prolactin)	Stimulates milk production by mammary glands
Pituitary (intermediate lobe)	Melanocyte-stimulating hormone (MSH)	Controls expansion of skin pigment cells in amphibia
Pituitary (posterior lobe)	See hypothalamus	
Placenta, chorionic	Gonadotrophins	Reinforce or substitute for LH
	Relaxin	Stimulates relaxing of pelvic ligaments
Stomach	Gastrin	Stimulates secretion of gastric juice
Testis	Testosterone	Initiates, maintains male secondary sexual characteristics
Thymus	Thymic hormone	Stimulates lymphoid tissue in development
Thyroid	Thyroxin	Stimulates oxidative metabolism; stimulates amphibian metamorphosis; inhibits TSH
	Calcitonin	Prevents excessive rise in blood calcium

ies that come up through the underlying muscle. There are two kinds, straight and coiled. The latter branch into tiny capillaries near the surface. The fundamental step in the menstrual breakdown is a shut-off of the coiled arteries, a fact established by J. E. Markee's clever observations. He implanted bits of monkey endometrium into the anterior chamber of the same animal's eye, just behind the clear cornea. The graft heals and grows. It responds to estrogenic hormone. When menstruation occurs in the uterus, it occurs at the same time in the eye graft, where it can be watched. The first step is the blanching of patches of tissue, as the coiled arteries contract. After a few hours, this phase is over. Blood flows again with renewed force. However, the capillaries and surrounding tissues have been damaged by the shortage of blood and give way, resulting in hemorrhage.

The deepest pits of the endometrial glands remain intact during menstruation, despite the sloughing of all the outer layers of the endometrium. In the first days after menstruation, new epithelium grows from the edges of the glands to cover the inner surface of the uterus. Then, under the influence of renewed estrogen production by the ovaries, the endometrial cycle begins all over again.

THE REPRODUCTIVE SYSTEM IN THE HUMAN MALE

The testes of the male (Fig. 15-10) produce billions of spermatozoa (sperm), whose structure and functions we have already considered (Chapter 7); they elaborate the androgens or male hormones, androsterone and testosterone, which control the development and maintenance of male characteristics. The production of these hormones is in turn controlled by the pituitary hormones FSH and LH. In the male, FSH and LH are released at the onset of puberty, and secretion continues throughout most of life. Both are necessary for continuous sperm production. In the male, release of the gonadotropic hormone is continuous, rather than cyclic as in the female. FSH causes proliferation of spermatozoa. LH stimulates the production of the androgens or male hormones, androsterone and testosterone, by the interstitial cells of the testis (Fig. 7-6).

SEX HORMONES AND HUMAN BEHAVIOR

We learned in Chapters 7 and 8 that development does not end at birth but continues throughout life. Nowhere is this statement illustrated more clearly than in the delayed action of the hypo-

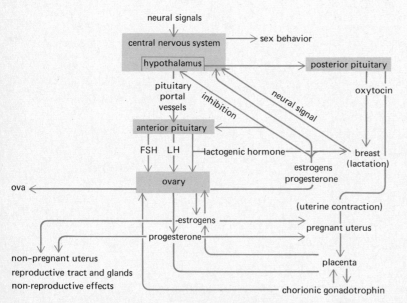

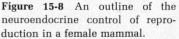

Figure 15-8 An outline of the neuroendocrine control of reproduction in a female mammal.

thalamus-pituitary-gonad axis. In humans the reproductive functions of the gonads do not begin until puberty, some 11 to 16 years after birth. We do not know what signals the release of FSH at puberty. We know only that once released it brings about maturation in both males and females.

In males, the second pituitary hormone, LH, is also released at puberty, which usually occurs somewhat later in males than in females. It is the release of testosterone, stimulated by LH, that profoundly influences the sexual maturation and behavior of the male at puberty. A growth spurt, production of hair, change in voice (caused by changes in the larynx) are all evidence of systemic changes in the male at puberty, as is his heightened aggressiveness. Puberty runs its course in about two years, after which the male becomes adjusted to the testosterone level which ordinarily remains rather constant over many years because of feedback controls on the pituitary-hypothalamus.

Puberty usually begins earlier in the female, at about 11 years. Under the influence of FSH, eggs begin to mature, and the female hormone ("the hormone of preparation of the uterus") estrogen is produced. Estrogen causes widespread systemic changes, resulting in the characteristic female form and behavior. The **vagina,** into which the male penis is inserted during copulation, and the **uterus** mature rapidly, the breasts grow and

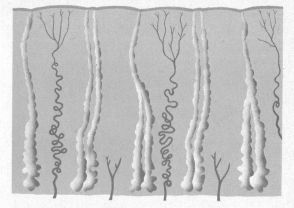

Figure 15-9   Diagram of the arteries of the uterus.

differentiate secretory cells for milk production, and the hips widen, providing an opening in the pelvic arch. In addition to estrogen small quantities of male hormones, androgens, are also produced in the female at puberty in the adrenal gland. These hormones act in initiating the appearance of pubic hair and in growth regulation. Adrenal tumors and other abnormal conditions may result in increased production of androgens, leading to disturbances in secondary sex characteristics.

In the female, LH, the hormone of gestation, is produced only after the first cycle is underway, with the production of the corpus luteum.

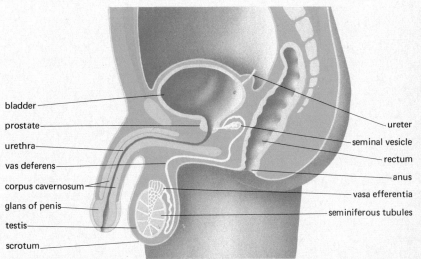

bladder

prostate

urethra

vas deferens

corpus cavernosum

glans of penis

testis

scrotum

ureter

seminal vesicle

rectum

anus

vasa efferentia

seminiferous tubules

Figure 15-10 Reproductive organs of the human male.

We have already observed that during ovulation a woman is ordinarily highly receptive to the male. After ovulation, when progesterone is produced (and the egg is not fertilized) and the menstrual period approaches, the woman is less receptive — increasingly irritable and aggressive. How much of this behavior can be attributed to progesterone is unclear, but its high level is surely a factor.

The emotional stress associated with the **menopause** is also well known. This period of a year or two at the end of a woman's reproductive life (in the absence of an outside source of hormones) may be attributed to the failure of the pituitary to secrete LH and to the subsequent dropping off of the ovary's ability to produce estrogen and progesterone. The withdrawal of these hormones, to which the body has been accustomed for 35 to 40 years, can cause physiological and psychological disturbances until a new balance is established.

MECHANISM OF SEX HORMONE ACTION

We have now seen that tissues associated with the reproductive process, such as the uterus and vagina, are unable to grow and function optimally without sex hormones. Either they have lost some

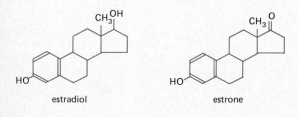

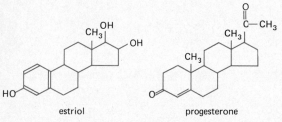

Figure 15-11 Estrogens (estradiol, estone and estriol) and progesterone.

vital capacity, which these steroid hormones restore, or they have gained an inhibitor or some mechanism of growth restraint, which the hormone relieves. The actual mechanism is still obscure, but enough is now known to indicate which directions may warrant closer scrutiny. The most promising lead has come through studies of the distribution of estrogens in the tissues of the body. These hormones circulate throughout the body, but only the "target" tissues — the uterus, vagina, and oviduct plus the interacting organs, anterior pituitary and hypothalamus — possess a striking affinity for them. This was learned by administering highly radioactive estrogens. The target cells contain specific estrogen-binding substances, generally believed to be proteins. Thus when an estrogen encounters the cell, it is coupled in some manner to a *receptor* protein. The hormone appears not to be metabolized or modified in any way. However, as a result of the binding, the protein may undergo a transformation in shape (conformation) which permits it and its attached hormone to move from the cell's cytoplasm into the nucleus. There the protein may be modified further and, interacting with the genes, regulate the synthesis of ribonucleic acids (leading in turn, as we know, to the synthesis of cellular protein and growth). Thus the mechanism has two steps; recent demonstrations that androgens function similarly led Jensen to comment on the uniformity of nature in handling sex hormones in the following rhyme:

> When hormones of gonadal extraction
> Dance through tissues of sexual attraction,
> It is not the fandango
> The twist or the tango,
> But the two-step that starts all the action.

HORMONES, RECEPTORS AND SECOND MESSENGERS

We have defined hormones, following the classic definition first advanced by Starling and Bayliss, as chemical agents which are released from one group of cells and travel via the bloodstream to affect one or more different groups of cells. Another definition focuses less on their mode of travel and more on their function; it states that hormones are information-transferring molecules, whose essential function is to transfer information from one set of cells to another. This definition is valuable for several reasons. (1) Since it includes neurotransmitter substances released from nerve endings, it emphasizes the similarities instead of the differences between the endocrine and nervous systems. (2) It is broad enough to include agents released from unicellular organisms; these agents may play a role similar to that of hormones in multicellular organisms and may also act similarly at the biochemical level. (3) It helps to distinguish hormones from other classes of biologically active compounds. Vitamins, for example, are concerned primarily with energy metabolism and not with the transfer of information.

Does this definition include all of the chemical agents included in the classic definition of hormones? It is not entirely clear that it does.

Today we usually recognize at least two very distinct types of hormones. These two types have not always been distinguished from one another, perhaps because they were included under a common definition. One type includes epinephrine, glucagon, insulin, gastrin, secretin, parathyroid hormone, calcitonin, and many of the hormones released from the anterior and posterior lobes of the pituitary gland. These and certain other hormones, together with the neurotransmitter substances released from nerve endings, seem clearly to play an *information-transferring role*. Cells respond to these hormones more or less rapidly, and the response is very often of short duration, or at least the magnitude of the response at any given instant is closely related to the amount of the hormone present.

The second class of hormones includes the steroid hormones, thyroid hormone, and at least one of the hormones produced by the anterior pituitary, namely growth hormone. These hormones play an information-transferring role in a sense, but differ in many ways from the hormones just mentioned. Cells respond to these hormones much less rapidly than to the others, and the changes produced by them may persist for long periods, regardless of the present concentration of the hormone. Indeed these differences have often been used, implicitly, if not always explicitly, to place these hormones in a separate category. They play an important *maintenance* role, and in their prolonged absence cells may become incapable of responding to other hormones. Consequently they have been referred to as *permissive* hormones. Since cells and tissues do not develop properly in their absence, they have also been called *developmental* hormones.

Biologists have designated the hormones of the first group as *messengers* and those of the second group as *maintenance engineers*. It is difficult to draw hard and fast lines, but in general despite some overlapping in functions, the distinction is a useful one.

We have already introduced the concept of the receptor, that portion of the cell which first interacts with the hormone. The interaction of hormone and receptor produces a further stimulus, which in turn effects cellular responses.

In discussing estrogens and androgens we were able to speak only of a receptor protein and to speculate about the consequences of hormone-receptor interaction. For many other hormones, however, we can be more precise. For all of the hormones listed in Table 15-2 the receptors are believed to be integral parts of one specific enzyme system, the enzyme being *adenyl cyclase*, which catalyzes the formation of cyclic AMP (adenosine 3′, 5′-phosphate) from ATP (Fig. 15-12). The hormone acts by inhibiting or stimulating cyclic AMP production. Cyclic AMP is now widely agreed to be a *second messenger*. The hormone might appropriately be regarded as a first messenger, with cyclic AMP as the second messenger. This basic concept is illustrated in

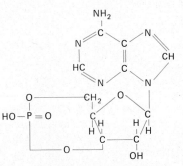

Figure 15-12 Adenosine 3′,5′-phosphate (cyclic-3′,5′-AMP, cyclic adenylic acid).

Figures 15-13 and 15-14. According to this concept the first messenger carries the required information to the cell, and the second messenger transfers this information to the cell's internal machinery. The end result of the increased level of cyclic AMP will depend entirely on the nature of the cell in which it occurs and on the prevailing conditions. How cyclic AMP acts to carry out this function is not clear.

Although cyclic AMP is the only second messenger recognized to date, it seems possible that other second messengers will be discovered. This seems especially likely in the case of certain hormones that are not known to be capable of stimulating adenyl cyclase.

PROSTAGLANDINS

The family of substances known as the prostaglandins resemble other known hormones in many of their effects; chemically, however, they are a quite different class of compounds.

In the early 1930s it was observed that substances in human semen caused uterine tissue to relax or contract, depending on whether the woman was fertile or sterile. Initially it was believed that the substances came from the prostate gland; hence the name, prostaglandins. However the prostaglandins in semen actually come from the seminal vesicle; hence the term is a misnomer. Prostaglandins were eventually found in many body fluids and tissues.

The prostaglandins are fatty acids, variants of a 20-carbon carboxylic acid incorporating a five-member cyclopentane ring (Fig. 15-15, top). They are synthesized in many tissues of the body from polyunsaturated fatty acids. The key to understanding prostaglandins and linking information about them to the rest of molecular biology is the fact that the main source of the precursors to prostaglandins is the phospholipids of the cell membrane. It seems that the cell membrane is a prime site of prostaglandin synthesis. These "hormones," if they can so be called, are present in very small quantities and are rapidly broken down. They are rarely concentrated, except in semen where there may be 100 micrograms per gram of wet weight, 100 times the usual amount in other tissues. The scarcity of prostaglandins resulted in a search for means of synthesizing them using precursors from plants or animals, including a fatty acid from an European herb, the borage plant, and a sea fan *Plexaura homomalla* found in coral reefs off the Florida coast.

Prostaglandins are metabolized very rapidly after intravenous injection, disappearing within a few minutes. They have a wide range of effects. One (PGE_2) lowers blood pressure; another (PGF_2-alpha) raises it. Both PGE_2 and PGF_2-alpha stimulate uterine contractions and are used to facilitate childbearing labor. Others (E_1 or E_2) inhibit gastric secretion in dogs and may prove useful in preventing peptic ulcers.

How can we relate the prostaglandins to cyclic AMP, the second messenger? They meet

Table 15-2   Hormones which utilize cyclic AMP as a second messenger

Hormone	Tissue
Catecholamines	Various tissues
Glucagon	Liver, pancreatic islets, and adipose tissue
ACTH	Adrenal cortex and adipose tissue
LH	Ovarian and testicular tissue
Vasopressin	Various epithelial tissues
Parathyroid hormone	Kidney and bone
TSH	Thyroid tissue
MSH	Frog skin
Prostaglandins	Various tissues
Histamine	Brain
Serotonin	*Fasicola hepatica*

most directly, it seems, in the membrane of the cell, which provides a common medium of communication for cells with other cells and with the battery of signals impinging upon them.

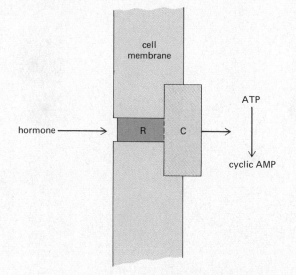

Figure 15-13 Possible model of the protein component of the membrane adenyl cyclase system. (R) regulatory subunit, facing the extracellular fluid; (C) catalytic subunit, with its active center toward the cell's interior.

CONCEPTION AND CONTRACEPTION —POPULATION CONTROL

Population research is the name we give to efforts to learn more about all the factors of human reproduction and how best to modify one or more of them so that conception may, at will, be harmlessly and effectively precluded as a possible consequence of coitus. Why? Simply because there are already more people on earth than presently utilized resources can properly care for, because unless the world is stricken with a great catastrophe, we shall soon have to apply our resources to a population of 10 billion, instead of today's 3.6 billion, and because Malthus was right when he observed, about 175 years ago, that "moral restraint is of dubious effectiveness."

What factors, what steps in the reproductive process can be modified? We have related most of the basic facts about the production of eggs and sperm. What remains to be considered are the steps leading to their union.

Coitus

In the male there is normally a supply of sperm in the testes and ducts, awaiting a signal from the nervous system. Sensory receptors (visual, touch, olfactory) are stimulated by interaction with a

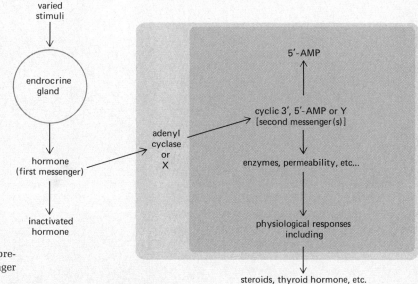

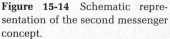

Figure 15-14 Schematic representation of the second messenger concept.

female. As a result of such stimuli in sexual fore-play, sperm move into the *epididymis* (Fig. 15-10) and *vas deferens*, the tubes and canal that connects to the urinary canal of the penis, re-spectively. The sperm remain inactive until they are discharged into the vagina during copula-tion, as a result of stimulation of sensory receptors in the head of the penis. At **orgasm**, a massive neuromuscular reaction, sperm move through the vas deferens into and out of the urethra; with them in the ejaculate flows the seminal fluid, provided by the seminal vesicles and *prostate gland*. This fluid contains substances which nourish and activate the sperm.

When sperm accumulate in the passages without discharge, muscle tension results in pain in the groin, which can be relieved by masturbation, a self-induced emission of sperm, or by spontaneous emission during sleep ("wet dream") as frequently occurs in adolescent males.

For the male the "essentials" are an erect penis, erection being the result of the filling of *cavernous bodies* with blood during foreplay and copulatory movements resulting in orgasm, with release of sperm. Although females gen-erally enjoy stimulation of the vaginal walls and the tip of the uterine cervix by the penis, the female orgasm results primarily from stimu-

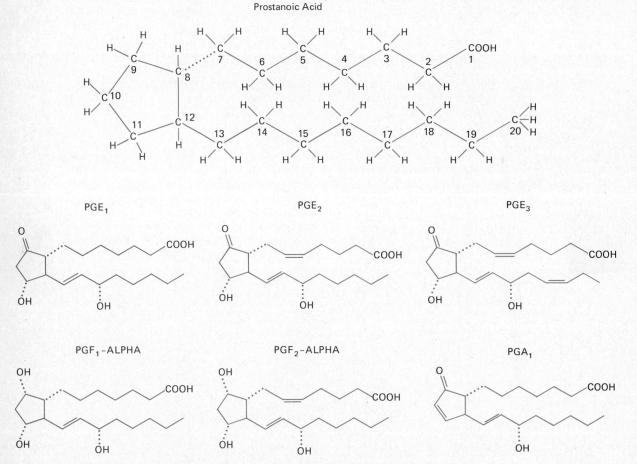

Figure 15-15   Some important prostaglandins. *Top:* prostanoic acid, the basic 20 carbon carboxylic acid; PGE and PGF, the primary prostaglandins. The PGE series have an oxygen atom (O) attached to the cyclopentane ring at site 9; PGF's have a hydroxyl (OH) at the same site. Hydrogen atoms are shown only in the top figure.

lation of the *clitoris* during movements of copulation. The clitoris is analogous to the *glans penis* of the male, becoming erect due to increased blood pressure. The female orgasm is similar to that of the male, with three exceptions: (1) it may be longer than in the male; (2) there may be multiple orgasms; and (3) there is no discharge.

Sperm, moving partly under their own power, but assisted by muscular movements of the uterus, traverse the cervix, the uterus, and the fallopian tube to reach the egg. If the male has a normal sperm count of 100 million or more (in an ejaculate of 3.5 milliliters), a few thousand will reach the egg. If the male has been hyperactive sexually (more than once every 24 hours) or inactive (less often than every 5 days), his sperm count may be too low, thereby reducing the probability of fertilization, which we have already examined (Chapter 7).

The "safe period"

There is evidence from many mammalian species, including humans, that eggs can be fertilized only while in the oviduct, some two to four days after their discharge from the ovary. We know, too, that in humans sperm cells can only survive about 36 hours in the female reproductive tract. Since ovulation usually occurs about midway between menstrual periods, it is sometimes thought that the fertile period can be safely charted, and that if copulation is avoided for five days preceding and five days following this midpoint, the possibility of fertilization can be eliminated. This is the theoretical basis of the so-called safe period or rhythm system of birth control. If all women had regular cycles, and things never happened out of turn, it would be an effective method. Irregularity and human error make it much less than certain. One of the "irregularities" stems from the discovery in humans of *reflex ovulation*. In the rabbit ovulation occurs as a consequence of copulation. Sensory receptors in the vagina signal the hypothalamus, which in turn triggers the pituitary to "inform" the ovary, and eggs are released. It was long thought that the process never occurred in humans. It now appears that copulation infrequently may start a cycle in the alternate ovary. Thus any calculation of a safe period is thrown

into doubt. While speaking of the "unsafeness" of safe periods, now is a good time to dispel the myth that pregnancy cannot occur during lactation. Ovulation may occur within a month after childbirth.

Birth control methods

The idea of **contraception** can be traced back some 5000 years. The first known medical text, which dates from about 1850 B.C. in Egypt, contains several prescriptions, one of them being the use of crocodile dung mixed with paste, probably intended for insertion into the vagina. Another consisted of plugging the vagina with honey and natron (sodium carbonate).

The desire and need to control fertility are one of contemporary man's primary goals. The fight to gain acceptance of this concept by the medical profession and the laws of the land is one of our most striking examples not only of how science contributes to the solution of societal problems but also of how social innovators influence the course of science.

Before we tell this fascinating story, let us review the available methods.

1. **Abstinence:** The complete elimination of sexual contact (*celibacy*) is even more difficult to achieve than the rhythm system, except for those rare instances where religious vows (and isolation) are strong enough to ensure separation of the sexes.

2. **Surgical ligature of the vas deferens or fallopian tubes** achieves infertility without affecting sexual drive. However, they are not "harmless" techniques for precluding conception. Although vasectomy is a simple office procedure, it is accepted by only about 40,000 American males annually. Sterilization of women requires hospitalization, but is nevertheless elected by about 100,000 women each year. The surgical procedures are being simplified; however, the problem is the very limited reversibility of the procedures in both males and females.

3. **Sperm blocking and killing** methods include a condom or rubber sheath worn over the penis, a rubber diaphragm, blocking the cervi-

cal opening, and spermicidal creams to kill sperm in the vagina. These methods may be used singly or in combination. Singly they may be 95 percent effective.

4. **Intrauterine devices** (IUDs), which are 98 to 99 percent effective, consist of a plastic or plastic plus metal (copper, zinc) coil inserted into the uterus by a physician. Normal fertility is restored upon removal of the loop. We do not know precisely *how* it works, but presumably it maintains the uterus in a "motile" state, thereby preventing implantation of the fertilized egg.

5. **The "Pill."** Estrogen and progesterone levels remain high during pregnancy, and ovulation is suppressed. The "Pill" contains a mixture of estrogens and progesterone in a balance designed to reproduce the hormonal level in the body during the period between ovulation and implantation (Fig. 15-16). If properly used, the Pill inhibits ovulation. When its use is discontinued, pregnancy may occur promptly, unless other methods are employed, for by regulating the cycle fertility is enhanced. Should pregnancy occur, owing to a lapse in using the Pill or its rare failure, its further use should be stopped at once, for it stimulates excessive production of male hor-

mone which may bring about the masculinization of female embryos.

There are other difficulties. The Pill is not medically safe enough for use by a substantial fraction of the population. For obscure reasons calcium balance is upset, leading to the formation of blood clots inside blood vessels (*thrombosis*). The incidence of thrombosis is high enough to preclude the use of the Pill by anyone with a history of vascular disease. In addition the relations between the pill and mammary cancer have not been clarified. Many physicians and patients find it unacceptable. Thus the search must be continued.

How the "Pill" was developed

The prospects for finding new methods are excellent, for we have several generations of research on which to build. The historical tracing of research leading to the Pill is shown in Figure 15-17. It began with the discovery of the existence of the male and female sex hormones in 1849 and 1896, respectively, with the description of the corpus luteum following shortly in 1898. From the 1920s through the 1940s the interactions of pituitary and gonads were described, and hor-

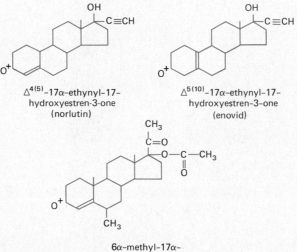

Figure 15-16 Synthetic progestins ("The Pill").

monal activities at all stages in the female cycle were determined.

At the same time, largely independent of the work on reproductive physiology, the field of steroid chemistry was being developed. Estrogen and progesterone were chemically purified and identified. However, the costs were prohibitive. For example, in the 1930s, in order to produce less than 300 milligrams of testosterone, a ton of bull testicles was processed. The ovaries of over 80,000 sows yielded 12 milligrams of estrogen. In 1940 the market price of progesterone was $200 per gram.

In the 1930s Marker developed methods for using plant steroids, compounds called "sapogenins" resembling cholesterol, which were derived from roots of sarsparilla, yucca, and the yam. He learned how to degrade sapogenin, convert it to progesterone, and progesterone to testosterone.

In the 1940s he went to Mexico and scouted sources of sapogenin-secreting plants, settling on the wild yam, *Dioscorea mexicana*. From these discoveries emerged the Mexican pharmaceutical firm Syntex S. A., the first to produce sex steroids from plants.

By about 1950, then, the stage was set *scientifically* for the development of the oral contraceptive pill. *But more than science was necessary.* The dangers of the "population explosion" and the importance of family planning had to be recognized by the public, medical practitioners, and lawmakers. This crusade was led by Margaret Sanger. Gregory Pincus, one of the pioneers along with John Rock in developing the Pill, stated that a 1951 visit by Sanger convinced him of the need to begin direct development of the Pill, which was in fact begun in 1952. Field trials were undertaken in Puerto Rico, Haiti, and Los Angeles, and in 1960 the first oral contraceptive was approved for general use.

We already noted that we are still concerned about the as-yet-unknown consequences of submitting a large population of healthy women to the Pill. The current methods are not anywhere near perfect. Among the innovative approaches now being studied are weekly or monthly pills for men or women, a small plastic implant placed under the skin that could last for years, thus rendering a specific, positive action necessary when pregnancy was desired, and techniques permitting reversible sterilization.

As Norman Borlaug, the botanist-nutritionist, stated in accepting the Nobel Peace Prize, "The green revolution has won a temporary success in man's war against hunger and deprivation; it has given man a breathing space. If fully implemented, the revolution can provide sufficient food for sustenance during the next three decades. But the frightening power of human reproduction must also be curbed; otherwise, the success of the green revolution will be ephemeral only."

STERILITY AND VENEREAL DISEASE

About one male out of every thirty is sterile, the most frequent cause being previous infection in the male genital ducts, although occasionally the seminiferous tubules of the testes may have been destroyed by mumps, typhus, or X-rays. Congenital defects may result in the production of abnormal sperm. The sperm may have two tails, two heads, or other less obvious abnormalities. These cannot fertilize an ovum, and when a large number of them occur in the ejaculate, infertility is almost certain.

About one female in fifteen is sterile. Although sterility may result from congenital defects or from dysfunction of the anterior pituitary (or other steps in the reproductive process), the most common cause is previous infection, blocking the oviducts, or enclosing the ovaries in scar tissue.

The most common infection causing female sterility, as is true also in the male, is *gonorrhea*. Venereal disease, including both gonorrhea and syphilis, is on the rise. There have been many positive consequences of the ready availability of IUDs and the Pill and of the changing attitudes throughout the world leading to increased sexual freedom. However, there has been at least one unfortunate consequence, the rapid spread of "VD," especially among young people who are reluctant to seek prompt, responsible medical care. The importance of immediate attention to these infections to avoid the prospect of permanent damage cannot be overemphasized.

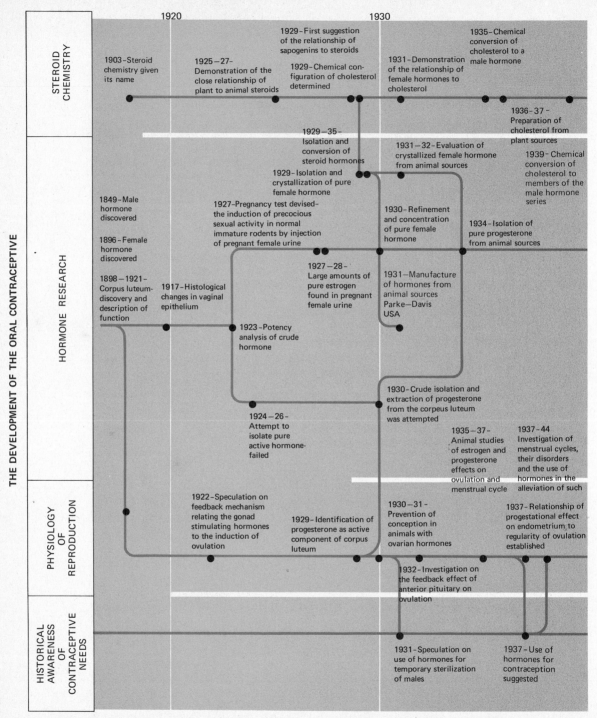

Figure 15-17   The development of the oral contraceptive.

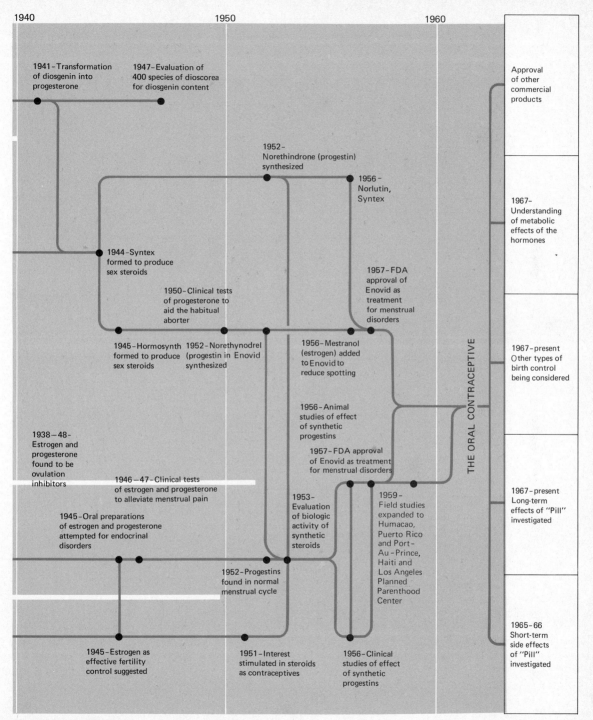

Figure 15-17 cont'd

HOMOSEXUALITY AND TRANSVESTISM

The general public's recent awakening to the reality of homosexuality has not brought anything approaching understanding. As early as 1953 the "Kinsey Report" indicated that almost 40 percent of adult American men and roughly 13 percent of adult American women have had overt homosexual contact to the point of orgasm. There are fewer overt "exclusive" homosexuals, the incidence being under 5 percent in both men and women. There are many forms of homosexuality; it is no more precise to speak of homosexuality than it is to speak of cancer, for in neither case is one dealing with a single condition.

It is especially important to distinguish between the purely erotic aspect of sexual behavior and the *gender role* (or preferred sex role), a term used to denote those aspects of attitudes and behavior which serve to disclose the person as having the status of boy or man, girl or woman.

Homosexuality may be *incidental*—adolescents get involved "for kicks"—or *situational*—in prisoners isolated from heterosexual possibilities. Some, but not all homosexuals display varying degrees of inversion of preferred sex role. Thus one can identify effeminate (or passive)

male homosexuals or masculine (or active) female homosexuals. The only common denominator in homosexuality is the use of, or preference for a sexual partner of the same somatic sex.

Transvestism ("cross-dressing") is the term used to refer broadly to those individuals who dress, or intensely desire to dress, in the clothes of the opposite biologic sex. Again, like homosexuality, transvestism is not a uniform behavioral condition, but occurs in varying degrees and with sundry implications.

We would not label a person cross-dressed for a masquerade party a transvestite. Only if the individual cross-dresses in search of *psychosexual comfort and satisfaction* is it to be considered transvestism. Many individuals who cross-dress are not homosexuals. Depending on life circumstances the psychologically inverted transvestite may attempt to live according to his psychologic orientation, that is, as a member of the opposite sex.

The term "transsexualist" has been used to distinguish transvestites who earnestly desire and often actively seek as a possible solution to their dilemma surgical removal or "correction" of what they consider to be inappropriate genital equipment. These individuals seek acceptance as a member of the opposite biologic sex, because it seems to them vastly more natural to behave and think like a member of the opposite biologic sex.

Table 15-3   Atypical sexual behavior and gender role inversion

---

Atypical sexual behavior not related to gender role inversion

1. Situational homosexuality
2. Incidental homosexuality
3. Neurotic (reparative) homosexuality
4. "Chaotic" sexuality (including homosexuality) in schizophrenics and mentally defective persons
5. Compulsive cross-dressing as an adjuvant to sexual performance
6. Fetishism
7. Exhibitionism
8. Impotence, frigidity

Atypical sexual behavior related to gender role inversion

1. Effeminate, "passive" homosexuality in males
2. Masculine, "aggressive" homosexuality in females
3. Transvestism; transsexualism in males or females

The human habit of labeling any given sex practice as *emotionally sick, abnormal,* or *perverted* is largely a judgment reflecting the consensus of attitudes about sexual behavior in the society. Flourishing societies exist where homosexuality is a normal and accepted part of the communities' sex life. Such societal differences appear to be the result of differing influences on social learning within the culture itself; genetic differences have never been found to be sufficient to account for the diversity of ethnic customs. It seems both inaccurate and unnecessary to regard all psychosexual deviations as manifestations of illness, unless the term is to be narrowly defined as any deviation from the cultural mode.

What evidence we now have tells us that "homosexuals are made, not born." There is no clear or compelling evidence of chromosomal or hormonal differences between homosexual and heterosexual populations. It appears that sexual behavior in humans, both typical and atypical, is the product of learning and experience in the social context. Psychosexually neutral at birth, an individual's gender role, including erotic orientation, becomes established during the process of growing up, as the result of a myriad of life experiences subtly imposed and governed by the culture of which he or she is a member.

The public's awareness of the existence of homosexuality has affected life throughout the world. We cannot be indifferent to homosexuality. Homosexuals live "against the grain"; when their behavior and life styles conflict with the cultural mode, their resulting emotional stress may be intense. Psychotherapeutic techniques can effect improvements in an individual's social adjustment and personal comfort. Looking toward the future, however, what are most needed are (1) public acceptance and a legal code that does not penalize homosexuality per se; and (2) greater emphasis on the detection of inversions in gender-role learning during a child's growing up years, for it is then, if ever, that remedial intervention can take place.

SUGGESTED READING LIST

Böving, B. G., "Anatomy of Reproduction," in *Obstetrics,* 13th edition, J. P. Greenhill (ed.). Saunders, Philadelphia, 1965, p. 3.

Burns, R. K. "Experimental Reversal of Sex in the Gonads of the Opossum, *Didelphys virginiana,*" *Proceedings of the National Academy of Sciences,* Vol. 41, p. 669 (1955).

———, "Role of Hormones in the Differentiation of Sex," in *Sex and Internal Secretions,* W. C. Young (ed.). Williams & Wilkins, Baltimore, p. 76.

Corner, G. W., *Ourselves Unborn.* Yale University Press, New Haven, Conn., 1944.

———, *The Hormones in Human Reproduction* (reprint). Atheneum, New York, 1963.

DeHaan, R. L., and H. Ursprung (eds.), *Organogenesis.* Holt, Rinehart and Winston, New York, 1965.

Etkin, W., and L. I. Gilbert (eds.), *Metamorphosis. A Problem in Developmental Biology,* Appleton-Century-Crofts, New York, 1968.

Goss, R. J. *Principles of Regeneration.* Academic Press, New York, 1969.

Jost, A., "Gonadal Hormones in the Sex Differentiation of the Mammalian Fetus," in *Organogenesis,* R. L. DeHaan and H. Ursprung (eds.). Holt, Rinehart and Winston, New York, 1965, p. 611.

Lillie, F. R., "The theory of the free-martin," *Science,* Vol. 43, p. 611 (1916). Reprinted in *Foundations of Experimental Embryology,* B. H. Willier and J. M. Oppenheimer (eds.). Prentice-Hall, Englewood Cliffs, New Jersey, 1964, p. 137.

Lloyd, C. W., *Human Reproduction and Sexual Behavior.* Lea and Febiger, Philadelphia, 1964.

Parkes, A. S., *Marshall's Physiology of Reproduction,* third edition. 3 volumes. Little Brown, Boston, Mass., 1956–1965.

Pike, J. E., "Prostaglandins," *Scientific American,* Vol. 225, p. 84, 1971.

Robison, G. A., R. W. Butcher, and E. W. Sutherland, *Cyclic AMP.* Academic Press, New York, 1971.

Schaffer, A. J., *Diseases of the Newborn,* second edition. W. B. Saunders Company, Philadelphia, 1965.

Short, R. V., "Reproduction," *Ann. Rev. Physiol.,* Vol. 29, p. 373, 1967.

Torrey, T. W., *Morphogenesis of the Vertebrates,* second edition. Wiley, New York, 1967.

CHAPTER 16

The Web of Life

For two hundred miles along the eastern shore of Lake Michigan, great dunes of sand rim the water's edge. The waves drive fresh sand up onto the beach, where it dries. Winds then drift it into shifting mounds that eventually become long dunes running parallel to the shore. At Sleeping Bear Point, near the southern edge of this vast system of dunes, the sand rises nearly 500 feet above the lake to form a horizontal ridge 5 miles long (Fig. 16-1).

If we were to walk inland from the shore, we would pass through a sequence of different communities of plants and animals, each adapted to the environmental conditions of its particular place. Together these communities form a **succession** from simple to increasingly complex organizational units in the *web of life*. Ecologists use the term **community** to refer to all of the populations of plants and animals that live together in the same area. As we shall see later these species are linked together in a system of functional relationships in which each kind of plant and animal has a particular role to play in the community.

Why is there an orderly sequence of different communities as one moves away from the lake, and why is it that a community near the edge of the lake contains relatively few species, whereas one a long way from the lake is far more complex and is dominated by large trees rather than low-growing grasses? We must look to the history of Lake Michigan for the answer to these questions.

TIME AND SUCCESSION

During the last few thousand years Lake Michigan has been steadily receding, so that its present shores are a long distance from where they once were. As the lake has grown smaller with the passage of time, it has left behind it old dunes which have become stabilized and have undergone a succession from simple dune communities to complex forests, while the younger dunes near the present shoreline show the early beginnings of this development. Thus by observing the sequence of communities from the shore inland, we can reconstruct the succession in time which led to the development of mature forests far from the present shore, where shifting sands were once the only feature on the landscape.

There are two types of ecological succession, **primary** and **secondary.** If a field that has been used to grow agricultural crops is abandoned, it will soon become overgrown with grasses, goldenrod, and other herbaceous plants. If this field was once a woodland, the weed fields will be invaded in a few years by shrubs such as blackberry, sumac, and hawthorn. In time the field will support a woodland of hickory, maple, oaks, or pines. This is not a haphazard process— it is an orderly sequence of change from simple to more complex communities, from a few species of grasses and herbs to a forest of trees, shrubs, and a variety of ground plants. For each geographic region, with its particular conditions of climate and soil, there is a characteristic sequence of communities. Because this succession takes place where vegetation was formerly established, and proceeds from a state in which a prior community of plants and animals was present, it is called **secondary succession.** Secondary succession occurs in areas that are disturbed by man or animals, or by natural forces such as fires and wind storms.

Primary succession begins on surfaces that have never before been colonized by plants or animals. Bare rocks on exposed sites are common locations for primary terrestrial succession. These rocks are first colonized by lichens which spread over the surface of the rock. When organic debris and soil begin to accumulate in crevices in the rock, other plants can take root. Mosses and grasses become established and, as wind and water and ice create deeper crevices with greater accumulations of soil, woody plants begin to take hold. Eventually deep soils may be formed, allowing the development of shrub communities and, finally, forests.

At the edge of Lake Michigan, between the water and the first sign of vegetation, is an expanse of bare sand that supports a small community of scavenger insects such as flies and beetles that feed entirely on organic debris washed ashore by the waves. Other insects may feed on these scavengers and on the invertebrates that burrow in the sand and also depend on organic debris for food. These animals are mostly transients that take advantage of a temporary abundance of organic debris, but retreat to other areas when storms sweep the beaches clean.

Figure 16-1 The Sleeping Bear sand dunes of Lake Michigan.

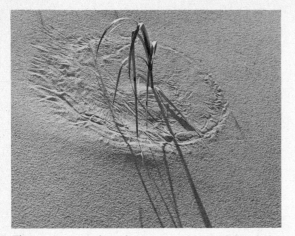

Figure 16-2 Early colonization of a sand beach by grass.

The first permanent community begins at the edge of the beach where grasses begin to take hold. The roots of the grasses and plants such as sand cherry bind the shifting sand and begin the process of stabilization of the substrate and soil formation (Fig. 16-3). As the leaves of these plants fall to the ground at the end of each growing season, they are eaten and broken down into small fragments by insects and other animals that live in the soil. This process produces a deposit

of organic material called *humus*, and also releases nutrients which the plants can use for further growth. When the humus has accumulated to a sufficient depth and the soil is enriched with nutrients, seeds of trees can germinate and grow.

The first trees to colonize the area immediately behind the grass-covered dunes are usually eastern cottonwoods. They are adapted to full sunlight and their seedlings can take root and grow even though there is no shade to reduce the soil temperature and conserve moisture. The cottonwoods, because of their large size and abundant leaves, provide even greater accumulations of humus in the soil as their leaves and branches fall off and decay. The droppings of insects, birds, and other animals also enrich the humus, and decay is hastened in the more humid atmosphere provided by the shelter of the trees. The trees break the force of the raw lake winds, they create shade, increase the moisture of the air, and promote the development of soft earth in place of shifting sands. Thus this first forest on the lake shore has reshaped the physical environment, paving the way for even greater changes to come.

Farther from the shore the cottonwoods gradually give way to pines, and a little farther inland the pines are replaced by oaks. Pine and oak seedlings require shade for their development.

Figure 16-3 Grasses and low-growing shrubs begin to bind and stabilize the shifting dunes.

The cottonwoods shield the young pine seedlings from too much sun, and the pines in turn offer more shade for young oaks. These changes in plant species are also accompanied by corresponding changes in the kinds of animals found in each of the communities. With the advent of trees, opportunities are quickly exploited by birds, squirrels, insects, and other animals that depend on trees in some way for food and shelter.

The final stage in this sequence is a forest of beech and maple and the various shrubs and ground plants associated with them (Fig. 16-4). Seedlings of beech and maple are adapted to shade and will not grow in open sunlight. On the other hand, a beech-maple forest is so densely shaded that young pines or oaks cannot take root and grow in it. In other words, we have reached the other end of the environmental spectrum from dune to forest. Within the environment of the beech-maple forest there is less wind, less light, more moisture, and much less evaporation than in any of the other communities between it and the open dunes. This forest is what ecologists call a **climax community.** Given the climate of the Lake Michigan region, no further development will take place beyond the stage of the beech-maple forest. It represents a stage of equilibrium with the physical environment. If not severely disturbed by drought, fire, or storms that blow down large numbers of trees and open the canopy to sun and wind, this forest will keep replacing itself with the same species year by year.

NATURE'S MACHINERY

Our observations along a transect in space have given us insights into the sequence of events that occurred during a few thousand years at a single point in which the present beech-maple climax now stands. So far our observations have been confined mainly to obvious features of the environment which are plainly visible—to a lake, the sand dunes at its edge, trees and shrubs, and a few conspicuous animals. But we have also suggested that these different communities are in some way organized into functional units, each with a special set of species and characteristics. Also the process of succession is an orderly pro-

gression from simple to complex communities, one community creating conditions that provide for the succession of the next. But in what ways are these natural groupings of plants and animals organized into well-defined systems? What conditions determine which plants or animals will occur in a particular environment and how they will interact to produce an orderly system?

Modern biology seeks to discover order in the enormous complexity of the living world and to determine how the systems at each level of biological organization have evolved and operate. **Ecology** is the study of the relationship between plants and animals and their environments, and ecologists are primarily concerned with the organization and functions of **populations, communities,** and **ecosystems.**

The **ecosystem** is the basic functional unit of ecology, and includes the nonliving, physical environment as well as the organisms that live in it. As we have already seen, organisms are dependent on the conditions of the physical environment, but may also modify that environ-

Figure 16-4 A mature deciduous forest represents the climax of dune succession.

ment by their presence. Each influences the properties of the other, and both are necessary for the maintenance of life. The study of ecosystems is in fact the study of the structure and function of nature. Once we understand how an ecosystem is constructed and how it operates, we will then have a greater understanding of how different subunits such as populations or communities, or particular environments such as oceans or lakes or forests, maintain themselves and contribute to the overall economy of the biosphere.

A community such as a beech-maple forest contains both **autotrophic** and **heterotrophic** organisms (see Chap. 9). For example, green plants (autotrophic) use light energy to manufacture organic substances from simple inorganic materials through the process of photosynthesis. These substances are then eaten by animals (heterotrophic) and the dead protoplasm of plants and animals is eventually broken down and decomposed by heterotrophic bacteria and fungi, releasing inorganic nutrients back into the system. These nutrients may then be used again by the green plants to manufacture more organic substance, completing a cycle in which the materials required for life pass through a sequence of inorganic and organic forms.

From this we can conclude that an ecosystem has four structural components: (1) **nonliving substances,** the basic chemical elements and compounds required by living organisms; (2) **producers,** usually green plants, which manufacture food; (3) **consumers,** animals that consume food manufactured by the producers or eat other animals; and (4) **decomposers** that break down the complex substances of dead protoplasm and release nutrients which the producers can reuse. Each of these components is essential to the total function of the system. Light and nutrients must be present in adequate amounts in order for the producers to carry on photosynthesis. Without the producers there would be no food for the consumers, and without the decomposers there would be a steady accumulation of undecayed organic debris which would overwhelm the environment, and the nutrients required by the producers would not be released from their organic forms.

THE CYCLE OF NUTRIENTS

A key to the proper functioning of an ecosystem is the movement of chemical elements through the more or less circular pathways that exist between organisms and the physical environment. These pathways are referred to as **biogeochemical cycles,** a term which indicates different positions in the cycle, from the organisms (*bio*), to the rocks, soil, air, or water (*geo*) of the physical environment. Living organisms require between 30 and 40 of the 90-odd elements that are known to occur in nature. Some of these elements, such as hydrogen, carbon, oxygen, and nitrogen, are required in large amounts, whereas others are needed in smaller quantities. Regardless of how small the quantity of an element that may be required, it may be essential to life.

Different chemical elements show characteristic biogeochemical cycles. Within the biosphere there are two general types of cycle: (1) a gaseous type, in which the major reservoir for the element is air, and (2) a sedimentary type, in which the environmental reservoir of the element is soil or rock. The carbon cycle is a gaseous type which is relatively perfect in that the element cycles freely between organisms and the environment and no major changes occur in the distribution of this element within the biosphere, even though there may be local shortages in particular environments. Phosphorus exhibits a less perfect, sedimentary type of cycle in which the element may become incorporated in rocks where it is unavailable to organisms for long periods of time, until the rocks are broken down into soil.

A critical question in man's relationship to natural ecosystems is the extent to which any of these different biogeochemical cycles can be modified and still continue to function. What would be the consequences of manipulating the environment in such a way, possibly through using pesticides, that an essential link in the biogeochemical cycle was blocked? Are we in danger of using some elements in such large quantities that we create critical shortages in the natural environment?

Let us consider two essential elements that illustrate both major types of biogeochemical

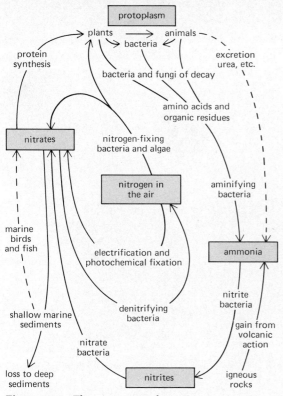

Figure 16-5 The nitrogen cycle.

steps, from nitrogen in the air or from nitrates to protein synthesis, require other energy sources, such as sunlight in photosynthesis.

Air contains 80 percent nitrogen and is the major reservoir for this element. Nitrogen enters the air through the action of denitrifying bacteria which break nitrates down into nitrogen. It is returned to the organic part of the cycle through the action of nitrogen-fixing bacteria and, to a far lesser extent, through electrification by lightning.

Nitrogen-fixing bacteria are associated with legumes such as clover and alfalfa, which is the reason these plants are used in crop rotation in agriculture. Certain species of bacteria have evolved that will grow only in small nodules that they have induced on the roots of the legumes (Fig. 16-6). The bacteria obtain carbohydrates from the plant and in the process fix atmospheric nitrogen, some of which is used by the plant and some excreted into the soil where it may be used by other plants. Thus by growing legumes in rotation with other crops, the fertility of a field can be maintained without the constant addition of nitrogen fertilizers.

This somewhat complex biogeochemical cycle is nearly perfect. The air provides a huge reservoir of nitrogen which can be made available by the action of nitrogen-fixing bacteria, and the system has few leaks through which the

cycles. Nitrogen (Fig. 16-5) exists in its organic form in protein in the tissues of plants and animals. When these organisms die, their protoplasm is acted upon by decomposer organisms, mainly bacteria, which break the nitrogen down into its inorganic forms through a series of definite steps. Each of the decomposer bacteria is specialized for its particular role in the process of decomposition. Certain bacteria and fungi produce the initial decay of protoplasm into amino acids and other organic residues. A second group of bacteria (amnifying) break the amino acids down into ammonia. Nitrite bacteria break the ammonia down into nitrites, and nitrate bacteria break the nitrites down into nitrates, which can then be used by green plants to synthesize protein. Each of these steps provides energy to the decomposers, allowing them to perform their essential functions in this system. The return

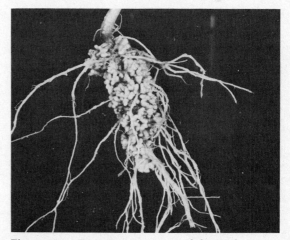

Figure 16-6 Bacteria-containing nodules on the root of a pea plant.

element is lost to the cycle. The only important loss occurs from nitrates which may be washed out of the soil by erosion. Nitrates that are eroded are carried by rivers into the sea, where they eventually become incorporated into deep sea sediments and are taken out of circulation for perhaps a few million years. But under natural conditions this loss is compensated for by additions from volcanic action in which nitrogen in the form of ammonia is produced from igneous rocks and is discharged into the atmosphere.

Let us examine this cycle in more detail to discover where it is vulnerable to human interference. Let us assume that the cash value of a crop such as corn is so great that it is more economical to add large amounts of commercial nitrogen to the soil and grow corn year after year, rather than rotate crops between corn and a legume which fertilizes the soil naturally. The nitrates that are artificially added to the soil will be subject to erosion, especially if they are applied in excess of the amounts that the plants can use, and will increase the rate of sedimentation in the seas. If this were to happen, we would have interfered with the natural biogeochemical cycle in such a way that the losses would be greater than the gains to the system. For how long could we continue this practice before we depleted the supply of available nitrogen, or seriously polluted our rivers with excess nitrogen carried as runoff from the land? There is evidence that the latter has already happened.

Let us consider another possibility. It is common practice in modern agriculture (see Chapter 17) to apply a wide variety of chemicals to crops to control weeds and animal pests. Insecticides such as DDT have been widely used throughout the world with no prior knowledge of what their side effects might be. They have also been carried by winds to places far from their original application. For example, DDT commonly occurs as fallout in remote areas in snow and rain from the atmosphere and has been found in the tissues of penguins. Consider what the consequences might have been if this, or any of the thousands of other chemicals that are constantly released by man into the environment,

were toxic to the specialized bacteria that are required for the breakdown of organic nitrogen.

The phosphorus cycle (Fig. 16-7) is simpler than the nitrogen cycle in that fewer components are involved. It illustrates the problem of a less perfect cycle, in which the great reservoir of the element is in rocks and other mineral deposits, rather than in air. Compared with nitrogen phosphorus is a relatively rare element. Although phosphorus is essential too for the growth of plants, its ratio to nitrogen in natural waters where it is available to organisms is only about 1 to 23.

Organic phosphorus in the tissues of dead plants and animals is broken down by bacteria to form dissolved phosphates, which then become available for synthesis into living protoplasm by plants. Some of the dissolved phosphates which are produced in this manner, and additional dissolved phosphates that enter the soil through erosion of phosphate rocks, are washed by rivers into the sea and become incor-

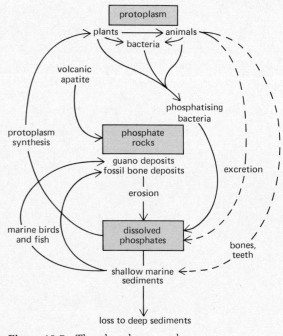

Figure 16-7 The phosphorus cycle.

porated into marine sediments in the bottom of the ocean. Much of this phosphate is lost to the system. To compensate for this loss of phosphorus from soil, phosphate rocks are mined in large quantities and added to the soil as fertilizer. But much of this phosphate is also lost through erosion, so that there is a continuing net loss of phosphorus from the cycle. The only way this lost phosphorus could ever be retrieved is by a major geological upheaval of the floor of the oceans, making these sediments available for mining. Agronomists and soil scientists tell us that the known reserves of phosphate rocks are so large that there is no reason for immediate concern, but if man must supplement the phosphorus cycle to keep it intact, this practice cannot go on forever.

LIMITING FACTORS

Nutrients such as oxygen, carbon dioxide, nitrogen, and phosphorus are needed in relately large quantities for the growth and productivity of an ecosystem. If any one of them is in short supply, the ecosystem will not function properly. However, there are other elements, called *trace elements*, that are required in only small amounts but are still essential. Boron is a trace element that may be very scarce in soils; although only minute quantities are needed for plant growth, plants cannot exist without it. If any essential constituent is in short supply, regardless of how little is needed, it will act as a **limiting factor** in the growth of an ecosystem.

Too much of a substance or condition can also be a limiting factor. An excess of nutrients such as nitrogen and phosphorus is often responsible for the pollution of lakes and rivers. They cause a breakdown in the normal function of an ecosystem by "overfertilization." When nitrates and phosphates in human sewage and phosphates in detergents are emptied into a lake faster than the ecosystem can process them, they often stimulate the growth of undesirable species of algae that are toxic to animals. The animals die and the biogeochemical cycle is broken, producing an essentially "dead" lake or river.

Physical conditions such as temperature or salinity may also be limiting. Every organism has a range of temperatures within which it can function. Heat is a limiting factor for trout in a mountain stream. If all other conditions are just right but the temperature is too high, the trout will die.

THE FLOW OF ENERGY

Theoretically, if there is no loss to a biogeochemical cycle through the deposition of an element in sediments where it is unavailable to the system, any element could continue to cycle indefinitely. Each atom of the element would be in constant circulation between its organic and inorganic forms. Energy, however, requires constant renewal in ecosystems and cannot be recycled.

Energy is the ability to do work. This ability is governed by two important laws of thermodynamics. The *first law of thermodynamics* states that energy may be transformed from one kind of energy to another, but is never created or destroyed. Energy may exist in a variety of forms such as light, chemical energy (in plant and animal protoplasm), or heat. When energy is changed from one of these forms to another, no new energy is created and none is destroyed. The sun's energy is used by green plants to produce organic compounds through the process of photosynthesis. Thus light energy is converted into chemical energy, which can then be used by the plants for their biological activities, and by animals that may, in turn, use the plants as a source of food and energy.

The *second law of thermodynamics* may be stated in several ways. Because no transformation of energy from one form to another is 100 percent efficient, some energy is converted into heat which is dispersed into the surrounding environment and is therefore not available to do more work. For example, the efficiency of an automobile engine depends on how much of the gasoline that is burned actually propels the car and how much is lost as heat from friction and radiation. If we place a warm object in a cool room, its heat is dispersed into the cooler room until it and the room are at the same temperature.

Thus we can state for the second law of thermodynamics that no spontaneous transformation of energy will occur without a degradation of the energy from a concentrated form to a less concentrated form. Although energy which is degraded during a chemical reaction or in combustion is not destroyed (because it still may exist as heat), it does represent a loss to the system if it can no longer be used to perform work.

How do these laws of thermodynamics apply to the structure and function of an ecosystem? We noted earlier in this chapter that green plants are the *primary producers* of an ecosystem. They convert energy from the sun and manufacture the food substances that supply the rest of the system. The efficiency of plants in converting available light energy to plant material is only about 1 to 5 percent. If green plants absorbed an average of 1500 calories of light energy per square meter of land area per day, which is a reasonable figure for the temperate regions of the earth, only about 15 calories would become food energy which would be available for animals that eat plants. The efficiency of animals in converting food to energy is about 10 percent (for some animals it is even less), so that the *primary consumers* would derive only about 1.5 calories from the 15 calories of energy in the plant food. The primary consumer might then be eaten by a predator, a *secondary consumer*, which would derive only about 0.15 calorie, and so on.

This sequence of energy transfers from plant to animal to another animal is called a **food chain.** For example, small single-celled algae are important primary producers in lakes. *Daphnia,* a small invertebrate called a "water flea," feeds on these green algae and is an important food source for small fish. A species of minnow which feeds on *Daphnia* might in turn support a larger fish, such as bass. The bass might be eaten by man, or by fish-eating birds such as ospreys or eagles.

In view of the large amount of energy that is lost during its transfer from one link to the next in this food chain, it is obvious that the amount of food remaining after two or three transfers is so small that it will not support large numbers of animals at the top of the food chain. Millions of *Daphnia* would be required to support one eagle,

which partly explains why large predators are never very abundant and why food chains are usually limited to three or four links.

The food chain described is an extremely simple one, with one species linked to the next in a linear sequence. In fact very few animals depend on only one species for their food; in addition to *Daphnia* a large variety of other invertebrate animals (Fig. 16-8) are found in lakes, feeding on algae and other plants, and forming the food base for several species of fish. When all of the feeding relationships in a natural community are described, the result is a complex food web rather than a simple food chain. It is important that animals be adapted to using more than one kind of food so that they can eat foods that are abundant when others are scarce.

A complex food web also provides greater stability for an ecosystem. The more the flow of energy in an ecosystem depends upon one species in a food chain, the more vulnerable the system is to disruption through the loss or scarcity of that species. In a simple food chain of green algae—*Daphnia*—minnows—bass—ospreys, a temporary scarcity of *Daphnia* would affect energy flow and the number of animals that could be supported all along the food chain. Unless other invertebrates were present as food for

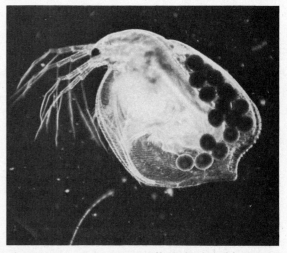

Figure 16-8   *Bosmina,* a small cladoceran like *Daphnia,* feeds on algae at the bottom of the food chain.

small fish, the result might be disastrous, especially toward the top of the food chain.

In order to consider the energy relationships of a total ecosystem, it is convenient to group the different organisms into **trophic levels** according to their functions as primary producers, primary consumers, secondary consumers, and so on. When this is done, we find that the trophic structure of an ecosystem consists of **pyramids of numbers, mass,** and **energy** (Fig. 16-9). At the base of the pyramid are the primary producers, the green plants. Above the base are the different consumer levels, from those that eat plants to the predators at the top of the pyramid.

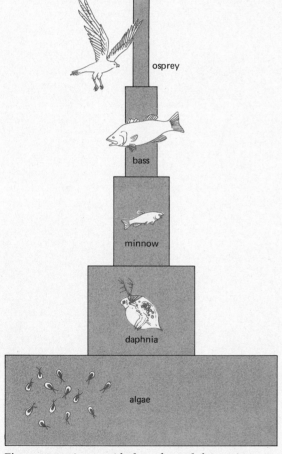

Figure 16-9 A pyramid of numbers of algae, minnows, bass, and osprey.

The fundamental reason for these pyramids is, of course, the loss of energy that occurs in the transfer of food between each trophic level. The pyramid of numbers depends also on the fact that predators usually eat animals smaller than themselves. Thus in addition to the large mass of food and energy that a predator requires, it must satisfy these needs with a very large *number* of individual prey. The number of trophic levels in an ecological pyramid is therefore limited by the size relationships between animals and the second law of thermodynamics.

We might ask ourselves how these principles of energy flow apply to man. What, for example, is the most favorable position to occupy in a food chain? Because of the loss of energy involved, short food chains are the most efficient. It is therefore advantageous to be as close to the base of the pyramid as possible. In the most heavily populated parts of the world, such as Southeast Asia, people eat plant products almost exclusively. More people can be supported on an acre of rice than on the same amount of land used to raise beef. Also most of man's domestic animals are herbivores, only one link away from producers. When man changed his culture from hunting to agriculture, he greatly improved his position in the economy of ecosystems, allowing support of larger human populations and the colonization of regions which would have previded only meager support for men who depended solely on hunting for their food.

THE IMPORTANCE OF PRIMARY PRODUCERS

It is clear, from the limitations that are placed on the trophic levels of an ecosystem by energy loss, that the number of species and the amount of life that can be supported depend, ultimately, on the amount of organic material manufactured by the primary producers. About half the solar energy that falls on the surface of the earth is outside the range of the spectrum of energy that can be used in photosynthesis. The efficiency of green plants in converting the remaining light energy into plant tissue depends on the limiting factors of the environment in which they live.

The rate of primary production is often expressed in terms of the dry weight of organic material produced per square meter of surface area of the earth per day.

Figure 16-10 illustrates the world distribution of primary production. What do these data tell us about the productivity of different kinds of plants and the effects of different limiting factors? First, the productivity of desert plants, which are adapted to dry conditions, is not much different from that of other plants, such as marine algae which are adapted to an aquatic environment. Second, different limiting factors may produce similar results. In other words, primary productivity is not governed by the kind of plant or the kind of environment as much as the availability of light and nutrients and the ability of ecosystems to use and recycle essential materials.

The limiting factor for deserts is obviously a shortage of water, while deep oceans are limited mainly by a shortage of nutrients. In each case the rate of primary productivity is below 0.5 gm/m²/day. Grasslands, shallow lakes, coastal waters, and most agriculture are intermediate, with rates of 1 to 3/gm/m²/day. The most productive natural environments are coral reefs, salt marshes, and estuaries, and forests along the flood plains of rivers. These communities, with rates of productivity from 10 to 20 gm/m²/day, compare favorably with the most intensive forms of agriculture, such as sugar cane crops in which

plant growth is continuous throughout the year. The highest rates of primary productivity are obtained when nutrients are abundant and physical factors such as light, temperature, and moisture allow long growing seasons for plants.

ECOSYSTEMS OF THE WORLD

So far in this chapter we have studied the general principles of succession, the structure and function of ecosystems, biogeochemical cycles, limiting factors, and the importance of primary producers as the basis of ecosystems. These principles apply not only to individual forests or ponds but also to the broadest patterns of life, the subdivisions of the biosphere that occur over large regions of the earth or are represented throughout the continents. Within the biosphere we can recognize nine major kinds of ecosystems: the seas, estuaries and seashores, rivers and streams, lakes and ponds, freshwater marshes, deserts, tundra, grasslands, and forests. Although there are different kinds of lakes or grasslands or forests, each of these major ecosystem types has certain common characteristics.

Sun, rain, wind, heat and cold, soil and topography interact to allow the development of a certain kind of plant life. These plants provide food and shelter for a characteristic sort of animal

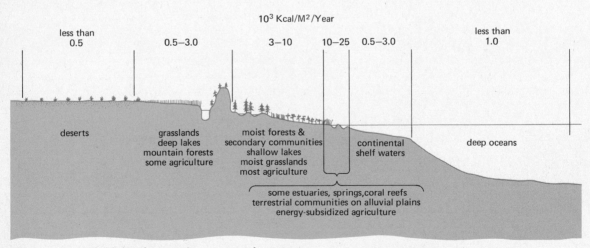

Figure 16-10   World distribution of primary productivity.

life. However, time and geographical isolation have allowed different species of plants and animals to evolve on different continents, and a forest or lake in North America will not contain the same species as its counterpart in Europe or Asia. Nevertheless a broad survey of each system will reveal fundamental similarities, with similar species fulfilling the same roles in the ecosystem. For example, grassland ecosystems typically support large grazing mammals that are adapted in various ways to life on open plains. In North America, this role was filled by bison and pronghorn antelope. In Australia, it is filled by kangaroos. In Africa, zebras and different antelopes are the principal herbivores in the grassland ecosystem. We speak of these forms as *ecological equivalents.*

The seas

The world's oceans occupy approximately 70 percent of the surface of the earth and make up the largest and most stable ecosystem in the biosphere. The most abundant producers are floating, microscopic green plants called *phytoplankton.* Phytoplankton occurs throughout the oceans, as deep as light can penetrate, but is most abundant in the waters over the continental shelf. The continental shelf is formed by the gradual slope of the bottom of the ocean away from land, before it begins its more abrupt decline into deeper waters. Because of the higher productivity of continental shelf waters, this is where most of the fish are taken that are harvested by man from the sea. The deep parts of the ocean are like vast deserts. There is very little productivity per unit area of the deep oceans, but because of their huge volume, their total contribution of energy flow to the biosphere is large.

Life in the sea is strongly influenced by waves, tides, temperature, and depth, but the most important limiting factors are light and nutrients. Much of the light falling on the ocean is reflected back into the air. The irregular surface created by waves increases the amount of reflection even further. The remaining light that penetrates the water is soon scattered and its energy absorbed by particles and the bodies of organisms in the water. Consequently, if we were to examine a column of the ocean, from surface to bottom, we would find only a very thin layer of water near the surface where light is adequate for photosynthesis. Everything in this column would depend upon this layer, the *photic zone,* for primary production.

When plants and animals in the photic zone die, their bodies descend to the bottom of the sea. The nutrients they contain become incorporated into the sediments on the ocean floor and are no longer available for recycling in the zone of productivity near the surface. Thus there is a continuous leak of important minerals in the biogeochemical cycles of the sea.

The supply of minerals in the oceans is constantly added to by materials that are eroded from the land and carried to the sea by rivers, but never in large enough amounts to sustain a high rate of productivity throughout the sea. Nutrients in the bottom sediments can only be returned to the biogeochemical cycle by upwellings of water from the floor of the ocean. Such upwellings do occur in a few places, one of which is off the coast of Peru, where commercial fishing fleets and huge populations of sea birds gather to harvest vast quantities of small fish that are spawned in these rich waters.

Although the deepest parts of the oceans do not provide much food, they are tremendously important to man and the economy of the biosphere. They are the major reservoirs of the world, storing up the runoff from the mountains and plains, so that water can be returned to the land as rainfall. They influence the climate of the land. Water has a great capacity to hold heat, and therefore moderates the temperatures of the land areas of the earth. Also since the total area of the sea contains more plant life than any other ecosystem, the seas are important determinants of the balance of oxygen and carbon dioxide in the atmosphere.

The sea was the first ecosystem on earth. This is where life originated and has flourished, producing a great array of organic diversity. The sea contains the smallest known plants and the largest animals, from microscopic algae to giant fish, squid, and whales. But in spite of its antiquity and importance, it remains one of the least known of our scientific frontiers.

Seashores and estuaries

At the junction of the oceans and the continents there is a narrow ribbon of ecosystems, the estuaries (Fig. 16-11), that are among the most productive of any on earth. An *estuary* is an area near the mouth of a river or a bay where salinity is intermediate between fresh water and the salinity of the oceans. Estuaries, beaches, and the rocky shores of seacoasts are continuously subjected to the action of tides that affect the plants and animals in this *intertidal zone.*

Tides are the key to the high productivity of these ecosystems. They keep nutrients in constant circulation and remove the waste products of metabolism quickly. Although physical factors, such as salinity and temperature, are far more variable in these environments, there is such an abundance of food that they are literally packed with life.

Organisms that live in the intertidal zones of salt marshes, rocky shores, coral reefs, and mud flats that are flooded at high tide but drain off at low tide have many adaptations that allow them to exploit the advantages that are offered by these environments. Barnacles and mussels have special means of attaching themselves firmly to rocks so that they can withstand the pounding of waves and not be dislodged. They open at high tide to receive the food they filter from the water. When the tide recedes, they close their valves and are protected from drying while they are exposed to the air. Fiddler crabs retire to burrows in mud during high tide, and emerge to forage for food when the tide is out. The lives of these animals are geared to the tides through "internal clocks" that regulate their daily activities. If they are removed from their natural habitats and placed in a constant laboratory environment, they will continue to show a rhythmic pattern of activity that corresponds to their tidal cycle.

The value of a salt marsh or a mud flat is not immediately apparent and we are often tempted to "improve" our environment by dredging a bay to make room for boats and marinas, or filling a salt marsh to provide sites for homes and factories. But these environments are extremely important sources of food and nutrients for deeper, offshore waters that supply most of the marine life we harvest. They act as breeding grounds and nurseries for our most desirable commercial and sport fish. The coastal marshes also act as a buffer against storms and hurricanes, dissipating the energy of waves before they reach higher ground. When we tamper with these ecosystems we violate important ecological principles that might have, in the long run, yielded the greatest benefits to man.

Figure 16-11 A barrier island, the North Carolina Outerbanks.

Rivers and streams

A river or stream is primarily a *transport system* between the land and the sea. Most of the nutrients in a river come from adjacent terrestrial ecosystems, and relatively little primary production occurs within the stream or river itself. The organisms that are characteristic of rivers and streams are especially adapted to feeding on organic detritus. Some insect larvae, such as caddis flies, attach themselves to rocks and spin nets which capture food particles from the flowing waters. In slow-moving rivers and pools the biological communities are much like those in lakes and ponds. If the load of organic material reaching a river in runoff from the land is not too great, it will be quickly converted into its inorganic constituents by an array of organisms adapted for this function. As we have seen already these inorganic nutrients will then be transported to the sea, where they will enter the biogeochemical cycles of that ecosystem.

Streams and rivers serve human society in many important ways. They provide water for drinking, irrigation, and industrial use. Rivers are used as disposal systems for sewage and industrial wastes, as well as for recreation. Boats and barges transport the raw materials and products of industry and society along the waterways of the world. In spite of the fact that they make up only a very small percentage of the earth's surface, man uses rivers and streams more intensively than any other natural ecosystem.

Before the industrial revolution our rivers could easily absorb the impact of man without damage to their normal functions. However, huge amounts of water used by modern society, and the heavy burdens of waste that are now emptied into these ecosystems, have placed them in a precarious state.

The two general categories of waste are those that consist of materials that are normally present in natural ecosystems and those that are not usually found in nature. In the first instance there are organisms in the ecosystem which are adapted to handle these products in reasonable amounts. Human sewage and fertilizers are in this category. As long as the sewage and fertilizers that enter a river do not exceed the river's capacity to use the breakdown products of these materials, the system will not be harmed. When the system is overloaded, pollution results.

Many of the chemicals in industrial wastes are in the second category. There are no organisms capable of breaking down pesticides and many industrial chemicals before they reach toxic concentrations, so these materials are poisons when they enter a natural ecosystem.

Figure 16-12   Lake Ann in the North Cascades, Washington.

Lakes and ponds

A limiting factor for lakes (Fig. 16-12) and ponds, as in the earlier case of the oceans, is the depth of light penetration into the water. Further zonation is found in the life forms of the primary producers. The edge of a lake or pond is often fringed with rooted aquatic plants such as cattails, reeds, and bulrushes whose leaves extend above the surface of the water. These *rooted emergents* are usually grasslike in appearance and occur in dense stands. In deeper water there is a zone of *rooted* plants with *floating* leaves. Water lilies are typical of this zone. As the photosynthetic surface is horizontal and is supported by the surface of the water, the leaves are typically large and round (Fig. 16-13). In even deeper water there is a zone of *rooted submerged* plants. The leaves of these plants tend to be thin and finely divided to facilitate exchange of nutrients between the leaves and the water. Finally, in deep water, the dominant primary producers are phytoplankton. These are mostly algae of different kinds.

The lifespan of a lake or pond depends upon its size, and especially its depth. Lakes gradually become filled with silt that washes into them and by accumulations or organic debris from the bodies of the plants and animals that live in them.

Figure 16-13 Water lillies with floating leaves supported by the surface of the water.

As the lake becomes shallower, rooted emergent vegetation will grow farther out into the lake and the shoreline will itself become constricted as terrestrial plants become established. The lake will gradually be reduced to a pond or marsh. Eventually this process of succession will lead to the elimination of the lake and the establishment of a forest where the lake once stood. The world's large, deep lakes are in no immediate danger, but the dynamic processes of succession are constantly altering smaller lakes and ponds.

Freshwater marshes

Freshwater marshes are more like estuaries than lakes. They are naturally fertile ecosystems in which changing water levels (due to variations in rainfall) and fire (during dry periods) clear out excess organic matter in much the same way that the tide flushes an estuary. When decayed organic material is allowed to accumulate over long periods of time, succession will lead (as we have already seen for lakes) to shrubs and eventually a forest.

Marshes are often regarded as a nuisance. They are a breeding ground for mosquitoes and other insect pests and they occupy areas that might otherwise be used for agriculture or building sites. In fact, however, they serve a useful function in helping to maintain proper water levels in surrounding areas and support animal populations that are valued by man.

Although the freshwater marshes of North America are usually left in their natural state, they are intensively managed in some countries as a source of food. Rice is a cultivated marsh grass that is the staple food of millions of people. In countries like Japan rice paddies are carefully drained and planted and reflooded each year to support not only rice but also fish that are introduced into the paddies to be harvested at the end of the season.

Deserts

A desert is a region where the annual rainfall is less than 10 inches, or where rain occurs only during short periods with long dry periods during the rest of the year. Deserts may therefore occur in cool regions, such as eastern Washington in the northwestern United States, or in warmer regions of the world such as the southwestern United States or the Sahara Desert of North Africa.

Life in a desert consists of strategies to overcome the limiting factors of high temperature and low moisture. Annual grasses survive by growing only when there is enough moisture and producing large numbers of seeds that will produce a new generation with the next rainfall. Desert shrubs have small, thick leaves that are shed during dry periods, allowing the plants to become dormant during periods of water stress. The most characteristic plants of the deserts are the *succulents*, the cactus plants of North and South America (Fig. 16-14) and the euphorbias of Africa. Their large, fleshy leaves and stems are reservoirs for water which they store within their tissues.

Desert animals solve the problem of temperature and moisture in somewhat similar ways—they adjust their activities to favorable periods and have evolved physiological adaptations that help them withstand the stress of a desert environment.

Most desert mammals are nocturnal. During the day they remain below ground in burrows that are relatively cool and moist and come out to forage at nightfall. Snakes and lizards would die in less than an hour if they were exposed to the full heat of the desert on its hottest days; they also live in burrows in the ground and adjust their activities to take advantage of favorable temperatures.

Camels, like cacti, store water in their bodies to sustain them during dry periods. Desert rodents do not have this ability, but they are physiologically adapted to conserve water by excreting highly concentrated urine.

Man has tried for thousands of years to conquer the deserts. With proper irrigations they may be turned into luxuriant agricultural areas, but a reclaimed desert is always dependent on its man-made water supply. One of the problems of desert agriculture is that mineral salts may accumulate at the surface of the ground and destroy its ability to produce crops. The abandoned remnants of ancient irrigation systems in the deserts of Africa and Asia are silent reminders that the desert ecosystem demands thorough understanding before man can live there.

Tundra

Arctic tundra rings the Northern Hemisphere between the polar icecaps and the most northern forests. Alpine tundras are similar and occur above treeline on high mountains. A tundra ecosystem is essentially a "physiological desert"; even though there is usually adequate rainfall, water is not available to support a large amount of biological activity because it is frozen most of the year. The tundra is like a frozen grassland.

The vegetation consists mainly of lichens, mosses, grasses, and grasslike sedges. The growing season is very short, but summer days are long in the Arctic and photosynthesis is almost continuous at this time of year. The plants that have been able to adapt to the tundra environment are extremely productive during the summer. Many of them start to grow and set flowers even before the warmth of spring releases them from their blanket of snow. They and the animals that live here are adapted to take the fullest possible advantage of the short growing season.

Relatively few species can survive the harshness of this environment in winter, but thousands of birds migrate north each spring to breed in the tundra and return south to more favorable climates in the fall. Even greater multitudes of insects spend the winter in a resting stage and

Figure 16-14 Desert plants in the Mohave desert of California.

flourish during the period of high productivity in the summer.

The permanent residents of the tundra are mostly mammals whose internal temperature controls protect them against the cold. The most abundant primary consumer is the lemming, a small rodent that lives on the low vegetation. The large herbivores are musk ox, caribou, and reindeer.

Many of the permanent residents of the tundra are nomadic, moving from place to place in search of food and favorable conditions for breeding. Caribou make long seasonal migrations, accompanied by their principal predators, the wolves. When lemmings are scarce, snowy owls, which depend on lemmings for food, often make long migrations south into Canada and the United States.

Grasslands

Grasslands occur in regions where the rainfall is intermediate between that in deserts and forest. In the temperate zones, this is usually between 10 and 30 inches of rainfall per year. In the tropics, grasslands may occur where the rainfall is as high as 60 inches, but there is a short wet season followed by a long dry season.

Grasslands occupy vast areas of North America, Africa, and Asia and about one-third of the United States is in the grassland ecosystem. Grasses do not live as long as trees and their

Figure 16-15   Bison, once common in North America, are characteristic animals of the grasslands.

leaves and stems decay quickly, so that there is a high rate of turnover of organic matter. This produces a great deal of humus, enriching the soil and making it particularly well suited for agriculture. Most of man's principal food crops, including wheat and corn, which are cultivated grasses, are grown in this ecosystem.

Fire is an important factor in the ecology of grasslands. Some species of plants found in the grasslands of the United States and Central Africa grow more vigorously when fires sweep the country periodically. Research has shown that in hot, dry regions the accumulated litter of dead plants gets so dry that bacteria and fungi cannot attack this material and break it down. Fire helps to decompose this litter, releasing minerals and increasing productivity. Indians in the western United States and Canada used to light prairie fires, probably to drive bison (Fig. 16-15). Although it was an unintentional benefit, this practice helped to maintain the health of the grassland ecosystem. But man has persistently misused grassland ecosystems through overgrazing and overcultivation, turning large areas into deserts.

Forests

A mature forest ecosystem is the result of a long, orderly process of succession from the first colonizers to the final climax. Nevertheless a variety of trees and shrubs will be found in any region, depending on the particular stage of succession that has been reached. There will also be local variations in soil type, moisture, and temperature that increase the diversity of plant species. An exposed, south-facing slope of a ridge will have a different vegetation from its north-facing slope where there is more moisture and less direct sunlight. Sheltered valleys, with less evaporation and more moisture, have a characteristic vegetation. For any given region, however, there is a dominant climax forest that is determined by the prevailing climate.

Forests at the edge of the tundra and in the coolest climates are dominated by evergreen conifers, usually spruce and fir (Fig. 16-16). Farther south in more temperature forests, deciduous forests predominate. They cover most of the eastern half of the United States but are also widespread in Europe and Eastern Asia. Pine trees are found in both coniferous and deciduous forests as a stage in succession leading to the mature ecosystem of the region.

Tropical forests may be either deciduous or evergreen, depending on the amount and seasonal distribution of rainfall. Where rainfall is abundant throughout the year, evergreens predominate, but if there is a prolonged dry season there will be a deciduous forest. A tropical rain forest is extremely diverse in its structure and the number of species it includes—there are probably more species of plants and insects in a few acres of tropical forest than in all of Europe.

ORDER IN DIVERSITY

This brief tour of the world's major ecosystems has revealed marked differences in the conditions for life in different environments. Natural selection has produced a bewildering array of organic forms and adaptations. But are there any common properties that are shared by these diverse ecosystems? As we have seen in earlier chapters and shall see throughout this book, order emerges from diversity when we examine the biological functions of a particular level of organization. All ecosystems have the same functions to perform and are subject to the same ecological laws, regardless of their physical and structural differences. Food must be manufactured and transferred from one set of organisms to the next. The organic matter that is produced must eventually be decomposed, and essential elements are then released to be recycled through the system. The laws that apply to energy flow and biogeochemical cycles in a desert apply equally to an ocean.

We can therefore conclude that the construction of a particular ecosystem is nature's response to the special set of limiting conditions found in that environment. The physical properties of an ocean set definite limits, but they also provide special opportunities for life in the sea. The availability of water, light, the chemical elements required for life, and the temperatures of an environment are the "ground rules" within which an ecosystem evolves to carry out its essential functions.

When man recognizes and accepts these ecological principles, he becomes better equipped to use the world's ecosystems in ways that supply his needs but are not destructive.

Figure 16-16   A Douglas fir forest in Washington.

SUGGESTED READING LIST

Bascom, W., "Beaches," *Scientific American,* August 1960.

Deevey, E. S., Jr., "Bogs," *Scientific American,* October 1958.

Elton, C., *The Ecology of Invasions by Plants and Animals.* Wiley, New York, 1958.

———, *Animal Ecology.* Methuen, London, 1966.

Odum, E. P., *Ecology.* Holt, Rinehart and Winston, New York, 1963.

Went, F. W., "The ecology of desert plants," *Scientific American,* April 1955.

Whittaker, R. H., *Communities and Ecosystems.* Macmillan, New York, 1970.

The Biosphere. A Scientific American Book. W. H. Freeman, San Francisco, 1970.

CHAPTER 17

Ecology and Man

A single bacterium may divide to produce two individuals, which in 20 minutes may again divide to produce four individuals. Each of these four bacteria may then divide again in 20 minutes to produce eight individuals, and so on. If the growth of a bacterial population were to continue at this rate without constraints, there would be a colony of bacteria one foot deep over the entire face of the earth in a day and a half. An hour later this colony would be over our heads.

This population projection is, of course, unreal—but why? We know in fact that all organisms, including man, have potentially high rates of growth, and the above illustration is actually based on the rate that a bacteria colony could grow if its environment were unlimited. The answer is that no population has an unlimited amount of space and resources available to it.

In Chapter 9 we considered an imaginary experiment in bacterial growth. If we look now at an actual case of population growth in a limited environment, we find that the pattern of growth again follows an S-shaped curve. Figure 17-1 shows the growth of a population of yeast cells in a laboratory culture. The culture was started with a few cells which multiplied slowly at first, then entered a phase of very rapid increase. As the population began to approach the limits of food and space in the culture, the rate of population growth slowed down and finally reached zero (when the number of births were equal to the number of deaths in the population). In this particular case the culture could support a population of about 665 yeast cells and this became the upper limit of population size.

We refer to this upper limit of population size as the **carrying capacity** of the environment for that particular organism. The population has entered into a balance between the resources of the environment (in this case food) and its own numbers. This is a fundamental property of all natural populations and environments and, since infinite population growth is impossible, a population has two alternatives available for the regulation of its numbers in relation to the ultimate carrying capacity of the environment: (1) the birth rate has to decrease or (2) the death rate has to increase, until these two values are equal.

HUMAN POPULATION GROWTH

How do these elementary laws of population growth apply to man? The human environment is extremely complex and our integration into natural systems is, to some extent, modified by man-made structures and social institutions. By discovering the uses of fire, by learning to make clothing and shelter, and being able to grow our own food, we have spread across the earth like no other species and can live in a greater range of environments than any other animal. In the pro-cess we have developed a somewhat detached view of nature—we tend to have the attitude that man is not an animal and exists apart from nature and natural laws. Is this in fact the case? Is the impact of our activities upon the natural environments of the world something we can disregard, and can the human population grow indefinitely?

In the year A.D. 1 there were about a quarter of a billion people living in the world (Table 17-1). Their numbers grew slowly, and it took approximately 1600 years for this population to double in size to half a billion. However, by 1830, only 200 years later, the population had doubled again to reach 1 billion people. By 1930, only 100 years later this time, the population of the world had doubled again to reach 2 billion. Today the world has 3.5 billion people spread across its continents and is now doubling at a rate of once every 37 years. Four babies are being born in the world every second. This is a net gain, or an excess of births over deaths, of 8000 new individuals every hour or 190,000 every day. The total of 70 million new people who are added to the world population every year is equivalent to the number of men, women, and children living in 20 cities the size of Chicago. This is enough people to add a city the size of New Orleans every three and a half days, or a new Los Angeles every two weeks. By the year 2000, if present trends continue, the population of the world will have reached 7 billion people (Fig. 17-2) and will double again to 14 billion 20 years later. The year 2000 seems a long way off, but most of the students who read this book will be alive in the year 2040 when there may be as many as 28 billion people crowded onto this planet.

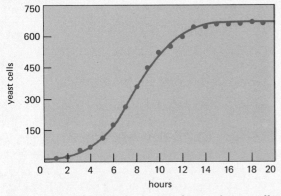

Figure 17-1 The growth of a population of yeast cells in a limited environment follows an S-shaped curve.

Table 17-1 History of world population growth

year (A.D.)	World population (Billions)	Years to double population
1	0.25	
1620	0.5	1600
1830	1.0	200
1930	2.0	100
1975	4.0	45
2000	7.0	20

Table 17-2 shows a comparison between population data for the world, the United States, and a selection of countries from Northern Europe, Latin America, Africa, and Asia. The rate of human population growth is measured by the difference between the number of births and deaths per thousand in the population each year. This is expressed as the mean annual growth rate (MAGR) according to the following formula:

$$MAGR (\%) = \frac{(births/1000) - (deaths/1000)}{10}$$

The average birth rate for the world in 1971 was 34 live births per thousand while the death rate was 14 per thousand. Substituting in the above formula, we have

$$MAGR = \frac{34 - 14}{10} = 2.0\%$$

as the growth rate of the world population. By comparison the growth rate of the United States population is 1.1 percent, in East Germany it is as low as 0.1 percent, and in Ecuador a very high 3.4 percent. The result is that the population of East Germany will double itself every 700 years, while the population of Ecuador will double in only 21 years.

DOUBLING TIME

The doubling time of a population is an extremely important factor. A population which is increasing at a rate of 1 percent per year will double itself in 70 years. Although the doubling time is derived from a rather complex equation, a rule of thumb is that the doubling time of a population may be found by dividing 70 years by the growth rate (Table 17-3). For example (Table 17-2), the

Table 17-2   Population data for selected countries for 1971.

	Population estimates mid-1969 (Millions)	Birth rate per 1000 population	Death rate per 1000 population	Current rate of population growth (MAGR)	Number of years to double population	Population under 15 years (percent)
World	3706	34	14	2.0	35	37
United States	207	18.2	9.3	1.1	58	30
Denmark	5.0	14.6	9.8	0.5	88	24
East Germany	16.2	14.0	14.3	0.1	233	22
Sweden	8.1	13.5	10.4	0.5	88	21
Colombia	22.1	44.0	11.0	3.4	21	47
Ecuador	6.3	45.0	11.0	3.4	21	48
Venezuela	11.1	41.0	8.0	3.4	21	46
Kenya	11.2	50.0	20.0	3.1	23	46
Uganda	8.8	43.0	18.0	2.6	27	41
Nigeria	56.5	50.0	25.0	2.6	27	43
India	570	42.0	17.0	2.6	27	41
Pakistan	142	50.0	18.0	3.3	21	45
Indonesia	125	47.0	19.0	2.9	24	42

world population growth rate is 2 percent per year. Thus the doubling time for the world population is 70 divided by 2, or 35 years.

The countries from Africa, Asia, and Latin America shown in Table 17-2 are doubling their populations every 21 to 27 years. All of the countries of Latin America together have an average

Table 17-3 Numerical relationship between mean annual growth rate (MAGR) and the doubling time of a population

MAGR (%)	Doubling time (Years)
0.25	280
0.5	140
1.0	70
2.0	35
3.0	23
4.0	18
5.0	14

growth rate of 3 percent—thus a doubling time of 23 years. This means that Latin American countries must double their food production, the number of homes and schools, water and electricity supplies, medical facilities, and all of the goods and services their people need every 23 years, just to maintain their present standard of living. Is their present standard of living in fact good enough, and do they have the resources to accomplish even this limited goal? These are questions that many countries will have to face in the next few years.

Another important fact is that the relationship between rate of increase and doubling time is curvilinear (Fig. 17-3). Thus if a country has a rate of increase of 3 percent and is able to achieve a 1 percent decrease in its growth rate, it will only gain 12 years in doubling time, compared with a country which lowers its rate of increase from 2 to 1 percent, and thereby gains 35 years. Thus the population problems facing countries such as

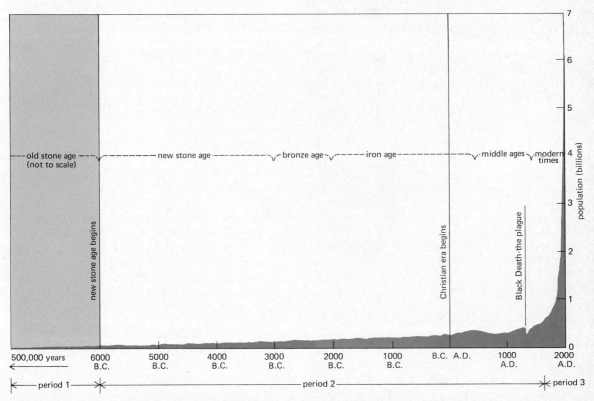

Figure 17-2   The growth of the human population shows its most dramatic increases in recent years.

those in Latin America, Africa, and Asia (Table 17-1) are further aggravated by the amount of decrease they would have to manage in their birth rates in order to realize any significant increases in their doubling times.

THE DEMOGRAPHIC TRANSITION

Ecologists and demographers are not only alarmed by the present rates of population growth that exist in most of the underdeveloped countries of the world but also by the requirements of a phenomenon known as the **demographic transition.** In order for a primitive or underdeveloped society to reach the condition of low death rates, relatively low birth rates, and the growth rates that are characteristic of modern countries such as the United States and most European countries, they must pass through the three stages shown below in Table 17-4.

Table 17-4   Stages in the demographic transition

Stage	Birth rate	Death rate	Growth rate
I	High	High	Low
II	High	Low	High
III	Low	Low	Low

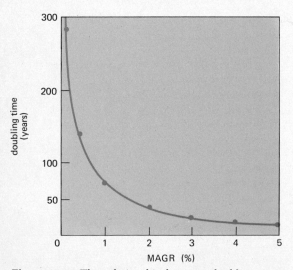

Figure 17-3   The relationship between doubling time and mean annual growth rate is curvilinear.

Advances in modern medicine have made it possible to control infectious diseases which were once important sources of human mortality, especially among infants and children. Measles, diphtheria, and smallpox were once important killers, but can now be controlled by immunization. We have therefore turned most of our attention in medical research to diseases of old age such as cancer and heart disease. We do not, however, have the same control over birth rates that we do over death rates. Consequently when a population moves from its initial condition (Stage I) of high birth rates and high death rates, the first to be reduced is the death rate. This produces a large difference between the birth rate and the death rate, and a corresponding increase in the rate of population growth. Latin American countries (Table 17-2) show this effect and are in Stage II of the demographic transition. Their death rates are as low as or, in some cases, lower than those in many more advanced countries, but their birth rates are still at 40 to 45 per thousand with rates of population increase of over 3 percent. The African countries shown in Table 17-2 have not yet made the transition from Stage I to Stage II. They have extremely high birth rates, but their death rates are about double what they could be if adequate medical services were available. Consequently when death rates are lowered in these countries, we can expect higher rates of population growth than those shown in the table.

A critical problem for the underdeveloped countries is that there has historically been a long time lag between a lowering of the death rate and a corresponding lowering of the birth rate that would bring a population to Stage III in the demographic transition. This fact is particularly alarming to demographers. What will be needed to bring these countries to the quality of life that can be provided by a modern society?

THE QUALITY OF HUMAN LIFE

How do these population statistics apply to people's lives? The consequences of overpopulation are already evident. Over 2 billion of the world's 3.7 billion people live lives that are dominated by shortages of food and water and inadequate resources of soil and forests. About 2250 calories a day are needed for a normal, active life. The average citizen of the United States receives 3200 calories daily, but more than half the people of the world receive less than the normal requirement and at least 10 to 15 percent are badly undernourished (Fig. 17-4). The seriousness of the world food problem can be seen in the fact that, if all of the food in the world were equally distributed and each person received an equal amount, we would all be undernourished.

Other nutrient deficiencies are even more serious. Vastly more people, perhaps as many as 1.5 billion, suffer from a shortage of nutrients of various kinds and are therefore "malnourished." Vitamin A deficiency frequently results in blindness. In India alone there are at least 1 million such cases. In East Pakistan 50,000 children are threatened every year with a lifetime of blindness due to a low intake of vitamin A.

It is difficult to establish the relationship among malnutrition, disease, and death. People who are chronically undernourished might not die of actual starvation, but they are far more susceptible to other diseases. Half the deaths in the developing countries occur among children under six years of age. In parts of Southeast Asia,

40 percent of the children die of some disease in their first four years of age. This proportion of deaths is not reached in most Western countries until the age of 60.

The social and political consequences of overpopulation are even more difficult to detect, but may be far more serious. In nations trying to support more people than they have resources for, subsistence levels are low and life is a struggle for existence. The hundreds of millions of people whose diets are deficient in proteins, minerals, and vitamins may not show exact symptoms of starvation, or even malnutrition, but they suffer from lowered efficiency and endurance and are not as productive as people with adequate diets. These people will find it difficult to build a modern nation. However, industrialized, economically advanced nations will also feel the impact of population pressures. The United States constitutes only about 6 percent of the world population, yet uses about 40 percent of the world's production of nonrenewable resources such as oil and minerals. Our standard of living depends, therefore, on events in other parts of the world and is derived in large part from the resources of other, less developed nations. Furthermore, because of the large amounts of resources that we use to maintain our present

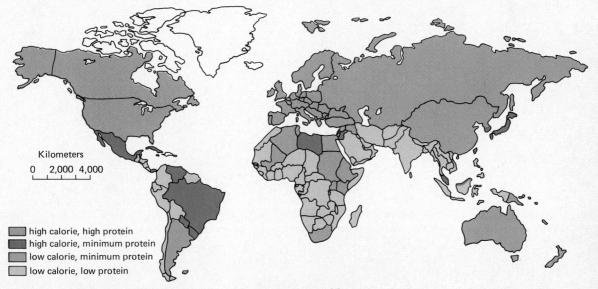

Kilometers

0   2,000 4,000

■ high calorie, high protein
■ high calorie, minimum protein
■ low calorie, minimum protein
■ low calorie, low protein

Figure 17-4 The geographical distribution of hunger in the world.

standard of living, we have far more impact on the environment than do less developed nations. An important conclusion to be drawn from these facts is that we are all part of one biosphere, in which no single individual and no nation is entirely self-sufficient.

HUNTERS, FOOD GATHERERS, AND AGRICULTURALISTS

The human population has been able to grow so rapidly because man has drastically changed his way of life in the last 12,000 years or so. Man was originally dependent on the food he could gather and the animals he could hunt in the forests and grasslands where he lived. But in time he learned to cultivate his own foods and to domesticate animals he could use for food and for hides that provided shelter and clothing. Archeologists in the Middle East have found evidence that 9000 years ago men had already learned to grow barley and wheat, and had tamed goats and perhaps other animals to live with them in their villages.

Today many of the plants and animals that we use for food and clothing have been altered so greatly by their continued association with man that they can no longer exist by themselves in the wild. We have selected and bred these plants and animals for various properties such as size and food quality at the expense of other characteristics which originally fitted them for survival in natural environments. Most of them have also been transplanted from their natural environments, often to far distant continents. The maize plant, which we call "corn," is an example. Maize originated in the New World but was carried to Europe, Asia, and Africa by explorers in the seventeenth century, where it is now a major food in those countries. It cannot reproduce in the wild. After thousands of years of artificial selection for large kernels and ears, man has created a plant in which the husks will not peel back and the kernels will not come off the cob without assistance.

In the course of becoming a food grower instead of a gatherer of wild foods, man has greatly simplified the ecosystems he lives in. Fields which originally contained hundreds of species of native plants and animals have been reduced to much simpler ecosystems containing only a few species, sometimes only one in large quantities, which benefit man. The western grasslands of the United States and Canada once supported vast herds of bison, antelope, deer, wild sheep, and a variety of other herbivores. These animals in turn supported a variety of predators, including the wolf. Today most of these species have disappeared or are very scarce, and fields of wheat or herds of cattle or sheep have taken their place. The midwestern United States was once covered with rich prairies and forests that contained a great variety of wildlife; today this vast area is dominated by only a few species, mainly corn, soybeans, and hogs.

In order to increase the efficiency of his agriculture, man has eliminated most of the plants and animals that are not agricultural products. However, after many years of study, ecologists have found that the stability of an ecosystem is highly dependent upon its diversity. The larger the number of plant and animal species in an ecosystem, the less likelihood there is of violent change; populations are less likely to increase drastically, thereby resulting in outbreaks of one species or the extinction of others, and the productivity of the system is greater over longer periods of time.

ECOSYSTEM DIVERSITY AND STABILITY

Unfortunately, when we reduce the diversity of an ecosystem, the first species to be eliminated are the predators and parasites that help to keep the system in balance. Thus species which would not be unusually abundant or a problem in complex, natural ecosystems suddenly become pests of agricultural crops. An attack of potato fungus or of witchweed on corn would not have serious consequences in natural environments, where plants of the same species are widely spaced and are interspersed with a variety of other plants which are resistant to infection. But by devoting vast acreages to one or only a few crops, man creates conditions which make it easier for pests to become a problem.

In addition to reducing the species diversity of ecosystems, we have also reduced genetic diversity in domestic plants and animals by selecting for strains with particular food qualities and high yields. In 1970 the corn crop of the United States was struck by southern corn blight, which severly reduced production of this important grain in many states. Corn blight is a fungus disease which is usually confined to a few southern states where it is of relatively minor importance. However, much of the corn which is planted in the United States consists of hybrid strains that are made with a particular gene (Texas male-sterile cytoplasm) that causes the male part of the corn plants to be sterile. One race of the corn blight fungus is specific and highly infectious for hybrids made with this gene. Warm, humid weather in August and in early September 1970 throughout most of the corn belt of the midwestern states allowed the disease to spread much farther than it normally occurs, as far as northern Minnesota. Because of the genetic uniformity of the corn that was planted and because of its high susceptibility to corn blight, the disease was able to spread rapidly through the corn belt causing heavy damage.

In natural environments that have not been altered by man, we also find that the simplest ecosystems are the least stable. We commonly speak of the "fragile arctic tundra." This is an extremely simple system which, because of its instability, is easily damaged but not easily repaired. It contains only a few species whose population sizes change greatly over short periods of time. The principal herbivore of the tundra is a small rodent, the lemming. Lemmings feed on the low-growing tundra vegetation and are the main food supply of a few species of predators such as the arctic fox and snowy owl. For reasons that are still not fully understood, lemmings have drastic population cycles, from extremely high numbers one year to almost complete disappearance the next year. This four-year cycle affects the whole tundra ecosystem. When lemming numbers are high, the foxes and snowy owls prosper and multiply, but when the lemmings disappear, most of the foxes die and the snowy owls either die or migrate south to southern Canada or the northern United States in search of food. Unlike the more complex ecosystems at more southern latitudes, where productivity is maintained at a fairly steady rate year after year, the arctic tundra cycles dramatically from years of extreme abundance to years of extreme scarcity.

CONTROL BY CHEMICALS

One of the ways in which we seek to improve the production of agricultural crops or to control diseases is through the application of a variety of chemicals that are designed to control weeds or kill animal pests. The effects of these *pesticides* are beneficial in some cases but not in others. The best known of these chemicals is DDT (dichloro-diphenyl-trichloro-ethane). DDT belongs to a family of chemicals called the *chlorinated hydro-carbons,* which also includes Aldrin, Chlordane, Endrin, and Heptachlor, all of which are widely used to control insect pests. Another important family of pesticides is the organic phosphates. The best known of these chemicals are Malathion and Parathion, both of which are commonly used as insecticides in gardens.

DDT has been used successfully to reduce typhus and malaria in many parts of the world and has probably saved the lives of millions of people, but, as is often the case in human affairs, what at first appears to be a good thing may be carried too far. Careful application of a pesticide to control a particular pest or disease-carrying organism may be extremely beneficial, but if the pesticide is broadcast over entire forests or ecosystems, or if it affects a large number of species, it may destroy many desirable plants or animals, as well as the pest species.

Several years ago the noted biologist Rachael Carson warned of the potential dangers of indiscriminate use of pesticides in her book *Silent Spring:* "... chemicals sprayed on croplands or forests or gardens lie long in soil, entering into living organisms, passing from one to another in a chain of poisoning and death. Or they may pass mysteriously by underground streams until they emerge and, through the alchemy of air and sunlight, combine into new forms that kill vegetation, sicken cattle, and work unknown harm on those who drink from once pure wells."

Modern chemical pesticides are so effective that only a few parts per million in the tissues of plants or animals may be lethal, but this is not the only hazard. Certain chemicals such as DDT are called "persistent pesticides," because they remain biologically active for years and represent a persistent threat to life. They are highly soluble in water and are therefore easily dispersed throughout ecosystems where they are readily absorbed by animals. But because they are less soluble in fat, they tend to accumulate in large amounts in fatty tissues. Repeated exposure to these chemicals allows more and more of the poison to be stored until the tissue contains large concentrations, or they may enter food chains where they become more concentrated at each step in the sequence. Small amounts of DDT in lake water will become incorporated into the tissues of the tiny invertebrate animals that are the food of small fish. As these fish must eat thousands of invertebrates to sustain themselves, they will tend to concentrate the DDT contained in the invertebrates into their own tissues. The small fish are eaten by larger fish, which in turn concentrate the pesticide even further. Predators, such as fish-eating birds at the top of the food chain, will therefore receive large doses of DDT with every fish they eat, and their own process of further concentration may prove lethal to them. It has been shown, for example, that the levels of DDT residues in fish-eating birds are 10 to 1000 times those of the fish on which they feed, and these birds, at the end of the food chain, may have concentrations of DDT in their tissues 1 million times greater than the original concentration in the water (Fig. 17-5).

An example of biological concentration of DDT occurred in Clear Lake, California. This lake is used for fishing and recreation, but there had been many complaints from users of the lake about dense clouds of small gnats. Because the larvae of these gnats live in the water, the lake was sprayed with DDT to kill the larvae before they reached the adult stage. An even distribution of the DDT throughout the lake would have produced a concentration of only 0.015 part per million (ppm) in the water. The first application of this pesticide seemed to be a success, with no apparent effects on other species. However, about five years later, large numbers of dead western grebes, fish-eating birds, were found along the shores of the lake. When the different organisms in the lake were analyzed for DDT, it was found that small invertebrates had concentrated the DDT to levels of 5 ppm in their tissues, the fish that lived on these invertebrates contained 10 ppm of DDT, larger fish had much larger amounts, and the grebes contained 1600 ppm of the pesticide.

It is through this mechanism of biological concentration along food chains that many species such as the peregrine falcon (Fig. 17-6), the bald eagle, and the osprey have become nearly extinct or are in serious danger. However, we are gradually learning that biological controls may be far better than chemical controls. If we

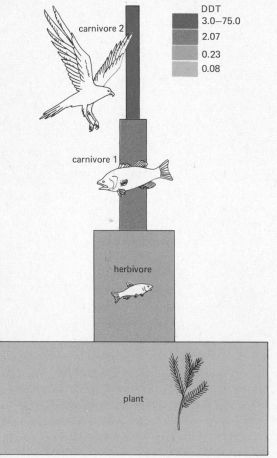

Figure 17-5 Concentration of DDT through a food chain.

can reconstruct ecosystems by reintroducing some of the predators or parasites we removed with pesticides, these may be far more effective and less harmful than the indiscriminate use of chemicals that are difficult to control. A particularly ingenious method of biological control was used against the screwworm fly, which used to cause $120 million worth of damage a year to cattle in the southeastern United States. Instead of using pesticides, male flies were attracted to traps where they were collected and later X-rayed to make them sterile. When these sterile flies were released, they mated with normal females who produced no offspring because the female mates only once during her lifetime. By releasing sterile males for several generations this pest was finally eradicated. This is a classic example of a biological control in which only the pest species are affected and other members of the natural community are undisturbed. This approach is being developed for a wide variety of pests.

From these examples we can draw the following general rule for pest control: nonspecific poisons that kill or harm many forms of life should only be used as a last resort. Nonspecific control agents should not be applied to the general environment or to complicated natural ecosystems where they may escape and destroy many desirable species as well as pests.

TOWARD A BALANCE

Although advances in agriculture are badly needed to increase the food supply of the earth and to raise the level of nutrition of its 3.7 billion people, it is obvious that food production cannot continue to keep up with the human population explosion. The curve of human population growth shows no sign of the S-shape (Fig. 17-1) that is characteristic of a population which is coming into balance with the carrying capacity of its environment. The human population is growing more like the bacteria colony we described earlier, at an ever accelerating rate. Unless man begins to limit his population size, the world's ecosystems will not be able to provide the food and other materials he requires and the ecosystems, themselves, will be in danger of irreparable damage.

The control of human populations is just as "natural" as the other controls that man exerts on the biological world for his own benefit. We have severely disturbed and modified large, natural ecosystems in order to increase food production, to raise better breeds of domestic animals, and to control pests and disease. This point can be illustrated with the cultivation of wheat. Wheat was originally grown in Egypt at least 7000 years ago and introduced into Virginia by English settlers during the seventeenth century. Wheat is not therefore a native plant to the New World. Wheat now covers millions of square miles of the midwest and the great plains of the United States and Canada, where forests and grasslands used to grow. In order to improve the production of wheat we have plowed the land, eliminated the natural plants that used to occupy the wheat fields, and we have had to apply hundreds of different chemicals to the environment in order to control the many pests and diseases to which wheat is so susceptible. It has also been necessary to add chemical fertilizers to compensate

Figure 17-6 The peregrine falcon is vanishing from North America and may soon become extinct.

for the loss of nutrients from the soil. These innovations which are having a heavy impact upon natural environments will not produce a better quality of life, however, unless human population size is limited.

As we have seen in Chapter 15, more effective methods of birth control are needed, but what is needed even more is the social and political commitment to use what means are already available. Human society must first recognize the nature of the balance between people and their environment, and then decide on courses of action which will bring about this balance.

Francis Bacon once wrote that "We cannot command nature except by obeying her." A great deal of the discussion in this chapter illustrates this maxim. Earlier in this chapter and in the introduction we raised the question of whether man can exist apart from nature or is subject like plants and other animals to natural laws. Every chapter of this book contributes to the answer to this question. The molecular basis of life applies to all life, including man's. The flow of energy and information in living systems, the principles of development, genetic control of development and characteristics, and food and energy requirements apply equally to all species, including man. Man is also an integral part of higher levels of organization such as ecosystems and the biosphere. The question therefore is not so much whether man is an animal—he is—but whether he will use his intellectual capacity, his cultural history, and his ability to analyze objectively his present condition and predict future events in order to preserve the essential components of the biosphere and ensure his own survival.

SUGGESTED READING LIST

Brown, L. R., "Human food production as a process of the biosphere," Scientific American, September 1970.

Cox, G. W., Readings in Conservation Ecology. Appleton-Century-Crofts, New York, 1969.

Deevey, E. S., Jr., "The human population," Scientific American, September 1960.

Ehrenfeld, D., Biological Conservation. Holt, Rinehart and Winston, New York, 1970.

Ehrlich, P. R., and A. H. Ehrlich, Population, Resources, Environment. W. H. Freeman, San Francisco, 1970.

Harte, J., and R. H. Socolow, Patient Earth. Holt, Rinehart and Winston, New York, 1971.

Huxley, J., "World population," Scientific American, March 1956.

Odum, E. P., Ecology. Holt, Rinehart and Winston, New York, 1963.

Woodwell, G. M., "The ecological effects of radiation," Scientific American, June 1963.

Part Four

Populations

The population is the
arena for evolutionary
change and the source
of biological diversity.

CHAPTER 18

Evolution in Action

Over a century ago English naturalists began to notice unexpected changes in the coloration of moths that lived in the woodlands surrounding many industrial areas. One of these species was the peppered moth, *Biston betularia*. Until then all of the peppered moths that had been collected were light gray or speckled, but in 1848 a black specimen was found near the industrial center of Manchester in northern England. There followed a rapid increase in black moths until by 1895 — less than 50 years later — about 98 percent of the peppered moth population in the Manchester area was dark-colored. Although it was not immediately evident at the time, English naturalists were witnessing evolution in action, starting almost ten years before Charles Darwin wrote his book *On the Origin of Species* and proposed the theory of natural selection in 1859.

INDUSTRIAL MELANISM

This rapid change in the predominant coloration of moths coincided with the dawn of the industrial revolution in England, and spread quickly to Europe and eventually into North America with the growth of large manufacturing centers and the pollution of the countryside by coal dust and soot. This change from light to dark forms of moths and other insects in industrial areas is now referred to as "industrial melanism." It is widespread among many species and many countries.

Figure 18-1 Air pollution from one industrial plant.

In addition to the peppered moth, seventy other British species were affected. More than forty species near the large manufacturing cities along the St. Lawrence River in Canada show industrial melanism. The proportion of black or dark-colored moths has reached 90 percent or more in populations near New York, Detroit, and Philadelphia and all of the species of moth in the Pittsburgh area have been affected. It has only recently been demonstrated scientifically that industrial melanism is the result of natural selection.

When scientists noted the correlation between pollution from soot and coal dust around industrial centers and the increasing prevalence of dark-colored moths, they postulated that the moths could have been affected in at least two ways. The simplest possible explanation, and the easiest to test, was that chemical pollutants in some way altered the physiology of the moths, causing an excess production of black (or melanic) pigments. But when this hypothesis was tested by exposing larval forms of the moths to various industrial chemicals, they developed into normal, light-colored adults and showed no evidence of industrial melanism. It was therefore evident that the change in coloration must have a genetic rather than a strictly physiological, or phenotypic, basis.

Two British geneticists, R. A. Fisher and E. B. Ford, then developed a theoretical explanation of industrial melanism. Animals are said to have "cryptic coloration" when their color and its pattern blend with their environmental background and make them inconspicuous to predators. A normal, light-colored peppered moth resting on the trunk of a tree that is covered with gray or light-green lichens is very difficult to see, but the same moth resting on a tree blackened with soot is quite conspicuous (Fig. 18-2). Therefore a particular coloration is only cryptic in certain circumstances, and a light color may be advantageous in one environment but disadvantageous in another. Fisher and Ford reasoned that a gene that produces a black phenotype would tend to be preserved and would spread through a population in industrial areas because of selection against its light-colored and more vulnerable allele.

The genetic theory advanced by Fisher and Ford rested partly on the assumption that the moths were subject to heavy predation, but it remained for another British geneticist, H. B. D. Kettlewell, to provide the experimental evidence that confirmed this theory. Kettlewell showed that the peppered moth is preyed upon by several kinds of birds and that the moths do depend upon their **cryptic coloration** for concealment from these predators. He released equal numbers of light and dark forms of the moth in an unpolluted woodland in Dorset, where the tree trunks are light colored and have abundant growths of lichens and where no black moths are normally found. The dark, melanistic forms were very conspicuous as they rested on the tree trunks, but the normal, light-colored moths were difficult to see. While Kettlewell watched with binoculars, birds ate a total of 190 moths, of which 164 were black and 26 were the light-colored form, even though equal numbers of each had been released.

This experiment was repeated near the industrial city of Birmingham, where the tree trunks are blackened with soot and about 90 percent of the moths naturally occurring in this area are black. When equal numbers of light and black moths were released, the birds were observed to eat 15 black and 43 of the light-colored moths. In another set of experiments Kettlewell

Table 18-1 Survival of light and dark phases of the peppered moth in an industrial area (Birmingham) and a non-industrial area (Dorset).

Locality	Number released (Dark)	(Light)	% recovered (Dark)	(Light)	Ratio (Dark:Light)
Birmingham	477	137	26.5	13.1	2:1
Dorset	473	496	6.0	12.5	1:2

marked black and light-colored moths with small dots of cellulose paint and released them in a woods. By later attracting them to a light at night and recapturing marked individuals, he was able to measure the relative survival of the two forms. When he released 477 black and 137 light-colored moths in the Birmingham woods, he recovered 27.5 percent of the black and only 13.1 percent of the light-colored form, showing that the latter form survived only about half as well as the black moths. When this experiment was repeated in the Dorset woods, the ratio was reversed: of 473 black and 496 light-colored moths released, 6 percent of the black and 12.5 percent of the light-colored forms were recaptured (Table 18-1).

Figure 18-2 Light and dark phases of the peppered moth on tree trunk covered with lichen (left) and soot (right).

The history of industrial melanism in the peppered moth is the most clearly documented case of evolutionary change in a natural population of animals. It illustrates how natural selection operates, giving one individual a selective advantage in relation to another; it also, however, illustrates another very important point. In Chapter 3 we discussed genetics with examples of inheritance in individuals and much of what we said about adaptation dealt with individual organisms. The example of the peppered moth shows, however, that the process of evolution is concerned with populations. Individual moths showed no evidence of change. When they emerged as adults, they either had the black or light-colored phenotype and remained this color for life. The dramatic evolutionary change that was observed was an increase in the proportion of black moths in the population with succeeding generations, as the selective agent, predation from birds, allowed relatively more black moths than light-colored ones to survive and reproduce. We must therefore shift our perspective to a larger focus in this and succeeding chapters in this section of the book, where we will be dealing with biological events whose effects are observed at the population, rather than at the individual level.

SOURCES OF VARIATION

It is evident that the process of evolution requires a continuous source of *heritable variation*, and that evolution is a population process rather than an individual process. As we learned in Chapter 3, Mendel presented experimental proof of the particulate nature of inheritance which led to the gene theory and provided a mechanism (which Darwin's theory lacked) for the transmission of inherited traits from one generation to the next. Mendel also established two laws of inheritance: (1) the *random segregation* of alleles during gamete formation and (2) the *independent assortment* of genes during gamete formation. We also learned about subsequent discoveries which provided an understanding of gene mutation and gene exchange (e.g., crossing over), which are the primary sources of all genetic variation in individuals. Thus *mutation* and *recombination* are

the two major sources of variation within populations upon which natural selection can act to produce adaptation and, eventually, evolution. Recombination is by far the most frequent source of variation in a population, but the new variations that are produced in this way are limited to the genes that exist in the population at that time. We refer to this total of all genes in a population at a given instant as the "gene pool." However, if there are to be new genetic themes beyond those that can be provided by recombination of the existing gene pool, there must also be mutation.

NO SELECTION, NO EVOLUTION

Let us first consider what would happen within a gene pool if there were no selection or random genetic events such as mutation. This has actually been done using a flock of chickens isolated in a large pen. Fifty roosters and 50 hens with black feathers made up half the stock; the other half consisted of 50 roosters and 50 hens with white feathers "splashed" with black. It was known that birds with black feathers are homozygous for one allele (pp) and white-splashed is homozygous for the other allele (qq). Heterozygotes (pq) are blue-gray. This gives us a single-gene model with two alleles at a given locus. The chickens were allowed to mate randomly with each other so that all three genotypes (pp, pq, qq) were produced.

As each bird had two genes for plumage color, there were 400 genes in the population, half of which were p and the other half q, so that the frequency of each gene was $1/2$. If we recall the diagram in Chapter 3 to illustrate the segregation of genes, we see that random matings in this population should produce first generation offspring in the ratio of $1pp:2pq:1qq$. What, if anything, has happened to the gene pool as a result? The proportion of the genes p and q has not changed, and the gene pool still contains $1/2 p$ and $1/2 q$. When genotypes occur in the proportion 1:2:1, the population is said to show genetic equilibrium. This is the **Hardy-Weinberg Law** which results from random mating. It is based on the simple binomial expansion of Mendelian inheritance, where

$$(p + q)^2 = 1 = p^2 + 2pq + q^2$$

We can predict the expected frequency of the three genotypes by substituting the gene frequencies in the equation:

$$\left.\begin{array}{lll} p^2 & = 0.50 \times 0.50 & = 0.25 \\ pq & = 2(0.50 \times 0.50) & = 0.50 \\ q^2 & = 0.50 \times 0.50 & = 0.25 \end{array}\right\} = 1.0$$

In this example we started with the simplest case, in which the gene frequencies were equal at the start of the experiment, but this is not necessary. If two alleles are not present in the population in the proportion of $p^2:2pq:q^2$, then one generation of random mating in the absence of selection will establish an equilibrium with $p^2:2pq:q^2$, and this will hold true for all subsequent generations. You can prove this for yourself by selecting any frequencies for the genotypes pp, pq, and qq which equal 1.0.

The Hardy-Weinberg Law of genetic equilibrium illustrates the conservative nature of heredity, which tends to favor the genetic *status quo*. It states that the proportion of genes in a population tends to remain constant, unless nonrandom forces such as selection alter the gene frequencies. The Hardy-Weinberg Law does not apply, however, when any of the following circumstances occur:

1. Individuals in the population do not have equal reproductive rates.
2. Mating does not occur at random.
3. Mutations occur and the back mutation rate (mutation back to the normal allele) is not the same as the mutation rate.
4. The population is so small that chance can influence the results of random mating in some significant way.

Let us look at an example of the second exception to the Hardy-Weinberg Law, which also shows how the processes of evolution can be studied in the laboratory. Normal, wild-type fruit flies have red eyes, as we have already learned. In 1909 a single white-eyed male *Drosophila melanogaster* appeared in a laboratory colony being cultured for genetic experiments. Additional white-eyed flies were bred from this original mutant to provide stocks for studies of the white-eyed trait. It was found that if normal males and females are kept together, random mating occurs and there is no apparent selection for particular males or females. If normal females are kept with white-eyed males, the flies will also mate normally, but if both normal and white-eyed males are present, nonrandom mating occurs. Given a choice of mates, normal females will reject white-eyed males and mate with normal males, so that eventually the mutant white-eyed gene is eliminated from the population.

SEQUENTIAL EVOLUTION

The Hardy-Weinberg Law is a valuable conceptual model which allows us to measure departures from genetic equilibrium, but natural environments are in a constant state of change and so many factors influence the lives of wild populations of plants and animals that genetic equilibrium is not likely to occur often in nature. Almost all natural populations undergo some evolutionary change from one generation to the next through modifications of the frequencies of genotypes in the gene pool. This process is called **sequential evolution.** For example, British geneticists conducted a 14-year study of changes in the frequency of two alleles which affect the pattern of spots in the wing of the scarlet tiger moth, *Panaxia dominula*. One allele (a_1) produces two light spots in the black anterior wing (Fig. 18-3), while the opposite allele (a_2) produces many light spots; heterozygous individuals have an intermediate number of spots. Figure 18-3 traces fluctuations in the frequency of the normal and two-spot alleles from 1939 to 1961 in a single population of the scarlet tiger moth. There is no indication of genetic equilibrium, and all available evidence indicated that this sequential type of evolution occurs with every generation in all natural populations.

Charles Darwin understood the importance of natural selection as a guiding force in evolution, even though he had not seen evolution in action when he formulated his theory and laid the foundations of evolutionary biology. He had read *Essay on Population* by Malthus, which contained the expression "struggle for existence" and the idea that a population in a limited environment will increase at a geometric rate and will multiply far beyond the food resources that the environment contains unless mortality keeps the

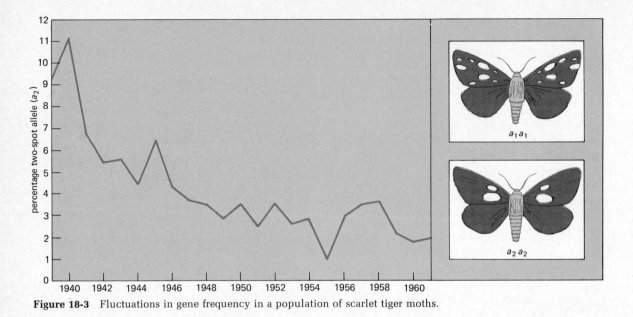

Figure 18-3 Fluctuations in gene frequency in a population of scarlet tiger moths.

population in check. During his travels on the *HMS Beagle* from 1831 to 1836, Darwin also observed that the past and present distributions and the morphological characteristics of plants and animals show orderly trends that are correlated with attributes of their environments. He was also aware, partly from his own selective breeding of fancy varieties of pigeons, that there is a wide array of individual variation in a population from which the "environment selects" those which are best fitted to contribute their qualities to the next generation. These observations placed differential reproduction and survival on a Malthusian background of organisms producing more offspring than can possibly survive, and provided the essential features of the theory of evolution by natural selection which can be summarized as follows:

1. Species are generally constant in population size.
2. Organisms produce far more offspring than can survive to become reproductive individuals in a stable population.
3. The production of excess offspring is balanced by a high mortality rate.
4. Individuals within a population are not identical and vary in their characteristics.
5. Some individuals are better fitted to the conditions of their environment than others, and the fittest will be most likely to survive and pass their characteristics on to the next generation.
6. Succeeding generations will tend to preserve and improve the adaptations contributed by their ancestors.

Thus Darwin concluded that advantageous variations, however small, which are inherited from one generation to the next, will be preserved and will accumulate in species populations, allowing them to adapt to the conditions of their environments. He also reasoned that with time and change in local environments, or with dispersal into new and different environments, species will gradually depart from their ancestral forms and alter their characteristics. Modern evolutionists refer to such changes in species over long periods of time as **phyletic evolution.** This is a concept we encounter again in later chapters when we consider the evolutionary origins of modern plants and animals.

The phrases "the struggle for existence" and "survival of the fittest" led to a popular misconception that natural selection is a continuous, physical battle among species. This view of nature was expressed by Tennyson in his poem *In Memoriam* as "Nature, red in tooth and claw." As we saw with the peppered moth, there was no battle or conflict among them. Those that were best adapted to the colors of their background survived to reproduce, and their genes contributed to the gene pool of the population. Bacteria that survive a dose of antibiotic, and continue to multiply until a resistant strain of bacteria is established, are the fittest to survive in this environment, but they are not responsible for the death of the other members of the population—the antibiotic is the selective agent. A tree may be more efficient than its neighbors and make better use of the sunlight, water, and nutrients that are available in an environment. It may flourish and produce a large crop of seeds, whereas its neighbors are less successful. Or the seeds from a particular tree may be better adapted to the soil and moisture conditions of an area, and will take root and grow where other seeds fail. Thus, far from "red in tooth and claw," natural selection is a subtle, continuous process that is measured by a more or less gradual change in the genetic constitution of a total population.

THE MODERN VIEW

Evolutionary biology has greatly matured in the twentieth century, largely because of the spectacular advances that have been made in our knowledge of genetics. We no longer need to list proofs for evolution, as Darwin and his supporters did when the theory was first proposed. Modern scientists accept the process of evolution as a fact, and the twentieth century emphasis has shifted to studies of the forces of evolution,

the various mechanisms and their actions, and to a greater understanding of the causes and effects of organic diversity. The principal defect in Darwin's theory was that he could not explain how variations in individuals and species occurred, or how they were transmitted from one generation to the next. His theory explained *why* evolution proceeds as it does, but not *how*. He knew that individual differences were in some way inherited, for there is no other way to account for natural selection. It remained, however, for modern genetics to provide the answers that Darwin could not supply.

Ironically, the mechanism of heredity was described by Mendel in 1865, 17 years before Darwin died and only 6 years after he published *On the Origin of Species*. But the results of Mendel's research were published in an obscure journal and were not rediscovered until 1900, so that Darwin was never aware that the clue to his puzzle had been found. When Mendel's principles of heredity and more recent findings in population genetics were combined with Darwin's theories, it became possible to explain the mechanism of evolution on the basis of the interactions between heredity and natural selection.

SUGGESTED READING LIST

Darwin, C., *The Origin of Species*. The New Library of World Literature, New York, 1958.

Dobzhansky, T., "The genetic basis of evolution," *Scientific American*, January 1950.

Hamilton, T. H., *Process and Pattern in Evolution*. Macmillan, New York, 1967.

Kettlewell, H. B. D., "Darwin's missing evidence," *Scientific American*, March 1959.

Savage, J. M., *Evolution*, second edition. Holt, Rinehart and Winston, New York, 1969.

Stebbins, G. L., *Process of Organic Evolution*, 2nd edition. Prentice-Hall, Englewood Cliffs, New Jersey, 1971.

CHAPTER 19

Races and Species

The broad course of evolution depends not only upon the continual adaptation of populations to changing environments, as in the process of sequential evolution described in the previous chapter, but also on the origin of new races and species. The factors responsible for the formation of races and species are considerably different from those involved in sequential microevolution, so that evolutionists no longer refer to a single origin of species. We now know that species may be formed in different ways.

An extremely important mechanism in speciation is **adaptive radiation** in which several species may be formed when different local populations of a single species invade different environments and become adapted to the special conditions of each. A classic example is Darwin's finches on the Galapagos Islands, which Charles Darwin observed in 1835 as chief naturalist on the famous voyage of the *H.M.S. Beagle*.

THE GALAPAGOS ARCHIPELAGO

The Galapagos archipelago lies across the equator, in the midst of the Pacific Ocean, 600 miles west of Ecuador and 1000 miles from Panama. The closest islands are Cocos and Malpelo, both about 600 miles northeast of the Galapagos. The Galapagos are therefore well isolated from the mainland of South America and the other Pacific islands.

The total Galapagos archipelago is roughly 150 miles square. The largest island (Fig. 19-1) is Albemarle, about 80 miles long with an elevation of 4000 feet. Several of the other islands are from 10 to 20 miles across and rise to heights of 2000 to 3000 feet, while there are many small, low-lying islands, some merely rock outcrops in the midst of the vast Pacific.

All of the Galapagos Islands are the crests of immense submarine volcanoes, some still active, which have risen above the surface of the water from almost 2 miles below. The low-lying islands and the lower elevations of the larger islands consist of barren outcrops of lava rock and a dry vegetation of scrub, thorn bushes, and cacti (Fig. 19-2). At higher elevations on some of the larger islands, moisture from fog and tropical rains has allowed the development of humid, tropical forests. At still higher elevations the trees give way to open country covered with grasses, ferns, and mosses.

Galapago is the Spanish word for "tortoise." The islands were well known to the early pirates of the Pacific, who stopped there to provision their ships with the 400-pound land tortoises (Fig. 19-3) for which the islands are famous. The giant tortoises were placed on their backs and lashed to the decks of the ships, where they remained alive for days, a ready source of fresh

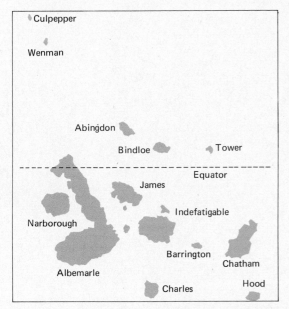

Figure 19-1 Map of the Galapagos archipelago.

Figure 19-2 View inside crater and cone on Narborough Island.

Figure 19-3   Abingdon land tortoise.

Figure 19-4   Galapagos land iguana.

meat for the sailors. But aside from the attraction of tortoise meat, the archipelago was a scene of utter desolation to most of its visitors and they spent as little time there as possible. Fitzroy, the Captain of the *Beagle*, wrote that the islands were "black dismal-looking heaps of broken lava, forming a shore fit for Pandemonium." But to Darwin, who viewed the archipelago through the eyes of a naturalist, it was "paradise."

The naturalist's paradise that Darwin instantly recognized was a unique fauna of birds, reptiles, and lizards (Fig. 19-4), unlike any other animals in the world, isolated on these remote islands but obviously related to similar species he had observed on the mainland of South America. He wrote

The natural history of these islands is eminently curious, and well deserves attention. Most of the organic productions are aboriginal creations, found nowhere else; there is even a difference between the inhabitants of different islands; yet all show a marked relationship with those of America, though separated from that continent by an open space of ocean, between five and six hundred miles in width. The archipelago is a little world within itself, or rather a satellite attached to America, whence it has derived a few stray colonists and has received the general character of its indigenous productions. Considering the small size of

these islands, we feel the more astonished at the number of their aboriginal beings, and at their confined range. Seeing every height crowned with its crater, and the boundaries of most of the lava-streams still distinct, we are led to believe that within a period geologically recent the unbroken ocean was here spread out. Hence, both in space and time, we seem to be brought somewhat near to that great fact—that mystery of mysteries—the first appearance of new beings on this earth.

DARWIN'S FINCHES

The finches of the Galapagos archipelago, which we now refer to as "Darwin's finches" (Fig. 19-5), have been studied intensively by modern evolutionists, following Darwin's original observations in 1835. They have confirmed his conclusion that the 13 species which now occur on the islands are descendants of a single, ancestral species which somehow managed to reach the Galapagos across the vast expanse of the Pacific from the mainland of South America (Fig. 19-6). When these birds arrived, probably blown across the ocean by a storm, they found a variety of foods and habitats to support them, but no other finch-like birds and few enemies.

Figure 19-5 Darwin's finches. (For names see Table 19-1.)

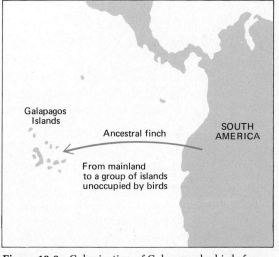

Figure 19-6 Colonization of Galapagos by birds from mainland South America.

Table 19-1 lists the 13 species of Darwin's finches that occur on the Galapagos. There are no established common names for these finches as there are for most birds, but the descriptive names indicate their roles in the habitats they occupy.

All of these finches are rather dull in appearance and habit, and their songs are also monotonous and unmusical. The eggs of the different species vary slightly in size, but are otherwise much alike, and there is very little difference in their breeding habits. The only really striking difference among the species is in the variety of size and shape of their beaks (see Fig. 19-5). There are small, finchlike beaks; heavy, finchlike beaks; parrotlike beaks; straight, wood-boring beaks; beaks which are curved for probing flowers; and slender beaks, like those of warblers.

The heavy, finchlike beaks of some of the ground-finches (*Geospiza magnirostris, G. fortis,*

Table 19-1   Species of Darwin's finches

Scientific name	Descriptive name (as numbered in Figure 19-5)
Geospiza magnirostris	Large ground-finch (1)
Geospiza fortis	Medium ground-finch (2)
Geospiza fuliginosa	Small ground-finch (3)
Geospiza difficilis	Sharp-beaked ground-finch (4)
Geospiza scandens	Cactus ground-finch (5)
Geospiza conirostris	Large cactus ground-finch (6)
Camarhynchus crassirostris	Vegetarian tree-finch (7)
Camarhynchus pauper	Large, insectivorous tree-finch on Charles Island (8)
Camaryhnchus psittacula	Large, insectivorous tree-finch (9)
Camarhynchus parvulus	Small insectivorous tree-finch (10)
Camarhynchus heliobates	Mangrove-finch (11)
Camarhynchus pallidus	Woodpecker-finch (12)
Certhidia olivacea	Warbler-finch (13)

G. fuliginosa) suggest a diet of seeds, which is in fact the principal food of these birds, even though they also feed on fruits and insects at certain times of the year. The long, decurved beak of the cactus ground-finch (*Geospiza scandens*) is adapted to flower-probing and nectar-feeding, and it feeds on the flowers of the prickly pear cactus, *Opuntia*. The woodpecker finch (*Camarhynchus pallidus*) has a stout, straight beak like a woodpecker and feeds mainly on insects. This bird has a most remarkable habit. When a woodpecker probes a branch or tree trunk for an insect, it inserts its long tongue into the crevice to retrieve its prey. The woodpecker finch lacks the long tongue of a true woodpecker, but achieves the same result by using a cactus spine or twig (Fig. 19-7) to probe into the crack, forcing the insect to the surface. When the insect emerges, the finch drops its twig and seizes the insect in its beak. This is one of the few recorded cases of the use of a tool by birds.

GEOGRAPHIC ISOLATION

Adaptive radiation requires isolation and vacant ecological niches. The first finches to arrive at the Galapagos found a variety of habitats and foods which they could exploit in the absence of competition from other, similar birds. These provided the vacant, ecological niches. With these opportunities available to them, all of the islands were soon inhabited by finches derived from the original, colonizing population. Also, because finches are relatively poor fliers, there was a strong tendency for the populations on different islands, once they were established, to remain isolated from each other. This produced a corresponding reduction in gene flow between the different populations. Furthermore, as we noted earlier, the islands of the Galapagos archipelago differ considerably in size, height, and vegetation, so that each tends to provide a particular set of habitat conditions. This combination of factors, different ecological niches, and genetic

Figure 19-7 Woodpecker finch probing for insects with cactus spine.

isolation, set the stage for natural selection to produce genetic divergence between populations on different islands. On each island sequential evolution, operating in the manner shown by the example of industrial melanism in the previous chapter, produced adaptation to a particular set of local conditions, leading to the formation of new races and, eventually, new species.

The importance of isolation is illustrated in Table 19-2, which shows the distribution and numbers of Darwin's finches on each of the islands. The largest islands, Albemarle, James, Narborough, and Indefatigable, tend to have the largest number of species, reflecting the number of different ecological niches available on the larger islands. However, the outer islands, such as Culpepper, Abingdon, Tower, and Chatham, have the largest number of endemic species. *Endemic species* are species that occur only in a particular locality and are therefore peculiar to that locality. Thus we may assume from the data in Table 19-2 that new races are formed more frequently on the outer than on the inner islands, and that the greater the isolation the greater the endemic differentiation and race formation.

When two populations are readily distinguishable from each other by some character such as color or body size or shape but can still inter-

Table 19-2 Distribution and numbers of species of Darwin's finches in the Galapagos archipelago

Island	Total species	Endemic species
Culpepper	4	2
Wenman	5	1
Abingdon	9	2
Bindloe	7	1
Narborough	9	0
Albemarle	10	1
Tower	4	3
Hood	3	2
Chatham	7	3
Charles	9	2
Jervis	9	0
James	10	0
Barrington	7	1
Duncan	9	0
Indefatigable	10	0
Seymour	8	0

breed, they are referred to as **races.** Races represent open genetic systems in that there is still some actual or potential gene exchange between different populations. However, when two populations have been isolated long enough to become reproductively isolated from each other, they are classified as **species.** Species are therefore closed genetic systems.

It is clear from the data in Table 19-2 that geographic isolation on the different islands was not perfect, even though it was enough to allow the formation of new species. Eventually members of different species found their way to other islands that were already occupied by populations of other finches. If differentiation had only proceeded to the level of races and the two populations could still interbreed, their differences would soon be swamped and they would have a common gene pool. However, if they had diverged sufficiently to be reproductively isolated and were not able to interbreed, they would not only remain distinct but there would also be a tendency for their differences to become even greater through the pressure of competition.

ISOLATING MECHANISMS

It may seem strange that closely related plants or animals could live together in the same environment and not mate together occasionally, but biologists have found that there are many kinds of **isolating mechanisms** which prevent hybridization between species. A list of isolating mechanisms is shown in Table 19-3. Some actually prevent mating from taking place, while others ensure that viable offspring will not be produced, even if mating does occur. In terms of heredity these mechanisms prevent gene flow between two species just as effectively as if they were geographically isolated on separate islands.

Some or all of these mechanisms probably occur among Darwin's finches, but let us consider another group of animals in which almost all of the mechanisms have been identified—the frogs and toads that occur in the vicinity of New Orleans, Louisiana.

Pig frogs and gopher frogs never mate because the former breed in deep water in large ponds, and the latter lay their eggs and fertilize

Table 19-3 Classification of isolating mechanisms

Mechanism	Type
Potential mates do not meet because of habitat or seasonal isolation	Ecological isolation
Potential mates meet but do not mate	Behavioral isolation
Copulation is attempted, but transfer of sperm does not take place	Mechanical isolation
Sperm transfer takes place, but egg is not fertilized	Gametic mortality
Egg is fertilized, but zygote dies	Zygotic mortality
Zygote is produced, but is unhealthy and does not reproduce	Hybrid inviability
F_1 hybrid is fully viable, but is partly or completely sterile, or produces an inviable F_2 offspring	Hybrid sterility

them in the shallows of small ponds and swamps (*ecological isolation*). Gray frogs and pine woods tree-frogs, on the other hand, frequently breed in the same ponds at the same time. However, the two species have different mating calls, and the females are attracted only to males of their own kind (*behavioral isolation*). This form of isolation is very common among animals and almost always depends on how the female responds to the mating display of the male. Differences in body size keep the oak toad and the Gulf Coast toad from mating (*mechanical isolation*). Male Gulf Coast toads are about twice the size of female oak toads and are more likely to eat them than mate with them; male oak toads, on the other hand, are too small to grasp and hold the Gulf Coast females.

Even when these mechanisms break down and mating does occur, others prevent healthy hybrids of two species from developing and reproducing. We can study these mechanisms in the laboratory by making artificial crosses between different species, either by creating special conditions which overcome the previously described barriers to reproduction or by artificial insemination. If a bullfrog is crossed with a gopher frog, the embryo begins to develop but never reaches the tadpole stage (*zygotic mortality*). Gulf Coast toads and Fowler's toads often breed in nature and can be crossed in the laboratory, but the offspring of a male Fowler's toad and female Gulf Coast toad is always a sterile male (*hybrid sterility*); crosses in the opposite direction, those between a female Fowler's toad

and male Gulf Coast toad, all die early in development (*zygotic mortality*).

Actually, in most species, more than one isolating mechanism may operate. This is especially well demonstrated in experiments with nearly identical species of *Drosophila*. Behavior patterns reduce the likelihood of mating, but if it does occur the offspring are not as strong and healthy as normal flies. Some eggs develop, others produce sterile young, and sometimes the offspring of the hybrids and the parental species are defective. In plants different habitats produce ecological isolation and reduce the chance of some crosses. Sometimes different flowering and pollinating seasons are involved, or variations in flower structure prevent fertilization. Weak or sterile hybrids often result if fertilization does take place.

We noted earlier that two species which have been formed in geographic isolation from one another may subsequently come into contact. This may occur because of geological changes in time which remove a previous physical barrier, or when geographical isolation is incomplete, as in the case of Darwin's finches on the Galapagos Islands. Secondary contacts among species are also caused by man. A species that would not be able to disperse over any great distance might be aided in its travels by human transportation, such as shipping or travel by planes. Seeds of plants and insects or their eggs probably are very frequent travelers this way, and are able to reach areas that, because of distance, would otherwise be inaccessible to them. Geographic isolation allows the genetic makeup of a population to change enough so that even if divergent populations begin living in the same area, genetic isolating mechanisms will maintain the separate identity of each species. Not only will the species remain distinct, they will also diverge further, for natural selection will increase the differences between them. In the course of sharing the same habitats, closely related species which resemble each other in structure and habitat are most likely to compete for food and places to live. The pressure of competition will therefore make it advantageous for each species to specialize along different lines.

CHARACTER DIVERGENCE

The tendency for two species to diverge from one another in one or more characters such as body size, color, or the size and shape of the beak is referred to as **character divergence** or **character displacement.** Character divergence is shown by two species of nuthatches which inhabit southeastern Europe and southwestern Asia. When the two species occur together in the same region, they differ more in color and in beak size than they do when a comparison is made between individuals from regions where they occur separately. Similarly, the sharp-beaked groundfinch, *Geospiza diffilis,* has several feeding roles on islands where there is a lack of competition (Wenman and Culpepper), but has been forced to specialize in its feeding when competition is present (Tower and the inner islands). Thus character divergence reduces the effects of competition through specialization and more efficient use of a particular resource. At the same time it increases genetic isolation by accentuating the differences between species.

There is one major exception to the general rule that divergent evolution begins with the combined effects of natural selection and geographic isolation. An additional mechanism has been identified in plants. Sometimes a plant population can become genetically or reproductively isolated and form a new species in a single generation, even though surrounded by members of its ancestral species. This isolation

comes about through major chromosome changes, self-fertilization, or asexual reproduction.

In some plants the number of chromosomes may increase spontaneously — doubling, tripling, or quadrupling, or even being multipled by six or eight times. Changes of this sort are often due to *nondisjunction,* which is the failure of chromosomes to separate during meiosis. The offspring produced following nondisjunction are immediately isolated from the parental gene pool because of the change in chromosome number. Hybrids of closely related species also tend to have faulty meiotic mechanisms, and will often produce individuals with higher than the normal number of chromosomes. If these plants then breed back to the original stock, the new chromosomal condition will become fixed.

Two other isolating mechanisms involve doing away with the genetic exchange that accompanies sexual reproduction. Self-fertilization, of course, cuts off genetic contact with other individuals, and can lead to an independent course of evolution. Many different asexual processes of reproduction are also known among plants and lower animals, all involving the development of unfertilized gametes, or propagation by vegetative means (Chapter 15). When a previously sexually reproducing population becomes asexual, the individuals in the population become separate, isolated gene pools with no other source of genetic diversity but mutation. They are denied the benefits of recombination, a major source of genetic diversity among sexually reproducing organisms.

It is clear that Darwin's finches have been successful in adapting to life in the Galapagos archipelago, and that they became far more diversified than their mainland ancestors. Darwin's finches were successful because they were able to meet the following three requirements necessary for an organism to occupy a new environment.

1. *Physical access.* The finches were able to reach the islands from the mainland of South America in sufficient numbers to colonize the new environment.

2. *Preadaptation.* They were sufficiently adapted to the new environment before they reached it

to be able to survive on the seeds and fruits of plants that were there when they arrived.

3. *Vacant ecological niches.* The lack of competition from other similar organisms permitted them to exploit the new environment and to evolve in several ways. Had representatives of all of the other South American land birds arrived at the Galapagos with them, the finches would have been just as limited in their adaptations as their mainland ancestors, and would not have been able to radiate into a variety of ecological roles.

We have chosen to illustrate *adaptive radiation* with the relatively simple example of Dar-

win's finches, but exactly the same process, magnified many times, produced the massive evolutionary radiations that covered the earth with its diversity of life. In later chapters there is further discussion about the history of life and evolution on earth and many diagrams like Figure 19-8. The divergent evolution treated in this chapter is the basis for such widespread radiations.

One feature of the adaptive radiation of Darwin's finches is that they were far enough advanced in evolution that they were strongly limited in their genetic potential. They could not, for example, become predatory mammals even though there were none on the Galapagos Islands,

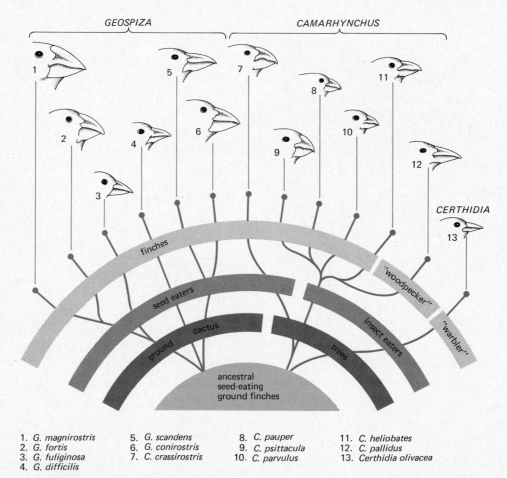

1. *G. magnirostris*
2. *G. fortis*
3. *G. fuliginosa*
4. *G. difficilis*

5. *G. scandens*
6. *G. conirostris*
7. *C. crassirostris*

8. *C. pauper*
9. *C. psittacula*
10. *C. parvulus*

11. *C. heliobates*
12. *C. pallidus*
13. *Certhidia olivacea*

Figure 19-8   Adaptive radiation in Darwin's finches.

so that this role in the community has gone un-occupied.

More significant in the broad course of evolution have been the radiations that involved opening whole new ways of life. One of the major landmarks in evolution came when the first vertebrates ventured ashore and began to exploit the land environment. Even then the three requirements of success that we have just outlined held true. These animals could not leave their normal environment in water until they could survive ashore, and they could not occupy terrestrial environments until plants had colonized the land and were there as a source of food.

But this is a part of the history of life which is described more fully later. Before we proceed to the broader spectrum of evolution, we must first complete the story of populations and how plants and animals survive long enough to reproduce and set the stage for the continuation of their sort of life. In later chapters we examine the interactions that occur between different kinds of plants and animals, to see how they divide up the environments of the earth and go about living together in integrated systems.

SUGGESTED READING LIST

Bowman, R. I., *The Galapagos.* University of California Press, Berkeley and Los Angeles, 1966.

Darwin, C., *The Origin of Species.* The New Library of World Literature, New York, 1958.

Eisley, L. C., "Charles Darwin," *Scientific American,* February 1956.

Hamilton, T. H., *Process and Pattern in Evolution.* Macmillan, New York, 1967.

Lack, D., "Darwin's Finches," *Scientific American,* April 1953.

Savage, J. M., *Evolution,* second edition. Holt, Rinehart and Winston, New York, 1969.

CHAPTER 20

The Origins of Life

Four and one-half billion years ago there was no life on earth. The planet's crust had probably just settled into place. Countless volcanoes spewed forth lava and steam. Giant thunderheads pelted the earth with a continual downpour of rain, while lightning crackled through the atmosphere.

The land was barren—hard rock and powdered grit. The seas were sterile, containing minerals and salts leached from the volcanic crust, but devoid of fish, seaweed, or even microscopic organisms. Today one cannot examine a pinch of dust or a drop of seawater without finding myriads of small living things. Then there were none.

But conditions favoring life were beginning to emerge. Lightning, solar radiation, the warmth of the still cooling earth—all provided energy for chemical reactions which set the stage for the slow march of evolution and the spread of life into virtually every habitable part of the planet's surface.

Such is the picture most scientists would paint of the origin of life. But the question may never be entirely answered. Life began between 3 and 4 billion years ago in an environment that will probably remain ill defined. Nevertheless the subject commands our interest, and progress in cosmology, geochemistry, and molecular biology, coupled with the space-age search for extraterrestrial life, has enabled us to develop plausible concepts.

ORIGIN OF THE SOLAR SYSTEM

There are many models of the origin of the solar system, none of which has won general acceptance. All theories start with a cloud of cosmic composition (cosmic "dust") called the solar nebulae condensing by gravity from the particles poured into space by exploding stars. According to this hypothesis, the center of this "cloud" became the sun, with peripheral condensations of solid particles, including heavy elements like iron oxides, becoming the planets. At the same time there must have been large amounts of hydrogen, the universe's most plentiful element, and helium as well. Since the inner planets have relatively little of these lighter elements, we assume they must have been carried outward, concentrating the heavier elements in the inner "proto-planets." This hypothesis suggests that the earth's atmosphere formed secondarily, as a result of the decomposition of solid compounds.

EVOLUTION OF THE EARTH'S ATMOSPHERE

The present atmosphere of the earth is *oxidizing*, with carbon present as CO_2, nitrogen as N_2, oxygen as O_2, and hydrogen as H_2O. But the *primitive* atmosphere was *reducing*, rich in hydrogen, present as H_2, but also forming simple compounds with the other common elements, carbon, nitrogen, and oxygen. In an excess of hydrogen carbon would be present as methane (CH_4), nitrogen might be found as N_2, and oxygen would form water. In the primitive oceans nitrogen would be found as ammonia, NH_3, or ammonium ion, NH_4^+. Finally, some carbon monoxide (CO) might have been formed, but it is highly reactive and might yield formic acid (in the oceans) and formaldehyde.

The hypothesis that the earth's primitive atmosphere contained hydrogen, ammonia, water, and methane is supported by studies of the atmospheres of our solar system's giant planets which have high gravities and have retained their original atmospheres. Jupiter, the largest planet, has hydrogen, methane, and ammonia in its atmosphere, but probably no carbon dioxide, nitro-

gen, or free oxygen. The earth, on the other hand, lost its hydrogen because the light gas was able to escape the relatively weak gravitational pull. The air we breathe today is about 80 percent nitrogen and 20 percent oxygen, with traces of carbon dioxide and other gases. How do we explain the transition from a reducing to an oxidizing atmosphere? First, we speculate that solar ultraviolet light penetrating to the earth's surface caused the dissociation of ammonia, methane, and water vapor, with the preferential escape of the lighter hydrogen. Later, as the first primitive green plants evolved, photosynthesis resulted in the accumulation of oxygen and ultraviolet-absorbing ozone.

PRIMITIVE EARTH— ITS TEMPERATURE AND ENERGY SOURCES

The temperature of the primitive earth is not known. We assume that it varied from the poles to the equator. We can, however, place some limits on temperature conditions. Very high temperatures result in the decomposition of organic compounds (lower temperatures lead to stability). Most organic syntheses that seem likely on primitive earth can take place at temperatures as low as 0°C. Locally high temperatures would have favored some reactions.

The ultimate energy sources were the sun, gravitational energy, and radioactivity. Sunlight is our greatest source of energy today (Table 20-1).

Table 20-1   Present sources of energy on the earth

Source	Energy $(cal/cm^{-2}/yr)^{-1}$
Total radiation from sun	260,000
Ultraviolet light	
$\quad \lambda < 2500$ Å	570
$\quad \lambda < 2000$ Å	85
$\quad \lambda < 1500$ Å	3.5
Electric discharges	4
Cosmic rays	0.0015
Radioactivity	0.8
Volcanoes	0.13

Gravitational energy results from the potential energy released when an expanded dust cloud forms a solid body, and this energy appears as heat in the interior of the earth. Some of this heat reaches the surface in lava from volcanoes.

THE EARLIEST FOSSILS

Geologists estimate that the earth was formed about 4.5×10^9 years ago, the estimate being based on the age of meteorites. The first hard-shelled animals appeared about 0.7 billion years ago, but recent discoveries reveal the presence of microfossils of bacteria and algae in rocks at least 3.1 billion years old. Life therefore must have started in the 1.4 billion-year period between 4.5×10^9 and 3.1×10^9 years ago (Table 20-2).

These new findings are remarkable, because of the difficulty in finding such fossils. Organisms without hard parts do not leave very clear indications of their remains. Moreover heat will destroy organic remains, and most very old rocks have been heated at some time or other. However, a few very old unheated rocks have been examined, a few sediments having escaped substantial alteration. Extensive deposits of black shale, black chert (a flintlike rock with a high silicon content), and other stratified sediments are scattered throughout the world in virtually

unmetamorphosed condition; the formations in the Lake Superior region of North America, in the Transvaal of South Africa, and in parts of western Australia are especially notable.

The oldest known group of sediments is located in the border region between the Republic of South Africa and Swaziland. It is known as the *Fig Tree* formation. The Fig Tree cherts contain traces of organic matter and a few microfossils (Fig. 20-1). Using a method based on the decay of radioactive strontium and rubidium, it seems probable that its age is at least 3.1 billion years.

Fig Tree chert contains microfossils of two kinds of organisms. One group a few microns long looks very much like modern bacteria. The second group, which contains forms as large as 20 microns, looks like modern blue-green algae. The existence of these two organisms, successful inhabitants of an aquatic environment more than 3 billion years ago, is evidence that the first *evolutionary threshold*—the transition from chemical evolution to organic evolution—had been safely crossed at some even earlier date. We now know that at least two living organisms appeared well before the first third of earth history had passed.

Was the alga-like Fig Tree organism photosynthetic? This question can be answered by measuring the abundance of two nonradioactive carbon isotopes, C_{12} and C_{13}. The carbon in CO_2

Table 20-2   Chronology of physical and biological evolution

Event	Approximate time (billions of years ago)
Origin of our solar system	Over 5.0
Formation of the earth with its present size and composition; formation of the earth's crust	4.5
Age of oldest rocks and minerals	3.6
Fig Tree microfossils—autotrophs—bacteria, unicellular blue-green algae	3.1
Gunflint microfossils—diversification—filamentous and other blue-green algae	1.9
Bitter Springs microfossils—eucaryotes—green algae—increasing O_2 in atmosphere	1.0
First hard-shelled animals	0.7
Age of the dinosaurs	0.151
Earliest appearance of man	0.001

in the earth's atmosphere normally consists of about 99 percent carbon 12 and 1 percent carbon 13. In the process of photosynthesis, however, plants tend to fix slightly more carbon 12 than carbon 13, so that plant tissues are even poorer in the heavier isotope. Measurements of the ratio of the two isotopes in Fig Tree organic material were made. The results indicate that the Fig Tree material is poor in carbon 13, and to a degree that is almost identical with the depletion in modern algae and other photosynthetic plants. This result supports the view that the alga-like Fig Tree organisms probably were photosynthetic.

Evidence of another great revolutionary threshold-crossing comes from North America. A remarkable outcropping of rocks along the shore of Lake Superior in western Ontario shows a sequence of sediments known as the *Gunflint Iron* formation. The rocks at the base of the Gunflint formation include beds of black chert. The Gunflint chert is a billion years or so younger than the Fig Tree chert, its age being estimated as 1.9×10^9 years. Most of the Gunflint fossils are three-dimensional, and many show exquisite anatomical detail because they have been infiltrated by silica and became preserved much as

a modern biological specimen is preserved by being embedded in plastic. The soft structure of the organism owes its preservation to the almost complete incompressibility of the silica matrix. There are many different kinds of algae-like organisms, ranging up to 30 microns long. Many resemble contemporary blue-green algae. One of the more unusual plants, called *Kakabekia*, looks like an umbrella.

The story of *Kakabekia* has had some interesting and continuing overtones. Late in 1964, quite independently (in fact unaware) of paleontological investigations of the Gunflint fossils, Sanford M. Siegel discovered a strange new form while he was studying soil microorganisms that can survive extreme atmospheric conditions. The organism defied identification or assignment to any known taxonomic category. Siegel set it aside as being an enigma, although he kept his preparations, drawings, and photomicrographs. A few months later a description of *Kakabekia* was published, and Siegel immediately noted a striking resemblance between this soil organism and some of the Gunflint specimens. Siegel's organism is very slow growing, contains no chlorophyll, and apparently has no nucleus; it may therefore be

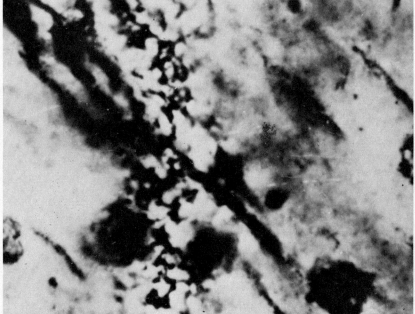

Figure 20-1 Three-billion-year-old alga-like fossils from the Fig Tree chert. Fossil bacteria are also found.

representative of a new group of procaryotic microorganisms. It was first found in ammonia-rich soil collected at Harlech Castle in Wales (actually at the site of an ancient latrine) and it has since been recognized in soils from Alaska and Iceland and recently in soil from the slopes of the volcano Haleakala in Hawaii. Whether or not it is related to the 2 billion-year-old *Kakabekia* is questionable, but the existence of two such bizarre forms is at least a remarkable evolutionary coincidence.

Evidence that many of the Gunflint organisms were photosynthetic comes from chemical studies: the C_{12}–C_{13} ratio already mentioned and the presence of two hydrocarbons, *pristane* and *phytane* which are "chemical fossils" probably derived from the breakdown of chlorophyll.

The variety of form (and therefore presumably of function) represented by the many kinds of Gunflint plants demonstrates that terrestrial life had crossed the second evolutionary threshold— the threshold of diversification—no less than 2 billion years ago.

The next great evolutionary event is recorded in fossils about 1 billion years old from the Bitter Springs cherts of Australia. These fossils include cells with nuclei, that is, green algae; indeed one series shows several of the cells of an alga at various stages of cell division and provides clear evidence of a nucleus.

Thus a billion years ago eucaryotic cells had arisen and along with them sexual reproduction. Soon thereafter—400 million years—the seas were swarming with highly differentiated aquatic plants and animals, evolved from primitive forebears. The late appearance of eucaryotic cells and sexual reproduction probably accounts for the failure of higher organisms to appear until about 3 billion years after the first simple organisms arose.

LABORATORY STUDIES

We can state our next question directly: Given plausible starting materials, is it possible to synthesize the major organic components of the genetic system (purine and pyrimidine bases, ribose and deoxyribose, and amino acids) under conditions presumed to be analogous to those on primitive earth?

The Russian biochemist, A. I. Oparin, first proposed the hypothesis of the hydrogen-rich atmosphere in his classic book, *The Origin of Life*, which laid the foundation for current thinking and research on this subject. The American chemist and Nobel laureate, Harold C. Urey, came to the same conclusion and determined to test a further hypothesis—that compounds associated with life might be formed if lightning or ultraviolet light acted on such an atmosphere.

It was one of Urey's students, Stanley L. Miller, then at the University of Chicago, who performed the first experiments. An electric discharge was passed through a mixture of ammonia (or nitrogen), methane, and water vapor above boiling water. At the end of the experiments the clear fluids were encrusted with tars, from which Miller identified a number of biologically important molecules, including the amino acids glycine, alanine, serine, aspartic acid, and glutamic acid. Aldehydes were formed in the discharge along with hydrogen cyanide, and these may have reacted with ammonia to give amino acids.

Since that time many similar experiments have demonstrated that other key molecules can be synthesized as well: the high energy phosphate, ATP, and purine and pyrimidine bases, the structural units of nucleic acids. For example, adenine is formed when ammonia, methane, and water are exposed to high energy electrons from a linear electron accelerator, simulating the shower of electrons on the earth from space. Adenine is also formed when a concentrated solution of hydrogen cyanide and ammonia are heated in water for several days.

At this point two general principles should be emphasized: (1) amino acids are not produced under strongly oxidizing conditions, thus arguing for the availability of H_2 in the primitive atmosphere and (2) other energy sources, for example, ultraviolet light, may replace the electric discharge.

The key building blocks of living tissue can be made from inanimate hydrogenous substances in the presence of a lightning storm, ultraviolet light, or other source of energy. These reducing mixtures produce intermediate molecules like hydrogen cyanide and 3-carbon molecules, which suffice as raw materials for synthesis of the biologically important purines and pyrimidines (e.g., cytosine and uracil) as well as amino acids.

Polymer synthesis

Thus far our case is plausible. On the primitive earth biologically important molecules could have collected in the oceans to make a "hot thin soup," as one biologist has called it. What *might* have occurred in this soup?

Efforts are underway in several laboratories to determine under what conditions amino acids may condense to form proteins, and purines and pyrimidines to form polynucleotides. Thus far there have been only modest successes. Random polypeptides ("proteinoids") have been formed by heating amino acid mixtures. Their relevance to the origins of life is not yet established. There have also been reports of the formation of nucleosides from bases and sugars, but these deserve more thorough exploration.

BIOLOGICAL ORGANIZATION

How could reactions proceed efficiently in a "primordial soup"? Some concentrating mechanism may have been required to speed up chemical reactions. At least three simple, plausible mechanisms could have been important: (1) evaporation, (2) freezing, and (3) adsorption of organic materials on clay minerals.

Another possibility is presented by a consideration of membranes and colloidal droplets (*coacervates*) which are capable of concentrating some materials by a process similar to selective absorption. Oparin pointed out that colloidal particles, jellylike semisolids such as certain proteins form, will break into smaller parts when they reach certain sizes. Each of these smaller units will then grow and divide again. In a very elementary way this resembles the events that we observe in cell division. Cells contain many molecules in colloidal suspension, and molecular tendencies may drive them to divide when they reach certain optimum sizes.

How do we discover a plausible route from the primordial soup to a self-replicating mutable system? How might structural order emerge? *Our discussion becomes increasingly speculative.* We know, for example, that polypeptides and simple proteins settle out of solutions in layers, or align themselves with one another and take on some of the ordered structure of crystals. Others will achieve even greater organization.

The organization of collagen offers an interesting model. This structural protein is widely distributed throughout the animal kingdom, accounting for up to 30 percent of the total protein in vertebrates. Many details of its structure are understood, and it can be readily isolated, purified, and identified.

The elementary molecule of collagen, called *tropocollagen,* is a long rod, composed of three helical polypeptide chains; in the body these rods are aggregated into fibrillar arrays (Fig. 20-2). Once synthesized, tropocollagen moves from the cell to the intercellular space. The manner in which this movement is accomplished is virtually unknown, but we do know something about the fate of tropocollagen outside the cell. Newly formed tropocollagen is soluble in cold salt solutions under physiologic conditions. As tropocollagen molecules progressively organize into collagen fibrils the protein becomes less and less soluble. The large bulk of collagen in an old animal is insoluble under ordinary conditions.

In considering aggregation of the basic units we must bear in mind the fact that when these identical molecules come together they fall into register—*order emerges.* That is, the self-ordering of macromolecules appears to be a major force in the origin of form. Furthermore the ordering requires a compatible environment. Although it is not yet established that changes in the extracellular microenvironment determine the aggregation patterns, it is clear that varying conditions do affect them. Even the use of different salt concentrations produces different structures of collagen (Fig. 20-3).

We know that combinations of proteins and fats form a structure resembling the cell membrane, thereby creating complex nonliving systems. The key feature of such organized systems may be their ability to remove and concentrate substances from the surrounding medium, making possible organized chemical reactions.

Suppose that a variety of cell-like aggregates did form from the atmospherically produced substances that washed into the sea. Some would attract new molecules more efficiently than others. One fact is clear, however; some of them, polypeptides and polynucleotides, had to be catalytic themselves or had the ability to form new catalysts. Polynucleotides, for example, would have served as templates for their own reproduction and also have organized amino acids into enzymes. However, there is no satisfactory hypothesis for the origin of the genetic code and the protein-synthesizing mechanism, nor are

there any plausible routes to the origin of cells.

We can speculate about the evolution of metabolic pathways. As essential nutrients disappeared from the primitive soup, a selective advantage would be gained by organisms capable of synthesizing essentials from potential precursors. Hence pathways could have evolved backwards. Each mutation would have been beneficial, making a new source of nutrients available.

Possibly the earlier pathways were fermentative, requiring no oxygen. Many microorganisms still use this method of securing energy. Yeasts, for instance, ferment sugar to alcohol and carbon dioxide. This process is not particularly efficient—only 1/35th as efficient as respiration.

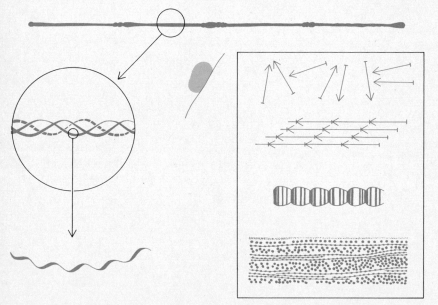

Figure 20-2 Collagen fine structure. The elongated tropocollagen molecule shown at top has an asymmetric fine structure roughly subdividing the molecule in quarters. It is compared with a 70S ribosome on a membrane sketched to scale below it. The enlargement of the section of the molecule to the left illustrates the three polypeptide chains, one of which, the dashed line, is different in amino acid composition from the other two. A further enlargement of a section of one of the chains illustrates the helical configuration of each of the polypeptides. In the box at right is a diagram of the manner in which tropocollagen units are postulated to aggregate, overlapping each other in a staggered array by about one quarter of their length, thereby giving rise to a collagen fibril with a characteristic repeating period. The fine structure within each of these periods would be a reflection of the fine details of asymmetry of the tropocollagen units arrayed in register. The lower portion of this block illustrates the manner in which collagen fibrils are found organized in plywood-like sheets in a variety of tissues.

The development of fermentation could have led to the end of life if no other changes had come about. The most advanced life forms would have used up all the available fermentable molecules. Without food they would have all died off. But a new process came into being. With the carbon dioxide resulting from fermentation some organisms were able to begin building the very molecules they needed to survive. Utilizing the energy of sunlight, they combined carbon dioxide and water to make sugar. This is the process we call **photosynthesis;** it made possible the great diversity of life on earth, both in the ocean and on land.

Oxygen is a waste product of aerobic photosynthesis. Now, for the first time, this life-giving element became available to living organisms. Previously, practically all of the free oxygen had been tied up in compounds of the earth's rocky crust. Organisms then evolved that were capable of utilizing oxygen. They developed mechanisms for respiration—a more efficient way to tap the energy potential of molecules and one in which waste products, carbon dioxide and water, are not toxic as are those of fermentation.

The combination of photosynthesis and respiration sets the groundwork for the evolution of life all over the planet. Photosynthesis produced more energy-storing foods than organisms needed to consume by respiration. There was a ready surplus of energy to be used by other organisms. In addition, photosynthesis changed the face of the earth. By producing oxygen it altered the atmosphere. This made possible much of the land life we know today.

Another series of events also occurred to make it possible for organisms to live ashore. The ultraviolet light which probably helped form early molecules would have been fatal to life. It can break down large molecules or critically alter their structure. Life could develop in the seas, however, because water absorbs the radiation.

Ultraviolet also breaks the bonds in water molecules. Free hydrogen molecules, so light they are not held long by the earth's gravity, escaped the early atmosphere. The remaining oxygen formed a 15 mile-high layer of activated ozone which turned out to be an excellent shield against ultraviolet and protects land life from most of the damaging effects of this radiation.

ORIGIN OF EUCARYOTES— SYMBIOSIS AND EVOLUTION

Some biologists now suggest that the organelles characteristic of the eucaryotic cell may once have been independent organisms that somehow came to live symbiotically inside larger host cells. It is not clear whether or not one of the host

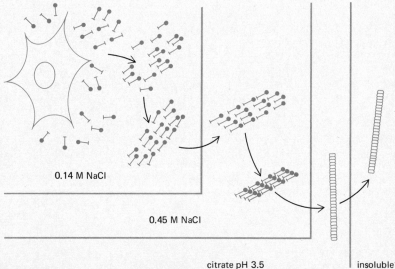

Figure 20-3 Diagram of hypothesis explaining the significance of the different extractable collagen fractions. Rodlike units represent tropocollagen molecules. Cold physiological saline extracts the most recently formed collagen; hypertonic salt solution extracts the same material plus older collagen in a more ordered state of aggregation; acid citrate buffer extracts all of the above plus some of the older collagen in the typical fibrillar form. The insoluble fibrils are older and their degree of crosslinking has prevented solubilization.

0.14 M NaCl

0.45 M NaCl

citrate pH 3.5   insoluble

cell's responses to the presence of such a guest was a regrouping into a nucleus of the genetic material formerly scattered throughout the host's cytoplasm. If symbiosis was indeed the first step toward evolution of the eucaryotic cell, it may be that certain of the Gunflint organisms show initial steps in this direction. A fascinating account of this possible transformation of various bacterial and algal organisms into components of the eucaryotic cell is given by Lynn Margulis of Boston University in a recent article. Ms. Margulis' argument rests on biological grounds; the increasingly detailed fossil record supports her thesis.

SPONTANEOUS GENERATION

Until the middle of the nineteenth century it was assumed that living organisms could arise spontaneously from decaying organic material. There were recipes for frogs and mice from garbage or mud. It was Redi, an Italian physician of the seventeenth century, who showed that maggots

Figure 20-4   The flasks used by Pasteur in his experiments on spontaneous generation.

would not appear in meat if insects were prevented from laying eggs in it. However, the belief persisted that bacteria could arise spontaneously. Bacteria could be found in boiled beef broth if the broth was simply left open to air. Pasteur's rigorous experiments in the nineteenth century showed that organisms would not grow in a broth if it was protected from contamination by bacteria or fungi in the air. He used a flask with a long swan-shaped neck which did not keep out air but trapped contaminants before they could enter the broth.

Among educated people belief in spontaneous generation died out, and scientists found it futile to inquire into the origins of life. Today we are again inquiring into our origins, recognizing that life reflects specific molecular combinations, and that these must have had a beginning.

The present theory is still a theory of spontaneous generation, but it does not try to explain how life might originate *now*. It does offer a plausible sequence of events by which life could have begun several billion years ago. It now seems quite reasonable to believe that life developed spontaneously from nonliving matter in the distant past. A similar development would be very unlikely to occur today for a very simple reason. The organisms populating the planet now are vastly more efficient at utilizing the raw materials of life than any prelife forms might be. These early forms simply could not compete with well-adapted, complex microorganisms, not to mention plants and animals.

THE SEARCH FOR EXTRATERRESTRIAL LIFE

In the foregoing discussion we have omitted one possibility—that life on earth was derived from preexisting life on another planet, somehow propelled to earth—the "panspermia hypothesis." If life on another planet were found identical to life on earth, this theory would have to be considered.

The search for extraterrestrial life is one of the primary scientific objectives of our space program. "Exobiology," as this new subject is named, has three goals: (1) to characterize other planetary environments; (2) to make remote observations of other planets, searching for life; and (3) to make direct observations.

The problem is not to decide whether a given environment may support terrestrial life but whether there might be forms of life adapted to conditions far different from those on earth. Little of biological interest is known about most planets. There is some ground for speculation that the lower strata of clouds of Venus might conceivably be habitats for organisms like those on earth. Principal interest centers on Mars, where there is a possibility of biologically interesting microenvironments.

The only life that might be detected beyond our solar system in the foreseeable future is intelligent life, as determined by a search for radio emissions. There have been two such searches, at the U. S. National Radio Astronomy Laboratory in 1960 and at the Shternberg Observatory in Moscow in the late 1960s, thus far without success. However, the search will continue.

Today the origin of life is viewed as a natural phenomenon, susceptible to corroboration. In fact it seems that the nature of atoms and molecules almost dictates that if conditions are suitable life should inevitably evolve. Possibly there are other planets in the universe capable of supporting life and where life has evolved as well as on our tiny way station in space.

In this chapter we have had to speculate about the nature of life before there were any fossils to document what really happened. In Chapter 22 we shall be able to deal with concrete evidence—the records left by living organisms in the stones of time.

SUGGESTED READING LIST

Barghoorn, E., "The oldest fossils," *Scientific American*, May 1971, p. 30.

Handler, P. (ed.), *Biology and the Future of Man.* Oxford University Press, New York, 1970. Chapter 5, "The Origins of Life," pp. 163–201.

Margulis, L., "Symbiosis and evolution," *Scientific American*, August 1971, p. 48.

Mueller, G., "Organic microspheres from the Precambrian of South-West Africa," *Nature*, Vol. 235, p. 90 (1972).

Wald, G., "The origin of life," *Scientific American*, August 1954, p. 44 (offprint 47).

CHAPTER 21

Cataloging Nature

Everywhere the earth is alive. Fish swim in the cold, lightless depths of the ocean abysses. Life goes on beneath the sunparched sands of the desert. Hardy lichens and mosses turn the banks of Arctic streams pale green during the short weeks of summer, then hold on in frozen, suspended existence for the rest of the year. Bacteria grow in mountain brooks and in scalding waters of natural hot springs. Animals roam the forests, plains, and rocky mountainsides, and soar high into the atmosphere. Plankton drift in oceans and lakes. Teeming populations of tiny organisms live between the particles of mud and sand on beaches and on the ocean floor. Living fish harbor populations of parasites. Dead fish harbor a multitude of bacteria and fungi.

Our planet is almost completely covered with living things, billions upon billions of them, as remarkable in their diversity as in their numbers. Approximately 1,200,000 different species of living animals and 400,000 species of living plants have been described by biologists in accepted scientific terms, and the list keeps growing each year as new animals and plants are discovered. Probably 4 to 5 million species exist today. In addition more than 500 million species have lived and died out since life began on earth. This is an almost incomprehensible diversity.

We have already begun to describe how this diversity arose (Chapter 20). Life appears to have

Figure 21-1 Homologous structures. The arm of man, the foreleg of a horse, the flipper of a whale, the wing of a bird and the front fin of a coelacanth provide different kinds of mobility. They are all **homologous** structures because they arise embryologically in the same way and have the same basic internal skeletal structure. Numbers refer to fingers in man.

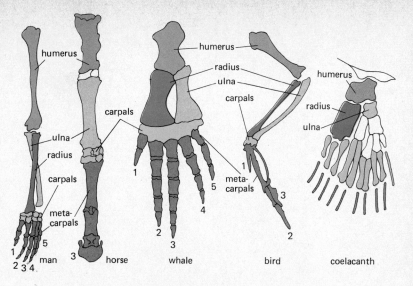

The wing of a housefly is **analogous** to the wing of a bird because it serves the same function but has a completely different embryonic origin and has no internal skeleton. Homology suggests a common ancestry, whereas analogy merely indicates a common function.

originated on earth only once. The diversity of life now found is due to evolution, to a continuous process of speciation—the process whereby one species yields several descendant species. Each major group of organisms—seed plants, insects, birds, mammals, primates, and so on— probably descended from one original *founder species*. Ninety-five percent of the organisms that lived on earth in past ages have no descendants alive today. The few that did leave descendants are responsible for the diversity we see around us.

UNDERSTANDING DIVERSITY

The field of biology that deals with this diversity is called **systematics.** It includes **taxonomy,** the cataloging or classification of animals and plants. Systematics also includes the study of evolutionary, ecological, and behavioral aspects of animal and plant diversity. There are three major objectives to the field of systematics:

1. To discover the relationships among species alive today and to trace the ancestors from which they descended. In other words, who is related to whom?
2. To catalog species of organisms in such a way that information about them can be retrieved easily, and made useful.
3. To determine how diversity increases and continues to arise. Why does it arise in some groups and not others, and in some places but not in others?

WHO IS RELATED TO WHOM— HOMOLOGY PROVIDES THE KEY

The nearest relatives of a species of organism are not always obvious, and the determination of relationships is one of the most challenging areas in systematics. In determining the relationships between organisms, a systematist looks for structures, developmental processes, and other features such as behavior that were inherited from a common ancestor. Such structures are said to be **homologous.** Homologous structures in different species have similar patterns of basic development and form, because of a common ancestor and similar genetic makeup.

Homologous parts in different organisms need not—in fact often do not—perform the same function. The human arm, a horse's front leg, the front flipper of a whale, and a bird's wing are all homologous (Fig. 21-1). They are located in similar parts of the body and have bones within them that are similar in basic arrangement, number, and shape. Intermediate forms between these animals and the common ancestors they shared can be found in the fossil records. In fact the fin of a now extinct ancestral fish related to the coelacanth, *Latimeria* (to be described in Chapter 22), apparently provided the basic patterns for the limbs of all these vertebrates. As the animals evolved, their limbs became adapted for different functions. Nevertheless, the essential similarities in the shape and arrangement of the bones confirm the evolutionary relationships, and demonstrate homology.

In many organisms, on the other hand, there are body parts that perform a similar function but have a different origin and structure. These parts are **analogous.** The wings of a bird and a housefly serve the same purpose, but are organized differently and originate in a different way during development. A bird's wing has certain bones that correspond to those of the human arm. A fly's wing is not a modification of a limb or a fin, but is a completely new structure which evolved independently of the limbs and develops as part of the insect's external skeleton. The wing of a bird is thus analogous to the wing of a housefly but homologous to the foreleg of a horse.

Another example of analogy is illustrated by the structural and functional similarities between the eyes of squid or octopus and those of mammals. Although similar in structure, the eyes of these mollusks and of mammals form differently during embryonic development. The eyes of a squid are homologous to those of an octopus, but they are both analogous to those of mammals.

Figure 21-2 Similarities in external form provide examples of convergent evolution. Left: dolphin (above) and shark (below). Right: cactus (above) and *Euphorbia* (below).

THE PROBLEM OF CONVERGENCE

How does a systematist decide what sorts of characteristics are homologous? The process is not always a simple one. However, let us consider an easy example. How should we classify a dolphin and a shark? These animals look somewhat alike. Both are streamlined, have fins, and are well adapted to life in the sea. They certainly share many features. However, evidence from fossils and other sources to be discussed presently shows that the external features which make a dolphin resemble a shark are relatively recent adaptations to aquatic life. Many major differences between them indicate that dolphins and sharks are not closely related. For example, the dolphin breathes with lungs, not gills; it has a four-chambered heart, mammary glands, and some hair. These are all characteristics of true mammals, showing that dolphins are much closer to cows than they are to sharks. A shark has gills, a two-chambered heart, and a skeleton made of cartilage, and is related to other cartilaginous fishes.

As another example let us try to classify certain desert plants. Some of the cacti found in the United States closely resemble the "corn-cob cactus" that grows in Africa (Fig. 21-2). Are they members of the same family? The answer is "No." Careful study puts the African form, euphorbia, in the same group with the leafy castor bean plant, the colorful poinsettia and the rubber tree. Both the euphorbias and the cacti have adapted in the same general way to desert life and therefore look similar at first glance. Closer examination, however, shows differences in the structure of their flowers, fruits, and seeds that indicate the plants are not closely related. Life in similar desert environments led to the evolution of similar adaptations in the stems and leaves of both groups. Both developed thick, fleshy stems with spongelike cells which store water. In both groups, the stems contain chlorophyll and do most of the photosynthesis. Most of them have reduced their leaves to spines which decrease water loss. Some of them also produce true leaves but they shed them during the dry season.

In addition to similar morphological adaptations to desert life, both euphorbias and cacti have evolved similar biochemical adaptations.

One of these adaptations is called **succulent metabolism.** Most plants use water and atmospheric carbon dioxide in photosynthesis and convert these into sugar. During this daylight process they give off oxygen. At night when plants respire they normally use oxygen and release carbon dioxide. But euphorbias and cacti produce instead *organic acids* as the waste products of respiration and do not release carbon dioxide. These organic acids are retained within the plant tissues until daylight and are then used by the plant instead of carbon dioxide in photosynthesis. Because of this built-in supply of carbon dioxide, euphorbias and cacti do not rely as heavily as other plants on atmospheric carbon dioxide. As a result it is not necessary for their stomates to open in the hot, dry daylight hours when there would be large water loss.

Even though euphorbias and the cacti share so many morphological and biochemical features, they are not closely related and evolved from ancestral leafy plants that were quite different from each other. These original differences can still be seen in many features of modern euphorbias and cacti, especially their flowers. If they are not blooming, an easy way to distin-

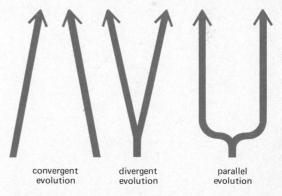

convergent   divergent   parallel
evolution   evolution   evolution

Figure 21-3 Three patterns of evolution. **Convergent evolution** occurs when two groups of unrelated organisms come to look like each other more and more as time passes. Such organisms usually live in the same kind of habitat to which they evolve similar adaptations. In **divergent evolution** one ancestral group splits into two groups which become less and less alike as time passes. In **parallel evolution** two related species evolve in similar ways for long periods of time. This usually occurs when the two species are exposed to similar selection pressures in their environments.

guish euphorbias from their cacti cousins is to wound the stem. If a milky substance flows from the wound, the plant is a euphorbia.

These two examples illustrate **convergent evolution.** They emphasize that similarity in structure and metabolism does not necessarily mean relationship. The environment may lead to the evolution of the same adaptations in unrelated groups. Apparently there are only a limited number of ways in which organisms can successfully meet the problems of their environment.

CLUES FROM REPRODUCTION AND EARLY DEVELOPMENT

How does a systematist distinguish between homologous structures reflecting a common ancestry and analogous structures reflecting adaptations to a common environmental problem? Usually systematists look for homologies among characteristics which do not change very much during the evolution of large groups of organisms. In flowering plants the parts involved in reproduction prove to be better indicators of evolutionary relationships than other organs. Leaves vary tremendously in their shape and we see considerable convergence. Witness the euphorbias and the cacti. In general, however, the number and arrangement of the parts of a flower, such as stamens and petals, and the position of the ovary are the same in large groups of plants. In fact the two great groups of flowering plants, the monocots and dicots, are distinguished by the number and arrangement of the parts of their flowers. In algae and fungi reproductive processes are also useful clues to relationships. Reproductive structures are so crucial to perpetuation of a species that there is little tolerance for change in them. Mutations which affect reproductive structures rarely survive.

In animals, embryonic structures and the processes of early development do not appear to change much during evolution. Consequently embryos and some larval forms are useful in analyzing relationships among major groups. For example, gill slits appear briefly during the embryonic life of all vertebrates (Fig. 21-4). Similarly the process of gastrulation occurs in one form or another in almost all animals, but not in plants.

Why do these embryonic structures appear to be insulated from evolutionary change? The reason becomes clear when one thinks about it for a moment. A mutation which affected early development would have little chance of being reproduced because it could affect the whole of development and produce massive effects. The resulting abnormal organisms would have little chance of surviving to reproductive age.

CLUES FROM THE ELECTRON MICROSCOPE, CHROMOSOMES, AND BEHAVIOR

Modern systematists look for homologies not only in morphological structures like the wings and legs of living animals and embryos and of fossils but also in many other features. Systematists are trained in many biological disciplines today, from electron microscopy to computer theory, from molecular biology to zoogeography, from organic chemistry to ethology. For instance, the analysis of ultrastructure has proved useful in identifying homologies. The flagella of bacteria have a different structure from the flagella of all other organisms. The presence of cell organelles, such as nuclei, enclosed by membranes, in all living organisms except bacteria and blue-green algae indicates a common ancestry for all animals, plants, fungi, and protozoans. The chloroplasts of all eucaryotic algae except for one group—the red algae—look similar to those of higher plants.

Studying the number and shape of chromosomes also provides useful clues to relationships, especially among plants. Analyses of instinctive behavior are also useful in uncovering relationships among animals. The cocoons of moths provide a morphological record of the caterpillar's behavior, for example (Fig. 14-1). So do the sounds of insects, frogs, and birds when analyzed by electronic techniques. The voice-prints of two crickets which look alike morphologically can tell the systematist that each cricket belongs to a separate species. Subsequent study shows that the populations of crickets do not interbreed in nature. Studies of courtship displays of birds and other animals are also useful.

CLUES FROM BIOCHEMISTRY— MOLECULAR TAXONOMY

Biochemical characters can also be used to establish relationships. The similarity in amino acid sequences of certain proteins in different species indicates homologies. Figure 21-5 illustrates homologies in the amino acid composition of the cytochrome C of various organisms. These homologies reflect, of course, homologies in the sequence of bases in the genes coding for cytochrome C in these different organisms.

Ultimately the degree of kinship among organisms may be established in terms of the per-

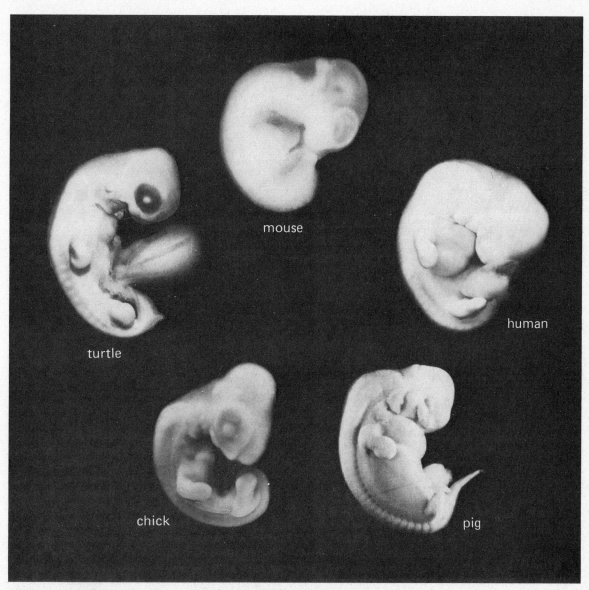

Figure 21-4 Embryos of various vertebrates look very much alike. At the stages shown here every one of them including the human embryo has gill slits and a tail. The reasons for these similarities are discussed in the text.

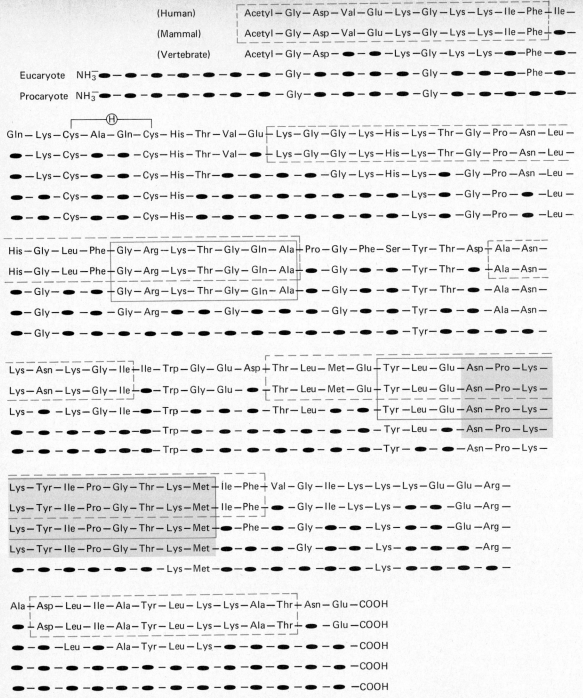

Figure 21-5 Homologous amino acid sequences of the protein, cytochrome C, in various organisms. The top sequence is for human cytochrome C. The second sequence denotes only those amino acids found in the specified position in all mammals examined so far, the third sequence those held in common by all vertebrates examined so far, and the fourth sequence those common to all eucaryotes examined so far. The last sequence stipulates only those amino acids found in the specified positions in both eucaryotes and procaryotes. The unspecified spaces may be filled by any of several amino acids. At least one long sequence of 11 amino acids is common to all eucaryotes, two long sequences to all vertebrates and five long sequences to all mammals.

centage of similar long sequences of bases in the DNA of various species. Chemical techniques have been devised to measure the percentage of long DNA sequences that the entire genomes of different organisms have in common. Figure 21-6 summarizes the results of one such analysis of the DNA of various vertebrates. The evolutionary continuity of vertebrates is suggested by their common long DNA sequences. The DNA of the baboon has the most long sequences in common with the Rhesus monkey. Similarly human and chimpanzee DNA closely resemble each other. But the very primitive primates known as lorises have long sequences of which only about half are identical with those of the Rhesus monkey.

The origins and relationships of many of the major groups of organisms such as the bacteria, many of the algae and fungi, and nematodes are still unknown to systematists. Biochemical taxonomy promises to provide important new information about their relationships.

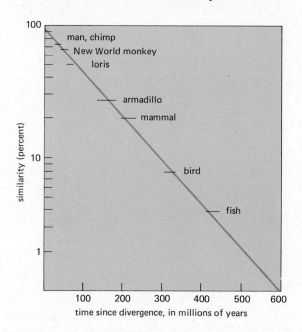

Figure 21-6 Relationship between similarity in polynucleotide sequences and time at which various groups of vertebrates diverged from each other. These results were obtained by a method of DNA-DNA hybridization, similar to the method discussed in Chapter 6. The vertical axis is logarithmic.

CLUES FROM NEWLY DISCOVERED ORGANISMS

In addition to the insights we gain from new techniques, important information also comes from newly discovered living organisms. It is surprising how often a significant specimen is uncovered that alters or strengthens our ideas about relationships. We already mentioned the coelacanth *Latimeria*. In 1944 representatives of a group of trees—the dawn redwoods—thought to be extinct for more than 10 million years were found in western China (Fig. 21-7). In 1952 an entirely new class of mollusks—the most primitive ever found—was dredged up from abyssal depths in the Pacific (Fig. 24-38). In 1955 a new subclass of crustaceans—the Cephalocarida—also the most primitive ever found, was discovered in the bottom mud and sand of Long Island Sound (Fig. 21-7).

Also the importance of key fossils in discovering relationships and origins can not be overestimated. They can show us the link leading from one form to another. One of the greatest discoveries was the finding of fossils of *Archaeopteryx* (Fig. 22-14), an animal which provided a link between reptiles and birds. As we have observed, fossils of blue-green algae have been found in rocks that may be more than 3 billion years old. These blue-greens look a great deal like present-day forms. Several recent discoveries of fossils of ancient primates which may be the ancestors of modern man have prompted intense interest throughout the world.

FRAGILE FAMILY TREES— PHYLOGENIES

When we attempt to classify a group of organisms, we must analyze many characteristics in order to determine what homologous structures they share with other organisms. Evolution results from the accumulation of many hereditary changes in populations. When we examine many characteristics and find only a few differences between two populations, we may conclude that the two populations are closely related to some ancestral population. But when we find that the differences between two populations are many and the homologies are few, then the organisms are more distantly related. Figure 21-8 summarizes homologous morphological and physiological features shared by several kinds of animals.

The picture of evolutionary relationships that results from the study of homology is called a **phylogeny.** It may be illustrated by a diagram called a **phylogenetic tree,** which is really a sort of family tree of the organisms. More complicated examples of phylogenies appear in Chapters 23 and 24 where we explore the diversity of organ-

isms. In an ideal phylogenetic tree the organisms are all descended from one ancestral population—the founder species. Such a phylogenetic tree is called **monophyletic.** But as we shall see in the following chapters, it is not always possible to decide this on the basis of the information available. Because so many different kinds of evidence must be weighed, the reconstruction of the evolutionary history of any group of organisms always entails considerable speculation. No

Figure 21-7 Newly discovered organisms which have provided clues to phylogenetic relationships and to evolutionary history. (Left) Dawn redwood, *Metosequoia glyptostroboides,* was one of the most abundant conifers in the Northern Hemisphere, and was found as far north as the Arctic Circle. As the climate cooled, its range diminished and it was thought to have become extinct. It was first described from fossils in 1941. Three years later a Chinese forester discovered a single huge *Metasequoia* tree in Szechuan Province. The people of the area had built a temple around its base. Subsequently, several thousand *Metasequoia* trees were discovered growing in remote valleys of Southwestern China, the last relics of the great dawn redwood forests of 25 million years ago. See fig. 22-2 for fossil branch of *Metasequoia* estimated to be 50 million years old. (Right) A crustacean belonging to the new subclass, *Cephalocarida.* These tiny blind shrimplike creatures which live obscurely in mud or silt were first discovered in 1955. The largest individual of this species is only about 4 mm long. They resemble primitive fossil crustaceans more closely than any living forms.

phylogenetic tree is a fixed scheme. Instead it is a working model and is merely accepted until better evidence becomes available. As one biologist put it: "phylogenetic trees have shallow roots and are easily blown over."

By grouping similar organisms on the basis of homologies and then preparing phylogenies, we can bring some order to the picture of life on earth. The 1½ million species of living organisms actually fall into comparatively few *phyla* (or *divisions* in the case of plants), each of which is a broad grouping of related organisms possessing certain major morphological and biochemical features in a combination not found in any other group. For example, all animals with jointed legs *and* external skeletons fall into a phylum of the animal kingdom (Arthropoda). Among plants, terrestrial plants with a vascular system *and* with leaves, roots and stems make up the separate division (Tracheophyta). Phylogeny thus provides us with a convenient framework for classification.

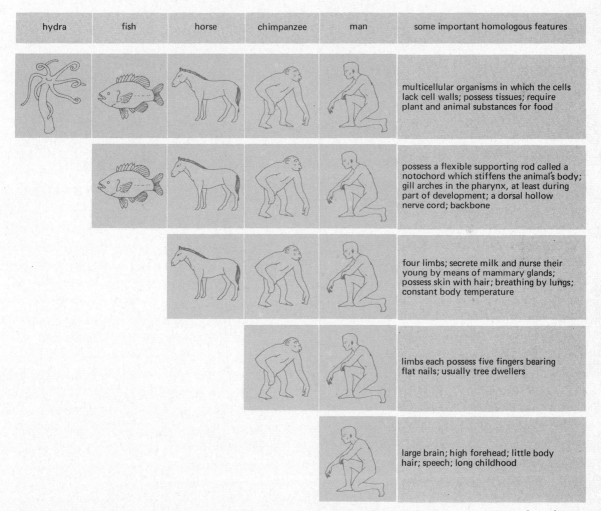

hydra	fish	horse	chimpanzee	man	some important homologous features
					multicellular organisms in which the cells lack cell walls; possess tissues; require plant and animal substances for food
					possess a flexible supporting rod called a notochord which stiffens the animal's body; gill arches in the pharynx, at least during part of development; a dorsal hollow nerve cord; backbone
					four limbs; secrete milk and nurse their young by means of mammary glands; possess skin with hair; breathing by lungs; constant body temperature
					limbs each possess five fingers bearing flat nails; usually tree dwellers
					large brain; high forehead; little body hair; speech; long childhood

Figure 21-8 Homologous features possessed by various kinds of animals. Man shares many important homologous features with chimpanzees, fewer with fish and still fewer with hydra.

HOW TO CATALOG ORGANISMS

How do we go about setting up a useful system of classification? There are three problems that must be solved: (1) what will the classification system be used for, (2) what is the reasoning behind the classification system, and (3) what is the technique of cataloging?

Biologists require a classification system which will enable them to retrieve information about particular species: all the information about a particular species of fruitfly, for example. They also want a classification system which will give morphological information to morphologists, physiological information to physiologists, biochemical information to biochemists, ecological information to ecologists, and so forth. A classification scheme based on phylogenetic relationships automatically takes into account information about every aspect of the organism it classifies. For example, when chemists discovered that a chemical extracted from a flowering

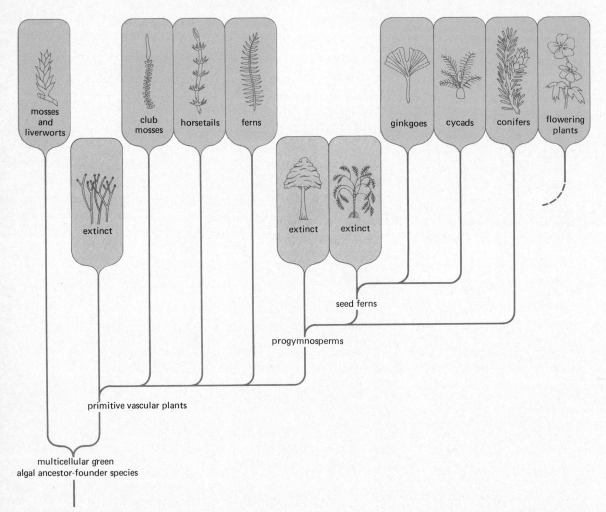

Figure 21-9 Phylogenetic tree describing possible evolutionary origin and relationships of conifers and of flowering plants from an ancestral population of green alga, the founder species. The origins of the conifers are clearer than those of the angiosperms.

plant called *Rauwolfia* was an effective tranquilizer, they immediately looked for other plants where similar chemicals might occur. Their first choices were plants which botanists believed were close phylogenetic relatives of *Rauwolfia*, because these plants were the ones most likely to produce the same or similar chemicals to those produced by *Rauwolfia*. From this analysis we see that a classification system based upon phylogenetic relationships will be useful to biologists doing many different things.

A VOCABULARY FOR CLASSIFICATION

The system employed in classifying organisms today is based on a method devised in 1758 by a great Swedish biologist, Carolus Linnaeus. Although Linnaeus' views of the origin of diversity were different from ours today, he attempted to divide organisms into groups which had certain morphological features in common. There was no thought at that time that similar groups of organisms were descendants of a common ancestor, the founder species. Linnaeus, however, was an astute observer of nature and many of the broader categories he established have not changed significantly in two centuries. The reason for this is that morphological characters are products of evolution and they can tell us much about evolutionary relationships. Organisms that share morphological features often are phylogenetically related, even though, as our examples have shown, convergence can complicate the picture.

The broadest general *category* of all is the **kingdom** (for example, animal or plant). The broadest category within the kingdom is the **phylum** in the case of animals or the **division** in the case of plants. Each phylum or division contains a group of related **classes;** each class contains a group of related **orders;** each order contains a group of related **families;** and each family contains a group of related **genera.** The final unit is, of course, the **species,** which is a population of organisms that share a common gene pool. Species may in turn be subdivided into more local interbreeding populations called **subspecies.** The members of each of these cate-

gories are probably the descendants of a common ancestor, and each of these categories is thus the product of evolution.

Table 21-1 shows how this system of classification applies to man and to white oak trees. When you know an organism's place in the sys-

Figure 21-10 Carolus Linnaeus was born in 1707 in southern Sweden. He became interested in botanical research, and at the age of 25, when still a student at the University of Uppsala, Sweden, he spent five months exploring Lapland for the Academy of Sciences in Sweden. (This portrait shows him in Lapland clothing.) Although not the first biologist to attempt to systematically organize plants and animals, he had a more far-reaching impact because he published so extensively and because he named everything in the living world around him. His invention of the binomial system of nomenclature and his passion for collecting and classifying led men from all over the world to send him plants and seeds. In addition he managed to dispatch collectors to all corners of the earth. Even Captain Cook had one of Linnaeus' collectors along with him. Linnaeus introduced order into the living world and laid a foundation for the work of Darwin and the other evolutionists of the 19th century.

Table 21-1   Classification of man and white oak

Man			White oak		
Category	Name	Key features	Category	Name	Key features
Kingdom	Animalia	Multicellular organisms which ingest plant and animal material for food.	Kingdom	Plantae	Organisms with rigid cell walls composed largely of cellulose; chloroplasts containing chlorophyll a.
Phylum	Chordata	Animals with a notochord, gill arches in the pharynx at some stage of development and a dorsal hollow nerve cord.	Division	Tracheophyta	Vascular plants; mostly terrestrial with a conducting system of xylem and phloem; sporophyte generation is dominant.
Class	Mammalia	Skin with hair, mammary glands which secrete milk to nourish young; constant body temperature; breathe by means of lungs; four-chambered heart.	Class	Angiospermae	Flowering plants; seeds enclosed in an ovary which is often fleshy.
Order	Primates	Five digits (finger or toes) on each limb with flat nails; hand with an opposable thumb; essentially tree-dwellers with eyes forward and binocular vision which permits judgment of distance; sense of smell reduced.	Subclass	Dicotyledonae	Embryo with two cotyledons; flower parts usually in 2s, 4s or 5s; net veins; vascular bundles in a single cylinder or "ring" in cross section.
			Order	Fagales	Trees with spike-like clusters (catkins) composed of flowers lacking certain parts (imperfect flowers); floral parts arise from the top of the ovary which is united with the sepals (inferior ovary).
Family	Hominidae	No tail or cheek pouches; two-legged, upright posture; hands and feet differently specialized; big toe not opposable, flat face.	Family	Fagaceae	Beech, oak, chestnut family; small staminate and pistillate flowers on same tree; fruit is a nut or an acorn; leaves undivided (simple), alternate and mostly deciduous.
Genus	Homo	Extremely large and complex brain; long period of childhood; speech; shorter arms and larger thumbs.			
Species	*Homo sapiens*	Scanty body hair; small front teeth and prominent chin; high forehead which houses frontal lobes of brain.	Genus	Quercus	Oaks; fruit is an acorn; buds clustered at tips of twigs.
			Species	*Quercus alba*	Leaves deciduous with rounded lobes; acorn set in a bowl-like cup covered with warty scales; shells are hairless inside; grows 80-100 feet tall, 3-4 feet in diameter with a wide spreading crown.

tem, you know a great deal about it. The descriptions indicate some of the important characteristics of the various categories.

Table 21-2 classifies dogs along with 7 other species. From the table you can see that the forms most similar to dogs are other members of the genus *Canis*, such as the wolf. As you read up the

hierarchy, the similarities become fewer. Dogs resemble other mammals such as man in some respects but not so much as they resemble other carnivores. Dogs also resemble other vertebrates such as the fishes but not as closely as they resemble other mammals or primates. The scheme of classification and its major terms were two

Table 21-2   Classification of eight species

Category	Dog	Wolf	Fox	Grizzly bear	Man	Atlantic salmon	Honey bee	White Oak
Kingdom	Animalia	Animalia	Animalia	Animalia	Animalia	Animalia	Animalia	Plantae
Phylum or Division	Chordata	Chordata	Chordata	Chordata	Chordata	Chordata	Arthropoda	Tracheophyta
Class	Mammalia	Mammalia	Mammalia	Mammalia	Mammalia	Osteichthyes	Insecta	Angiospermae
Order	Carnivora	Carnivora	Carnivora	Carnivora	Primates	Salmoniformes	Hymenoptera	Fagales
Family	Canidae	Canidae	Canidae	Ursidae	Hominidae	Salmonidae	Apidae	Fagaceae
Genus	Canis	Canis	Vulpes	Ursus	Homo	Salmo	Apis	Quercus
Species	Canis familiaris	Canis lupus	Vulpes fulva	Ursus horribilis	Homo sapiens	Salmo salar	Apis mellifera	Quercus alba

important contributions Linnaeus made to taxonomy.

Even more important was the method he provided for naming species. This enabled biologists to label every species in a simple, unambiguous way, no matter where it appeared on earth and what its local name happened to be. It may sound unfamiliar to call a mountain lion *Felis concolor.* When scientists use that name, however, they know just what animal they are referring to. Perhaps in certain parts of the western United States *Felis concolor* is called a mountain lion. In South America it may be a cougar or a puma, or even a panther. In Africa, though, "panther" refers to *Felis pardus,* the spotted leopard. The spotted cat of North and South America, on the other hand, is *Felis onca,* the jaguar, which is also occasionally known as a panther. So it is obvious that local names may not be much help, and in fact often confuse the issue. Add to that the many language differences around the world, and the necessity for standardization in designating species becomes clear.

The two-name, or **binomial** system devised by Linnaeus indicates an organism's genus in the first word, which is capitalized. The second word, ordinarily written without capitals, designates the species in that genus. *Homo* refers to the genus which includes man and *sapiens* to his species. Occasionally a subspecies is recognized, and a third, uncapitalized word is added: *Homo sapiens neanderthalensis* (Neanderthal man), for

instance, and *Homo sapiens sapiens* (modern man).

This simple system is used today for all organisms. Even when a genus or species name must be coined, perhaps to give credit to its discoverer (for example, *Attacus edwardsii*) or to a benefactor who supported an expedition (for example, *Rothschildia orizaba*), the term is latinized. Linnaeus wrote in Latin, as did most scholars of his time, and that language has persisted in this branch of science. No matter what language a scientist is writing in today—English, Russian, or Japanese—international convention calls for a Latin name to designate the name of the species. This guarantees that each scientist will know exactly which species another scientist is discussing.

SOME PROBLEMS OF SYSTEMATICS

During the last few years the whole field of systematics has undergone renewed and exciting growth. Systematists have become interested not only in how organisms are related to one another but also in the mechanism of evolution itself and the rate at which it continues to go on. Consider the following challenging and perplexing problems.

Why are there so many species in the tropics? A tropical forest, like that shown in Figure 21–11, contains many species, each with relatively few individuals. This contrasts to a nontropical forest which contains many individuals of relatively few species. This is also true of organisms in the oceans. A tropical reef is ten times as rich in species as a rocky reef in cool water.

Or how about the following: Why do certain groups remain virtually unchanged for hundreds of millions of years, whereas others simultaneously undergo radical changes, seemingly in the same environment? Why do some groups suddenly and almost explosively diversify and appear to occupy simultaneously many new niches—adaptive radiation—whereas other groups remain within their ancestral niche?

Another unsolved problem is the process of *extinction*. More than 95 percent of all evolutionary lines that once existed are now extinct and have left no descendants. Why did entire groups such as the dinosaurs succumb? Big ones, small ones, herbivores, carnivores, running forms, and slow-moving swamp dwellers all died out, and no one has so far advanced a convincing theory to explain why it happened. Why did natural selection fail to prevent them from becoming extinct?

In order to answer these questions the systematist uses many different tools, some of which we have mentioned. But in addition, he commonly looks at organisms somewhat differently from other biologists. Whereas many biologists who work at the cellular or molecular levels view organisms in terms of chemistry and physics, the systematist looks at organisms as the products of individual genetic programs carefully regulated by hundreds of millions of years of natural selection. Each living species is the product of its evolutionary history. This past history of the species is a key to understanding its present status and perhaps for predicting its future. For the systematist each living species is a success story, for it has persisted and today living species represent the descendants of fewer than 5 percent of the species which ever existed.

At this time we know very few of these success stories. For more than 90 percent of the 1½ million known species of animals and plants on earth, nothing more is known than a name and a few diagnostic characters. Today many major groups of organisms are "orphaned" and not one scientist in the whole world is studying them. There remain major groups to be studied, even new environments to be explored. For example, within the last few years scientists discovered a new environment teeming with organisms which live between the particles of mud and sand on beaches and on the ocean floor. This new environment contains organisms considerably bigger than microorganisms and contains many previously unknown types of organisms, including new orders and classes.

Figure 21-11 A tropical rain forest in the Caroline Islands. Many more species are present here than in a temperate forest.

SYSTEMATICS IN THE SERVICE OF MAN

The study of diversity is of great practical value for man. It enables us to recognize and control noxious organisms like the corn borer and the boll weevil. It permits us to harness beneficial forms, such as releasing ladybird beetles to control citrus scale insects and releasing fungal spores to control Japanese beetles. Yet we have barely made use of systematics. Nowadays we fish for only a few species. What other species of fish may provide us with food? Most of the shrimp catch of the United States consists of only a few species. How many other crustaceans are available and suitable for food? We are so ignorant about the properties of most organisms that we are unable to decide which ones may be useful for man. Who would have suspected that molds would be of great medical value until penicillin, aureomycin, and other antibiotics were discovered? What medically important chemicals are waiting to be discovered within the many organisms that have not yet been described? Recently a powerful insecticide was isolated from balsam fir trees, and a substance which was found to inhibit the growth of tumors was extracted from puff balls.

We had best do a great deal of this describing soon, because the rapid growth of human populations is rapidly destroying the natural habitats of most organisms. The tropical rain forest is the richest and the least explored natural habitat in the world in regard to species of animals and plants. Yet by the end of this century there will be no undisturbed tropical rain forest anywhere in the world. One authoritative estimate is that 80 percent of the world's organisms will become extinct during the next century. Most of the hundreds of thousands of species of mites and of ichneumonid wasps will never be described, nor the many species of flowering plants, molds and Rickettsiae of very high medical, scientific, and economic importance.

In addition to its practical value, systematics is important to biologists in most other fields. For somewhere in nature there are organisms ideally suited to help biologists solve almost any problem in biology. *Drosophila*, bread mold, and bacteria are widely used in attacking genetic problems. Organisms as diverse as sea urchins and frogs have taught us about the development of complex multicellular organisms from single cells. Green algae provide us with clues to the evolution of sexual reproduction and the origin of higher plants. Plants of all sorts have been investigated to solve riddles of photosynthesis (green algae and euglenas), development (carrots and tobacco), and genetics (corn). A great deal of our knowledge of regeneration comes from studies of hydra, planaria, and newts. The best studied nerves in all the animal kingdom are the giant axons of the squid. Nudibranch mollusks have proven ideal for studies which correlate the behavior of an organism with the input and output of individual nerve cells. For studies of the organization of simple nervous systems composed of less than 100 cells, nematodes are being used. Many inquiries into how muscles work have employed insect flight muscles, grasshopper jumping muscles, and the muscles of barnacles. For studies of kidney function, kangaroo rats and sharks are often chosen. For the analysis of behavior, social insects, fruit flies, birds, and primates—to name only a few—are selected.

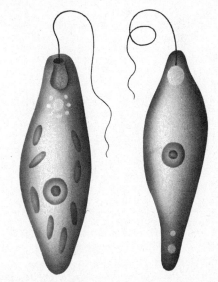

Figure 21-12 Plant or animal? The unicellular flagellate *Euglena* (left) has chloroplasts and photosynthesizes. A similar-looking flagellate, *Astasia* (right), absorbs decayed organisms. When grown in the dark *Euglena* feeds like *Astasia*.

THE KINGDOMS OF ORGANISMS

Chapters 23 and 24 provide a brief survey of the major groups of present-day organisms, their role in the economy of nature, and their importance to man. We shall also try to identify their ancestry and examine some of the factors which shaped their evolutionary history. But before we do this it is helpful to take an overall look at the several kingdoms of organisms that make up the living world.

For a very long time biologists divided organisms among two great kingdoms, one for plants and the other for animals. Animals moved, ate things, and had a definite size. Plants did not move, manufactured their own food, and seemed to grow almost indefinitely. There was little difficulty in classifying most familiar organisms. Daffodils and Sequoia trees are definitely plants, and chickens and earthworms are definitely animals. However, organisms like the unicellular flagellate, *Euglena*, proved to be much harder to categorize. *Euglena* has chloro-plasts and makes its own food by photosynthesis. But it also moves. There are other flagellates, one called *Astasia*, which are virtually identical with *Euglena* and are obviously closely related to it, which do not have chloroplasts. Instead *Astasia* gets its food by absorbing the breakdown products of dead organisms. To make matters even more interesting, if we grow *Euglena* in the dark where its chlorophyll cannot function, then it too feeds like *Astasia*. Is *Euglena* a plant or an animal? Have we transformed it from a "plant" to "animal" by growing it in the dark? Obviously two kingdoms, plant and animal, cannot accommodate certain organisms. There are many other problems of this sort which require a broader classification than two kingdoms.

Earlier in this chapter we indicated that one of the main aims of a system of classification is to group together organisms that share a common ancestry. Unfortunately we have few useful clues to the origins and relations among most great groups of organisms, notably the fungi and most of the algae. For this reason biologists have

Table 21-3   Three classifications of the kingdom of organisms. The most difficult groups to classify are in shaded boxes.

Three kingdoms	Four kingdoms	Five kingdoms (Whittaker)
MONERA Bacteria Blue-green algae	MONERA Bacteria Blue-green algae	MONERA Bacteria Blue-green algae
	PROTISTA Protozoa All algae Slime molds True fungi	PROTISTA Protozoa Diatoms Dinoflagellates Euglenas Slime molds
		FUNGI True fungi
PLANTAE All algae Slime molds True fungi Bryophytes Tracheophytes	PLANTAE Bryophytes Tracheophytes	PLANTAE Red algae Brown algae Green algae Bryophytes Tracheophytes
ANIMALIA Protozoa Multicellular animals	ANIMALIA Multicellular animals	ANIMALIA Multicellular animals

not reached agreement in which kingdoms to place these organisms. Three schemes which have attracted considerable favor among biologists are summarized in Table 21-3. The scheme recently devised by Richard Whittaker has a number of attractive features which are worth thinking about as you read the following three chapters and learn more about the diversity of organisms.

Whittaker recognizes five kingdoms of organisms: Monera, Protista, Plantae, Fungi, and Animalia. The **Monera** are the simplest known organisms and presumably resemble the early organisms from which the other kingdoms evolved. It includes all of the procaryotic organisms, the *bacteria* and *blue-green algae*. They have no membrane-bound cellular organelles, no microtubules, simple flagella, and their genetic material is a single circular molecule of DNA not linked with proteins. Because all procaryotes have many features in common, it is likely that they originated from some original founder cell population and are a monophyletic group.

The other four kingdoms are composed entirely of eucaryotes. They have a definite nucleus bounded by a double unit membrane that contains chromosomes made up of both DNA and proteins. These chromosomes divide and segregate regularly by mitosis. They have spindle fibers composed of microtubules, complex cilia, and flagella (with 9-plus-2 microtubules), and mitochondria and vacuoles bounded by a single unit membrane. Eucaryotes also have two other important features which are not found in procaryotes: *sexual reproduction* and *multicellularity*. Sexual reproduction is characteristic of all eucaryotes and was probably a characteristic of the common founding ancestor of the eucaryotes. Multicellularity evolved independently in several different lines of eucaryotes, but is not found in any procaryotes. The cells of procaryotes occasionally remain together after cell division and form filaments or sometimes a solid mass of cells. But the individual cells have few protoplasmic connections among themselves and as a consequence there is little overall integration of the filament or mass. In contrast the cells of multicellular eucaryotes communicate with each other in various ways. In multicellular eucaryotic plants there are pores in the cell walls through which protoplasmic processes, the plasmodesmata, connect the cells. In multicellular animals the cell membranes are in direct contact with each other, permitting ready communication.

The **Protista** include a variety of organisms composed of a single eucaryotic cell or a colony of similar eucaryotic cells. It includes both plant-like and animal-like simple organisms. Among these are the protozoans which ingest or absorb their food and are usually called animals, as well as several groups of unicellular algae which manufacture their own food by photosynthesis and are commonly called plants. All protists appear to share a common ancestry in the primordial eucaryotic cell which evolved from the procaryotes. However, the protists have evolved in several divergent directions and have given rise to several lines of algae and protozoans.

The kingdoms Plantae, Fungi, and Animalia consist of multicellular eucaryotic organisms which during the course of evolution have come to specialize in three different sorts of nutrition: *photosynthesis, absorption,* and *ingestion.*

Members of the kingdom **Plantae** have specialized in photosynthesis and include multicellular organisms composed of eucaryotic cells with cell walls and photosynthetic pigments in plastids. Some closely related unicellular organisms and a few related organisms without chlorophyll are also included among the plants. Plants appear to have evolved from several different ancestral populations of protists.

The **Fungi** lack photosynthetic pigments and feed by growing *through* their food, absorbing nutrients as they go. They are usually composed of filaments containing many nuclei, and the filaments may or may not be divided by crosswalls. Most biologists will continue to refer to fungi as plants and botanists will continue to study them, but they have very little in common with green plants and clearly represent a separate evolutionary line. Whether they arose from one or several ancestral populations of protists is unknown.

Members of the kingdom **Animalia** are multicellular organisms with eucaryotic cells which lack cell walls, plastids, and photosynthetic pigments. They feed mainly by ingestion and digestion and move about by means of contractile fibrils. The animals appear to have evolved from two ancestral protists, one giving rise to the sponges and the other to all the other multi-cellular animals.

The **viruses** do not fit easily into any of these categories and there is a real question of whether they should be regarded as living things. They differ fundamentally from all cellular organisms in having either RNA or DNA and in being parasites inside of other cells. For this reason they are treated as a separate group.

These five kingdoms and the viruses provide a useful framework in which to classify the organisms we are about to consider. However, like all phylogenetic schemes it is only a working model and must be revised as new facts are discovered. Viruses and Monera have already been treated in Chapter 9; the Protista are treated in Chapters 23 (algae) and 24 (protozoans), Plantae and Fungi are considered in Chapter 23, and Animalia in Chapter 24. This arrangement is a convenient one, and although it does not reflect phylogeny perfectly, it does reflect the fact that research on groups of organisms and scientific publications about them are usually separated as we have separated the chapters. If you wish to learn about a group of viruses or bacteria, ask a microbiologist or read a microbiological journal (Chapter 9); for algae, fungi, and higher plants, go to a botanist or a botanical journal (Chapter 23); for protozoans and multicellular animals, go to a zoologist or a zoological journal (Chapter 24).

SUGGESTED READING LIST

Barghoorn, E. S., "The Oldest Fossils," *Scientific American*, May, 1971 (Offprint 895).

Dayhoff, M. O., "Computer Analysis of Protein Evolution," *Scientific American*, July, 1969 (Offprint 1148).

Dickerson, R. E., "The Structure and History of an Ancient Protein," *Scientific American*, April, 1972.

Hanson, E. D., *Animal Diversity* (second edition). Prentice-Hall, Englewood Cliffs, New Jersey, 1964.

Jensen, W. A., and F. B. Salisbury, *Botany: An Ecological Approach*. Wadsworth Publishing Company, Inc., Belmont, California, 1972.

Margulis, L., "Symbiosis and Evolution," *Scientific American*, August, 1971 (Offprint 1230).

Mayr, E., *Principles of Systematic Zoology*. McGraw-Hill Book Company, New York, 1969.

Salthe, S. N., *Evolutionary Biology*. Holt, Rinehart and Winston, Inc., New York, 1972.

Solbrig, O. T., *Principles and Methods of Plant Biosystematics*. Macmillan, New York, 1970.

Whittaker, R. H., "New Concepts of Kingdoms of Organisms," *Science*, Vol. 163, 150-160, 1969.

CHAPTER 22

Unfolding Diversity

Over a million years ago a dinosaur walked across a lake bed in Texas, leaving its footprints embedded in the mud. The lake bed dried and the mud eventually turned to stone, preserving these footprints as fossils (Fig. 22-1). Other fossils are the bones or shells of animals whose

Figure 22-1 Fossil footprints of a dinosaur.

553

remains settled into mud or were covered by sand which later turned to rock, or the impressions of leaves in soft clay (Fig. 22-2) that hardened with time, or insects trapped in sticky pine sap that later turned into amber, or the entire skeletons of larger animals that fell into pits of oil that turned into tar and preserved them for later study.

In Chapter 20 we examined the origins of life from its organic beginnings to the conditions that not only sustain today's living systems, but made it almost inevitable that life would arise. Once life came into being, it proliferated into the millions of species of plants and animals that today cover the earth's surface.

THE FOSSIL RECORD

Nothing is directly known about the origin of life. The first organisms were undoubtedly extremely small and soft-bodied and could not have left a fossil record, so that we have to reconstruct the origin of life from what we know about living organisms. However, much of the later

history of life on earth is written in fossils which give us direct evidence of procaryotes, plants and animals that existed millions of years ago.

Charles Darwin found some of his first clues to the pattern of evolution from fossils he discovered in South Africa. These were the remains of large, extinct animals that were clearly related in form and structure to smaller species which still live on that continent. For example, one of his finds was a huge animal covered with armor which, except for its size, looked like the armadillo (Fig. 22-3) that still lives today in the southwestern United States and South America. There was too much similarity in the details of the structures of extinct and living animals for Darwin to believe that they were unrelated, and the fossil record became one more strong piece of evidence for his theory of evolution. Since then, as we see later in this chapter, scientists have discovered fossils in all parts of the world which have enabled them to reconstruct the story of past life and its relation to organisms living today.

However, it takes more than just fossils themselves to put together the story of the past. To reconstruct history we must provide dates. In the case of the fossil record, if we cannot date the existence of a species exactly, we must at least be able to show the relationship between different species on some sort of time scale. We need to know the order in which different forms followed one another, and which forms lived at the same time. Originally scientists had to draw such conclusions from the approximate ages of the rocks in which the fossils were found. Some rocks were obviously older than others, because they were

Figure 22-2   Fossil impressions of leaves of trees.

Figure 22-3   A modern armadillo looks much like a small version of a fossil found by Darwin in South Africa.

more deeply buried in the sediments covering the earth. From this kind of information it was possible to establish, for instance, that amphibians appeared on the earth before reptiles, and that reptiles were followed by mammals, but not how many years ago these events occurred.

In recent years better dating methods have been developed. These depend on the disintegration of radioactive elements which decay at known rates, so that the amount of the decay product compared with the amount of the original element provides a fairly accurate estimate of the age of the sample. For example, uranium disintegrates to form an *isotope* of lead that stays in the rock with what is left of the uranium. We can therefore date a rock by comparing the amounts of uranium and lead by the following formula:

$$\frac{\text{grams of lead}}{\text{grams of uranium}} \times 7.6^8 = \text{age in years}$$

Potassium is more common than uranium in fossil-bearing rocks and disintegrates into the inert gas argon. Also plant and animal tissues contain carbon. One isotope of carbon, carbon-14, has a half-life of 5568 years and can be used to date fossil wood or bone.

Although these "atomic clocks" have enabled scientists to establish a better calendar of life on earth than ever before, these methods of dating by radioactivity give only useful approximations of the exact age, and a precise calendar is not yet complete. Nevertheless the broad outline of events in the history of the planet earth is fairly clear. There were major time periods during which certain key events took place in the evolution of organisms and in changes in the environment, although it is not yet possible to date these events precisely, and it is more convenient to refer them to the geological time scale, rather than to actual dates (Fig. 22-4).

TIMESPANS OF THE EARTH'S HISTORY

Scientists divide the earth's history into four major timespans called *eras*. The earliest era was the Pre-Cambrian, which covers the events from the beginning of the earth to about 600 million years ago. Until recently it was believed that the Pre-Cambrian was the era before life began, but as we have seen, recent methods of detecting and identifying fossils have provided evidence that bacteria and algae existed even then.

The *Paleozoic* era covers the timespan from approximately 600 million to 230 million years ago. This era is subdivided into seven *periods* which correspond to important geological events during this era. The name of the era means "ancient life" and is marked by the appearance of the earliest known fishes, the first land plants, and the first amphibians, insects, and reptiles.

The *Mesozoic* era (meaning "middle life") extended from 230 to 70 million years ago. This era contains three periods and is also called "The Age of Reptiles," for this was when the great reptiles, the dinosaurs, flourished and became extinct. It is also the era when the first birds and mammals appeared on earth.

The *Cenozoic* era ("recent life") began 70 million years ago and extends to the present. It contains two periods, subdivided into seven epochs, and is called "The Age of Mammals." During this era the mammals expanded tremendously in numbers of species and developed into the modern forms we know today, the apes evolved from primitive, primate ancestors, and man appeared on the scene.

As each of the last three eras began, there were major changes in the climate of large parts of the earth and, in many cases, in the geology of the earth. The change from the Pre-Cambrian to the Paleozoic marked a transition from a cold climate, when glaciers were widespread, to much warmer temperatures which allowed the evolution of the many new forms which appeared during the Paleozoic. At the end of the Paleozoic era warm climates gave way to cooler ones, and again glacial ice sheets covered large parts of the continents. This was also an era of mountain building, when rising lands provided new and varied environments for the animals of the Mesozoic. This was when the Appalachian range in the eastern United States was formed.

Tropical climates characterized much of the Mesozoic era. They allowed the dinosaurs to flourish, but finally gave way to the climatic zones still prevalent today. At the end of the

relative durations of major geologic intervals	era	period	epoch	duration in millions of years (approx.)	millions of years ago (approx.)	
Cenozoic	Cenozoic	Quaternary	Recent	approx. last 5,000 years	0	
			Pleistocene	2.5	2.5	
Mesozoic		Tertiary	Pliocene	4.5	7	
			Miocene	19	26	50
			Oligocene	12	38	
Paleozoic			Eocene	16	54	
			Paleocene	11	65	
	Mesozoic	Cretaceous		71		100
					136	150
		Jurassic		54		
					190	200
		Triassic		35		
					225	
	Paleozoic	Permian		55		250
					280	
		Carboniferous Pennsylvanian		45		300
					325	
		Carboniferous Mississippian		20	345	350
		Devonian		50		
					395	400
		Silurian		35		
					430	450
Precambrian		Ordovician		70		
					500	500
		Cambrian		70		550
					570	
		Precambrian		4,030		
					4,600	

formation of Earth's crust about 4,600 million years ago

Figure 22-4 The geological time scale.

Mesozoic and the start of the Cenozoic, there was another great change in the land. More mountains took shape, including the Rocky Mountains, the Andes, the Alps, and the Himalayas.

EARLY LIFE

The first organisms on earth lived in oceans and lacked shells or bones or other hard parts that would be preserved as fossils. Compounds of hydrogen and carbon, called *hydrocarbons*, are synthesized only by living organisms and have been found in rocks over 3 billion years old. It is very difficult to prove, however, that organic matter in an ancient rock is the same age as the rock. Petroleum moves underground, and water containing organic compounds percolates through rocks. Rock may also pick up contaminants after it has been unearthed. Oil from printer's ink has been absorbed by rocks that were wrapped in old newspapers. Nevertheless, when we account for all the different possible sources of error, we must conclude that some 3-billion-year-old molecular fossils are real.

There are still earlier signs of possible life in the Pre-Cambrian era. Geochemists have found remnants of organisms that resemble one-celled algae that are estimated to be over 3 billion years old. We also believe that some limestone reefs formed in the Pre-Cambrian era were started by algae. These reefs were not formed in layers like most sediments but were deposited in concentric circles similar to those formed by modern blue-green algae. One of these ancient limestone reefs has been estimated to be 2.3 billion years old.

The first signs of animal life appear much later in the Pre-Cambrian era, perhaps 750 to 600 million years ago. They are impressions in rock that resemble the soft bodies of jellyfish and marine worms. In this era, then, we already see the division of primitive forms that gave rise first to microorganisms and then to plants and the animal kingdom.

Just how this division of kingdoms occurred we shall probably never know, although there is good evidence that organisms resembling today's one-celled flagellates might have been involved. Flagellates have certain characteristics typical of both plants and animals. Accordingly, they might be transitional organisms, not entirely plantlike and not entirely animal-like but able to combine some of the characteristics of both. Some of the organisms that are discussed in later chapters, such as sponges among the animals and algae among the plants, have features similar to those found in flagellates.

We have already noted that some modern flagellates are able to carry out photosynthesis, whereas others, seemingly identical, cannot. Experiments indicate that the nonphotosynthetic flagellates probably evolved from photosynthetic ones and survived by finding ways of obtaining the foods they can no longer manufacture. This gives us reason to believe that animals might be ultimately traced back to a primitive, one-celled ancestor that was once able to manufacture carbohydrates like a plant. Furthermore we know that chloroplasts and mitochondria contain DNA — their "own" DNA — which is different from that in the nucleus of the cell. As we proposed earlier, these organelles might have originated as **symbiotic** organisms, contributing to the welfare of other cells and benefiting from their hosts as well.

Pre-Cambrian history will probably be rewritten many times in the next few years, as new experiments give us more information and as new techniques are developed for dating and interpreting fossils. For a long time it was assumed that life became plentiful only in the Paleozoic era, as though this was a timespan when the conditions for life were rather marginal, but today we know that primitive plant forms were plentiful nearly 2 billion years ago and were in existence well before that. If the estimates of early life are confirmed by further research, we shall have to alter considerably our ideas about this era and the history of life. The boundaries are gradually being pushed back farther into time, nearer the time when the earth was formed — something on the order of 4.5 billion years ago.

THE PALEOZOIC ERA

The Paleozoic era dawned with the end of the Pre-Cambrian glaciers and a gradual warming of climates approximately 600 million years ago, but it was a rather harsh dawn by present standards. The land was bare and lifeless and only the seas showed signs of what was to come. For at least 100 million years, during the period called the Cambrian, and perhaps for another 75 million years of the Ordovician period, life was probably confined to the oceans, but it contained the potential for considerable development of new forms. Flagellates suggest this and so do the algae, which today have many diverse ways of life. They have characteristics which suggest the origins of important features of higher organisms, such as the specialization of cells, sexual and asexual reproduction, the spontaneous movement of animal cells, and the system of vessels through which vascular plants conduct fluids to every part of the organism. These characteristics were necessary for the next major step in evolution, when plants led the move to *terrestrial life*. Once plants established a foothold on land, animals followed, and by the end of the Paleozoic era, 370 million years after it began, giant forests covered many parts of the land and amphibians and reptiles were abundant.

The appearance of marine invertebrates

Animals first became really plentiful in the fossil record during the Cambrian period. Paleontologists have found traces of nearly every major invertebrate group except one in the rocks of this period. Among the animal phyla that were common were *sponges*; *coelenterates*, ancient relatives of the jellyfish and corals; *brachiopods*, which resemble mollusks but are sufficiently different to be in a separate phylum; and *arthropods*, a phylum that includes today's lobsters, crabs,

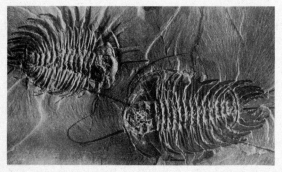

Figure 22-5   Fossil trilobites from the Cambrian period.

and insects. An arthropod which is now extinct but was prominent in the Cambrian and Ordovician seas was the *trilobite* (Fig. 22-5), some of which were a foot long.

It sometimes sounds as though animals suddenly appeared in great numbers during the Cambrian period, in an "explosion" of animal life, when there was actually a very slow multiplication of new forms. Even though the Cambrian is only one period in the Paleozoic era, it covered 100 million years, which is more than the span of the entire Cenozoic era, the era of recent life. During this 100 million years new animals appeared only gradually in a long, slow process of evolution.

The Cambrian and Ordovician periods teach an important lesson about evolution; all of the species known to exist by Ordovician times are extinct today, only one genus has managed to survive from this period, and yet all the phyla still exist. In other words, the broad life types represented in these phyla have proven successful, while the genera and species within them have been replaced by new forms better adapted to today's environments. The phenomenon of *replacement* occurs over and over again, among plants and animals alike, in the fossil record.

The first land plants

There are some clues that indicate that plants might have begun living on land in Cambrian times, but they are not entirely convincing. The fossil evidence for this possibility consists of deposits of plant spores, but because of their size, spores are easily carried by wind and other agents far from where they originated, and botanists generally prefer to say that the first truly terrestrial plants can be identified from the third Paleozoic period, the Silurian. These were leafless plants that had the vascular conducting system typical of the most advanced plant phylum. This phylum produced ferns and all the seed plants—trees, shrubs, and the flowering plants, including grasses.

Green plants also evolved in another direction to produce the mosses and liverworts which lack a vascular structure. Fossils from this phylum are found in the Devonian period, which followed the Silurian, but because they lack the tougher tissues of vascular plants, they decay more easily, and earlier forms may not have been preserved as fossils.

Botanists do not know whether there was a single plant migration ashore or whether there was more than one evolution of land plants from their aquatic ancestors, but they generally agree that the ancestral water plants that produced land forms were most likely green algae. Chemical studies provide the strongest evidence for this belief. As we learned in earlier chapters land plants and green algae both depend on the chlorophylls a and b in photosynthesis and have similar proportions of other pigments as well. Also they both store the products of photosynthesis as starch.

Primitive landscapes

The Carboniferous period (often subdivided into the Mississippian and Pennsylvanian) produced the giant swamps and forests that eventually formed many of the earth's coal deposits, including those in the Appalachian region of the United States. These forests were not like today's forests (Fig. 22-6). The trees were giant *club mosses*, which are not true mosses but vascular plants, *mammoth ferns* and *seed ferns*, and primitive *conifers* that were related to today's evergreen trees. These primitive seed ferns and conifers included some of the first-known seed plants.

Another prominent plant during the Carboniferous was the *giant horsetail* which grew to heights of 30 to 40 feet. Today's horsetails, which grow only 1 to 3 feet high, are often referred to as "living fossils" because of their relationship and close resemblance to their Carboniferous ancestors.

The *ginkgo* tree is another so-called living fossil. The first ginkgos appeared in the Permian period, just before the end of the Paleozoic, and by the middle of the Mesozoic era they had spread around the world. Although they were once represented by several genera and species, today there is only one species, called the "maidenhair tree," which is common in the United States as an ornamental shade tree.

Animal life on land

If animals are to evolve from an aquatic existence to life on land, what will their requirements be? Obviously they would require food, which the earlier evolution of land plants provided, but an aquatic environment has several advantages which are not found on land, so that a terrestrial animal would also have to introduce several new adaptations. Because water is a dense medium, it provides support for the bodies of large animals which would otherwise have to develop strong, internal skeletons or a rigid outer shell in order to move about on land. This, of course, is why an animal such as the blue whale can grow to such tremendous size. Water temperature is also more stable than air temperature, so that

Figure 22-6 A Carboniferous landscape of primitive plants.

land animals must either remain inactive when it is too hot or too cold or evolve a system which will allow them to regulate their internal body temperatures independently of the air temperatures that surround them.

However, the most critical problem in the evolution of land animals was that reproduction requires a liquid medium. Sperm cannot be transferred from male to female, eggs cannot carry on the processes of embryonic development, and the embryo itself cannot exist outside a liquid medium. Female fish, for example, deposit their eggs on the bottom of a lake or stream, the male swims over the eggs and releases sperm where it will reach the eggs and fertilize them, and the developing embryo is constantly bathed in the water of its environment where it can respire and develop. Land animals had to evolve special adaptations that would provide similar conditions. Two of these were internal fertilization and eggs protected by shells which contained a liquid medium and allowed respiration through the shells.

Hard-shelled aquatic arthropods were partly preadapted to life on land, in that their shells provided the body support they required. It is not surprising, therefore, that there is evidence of a fossil scorpion in the Silurian that suggests this first migration. It resembles living scorpions so much that it might have been able to live on land,

but it also had many features of its aquatic ancestors, and paleontologists are not certain that it was entirely independent of water. It is in the following period, the Devonian, that we find better evidence of several arthropods becoming land dwellers. This group includes mites, which are ancestors of the spiders, and a form that apparently led to the insects.

Earlier, in the Ordovician period, there were other changes occurring that led to the second animal migration ashore. Fossils from that period show the first evidence of vertebrate animals with internal skeletons. These were jawless fish called *agnatha*. The agnaths are extremely important, for there is a clear evolutionary pathway from these fish to amphibians and all other terrestrial vertebrates. Lampreys and hagfishes are the only agnaths living today.

The age of fishes

The fourth period of the Paleozoic era, the Devonian, is often called "The Age of Fishes"; this was the great period of fish evolution. When the Devonian period began, there were not only agnaths but also *placoderms*, fish with bony plates embedded in the skin. They were the first fishes with jaws. This adaptation was far more successful and provided more opportunities for different feeding habits than was possible among the jaw-

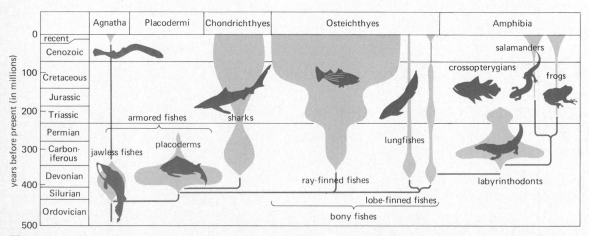

Figure 22-7   The evolutionary history of fishes. The dashed lines indicate the probable relations between groups and the light-colored areas show the relative abundance of each group. (Modified from "The Coelacanth" by Jacques Millot. Copyright © December 1955 by Scientific American, Inc. All rights reserved.)

less fishes; hence the placoderms expanded for millions of years while their less successful competitors, the agnaths, declined. By the end of the Devonian, the agnaths were nearly extinct, except for a few species which led to the two forms that still exist today.

Meanwhile the later placoderms proved to be less successful than two other important groups which had evolved from early Devonian placoderms; these were the *cartilaginous fish*, with a skeleton of flexible cartilage rather than bone, and the *bony fish*. The cartilaginous fish of the Devonian led to modern sharks and rays and their relatives, while the bony fish include most of the fish that are common today, such as trout, tuna, catfish, perch, bass, and eels. The placoderms were nearly extinct by the end of the period and died out entirely before the end of the Paleozoic era, to be succeeded by the cartilaginous fishes

and bony fishes. Thus we see two large-scale replacements of fishes during the Devonian—first the placoderms succeeded the agnaths and were succeeded in turn by the cartilaginous and bony fishes (Fig. 22-7).

An important class of bony fish that had its origins in the Devonian was the *lobe-finned fish* (Fig. 22-7). These fish had large, fleshy, lobe-shaped fins with considerable skeletal development within the lobes (Fig. 22-8). They lived in large numbers in shallow water in pools which frequently became dry. The lobe fins, with their internal skeletal support, allowed them to leave a dry pool and crawl across land to find new water. Until recently it was thought that the lobe-finned fish had become extinct, but in 1938 a specimen of this fish, called *Latimeria* (Fig. 22-9), was caught by a fisherman on the southeast coast of Africa. Several other specimens have since been taken in this region and paleontologists have been able to study the anatomy of these fish in much greater detail than they could previously with only fossils to work from.

Surprisingly enough early bony fish had lungs as well as gills for breathing. Although these lungs no longer function for breathing in most modern fish and are not well developed, they would have helped early forms survive in water that was low in oxygen and they probably could have survived for short periods out of water in times of drought. Today's lungfish, for example, survive in polluted waters by coming to the surface periodically to gulp air in order to supplement the oxygen they take in by their gills. This adaptation provided one more step in the requirement of a terrestrial animal.

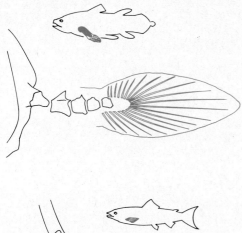

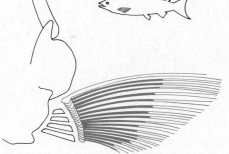

Figure 22-8 Arrangement of fin bones of typical lobe-finned fish (top) and a modern ray-finned fish (bottom). The fins and skeletal structure of the lobe-finned fish led to the development of the limbs of land-dwelling amphibians. (Modified from Thomson, 1966.)

Figure 22-9 Living lobe-finned fish, *Latimeria*, caught off the coast of South Africa.

Early amphibians

Before the end of Devonian period some lobe-finned fish had evolved into primitive *amphibians* (Fig. 22-10) which evolved legs and feet from primitive lobed-finned fish and had lungs and an internal skeleton. They could therefore move about freely on land, but were restricted in two ways: like today's amphibians, the frogs and toads and salamanders, they had soft skins which forced them to stay in relatively moist environments and they had not developed shelled eggs. They had to return to water for reproduction. Amphibians are therefore only a transitional form between aquatic and terrestrial forms, but they represent a great step in the sequence of events leading to the evolution of true terrestrial vertebrates, from reptiles to birds, and thus to mammals.

The rise of reptiles

Amphibians multiplied in numbers and kinds during the Carboniferous period, but, while they were dominating the land, the *reptiles* were becoming even more successful. The reptiles in-

Figure 22-10 Primitive amphibian descendants of the lobe-finned fish were able to invade the land.

troduced two extremely important adaptations; they had dry, scaly skins which were much better at retaining body moisture and they developed eggs with shells. Shelled eggs allowed reptiles to lay their eggs anywhere on land, thus freeing them entirely from aquatic environments, and the reptiles therefore completed the change to life on land which the amphibians had begun. Their fossils are found in the Carboniferous period and became plentiful in the next (and last) period of the Paleozoic era, the Permian. By the Permian period reptiles had already become more numerous than the amphibians and had also begun to branch into two different lines. One of these led to a great expansion of reptilian forms during the Mesozoic era, and the other led to even more abundant forms which gave rise to the mammals.

During the 370 million years of the Paleozoic era, life first conquered the seas then invaded and conquered the land, but late in the Permian the land was beginning to change. It rose in many places, pushing up mountain ranges like the Appalachians so that there were fewer jungles, swamps, and shallow, inland seas. There was much more dry land, a great deal of it on high elevations far removed from water, and new environments were being formed which would provide opportunities for new forms of life and reduce the potential for some of the older forms.

THE MESOZOIC ERA

During the first period of the Mesozoic era there were also great changes in the forests. Club mosses, seed ferns, and true ferns declined in importance, with some becoming extinct, while the ginkgos, conifers, and two groups of plants resembling palms but unrelated (*cycadeoids* and *cycads*) began to dominate the landscape.

The flowering plants

During this period, the Triassic—which began 230 million years ago—flowering plants may have first evolved, as some fossil impressions in rock resemble palm leaves and palms are true flowering plants. However, it is not until the final period of this era, the Cretaceous, that there was a noticeable increase in the number and spread of flowering plants. One possible reason for the lack of early fossils of these plants is that the new developments in plant evolution occurred in the highlands where opportunities for fossil formation were fewer than in marshy lowlands where leaves and other plant structures could become embedded in mud that would later turn to rock.

However, by the late Cretaceous, the flowering plants were modern in form and had become the dominant vascular plants they are today. Most of the subsequent changes among plants involved only geographic shifts brought about by changes in the climate, and all of the important groups of plants present in the last part of the Mesozoic era still survive today.

The expansion of flowering plants was paralleled by a great expansion of insect forms (Fig. 24-50). Most of today's insect families and many genera had appeared by the end of the Cretaceous period. Insects play an important role in the cross-fertilization of plants—an important source of genetic diversity. There was undoubtedly a complementary evolution of flowering plants and insects that accounted for the great development of both kinds of organisms during this period.

The age of reptiles

There were also great changes in the evolution of other land animals during the Mesozoic era. Amphibians declined sharply during the transition from the Permian to the Triassic period, and the only survivors were the forms which gave rise to the small amphibians that exist today. The reptiles also declined at first, but, because of their better adaptations to a land environment, they then began an expansion of forms that carried on through the Mesozoic, giving rise to its title as "The Age of Reptiles."

Our popular impression of dinosaurs is that they were all huge, but they were not. They ranged from small creatures about the size of a chicken to *Brontosaurus* (Fig. 22-11), the largest land animal of all times; it was over 80 feet long and weighed more than 25 tons. The dinosaurs reached their peak during the Cretaceous. Some were herbivores and others were predators, including some that preyed upon other dinosaurs.

The ancestral line that produced the dinosaurs also led to four existing forms of reptiles: *turtles* and *crocodiles*, both from the Triassic, *lizards* from the Jurassic, and *snakes*, which first appeared in the Cretaceous period. These are the only reptiles that have survived from the Mesozoic. Reptilian evolution also produced several kinds of marine animals, including the *ichthyosaurs*, which may have looked superficially much like dolphins or porpoises and represented a return adaptation to the sea (Fig. 22-12). Some reptiles also took to the air, producing the now extinct flying reptiles, the *pterosaurs* (Fig. 22-13), and eventually, modern *birds*.

The first birds

The earliest bird fossil is that of the Jurassic *Archaeopteryx* (Fig. 22-14), clearly a transitional form with both reptilian and bird characteristics. It had teeth and a long, reptilian tail, but the fossil also shows distinct impressions of feathers. This specimen is a particularly fortunate find, as

Figure 22-11 The dinosaur *Brontosaurus* was the largest land animal of all time.

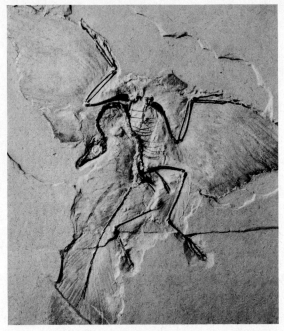

Figure 22-12 Aquatic reptiles from the Mesozoic era. The plesiosaurs (top) reached lengths of 35 feet, and the ichthyosaurs (bottom) were about 10 feet long.

Figure 22-14 The earliest bird fossil is of the Jurassic *Archaeopteryx.*

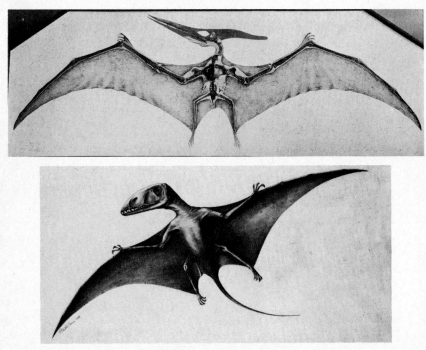

Figure 22-13 Pterosaurs had a wingspread of about 20 feet. A complete Cretaceous fossil is shown with a restoration of its body outline (top) and the probable appearance (bottom) of a living pterosaur.

feathers seldom fossilize. Birds continued to evolve in the Cretaceous and by the close of the Mesozoic era had many of their modern characteristics.

Early mammals

By the middle of the Jurassic another important evolutionary sequence was well under way. Few of the mammal-like reptiles survived the environmental changes that occurred between the Paleozoic and Mesozoic eras, but those that did provided the basis for mammalian evolution (Fig. 22-15). For 100 million years these mammals remained quite inconspicuous, but during their relatively minor position in the Mesozoic era they evolved many of the important adaptations that would allow them to flourish later. They developed an insulating coat of hair and internal temperature controls that eventually made them better adapted to changing temperatures than reptiles, and they also evolved a reproductive system in which the embryo developed within the body of the mother, rather than in unprotected eggs. These and other developments led finally to domination of the earth by mammals and a comparatively minor role for reptiles.

As the Mesozoic era came to a close there were vast, devastating changes taking place in the environment. The great mountain ranges of the North American continent came into being at this time. Many of the animals of the Mesozoic became extinct, while others persisted without significant change. The marine reptiles disappeared, but the fishes they fed upon remained relatively unchanged. But as the Cenozoic era began, the animal forms most prominent until then no longer existed.

THE CENOZOIC ERA

There are no dinosaur fossils in the deposits of the Cenozoic era. They disappeared from the fossil record very quickly. What can we assume from this? It seems unlikely that this was a case of the gradual replacement of the dinosaurs by another group such as the mammals, in the manner of the placoderms replacing the agnaths or the modern fishes replacing the placoderms during the Paleozoic. This dramatic disappearance of the dinosaurs was more likely due to the large-scale changes that occurred in the environment. Perhaps a cooling of the climate eliminated much of the lush vegetation these large herbivores ate or created temperatures to which they could not adapt, or several other factors may have been involved. However, whatever the cause of the extinction of so many animal forms, it was an extremely important evolutionary event.

The age of mammals

The Paleocene epoch, the first in this era, shows that mammals were abundant, but there were very few kinds. However, within the 15 million years of the Paleocene (beginning about 70 million years ago) the Age of Mammals was well under way. There were plant and meat eaters of several sizes and shapes, including ancestors of the rodents and monkeys, and when the environments and ways of life that had been occupied by the reptiles became empty, the mammals radiated quickly into these vacant niches and exploited the new opportunities.

In spite of their common ancestry in the distant past, the mammals and reptiles had diverged into different forms and modes of life, until the disappearance of the reptiles at the end of the Mesozoic. Now, with most of the reptiles gone and the rapid evolution of new mammal forms, *convergent evolution* could occur. For example, the reptilian ichthyosaurs that reinvaded the marine environment evolved a streamlined body and appendages like that of a shark, an obvious adaptation for fast movement in water. The present-day porpoises, which are mammals, have this same kind of shape, although they did not evolve until this era. In other words, convergent evolution occurs when unrelated animals evolve similar adaptations for a common way of life.

The evolution of the mammals also demonstrated another evolutionary phenomenon—that of expansion into new environments. Mammals not only occupied all of the environments vacated by the reptiles but also expanded into environments the reptiles had never been able to exploit. For instance, horses began life as small forest animals, but in the evolution of the modern horse

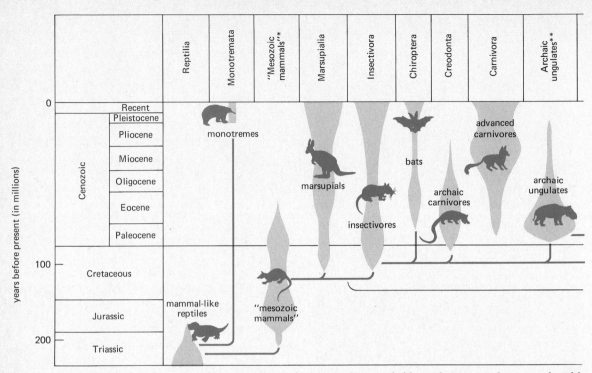

Figure 22-15 Evolutionary history of the mammals. Dashed lines show probable evolutionary relations and width

there were new adaptations in size, the structure of the hoof, and running ability which permitted them to live on the open plains. They were joined in this environment by a variety of mammals, such as antelope and others which also became adapted to this way of life.

By the Eocene epoch, it seems likely that most, if not all, of the basic groups of mammals had evolved, and as this period progressed modern forms replaced more primitive Paleocene representatives within the major mammal groups. The modern groups were generally dominant by the middle of the third epoch, the Oligocene. Some of these mammals might appear somewhat unusual in comparison with today's mammals, but all would be recognizable as relatives.

Modern mammals

Modern mammals are classified into three categories. One group, the *monotremes,* includes only two forms, the *duckbill platypus* and *echidna* (Fig. 22-16). These relatively primitive

mammals lay eggs like reptiles, but have mammary glands and suckle their young, a mammalian characteristic. The second group, the *marsupials,* or pouched mammals, carry their young in a pouch on the mother's body, where the young are nursed and protected until they have developed enough for a more independent existence. This group includes the Australian kangaroos and the American opossum among others. The largest group consists of the *placental mammals.* As we have learned, during pregnancy the embryo is nourished within the uterus of the mother through a special structure, the *placenta.* The placental mammals have been the dominant vertebrates throughout the Cenozoic.

In the following chapters we shall survey and classify the different forms of plant and animal life and the origins of man more completely. A fuller picture will be obtained of the variety of forms that evolved during the evolutionary sequences just described. But this brief recounting of the unfolding of life's diversity may have made the process of evolution sound far more direct

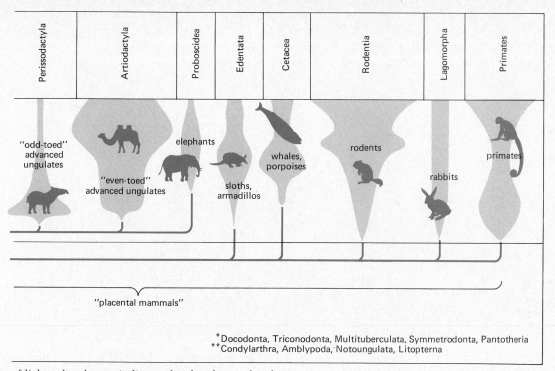

of light-colored areas indicates the abundance of each group.

*Docodonta, Triconodonta, Multituberculata, Symmetrodonta, Pantotheria
**Condylarthra, Amblypoda, Notoungulata, Litopterna

than it actually was. It is true that the fossil record and our studies of the anatomy of modern plants and animals allow us to trace the lineage of living things back in time. We can describe the origins of many kinds of mammals from an advanced reptilian form, and then, for example, trace the reptiles back to their amphibian ancestors, but this evolutionary sequence could not have been predicted in advance, and looking back at it from our perspective of today reveals many surprises.

For example, without the fossil record there would be no way of knowing that dinosaurs ever existed, for there are no animals like them among today's reptiles, but they did exist, along with many other extinct lines of plants and animals. There were plants and animals that appeared for relatively short times and died out. Others lasted

Figure 22-16 The monotremes (egg-laying mammals) are represented by only two living forms: the duckbill platypus (above) and the spiny anteater (right).

millions of years and seemed very successful, but then became extinct. Some forms, such as the mammals, barely survived for a long time and then became the dominant forms on earth. Still others, like the shark, have lasted virtually unchanged for even longer periods of time and are still present today.

In other words, the path of evolution was filled with many dead ends, twists and turns, and occasional detours. It was, in a certain sense, a massive organic experiment, like pioneers starting out across a continent with no idea of where they were going. New environments were explored with different adaptations, some successful and some not, until the survivors reached their unpredictable destination. Evolution has in fact followed a complicated, tortuous series of paths across the map of time.

SUGGESTED READING LIST

Barghoorn, E. S., "The oldest fossils," *Scientific American*, May 1971.

Brues, C. T., "Insects in amber," *Scientific American*, November 1951.

Deevey, E. S., Jr., "Living records of the ice age," *Scientific American*, May 1949.

Delevoryas, T., *Plant Diversification*. Holt, Rinehart and Winston, New York, 1966.

Fingerman, M., *Animal Diversity*. Holt, Rinehart and Winston, New York, 1969.

Fuller, H. J., Z. B. Carothers, W. W. Payne, and M. K. Balbach, *The Plant World*. Holt, Rinehart and Winston, New York, 1972.

Glaessner, M. F. "Pre-Cambrian animals," *Scientific American*, March 1961.

LaPorte, L. F., *Ancient Environments*. Prentice-Hall, Englewood Cliffs, New Jersey, 1968.

McAlester, A. L., *The History of Life*. Prentice-Hall, Englewood Cliffs, New Jersey, 1968.

Millot, J. "The coelacanth," *Scientific American*, December 1955.

Part Five

The Diversity
of Life

Our aim is to examine the immense diversity
of life on earth, and to explore the role of
different groups in the economy of nature,
and their special importance to man.

Plants

About 250,000 species of flowering plants and conifers are now known, along with 100,000 species of fungi and thousands of species of algae, ferns, mosses, and other plants. They include unicellular yeasts living in beer, red algae living at ocean depths, yellow diatoms coating the undersides of blue whales, 4000-year-old bristle cone pines on mountain peaks, fungi living on human skin, and hardy lichens living on bare rocks in the Arctic. In this chapter we examine the diversity of plant life, the places where different kinds of plants are found, the role of different plants in the economy of nature, and their special importance to man. We also try to discover the origin of this diversity as revealed in the major trends of plant evolution and the phylogenetic relations of the various plant groups.

As might be expected many useful clues to plant relationships have come from comparing the structures of different kinds of plants. Some information has also come from the fossil record. But unfortunately, although there are many excellent fossils of more recently evolved plants such as conifers, the fossil record is far less satisfactory for more primitive groups of plants such as algae, which lack vascular tissue with the tough supporting substances that become fossilized. A few good fossils, like the green algae found in the Bitter Springs cherts of Australia (Chapter 20), give assurance that unicellular eucaryotic green algae existed a billion years ago, but provide few clues to whom the green algae or any other algae are related. For this reason botanists have turned to other kinds of information to help them unravel plant relationships, such as the chemical structure of the pigments found in different plants, differences in ultrastructure, metabolic processes, life histories, and reproductive processes.

Using all of this information botanists have characterized the major groups of plants and uncovered major evolutionary trends. They have also obtained a good idea of the phylogenetic origin of certain recently evolved groups, such as the conifers. However, challenging questions remain. Even today we do not know from which ancestors flowering plants derived. Nor do we have a clear picture of the relationships and origins of various divisions of algae and classes of fungi. One attempt at a phylogeny for the major plant divisions is outlined in Figure 23-1. This phylogenetic scheme has three central features. These features should be borne in mind as we analyze the diversity of plants in the rest of this chapter.

1. Fungi and green plants are separate kingdoms that evolved independently from distinct groups of single-celled eucaryote ancestors.
2. Among the green plants, each of the algal divisions evolved independently by two distinct processes. The first process was *symbiosis* (see Chapter 20) in which a single-celled, non-photosynthetic, protozoan-like eucaryote developed a symbiotic relationship with a photosynthetic procaryote to produce a new photosynthetic symbiotic eucaryotic organism. This symbiotic organism was able to evolve further by the second process, evolution by *mutation, recombination,* and *natural selection.* Thus each of the algal divisions started from the merger of two separate evolutionary lines.
3. One division of algae, the green algae, gave rise to mosses and to vascular plants.

Hopefully some of the biochemical tools mentioned in Chapter 21, such as protein sequence determination and nucleic acid hybridization, will permit us to check some of the phylogenetic relationships suggested in Figure 23-1. In addition to these three features of plant evolution which are evident in the figure, three further features are noteworthy.

4. Multicellularity evolved independently at least once in several divisions of the algae (it arose several times in the green algae).
5. There have been numerous adaptations among vascular plants for life on land. The degree to which these adaptations have evolved enables us to distinguish major groups.
6. There have been characteristic evolutionary changes in life cycles and methods of reproduction among many groups of green plants. In general the majority of simple green plants are haploid for most of their life cycles, whereas the majority of complex green plants are diploid for most of their life cycles. This evolutionary tendency is evident among several divisions of algae and among the vascular plants. Since we shall encounter these differences in life cycles and reproduction several times in the following pages, it is important to understand at the outset some basic features of plant reproduction.

THE "DOUBLE LIFE" OF A PLANT— ALTERNATION OF GENERATIONS

In Chapters 7 and 15 when we discussed the development and reproduction of organisms, we learned that the life cycle of almost all plants and animals consists of a phase in which the cells of the organisms are haploid and a phase in which the cells are diploid. Haploid cells have one member of each pair of homologous chromosomes, whereas diploid cells have both members of the pair. Figure 23-2 diagrams the principal types of life cycles found in both animals and plants and indicates the proportion of the life cycle spent in the haploid and the diploid states.

In animals (Fig. 23-2a) the haploid phase consists of the gametes, the egg, and the sperm. The egg and the sperm fuse to form the diploid zygote —the fertilized egg cell. In almost all multicellular animals this initial cell, the zygote, divides mitotically to produce an animal composed of diploid cells. Certain cells in this diploid animal then undergo meiosis and produce haploid gametes. Thus the life cycle is ready to begin again.

There are several important differences between the animal life cycle and the life cycles

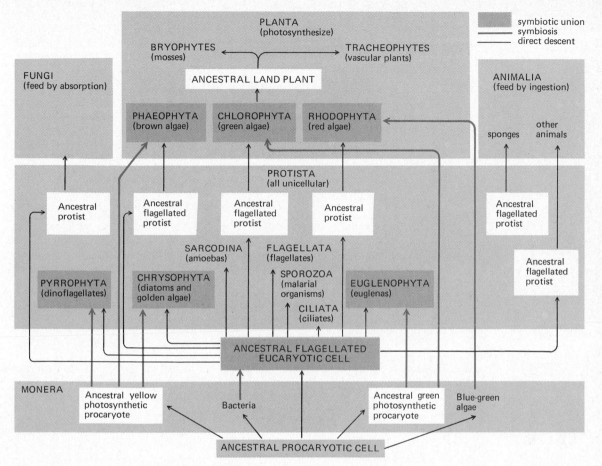

Figure 23-1 A possible phylogeny of the major divisions of plants showing their relations to other organisms. A black line indicates a *line of direct descent*. A brown line indicates the establishment of a *symbiotic relationship*. The founder species for all living organisms was a line of *ancestral procaryotic cells* which gave rise to the bacteria, the photosynthetic blue-green algae and two other groups of photosynthetic procaryotes called "yellow" and "green procaryotes." The three photosynthetic procaryotes differed in the chlorophylls and other pigments they possessed.

Subsequently two evolutionary processes occurred. One was evolution by mutation, recombination and natural selection and the second was evolution by symbiosis. The first symbiosis led to the formation of an *ancestral eucaryotic cell*. The first symbionts later became the mitochondria of this cell. The descendants of this first eucaryotic cell evolved further by mutation, recombination and natural selection and gave rise to various *protists*. Some of these protists formed symbiotic unions with a yellow photosynthetic procaryote and gave rise to three divisions of algae: the *brown algae*, the *dinoflagellates* and the *diatoms*. The ancestral yellow procaryotes evolved into the chloroplasts of these algae. Another ancestral protist formed a symbiotic union with a blue-green alga and gave rise to another division, the *red algae*. A euglena-like protist formed a symbiotic union with a green procaryote and gave rise to the photosynthetic *euglenas*. Finally another flagellated protist entered into a symbiotic association with a green procaryote and evolved into the *green algae*. These in turn gave rise to an ancestral land plant from which the vascular plants and mosses both evolved.

The yellow and green procaryotes which entered into these ancient symbioses no longer exist as free-living organisms. Blue-green algae and bacteria continue to have free-living descendants. The fungi probably arose from some ancient motile protist. Animals also arose from ancient protists. This scheme is based on one developed by Peter Raven.

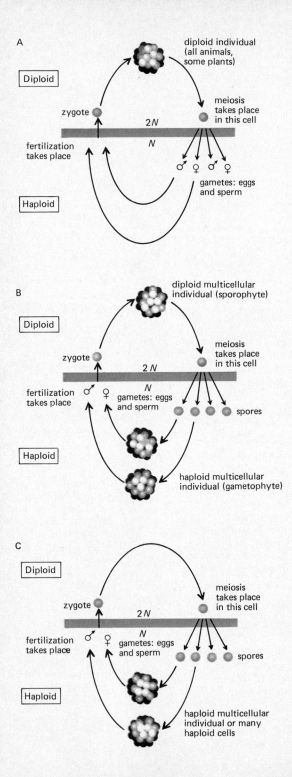

we find in most plants. For one thing, in most plants after meiosis the haploid cells usually do not transform immediately into gametes. Instead they divide mitotically to produce a haploid multicellular plant. Some of the cells in this haploid plant then differentiate into gametes. This gamete-producing haploid plant is called a **gametophyte.** The gametes formed by the gametophyte fuse to form a diploid zygote, as with animal cells. This zygote may develop in one of two ways. In some groups of plants (Fig. 23-2b) the zygote divides mitotically to produce a diploid, multicellular plant. However, this diploid multicellular plant, unlike the diploid animal, does not usually produce gametes. Instead certain cells of the diploid phase undergo meiosis to form haploid cells called spores. The haploid spores divide mitotically and grow to produce the haploid plant stage, the gametophyte, in which certain cells differentiate into gametes.

In other groups of plants (Fig. 23-2c) the diploid zygote instead of dividing mitotically undergoes meiosis immediately to form haploid spores. These spores then divide mitotically to form the haploid gametophyte phase of plant life.

Figure 23-2 Outlines of the main kinds of life cycles found in plants and animals. The *haploid phase* N. of the life cycle occurs below the horizontal bar. The *diploid phase* 2N. occurs above the bar. Each meiotic division yields four haploid products which are denoted by brown arrows. A. The diploid individual forms gametes by meiosis and these fuse to form a diploid *zygote*. The zygote divides to produce another diploid individual. Almost all animals, a few fungi and a few brown algae have such a life cycle. B. The diploid individual forms haploid cells called *spores* by meiosis. These spores undergo mitotic divisions to form multicellular individuals. These in turn produce cells which differentiate into gametes which fuse to form a zygote. The zygote develops into a diploid individual. Bryophytes, vascular plants and many algae have this kind of life cycle which is known as *alternation of generations.* C. The diploid individual is limited to the zygote. The zygote divides by meiosis to produce haploid cells called spores. These spores divide mitotically to produce more haploid cells or a multicellular haploid individual. Eventually some of the haploid cells differentiate into gametes. *Chlamydomonas, Ulothrix* and *Oedogonium,* as well as a number of other kinds of algae, have a similar sort of life cycle.

It is important to recognize the basic difference between a gamete and a spore. A **gamete** is a haploid cell that undergoes fusion to form a zygote. A **spore** is a haploid cell that undergoes mitosis. Sometimes the daughter cells remain attached to form a *multicellular* haploid gametophyte. In other cases each of the daughter cells becomes a *unicellular* haploid gametophyte. Eventually some of the haploid cells in the gametophyte differentiate into gametes.

In the two examples diagrammed in Figure 23-2b and c, the diploid phase of the plant produces spores. For this reason the diploid phase is called the **sporophyte,** a plant that produces spores. There is no sporophyte stage in the animal world. Diploid animal cells produce haploid gametes directly, not haploid spores.

Although the development of a multicellular haploid stage is rare in animals, it is almost universal among multicellular green plants. A green plant leads a "double life." Its life cycle alternates between the haploid, gamete-producing stage and a diploid, spore-producing stage. This shifting of phases is known as the **alternation of generations** or *alternation of phases.* Since both the fusion of gametes, which occurs in the haploid gametophyte generation, and meiosis, which occurs in the diploid sporophyte generation, are essential features of sexual reproduction, both gametophyte and sporophyte are sexual generations.

In most simple algae, such as the green algae, the haploid gametophyte is the prominent phase of the life cycle. It is the plant one sees, whereas the diploid phase of the life cycle is limited to the single-celled zygote. In more complex algae such as many seaweeds, the diploid sporophyte also becomes very prominent and and in some cases the haploid gametophyte is much reduced. The same tendency is seen in the vascular plants in which the diploid sporophyte is the prominent stage. Although the gametophyte in these plants is multicellular, it is usually microscopic. Accordingly, when we look at an elm tree, a corn plant, or a fern we are seeing the diploid sporophyte phase of the plant's life cycle.

When reading further about plants, the terms *thallophytes* and *embryophytes* are likely to appear. Many botanists combine the algae and the fungi in a group called the Thallophyta, and the mosses and their allies and the vascular plants in a group called the Embryophyta. These groups include divisions which are not closely related phylogenetically, but the terms are often used for convenience to distinguish simple plants from more complex plants. An outline of the classification of plants is given in the Appendix.

ALGAE—THE GRASS OF MANY WATERS

Algae are extremely diverse and plentiful organisms—more than 20,000 species have been described. They flourish in a wide range of habitats all over the world, but are principally found in water. Some are microscopic and unicellular, whereas others, such as giant kelps, are more than a 100 feet long, consist of billions of cells, and weigh hundreds of pounds. Except for these large forms, the body of a multicellular alga is quite simple compared to that of a vascular plant and they do not have the array of different kinds of tissues seen in vascular plants. Most multicellular algae consist of simple filaments or plates of cells. Also, their reproductive structures are almost always single cells, not multicellular structures like those found in mosses or vascular plants.

Cell division in algae differs from cell division in mosses and vascular plants. When the cells of most algae divide, the cell membranes usually pinch inward from the edge of the cell. The same process occurs in animals, protozoans and fungi. In contrast, when the cells of mosses and vascular plants divide, they synthesize a cellulose wall which grows outward from the interior of the cell to the margin.

Virtually all algae are photosynthetic and they play a major ecological role as the primary producers of most of the organic matter in oceans and lakes and all of the oxygen. They are the world's largest crop and produce at least five times as much living material each year as all of the land vegetation put together, including cultivated plants. They provide food, either directly or indirectly, for all aquatic animal life.

Recently scientists have been devoting more and more time to their study, because algae seem to be a major target of the pollution of fresh water. They are key indicators of chemical and thermal pollution; when the chemistry and temperature of the water change, the numbers and kinds of algae change. They can provide us with a warning that an environmental problem is about to appear. One such problem is the release of organic wastes into rivers and lakes, which provide nutrition for algae which then grow rapidly. During the day when photosynthesis takes place, oxygen is produced and carbon dioxide is used up from the water, and both animals and plants thrive. At night, however, when photosynthesis stops, the algae use up the oxygen in the water to such a degree that fish and other organisms die.

The algae are classified in six divisions, principally on the basis of the pigments they possess. Table 23-1 summarizes some of their characteristics. Like all photosynthetic organisms except bacteria, they contain chlorophyll a. Each division, however, has additional pigments which are characteristic of it alone. Each division also has characteristic chemicals in its cell walls (four of them contain cellulose) and characteristic food storage products. These biochemical features along with morphological features, such as the presence or absence of flagella, the structure and arrangement of flagella, and the structure of their chloroplasts, provide the principal basis for their classification. Without biochemical markers it would be difficult to classify many of the algae for there has been considerable parallel evolution. Most of the divisions, for instance, have independently evolved complex forms of sexual reproduction, and several have evolved complex branching patterns and leaflike structures.

Three of the six divisions are entirely unicellular, two include both unicellular and multicellular forms, and one is exclusively multicellular. As we have noted, the six divisions are probably not related and evolved independently by symbiosis with various photosynthetic procaryotes.

Green algae—Division Chlorophyta

Anyone knows what **green algae** look like if he has ever noticed greenish scum on a freshwater pond or found "sea lettuce" on an ocean beach.

The "moss" on a tree bark is usually a green alga. So is the giant unicellular alga *Acetabularia*, discussed in Chapter 8, which has been used extensively for nuclear transplantation studies. These are just four representatives of the 7000 species of the division *Chlorophyta*. They are the largest and most diverse of the algae. Green algae may be free-living single cells, colonies of similar cells which show little specialization, or true multicellular organisms in which groups of cells take on specific vegetative or reproductive roles. They are commonly found in fresh water, but marine species are abundant. They also occur widely in the soil, actively growing after rainfalls. A number live symbiotically among the tissues or within cells of other organisms.

The green algae share three significant features with all of the higher green plants, which suggest that they probably gave rise at various times to the major groups of green land plants.

1. They have cell walls containing cellulose.
2. Their food is stored in the form of starch.
3. They contain, in addition to chlorophyll a, a closely related compound, chlorophyll b.

None of the other algae share all of these features. The green algae also show evolutionary tendencies that resemble some of the evolutionary steps that were probably taken by the ancient ancestors of modern land plants. Indeed most botanists believe that among the green algae lie clues to several major steps in the evolution of higher plants, particularly the origins of *sexual reproduction* and of *multicellularity*. Some of these clues are provided by three representative green algae we now consider (Fig. 23-4).

Chlamydomonas is a genus of modern green algae, the species of which closely resemble unicellular flagellate protozoans.

Ulothrix is a genus of threadlike filamentous algae that shows some signs of multicellularity and specialization.

Oedogonium is a genus of filamentous algae with advanced sexual reproduction.

Chlamydomonas is a common single-celled inhabitant of stagnant freshwater pools, about 25 microns long. It generally has two flagella, a rigid cellulose cell wall, and a nucleus lying in

Table 23-1 Some Characteristics of the Six Divisons of Eucaryotic Algae

Phylum	Major pigments	Major food storage product	Cell wall component	Flagella[1]	Organization	Habitat
Chlorophyta (green algae)	Chlorophyll *a* and *b*, carotenoids	Starch	Cellulose	Apical, 2 or more, equal, whiplash	Unicellular and multicellular	Freshwater and marine
Euglenophyta (euglenas)	*a*, *b*, carotenoids	Paramylon (a polysaccharide)	No cell wall, pellicle made of protein	Apical, 1 to 3, unequal, tinsel	Unicellular	Mostly freshwater
Pyrrophyta (dinoflagellates)	*a*, *c*, carotenoids	Starch	Cellulose	Lateral, 2, unequal, whiplash or tinsel	Unicellular	Freshwater and marine
Chrysophyta (diatoms and golden algae)	*a*, many have *c*, carotenoids, fucoxanthin	Leucosin (a glucose polymer, similar to starch); oil	Pectin plus silica	Apical, 1 or 2, equal or unequal, whiplash or tinsel	Unicellular	Mostly marine
Phaeophyta (brown algae)	*a*, *c*, carotenoids, fucoxanthin	Laminarin (a glucose polymer, similar to starch)	Cellulose plus polysaccharides called algins	Lateral, 2, unequal, tinsel; (in reproductive cells only)	Multicellular	Mostly marine
Rhodophyta (red algae)	*a*, carotenoids, phycobilins	Floridean starch (a glucose polymer, similar to starch)	Cellulose plus pectin	None	Unicellular and multicellular	Mostly marine, some freshwater

[1]There appear to be two sorts of flagella. When the outer membrane of the flagellum is smooth it is referred to as a *whiplash* flagellum. If it has thin hairs along its length, it is referred to as a *tinsel* flagellum.

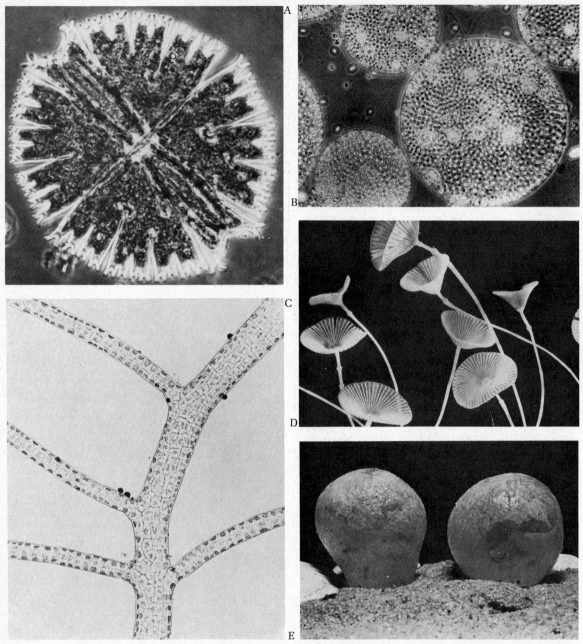

Figure 23-3 Green algae. These photographs give some idea of the diversity of form of green algae. These include freshwater forms like (A) the unicellular *Micrasterias* and spherical colonial forms such as (B) *Volvox.* Marine green algae include (C) the filamentous *Enteromorpha* as well as (D) giant unicellular forms such as *Acetabularia* which have been used extensively in studies of nuclear cytoplasmic interactions. Another impressive marine green algae is (E) *Valonia.* Each *Valonia,* which may be as large as a hen's egg, is a single multinucleate cell—among the largest cells known.

the depression of a single, large cup-shaped chloroplast. It may resemble the ancestral organisms from which all green land plants arose.

Chlamydomonas spends most of its life in the haploid phase and reproduces asexually.[1] It also has one of the most primitive kinds of sexual reproduction. Both its sexual and asexual cycles begin the same way, with mitotic division of the haploid parent cell. When reproducing sexually, the daughter cells are liberated and function as gametes. Two of them fuse to form a diploid zygote. In most species of *Chlamydomonas*, the zygote loses its flagella, sinks to the bottom, and secretes a protective wall which enables it to survive unfavorable conditions, such as cold or drying up of the pond. As soon as environmental conditions again become favorable, the zygote is activated and divides by meiosis to produce four haploid cells, each of which quickly matures, develops flagella, and becomes a haploid unicellular *Chlamydomonas*.

When reproducing asexually, the daughter cells which are liberated simply swim away and grow into new individuals. There seems to be little morphological difference between cells that swim away and those that act as gametes. Nor is there any detectable difference between the gametes; we cannot yet call one a sperm and the other an egg, for they are morphologically identical. This lack of specialization of gametes into morphologically distinct cells is probably a primitive condition. We may be observing in *Chlamydomonas* a process akin to that which marked the origin of sexual reproduction in plants. It is also noteworthy that the diploid phase of its life cycle is brief and limited to the zygote, whereas the haploid phase is the prominent phase. This also seems to be like the ancestral condition.

Ulothrix is also a simple alga which consists of unbranched filaments of small haploid cells and is found mainly in fresh water. The fila-

[1] The term *vegetative reproduction* is often used by botanists to denote *asexual reproduction* in plants.

Figure 23-4 (Right) Life cycles of three species of green algae: A. *Chlamydomonas*, B. *Ulothrix* and C. *Oedogonium*.

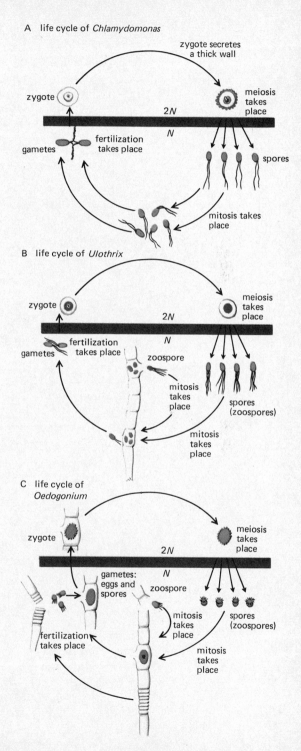

ment is commonly attached to a rock by a specialized cell called a *holdfast*. This makes *Ulothrix* more than a colony of similar cells. It is multicellular and shows cellular specialization.

Ulothrix has two methods of asexual reproduction. Fragments of the filament can break off and thereby form new plants. But the cells (with the exception of the holdfast) can also divide mitotically to produce flagellated spores called *zoospores*, each with four flagella, which give rise to an entire plant. In addition, *Ulothrix* can produce gametes and can reproduce sexually. These resemble zoospores, but are smaller and have but two flagella. The morphological differences between zoospores and gametes represent an evolutionary advance. The two gametes are, however, identical as in *Chlamydomonas*.

The third green alga we examine is *Oedogonium*. Species of *Oedogonium* are widely distributed and grow as unbranched filaments which are anchored by a holdfast when young but may float free in the water in large masses when mature. *Oedogonium* shows two significant differences from *Ulothrix* in sexual reproduction. First, the gametes are no longer identical. There is a large egg and there are small free-swimming sperm. Second, the egg and the sperm are produced by specialized cells in the filament. *Oedogonium*, like *Ulothrix*, can also reproduce asexually by fragmentation and through zoospores.

Chlamydomonas and Ulothrix illustrate the most primitive method of sexual reproduction. These plants release similar gametes which then fuse in the water. In *Oedogonium* there are two new features that were necessary steps in the evolution which led to the reproductive processes now observed in higher plants — male and female gametes and cells specialized to produce male and female gametes.

This does not mean that these three modern green algae are ancestors of more advanced plants. However, their life cycles very likely illustrate stages that took place as higher plants evolved from green algae. Accordingly we conclude that these plants which are living today probably illustrate the kinds of events that occurred in the evolution of both multicellularity and sexuality. These steps also occurred in other algal divisions through parallel evolution.

Love thy neighbor — Algal symbioses

One noteworthy feature of several sorts of green algae is the ease with which they form symbiotic relationships of various kinds with other organisms. Many fungi have green algae growing closely with them. The symbiotic organism composed of the alga and the fungus is called a *lichen*, which we discuss later in connection with fungi. In addition many green algae actually live inside the cells of other organisms, usually within a vacuole. More than a hundred different genera of invertebrates in eight different phyla have green algal symbionts growing within their cells. A familiar example is the green hydra, *Chlorohydra viridissima*, which has a species of the green alga *Chlorella* growing within its cells. The algae secrete a sugar which the hydra uses for nourishment. In some animals the algae establish an even more intimate relationship with their host cells. Inside certain cells of mollusks called sea slugs are found the chloroplasts of certain algae which the slug has eaten, still "alive" and photosynthesizing (Fig. 23-5).

Plentiful food — Divisions Chrysophyta and Pyrrophyta

The most well-known members of the division *Chrysophyta* are the **diatoms** found in both oceans and fresh water. By weight they are among the most abundant organisms in the oceans. In some locations 2 million diatoms can be found in a single gallon of seawater. They are one of the primary producers of organic matter and oxygen in the sea and are key organisms in the major food webs of the ocean. They are truly the grasses of oceans and lakes on which animals graze.

Most diatoms exist as plankton, floating here and there with the tides and currents. Some attach themselves to plants and other objects. For instance, colonies of diatoms grow on the underside of the blue whale. Their yellow color is so noticeable that sometimes the animal is called the sulfur bottom whale.

Diatoms may be unicellular or colonial and are all microscopic. Each is enclosed in a cell wall of two halves that fit together like a box and its lid (Fig. 23-6). These walls consist almost entirely of glass-like silica. Each of the 10,000 species described so far produces silica walls or shells with a specific pattern. Diatoms store their food in a special carbohydrate reserve that differs from starch or glycogen, and also as oil droplets. Unlike almost all other algae their vegetative cells are ordinarily diploid. The zygote differentiates and functions as a diploid unicellular plant. Contrast this with *Chlamydomonas* in which the zygote remains diploid for only a short period and immediately undergoes meiosis.

The fossil record of the diatoms shows that they were present in the Mesozoic era. The silica shells of some of these fossil diatoms appear to be identical in size and pattern with the shells of some living diatoms. This suggests that some species have not changed during the last 200 million years, which indicates how well adapted diatoms have been to their environments. When the diatoms die, their hard shells settle to the bottom of the ocean to form vast deposits that may be more than 1000 feet thick. These ancient deposits are now mined extensively and *diatomaceous earth* is used to filter impurities from gasoline, in heat and sound insulation, and in many other ways. One cubic inch of diatomaceous earth may contain 75 million individuals.

Some scientists believe that the dead remains of diatoms and other marine plankton gave rise to natural gas and petroleum.

Another group of algae in the division Chrysophyta is the **golden algae.** They occur as single cells, in filaments or in colonies. Until recently, golden algae were thought to be mainly freshwater forms and attracted little attention. Within the last few years, however, marine biologists have discovered that the oceanic plankton is filled with them. They had been overlooked for years because they were so small that they passed through the fine nets used to collect plankton. By weight, the golden algae may be the most abundant organisms in the oceans and the major food producers in the world!

As runners-up to the golden algae and the diatoms the largest producers of organic matter in the oceans are the **dinoflagellates.** These single-celled algae belong to the division *Pyrrophyta*. They are chiefly marine, but some species occur in fresh water. The surface of each dinoflagellate has two characteristic grooves set at right angles to each other (Fig. 23-7). One flagellum extends from each groove. The entire cell is often armored with cellulose plates that are completely enclosed by the cell membrane.

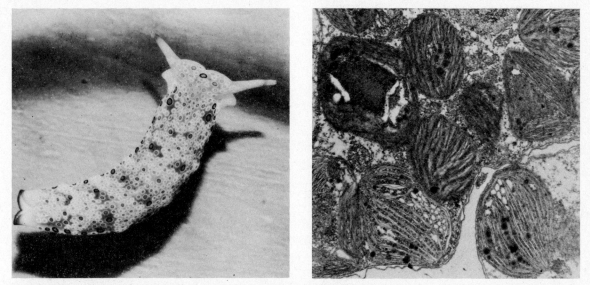

Figure 23-5 Algal symbioses. The photograph (left) shows a sea slug, *Placobranchus ianthobapsus*, in whose cells are found living chloroplasts of a green alga. The slug is about an inch long. The electron micrograph (right) shows the chloroplasts of the alga in the digestive cells of the slug.

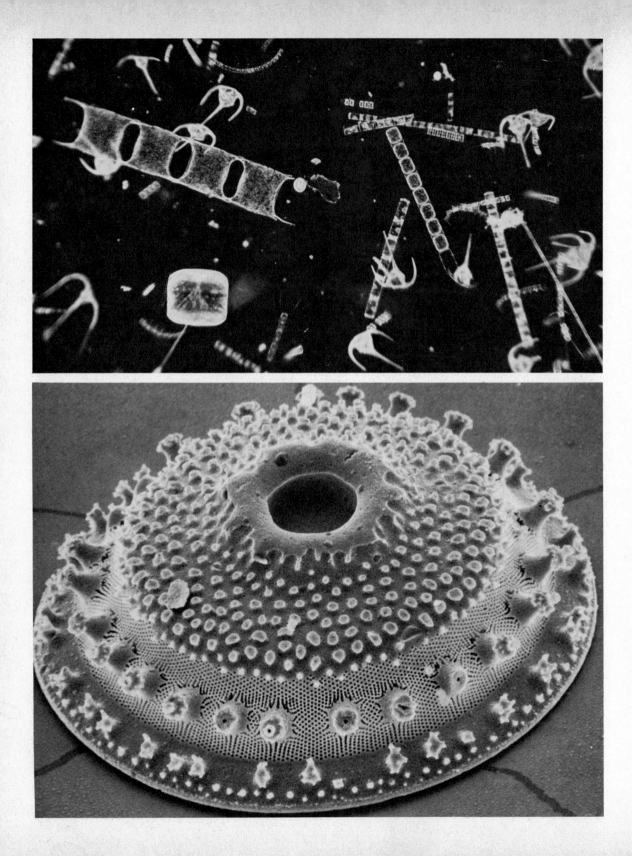

Figure 23-6 (Opposite) Diatoms are unicellular algae which form a shell consisting of two halves or valves. Each of the various species illustrated in this photograph of living phytoplankton (upper left) has its own characteristic form and species-specific pattern of perforations in the shell. This is seen clearly in the scanning electron micrograph (lower left). These patterns are so precise and symmetrical that lens makers used them to test for resolution, astigmatism and other imperfections. The drawing (right) illustrates how diatoms reproduce by cell division. The two valves of a diatom shell fit together like a "pill box" with one valve forming an overlapping "lid" and the other forming the "bottom." When the diatom divides, the old valve always forms the top lid of the box and the new valve fits inside it. As a result the cell receiving the upper valve is the same size as the parent, whereas the daughter cell is smaller. In some forms the shells are expandable and grow to full size. In others the continued decrease in size stops when sexual reproduction occurs.

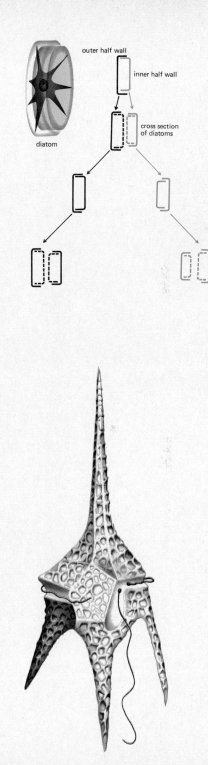

Figure 23-7 (Right) *Ceratium* has armor plates with the two deep grooves, each containing a flagellum, possessed by all dinoflagellates. A short flagellum lies in the horizontal groove and a longer one trails along the vertical groove. The anchor-shaped organisms in 23-6 are *Ceratium*. Dinoflagellates like *Ceratium* have both procaryotic and eucaryotic features. Their chromosomes lack histones and occur in bead-like chains. However, they have a nuclear envelope.

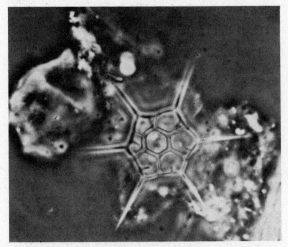

Figure 23-8 The golden alga *Distephanus*. Golden algae have silica plates or scales which are homologous to the valves of diatoms.

Because the dinoflagellates possess both plant-like characteristics, such as a cellulose wall and chlorophyll, and animal-like characteristics, such as motility, they are claimed by both botanists and zoologists. They reproduce by a unique type of mitosis that appears to combine features of both procaryotes and eucaryotes. It has been suggested that the dinoflagellates are a kind of missing link that illustrates the evolution from the procaryote to the eucaryote nucleus.

Dinoflagellates are ordinarily inconspicuous, but occasionally they put on a great display. One dinoflagellate called *Noctiluca* is about 2 millimeters long and is *bioluminescent,* converting chemical energy into flashes of light. The luminescent wakes of ships at night and the underwater fireworks seen by skin divers are caused by *Noctiluca* and its relatives. *Gonyaulax* and some other dinoflagellates which contain a red pigment may become so plentiful that they turn the ocean red, producing the "red tides" sometimes reported along the seashore. A toxin released by these dinoflagellates kills fish by poisoning their nervous system. It also acts on other vertebrates including man. A mere 0.2 gram can kill a million mice in 15 minutes. Oysters and clams become poisonous after feeding on species of *Gonyaulax* that are most abundant in the "r-less" months (May, June, July, August).

Dinoflagellates also play an important geological role. Certain species inhabit the tissues of reef-building corals and appear to accelerate the growth of the reef. Other dinoflagellates live as symbionts inside the cells of protozoans, sponges, flatworms, and mollusks.

Another division of algae, the *Euglenophyta,* includes unicellular organisms, the *euglenas,* which also show plant-like and animal-like characteristics. Unlike other algae they do not have a cell wall. Instead, the cell is enveloped by a cell membrane inside of which are a series of flexible strips composed of proteins. This proteinaceous *pellicle,* as it is called, also occurs in flagellate protozoans (see Chapter 24). Euglenas also have an organ called a *contractile vacuole* which occurs in most protozoans. Indeed euglenas look like protozoans with chloroplasts. This is probably what they are: flagellate protozoans which formed a symbiotic relationship with the same group of green procaryotes that formed the chloroplasts of the green algae.

Seaweeds—Divisions Phaeophyta and Rhodophyta

Most of the seaweeds encountered on the ocean shore are brown algae and red algae. The **brown algae** comprise the division *Phaeophyta.* Of the 1500 known species almost all are marine and all are multicellular. Common members of the division include various species of *Fucus,* also known as rockweed, which are found all over the world. Another common group are the *kelps* that grow along both the east and west coasts of North America. These are the largest of all algae, with stalks more than 100 feet long bearing broad 50-foot fronds. In some places kelps are so dense that they form a true underwater jungle. Most brown algae are attached to the bottom, but a few live as free-floating plants. One species of *Sargassum* floats free in the Sargasso Sea, covering some 2½ million square miles of ocean between the West Indies and North Africa. This floating forest of brown algae shelters a rich and varied marine life. Stories of ships being trapped in these masses of seaweeds are mythical, but sailors avoided the Sargasso Sea because of its dead calms.

The color of brown algae comes from a characteristic pigment called *fucoxanthin* which is present along with chlorophyll *a* and another chlorophyll called chlorophyll *c.* Brown algae reproduce either asexually or sexually and usually have a multicellular haploid gametophyte generation which alternates with a multicellular diploid sporophyte generation. In a few species such as *Fucus* the haploid gametophyte is reduced to the point that the only haploid cells in the entire life cycle are the gametes themselves (Fig. 23-10). In this sense *Fucus* has a life cycle just like that of animals but a very peculiar one for a plant. (The early embryology of *Fucus* was discussed in Chapter 8).

Many tissues of some of the large brown algae such as kelps are much more specialized than the tissues of other algae. For example, kelps have vascular tissue which conducts sugars from the upper parts of the plant that are

Figure 23-9 Brown algae. A. Sea palms (*Postelsia palmaeformis*) off the California coast thrive where the surf beats hard. When struck by the waves, the elastic stalk is bent almost horizontal but regains its upright position as soon as the water recedes. B. A large kelp (*Laminaria saccharina*) found off the coast of Maine. C. Rockweed (*Fucus vesiculosus*) forms a dense mat on many rocky shores. At low tide it is exposed to the air and sun but it is fairly drought-resistant. The blades contain air-filled bladders which cause them to move upward where there is more light.

well illuminated to the poorly illuminated parts far below the surface of the sea. They also have a meristematic region similar to the cambium of higher vascular plants. Apparently by convergent evolution the kelps evolved some structures similar to those in vascular plants.

Kelps are of some economic importance because they concentrate many minerals from seawater. In some parts of the world they are harvested as sources of minerals for fertilizer, or as a food for farm animals or man. In addition, kelps are extracted to obtain *algin*, a chemical present in their cell walls which is widely used to impart smoothness to ice cream and other foods.

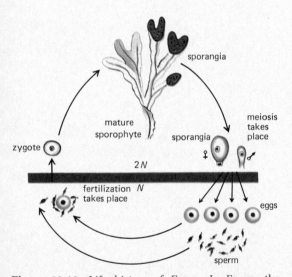

Figure 23-10 Life history of *Fucus*. In *Fucus*, the zygote develops into a conspicuous diploid multicellular plant. At the tips of the fronds are swollen hollow chambers which contain male or female *sporangia*. Meiosis occurs in these sporangia and the male sporangia give rise to sperm and the female sporangia to eggs. The gametes are released into the water where they fuse to form a zygote. Gamete release appears to be correlated with the tides. Unlike other plant life cycles, there is no gametophyte stage except for the gametes themselves. Incidentally, the common species of *Fucus* found on the Atlantic coast is *dioecious*—that is, plants are either male or female. However, the Pacific coast species are *monoecious* and both eggs and sperm are produced by a single plant.

The **red algae** belong to the division *Rhodophyta*. All but a few of the 4000 known species are multicellular and practically all of them are marine. Red algae are usually small; only a few grow longer than 3 feet. They are attached to the bottom by holdfasts and none are free floating. In addition to the chlorophyll *a*, red algae contain several accessory pigments, including red pigments called *phycobilins* which make many of them red or pink.

Many red algae are found in shallow water, but some live at depths far greater than any other algae, more than 600 feet. The accessory pigments they contain enable them to do this. At the depths at which these red algae grow, the wavelengths of light which are absorbed by chlorophyll *a* do not penetrate. However, blue and green light do penetrate and these are absorbed by the accessory pigments the algae contain. These pigments then transfer the energy to chlorophyll *a* for photosynthesis. By this means red algae can occupy an environment from which most other plants are excluded.

In addition to their characteristic pigments, phycobilins, red algae have several other novel features which distinguish them from all other algae. They have simple chloroplasts and no flagellated cells at any stage in their life history. Sexual reproduction occurs, but the process is unusual: although the gametes differentiate into eggs and sperm, the sperm have no flagella and are carried to the female by water currents. Interestingly both phycobilins and simple chloroplasts are also found in blue-green algae. Also, blue-green algae have the same kind of storage carbohydrate and cell walls as do red algae and also lack flagella. These observations support the idea mentioned earlier that the red algae originated from the symbiotic union between some procaryotic blue-green alga and some primordial eucaryotic protist.

Many red algae extract calcium from seawater and deposit it as calcium carbonate, or limestone, in or around their cell walls. Numerous *coral reefs* might more appropriately be called *algal reefs*, for the red algae have often contributed more to reef formation than the coral animals. They have been active reef builders in the past as well as in present times.

Red algae are major sources of food in the Far East. One of them, *Porphyra* ("nori"), has been cultivated for hundreds of years. In Japan alone, there are nearly a hundred thousand "nori fisherman". However, neither the red algae nor the brown algae have much nutritive value, because humans and most other animals do not have the enzymes to digest the carbohydrates found in seaweeds. Red algae are also the source of *agar*, which is a dried gelatinous extract obtained from various red algae. Agar is an essential ingredient of media upon which bacteria and fungi are grown in laboratories and no one has yet found a suitable substitute for it.

Figure 23-11   Red algae. A. *Ceramium rubrum* is a common red alga. B. *Porolithon crasspedium* is a typical atoll reef alga of the tropical Pacific. C. *Chondrus crispus* or Irish moss is an important source of agar. D. The electron micrograph is of a one-celled red alga, *Porphyridium cruentum*. Its single large chloroplast with parallel lamellae closely resembles the simple chloroplasts of blue-green algae and provides evidence for the symbiotic origin of red algae from a union of a protist with a blue-green alga.

FUNGI—THE FILAMENTOUS WAY OF LIFE

Fungi are an extremely large group of organisms that are of great ecological and practical importance. They are as different from green plants as they are from animals and most biologists place them in their own kingdom. Together with bacteria they are the major agents of decay which render the remains of dead organisms available to green plants and return essential materials to food chains. The top few inches of an acre of fertile soil may contain tons of fungi and bacteria.

Fungi are tremendously diverse, numbering on the order of 100,000 species. More than 1000 new species are described each year, but probably at least a half million are yet to be described. Many of them are familiar. They include complex multicellular forms, such as mushrooms, the mold on decaying bread and other food, mildew on rose plants, dry-rot in the wooden beams of a house, and the organisms that cause athlete's foot.

All fungi except those which are single-celled are composed of branching tubular filaments called **hyphae,** a mass of which is called a **mycelium.** In some fungi the hyphae have no crosswalls and contain many nuclei. In other fungi the hyphae have crosswalls called **septae** which divide the filaments. These walls are perforated and cytoplasm streams freely along the hyphae. Thus although fungi themselves do not move about, the protoplasm inside their hyphae flows. Usually the hyphae have walls composed mostly of *chitin* instead of cellulose. Chitin, as may be recalled, is a molecule also found in the exoskeleton of arthropods. The hyphae may be packed together to form a complex three-dimensional plant body like a mushroom. However, there is little differentiation of any tissues of the body except the reproductive tissues.

Growth takes place at the tips of the hyphae and is usually extremely rapid. Within a few hours a fungus colony may produce hundreds of feet of new hyphae. This rapid growth is possible because the entire hypha is able to synthesize proteins, all of which stream into the growing tip. This rapid growth of the hyphae is a key part of the way fungi obtain food. Fungi have no chlorophyll and do not manufacture their own food. Instead they live embedded in their food supply through which they send their hyphal filaments. By means of these hyphae, fungi feed on dead, decaying, or nonliving organic matter, or they invade the tissues or cells of living organ-

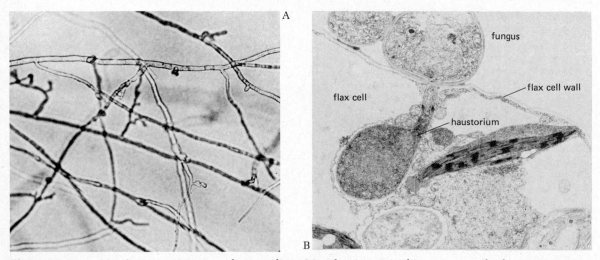

Figure 23-12   A. Mycelium growing on a culture medium. Most fungi grow in this manner on dead organic matter. B. Some fungi, like the parasitic fungus *Melampsora lini,* attack living cells. In this case a specialized fungal hypha called a haustorium is growing into a cell of a flax leaf. The fungal hypha breaks through the cell wall of the flax cell but does not rupture the plasma membrane.

isms and live as parasites. In the process of feeding, most fungal hyphae secrete digestive enzymes onto their food to digest it and absorb the products of this *extracellular digestion.* Fungi have evolved a remarkable number of different mechanisms which enable them to exploit various food sources in their environment. In the case of some parasites the hyphae penetrate into the cells of the organisms upon which they feed.

Fungi cause many diseases—in man, in other animals, and in plants. These range from ringworm to wheat rusts. Farmers, nurserymen, and lumbermen in the United States pay hundreds of millions of dollars each year to combat fungi that attack plants. Fungi cause decay and other diseases which claim more than 20 billion board feet of lumber annually— "enough lumber to construct a wooden sidewalk a mile wide from New York to San Francisco." However, there are numerous beneficial fungi too, such as the yeasts used for brewing beer, baking bread and making cheeses, and the molds which produce the antibiotic penicillin (Chapter 9). It is worth noting that fungi are capable of making many sorts of exotic molecules besides penicillin and can carry out almost any conceivable chemical transformation. Some mushrooms produce polyacetylenes which are explosive. Others make compounds so energy-rich they give off light. These form fantastic patterns of color and designs on jungle floors. In the future it is to be expected that fungi will be used more and more in laboratories and in industry to perform complex chemical changes which are difficult for chemists but easy for fungi. Pharmaceutical companies continuously test chemicals produced by fungi in their search for new drugs.

Fungi reproduce by means of *spores* which germinate to produce a new mycelium. Most of them reproduce both sexually and asexually, and some have complex life cycles. The mechanism of sexual reproduction is an important feature by which we can distinguish groups of fungi. The fungi are placed in a single division, *Eumycota,* with five classes. *Oomycetes, Zygomycetes, Ascomycetes,* and *Basidiomycetes* each have characteristic hyphae and reproductive structures. The fifth class, the *Fungi Imperfecti,* includes more than 25,000 species in which sexual reproduction

either does not occur or has not been discovered (they are called imperfect for this reason). Most can be placed in one of the other classes once their life cycle is known.

Water molds and bread molds— Classes Oomycetes and Zygomycetes

Several classes of fungi superficially resemble algae. For many years they were classified together as *Phycomycetes,* which means "algal fungi," but it is now clear that they belong to distinct classes. All of them have *hyphae without crosswalls,* whereas in the other classes of fungi the hyphae do have crosswalls. They look very much like filaments of algae without chlorophyll.

The fungi which look most like algae, and are in fact aquatic, are the *water molds* in the class Oomycetes. Like many algae, water molds and other Oomycetes produce motile flagellated *zoospores.* Also, their cell walls are primarily composed of cellulose, whereas in most fungi the walls are mainly chitin. Water molds are easy to find. Place a dead insect or a small dead fish in a jar of pond water. Within a few days a white furry mass of hyphae will cover the material. They are common parasites of tropical fish.

Not all Oomycetes are aquatic. A notorious terrestrial representative is the *late blight fungus* which caused the potato famine in Ireland and the rest of Europe during 1845–1847. More than a million people starved in Ireland, and a third of the population emigrated to escape. Even today, despite potent fungicides, late blight of potato remains a serious disease and causes heavy economic loss. Yearly losses from all plant diseases in the United States exceed a half billion dollars.

Another class of fungi which superficially resemble algae are the Zygomycetes. They are all terrestrial and have developed numerous adaptations for life ashore. Instead of swimming zoospores these fungi (and all the other fungi as well) produce small, light, airborne spores. The air is filled with vast numbers of these fungal spores which remain alive for many months and can survive extreme temperature and drought. Each spore may give rise to a whole plant. This mass production ensures that at least a few will be carried by air currents or animals to places where they can germinate and obtain nourishment, whether this is a rotting log, a piece of bread, or even a sheet of paper. For this reason it is very difficult to maintain sterile conditions in hospitals, laboratories, or at home.

The best known member of the Zygomycetes is the black bread mold, *Rhizopus* (Fig. 23-13). This organism forms a cottony growth on bread, fruit, and other food. It can damage many fresh fruits and vegetables if they are not properly handled during storage or transportation to market. Its spores are so abundant and so easily spread that chemicals like calcium propionate are often added to bread dough to prevent its growth. *Rhizopus* resembles the majority of the other Zygomycetes in that it spends most of its life in the haploid condition. Sexual reproduction takes place between the hyphae of two different *mating strains* (usually designated + and −). When both mating strains are present, hormones are secreted causing short branches from hyphae of the two different strains to meet and their tips to fuse. One nucleus from each hypha functions as a gamete and the nuclei fuse to form a zygote.

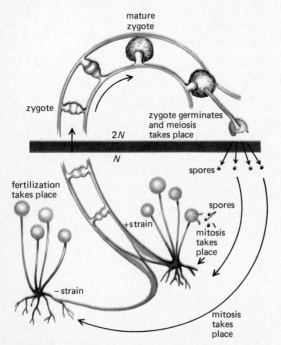

Figure 23-13 Life cycle of *Rhizopus*.

This undergoes meiosis to form haploid hyphae which produce spores. These spores grow into hyphae which produce more spores and the cycle continues. Thus the zygote is the only diploid phase in the entire life history.

Sac fungi—Class Ascomycetes

Sac fungi are very common and more than 30,000 species have been described. They cause Dutch Elm disease, chestnut blight, peach leaf curl, apple scab, and other damaging diseases in ornamental and food plants. But they also include many useful organisms. Yeasts used in brewing beer, fermenting wine, and baking bread are single-celled sac fungi. Delicious truffles and morels also are sac fungi.

A characteristic feature of all sac fungi is a saclike enlargement, called an **ascus** (plural, *asci*), at the end of each hypha which is formed during sexual reproduction. In most species the hyphae have crosswalls.

A typical sac fungus is haploid for most of its life cycle (Fig. 23-14). During sexual reproduction the hyphae of one strain, called a + strain, develop large female organs. Hyphae of another strain, called a − strain, develop smaller male organs. A finger-like tube grows from the female organ to the male organ. Nuclei from the male hypha migrate through this tube into the female organ and pair with the female nuclei, but do not fuse with them. The female organ ends up with pairs of male and female nuclei. This strange structure gives rise to hyphae in which each cell has two nuclei not fused but merely paired. Finally the male and female nuclei in the cell at the tip of each hypha—the ascus—fuse to form a diploid zygote. The zygote divides by meiosis and forms four haploid cells. Each of these then divides once by mitosis. Each ascus thus produces a total of eight haploid cells known as *ascospores*, each of which can grow into a new haploid plant. As was true for *Rhizopus*, the diploid phase is very brief and involves only a single cell. However, a new phase in which male and female nuclei remain paired but unfused has appeared. This phase is longer than the diploid phase, but is still only a small part of the entire life cycle. Although cells with paired nuclei of two different genotypes—*heterocaryons*—do not normally occur in animal cells (except during fertilization), as we described in Chapter 8, scientists have been able to produce them experimentally by fusing animal cells in tissue culture.

Sac fungi also reproduce asexually by special haploid spores called **conidia** which are produced in chains at the end of haploid hyphae. Each conidium grows into a haploid plant.

Recall that geneticists have used the sac fungus *Neurospora* extensively in fundamental investigations of the relationships between genes and enzymes. These studies led to the idea that genes code for specific proteins. *Neurospora* is especially suited for such work because it is haploid for most of its life. In a haploid organism the effects of a single gene can be observed, whereas they might be masked by an allele in a diploid organism.

Lichens—Collective security

Certain Ascomycetes, and occasionally other fungi, often develop a symbiotic arrangement with green or blue-green algae. This novel arrangement produces plants called *lichens*. Both partners in a lichen contribute to the welfare of the collective plant. The alga obtains water and minerals from the fungus; the fungus in turn draws upon the food manufactured by the photosynthetic alga. The algal partner can survive in some habitats without the fungus, but the fungus is usually completely dependent on the alga and is never found living independently in nature.

A lichen is a plant in which the whole is much more than just the sum of its parts. It differs in appearance and in physiology from its component alga and fungus. In most cases which have been studied, the fungal partner determines the form of the lichen. More than 17,000 different combinations of algae and fungi are known which form lichens. Since each combination reproduces itself consistently, it is called a species.

The most familiar lichens are the flat, scaly forms growing on rocks. These plants cause the stones on which they live to decompose. This is the first step in soil formation. Lichens are extremely hardy plants. They withstand drying and freezing extremely well. Reindeer "moss," for instance, withstands the drying chill of the

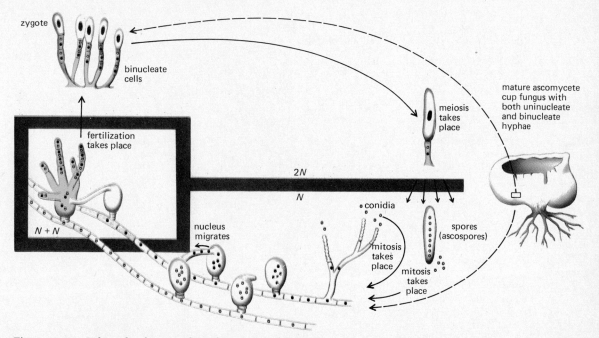

Figure 23-14 Life cycle of a typical sac fungus. For details see text.

Figure 23-15   Lichens. (Left) A foliose or leafy lichen (*Parmelia*) growing on a rock. (Right) Reindeer "moss" (*Cladonia rangiferina*), a fruticose or shrubby lichen that forms patches on the soil.

Arctic and provides food for caribou, reindeer and musk ox. When a lichen is dry, photosynthesis ceases. But when it is wetted by rain it behaves like a blotter and absorbs up to 35 times its own weight in water in a short time and starts to photosynthesize. Although lichens withstand harsh climates and frequent drying out, they are among the most sensitive organisms to atmospheric pollutants, particularly the sulfur dioxide in smog. For great distances around all great cities of the world there are "lichen deserts" in which the former lichen population has disappeared.

There are a number of other intimate associations between fungi and other organisms. We have already discussed (Chapter 10) the thin filaments of fungi known as *mycorrhiza* which increase the water-absorbing capacity of the roots of many trees and crop plants. Another example is found in orchid embryos which do not develop in nature unless certain fungal filaments invade their cells. The fungus provides sugar and controls the acidity of the soil during germination of the orchid seeds.

Club fungi—Class Basidiomycetes

Edible mushrooms, poisonous toadstools, and bracket fungi on the bark of trees are members of an extraordinarily widespread and numerous class of fungi, the club fungi. Equally plentiful are the rusts and smuts that attack valuable grain crops, such as barley, corn, oats, and wheat. Altogether more than 25,000 species of club fungi have been described.

Figure 23-16 Club fungi. A. The field mushroom, *Agaricus*, one of the common edible mushrooms. The gills which bear spores fan out under the cap of the mushroom. B. The destroying angel, *Amanita phalloides*. This species and other members of the genus *Amanita* are responsible for about 90 percent of the mushroom poisoning in North America. In the case of *Amanita phalloides* the symptoms of poisoning appear 6-15 hours after eating when the toxins have been well absorbed. As a result as many as half of those eating the fungus die. Other species of *Amanita* may produce symptoms sooner. The panther fungus *Amanita pantherina* which is the most poisonous fungus in the Pacific northwest may cause symptoms in 15 minutes to 3 hours. Some Amanitas are edible but experts suggest that they all be avoided to eliminate any chance of wrong identification. C. Species of wood-rotting bracket fungi on a dead tree. These fungi usually form overlapping or fan-shaped shelves or brackets on dead trunks or limbs of trees. However, they may invade the trunks of living trees and turn the wood into a charcoal-like mass. D. Puffballs like this one are familiar woodland fungi. The group includes *Calvatia maxima*, one of the largest of all fungi, individuals of which may weigh as much as ten pounds.

Club fungi get their name from a club-shaped cell found at the tip of some filaments. They resemble sac fungi in having crosswalls in their hyphae. The structure and reproduction of a typical club fungus is shown in Figure 23-17. Like the sac fungi they have no flagellate sperm. But unlike all other fungi both the haploid and diploid phases are brief. During most of its life each cell in a club fungus contains two nuclei, one male and one female, but they are not fused, simply paired. The fleshy parts of mushrooms are made up of masses of filaments, composed of cells with such paired nuclei.

We have already met cells with paired nuclei in the life cycle of the sac fungi. The advantages of paired nuclei to the fungi are similar to those of the diploid condition. It permits the storage of more genetic information and it enables genetically different nuclei to interact.

Eventually the paired nuclei in the club-shaped cells fuse to form zygotes. Each zygote undergoes meiosis and produces four haploid

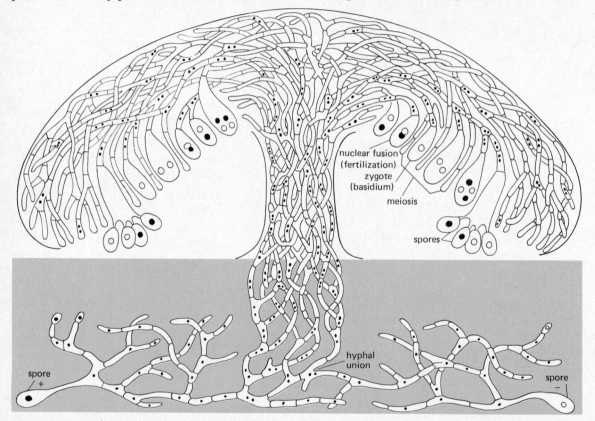

Figure 23-17 This famous diagram by L. W. Sharp illustrates the structure of a typical club fungus, a mushroom, and also outlines their basic reproductive processes. Let us start underground with uninucleate haploid hyphae of two different strains, a + or male strain and a − or female strain. These hyphae fuse to give rise to a hypha in which each cell contains one male and one female nucleus. This binucleate hypha grows rapidly and gives rise to the part of the mushroom which grows above ground. The stalk and the cap are composed of these binucleate hyphae packed together. These paired nuclei fuse in certain cells on the lower surface of the cap, each of which forms a zygote called a *basidium* (plural, *basidia*). Meiosis occurs in these basidia and haploid spores are produced which give rise to uninucleate haploid hyphae and the cycle begins again. If you compare this figure with Figure 23-14 you will notice that the binucleate stage, which is very brief in a sac fungus, represents most of the life cycle of a club fungus. (From *Fundamentals of Cytology* by L. W. Sharp. Copyright 1943 by McGraw-Hill Book Company, Inc. Used with permission of McGraw-Hill Book Company.)

spores. In most mushrooms the club-shaped cells which bear these spores are attached to thin flaps of tissue called *gills*. These fan out under the cap of the mushroom.

The growth of these special cells and their spores may go on for several days, and the number of spores released may be astronomical. According to one estimate as many as 10 billion spores may be formed by a single mushroom of the species grown commercially, that is, the meadow or field mushroom. Mushrooms not only produce vast numbers of spores, but they also grow extremely rapidly.

The most important club fungi from a practical point of view are the *rusts*. They are major parasites on numerous agricultural, forest, and ornamental plants. The rusts' life cycles are extremely complicated. They may involve as many as five different kinds of spores that follow one another in sequence. Control of these fungi is a major economic problem. One way to reduce damage is the breeding of varieties and hybrids of economic plants that are resistant to infection. Unfortunately, however, new races of rusts often appear which overcome the resistance of the new plant breeds.

Sexless fungi—Class Fungi Imperfecti

This class includes a large number of microscopic fungi whose life cycles, particularly the sexual stages, are not known. Most of them appear to be *Ascomycetes*. They do not form conspicuous fruiting bodies but reproduce by air-borne asexual spores. A number are of great economic importance. *Penicillium camembertii* and *Penicillium roquefortii* give well known cheeses their special flavor and smell. Another species of *Penicillium* produces the *antibiotic* penicillin which kills many bacteria and cures many bacterial diseases.

Imperfect fungi also cause many skin infections, such as athlete's foot and ringworm, as well as many plant diseases. Molds that spot clothes and mildew walls are also included in this group.

The evolutionary relationship of the five classes of fungi to one another and to other organisms continues to be debated. It appears likely that the first fungi were motile unicellular eucaryotes that evolved into filamentous fungi like the Oomycetes with many nuclei but no crosswalls. These invaded the land and evolved the habit of producing airborne spores. These early land dwellers evolved into the Ascomycetes, which in turn gave rise to Basidiomycetes.

Slime molds—Division Gymnomycota

Before leaving the fungi, mention should be made of a bizarre group of amoeba-like organisms, often classified with the fungi, called *slime molds*. These organisms resemble fungi in that they are heterotrophic and reproduce by forming spores. However, unlike the fungi these spores do not form hyhae when they germinate. Also unlike fungi, slime molds have no cell walls during most of their life cycle. They ingest their food, largely yeasts and bacteria, like an animal instead of absorbing predigested food like a fungus. Their evolutionary origins and relations are unknown. They are placed in a separate division, the Gymnomycota, but it is not clear to which kingdom they should be assigned. Perhaps they evolved from some ancient protozoan. There are several different groups of slime molds totaling about 500 species, but the best studied are the cellular slime molds.

Dictyostelium discoideum is a *cellular slime mold* (Fig. 23-18). At one stage of its life it exists as free-living haploid amoebae. These individuals move about independently, feed, and divide mitotically to produce more amoebae. When the population density reaches a certain point, changes occur in some of the amoebae and they begin to secrete *cyclic adenosine monophosphate*. (This is the very same cyclic AMP discussed in Chapter 15 which acts as a "second messenger" within the cells of vertebrates following stimulation by various hormones.) In response to this stimulus, the amoebae swarm together and form a multicellular haploid organism which looks like a slug. The slug differentiates to form a stalked fruiting body which produces haploid spores. When each spore germinates, it grows into a haploid amoeba and the life cycle begins again. Although sexual reproduction in these slime molds has not been well studied, there are some reports of free-living amoebae functioning as gametes and fusing to form a diploid zygote.

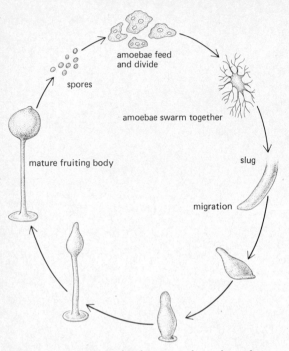

Figure 23-18 Life cycle of *Dictyostelium discoideum*.

INVADING THE LAND

Two major divisions of green plants have been successful on land, the mosses and liverworts or *bryophytes* and the vascular plants or *tracheophytes*. They appear to have evolved from some ancestral green alga which invaded the land some 500 million years ago. Like the green algae both bryophytes and vascular plants contain chlorophyll *a* and chlorophyll *b*. The nongreen pigments in their cells, carotene and xanthophyll, occur in about the same proportions as in green algae. They store the products of photosynthesis as starch and have cell walls of cellulose. When flagella occur, they are found only on sperm.

The evolutionary history of bryophytes and vascular plants can be understood best in terms of various adaptations they evolved to cope with the special problems of life on land. Three major problems that they solved are the following:

1. Preventing the plant from drying out while at the same time permitting carbon dioxide uptake. Some bryophytes and all vascular plants have evolved a *cuticle* on their leaves which

cuts down on water loss, and *stomates* which regulate gas exchange.

2. Preventing the gametes from being exposed to air where they would quickly dry out. Bryophytes and vascular plants have evolved *multicellular sex organs* covered by a jacket of nondividing cells which keeps the gametes from drying out. These organs are very different from the single-celled sex organs without jackets found in most algae.

3. Protecting the young embryo from drying out. Bryophytes and vascular plants have evolved mechanisms for *internal fertilization* and for the *development of the early embryo within the female sex organ*. The young embryo receives food and water from the parent. This pattern is similar in some respects to the way young develop in placental mammals.

These three sets of adaptations were evolved by the green algal ancestor of both bryophytes and vascular plants when it first invaded the land. In addition both bryophytes and vascular plants each evolved numerous other adaptations for life on land.

MOSSES AND LIVERWORTS— DIVISION BRYOPHYTA

The bryophytes, the mosses and liverworts, are the simplest green land plants. Compared to most vascular plants they seem quite inconspicuous, but they are still a fairly numerous and widespread group comprising more than 30,000 species. Individual plants are small and usually grow together in large numbers as a mat. These mats are commonly found in moist, shady locations in which the shade may be too deep for most other green plants. A few species grow in arid habitats and a few others, like *Sphagnum*, are aquatic. When droughts occur, they may survive the way lichens do, shriveling up as they dry out and not photosynthesizing again until water is available once more.

The bryophytes are rather like amphibious animals, because they never become completely independent of a watery environment. Two ancestral ties bind them to the water. First, they have *flagellated sperm cells* that must swim or be washed by moisture to the egg in the female sex organ. Second, they have *no vascular tissue* to transport fluids along distances. Lacking this capacity as well as the supporting tissues found in vascular plants, they have remained small.

Bryophytes do not have leaves or roots like those in vascular plants. Mosses, for example, have short, slender stems with leaflike structures radiating out in all directions. The stem is an-

Figure 23-19 *Bryophytes.* A. Moss plants usually grow in groups or mats and are rarely found singly. Here is a clump of urn moss, *Physcomitrium.* The leafy plants are the gametophytes. The long stalks with capsules on their ends that extend from the gametophytes are the sporophytes. Spores produced in the capsules germinate to form a new gametophyte generation. B. The spores germinate into a threadlike young gametophyte called a protonema which resembles a filamentous green alga. C. A liverwort, *Marchantia.* The flat scaly organism is the gametophyte. The cup-like structures on the surface of the gametophyte produce special cells called *gemmae* which can grow into new gametophytes. Sporophytes are not shown. D. Sphagnum or peat moss forms a dense mat in bogs.

chored to the soil or rock by threadlike filaments called *rhizoids*. In most bryophytes the rhizoids only serve to anchor the plant, and water and minerals are absorbed directly through the leaves and the stem. The stems of most mosses have no water-conducting tissues like the stems of vascular plants.

Bryophytes and vascular plants also differ in the relative development of their haploid gametophyte and diploid sporophyte phases (Fig. 23-20). In bryophytes the gametophyte generation is the prominent generation and the sporophyte is much reduced. In vascular plants the sporophyte is the conspicuous generation and the gametophyte is much reduced. The gametophyte of a moss is the moss we see. At the tips of its leafy stems are male and female sex organs that produce gametes. The flagellated sperm swim to the egg cells in the female sex organs. Following fertilization the zygote divides by

mitosis. The result is a diploid sporophyte that takes the form of a stalk with a *capsule* or *sporangium* at its upper end. The stalk and capsule of the sporophyte contain chloroplasts and manufacture most of their own food supply. The base of the stalk, however, is buried in the tissue of the leafy gametophyte plant and obtains water and minerals from the gametophyte. Interestingly the sporophyte has stomates just like the vascular plants, a fact which supports the idea that bryophytes and vascular plants had a common ancestor.

Meiosis takes place within the diploid sporangium and haploid spores are formed. These spores spill into the wind. When a spore germinates it produces a threadlike young gametophyte which looks very much like a filament of green alga. Small buds of cells develop on the young gametophyte and grow into a typical gametophyte moss plant.

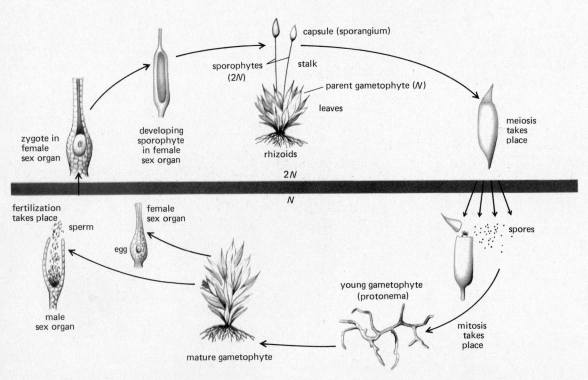

Figure 23-20 Life cycle of a moss.

Liverworts are also bryophytes. They look much like mosses, but are often scaly, growing more or less flat against the ground, the bark of a tree, or rotting wood. Although they are rarely abundant or conspicuous, they are found in all parts of the world.

Like the lichens, bryophytes are pioneers and literally "break new ground" by attaching themselves to rocks. Soil and organic material collect among their filaments and begin to form the sort of soil that vascular plants favor. The richest bryophyte vegetation in North America is found in humid areas along the Pacific Coast and in the high rainfall areas of the southern Appalachian Mountains. Deep masses of plants often cover the cliffs, forest floors, and even festoon the trees.

Large accumulations of one genus of moss,

Sphagnum, are widespread in bogs throughout the world. *Sphagnum* makes bogs extremely acid by releasing hydrogen ions and absorbing cations. It also produces antibiotic materials which kill microorganisms. As a result decay goes on very slowly in bogs and the old dead gametophytes of the *Sphagnum* accumulate to many meters in thickness. Such accumulations are called *peat.* It is often cut into blocks, dried, and used as fuel. Some bogs are thousands of years old and contain the remains of all sorts of organisms. Occasionally the body of a large animal or a human is recovered, well preserved, although dead for thousands of years. Botanists can learn a great deal about the changes in vegetation over the centuries by examining the pollen grains that have blown into bogs and been preserved.

Figure 23-21   Iron Age man recovered from a peat bog in Denmark, where he had died more than 2000 years ago.

VASCULAR PLANTS
—DIVISION TRACHEOPHYTA

The vascular plants are truly terrestrial and have evolved many adaptations to life ashore. Their present diversity can be understood in terms of these adaptations which enabled them to solve four additional problems which bryophytes failed to solve.

1. They absorb and conduct water and minerals from parts of the plant which have access to water to other parts, and conduct products of photosynthesis from photosynthesizing tissues to parts of the plant which may not be exposed to light. As we have learned, in fairly simple plants, such as most algae, fungi, and mosses, water, minerals, and food materials are passed along from cell to cell by osmosis and diffusion. This is fine for short-distance transport. But if water had to diffuse from cell to cell up a pine tree and sugar had to diffuse from cell to cell down the tree, then the leaves would soon die of "thirst" and the roots would soon die of "hunger." Vascular plants evolved roots which absorb both water and minerals. They also evolved a vascular system of *xylem* or *wood* which carries water and minerals upward in the plant, and of *phloem* which transports food from one part of the plant to another. In addition they have developed remarkable photosynthetic organs—leaves— which have their vascular system connected to the vascular system of the stem.

2. They support large structures without the buoyant effects of water. The stem is the evolutionary adaptation that copes with this problem. In many vascular plants the stem is rigid and woody so that the plant may grow to considerable height. Vascular plants also have an efficient anchoring system for the stem.

3. They are less dependent on water for reproduction than are the bryophytes. Primitive vascular plants still require some water through which the flagellated sperm may swim to the egg. But most seed plants need no water at all for reproduction.

4. They can be dispersed with little or no water.

Many adaptations for dispersal were evolved by different vascular plants culminating in seeds, flowers, and fruits.

These adaptations all occurred in the diploid sporophyte generation which is free-living in vascular plants. The gametophyte generation of vascular plants became much reduced and generally has no vascular tissue. In flowering plants the gametophyte finally ended up as a microscopic, well-protected parasite in the sporophyte.

The extent to which vascular plants conquered the land is extraordinary and today they are the most numerous and widespread plants. In the 400 million years since the first members of the division appeared, they have managed to invade almost all environments. They grow in Greenland only 7° in latitude from the North Pole, and in the Himalayas at an altitude of 18,500 feet. They occupy virtually every sort of habitat except the most extreme deserts and regions covered by permanent ice and snow. Most are terrestrial, although a few inhabit fresh water and fewer, such as eel grass, are marine. A small number are parasites either on other vascular plants (for example, dodder, *Cuscuta*) or on fungi (for example, Indian pipe, *Monotropa*), but most are photosynthetic. They range in size from the tiny water meal (*Wolffia*), which is at most 1.5 millimeters long, to the giant redwoods, *Sequoia sempervirens*, of coastal California and southwestern Oregon which are more than 60,000 times as long. These trees attain heights of 385 feet and may be more than 35 feet in diameter. The importance of vascular plants is incalculable, for they are the primary source of energy for all terrestrial ecosystems and their inhabitants, including man.

Botanists are still not certain just how the five major subdivisions of vascular plants originated. By Devonian times, 400 million years ago, all these subdivisions existed:

1. Subdivision Psilopsida (the psilopsids, all extinct)
2. Subdivision Lycopsida (the lycopsids)
3. Subdivision Sphenopsida (the horsetails)
4. Subdivision Pteropsida (the ferns)
5. Subdivision Spermopsida (the seed plants)

Figure 23-22 Some unusual vascular plants: A. The lesser duckweed, *Lemna minor*, one of the smallest vascular plants. B. Indian pipe, *Monotropa uniflora*, a nonphotosynthetic parasitic vascular plant which uses fungi to connect it with a photosynthetic plant from which it gains nourishment. C. Dodder, *Cuscata gronovii*, twining about the stem of another plant. Dodder is a parasitic flowering plant which sends outgrowths into the phloem of its host to absorb nutrients. D. Eel grass, *Zostera marina*, one of the few marine vascular plants. E. Coast redwood, *Sequoia sempervirens*, the largest vascular plant.

Most scientists believe that these groups shared a common ancestor because they all have similar water-conducting systems.

The first four subdivisions are important largely for what they tell us about the evolution of land plants. Each was successful in covering the land with vegetation at some period in the earth's history. But eventually all of them were almost completely supplanted by the seed plants. Let us examine the evolutionary history of the vascular plants and explore the strategy of their conquests of land.

Primitive vascular plants—Subdivision Psilopsida

The oldest undisputed remains of vascular plants have been found in deposits from the Silurian period, about 425 million years ago. They belong to members of the subdivision Psilopsida and are the earliest known plants that left the water to live on land. The psilopsids were starkly simple plants (Fig. 23-23A). They had no roots, only an extension of the stem to which were attached multicellular rhizoids. They were leafless, but part of the stem above ground was green and carried out photosynthesis. Their most significant feature was the presence of both xylem and phloem (Fig. 23-23B), indicating that the plants were tracheophytes. The stem was sufficiently woody to help them to stand upright.

Although no psilopsids persist today, several primitive ferns resemble them closely. The most well known is *Psilotum nudum*, the whisk fern, which is among the most primitive of living vascular plants. The prominent stage of *Psilotum* is the diploid sporophyte which has neither roots nor leaves (Fig. 23-23C). Photosynthesis is carried out by cells lining the outside of the stem. Some of the stems end in **sporangia** which produce haploid spores. These spores develop into small, colorless gametophyte plants that grow underground parasitically with a fungus. The gametophyte produces both eggs and motile sperm. After fertilization the diploid zygote develops into a sporophyte plant and the life cycle begins again. This probably resembles the life cycle of the ancient psilopsids.

Club mosses—Subdivision Lycopsida

Fossil evidence indicates that lycopods lived in the lower Devonian period. They show an evolutionary advance over the psilopsids. In addition to the vascular tissues in their stems, these early lycopods had true leaves and also true roots. The leaves probably arose as outgrowths of the stem, whereas the roots may have evolved from branches. Interestingly, in lycopod embryos, the roots are the last structures to develop, a fact which is consistent with their late evolutionary appearance.

The most familiar living lycopods are the so-called club mosses or ground pines, *Lycopodium*, which are neither mosses nor pines (Fig. 23-24A). They are often sold for Christmas greenery. During the Carboniferous period, between 280 and 310 million years ago, this subdivision produced giant forests with trees as tall as 130 feet and with trunks measuring 3 feet across. These trees later became extinct and the 1000 species of lycopods surviving today probably descended from their smaller relatives.

As was the case with psilopsids the conspicuous club moss or other lycopod is the sporophyte phase. Certain of the leaves called **sporophylls** were modified to produce haploid spores which develop into small gametophytes that usually grow underground, parasitic on a fungus. The rest of the life cycle resembles that of *Psilotum* except for the development of a root by the embryo.

Horsetails—Subdivision Sphenopsida

The horsetails first turn up in the fossil record some 400 million years ago, during the Devonian period. By Carboniferous times some species were among the dominant forest trees (Fig. 22-6), and their dead bodies form much of the coal we use today. Since then they have declined. Today's forms are restricted to a single genus, *Equisetum*, which contains only 24 species (Fig. 23-24B). They are often found in moist or damp places, along the edge of woods or by streams. They are easily recognized because of their jointed stems and coarse texture. The epidermal cells on their stems

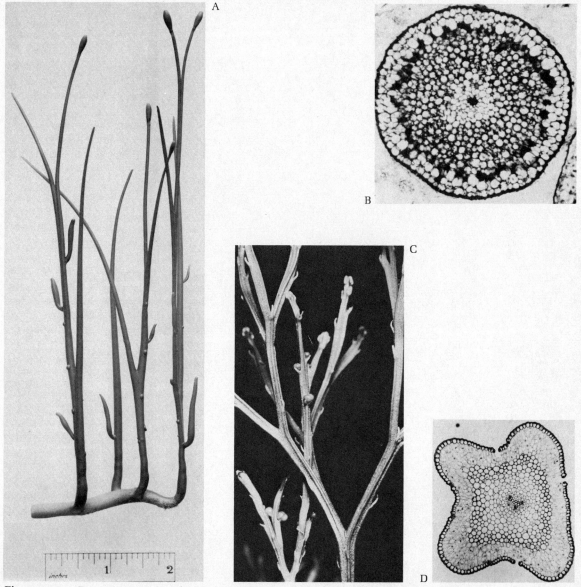

Figure 23-23 Primitive vascular plants. A. A reconstruction of *Rhynia Gwynne-Vaughni*, an extremely primitive leafless, land plant from the Silurian. Excellent fossil remains of this plant reveal not only its general form, but also the cellular organization of the stem. B. Cross section of stem of *Rhynia* revealing a solid rod of xylem and phloem in the center. C. *Psilotum nudum* is one of the most primitive of present-day vascular plants. It is found on humus and logs in tropical and subtropical regions and ranges north to South Carolina. Because it lacks both roots and leaves and has many primitive features, *Psilotum* and a close relative *Tmesipteris* were long regarded as the only living representatives of the Psilopsida which otherwise became extinct at the end of the Devonian period. Recent studies, however, indicate that they are probably primitive ferns. Hence all the Psilopsida are extinct. D. A cross section of the stem of *Psilotum*. The vascular tissue is clearly evident in the center of the stem.

Figure 23-24   A. *Club moss.* The running ground pine, *Lycopodium complanatum* is commonly used for Christmas decorations. It is easy to identify by the rounded branching pattern of the upright stems and the striking clusters of one to four long yellowish cones on each stem. It usually grows no higher than 10 inches. B, C *Horsetail.* The field horsetail, *Equisetum arvense,* is the most familiar species in North America and is usually 8 to 20 inches tall. The hollow sterile stems (B) are green and have whorls of branches. In early spring pink fertile stems with cones appear (C). This and other species of horsetails are poisonous to livestock.

have tough silica deposits which make them indigestible to most herbivores. In colonial and frontier days they were used to clean pots and pans, and they are still called "scouring rushes." In the United States horsetails grow to a height of 3 feet, but those in tropical forests may be more than ten times as tall.

The fundamental structure of the sphenopsids has changed very little since the early part of their evolution. Their vascular tissues are arranged in a ring around a hollow stem. The stem is jointed and leaves radiate in whorls from the joints. These leaves, like those of lycopods, apparently arose as outgrowths of the outer tissues of the stem. The sphenopsids have effective roots which sometimes penetrate as far down as six feet below the surface of the soil. *Equisetum* has a life cycle similar to that of a club moss. This cycle involves a small gametophyte stage and motile sperm. However, the gametophyte of *Equisetum* lives above gound; moreover, it is photosynthetic, and does not depend upon a fungus.

Ferns — Subdivision Pteropsida

The three groups of vascular plants just described were prominent in the Carboniferous period and then declined. A fourth group, the ferns, was also important then, and has managed to persist even in the face of tremendous competition from seed plants. Today ferns still include about 11,000 living species. They are found as far north as the Arctic Circle, but are the most plentiful in the tropics, where they grow as large as trees (Fig. 23-25).

Except for a few unusual forms like *Psilotum*, ferns are quite advanced structurally. They have a well-developed vascular system, with roots, stems, and usually large, feathery leaves. The leaves probably evolved from flat, webbed branches. The stems of most of the ferns found in temperate climates are underground. The only part of the plant normally seen is the leaves.

Ferns have a life cycle like that found in the primitive vascular plants we have already described (Fig. 23-26). The leafy fern is the diploid sporophyte. Spores are produced in sporangia

Figure 23-25 Ferns. A. The bracken fern, *Pteridium*. B. Growth of fern fronds or "fiddleheads" of the Christmas fern, *Polystichum achrostichoides*. C. A tree fern, *Cyathea*, in a Puerto Rico rain forest. D. On the undersides of some fern leaves are found clusters of sporangia, called *sori*. This shield fern, *Dryopteris spinalosa*, is covered with sori. E. This higher magnification view of the sori of a staghorn fern shows clusters of granules, which are the individual sporangia in which spores are formed.

which grow on the underside of some leaves (sporophylls). Each spore germinates and develops into a tiny, free-living green gametophyte which is about 5 millimeters long and often heart-shaped. This gametophyte has both male and female sex organs. The sperm are flagellated and need water to swim to the female sex organ and fertilize the eggs. Each zygote develops into a young green sporophyte, absorbing food and water from the gametophyte before becoming independent.

The differences between the diploid sporophyte fern and its haploid gametophyte stage are so striking that they have aroused a great deal of scientific curiosity. One question botanists asked was whether the differences depend in part on the chromosome number of each phase. The answer

was provided by causing a haploid gametophyte to reproduce vegetatively. The result was a leafy haploid plant that looked just like a normal sporophyte. Similarly, in several species, diploid heart-shaped plants that look like gametophytes develop from the edges of the sporophyte leaves. This evidence leads us to believe that the *ploidy* or number of full chromosome complements does not determine the growth pattern of either the gametophyte or the sporophyte.

One might wonder why the ferns have survived in fairly large numbers, whereas the other two groups of primitive plants have almost disappeared. One reason may be that the ferns have much larger leaf areas exposed to sunlight than the lycopods or sphenopsids. Fern leaves are extremely large in proportion to the plant's weight.

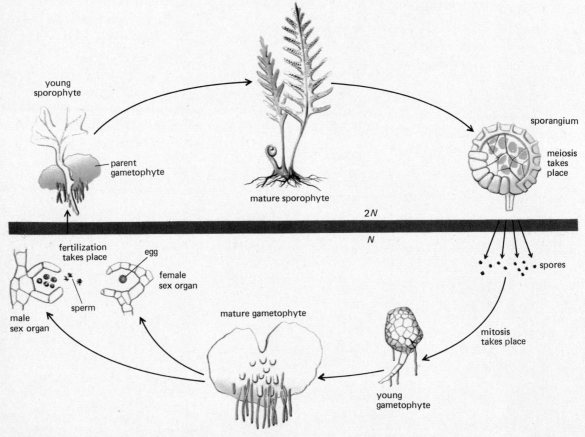

Figure 23-26 Life cycle of a fern.

This probably provided a slight advantage that enabled ferns, like the mosses, to occupy shaded areas in which other plants could not survive.

Seed plants—Subdivision Spermopsida

The seed plants include the vast majority of green plants, and among them are the ones we know best. They are also the ones best adapted for life on land. Almost all of the green land plants we have considered until now are partly "amphibious." They require some water for fertilization. In addition their gametophyte is not hardy enough to withstand the drastic climatic changes which may occur on land. The seed plants are much better adapted for reproducing in a dry environment and for withstanding drastic changes of climate. Two adaptations have been keys to their evolutionary success: (1) pollen and (2) seeds. We discussed both of these in Chapter 7 in connection with the development of a flowering plant. Now let us examine how these two reproductive features function as adaptations for life on land.

In all seed plants the sporophyte is dominant. The gametophyte is extremely small, is not photosynthetic, and lives as a parasite on the sporophyte. *Pollen* and *seeds* appear to have evolved following the development of two types of haploid spores by the sporophyte. One kind of spore called a **megaspore** gives rise to female gametophytes. These remain enclosed within the tissues of the sporophyte. Another kind of spore, a **microspore,** produces male gametophytes called *pollen grains.* Pollen grains are released from the parent plant and transported to the female gametophyte by wind, insects, or other means to bring about fertilization. This behavior of seed plants contrasts with most ferns and most other primitive vascular plants that have only one kind of spore that usually produces an independent, free-living gametophyte bearing both male and female sex organs.

Pollen grains eliminated the need for free-swimming sperm which require water for fertilization. Unlike the fragile gametophytes of ferns and other primitive vascular plants, pollen grains are extremely hardy and can float in the air for long periods without drying up. When a pollen grain lands next to a female gametophyte, the pollen grain completes its development into a tiny male gametophyte containing a few nuclei and usually two sperm. In this process the pollen grain develops a tubular extension, the **pollen tube,** which grows toward the egg in the female gametophyte. The pollen tube then discharges its sperm and one of these fertilizes the egg. This adaptation freed seed plants from requiring water for sperm transfer.

Seeds protect the embryo from dryness and other unfavorable conditions. At the same time they promote the widespread distribution of the plant. A mature seed usually consists of the embryo of the sporophyte generation along with food reserves, all packaged inside a protective seed coat from the parent plant.

The female gametophytes of seed plants are inconspicuous and completely parasitic, and live and produce eggs within the tissues of the sporophyte parent plant. Here they are relatively protected from unfavorable environmental conditions. As we noted in Chapter 7 the female gametophyte in flowering plants consists of only eight haploid nuclei organized into seven cells. When an egg is fertilized by a sperm brought to the female gametophyte by a pollen grain, the zygote and the rest of the female gametophyte remain embedded in the tissues of the parent plant. Before the embryo is freed from the parent plant it is provided with a ready supply of food and a tough outer seed coat. When the parent plant releases the seeds, most of them are in suspended animation. Such a dormant seed will not continue its development until moisture, temperature, light, and oxygen are available in the proper proportions for it to germinate. Then it draws upon its food reserves and grows until it can manufacture its own food photosynthetically.

Importance of Seeds

Seeds themselves have played a large role in the widespread distribution of land plants. They can be carried vast distances by the wind or by animals. Many seeds can survive conditions that would quickly kill their parent plants, such as drought, extreme heat and cold, and permanent darkness. They are equipped with mechanisms to keep them from germinating until conditions are favorable for growth. Seeds like those of lettuce with small food reserves often will not germinate unless exposed to fairly bright sunlight; if they germinated in the shade of other vegetation, they might not be able to grow enough. Many desert plants require a thorough soaking to remove a growth-inhibiting substance that keeps them from germinating unless there has been enough rain to support proper growth.

Some seeds, like those of poplar and orchids, live for only a few days. The seeds of many crop plants and of many weeds live for about 30 years. A few seeds may live even longer. Several years ago seeds of the Oriental lotus were found by a Japanese botanist in a dried lake bed in Manchuria. Radiocarbon dating showed that they were at least 800 years old. When the seed coats were filed to let water seep through, all of the seeds germinated. More recently seeds of an Arctic lupine were found in the Yukon in a lemming burrow which was in permanently frozen silt from the Pleistocene age. The seeds were at least 10,000 years old, yet they germinated!

Seeds vary tremendously in size. The smallest are found in orchids where nearly 4 million dustlike seeds have been counted in a single fruit. Probably the largest seed, that of the double coconut, Coco-de-mer, may weigh 40 pounds!

In addition to seeds and pollen, other evolutionary changes in the seed plants favored life on land. In particular, the vascular system and the other tissues of the root, stem, and leaves underwent extensive development.

The fossil record reveals that the first seed plants appeared 365 million years ago in the upper Devonian period. Their seeds developed on leaflike structures, called *sporophylls,* or on branches, but were not enclosed. Ancient and modern plants of this sort, with comparatively unprotected seeds, are known as **gymnosperms.** Their name comes from two Greek words for "naked seeds." Later on plants evolved that had flowers, and their seeds were enclosed in an ovary which probably was formed by the sporophylls. The flowering plants are therefore called **angiosperms** ("containered seeds") because of their enclosed seeds. The angiosperms are generally thought to have a common ancestor and appear to be closely related phylogenetically. The gymnosperms, in contrast, include several different groups of plants which evolved at different times from various groups of ferns and are not closely related phylogenetically. They all have certain common features, however, and for convenience we can examine them together.

CONIFERS AND THEIR RELATIVES — THE GYMNOSPERMS

The probable ancestors of some of the gymnosperms belong to a group called the **progymnosperms.** Originally these plants were only known from impressions of their leaves found in Upper Devonian rocks, and they seemed to be ferns, for they had spores which were shed. In 1960, however, a fossil was found that showed a leaf attached to a stem that was unquestionably coniferlike. The leaf belonged to a tree, 50 feet tall. Because they shed their spores, the progymnosperms were not seed plants, but they appear to be a step along the way.

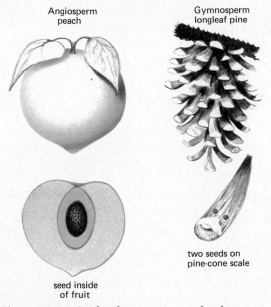

Angiosperm peach

Gymnosperm longleaf pine

two seeds on pine-cone scale

seed inside of fruit

Figure 23-27 Seeds of gymnosperms develop on an exposed surface, as on pine-cone scales. The enclosed seeds of angiosperms are surrounded by the wall of the ovary that develops into a fruit as in a peach.

Many experts in botanical fossils believe that these ancient plants gave rise to earlier true seed plants, the **seed ferns,** and to the **conifers.** The seed ferns are an order of extinct gymnosperms that had leaves which bore seeds, not naked spores. A major part of the coal forests so prominent in late Carboniferous times was composed of seed ferns. Another group of gymnosperms, the conifers, lived along with the seed ferns in the late Carboniferous period. The conifers did not die out, however, and are still an abundant part of the earth's vegetation today.

Conifers are certainly the best known gymnosperms. This order contains about 450 living species, including such common trees as fir, hemlock, juniper, pine, and spruce. Also in the order are some of the largest living things, the sequoias, and the oldest living things, the bristlecone pines and the Montezuma bald cypress, some of which are estimated to be nearly 5000 years old. Dense forests of conifers are found in most of the temperate regions of the world, and they furnish about 75 percent of the world's lumber and most of its pulpwood. These trees grow most abundantly in comparatively hostile environments, where it is cold, windy, or the soil is poor. In more hospitable areas angiosperm forests are far more common. Like all other gymnosperms, conifers are woody trees or shrubs and all are perennial.

The leaves of most conifers are small needles or scales. Practically all of these plants are evergreen and keep their leaves for two or more years. A few shed their leaves at the end of the growing season, for instance, the larch and the bald cypress. Conifers get their name from their characteristic **cones** which serve as reproductive organs. Each cone is a short stem covered with a spiral cluster of **cone scales** which evolved from flattened lateral branches.

The pine trees (genus *Pinus*) are an ancient group and have been traced back some 170 million years to an uncertain origin in northern Asiatic Russia. Their life cycle is typical of conifers and similar to that of most other gymnosperms (Fig. 23-30). A pine tree has two kinds of cones — small *pollen-bearing cones* and large *seed-bearing cones.* The scales on seed cones produce *megaspores,* whereas the scales on pollen cones produce *microspores.* Each megaspore develops into a tiny female gametophyte composed of several thousand cells. Both the megaspore and the female gametophytes remain embedded in the seed cone scale. The female gametophyte bears the eggs and in some cases accumulates all the food reserves around the eggs *before* fertilization and seed production occur. The microspores in the pollen cones, unlike the megaspores in the seed cone, are eventually released. This occurs after they have partially germinated and developed into pollen grains. Millions of pollen grains are produced by each tree in the spring. The few pollen grains that happen to land on seed cones complete their development and end up as tiny male gametophytes containing only six cells, two of which are sperm. This is a far cry from the multicellular gametophytes of ferns. One of the sperms fertilizes the egg and the resulting zygote develops into an embryo. Finally the pine seed is shed. It consists of the embryo of the new sporophyte pine tree, stored food material produced by the female gametophyte in part before fertilization, and a seed coat from the parent plant.

Figure 23-28   A. Reconstruction of an Upper Devonian progymnosperm, *Archaeopteris,* about 50 feet tall. B. Reconstruction of a Carboniferous "seed fern," *Medullosa noei,* about 15 feet tall.

Figure 23-29 Conifers are mainly trees but include a few shrubs. Their distinctive reproductive structure, the cone, is of two types; one forms pollen whereas the other forms eggs and finally seeds. The cones are usually conspicuous as in the Eastern white pine but in yews they are fleshy structures. A. Bristlecone pines, *Pinus arisata*, like this one from the mountains of eastern Nevada, may get to be 4900 years old. The chief living part of the tree is the treelike shoot at right. B. Bald cypress, *Taxodium distichum*, sheds its leaves in the fall. It thrives in the swamps of the southeastern United States. Its roots take in oxygen by means of conical "knees" which extend above the water. C. Pollen cones of loblolly pine (*Pinus taeda*). D. Japanese yew, *Taxus*, showing fleshy cup that bears seeds.

As a group, the conifers are significant in numbers of individuals and they are of immense value as timber trees. But in an evolutionary sense they are probably past their prime. Many living species appear to be nearly extinct and are limited to a few locations. Among the most remarkable of these is the dawn redwood *Metasequoia*, which until 1944 was known only from fossils and presumed to be extinct for millions of years. Then unexpectedly a stand of living *Metasequoia* trees was found in a remote part of western China (Fig. 21-7). This remarkable survivor is now being cultivated in several parts of the United States as an ornamental tree.

Perhaps the best known gymnosperm besides the conifers is the *ginkgo,* or the maidenhair tree, first appearing in Permian fossils about 280 million years old. Once represented by many genera and species, there is now only one surviving species. Nevertheless ginkgos are common in the United States because they are popular shade trees. The ginkgos retain a primitive feature lost in most other living gymnosperms and in all angiosperms—their pollen tubes produce large *motile* sperm that fertilize the eggs. This same feature is found in only one other gymnosperm group, the *cycads,* a small group of palmlike plants common in the tropics (Fig. 23-31).

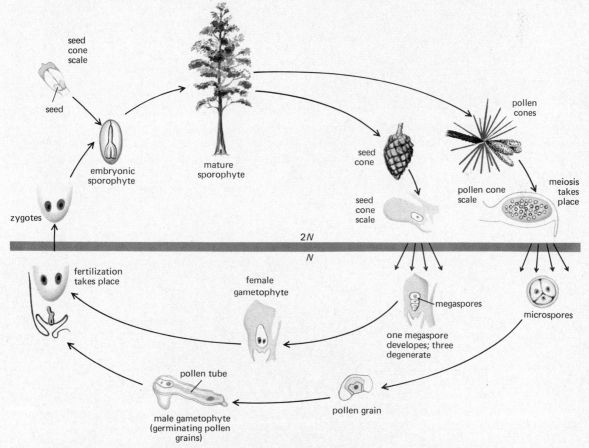

Figure 23-30 Life cycle of a pine tree. A pine tree is the mature sporophyte plant. The gametophytes are much reduced structures and are enclosed within the tissue of the sporophyte. After fertilization the female gametophyte with its enclosed zygote and a tough seed coat produced by the parent sporophyte becomes the seed which forms on the cone scale.

Figure 23-31 Various kinds of gymnosperms. A. Leaves of the ginkgo or maidenhair tree (*Gingko biloba*), the lone surviving species of an ancient group that once formed dense forests in North America. Because it is remarkably resistant to smog it is a widely planted street and shade tree reaching a height of 50–80 feet. Its bilobed leaves have a distinctive shape. B. Cycads are ancient trees that have persisted since the time of the dinosaurs. They all have large, fern-like leaves that form a crown at the top of the stem. *Zamia floridana* shown in this photograph is a common cycad of the Florida pinelands. Most of its stem is underground. Indians and early settlers ground their underground stems into a flour. It is poisonous when fresh but the poison is water-soluble and was washed out while the stems were being ground. C. *Weltwitschia mirabilis* is one of the most unusual gymnosperms. It produces only two long ribbon-shaped leaves during its entire lifetime which may last a hundred years. The distal ends of the leaves break off as the plant grows. It is found only in the Kaokoveld of South Africa. It has a long tuberous root that grows deep into the desert soil and when grown in the greenhouse may require several pots, piled one on top of the other. D. Ephedras like the one in this photograph are shrubby plants 2 to 6 feet tall, which are fairly common in the southwestern deserts. They have jointed green stems and small, scale-like leaves. Indians made a medicinal drink called Mormon tea from the plants.

FLOWERING PLANTS—
CLASS ANGIOSPERMAE

The flowering plants appear in the fossil record rather suddenly in the Cretaceous. Their sudden appearance has puzzled botanists and, as mentioned in Chapter 22, has led to the suggestion that angiosperms existed earlier but lived in upland environments in which fossils were rarely preserved. Their ancestors are not yet known, but may have been seed ferns. The first angiosperms of which we have a record were trees and must have been perennials. However, in the 135 million years or so since they first appeared in the fossil record, angiosperms have diverged tremendously from these ancestral trees in growth habits and structure. To be sure there are still many angiosperm trees—oaks, palms, and magnolias. But today there are also numerous nonwoody angiosperms like weeds and grasses which cover most of the ground of the earth. Many of these are annuals, bearing seeds and dying in one growing season. All of these herbaceous plants are thought to have evolved from the early woody angiosperm.

Approximately 250,000 different species of angiosperms have already been described. But there are still many more to be discovered. In the tropics there is an incredibly large number of different kinds of flowering plants. Many of these are so scattered that only one or two specimens of a particular tree species may occur in the same square mile. To find particular kinds of trees botanists may have to hire a local "tree finder" who knows where individual trees are located. So numerous are the different species of angiosperms that botanists believe that there may be in remote tropical areas more than 200,000 species of angiosperms not yet described.

Secrets of success

Angiosperms dominate the earth's landscape today. Interestingly they became the dominant plants soon after they first appeared, displacing most groups of gymnosperms in almost every location. Moreover they had the bewildering capacity to adapt to a much wider range of environmental conditions than gymnosperms and occupied habitats to which gymnosperms were not adapted. One such habitat is the tundra, another is the desert, and a third is fresh water.

What adaptations enabled them to win the competition with the gymnosperms? We can list at least eight important factors.

1. The development of flowers which permitted improved pollination.
2. The formation of fruits.
3. Double fertilization.
4. Development of nonwoody plants such as grasses.
5. Development of rapid life cycles, as in annual plants such as marigolds.
6. Better water- and mineral-conducting systems, larger leaves, and faster growth rates.
7. Development of resistance to drought and cold.
8. Greater genetic flexibility.

Since several of these adaptations involve reproduction, let us take another look at the basic features of angiosperm reproduction which were described briefly in Chapter 7.

What is a flower?

The flower of an angiosperm, like the cone of a gymnosperm, is the site of reproduction, and consists of a short length of stem with modified leaves attached to it. But the angiosperm flower has a very different structure from the cone of a gymnosperm. Figure 23-33 shows a diagram of a flower. The enlarged end of the stem on which the flower develops is called a **receptacle.** Four sets of modified leaves develop from the receptacle. The **sepals** which are outermost are usually green and leaflike in structure and enclose and protect all the other parts of the flower bud. The next ring of modified leaves are the **petals,** which are also usually shaped somewhat like leaves and often brightly colored and advertise the flower to prospective pollinators.

Inside the petals are the **stamens,** which are the male reproductive organs. They are the sporophylls that produce the microspores that develop into pollen grains. Each stamen usually consists of a slender stalk, the **filament,** bearing at its end a boxlike structure, the **anther.** The pollen grains, when ripe, are released from within the anther.

Figure 23-32   Various angiosperms. A. *Impatiens*, a dicot in flower. B. Yellow lady's slipper orchid, a monocot in flower. C. *Amorphophallus titanum*, a close relative of the calla lily, has one of the largest flowers of any of the flowering plants. This photograph shows Hugo de Vries, the Dutch botanist who introduced the concept of genetic mutations, standing next to a splendid specimen of *Amorphophallus*. D. The largest living angiosperm is *Eucalyptus regnans*, which may be 250 feet tall and 60 feet in diameter at the base.

They are the male gametophyte, usually composed of only three cells, two of which are sperm.

In the center of the flower are the **carpels,** which are the female reproductive organs. Each carpel consists of a **stigma** at the top on which the pollen lands, a slender stalk called a **style,** through which the pollen tube grows, and at the base a hollow ovary. (Some botanists refer to an individual carpel or fused carpels as a *pistil.*) Carpels are the sporophylls which produce the megaspores. These megaspores are produced in sporangia called **ovules** attached to the wall of the ovary. The megaspores develop into the female gametophyte or **embryo sac,** which is composed of eight haploid nuclei organized into seven cells. One of these cells is larger than the others and contains two nuclei, called **polar nuclei.** Another cell will act as the **egg cell.** Compared to the multicellular female gametophyte of a gymnosperm, which contains several thousand cells, the seven-celled female gametophyte of an angiosperm is streamlined down to bare essentials.

When the pollen grain of a flowering plant lands on the stigma and germinates, its *pollen tube* grows down through the tissues of the stigma and style and enters the ovary. The tip of the pollen tube reaches the female gametophyte and discharges its two sperm. One of the sperm fertilizes the egg cell to form a diploid zygote which develops into an embryo sporophyte. While all this is going on, the two polar nuclei of the female gametophyte fuse together to form a **fusion nucleus.** The second sperm from the pollen tube usually unites with the fusion nucleus. This second union usually produces a triploid tissue with three times the number of chromosomes as haploid tissue. This triploid tissue serves as source of food for the embryo in many angiosperm seeds and is called **endosperm.**

After fertilization, most of the flower falls away, leaving the ovary with its ovules, each of which contains a fertilized egg. Each ovule develops into a seed which consists of a seed coat, stored food in the form of endosperm, and the embryo. The ovary which envelops the seeds usually enlarges greatly to form a *fruit.*

With this picture of angiosperm reproduction in mind let us examine the adaptive significance of several angiosperm features.

Teaming up with insects— Improved pollination

Angiosperms have evolved a more effective way for transporting pollen from the male to the female parts of plants. Gymnosperms appeared before many land animals had evolved, particularly insects and birds. Accordingly they had to depend on the wind to carry pollen to its destination, and they continue this practice today. But the angiosperms evolved at the same time that insects, birds, and mammals did, and were able to take advantage of animals as pollinators. Most flowering plants, in fact, depend on animal pollination, and their evolution was greatly influenced by animals, particularly the insects.

Pollination with the help of animals is usually more efficient than wind pollination. The reason is a simple matter of statistics. Anyone who has ever seen the yellowish, sulfur-colored pollen given off by pine trees will recognize that a great deal of it goes to waste. It is so plentiful that it can cover a lake surface or form dusty windrows ashore. Flowering plants need not be so prodigal with their pollen. By enticing insects with flowers containing nectar or pollen, or with flowers whose color or shape resembles female insects, prey, or enemies, the flowering plants get free

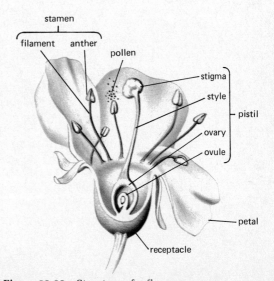

Figure 23-33 Structure of a flower.

transportation for their sperm. A honeybee, for example, will not drop pollen from the pollen baskets on its legs, but always has enough on its body to leave some of the sticky pollen on the stigmas of the flowers it visits.

Insects are by far the most important pollinators. They visit flowers to obtain food, either nectar or pollen. Insects are attracted to flowers chiefly by color and scent. In the course of evolution flowers have evolved showy colors, scents, nectars, and other structures which attract insects and other animals (principally birds) to them. They have evolved sticky pollen such as we find in apple and peach blossoms, which clings to insects.

Often a particular species of flower will be pollinated by only one insect species. The nectar of red clover, for example, is located nearly half an inch from the mouth of the flower. It can be reached only by bumblebees, which have comparatively long tongues. When settlers first planted red clover in Australia, it failed to go to seed until they had imported bumblebees too. Figs provide another example. Attempts to grow figs in the United States failed completely until a tiny wasp, previously not known to be necessary for the pollination of fig flowers, was imported. A more exotic example involves an orchid from Madagascar. Its nectar is at the bottom of a thin tube nearly a foot long. When this plant

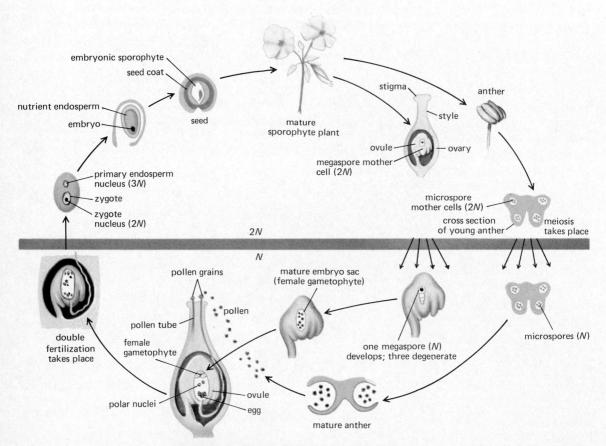

Figure 23-34 Life cycle of an angiosperm. For details see text. Note the megaspore and microspore mother cells which give rise to megaspores and pollen grains respectively.

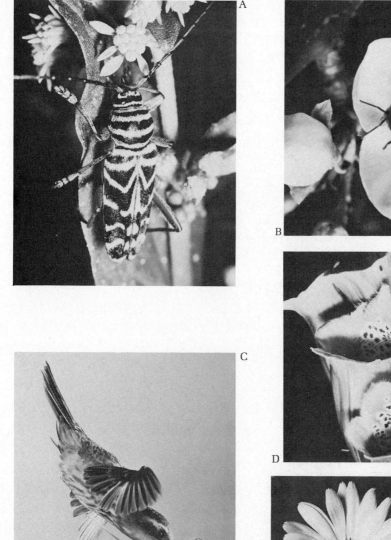

Figure 23-35 Pollination by insects and other animals. A. Beetle crawling among goldenrod flowers. B. Honeybee foraging in a pear flower. Note the filled pollen basket on the left hind leg. C. Brazilian hermit hummingbird pollinating a flower. D. "Honey guides" are conspicuous markings which indicate the location of the nectar supply, as in this foxglove. E. *Coreopsis* has honey guides which are not evident in visible light (left) but are evident in ultraviolet light (right). Insects see in the UV and can see these honey guides. Flowers normally pollinated by insects have honey guides visible in the UV whereas flowers pollinated by birds and bats do not.

was first brought to England, Charles Darwin predicted that a moth would someday be discovered with a tongue long enough to reach the nectar. This apparently caused some furor and much laughter. But about 40 years later the moth was discovered by a missionary. It has a wing span of about 6 inches and a tongue about a foot long. Darwin's prediction is taken note of in its name, *Xanthopan morgani praedicta!*

There are thousands of other examples of ways in which the evolution of flowers has been affected by insects. But the influences have not all been one way. Insects in their turn have often evolved elegant adaptations that benefit plants. The dense hairs in the body of a honeybee, for instance, greatly increase the amount of pollen the bee picks up from plants. It is quite clear that insects and flowering plants have evolved to-

gether; their evolution and their present lives are interdependent. We cannot understand the evolution of one of these groups without considering the other.

Wind-pollinated flowers are thought to have evolved from insect-pollinated flowers. These flowers are often small. They usually have a light and airy pollen and feathery stigmas that are efficient in collecting such pollen from the air. This is true of many grasses. Many people develop an allergic reaction to wind-borne pollen, called *hay fever,* and become amateur experts in identifying the cause of their misery. From March to May it is tree pollen and from May to July it is grass pollen. August and September are pollination times for ragweeds and sagebrushes. Plants like goldenrod which are pollinated by insects cause little hay fever.

Figure 23-36 Wind-pollinated flowers often cause hay fever. (Left) Great ragweed (*Ambrosia trifida*) showing feathery stigmas which collect air-borne pollen. (Right) Flowers of paper birch (*Betula papyrifera*).

Fruit peddling—Improved distribution of seeds

Angiosperms have also evolved special mechanisms for promoting the distribution of their seeds. **Fruits** are one such mechanism. Fruits usually develop only in flowers in which the egg has been fertilized. Their formation is initiated by hormones brought in by the pollen tube and, as we mentioned in Chapter 10, is further stimulated by hormones from the developing embryo. Fruits consist of the ovary and often some other structures, as well as the seeds. The attractiveness of fruits to various animals ensures the dispersal of seeds. An animal will eat the fruit, digesting the soft outer coat, and then pass the seeds through its intestinal tract, depositing it in some other place with a neat coating of fertilizer around it.

Unripe fruits are often green and hidden among the leaves, somewhat concealed from birds, mammals, and insects. They often taste disagreeable and thus discourage animals from eating them before the seeds within are ripe. When the fruit ripens, its changing color provides a signal that the fruit is ready to eat and that its seeds are ripe and ready to be dispersed. One of the commonest colors of ripe fruit is red. Although red is clearly visible to mammals and birds, it is not visible to insects. Hence birds and mammals will usually get first crack at a ripe red fruit before the insects see it. Since most seeds are too large to be dispersed by insects, the color red in ripe fruits is obviously adaptive.

Many other mechanisms of seed distribution have been evolved, for example, fruits which stick to fur and feathers such as beggar's tick, or those carried by the wind such as dandelion.

Plant parenthood—Double fertilization

In an angiosperm the endosperm which provides food for the embryo will not develop within the ovary until fertilization actually occurs and a zygote exists. In this respect the flowering plants resemble placental mammals whose young receive virtually all of their nourishment from the body of the mother *after* fertilization. In many gymnosperms, on the other hand, much of the food reserves are built up in the female gameto-

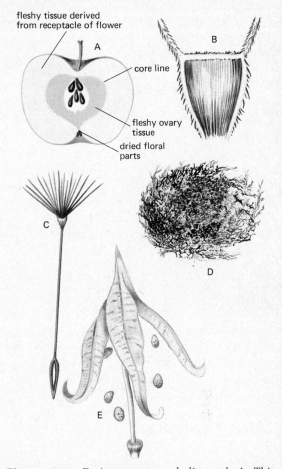

Figure 23-37 Fruit structure and dispersal. A. This apple, which has been cut longitudinally, illustrates the important parts of a fruit. Most of the edible portion develops from the floral tube of the flower. The seeds and the surrounding ovary wall comprise the core of the apple. Seeds of fleshy fruits such as apples are usually dispersed by vertebrates that eat the fruits and eliminate the seeds after they have carried them for some distance. B. Beggar's tick (*Bidens*) is an example of a fruit which adheres to fur or feathers. C. Dandelion (*Taraxacum*) is a fruit which is displaced by the wind. D. In tumbleweeds (*Salsola*) the fruiting portion or the whole plant breaks from the stem and is blown about the countryside, scattering seeds as it goes. E. In some species such as touch-me-not (*Impatiens*), parts of the fruit develop a high turgor pressure. When ripe the fruit will split open at the slightest touch and throw its seeds far from the parent plant.

phyte willy-nilly *before* fertilization. They consist not of triploid tissue but of haploid tissue of the female gametophyte itself. When fertilization occurs in a gymnosperm, a single sperm fertilizes the egg and the other sperm in the pollen tube dies. If the egg is not fertilized, the food in the gametophyte is wasted. In this respect some gymnosperms can be compared to reptiles or birds which store food in their eggs before fertilization. Angiosperms are thus more efficient.

Nonwoody plants

A great advance in the angiosperm was the development of *nonwoody* or *herbaceous plants* like grasses and weeds. This enabled plants to complete their life cycle more rapidly. The extremes in these short life cycles are, of course, **annuals** which complete their life cycles from germination through seed production in a single growing season. Gymnosperms, in contrast, are without exception **perennials**—their plant bodies persist for many years. The streamlined life cycle of the annual angiosperm confers several advantages. It enables plants to take over an area such as a newly burned-over forest, complete a generation in a season, and leave seeds for the next season. It also enables plants to continue from year to year in harsh climates. Even though a season may be too cold, hot, or dry for the parent plant, the seeds survive. A final advantage of tremendous importance in adaptation is that rapid life cycles permit more generations in a given period of time. This results in much more genetic recombination and thus more rapid evolution than in plants with long life cycles.

Some nonwoody plants evolved other mechanisms which enabled them to survive harsh climates. In these plants the part of the plant which is above ground dies back each winter. The part below ground remains alive. In the spring, growth resumes from **underground stems** which have survived the winter. A potato *tuber* is an example of such an underground stem. So are the extensive *rhizomes* of many of the weedy grasses that are familiar to anyone who has tried to dig them up. The development of these underground stems, which renew growth each spring and reproduce a plant above ground, has enabled non-woody angiosperms to develop in temperate and even Arctic environments. Most of the plants of the tundra are nonwoody angiosperms.

The fast life of a flowering plant

The rapid life cycle of angiosperms requires rapid growth. In fact most angiosperms grow far more rapidly than gymnosperms. Angiosperms appear to do everything faster than gymnosperms. The time between pollination and fertilization in a barley plant is less than an hour. In a pine tree it takes a year. It has been claimed that on a warm summer day in Iowa one can hear corn grow as the leaves brush by each other. Some bamboos grow 3 feet in one day.

Rapid growth requires tremendous photosynthesis and this in turn requires large leaves. But large leaves require considerable water because, as we pointed out in Chapter 10, 95 percent of the water lost by a plant is lost by the leaves. Almost all of the angiosperms have evolved more effective water- and mineral-conducting mechanisms than the *tracheids*, which conduct water and minerals in gymnosperms. These are the *vessels* which were described in Chapter 10. Vessels are tubes made of structures arranged end to end in which the walls have dissolved away at places where cells touch. They look like a series of drain tiles and permit tremendously rapid conduction of water and minerals. The evolution of this rapid water transport system permitted angiosperms to develop much larger leaves than gymnosperms, and consequently permitted more photosynthesis and more rapid growth.

The development of resistance to drought and cold

Angiosperms have evolved a number of adaptations to resist drought and cold. We have already considered tough resistant seeds. In addition many groups of angiosperms became **deciduous,** that is, they shed their leaves at times of the year when growth is not possible. This occurred in the north, where certain parts of the year were so cold that no water was available for growth. It also happened in savannas—tropical areas that have periodic drought. A few gymnosperms like larch are also deciduous, but the habit is largely an angiosperm feature which has enabled them to invade both dry and cold areas. Another adaptation of angiosperms to resist drought are **taproots** which permit them to tap deep sources of water and occupy relatively arid places.

Genetic flexibility

In addition to these numerous adaptations it appears that angiosperms have greater genetic flexibility and the capacity to respond to selection pressure than do gymnosperms. The leaves of most gymnosperms are homogenous compared to the diverse leaves of angiosperms which show tremendous adaptations to the physical environment. Consider angiosperms growing in dry areas. They have evolved thick waxy cuticles on their leaves and they may reduce the size of their leaves, which reduces water loss. Compare the $1/2$ inch leaf of a desert creosote bush with a 3 foot by 12 foot banana leaf!

The diversity of leaves similarly reflects the remarkable ability of angiosperms to adapt to adverse physical environments. In contrast, the diversity of flowers which we discussed in connection with pollination reflects the ability of angiosperms to adapt to the biological environment — the presence of insects and other animals which are able to pollinate them.

This genetic flexibility is not uniform among the angiosperms. Some families like the cactuses and the waterlilies have become entirely adapted to specific environments. Others such as the sunflower family include weeds, trees, shrubs, vines, and cactuslike succulents which are of worldwide distribution, from tropical rainforests to arctic tundra to arid deserts. Scientists still do not know what confers this great genetic flexibility on many groups of flowering plants. Part of this variability may come from their ability to increase their chromosome number and become *polyploid*. This permitted all sorts of additional variability to evolve. About half of the species of flowering plants owe their origin to an increase in chromosome number. Only a few gymnosperms are polyploid. Another source of genetic variability has been the evolution of *mechanisms which prevent self-pollination*. This is particularly important in the evolution of long-lived angiosperms such as forest trees, and probably has been important in gymnosperm evolution.

All of these factors taken together seem to have given the flowering plants significant selective advantages over the gymnosperms. These advantages must have accounted for a large part of the angiosperms' evolutionary success. Let us look briefly at the major groups of angiosperms.

Two subclasses — Dicots and monocots

Angiosperms are commonly divided into two subclasses, the Dicotyledonae and the Monocotyledonae. Botanists are not in thorough agreement on the matter, but most believe that monocots evolved from a primitive buttercup-like dicot soon after the flowering plants first appeared. Several features enable us to tell the two types apart. They get their names from differences in embryo structure. The bulk of an angiosperm embryo is made up of one or two leaflike units which may be packed with food. These seed leaves are called **cotyledons.** A monocot embryo such as we find in a grain of corn has one cotyledon. A pinto bean, on the other hand, has two cotyledons and is a dicot.

There are other distinguishing characteristics for each subclass. In *monocots* the veins of the leaves are usually parallel or nearly so. Their stems usually have bundles of vascular tissue scattered throughout. Ordinarily they do not have a cambium and secondary growth so that they do not usually become woody. The flower parts such as petals occur in multiples of three. Their root system is usually fibrous and lacks a principal root. In *dicots*, the veins of the leaves usually have many branches from a central vein instead of being parallel. The conducting vessels in dicot stems form a cylinder; they usually have a cambium and secondary growth and may become woody. The flower parts come in twos, fours or usually fives. They ordinarily have a large primary root with branch roots growing from it.

Among the *monocots* are such familiar plants as grasses and rushes; lilies, yuccas, irises, amaryllids, and orchids; palms, cattails, and skunk cabbages; and many of the pond weeds and all of the few dozen marine angiosperms. There are about 60,000 species of monocots. Most trees and bushes with broad leaves are *dicots*, as are herbaceous, soft-stemmed plants such as buttercups, carnations, poppies, mustard, roses, peas, cotton, cacti, parsley, blueberry, mint, potatoes, snapdragons, marihuana, and sunflowers. About 190,000 species of dicots have been described.

THE IMPORTANCE OF FLOWERING PLANTS TO MAN

It is difficult to overemphasize the importance of flowering plants to man. Humanity depends largely on them for survival. In the last 10,000 years the human population has increased from less than 10 million to more than 3.7 billion, largely as a consequence of the development of agriculture and the cultivation of angiosperms.

If some incurable disease were to destroy only one family of flowering plants, the grasses, human life as we know it would come to an end. The grasses include all of the cereal grain plants, and we are so dependent on them that without them our population would be reduced to a minute percentage of its present size.

Most of our important angiosperm food and fiber plants are annuals, and have been cultivated by man since prehistoric times. A number of them were first cultivated by the Indians of North and South America and were not introduced to the rest of the world until after Columbus. Today 42% by weight of the world's major food crops is composed of three crops first grown by Indians: potatoes, corn and peanuts. Lima, kidney, string and navy beans, tomatoes, avocados, chili peppers, pineapples, and sweet potatoes were also first cultivated by the native Americans. Corn was particularly important to the pioneers who learned from Indians about

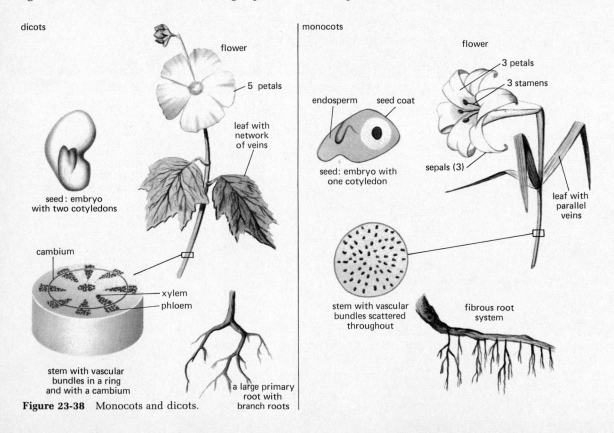

Figure 23-38 Monocots and dicots.

the many kinds of corn and their uses. The Indians grew every kind of corn: sweet, dent, flour, flint and popcorn and they made hominy, succotash, corn bread and cornmeal mush. In the Southwest and in Mexico they made tortillas and tamales. The pioneers became corn enthusiasts and from the beginning it outranked all European grains as the crop of choice. When money was scarce in Massachusetts Bay, corn was made legal tender. It is easy to see Robert P. Tristam Coffin's meaning in his poem "America is Corn":

> This sweetness built our beauty
> These kernels made us great.
> The corn ear is the Great Seal
> Of this American state.

The Indians knew of hybridization and planted different varieties of corn far enough apart to prevent undesired mixtures.

Few new species of food and fiber plants have been domesticated in modern times, but man continues to select new strains with special features. New varieties of corn have been selected which are rich in *lysine,* an amino acid which is necessary in human nutrition but is present in only small amounts in most plants. Proteins from plants such as lysine-rich corn may be able to substitute for animal protein in human diets and improve nutrition, particularly in countries which lack adequate sources of protein.

Flowering plants also provide us with numerous drugs which have physiological effects on man. These include nicotine from *Nicotiana tabacum,* caffeine from *Coffea arabica,* tetrahydrocannabinol from marihuana, *Cannabis sativa,* tranquilizers from *Rauwolfia,* quinine from *Cinchona.* It is worth inquiring how plants came to synthesize so many chemicals which affect human behavior. The reason is not hard to find: these chemicals are *defense mechanisms* for the plants that produce them. Grazing mammals avoid eating such plants because of their physiological effects. Similarly, many insects avoid eating certain plants because they contain materials which are repellent or noxious to them. Groups of plants which evolved the ability to manufacture such chemicals have a biochemical shield which protects them from most grazing mammals and from many insects. Not surprisingly, most chemicals which have physiological effects on grazing mammals also affect humans.

The use of flowering plants for medicinal purposes was well known to American Indians. More than 200 drugs derived from flowering plants that are part of the official drug com-

Figure 23-39   Eastern Woodlands Indians. Painting by A. A. Jansson.

pendia of the United States were used by the native Americans. These include quinine, sassafras, and witch hazel. Tobacco is, of course, one of the Indians' most famous contributions. For a short time it was thought to have healing properties. But, in light of current information, it looks more like the Indians' revenge for the various gifts they received from the white men such as cholera, smallpox, tuberculosis, and measles.

In addition to their role in providing food, shelter, and drugs, flowering plants play a crucial ecological role in preventing erosion. Their extensive root systems stabilize soil. When soil-retaining plants cease to grow because of poor agricultural practice, drought, or fires, vital topsoil is lost to wind and water, and the land is rendered useless for agriculture, recreation, or as watershed. Proper management of our environment requires thorough understanding of the interactions of angiosperms in various ecosystems, particularly the tropics.

Science also owes a great deal to the angiosperms for what they have revealed about life's processes. Gregor Mendel first gained insight into the hereditary action of genes by experimenting with the garden pea (Chapter 3). Today corn is widely used in basic studies of genetics. Tissues and organs of certain flowering plants grown in culture have given us important clues about the nature of embryonic development.

In the future, angiosperms will be studied more and more intensively as part of our efforts to meet the food demands of an increasing world population. We may expect advances in agricultural practice to help alleviate the famines which are approaching. But this new knowledge about angiosperms and their growth can help improve man's condition on earth only if the growth of human populations throughout the world ceases as soon as possible.

SUGGESTED READING LIST

Ahmadjian, V., "The Fungi of Lichens," *Scientific American*, February, 1963.
———, *The Lichen Symbioses*. Blaisdell Publishing Company, Waltham, Massachusetts, 1967.
Baker, H. G., *Plants and Civilization*. Wadsworth Publishing Company, Inc., Belmont, California, 1965.

Bell, P. R., and C. L. F. Woodcock, *The Diversity of Green Plants*. Addison-Wesley Publishing Co., Reading, Massachusetts, 1968.
Billings, W. D., *Plants and the Ecosystem*. Wadsworth Publishing Company, Inc., Belmont, California, 1966.
Bonner, J. T., *The Cellular Slime Molds*. Princeton University Press, Princeton, New Jersey, 1959.
Cobb, W. C., and J. S. Niederhauser, "The Late Blight of Potatoes," *Scientific American*, May, 1959 (Offprint 109).
Corner, E. J. H., *The Life of Plants*. Mentor Books, New American Library, Inc., New York, 1968.
Cronquist, A., *The Evolution and Classifications of Flowering Plants*. Houghton-Mifflin Co., New York, 1968.
Dawson, E. Y., *Marine Botany: An Introduction*. Holt, Rinehart and Winston, Inc., New York, 1966.
Delevoryas, T., *Plant Diversification*. Holt, Rinehart and Winston, Inc., New York, 1966.
Ehrlich, P. R., and P. H. Raven, "Butterflies and Plants," *Scientific American*, June, 1967 (Offprint 1076).
Jaques, H. E., *Plant Families. How to Know Them*. William C. Brown Company, Publishers, Dubuque, Iowa, 1949.
Jensen, W. A., and F. B. Salisbury, *Botany: An Ecological Approach*. Wadsworth Publishing Company, Inc., Belmont, California, 1972.
Lamb, I. M., "Lichens," *Scientific American*, October, 1959 (Offprint 111).
Large, E. C., *The Advance of the Fungi*. Dover Publications, Inc., New York, 1962.
Margulis, L., *Origin of Eukaryotic Cells*. Yale University Press, New Haven, Connecticut, 1970.
Raff, R. A., and H. R. Mahler, "The Non-Symbiotic Origin of Mitochondria," *Science*, Vol. 177, 575-582, 1972.
Raven, P. H., "A Multiple Origin for Plastids and Mitochondria," *Science*, Vol. 169, 641-646, 1970.
———, and H. Curtis, *Biology of Plants*. Worth Publishers, Inc., New York, 1971.
Scagel, R. F., R. J. Bandoni, G. E. Rouse, W. B. Schofield, J. R. Stein, and T. M. C. Taylor, *An Evolutionary Survey of the Plant Kingdom*. Wadsworth Publishing Company, Inc., Belmont, California, 1965.
———, *Plant Diversity: An Evolutionary Approach*. Wadsworth Publishing Company, Inc., Belmont, California, 1969.
Watson, E. V., *The Structure and Life of Bryophytes*. Hutchinson and Co., Ltd., London, 1964.
Wilson, C. L., W. E. Loomis, and T. A. Steeves, *Botany* (fifth edition). Holt, Rinehart and Winston, New York, 1971.

CHAPTER 24

Animals

Some 1,200,000 species of multicellular animals have been described. These include about 900,000 species of arthropods, 80,000 mollusks, 40,000 vertebrates and other chordates, 12,000 roundworms, 10,000 coelenterates, 7000 annelids, 7000 flatworms, 6500 echinoderms, 4500 sponges. In addition fully 50,000 single-celled protozoans are known. These ten major phyla contain more than 99 percent of the known living species of animals. The remaining 1 percent of living species belong to about 20 smaller phyla. They live everywhere one finds plants, except in very hot springs. Apparently in the course of geological time, wherever plants went animals were not far behind. Animals also live in some places in which no plants except a few fungi live, such as the bottom muds of ocean deeps. Thousands of species of animals spend their lives deep within the bowels of other animals, an ecological niche that even the enterprising fungi rarely invade. This chapter examines the immense diversity of animal life and habitat, the role of different groups of animals in the economy of nature, and their special importance to man.

In addition we shall try to discover possible evolutionary patterns among living groups of animals—who is related to whom. We shall also identify some of the factors that shaped the evolutionary history of various groups of animals. For example, what factors in their evolutionary history have resulted in all insects being small? Why have certain groups of animals like horseshoe crabs remained virtually unchanged for hundreds of millions of years, whereas others, such as the bony fishes, are undergoing rapid evolution even today? Finally we shall try to uncover the adaptive significance of certain major features of animals. What is the advantage of having segmented body parts like an earthworm, an insect, or a man? (Your nervous system is clearly segmented.) What is the adaptive significance of a coelom or body cavity like the one in which our own internal organs are suspended?

Our best clues to kinships among different groups of animals have come from comparing the structures of different kinds of living animals and seeking homologies. This is not always simple. As we already learned, most—and perhaps all—of the animal phyla arose in the sea. However, there has been an immense amount of **adaptive radiation,** and some phyla and classes of animals have many species occupying all of the major ecological niches, all well adapted. Mammals are a good example: they occupy land, sea, air, underground, treetops, apartment houses, Arctic ice floes and deserts. They have invaded the water at least three times (whales, sea cows and seals). In the process of adapting to these different environments modern mammals have become distinct from their ancestors. What did their ancestors look like? Clues are provided by the fossil record which tells us that the ancestors of modern mammals were ancient reptiles which lived some 200,000,000 years ago. Unfortunately the fossil record lets us down when we try to search out the origins of the phylum to which the mammals belong, the chordates, or of any of the other major animal phyla. Apparently the earliest members of each animal phylum were extremely ancient, soft-bodied creatures that have not yet been found as fossils. When the animal fossil record began in the Cambrian period, almost all of the major animal phyla had already evolved and were distinct.

Because of this lack of early fossils, zoologists have turned to other kinds of information to provide clues to phylogeny, especially the structure of embryos and of immature or larval forms. Embryonic structures appear to be insulated from most evolutionary changes, and the major features of an organism which form early in em-

bryonic development do not change much during evolution. For this reason we can obtain some useful clues about phylogenetic relations among major groups of animals by comparing embryos and immature or larval forms. Animals in which the early embryonic development is similar are probably related.

In addition to embryology zoologists have used all sorts of anatomical, cytological, biochemical, and behavioral information to search out the course of animal evolution and to characterize major living groups. Certain major features of this phylogeny are widely accepted—for

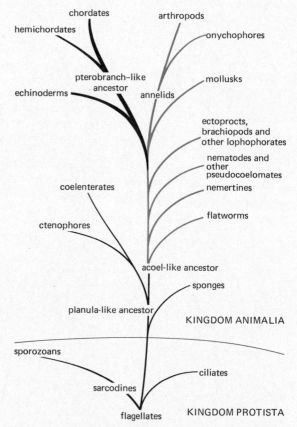

Figure 24-1 Possible evolutionary relationships among the major animal phyla. Two major evolutionary lines are indicated by different colors. The protostomes are colored brown whereas the deuterostomes are shown in black. These evolutionary relationships will be discussed in this chapter.

example, that arthropods descended from annelid worms. However, other parts are much less certain and are hotly disputed, such as the origin of vertebrates. One attempt at a phylogeny for the major animal phyla is outlined in Figure 24-1. Like all phylogenetic trees it is a working model and we accept it only until better evidence becomes available. It has four features which should be kept in mind as we analyze the diversity of animals in the rest of the chapter:

1. Protozoa and multicellular animals are placed in separate kingdoms.
2. The protozoans, which are largely single-celled, mobile heterotrophs, belong to the kingdom *Protista* along with most unicellular algae. Multicellular animals which are mobile heterotrophs are members of the kingdom *Animalia*.
3. One group of protozoans, the *flagellates*, evolved from the common ancestor of all the eucaryotes and gave rise to the other groups of protozoans.
4. The flagellates also evolved at least twice into multicellular animals. Once they gave rise to sponges and the second time to the common multicellular ancestor of the rest of the animal kingdom.

MAJOR TRENDS IN ANIMAL EVOLUTION

A brief look at the major trends and innovations that have occurred in animal evolution provides a road map for our journey ahead through the animal kingdom. The central evolutionary trends which we shall encounter are the following:

1. Multicellularity.
2. Division of labor among cells with specialized cells performing special functions such as muscle cells and reproductive cells.
3. Tissues—groups of cells performing specific functions, such as muscles and nerves.
4. Organs and organ systems—groups of two or more tissues performing specialized tasks, usually combined in an organized fashion to make up an organ such as a bone or a heart. Groups of organs usually function together to comprise an organ system such as the skeletal system and the excretory system.
5. Bilateral symmetry—the development of a right and a left side, a front and a back side.
6. A tubular gut with a mouth and an anus.
7. A body cavity or coelom.
8. Segmentation and appendages.
9. Exoskeletons or endoskeletons and jointed appendages.
10. Complex nervous systems with brains and sense organs concentrated in the anterior end of the animal.
11. Behavior patterns that can be modified by experience.
12. Increased specialization of the brain ultimately yielding the ability to abstract.

These trends describe a successive increase in the complexity of animals during the course of evolution. In many cases, the trends appeared independently in unrelated groups. For example, segmentation arose independently in annelids and vertebrates, and large brains arose independently in mollusks and vertebrates. The greater complexity of such advanced animals enables them to live under conditions that would have been impossible for their simpler ancestors.

An outline of the classification of animals is given in the appendix.

PHYLUM PROTOZOA

Protozoans include the simplest eucaryotic organisms. Their cell structure has all the features of a typical eucaryotic cell such as a nucleus and other organelles surrounded by a double unit membrane, mitochondria, vacuoles, microtubules, and chromosomes composed of both DNA and proteins. Their key feature is that they are all single-celled organisms. Most are solitary, but occasionally colonies of similar cells form. Since all the functions of an entire organism are performed within the confines of one cell, it is not surprising to discover that some protozoans have become very complex cells indeed, far more complex than any cell of a multicellular animal. No cell in a man is as complex in its structure and function as a single-celled *Paramecium* (Fig.

24-2). Although protozoans are the simplest eucaryotic organisms, they are not the simplest eucaryotic cells. They range in size from a 2 micrometer long protozoan called *Plasmodium*, which causes malaria and which occupies only a small part of a human red blood cell, to colonial protozoans 3 to 4 centimeters long. Most are about 100 to 300 micrometers long, that is, just barely visible to the eye.

Although the name "Protozoa" means "first animals," the idea behind it should not be taken too seriously. Admittedly, most protozoans are simple and show definite animal-like characteristics. They move about actively, scouring their surroundings for food, which they take into their body in small pieces the way most animals do. They are *mobile heterotrophs*. On the other hand, some single-celled organisms (often called protozoans) such as the euglena-like flagellates have chloroplasts and engage in photosynthesis and in this respect resemble green plants. Such flagellates are *mobile autotrophs*. The green plantlike forms were considered along with the algae in Chapter 23. The heterotrophic protozoans are considered in this chapter. As we indicated earlier, many biologists place the mobile hetero-

trophic protozoans and most mobile autotrophic unicellular algae in their own kingdom, the Protista.

Protozoans have no special organs for respiration or excretion and all of the exchange of gases and of waste products takes place across the cell membrane. Those that live in fresh water have special structures called **contractile vacuoles** that pump out the excess water that enters their bodies (it enters because the protozoan has a higher osmotic pressure than the water around it). These vacuoles become filled with water and periodically collapse, releasing their fluid contents to the outside. All types of nutrition occur in protozoans. Some are autotrophs and were already considered along with the algae in Chapter 23. Others absorb dissolved and digested food in the same way as do fungi and many bacteria. Still others have special feeding structures and ingest particles of food or entire organisms which they digest within a **food vacuole.** Some protozoans have complex skeletons and most have special structures for movement. They reproduce by the same kind of mitosis as the cells of higher plants and animals. In addition to dividing asexually by mitosis, some protozoans

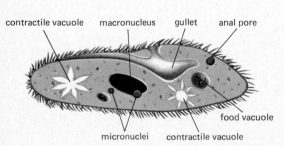

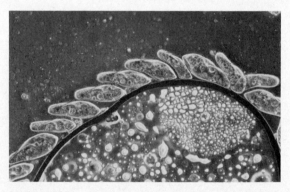

Figure 24-2 *Paramecium* (left) is a single-celled organism. The flower-like structures are contractile vacuoles which pump out excess water. Food particles are driven into the gullet by the beating cilia. The food particles are enveloped by a food vacuole into which digestive enzymes are secreted. The food is digested and the products of digestion are absorbed into the general cytoplasm. The food vacuole then moves toward the anal pore to which it attaches and expels digestive wastes. The body of a paramecium is completely covered by about 3000 cilia, although only a few are shown here. *Paramecium* and other ciliated protozoans differ from all other cells in possessing a large macronucleus and one or more smaller micronuclei. Besides these special organs *Paramecium* contains the mitochondria, endoplasmic reticulum and other organelles found in most eucaryotic cells.

Photograph at right shows paramecia clustered about an air bubble. The high oxygen content of the water next to the bubble attracts these protozoans.

have complex reproductive cycles in which they undergo meiosis and give rise to haploid gametes. Many others like the common amoeba have only asexual reproduction.

Protozoans are found wherever there is moisture, in soil, oceans, fresh water or in the moist interiors of other animals. This, however, does not mean that they die when moisture disappears. Although they require moisture for active life, most freshwater protozoans can survive extremely unfavorable conditions by secreting around themselves a hard, impervious, protective shell called a **cyst.** When the pond in which an amoeba lives dries up, when food gets low, or temperatures get too high or too low, the amoeba forms a cyst. Its metabolism drops extremely low within the cyst and it stays in that resting state until environmental conditions improve, whereupon it breaks out of the cyst and begins an active life again. The amoeba remains ready to encyst again if unfavorable conditions arise. The cysts of many protozoans can resist drying, boiling, freezing and the like.

Interestingly, unlike most larger animals, protozoans are cosmopolitan—the same species may be found on every continent. The reason is the ease with which these tiny organisms can spread from place to place. The movements of larger animals are commonly limited by various geographical barriers such as oceans and deserts, but protozoans are swept along by rivers and streams and their cysts are blown along by the wind. Mud on the feet and bills of shore birds and on the bodies of insects also disperses protozoans from place to place.

Their importance to man is incalculable. Hundreds of millions of human beings suffer from serious diseases caused by protozoans, such as malaria, sleeping sickness and amoebic dysentery. The rest of the animal kingdom is prey to protozoans that live in their bloodstreams and invade their tissues. In addition to leaving their mark on man and other animals, protozoans have left a gigantic mark on the geology of the earth. As the shells of marine protozoans called foraminiferans and radiolarians settle on the ocean bottom, they form mineral layers. The White Cliffs of Dover are the remains of foraminiferans which were raised when the land mass of the British Isles was formed.

So far about 50,000 species of protozoans have been described and these are traditionally divided into four classes principally on the basis of their means of locomotion. Members of the class *Flagellata* move about by means of their whiplike flagella. Those in the class *Sarcodina* extend forward protoplasmic processes called **pseudopods** or false feet, then pull themselves about with them. Those in the class *Sporozoa* are all parasitic and are unable to move themselves independently during much of their life. Members of the class *Ciliata* move by rhythmically beating the hairlike cilia which cover most or all of their body.

The relationships among these four classes is far from clear. The flagellates and the sarcodines share some features and are often placed in the same subphylum. Each of the other classes is commonly placed in its own subphylum.

Simplest protozoans—Class Flagellata

This class is sometimes called Mastigophora as well as Flagellata. Both names mean the same thing, "whip-bearers," but the former comes from Greek instead of Latin. Unlike amoebas, most flagellates have a fixed body form made possible by their stiff outer covering or **pellicle** (Fig. 24-3). As we noted earlier, their flagella have the typical structure characteristic of the flagella and cilia of all higher plants and animals and of other protozoans. Most of them have one or two flagella, but some have many. They use their flagella like airplane propellers that pull the animal forward. The flagellum may also act as a sensory organelle and stimulate some flagellates to activity when they contact prey. Most species reproduce asexually by cell division.

There is good reason to think that the common ancestor of all eucaryotes was a flagellate and that different groups of ancestral flagellates gave rise to some (and possibly all) of the other groups of protozoans, to various groups of algae, to multicellular plants, to multicellular animals, and to the sponges (although this is not the only alternative). Today's flagellates are, of course,

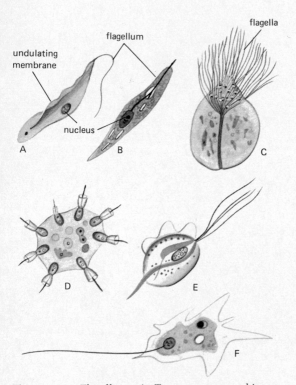

Figure 24-3 Flagellates. A. *Trypanosoma gambiense*, the cause of a major type of African sleeping sickness. The flagellum is attached to the cell by a thin membranous structure. This undulating membrane, as the flagellum and membrane are called, is used in locomotion. B. *Heteronema acus*, a free-living euglena-like flagellate. C. *Calonympha grassii*, a symbiotic flagellate that lives in the gut of termites. These flagellates aid in the digestion of the cellulose eaten by the termites. D. *Protospongia haeckeli*, a colony of flagellate cells which closely resemble the collar cells of sponges (Fig. 24-12). The individual flagellate cells are embedded in a gelatinous mass. The surface of the collar streams downward toward the cell surface. This enables the collar to act like a continuously moving flypaper that brings trapped food down to the cell surface, where it is engulfed by a food vacuole. E. *Trichomonas vaginalis*, a flagellate parasitic in the human vagina and male reproductive tract. The organism has four flagella and an undulating membrane. F. *Mastigamoeba* is particularly interesting because it has a permanent flagellum but utilizes pseudopods for feeding.

different from their ancient relatives. But they probably show fewer changes from them than any other living protozoan. Among the evidence of the relationship of flagellates to other organisms is the fact that many of the algae have cells which can hardly be distinguished from flagellates, and flagellated body cells are common in sponges and coelenterates. Even in higher organisms like man, the sperm, which is a brief haploid phase of human life, resembles a modified flagellate, and in those higher plants such as ferns which have mobile sperm, kinship with flagellates is evident.

Some of the animal-like heterotrophic flagellates are free-living and are found in both fresh and salt water, and moist soil. But most heterotrophic flagellates live within the bodies of higher animals and plants. An example is the flagellates which live inside the gut of termites and aid in the digestion of the cellulose eaten by the termites. Both the flagellate and the termite gain from this kind of relationship, which is called **mutualism.** In other relationships between flagellates and other organisms, only one partner, the flagellate, benefits; such relationships are called **parasitism.** African sleeping sickness is caused by two species of parasitic flagellate called *Trypanosoma gambiense* and *Trypanosoma rhodesiense*. Other members of this same genus cause numerous other diseases in man and animals. The trypanosomes are carried from animal to animal by bloodsucking flies called tsetse flies. Large parts of Africa will continue to be inhospitable to man and domestic animals until the flies or the disease can be controlled.

Amoebas and their relatives— Class Sarcodina

All sarcodines have flowing extensions of the body called pseudopods with which they capture food. The most famous organism in this class is the *amoeba*, but there are thousands of other species in both oceans, fresh water, and soil, and a few live inside other organisms. Some sarcodines, like amoeba, are naked, but most of them produce a shell of some sort around themselves. These shell-producing forms include the Foraminifera and Radiolaria which we mentioned earlier in this chapter. Modern sarcodines appear to have evolved from flagellates which developed pseudopods which helped them in feeding. Certain forms still alive today suggest some of the intermediate stages which may have occurred (Fig. 24-3). *Mastigamoeba*, for example, has pseudopods just like an amoeba, but it also has a long flagellum. It usually is classified among the Flagellata, but it could just as logically be placed in the Sarcodina. Another interesting form called *Naegleria* develops pseudopods or flagella depending on the salt concentration in the environment. There are also numerous sarcodines which produce flagellated gametes. These observations are evidence of close relationship between flagellates and sarcodines. However, as the sarcodines evolved, they exploited the new development of pseudopods to its fullest. Pseudopods not only became an important means of engulfing food for the whole class, but also provided a new method of movement. Most sarcodines use their pseudopods to creep slowly over the bottom of the ocean or a pond or crawl about over the surface of submerged vegetation.

Amoebas have flexible surfaces (Fig. 24-5). When the pseudopods extend outward, the cell cytoplasm flows into them. Food such as a bacterium, a *Paramecium*, or perhaps a small nematode is surrounded by pseudopods and water, which form a food vacuole lined with plasma membrane. Food is digested within this vacuole and whatever cannot be digested is later expelled. Most sarcodines ingest small particles of food, the

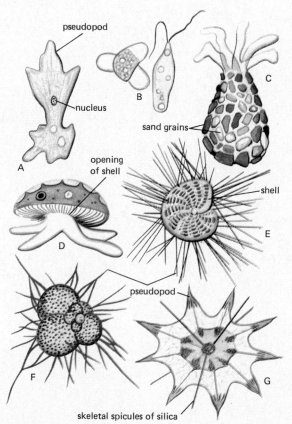

Figure 24-4 Sarcodines. A. *Amoeba* uses pseudopods for both locomotion and feeding. B. *Naegleria* develops pseudopods or flagella under different environmental conditions. C. *Difflugia* is a shelled amoeba which embeds sand grains in its shell. D. *Arcella* is another shelled amoeba. E. *Elphidium* is a foraminiferan. Its fine pseudopods protrude through perforations in its shell. F. *Globigerina* is a common foraminiferan whose remains form massive deposits called globigerina ooze. G. *Acanthometra* is a radiolarian. Note the spikes made of silica, each of which is covered with a pseudopod.

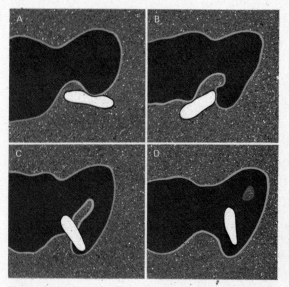

Figure 24-5 An amoeba captures a *Paramecium*. The amoeba detects its prey, moves toward it and surrounds it with a pseudopod. It forms a food vacuole lined with plasma membrane around the *Paramecium* and digests it. See also Figure 11-19.

way amoeba does and the way most animals do. But a few forms that live inside of other organisms absorb dissolved nutrients through their body surface. One such form is the amoeba which causes amoebic dysentery in man. The dysentery amoeba lives in the large intestine where it feeds upon the living cells and tissues, leading to the formation of abscesses and bleeding ulcers. Some amoebas, especially freshwater forms, contain contractile vacuoles which serve to pump out excess fluids.

The most spectacular looking and ecologically important of the sarcodines are the **foraminiferans,** which secrete shells composed of calcium carbonate, and **radiolarians,** which make shells composed of silica. Their pseudopods, which are usually thin and pointed, move in and out of holes in their shells and serve as a trap for food. When a diatom, nematode, or protozoan touches one of these thin pseudopods, it becomes stuck in an adhesive secretion that coats the pseudopod, and is engulfed into a food vacuole. Both foraminiferans and radiolarians occur in spectacular abundance in the plankton of the oceans. When they die, their shells sink to the ocean bottom and form a major part of the mud and ooze. They are often fossilized and as we mentioned earlier, much of the limestone, chalk and flint now present on earth was formed from deposits of foraminiferan or radiolarian shells. They are the only protozoans that have an extensive fossil record. Their fossils have been extremely well studied and provide petroleum geologists a convenient way of identifying the age of various formations. Some radiolarians have been found in the pre-Cambrian era and foraminiferans first appeared in the Cambrian. The largest single-celled protozoan ever discovered is a fossil foraminiferan called *Nummulites* which was 19 centimeters across. How what looks like a featureless blob of protoplasm can secrete a structure as intricate as one of the shells depicted in Figure 24-6 remains to be discovered. But they clearly do, and furthermore the structure of the shell is characteristic of each species and fixed by heredity.

Malarial organisms and their kin — Class Sporozoa

Sporozoans differ from the other protozoans in being motile only during a brief part of their life cycle. They are all parasites and live most of their lives inside another organism, feeding on its

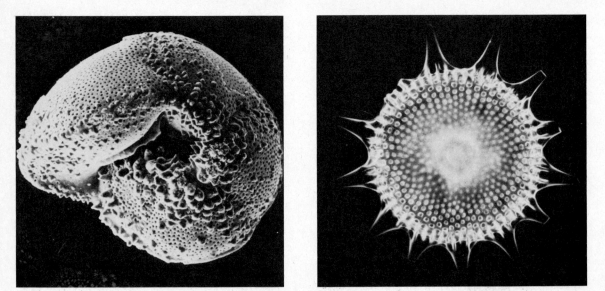

Figure 24-6 (Left.) Scanning electron micrograph of a foraminiferan, *Globorotalia trunculatulinoides* (X 200). (Right.) Finely sculptured shell of a radiolarian, *Aulacantha* (X 450).

tissues. The smallest sporozoans, only a few microns long, live inside the cells of other animals. The largest sporozoans may reach 10 mm in length and live in the body or gut cavities of various animals. They infect every major phylum including the Protozoa. Many sporozoans have complex reproductive cycles and spend part of their lives in several different organisms.

Most sporozoans reproduce asexually by forming **spores.** In spore formation, the nucleus of the protozoan divides mitotically many times to form a number of nuclei. The protozoan then splits up into a number of offspring corresponding to the number of nuclei, a process called sporulation. In some species the offspring are enclosed in a resistant wall and form a spore which resembles the cysts of other protozoans. In other species they remain naked. Spore production ensures infection from one host to another and continuation of the species.

The sporozoan which causes malaria in man belongs to the genus *Plasmodium* (Fig. 24-7). These sporozoans divide their life between *Anopheles* mosquitoes and man. When an infected mosquito bites a man, sporozoans from the mosquito enter the man's bloodstream and invade the liver cells. They divide several times into a dozen or more naked spores and then reenter

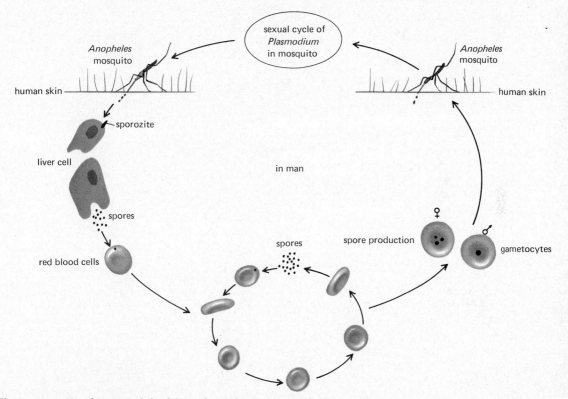

Figure 24-7 Key features of the life cycle of the malaria parasite *Plasmodium*. The female *Anopheles* mosquito injects a minute spindle-shaped form of the parasite called a *sporozoite* into the bloodstream of a man. The sporozoite finds its way to the liver where it feeds, enlarges and produces many naked spores. These spores enter red blood cells, grow and undergo rapid division into more spores which infect more red blood cells and the process repeats itself over and over until billions of red blood cells are infected. After a time, some of the spores in the red blood cells become female gametocytes and others become male gametocytes. When these cells are ingested by a mosquito, the sexual phase of the parasite's life cycle takes place in the tissues of the mosquito and results in more sporozoites capable of infecting man.

the bloodstream. Once in the bloodstream they enter the red blood cells, multiplying again. Periodic chills, fever and sweating occur when the billions of red blood cells rupture and release

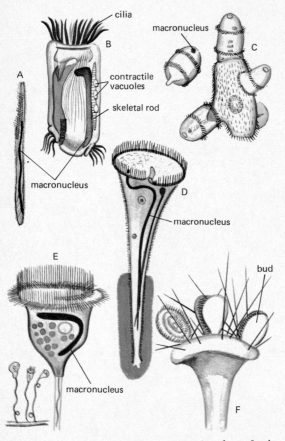

Figure 24-8 Ciliates. A. *Spirostomum* is a large fresh water ciliate. B. *Cycloposthium bipalmatum* is a parasitic ciliate found in the gut of horses. Its cilia are stuck together like paddles. C. *Didinium* can open its mouth almost as wide as its body. Here we see several of these barrel-shaped ciliates attacking a paramecium. D. *Stentor mulleri* lives in a gelatinous tube which it secretes. The large cilia around the "funnel" of the animal sets up a current of water that brings food to the mouth. E. *Vorticella* has a stalk which quickly contracts when the animal is stimulated. F. *Ephelota gigantea* belongs to a group of ciliates called suctorians. Here we see it giving off buds. The buds, which have cilia, will eventually metamorphose into non-motile adults which are attached by a stalk.

more parasites as well as poisons produced by them. Another human being may become infected when a mosquito bites the first person, picks up the parasite from him, and then bites another individual. Despite extensive efforts to control *Anopheles* mosquitoes and to improve chemotherapy, malaria still remains mankind's most important microbial disease. It weakens, kills and impoverishes hundreds of millions of people throughout the world.

The relation of sporozoans to other groups of protozoans is not entirely clear, but the life cycle of *Plasmodium* suggests their evolutionary origin: *Plasmodium* goes through an amoeboid stage within the red blood cells and produces flagellated male gametes in the mosquito. This suggests that sporozoans arose from the sarcodines, which in turn arose from flagellates.

The ciliates—Class Ciliata

Ciliates get their name from the fine cilia which cover them. These cilia have the same structure as flagella, but are usually more numerous and much shorter. More than 6000 species of ciliates have been described and they are found in both oceans and fresh water (Fig. 24-8). A few live inside other organisms. Ciliates have the most complex structure of any protozoan, including many of the structural features which evolved in multicellular organisms, such as a mouth opening and an anal pore. All of these structures evolved within a single cell through the specialization of cell organelles.

Ciliates differ from all other members of the animal kingdom in having two quite different types of nuclei: a single large **macronucleus** and one or more small **micronuclei** (Fig. 24-2). The macronucleus controls the metabolism of the cell. The micronuclei are involved only with reproduction and give rise to the macronuclei. When a ciliate divides asexually, the micronucleus divides by mitosis, but the macronucleus usually constricts in half (Fig. 24-9a). Many ciliates have an interesting sexual process called **conjugation** in which two individuals join together and exchange a haploid set of chromosomes from their micronuclei (Fig. 24-9b). Each conjugation results in the formation of four new ciliates.

Paramecium, shown in Figure 24-2, is a typical freshwater ciliate and is found in almost every pond, stream, lake, and river on earth. Its cilia move in coordination like oars and propel it through the water. A network of fine filaments connects the cilia and may coordinate their movement like a primitive nervous system. (Although *Paramecium* is not particularly speedy, some ciliates can oar their way through the water at 2 millimeters per second, a pace that would take them the length of an olympic-size swimming pool in three hours, which is not bad if one

is less than ½ millimeter long.) On one side of the *Paramecium* there is an oral groove which serves as the mouth. Cilia line the gullet and sweep the food inward. Food is digested within a vacuole, the remains being excreted through the anal pore. There are also contractile vacuoles like those found in amoebas.

Paramecia and many other ciliates possess explosive thread-like darts called *trichocysts* which they can discharge in a few thousandths of a second (Fig. 24-10). In some ciliates, trichocysts are used in defense or in capturing prey,

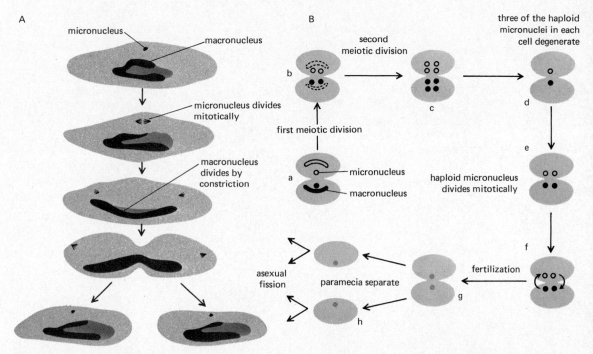

Figure 24-9 A. Asexual reproduction in *Paramecium.* The micronucleus divides mitotically, forming a spindle inside the nuclear membrane. The macronucleus divides by constriction. B. Sexual reproduction by conjugation in *Paramecium.* A species of *Paramecium* may have several mating types. When two mating types are mixed, conjugation occurs. An individual of one mating type becomes attached to an individual of another mating type. The large macronucleus in each cell breaks up and disappears and seems to play no role in conjugation. The micronucleus undergoes two meiotic divisions (a, b, c) to produce cells with four haploid micronuclei. Three of these four haploid micronuclei in each cell then degenerate (d). The haploid micronucleus which remains in each cell then divides mitotically to produce cells each with two haploid micronuclei (e). The cells now exchange one of their micronuclei (f) and fertilization occurs as the migrating micronucleus from one cell fuses with the resident micronucleus of the other to produce a zygote micronucleus in each cell (g). The zygote micronucleus is diploid, containing a haploid set of chromosomes from each of the two conjugating organisms. The cells now separate (h) and each individual undergoes a series of mitotic nuclear and cell division during which the macronucleus is reformed and four new paramecia are formed. These last events are not illustrated.

and in others they are used to anchor the animal when it feeds. In paramecia and other ciliates, the trichocysts, the cilia, and sometimes stiff plates which form a sort of skeleton are all contained in the outermost layer of the animal, the pellicle. This pellicle is an extremely complex living structure, far more elaborate than the simple plasma membrane of an amoeba.

When paramecia and most other ciliates reproduce asexually, they simply split in half. But when some ciliates divide asexually, one of the daughter cells is very small and arises as a tiny bud from the body wall. When this bud is released, it often looks very different from the adult it came from. After some hours the bud transforms into an adult. Here in a single-celled organism we have a "metamorphosis" (Fig. 24-8F).

From this description of the structure and functioning of ciliates it certainly appears that a ciliate has become just about as complex as it can without becoming multicellular. It seems to have

reached the limit of the number of specialized functions a single cell can carry out even with several nuclei. Ciliates in this sense are an evolutionary blind alley and are probably not on the main line of development of any higher organisms. To reach a higher level of organization, an organism must become multicellular and have different cell types available to carry out special functions. The most primitive and least complex of such multicellular organisms are the sponges which are considered next.

SPONGES, FILTERS OF THE OCEAN— PHYLUM PORIFERA

Sponges are a large group with more than 5000 living species. They are all aquatic and are abundant in every sea from the equator to the poles. Most of them prefer shallow water, but a few inhabit ocean depths and some live in fresh water. Some are as small as thimbles, others as big as bathtubs. They have been used by man since antiquity and sponge diving is still a thriving occupation in warm shallow waters near Greece, Florida, and Australia. A living bath sponge looks like a shiny piece of raw liver, but when dried and cleaned its skeleton is revealed as the familiar sponge we use for cleaning. The ancient Greeks used sponge skeletons to pad their armor. Romans used them to paint and polish. Even today most professional car washers and window cleaners agree that no man-made sponge can beat a good natural bath sponge for cleaning.

Sponges are all built on the same basic design—a hollow tube anchored to the bottom. They seem unresponsive and plantlike, but if some colored particles are added to the water near a sponge, considerable unsuspected activity is revealed. A steady stream of water flows out from one or more large openings at the top of the tube. The water enters the body of the sponge through thousands of microscopic pores which riddle the walls of the tube. These pores give sponges their name *Porifera*, which means "pore bearers." The sponge lives out its life as an animated filter, straining out diatoms, protozoa, and other microscopic bits of food from the water current that

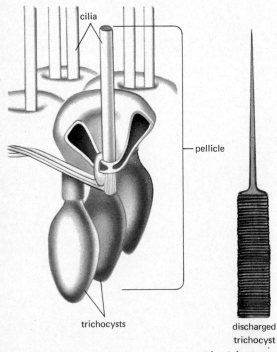

Figure 24-10 Pellicle of *Paramecium* with trichocysts. Discharged trichocyst at right.

flows constantly through its body. For each ounce of weight they may have to filter as much as a ton of water. Fortunately, sponges are remarkably good at filtering, and some filter more than 1000 times their own volume of seawater each day.

Sponges are multicellular and are composed

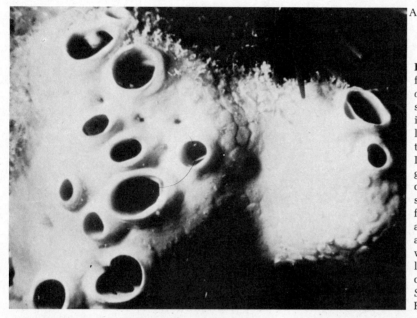

A

B

C

Figure 24-11 Sponges. A. California loggerhead sponge, *Spheciospongia confoederata*. This sponge is found typically on pilings and submerged rocks below low tide. It is closely related to the world's largest sponge, a West Indies species. A single large loggerhead sponge was once found to contain more than 16,000 small shrimp. B. A cluster of three different kinds of sponges growing on a frond of seaweed. The opening at the top of each sponge through which water flows out, the osculum, is clearly visible. C. A hermit crab carries a concealing sponge, *Suberites*, on its back. (Photo by Ralph Buchsbaum.)

of several different kinds of cells, but they are far simpler than other multicellular organisms. They have no mouth, no digestive cavity, no nervous system, no circulatory system. Indeed they have no organs of any kind. Even their tissues are poorly developed and show no real division of labor. Most of their cells do several jobs. For instance, one particular kind of cell called a collar cell pumps water, captures food, digests food, and also forms sperm and eggs.

The body wall of a sponge contains three layers (Fig. 24-12). The *outer* layer of cells forms a simple sort of epidermis. Next is a gelatinous *middle* layer through which amoeba-like cells wander. These wandering cells can form other kinds of cells when needed and also make the sponge's *internal skeleton*. This skeleton may be made of protein fibers like those we associate with bath sponges, or beautifully-shaped crystalline needles called **spicules,** or both. The *inner* layer of the tubular body consists of **collar cells.** The flagella within these cells sweep water through the sponge and out of the opening at the top. Their high collars are sticky and trap food particles which are drawn down the outer surface of the collar and engulfed by the cell body. The food is digested and shared by diffusion with the other cells in the sponge. In addition, digested food is carried by amoeba-like cells from the collar cells to epidermal cells and other nonfeeding types. Because all of the digestive processes of sponges are carried out within these single cells, sponges eat only microscopic particles from the sea water that is swept through their pores. All three layers are perforated by thousands of pores. Each pore is surrounded by a single cell shaped like a tube called a **pore cell** through which water enters. These pores and the large opening at the top of the tube through which water leaves are surrounded by spindle-shaped cells that can contract. If toxic substances are present in the water, these cells contract and cut down the flow of water through the sponge. Since no nerves are present, these contractile cells must respond directly to stimuli in the environment.

Sponges have no permanent sex organs, but they still manage to reproduce sexually. Eggs and sperm are formed from both amoeba-like cells and from collar cells. The larvae are ciliated and

free-swimming before they settle down on the bottom. Sponges also reproduce asexually by budding and by forming balls of amoeba-like cells called **gemmules** (Fig. 24-12B). These balls of cells secrete a tough outer coat and can survive

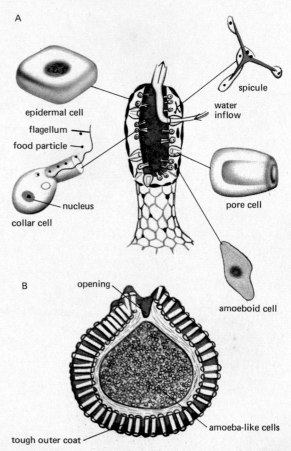

Figure 24-12   A. Part of a colony of sponges. The upper part is cut away to show the basic arrangement of the cells. Note that the collar cells look and function very much like the flagellated cells of *Protospongia* (Fig. 24-3). The surface of the collar streams down toward the cell surface where trapped food particles are engulfed by a food vacuole. B. Gemmules are balls of amoeba-like cells covered with a protective outer coat. They are produced by freshwater sponges and by some marine sponges, and enable the sponge to survive unfavorable conditions such as winter or drying out. When conditions again become suitable for sponge growth, the amoeba-like cells flow out of an opening at one end and form a new sponge.

freezing and drying. This is especially important to freshwater sponges in which the gemmules that survive the winter form new sponges.

In 1907 a marine biologist, H. V. Wilson, demonstrated that sponges have a fantastic capacity to reorganize themselves. He squeezed a living sponge through silk cloth to break it up into small clumps of cells. Within a short time the separated cells had reaggregated to form many small new sponges. Wilson also found that when he mixed the cells of two different species of sponges together, the cells reaggregated only with other cells of the same species. In some way the cells had an affinity for one another. These experiments which we discussed in Chapter 8 have provided some important leads in learning more about the interactions between cells when an embryo is taking shape and how cells recognize one another during development.

Sponges appear to have few enemies; apparently their skeletons make them distasteful or inedible to most predators. Other animals sometimes use the distastefulness of sponges for their own protection. Some species of crabs "plant" a sponge on their backs by holding it with their clawed hind legs until it has become attached.

Although sponges appear to represent a first step in evolution toward higher animal forms, they are considered an offshoot of the main course of animal evolution. There is no evidence that the sponges have ever given rise to any other group of animals. Fossils from Precambrian seas show that sponges have existed since life was in its early stages. They have not, however, advanced much for hundreds of millions of years. Although modern forms can pass more water through their cells than ancestral sponges, they have kept fundamentally the same way of life their ancestors had more than a half billion years ago. It is obviously a successful way of life, for they are an abundant and widespread phylum.

Some clues to the origins of the sponges are found among the flagellate protozoa. Their collar cells resemble certain free-living flagellate protozoans. Also the larvae of sponges resemble some hollow free-swimming colonial green algae such as *Volvox.* Sponges probably evolved from free-swimming, hollow colonies of flagellates. Because their development is so different from that of any other multicellular animal, it appears likely that they arose independently from the rest of the animal kingdom. They stand as living evidence that multicellular animals probably evolved at least twice from flagellates.

HYDRAS, JELLYFISHES, CORALS— PHYLUM COELENTERATA

The coelenterates live up to their name, which is derived from two Greek words meaning "hollow gut." Their digestive system is a hollow sac with one opening that serves as both a mouth and an anus. Although this is a primitive sort of an arrangement, it serves these animals well. They make up a diverse and widespread phylum with 9000 living species, including relatively familiar creatures such as jellyfish, sea anemones and corals among the marine forms, and the freshwater hydras mentioned so frequently in earlier chapters. They range in size from tiny hydra-like creatures, a millimeter long, to monstrous orange and blue jellyfish called *Cyanea,* which are found in the North Atlantic. The largest of these chilling menaces on record had a disk 12 feet in diameter and trailed tentacles 100 feet long.

Most coelenterates have the same basic body plan as hydra and are *radially symmetrical.* This has advantages for an animal that is attached to the bottom and does not move much, because it enables it to receive input from its environment in all directions. Some sponges are also radially symmetrical. Only one other major group—the echinoderms, many of which are attached to the bottom—is radially symmetrical.

Other features that typify the coelenterates are the presence of three layers of cells in their bodies, tentacles, and characteristic stinging cells, which produce microscopic harpoons called **nematocysts.** Zoologists once believed that the body wall of coelenterates was made of only two cell layers—an outer epidermis or *ectoderm* and an inner gastrodermis or *endoderm.* But most zoologists now consider the jellylike middle layer between the epidermis and the gastrodermis to be a true mesoderm. This layer, called the **mesoglea,** contains some wandering amoeba-like cells and a few connective tissue cells. It is comparable to, but much less developed than, the mesoderm found in higher animals.

Unlike sponges, coelenterates have some real tissues and more division of labor among the cells. Certain functions are concentrated in particular regions. Thus digestion is principally dele-

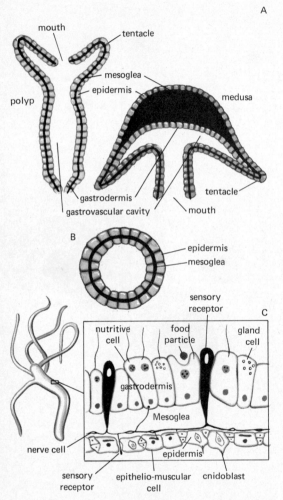

Figure 24-13 Coelenterate structure. A. Adult coelenterates are radially symmetrical and have two basic body forms, the tree-like polyp and the bowl-shaped medusa. The medusa resembles a flattened polyp turned upside down. B. Cross section of an adult coelenterate. C. Enlarged longitudinal section of part of body wall of hydra showing three layers of cell and several cell types.

gated to cells lining the gut. However, this division of labor is not complete, and certain cells carry out several important functions simultaneously. For instance, as we noted in Chapter 12, certain epidermal and gut cells also serve as muscle cells. Although muscular activity is carried on as a sideline by certain cells, nervous activity seems to be a full time job and coelenterates have perfectly good nerve cells. A primitive feature of these nerve cells which we discussed in Chapter 13 was that the synapses between them are not polarized, and impulses can go from cell to cell in either direction.

A good look at a jellyfish jet-propelling itself along, or a hydra feeding on brine shrimp reveals that the coelenterate nervous system, although simple, permits true coordinated activity. In this respect coelenterates are much further advanced than sponges. They have no organs such as hearts, brains, kidneys or gonads, which are made up of several tissues. They carry out all of their activities by parceling out the labors of the animal as a whole among individual tissues.

Earlier we noted that the epidermis and especially the tentacles, which are extensions of the body wall, are equipped with nematocysts loaded with poison, usually a nerve poison (Fig. 11-7). They are formed inside special cells that are found in the epidermis of all coelenterates and of one member of a closely related phylum of animals called *ctenophores*. Coelenterates use nematocysts to paralyze or otherwise immobilize the crustaceans and fish on which they feed. When a prey animal brushes against a tentacle, the stinging cells discharge nematocysts. This discharge is not controlled by the coelenterate. Instead the stinging cell seems to respond directly to mechanical stimuli and to chemicals in the prey. The advantage of this system is obvious: by requiring both a chemical and a mechanical trigger, they do not waste nematocysts on rocks, seaweeds, and so on. It is these nematocysts which make some coelenterates unpleasant and others, such as the Portuguese man-of-war, a menace to swimmers.

Coelenterates come in two basic forms, either a treelike **polyp** or a bellshaped **medusa** or jellyfish (Fig. 24-13). We are already familiar with polyps from our discussions of hydra. They are essentially long cylinders; one end is usually attached to a firm surface and the other end has the tentacles and mouth opening. The medusa is flatter and broader, like an inverted bowl, and is usually free-swimming. The two forms may even occur within the same species. The occurrence of different kinds of individuals within a species is called **polymorphism** (meaning many forms), and is a widespread phenomenon in animals.

We now review the life cycle of a typical coelenterate, considered in Chapter 15 (Fig. 24-14). The eggs and sperm are produced by a medusa and the zygote develops into a tiny ciliated larva called a **planula.** This larva settles down and develops into the next developmental stage, the polyp. The polyp then buds off medusae by an asexual process. The medusae swim away, mature, produce eggs and sperm, and the

cycle is complete. We therefore see an alternation between a sexual form, the medusa, which produces gametes, and an asexual form, the polyp, which buds off medusae.

Classes of coelenterates

There are three classes of coelenterates. The **Hydrozoa** include the freshwater hydras that were discussed earlier, simple marine hydroids like *Tubularia* and *Obelia,* as well as complex forms such as the *Portuguese man-of-war* (Fig. 24-15). The latter represents a group in which the organism is made up of colonies of polyps, each differentiated to perform specialized tasks, such as making a float to hold up the organism, capturing food, digesting it, or reproducing the organism. Both polyps and medusae are found among hydrozoans, but polyps predominate.

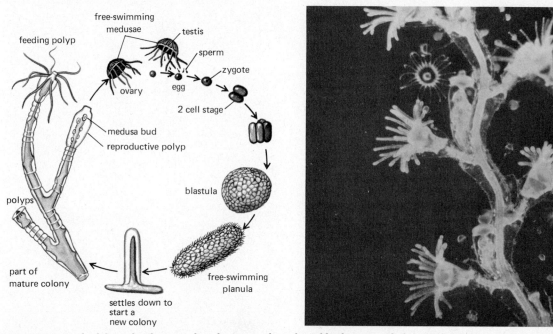

Figure 24-14 The life cycle of a typical coelenterate, the colonial hydrozoan, *Obelia.* In *Obelia* the predominant stage is the polyp which by asexual budding forms a branched colony of many polyps. Each polyp lives in a protective sac and is attached to other members of the colony by an extension of its body. Some of these are feeding polyps with tentacles. Others are reproductive polyps which bud off tiny free-swimming medusae. The medusae are either male or female and produce either sperm or eggs. The resulting zygote develops into a ciliated planula larva. In *Obelia,* the free-swimming medusae disperse the species and are the sexual stage. The polyps are the asexual stage. Both stages are diploid. The only haploid cells are the sperm and eggs themselves. Photograph shows extended polyps and newly liberated medusa. Notice the suctorian protozoan, *Ephelota* (Fig. 24-8), growing on the *Obelia.*

Figure 24-15   Hydrozoan polyps and medusa. A. A colony of the hydroid *Tubularia larynx*. B. Minute medusa of a hydrozoan, *Podocoryne carnea*. C. The Portuguese man-of-war, *Physalia physalis*, is a colonial hydrozoan capable of capturing fish in its trailing tentacles which bear powerful nematocysts. D. *Calpasoma dactyloptera*. The adult of this minute freshwater hydroid is only 200 to 300 microns long. Its tentacles are each composed of a single epithelio-muscular cell which has several channels, each containing a cnidoblast.

Figure 24-16 Scyphozoan medusae. (Left) *Chrysaora hyoscella*, an open ocean jellyfish. (Right) A few jellyfish do not swim free but live fastened by the top of their bell to seaweeds or stones. Two species of stalked jellyfish are here attached to a red alga. On the left is *Haliclystus auricula* and on the right is *Craterolophus convolvulus*. *Haliclystus* is less than 1 inch in diameter.

Their life cycle is usually like the typical one described. In some species, however, such as the freshwater hydras, this life cycle has been modified during the course of evolution so that the medusa stage has been lost completely and the polyp itself produces gametes.

The class **Scyphozoa** includes most of the larger jellyfishes. Many of these have the typical life cycle, but the polyp stage is usually very brief; the medusa is the most common form (Fig. 24-17). We have already mentioned giant jellyfish such as *Cyanea* which float on the open ocean trailing long tentacles bristling with stinging cells. Although these jellyfish are menaces to many animals, some have found them to be a haven from predators, oases of refuge in the hostile sea. Under the slowly pulsing bell of any

Figure 24-17 (Right) Life cycle of a scyphozoan, the jellyfish *Aurelia*. In *Aurelia* the medusa is the predominant body form. The polyp is a small brief stage that actively buds off young medusae in a manner which one zoologist likened to "the way slices might be removed from an ever-growing loaf of bread." These young medusae mature to form large jellyfish about 6 inches in diameter which produce either sperm or eggs. The zygote becomes a planula larva which settles down to become a polyp. As in *Obelia*, the medusa is the sexual stage and the polyp is the asexual state. Both sexual and asexual stages are diploid.

large jellyfish one is likely to find a variety of small animals, including young whiting, cod, and haddock, enjoying considerable protection from the jellyfish. Just how they themselves avoid destruction is not yet known. **Protective associations** of this sort are common in nature and

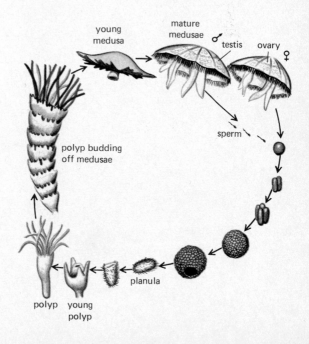

involve animals in almost every phylum.

Members of the class **Anthozoa** ("flower animals") have only the polyp stage and include the sea anemones and the corals (Fig. 24-18). They are considered the most advanced of the coelenterates and have a much more complex

A

B

C

Figure 24-18   Various anthozoans. A. A sea anemone, *Aptasia*. On the left is a cluster of another anemone, *Zoanthus*. B. *Stoichactis*, like several other large anemones, often has a small clownfish living among its tentacles. The fish, *Amphiprion percula*, is immune to the nematocysts and is thought to drive away other fish and sea turtles which might feed on the anemone. C. Although most corals are found only in warm waters, *Astrangia danae* can live in the cold waters off Cape Cod, Massachusetts. This colony consists of about a dozen individuals about ½ inch high. New individuals are produced by budding.

body structure than a simple hydra. Sea anemones are solitary and usually much larger than the polyps of hydrozoans. The giant anemone of the Great Barrier reef, *Stoichactis*, may be more than a yard in diameter. Most corals and other anthozoans form colonies of thousands of individuals. Some corals have soft bodies stiffened with scattered spicules, but the reef corals secrete a hard skeleton of calcium carbonate or limestone. Dense colonies of such reef corals occur in warm shallow seas and reefs may form as a result of the accumulation of layer upon layer of coral skeletons. Because of this ability to form reefs, corals are ecologically the most important of all coelenterates. A reef 150 feet deep may form in the course of 1000 years. The greatest of these reefs is the Great Barrier reef which runs for 1200 miles along the Australian Coast. Coral reefs are of special interest to geologists because ancient reefs that are now part of the continental land masses are important reservoirs of petroleum.

Coral reefs have the most abundant and varied array of animal life of any natural habitat. They are the skin diver's paradise. Until you explore a coral reef it is hard to take coelenterates seriously. There towering growths of colonial polyps and huge banks of reef building corals cover almost every square inch of the ocean floor. They replace the plants and, as a well known naturalist wrote, "They dominate the lives of the other invertebrates and even of the fishes, as the trees in a forest dominate the other plants and animals." Recently giant starfish have begun to devastate the great tropical coral reefs of the Pacific Ocean and many reefs may soon disappear. This could have catastrophic effects on fisheries and denote economic disaster for the small isles and coral atolls of Oceania. We shall return to this subject again when we discuss starfish and their kin.

Coelenterates have been extensively used in research because they are so simple. Their nervous system, stripped down to bare essentials, provides a useful model for studies of more complex nervous systems. Their developmental processes are also being actively studied, particularly their phenomenal capacity to regenerate a whole animal from a tiny tissue fragment containing a few hundred cells or less.

Related to the coelenterates are the **ctenophores** or comb jellies that are common in coastal waters during the summer (Fig. 24-19). They share several features with coelenterates including radial symmetry. At least one species has nematocysts. They are usually placed in a separate phylum, the *Ctenophora*, which comes from the Greek words for "comb bearers." Their most characteristic feature is eight vertical rows of plates of fused cilia, called *comb plates*, which move the animal about. These comb plates are iridescent in the sun, and a swarm of ctenophores swimming together is an unforgettable sight.

Ancestors and descendants of the coelenterates

Like the sponges, the coelenterates have not progressed greatly over hundreds of millions of years. All of the classes present today were already present 500 million years ago. It is widely believed that the ancestors of coelenterates were

Figure 24-19 Ctenophores. Sea gooseberries *(Pleurobrachia pileus)* are typical comb jellies and possess the two long tentacles used for food capture and the rows of comb plates for swimming.

colonial flagellates which gave rise to a ciliated planula larva of the sort found in most modern coelenterates (Fig. 24-20). This type of primordial ciliated larva is considered ancestral to all modern coelenterates. Furthermore, since the planula larva of modern coelenterates resembles a primitive flatworm called an *acoel*, many specialists believe that the coelenterates gave rise to ancient flatworms and to all higher groups.

That great student of invertebrates, Libbie Hyman, suggested that if an ancestral ciliated larval stock took up life on the ocean bottom and embraced the habit of creeping, this would have led to the development of a head-end and tail-end as well as an up-side (dorsal) surface and a downside (ventral) surface. This would, of course, result in *bilateral symmetry* in which there are two halves—right and left—that are almost mirror images. Such bilateral symmetry is found in all of the higher invertebrates except the echinoderms, and was an important advance. It paved the way for highly mobile organisms. The ventral surface was applied to the bottom mud, or rocks, or plants and became involved in processes like crawling. On the principle that "it's better to see where you are going than where you've been,"

the leading anterior end became the center for various sense organs and their accompanying nerves, and ultimately a brain and a head.

FLATWORMS—PHYLUM PLATYHELMINTHES

This phylum includes what most biologists consider to be the simplest and probably most primitive bilaterally symmetrical animals. They are of great scientific interest because of their simplicity and because among them are found the most insidious disease organisms known to man, the schistosomes. These parasitic flatworms cause misery and death to tens of millions of human beings each year.

As their name implies, almost all members of this phylum are extremely flat. They range in size from microscopic to 70 feet. Like coelenterates their bodies are composed of three layers, ectoderm, mesoderm and endoderm. But unlike that of the coelenterates, the mesoderm is well developed and contains many cells and fills solidly the space between the gut and the body wall. The fact that there is no body cavity or coelom between the gut and body wall is of considerable evolutionary significance, as we shall see.

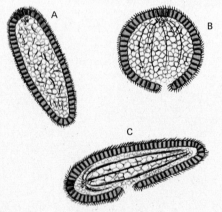

Figure 24-20 Possible origin of bilaterally symmetrical organisms like flatworms from the ciliated planula larvae of some ancient coelenterate. A. Planula larva without a mouth. B. Mouth appears and organism shortens. C. The body elongates and sense organs and the nervous system shift to what will become the head end. This form looks much like a primitive type of flatworm called an acoel.

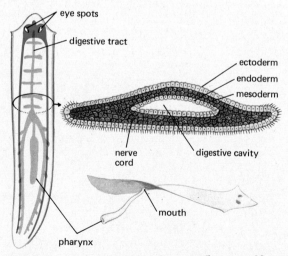

Figure 24-21 Structure of a planarian flatworm. Note the well developed mesoderm between the gut and body wall.

A big evolutionary advance over the coelenterates is the presence of several real *organs* performing specialized tasks. Organs for reproduction and excretion, as well as muscles, are particularly well developed. The development of an extensive mesoderm in the flatworms apparently led to an increased division of labor, and this in turn allowed the development of these specialized organs. There is also a simple central nervous system with a brain. However, flatworms have no skeleton or appendanges and there are no special organs for circulation or respiration. Absorbed nutrients and oxygen are distributed within the body by diffusion from cell to cell.

There are about 7000 species of flatworms divided into three classes, the free-living **Turbellaria** and the parasitic **Trematoda** and **Cestoda.** In adapting to a specialized life the parasitic classes have lost many of the features that typify the phylum, and therefore the Turbellaria hold the most evolutionary interest.

Planarians—Class Turbellaria

The best-known turbellarians are the *planarians*, which range in size from microscopic to several feet in length. At least part of the body is covered with cilia. They have an extensible pharynx in the middle of their underside with which they suck in food to their mouth. Their mouth opens into a *gastrovascular cavity.* Most of the planarians are carnivorous. Like coelenterates, they have no anus, and wastes are expelled through the mouth and the pharynx.

A familiar species, *Dugesia tigrina,* lives in many streams and ponds. The adult is about 2 centimeters long and gray in color. The one-opening digestive system is still primitive, but has many branches for carrying food throughout the body. They have well-developed reproductive organs, not just single cells that produce eggs and sperm. They also have excretory organs called flame cells and well-developed muscles. Their cilia propel them, but they use their muscles to wriggle and squirm along as well. There are also two eyespots near the worm's head; although they do not form images, they are light-sensitive. A nerve cord runs down each side of the worm.

The two cords have cross-connections between them, like a ladder, and join at the front end to form a rudimentary brain. The nerve cords are located ventrally just like the nerve cords of most major groups of invertebrates. Simple as this nervous system seems, as we mentioned in Chapter 13 it has many of the features characteristic of the nervous system of higher animals and can provide us with insights into the evolutionary origins of nervous systems.

The most primitive turbellarians are small marine worms called the **acoels.** These simple creatures do not even have a gut. The food they ingest is simply engulfed by *phagocytic cells* in their endoderm. Acoels are believed by many to have evolved from ciliated coelenterate larvae, which they closely resemble, and are thought to be the common ancestor of the more complex turbellarians and of the other metazoans (Fig. 24-20C). The turbellarians are probably the ancestors of the two parasitic classes that we consider next. These parasites have undergone immense modifications in adapting for "indoor existence." Since parasitism is so widespread a habit among animals, it is worth mentioning briefly some of the things that happen to all parasites.

The road to parasitism

A parasitic animal lives symbiotically in or on its host for at least part of its life cycle and benefits at the expense of its host. Parasites feed repeatedly on the bodies of their hosts but do not destroy them as a predator would its prey. In fact, even though parasites often harm their hosts, the parasites and the host sometimes reach a state of balance. This state of balance is of great value to the parasite because it cannot live with a dead host.

The greatest problem a parasite faces is "house hunting." Since it lives in some organ of a host animal, when the host finally dies, the parasite has to find a new host. It is like a fish having to find a new lake to live in every generation. Parasites respond to this problem by producing large numbers of eggs. The chance that any one egg or larva will reach the right kind of host may be only one chance in a 100,000, but if

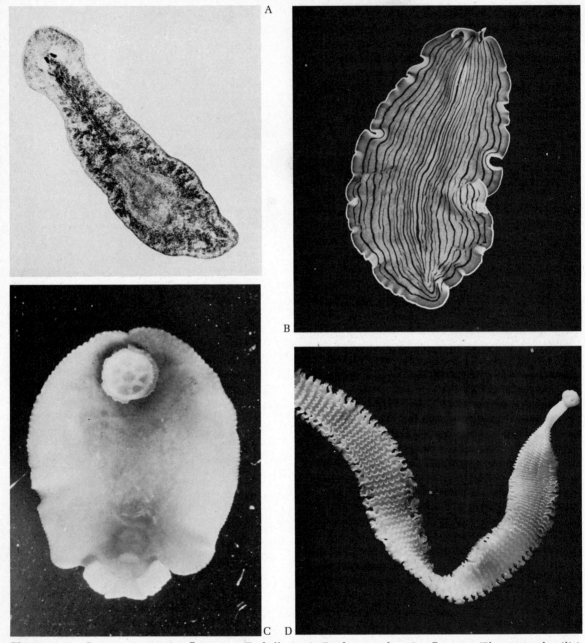

Figure 24-22 Some representative flatworms. *Turbellaria:* A. Freshwater planarian flatworm *(Phagocata glossilis)* and B. marine polyclad flatworm *(Prostheceraeus vittatus).* *Trematoda:* C. This fluke was found attached to the gill surface of a broadbill swordfish. This view reveals the anterior mouth and the posterior sucker disc which is divided into seven sections. *Cestoda:* D. A tapeworm found in an African rhinoceros. The segmentation of the animal and the head which bears suckers and hooks are clearly visible.

a parasite lays a million eggs it is reasonably certain that a few will make it. For this reason they all reproduce in great quantity.

Parasites have become extremely well adapted for their environment and have lost many structures commonly found in their free-living relatives. Like cave animals, internal parasites rarely have eyes. Apparently not having eyes is no disadvantage if one always lives in the dark, so there have been no selective pressures for eyes. Let us look at the two great classes of parasitic flatworms, the flukes and tapeworms.

Flukes—Class Trematoda

Flukes are mainly internal parasites of higher animals. They range in length from a few millimeters to several centimeters. Both they and the tapeworms have lost the soft, ciliated epidermis of the free-living flatworms. Instead, they have developed a hard outer cuticle that protects them from the body fluids and antibodies of their hosts. Most flukes have complex life cycles involving several hosts, one of which is almost always a snail. A common one in some parts of the world is the *Chinese liver fluke*, whose life cycle involves three hosts. A human being can pick up the organism by eating raw or undercooked fish, usually a carp or a trout, which contains larvae of the parasite. The immature flukes enter man's liver, living there and causing extensive damage, and producing eggs which leave his body with the feces. If some of the fecal material reaches fresh water, the fluke eggs hatch into larvae and some of them infect snails. They eventually reach the snail's liver. Here the fluke larvae reproduce asexually and produce another larval stage. A single infected snail may release as many as 200,000 larvae. This second larval stage leaves the snail and burrows through the skin of a fish. The larvae form cysts in the fish's muscles. The cycle starts all over again when a human being consumes the raw fish. All three hosts—snail, fish, and man—are needed to complete the reproductive cycle of this fluke. Such a pattern involving several hosts is common among many flukes.

Flukes are serious disease-producers in man and domestic animals. The ones taking the great-

est toll in human life belong to the genus *Schistosoma*. The life cycle of schistosomes is simpler and more insidious than that of the Chinese liver fluke. The larvae develop in freshwater snails and emerge from the snails into the water. They enter the human body either in drinking water or by boring directly into the skin. They then migrate into large blood vessels and cause local hemorrhages as they lay their spined eggs. The misery they cause is astonishing. Today more than 100 million human beings in tropical countries have schistosomes in their blood and are doomed to a wasting and debilitating disease. Wherever schistosomes are present it is dangerous to wade or bathe in fresh water and hazardous to drink untreated water. Each year millions of Asians standing barelegged in flooded rice fields pick up the disease as they plant their rice. In the Middle East the construction of new dams which supply irrigation canals, which we usually think of as progress, has created new breeding places for snails and *schistosomiasis* is on the rise. The

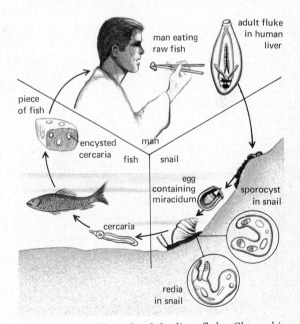

Figure 24-23 Life cycle of the liver fluke, *Clonorchis sinensis*. For details see text. The several different larval stages have specific names that are used in the figure.

disease will only be reduced when we control the snail populations, improve sanitation, and employ new drugs or perhaps immunization to kill the flukes in man.

Tapeworms—Class Cestoda

Adult tapeworms are all parasites of vertebrates, living almost always in the intestine. They have become so well adapted to their parasitic life that they no longer have a digestive system, not even a mouth. They absorb digested food from their host directly through the body wall. The head has suckers and hooks that enable the animal to hold on to the host. The extremely long ribbonlike body is composed of many segments.

Each segment in a tapeworm is really a reproductive apparatus, complete with male and female sex organs. When the segment matures it contains thousands of fertilized eggs. Mature segments, laden with fertile eggs, are continuously dropped off the tail end of the worm and leave the host animal with the feces. New ones are formed by a growth region near the "neck" of the worm. The tapeworm is clearly an impressive reproductive machine that turns out egg packages on an assembly line basis one after the other. In man tapeworms have been known to reach a length of 70 feet and may contain hundreds of segments. The segmentation of the tapeworm appears to be a special adaptation for the high reproductive potential required of almost all parasites. Free-living relatives of the tapeworms, such as the planarians, have no more than the usual number of reproductive organs and are not segmented.

The tapeworm most commonly encountered (or acquired) is the beef tapeworm. Adults live in the human alimentary canal. Eggs pass out in the feces. One beef tapeworm in a man may shed 800,000 eggs per day. When these eggs are picked up by a cow they develop into larvae which come to rest in the muscles of the cow. When man eats the beef raw or rare, the larvae hatch in his intestine and grow into adult tapeworms. Nowadays beef tapeworms are rare in urban communities. They can be avoided by never eating beef rare. Although unpleasant, they are not such serious disease producers as the flukes.

RIBBON WORMS—PHYLUM NEMERTINEA

This phylum is composed of slender unsegmented worms that have long, extensible proboscises for capturing food. The proboscis lies above the mouth and can be extended a distance sometimes equal to the length of the worm. They are all carnivorous and usually feed upon annelid worms by extending the mucus-covered proboscis out of their body, entangling their victim. They then pull it back into the mouth and suck it in whole. They are commonly found on the ocean bottom living between tidemarks, under

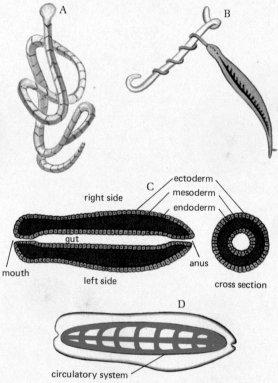

Figure 24-24 Nemertines. A. Pacific coast marine nemertine *Tubulanus capistratus*. B. Freshwater nemertine *Prostoma* capturing an annelid worm with its proboscis. In some nemertines the proboscis is armed with a sharp barb. C. Diagram showing three body layers of a nemertine and the digestive system with mouth and anus. D. Diagram of the circulatory system of a nemertine.

stones or seaweed, or coiled up in burrows in muddy sand. Most are only a few inches long, but one common New England species, *Cerebratulus*, reaches lengths of more than 6 feet.

The ribbon worms appear to be an offshoot of the free-living flatworms with whom they share several features, such as bilateral symmetry, a ciliated epidermis, the absence of a coelom and segmentation, similar excretory and muscular systems, and similar embryonic development. However, they show two significant evolutionary advances over the flatworms: a digestive system with two openings—a mouth and an anus, and a circulatory system for transporting substances from one part of the body to another. Food is taken into the mouth, gradually digested as it passes through the gut, and wastes are discharged through the anus. The adaptive value of this one-way traffic is that partially-digested and indigestible particles are not mixed. In addition the gut itself can readily evolve specializations along its length for digestion and can add or remove substances at certain points. This happens in more advanced ribbon worms and in higher animals.

The circulatory system is very primitive, with three large vessels running the length of the animal; they are connected at the ends. The animal has no heart, and blood flows when the animal's muscles contract as it moves.

The significance of ribbon worms to man is evolutionary. They show us the beginnings of two important adaptations that have been extensively exploited by other organisms.

TWO ROUTES TO ONE-WAY TRAFFIC

At this point it becomes useful to look for a moment at the embryology of representatives of several of the phyla for clues to their kinship. Recall from Chapter 7 that during embryonic development the primitive embryonic gut or archenteron has one opening to the outside, the blastopore. In hydras and flatworms the situation is similar and the gut has one opening. But in nemertines and all other animals the gut has two openings—a mouth and an anus. Does the blastopore become the mouth or the anus?

In one large group of animals that embraces the nemertines, nematodes, annelids, mollusks, and arthropods, the embryonic blastopore becomes the mouth, and the anus is a new opening. These phyla are often called the **protostomes** (Greek for "first mouth"). In the other phyla, notably the echinoderms and chordates, the mouth is a new opening and the embryonic blastopore becomes the anus. These phyla are called the **deuterostomes** (Greek for "late mouth"). In both cases the result is one-way traffic through the gut. It appears, however, that each of these great groups of phyla developed one-way traffic independently. Apparently basic differences in embryonic development led to the appearance early in the Precambrian era of these two major evolutionary pathways of the animal kingdom.

These two major groups also differ in other features of their early embryology which are summarized in Figure 24-25. A notable difference is in the origin of the **mesoderm**. In coelenterates what little mesoderm there is comes from ectodermal cells which wander inward. In protostomes most of the mesoderm is derived from a few cells located near the blastopore. In most of the deuterostomes (except for the vertebrates) the mesoderm arises from pouches that are pinched off from the embryonic gut wall.

Another important difference between protostomes and deuterostomes concerns the formation of the **coelom,** which is the cavity between the gut and the body wall. If you were to cut a slice through an earthworm, you would see that the animal really consists of a *tube within a tube*. The inner tube is the gut which is encased by muscle; the outer one is the muscular body wall. Between the tubes is the fluid-filled body cavity or coelom. It is lined with muscle which is mesodermal tissue. If you were to cut a slice through a nemertine worm, you would find the space between the body wall and the gut filled with mesoderm and you would not find a coelom. With the exception of roundworms and their relatives, which we consider in a moment, all of the other major animal phyla we discuss from now on have a true coelom.

What is the adaptive significance of the coelom? In all phyla that possess it, it provides a cushioned area of the body for many of the vital

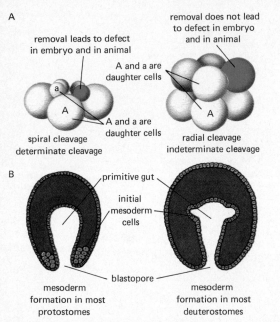

Figure 24-25 Differences in the embryonic development of protostomes and deuterostomes. A. *Spiral and radial cleavage*. In members of the protostome phyla, such as the annelids, after fertilization the cell divisions of the zygote orient the resultant cells into a definite spiral pattern, called spiral cleavage. In members of the deuterostome phyla, such as sea urchins, cleavage gives rise to cells stacked above one another, or radial cleavage. *Determinate and indeterminate cleavage*. In protostomes, the commitment by cells to differentiate in a particular direction often begins with the first cleavage after fertilization. Even in the two-celled stage, each cell is destined to give rise to different tissues and organs of the adult. If a particular cell is removed from the embryo, then the animal which develops usually lacks a certain part. For instance, if you remove a particular cell from an early cleavage embryo of a snail or an annelid worm, the larva will be missing a specific structure. Most of the cells seem to have a fixed fate. This is called determinate cleavage and is characteristic of protostomes. In deuterostomes the cells formed by the first few cleavages of the fertilized egg do not become committed as quickly and retain the ability to develop into complete adults if separated. For instance, if you remove a cell from an early cleavage embryo of a starfish or a frog, a normal embryo and adult will still develop. This is called indeterminate cleavage. B. *Origin of mesoderm*. For details see text.

organs that protrude into it. It has fluid in it that makes it possible for internal organs to fill, to empty and to slide past one another. The long, coiled human intestine and the expanding and contracting human lung are examples of organs occupying the coelom. For animals that move rapidly, protective cushioning is clearly an advantage. In some animals such as earthworms, the coelom is filled with fluid and functions as a hydraulic skeleton of the sort we described in Chapter 12 which enables the animal to move vigorously. In many organisms the coelom also serves as a place where sperm and eggs can unite, thus increasing the efficiency of fertilization.

Because of differences in the way they form mesoderm, protostomes and deuterostomes form the coelom in a different manner. However, in both cases it is a body cavity lined with mesoderm. In protostomes the cells that give rise to the mesoderm produce a solid mass of mesoderm tissue between the ectoderm and the endoderm. This mass of mesoderm splits down the middle and the cavity becomes the coelom. In the deuterostomes the coelom arises differently as a cavity in the mesodermal pouches that form in the gut wall of the embryo.

These many differences in early development between protostomes and deuterostomes indicate that they represent two independent lines of evolution. On the other hand, the many similarities in early development of the several protostome phyla indicate that all the protostomes probably had a common ancestor. The same is

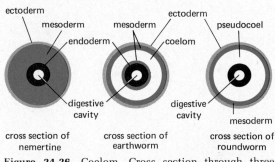

Figure 24-26 Coelom. Cross section through three different kinds of animals to illustrate basic body plans of an animal without a coelom (nemertine), with a coelom (earthworm) and with a pseudocoel or false coelom (roundworm).

true of the deuterostomes. Figure 24-1 summarizes a possible phylogenetic scheme. Other schemes are being seriously considered at this time. This is clearly one of the places where some of the techniques of molecular taxonomy we discussed in Chapter 21 might enable us to confirm or reject various hypotheses about possible phylogenetic relationships.

ROUNDWORMS AND OTHERS— PHYLUM ASCHELMINTHES

Most of us have heard of roundworms even if we have never seen any of them, for the phylum includes several common parasites found in man—hookworms, pinworms, and the trichina worms responsible for trichinosis. Actually, only about 50 of the 10,000 described species of roundworms afflict man. The rest live almost anywhere—as parasites in other animals and in plants or free-living in the soil and in salt- and freshwater bottom deposits. Most nematodes are free-living. Dig up a shovelful of earth almost anywhere and you will turn up thousands of them, including probably several that have never been classified. One decomposing apple may contain nearly 100,000 roundworms belonging to a number of different species. One forkful of undercooked pork may contain 100,000 cysts of the trichina worm, each enclosing a young worm capable of producing more than 1000 offspring. The diets of roundworms are as different as their habitats.

All species of higher plants and animals have at least one species of nematode parasite. At this present moment about half of the readers of this book are hosts to either adults or eggs or larval stages of these worms. Nematodes, which are parasitic in plants, destroy more than one-tenth of the crops grown by American farmers. By sheer weight of numbers they even vie with insects. They are not to be confused with earthworms, which are segmented worms in another phylum. Roundworms make up the class **Nematoda,** the most plentiful and important of the five classes in the phylum **Aschelminthes** and the only class we shall consider. Nematodes range in length from about 0.1 to 1 millimeter for most free-living

forms up to 2 meters in the case of some parasites. They look like the *Enterobius* in Figure 24-27. Interestingly, no matter whether nematodes are free-living or parasites of plants or of animals, they all look very much alike, even in their larval stages. This is in sharp contrast with most other animals in which adaptations to different environments are usually accompanied by conspicuous changes in structure.

Like the ribbon worms, all members of the phylum Aschelminthes have a three-layered body and a complete digestive tract with a mouth and an anus. But a significant structural innovation is a cavity that lies between the gut and the outer body wall. The body wall is lined with a layer of mesoderm. Hence the cavity is partly lined by mesoderm and partly by the endoderm of the gut. It is called a **pseudocoel** or false coelom to distinguish it from a *true coelom,* which is completely lined by mesoderm and found in all of the other major animal phyla that we discuss from now on. The pseudocoel of the nematodes and other aschelminths provides space for the reproductive organs; the other organs are contained in the body wall itself.

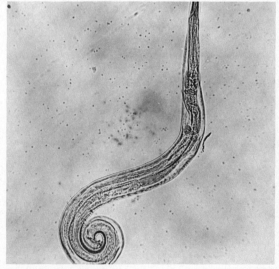

Figure 24-27 Nematode. *Enterobius vermicularis,* the familiar "pin worm," is parasitic in the intestines of man. This animal and all other nematodes have a similar structure. In this specimen the contents of the body can be seen through the transparent body wall. The mouth is at the top.

Movement is not one of the specialties of nematodes. First, they have no cilia. Second, they have no circular muscles, only longitudinal muscles. A few manage to swim moderately well, but most are quite inactive. Because nematodes almost always live in their food source as do bacteria or tapeworms, this inability to move well is not much of a disadvantage. Another feature of nematodes worth noting is that they are all covered with a thick cuticle which they shed as they grow, like the arthropods.

One of the striking features of the nematodes and of the other aschelminths is that many organs of the adult body contain a fixed number of cells. For example, in a certain species of nematode there are 120 epidermal cells, 200 nerve cells, and 172 gut cells. This feature makes them excellent objects for studying development and neurophysiology. The fact that all members of a given species have a constant and small number of nerve cells makes nematodes ideal organisms in which to analyze the basic architecture of the nervous system. Another novel feature of many nematode muscle cells is that the muscle cell sends out long branches that make contact with a nerve fiber. In most other animals it is the nerve fiber that branches to make contact with an unbranched muscle cell.

The most dangerous nematode you are likely to meet is the trichina worm which is found in rats, in pigs and in man, where they cause the disease of trichinosis. The adult worms are 3 to 4 millimeters long and live in the small intestine where they do little harm. However, after they mate the female burrows out into the blood with a load of more than 1000 fertilized eggs. She keeps these inside her uterus until they hatch. The young worms, about 0.1 millimeter long, are carried to all parts of the body by the blood and lymph. They then bore out of the bloodstream into muscles, particularly the tongue, eyes, ribs, and diaphragm. There they grow further and finally a cyst forms around each worm and becomes calcified. The worm, now entombed, stops growing until the cysts are eaten by a new host. The cysts break down in the gut, the worms escape, grow, mate, the females burrow out again, and the cycle is repeated. Most of the damage is done when these millions of larvae bore through the body almost simultaneously. Anemia, fever, and excruciating muscular pain occur; there may be permanent damage to muscles and even death.

An ounce of infected pork in the middle of a large piece of meat, where the heat does not easily reach, may contain 80,000 cysts which will hatch out in the gut. About half of these will be females, each of which will produce more than 1000 larvae. Thus eating that ounce of infected pork can result in 50,000,000 trichina larvae burrowing through a man's blood vessels and muscles. The most common reason for meat inspectors rejecting pork as unfit for human consumption is the presence of trichina worms. Just as in the case of tapeworms, if the meat is thoroughly cooked, infection can be avoided.

The kinships of the Aschelminthes are unknown. Perhaps they originated among the turbellarian flatworms or perhaps from degenerate annelids. They are an interesting side branch on the main line of invertebrate evolution.

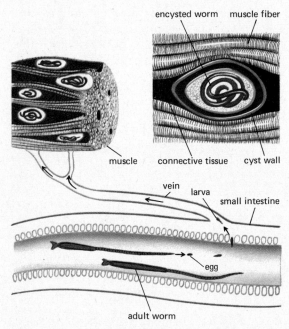

Figure 24-28 Life cycle of the trichina worm, *Trichinella spiralis*, which causes trichinosis.

EARTHWORMS, LEECHES, AND MARINE WORMS— PHYLUM ANNELIDA

The most familiar annelids are earthworms. The name of the phylum **Annelida** is derived from a Latin word meaning "little ring"; the ring-shaped segments that make up these animals are very apparent. They are all soft-bodied worms and are found in the oceans, fresh water and soil. About 9000 species have been described. All have well-developed organs, most of which are segmented. We find excretory organs, gonads, nerve ganglia, muscles, and blood vessels repeated in many segments. Only the digestive system is unsegmented. Most annelids have appendages on the segments which bear stiff bristles called **setae.** They move their appendages by means of muscles, but their appendages are not jointed. Earthworms have no appendages, but they do have setae. Furthermore these setae have muscles, and earthworms use the setae in locomotion, as described in Chapter 11. The basic body plan of an earthworm is illustrated in Figure 24-29.

Annelids do not have a rigid skeleton. Instead their coelom is filled with fluid and serves as a *hydraulic* skeleton, as described in Chapter 11, and transmits pressure from one part of the worm's body to another. When any muscle in the body wall of an annelid worm contracts, it presses the body wall against the fluid in the coelom. This in turn causes the stretching of any relaxed muscles. By this means annelids can

Figure 24-29 Anatomy of an earthworm, illustrating the annelid body plan. A. *The digestive tract and circulatory system.* The mouth opens into a highly muscular pharynx which sucks soil containing decaying vegetation into the gut. This material is stored in the crop and pulverized in the gizzard with the aid of the sand grains in the soil. The rest of the gut consists of a long intestine where food is digested and absorbed. The circulatory system is composed of three longitudinal blood vessels which run from head to tail, a dorsal vessel and two ventral vessels. The dorsal vessel receives blood from smaller vessels in each segment. It is extremely muscular and pumps the blood forward. Five muscular pumping vessels, the hearts, propel blood downward to the ventral vessels from which it flows back to the posterior segments. The dorsal vessels and the hearts have valves that prevent backflow, and as a result the blood flows in a circuit. B. *The nervous system* consists of two ventral nerve cords which are fused throughout most of the body. This fused nerve cord divides near the pharynx and fuses again to form the earthworm's brain. C. Cross section revealing the large coelom in which numerous organs are suspended. The excretory organs, called nephridia, are also visible. Each of these simple kidneys consists of a long tubule lined with cilia. One end is funnel-shaped and opens into the coelom. The other end of the tubule opens to the outside via an excretory pore. Coelomic fluid is drawn into the tubule by the action of the cilia. As the fluid passes down the tubule various useful materials such as salts and sugars are reabsorbed by the cells of the tubule and returned to the coelom.

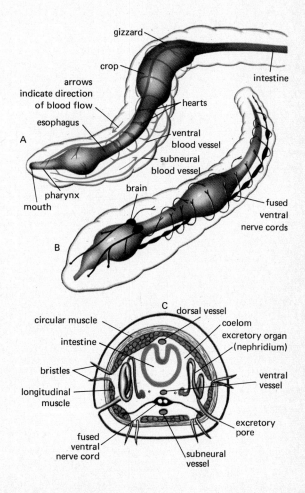

crawl and burrow far faster and more efficiently than ribbon worms which have no hydraulic skeletons.

The circulatory systems of annelids are especially well-developed, and they exchange oxygen and carbon dioxide through minute blood vessels in the appendages or at the surface of the body.

There are three classes in the phylum Annelida. The class **Oligochaeta** includes the earthworms and most freshwater annelids. Earthworms have an unusual way of life: They simply eat soil. Their digestive system absorbs decaying organic material, and whatever cannot be digested passes out the other end. Because of this,

Figure 24-30   Representatives of three classes of annelids. A. *Oligochaetes*. Earthworms like *Lumbricus terrestris* are the most familiar annelids and have bodies composed of many separate segments. They are hermaphrodites and mating involves the exchange of sperm between the two partners. The setae are not visible in the photograph but are evident in Fig. 24-29. B. *Polychaetes* are the most abundant and typical members of the phylum Annelida. The simplest polychaetes are free-swimming worms like this clamworm, *Nereis*, which has paddle-like appendages on each segment bearing setae. These appendages are used for swimming, but they also serve as gills because they have a large surface area and a rich blood supply. They are carnivorous and have stout jaws. At night *Nereis* swims or creeps over the bottom and hunts for dead mollusks and other bits of food. C. *Leeches*. *Haemopsis grandis* is a leech which feeds on the blood of various aquatic vertebrates. One of its suckers is clearly evident.

they play an important role in mixing layers of soil, aerating it, and reducing its acidity with their secretions, thereby improving its agricultural value. Although most earthworms are of modest size, giant earthworms 10 feet long and 1 inch in diameter occur in Australia.

A second class, the **Hirudinea,** includes the leeches. Many of these are bloodsucking and feed on vertebrates. Others feed on the blood of insects and snails, and still others are more typical carnivores, eating earthworms, slugs and insect larvae. Their distinctive features are the presence of two muscular suckers, one in front and one in back. When a leech feeds it selects a thin area of skin—an ankle rather than the sole of a foot—and attaches itself with its posterior sucker. It then applies its front sucker tightly to the skin, slits open the skin with its jaws, and secretes a chemical into the wound that prevents the blood from coagulating. It then feeds for a time, usually gorging itself. Then it drops off and may not feed again for months.

Earthworms and leeches, although the most familiar annelids, are not the most numerous. Most annelids live in the sea and belong to the third class called the **Polychaeta.** The clamworm or ragworm *Nereis,* illustrated in Figure 24-30, is a fairly typical polychaete. Most of its segments bear a pair of flaplike appendages that have setae. These basic features of polychaetes are used along with wriggling movements of the body in swimming, crawling, and burrowing. Many burrow in sand or mud or beneath rocks. Others live within tubes they themselves manufacture. Some of these display exquisite plumes that function in feeding and in respiration. Others live completely hidden within their tubes. One of the most interesting of these hidden forms is *Chaetopterus,* the parchment worm.

Chaetopterus secretes a parchmentlike U-shaped tube in the mud and lives in this tube. It uses its paddle-shaped appendages to create a current of water through its tube. At the same time it secretes a thimble-shaped bag of mucus in which it catches food particles brought in by the water currents. About every 15 minutes it eats the mucus bag full of food and makes a new one. The bag is so efficient in capturing food particles that there is evidence that it may even trap large protein molecules. The diligence of this little worm has been exploited by other animals. Inside *Chaetopterus* tubes one commonly finds tiny crabs called pea crabs and porcelain crabs that feed by filtering the water current maintained by the worm. They also get their oxygen from the water current. This kind of food-sharing association between two different animals, widespread in nature, is called **commensalism.** In the case of *Chaetopterus* and the crabs, one of the partners, the crab, clearly gets the better end of the deal, but neither is harmed by the partnership.

An interesting difference between marine polychaetes such as *Chaetopterus* and freshwater oligochaetes concerns the size and number of eggs they produce. Marine polychaetes lay thousands of tiny eggs. Freshwater oligochaetes lay a much smaller number of large eggs rich in yolk. Other groups of animals also show a similar correlation between the type of environment they inhabit and the size and numbers of the eggs they produce. By and large, most marine annelids, mollusks, crustaceans, echinoderms, and other animals produce large numbers of small eggs. Sperm and eggs are released into the sea water and fertilization usually takes place externally. This is true of polychaetes. In contrast,

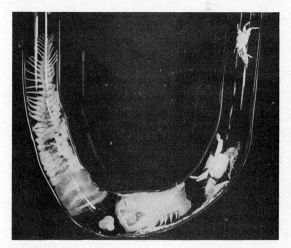

Figure 24-31 *Chaetopterus variopedatus* lives in a U-shaped tube which it secretes in the mud. It shares its tube with two tiny crabs, a pea crab, *Pinnixa chaetopterana* (top), and a porcelain crab, *Polyonyx macrocheles* (bottom).

freshwater animals produce relatively few large eggs rich in yolk and fertilization is commonly internal. Freshwater oligochaetes copulate and sperm is transferred from the body of one individual to the body of another. There are also other common differences between marine and freshwater animals. When a marine animal hatches, it is usually a tiny weak ciliated larva which drifts in the surface waters and feeds on plankton. But, when a freshwater animal hatches, it is able to swim and forage for itself. If it lives in a river, it is not at the mercy of the current.

Terrestrial oligochaetes such as the earthworm have followed the habits of their freshwater oligochaete ancestors. Like almost all terrestrial animals they have internal fertilization, and produce a few large eggs, rich in yolk. The young earthworm passes through the most vulnerable stage of its life protected by its eggshell. It hatches as an immature adult, fully able to fend for itself. Similar reproductive adaptations are required of all terrestrial animals unless they return to the water to reproduce.

Ecologically, the annelids are quite significant. In addition to being key parts of food webs, they physically modify the habitats in which they live. Tube-building polychaetes may stabilize drifting marine bottoms and so transform their habitat. We have already mentioned the soil-mixing activities of earthworms. In addition earthworms appear to accumulate certain pesticides and may provide us with advance warning about certain kinds of pollution such as excessive DDT.

The fossil evidence suggests that the annelids evolved from flatworm ancestors in Precambrian times. Fossils of well-developed polychaetes have been found in Cambrian rocks, and at least one Precambrian fossil annelid is known that antedates the advent of arthropods. Furthermore, fossils resembling worm tracks and burrows are quite common before the Cambrian period. The polychaetes probably gave rise to the oligochaetes and these in turn to the more specialized leeches. It is likely that earthworms evolved during the Cretaceous period along with the flowering plants and the insects. The terrestrial soils of today evolved along with the seasonal and deciduous vegetation of the angio-sperms and along with the earthworms. As one distinguished zoologist put it, "Angiosperms, earthworms and humus all have exactly the same geological antiquity."

Did ancient annelids give rise to any other groups? One clue comes from their embryology. Annelids all have the embryonic characteristics, such as spiral cleavage, of the other protostomes. In addition they have a distinctively-shaped ciliated larva called a **trochophore.** Almost identical trochophore larvae are found in mollusks, suggesting that the two phyla had a common ancestor. The ancestral annelids also gave rise to the arthropods, as we see presently.

THE SIGNIFICANCE OF SEGMENTATION

Segmentation apparently had considerable adaptive significance and was passed on by the annelids to the arthropods. It also evolved independently in tapeworms and in deuterostomes and reached a high state of development in the vertebrates. What is the adaptive significance of segmentation? Or, more properly, what *was* the adaptive significance of segmentation when it first appeared early in the evolutionary history of these animals?

In tapeworms, segmentation is clearly a special reproductive adaptation to a parasitic way

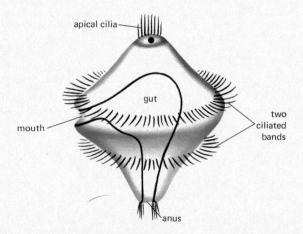

Figure 24-32   Trochophore larva of an annelid. Almost identical trochophore larvae are found in mollusks.

of life. In vertebrates, an excellent fossil record shows that ancestral vertebrates, fish, had their vertebrae and muscles arranged segmentally, permitting the undulating movements necessary for swimming. By such movements, water is pushed back and the body is thrust forward. Segmentation thus arose in vertebrates as an adaptation to locomotion.

What is the situation in annelids? Unfortunately we do not know how the ancestors of modern annelids moved. However, the fossil record tells us that Cambrian polychaetes looked much like present-day ones, many of which move by the appendages they have on their segments. Moreover, the only Precambrian annelid known also had appendages of sorts on its segments. Perhaps segmentation arose in connection with the development of appendages used for swimming, or perhaps it evolved as an adaptation for burrowing. No unsegmented worm can burrow as effectively as an annelid. In any case annelid segmentation was an innovation which had enough adaptive value to persist, and was passed on by the annelids to the arthropods. However, close relatives of the ancient annelids, the mollusks, either never became segmented or abandoned segmentation. Yet, as we shall now see, they have done very well indeed.

SNAILS, CLAMS, AND SQUIDS—PHYLUM MOLLUSCA

Clams, scallops, mussels, oysters, limpets, conches, whelks, snails, slugs, abalones, squids, and octopods: all of these and many others make up this large and diverse phylum which includes more than 80,000 living species and 35,000 extinct forms. They penetrate all habitats: oceanic deeps, mudflats, coral reefs, the tops of the tallest topical trees, deserts, underground, lakes, and rivers. They are found at the snow line of the Himalayas and at oceanic depths of 3 miles where the pressure is 4 tons per square inch. They may become hidden as parasites inside other animals. They range in size from tiny snails, the size of a pinhead, to giant squids of the genus *Architeuthis* which roam the North Atlantic. The largest one on record weighed about two tons and was about 60 feet long, including the tentacles. Beaks of even larger specimens, found in the stomachs of sperm whales, show that they may reach a length of 140 feet.

Although the various classes of mollusks are strikingly different from one another in outward appearance and behavior (compare a clam and a squid, for instance), they all have a similar body plan and many common features. Chief among

Figure 24-33 Fossil annelids. A. A Cambrian polychaete *(Canadia setigera)* which has appendages and looks very much like present day polychaetes. B. A Precambrian annelid from Australia *(Spriggina)* which appears to have appendages on its segments.

these is the presence of a large muscular **foot.** It is this foot on which a snail creeps along or that sometimes projects between the shells of a clam. Its action resembles that of the human tongue in the variety of effective shapes it can assume. Clams use the foot for creeping and for digging. In squids and octopods the foot has been modified into separate arms that surround the mouth.

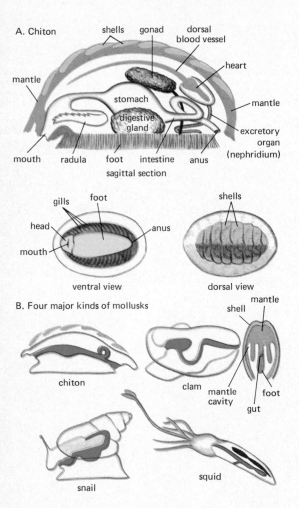

Figure 24-34 Molluskan body plans. A. All mollusks have a basic organization similar to this chiton. Virtually all that is left of the coelom is the pericardial cavity which surrounds the heart. B. The molluskan body plan has been modified in various groups. Here are representatives of the four major classes of mollusks.

Almost all mollusks are unsegmented, their bodies are soft ("mollis" is the Latin word for soft), and are covered by a heavy fold of tissue called a **mantle,** which usually secretes a hard calcium-containing **shell.** In squids and octopods the shell is internal or missing completely. Mollusks have well-developed organs for excretion, hearts for pumping blood, a digestive system with a mouth and anus, and complex respiratory organs. Almost all of the aquatic forms have gills, and the land snails have modified their mantle into a lung of sorts. Their nervous systems and sense organs vary from the simplicity of a clam to the great complexity of squid.

Almost all mollusks except clams and their relatives have a ribbonlike feeding device called a **radula** that is located just inside the mouth. It is covered with horny teeth and can be moved back and forth like a file about 40 times a minute. In addition to rasping off food, the radula brings food back into the gut like a conveyer belt. Although some mollusks feed on large plants and animals and some are filter feeders, most of them make their living by scraping food from a surface with a radula. Many zoologists believe that the ancestral mollusk from which modern mollusks descended crawled about over rocks and other hard surfaces in shallow water in the Precambrian oceans, scraping off food with its radula. Its feeding habits were probably like those of present-day chitons, considered next, except that it was more active and had a distinct head.

Most mollusks reproduce sexually; an individual is either a male or a female. However, like the earthworms, some mollusks are **hermaphroditic,** which means that a single animal has both male and female reproductive organs at the same time. A few other mollusks start their life as males, but when they get older transform into females.

The development of mollusks is like that of other protostomes, and has already been considered. They have a trochophore larva like the annelids. The only unusual feature of their development is that their coelom is quite small. Because of their wide diversity and great internal complexity, they must be recognized, along with the arthropods and chordates, as one of the most advanced phyla.

Chitons—Class Amphineura

The simplest living mollusks likely to be encountered are called chitons or sea cradles (Class **Amphineura**). Their most characteristic feature is a shell made up of eight overlapping plates. They lack a distinct head. Chitons are all marine and are found largely in the intertidal zone attached to rocks and shells by a broad oval foot. When dislodged from their anchorage, they curl up like an armadillo. Although most are only one or two inches long, some of the Pacific Coast chitons are a foot long, and were used as food by American Indians.

Chitons feed on algae that they scrape off with their radula from the rocks and shells to which they attach. Some chitons forage each night near their home base to which they return at daybreak. Others never move from their home base and feed on bits of seaweed deposited near them by the waves. Their lack of a head is probably related to their inactive way of life.

Snails—Class Gastropoda

The largest class of living mollusks is the **Gastropoda,** so named because they seem to move on their stomachs. This group includes the land snails, one of the few groups of truly terrestrial invertebrates as well as all the other forms with spiral shells. The common garden slugs are gastropods that have lost their shells. Their snailhood is still evident internally, however, for their organs are twisted as if they still had to fit in the shell. Gastropods have invaded almost every environment. Marine forms are the most common and usually creep about on the bottom, but there are also some marvelous free-swimming forms called pteropods ("wing footed") in which the foot has become transformed into two broad, thin winglike organs with which they swim at or near the surface of the sea.

The food habits of gastropods are diverse. They include scavengers, herbivores, carnivores, and some filter feeders. There are even some parasitic gastropods which live in the body of sea cucumbers. These bizarre creatures have lost their shells and digestive tracts, and absorb food through the body wall. In most gastropods the radula is used in feeding for scraping and grinding. In others, however, it has become modified for different feeding activities. Some snails bore through clam or oyster shells with the aid of the radula. When the shell is pierced, they lick the flesh from their helpless victims. One group of carnivorous tropical snails, the cone shells, use their radula as a weapon. They feed mainly on fish and marine annelids. They can shoot out the radula with great speed and force. The hollow teeth are packed with a powerful nerve poison. Native inhabitants of places like New Guinea have a wholesome dread of the bite of a cone, for it can kill a man.

The silvery trails of mucus that snails and slugs leave are familiar sights to anyone who has a garden. The trail is secreted by glands in the front of the foot and the animal skids along it, propelled by movements of the foot. Some gastropods spin mucous threads which they put to unusual use. For example, certain slugs use them in their courtship. After executing a "wedding dance," a pair of these slugs spins a thread. Holding on to their thread, they drop free from some overhanging support and mate. After mating, they climb back up the thread (Fig. 24-35D).

There are numerous species of snails in diverse environments. But for many species there is only a very limited environment in which they are found living and reproducing. Like the finches of the Galapagos Islands discussed in Chapter 19, some species of Hawaiian tree snails are confined to a single island, sometimes to a single valley. Other species that live in the ocean are limited to particular habitats, such as certain depths, temperatures, and bottom conditions. If an expert on mollusks is given a great many snail and other mollusk shells collected from a particular part of some ocean, he can readily give a picture of the habitat and name the general area from which the collection came. With equal ease the geologist or paleontologist can tell a great deal about the environment of some ancient sea from a collection of fossil mollusk shells. He can reconstruct a picture of the bottom and can determine facts such as the temperature and salinity of the covering sea and the amount of movement to which the sea was subjected. For this reason, the fossil shells of snails

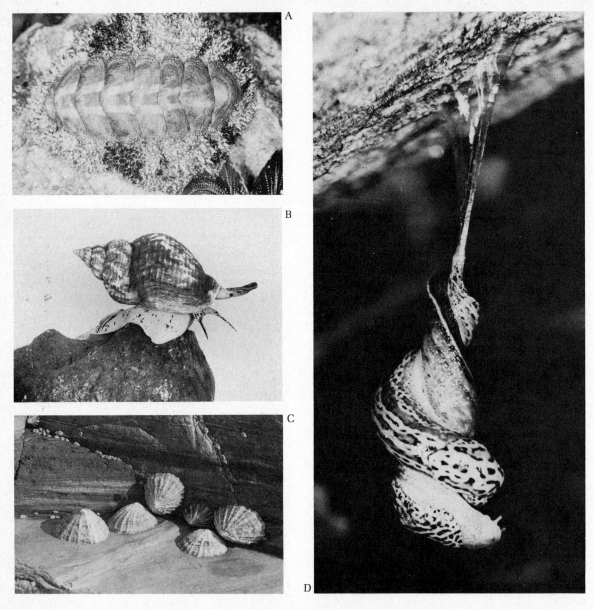

Figure 24-35   A. A chiton attached to a rock. A group of mussels are at the lower right. *Various gastropods:* B. A carnivorous snail, *Buccinum undatum* moving along a rock surface by means of its huge foot. The head with its two tentacles are visible. At the tip of each is a simple eye. Above the tentacles is the siphon tube which brings sea water into the mantle cavity which contains the gills. C. A limpet, *Patella vulgata*, on a rock. D. Land slugs are shell-less snails. They are hermaphroditic; that is, they have both a male and a female set of reproductive organs so that self-fertilization is possible. However, they usually cross-fertilize. Here is a pair of mating slugs (*Limax*) which have lowered themselves from a tree trunk by means of a mucous cord. They are entwined together with their genital apertures opposed and are exchanging sperm. Another group of shell-less snails called nudibranchs (meaning naked gill) are marine. One of them is illustrated in Figure 23-5.

and other mollusks are widely used by petroleum geologists to determine the climate and other features of ancient environments and the age of geological formations.

Snails and their allies are of major ecological importance. In some intertidal regions hundreds of snails are found in every square yard. They provide food for many fish and aquatic birds and occasionally for man. In some areas land snails and slugs are serious agricultural pests and cause extensive destruction of crops. They are also of major importance to man because certain snails harbor immature flukes and are links in complex disease chains.

Bivalves—Class Pelecypoda

Clams, scallops, mussels, oysters, and other mollusks with two hinged shells fall in the second largest class of mollusks, the **Pelecypoda.** The shells are sometimes called valves, so these animals are known as bivalves. Even less active than their snail relatives, they often rest on the ocean bottom or become attached to rocks. They secure food by filtering fine material from the water. Their large gills are used for both respiration and food gathering. The gills are covered with cilia that create a water current and carry in food particles (Fig. 12-2). The ciliated gills and surfaces around the mouth also serve to sort out particles by size. They reject large particles like sand grains but allow smaller microorganisms to pass into the mouth. This effectively separates inedible particles. Earthworms and burrowing polychaetes ingest a good deal of useless inorganic matter in order to obtain the small amount of nutrient that is also present. Clams appear to be much more efficient and ingest mostly edible particles. Since most bivalves are filter feeders, it is not surprising to discover that they do not have radulas, nor a head. As was true for the chitons, the lack of a head probably is typical of their sedentary life.

Members of this group of mollusks have been highly prized as food since antiquity. In many parts of the world oysters and mussels are grown as a commercial crop. Oystermen try to provide an attractive bottom for young oysters to grow on and to prevent overcrowding. They also try to reduce the numbers of starfish and other organisms which feed on growing oysters. Oysters are also cultured for the production of pearls.

Squid and octopods—Class Cephalopoda

The largest, most active mollusks are in the class **Cephalopoda,** which includes squids which have ten arms around their mouth and octopods which have eight. The name suggests their structure— the typical foot of the phylum is located in the head region. Part of the foot forms a siphon through which water can be forcibly expelled, providing for jet-propelled locomotion. The rest of the foot has been subdivided into separate arms that move freely by well-controlled muscular contractions. These arm movements require precise regulation, which is provided by a well-developed nervous system. The brain of an octopod is larger and more complicated in structure and function than that of any other animals except the vertebrates.

Cephalopods show extremely complex behavior. Their most important sense organs in this behavior are their eyes, which are large, camera-type, and image-forming. They have a cornea, iris diaphragm, movable lens and retina. Their eyes work just the way ours do except that the image is not inverted. Consider, for example, some behavior of octopods. They lay masses of rather large eggs from which miniature adults will hatch. They often have elaborate rituals of parental care. For instance, they return regularly to the egg mass, wash it with water, and wipe each egg with the tips of the tentacles. Octopods also have elaborate courtship behavior which involves tactile caresses and color changes. The male produces a packet of sperm called a spermatophore or sperm bag. He picks up this spermatophore with a special arm called the *hectocotylus*. He then usually thrusts the arm deep into the female's mantle cavity and presses the spermatophore into her genital opening. In a few species of octopods the hectocotylus becomes detached from the male near the end of courtship. The arm continues to contract its muscles and moves within the mantle cavity of the female for some time. It finally attaches by its suckers near the genital pore and pushes in the spermatophore.

A

B

C

Figure 24-36 Various pelecypods. A. Queen scallops, *Chlamys opercularis,* swim by rapidly opening and closing their shell valves, producing a jet-propulsion effect. They are able to detect predators by means of chemical receptors. Here they are escaping from a starfish. At the edge of the valves of the scallop are lobes of the mantle which bear long tentacles and eyes (dark spots). Deep within the mantle cavity are the broad plates of the gills which are covered with cilia and enlarged for food collection. B. The common edible mussel, *Mytilus edulis* attached to a rock by fine threads which it secretes. (Photo by Ralph Buchsbaum.) C. Giant clam, *Tridacna gigas,* is the largest bivalve ever evolved. It may exceed 4 feet in length and has been reported to trap divers between its shell valves. Like its smaller relatives in the other photographs, it is a filter feeder. Here is a model of a *Tridacna* shell with a coral growth.

In addition to these elaborate innate behavior patterns, cephalopods are also capable of
learning and of memory. J. Z. Young and his colleagues trained an octopus by electric shocks not
to attack a crab whenever a white square was
placed in its aquarium. In the absence of the
white square the octopus attacked the crab which
is one of its normal foods. An octopus could remember this for at least two weeks. However, if a
specific part of the brain was removed, it remembered for only about two hours. Furthermore, if

this region was removed from the brain of a
trained octopus, it forgot what it was trained to
do. This result suggests that the establishment of
memory involves both short-term and long-term
changes in the brain. This situation is similar to
human memory (discussed in Chapter 14), where
both short-term and long-term processes are involved.

The chambered nautilus is also a cephalopod.
Unlike squids and octopods it has retained an
external shell. It also has many more arms. It is a

Figure 24-37   Cephalopods.
(Right) Common octopus, *Octopus vulgaris,* clearly showing
its eight tentacles armed with
suckers. Its bag-like body
opens on the underside into a
mantle cavity containing gills.
(Below) American squid, *Loligo pealii,* is about 10 inches
long. It is found in great
schools from Massachusetts
Bay to Cape Hatteras and
serves as food for fishes. A
third kind of cephalopod,
called a nautilus, retains its
shell and is illustrated in Figure 11-8C.

descendant of the first cephalopods, which lived in Paleozoic seas. For reasons not yet understood, the cephalopods have not evolved a single fresh-water representative.

Evolutionary origins of mollusks

The evolutionary origins of the mollusks are unclear. They share many key characteristics with polychaete annelids, such as a trochophore larva, spiral cleavage, and annelid-type coelom formation. These facts make it highly likely that mollusks share the same ancestry with annelids. However, it is uncertain whether the ancestors of mollusks were segmented. The fact that chitons have a shell which is in eight parts seems to indicate that mollusks were originally segmented. In chiton embryos, however, the eight parts of shell develop from a single shell gland and not from eight separate shell glands as might be expected in a truly segmented animal.

In 1952 evidence that mollusks might have a segmented ancestor dramatically appeared in the form of a true living fossil. Ten living specimens of a class of mollusks which had been thought to be extinct for 400 million years were dredged from a trench 2 miles deep off the Pacific Coast of Nicaragua. Four other specimens were dredged up in 1958 from an even deeper trench near Chile.

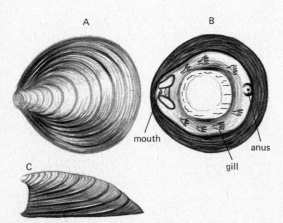

Figure 24-38 *Neopilina* is an extremely primitive mollusk. A. Dorsal view. B. Ventral view. C. Lateral view.

The mollusk was called **Neopilina,** and it belongs to a class of mollusks known as the *Monoplacophora. Neopilina* is about 1 inch long and looks like a cross between a chiton and a common marine gastropod called a limpet. It has a single shell like a limpet, but its underside reveals five pairs of gills and eight pairs of muscles attached to the shell. For a time *Neopilina* was viewed as a sort of missing link between mollusks and annelids. More recent studies, however, indicate that its segmentation like that of chitons was acquired secondarily and is not a primitive feature. The facts are still incomplete, but the weight of evidence indicates that the ancestors of the mollusks were never segmented.

A LINK TO THE ARTHROPODS—THE PHYLUM ONYCHOPHORA

This phylum is so small it might well be skipped over in a brief survey if it were not for the unusual combination of characters the animals have. **Onychophores** are small, wormlike creatures (often called walking worms) that live in the tropics. Their skin is not well waterproofed and they live only in damp places where there is little danger of dying of desiccation. Like most land animals that are not as well adapted to life on land as insects and vertebrates, they avoid the light and remain hidden most of the time. They can be found in decaying leaves, rotting logs, beneath stones, or along stream banks. They feed on both animal and plant material. They capture small animals and defend themselves by squirting out a sticky material. Some onychophores can spit almost 2 feet and these animals have been termed "the champion spitters of the animal kingdom, gram for gram."

In many ways onychophores resemble annelid worms. They have soft segmented bodies with muscles arranged like annelids. They also have annelid-like excretory organs repeated in many segments. In other ways they are like arthropods. They have the beginning of a hard, outer coating or cuticle containing chitin, which in the arthropods serves as an exoskeleton. Each segment has a pair of short walking legs, tipped with claws similar to those of insects. The legs, however, are unjointed. Also, the onychophores breathe like land arthropods. They have thin tracheal tubes through which air diffuses directly to the tissues. Development of the fertilized eggs of onychophores and arthropods is similar.

Another feature shared by onychophores with all annelids and several arthropods is continued dependence on a hydrostatic skeleton for many functions. Examination of the muscles of the leg of an onychophore reveals no mechanism for extending the leg. There are flexor muscles for raising the leg but no extensors. The onychophore must instead blow its legs out with hydrostatic pressure generated by contractions of its annelid-like body wall muscles. A similar mechanism is found in some arthropods such as centipedes, millipedes, and spiders which also lack extensor muscles in most of their leg segments.

The combination of annelid and arthropod characters which onychophores possess suggests that they are an evolutionary offshoot of the line leading to the arthropods from an ancient annelid-like ancestor. Their fossil record goes back to the Cambrian period (Fig. 24-39). The fossil onychophores were marine but looked very much like modern species. They probably adapted to life along the shorelines and spread to humid land environments.

A brief lesson in animal geography

The geographical distribution of onychophores is unusual. Species within the same genus are often widely separated. For instance, the closest relative of a South African onychophore is found in Chile and that of a Congolese onychophore is found in islands of the Caribbean. What accounts for this odd distribution? In seeking an explanation, it must be borne in mind that onychophores are delicate creatures and it is unlikely that they would be spread from one remote place to another by other animals, by air or by ocean. This rules out any overseas invasion of South Africa by Chilean onychophores or vice versa. But what about an overland invasion? The peculiar distribution of onychophores suggests that at some time in the past land bridges connected Africa and South America. In those ancient days South African and Chilean onychophores were one interbreeding population or species, and the Congolese and Caribbean onychophores were another species. Subsequently the land bridges disappeared, probably as the continents drifted apart, and the onychophores of South America and of South Africa, no longer able to interbreed, gradually became different. By analyzing the geo-

Figure 24-39 Onychophores. Upper: *Peripatus* is a modern onychophore found in the tropics. It has many features of both annelids and arthropods. (Courtesy of Carolina Biological Supply Co.) Lower: *Aysheaia pedunculata* is a fossil marine onychophore from the Cambrian.

graphical distribution of various groups of extinct and contemporary animals, animal geographers and geologists can learn about the probable movements of land masses in ancient times.

INSECTS, SPIDERS, AND CRUSTACEANS— PHYLUM ARTHROPODA

Almost everywhere on earth there are arthropods. They fly thousands of feet in the air or thrive thousands of feet below the surface of the ocean. They are denizens of deserts, tropical jungles, and Arctic regions. Many live deep within the bowels of other animals as parasites which have become so modified that they resemble a tumorous growth rather than an animal. A few live in hot springs, where temperatures may get as high as 47°C, close to the upper limit for any metazoan. The 1,000,000 described species outnumber those of all the other animals, plants, and protists together, and there is probably an equal number yet to be discovered. Arthropods are also vast in numbers. A swarm of locusts can darken the sky

and blot out the sun like a black cloud; minute crustaceans in the oceans total billions upon billions. Indeed the single most abundant species of animal, both in numbers of individuals and in fraction of total animal biomass in the whole world, is a tiny marine crustacean called *Calanus finmarchicus* or one of its close relatives.

Arthropods are obviously one of the most successful groups of animals alive today. What is of equal interest is that they have always been a dominant and successful group since they first appeared 550 million years ago. Soon after arthropods appeared on the scene in the early Cambrian seas, the annelids, which had been the dominant form, apparently decreased in numbers and were displaced by the arthropods. Most of the annelids retreated to a sheltered life on the ocean bottom, in the mud or in burrows, or to occasional forays at night—a mode of life most of them retain today. The arthropods were almost certainly the cause of their downfall. What accounts for this initial success of the arthropods and their continued success today?

To the annelid characters they inherited, arthropods have added during their evolution two

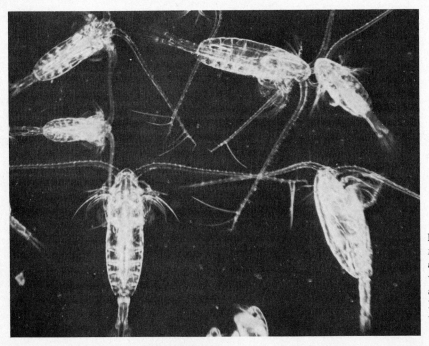

Figure 24-40 *Calanus finmarchicus*, a marine copepod about the size of a grain of wheat, is probably the most abundant animal in the ocean. It makes daily migrations from the ocean surface to depths of 1500 feet.

important features with great adaptive value: *jointed legs* and a hard and almost impermeable outer skeleton or *exoskeleton*. They are the only animals besides the vertebrates that have jointed legs; in fact the name "arthro-poda" means "jointed feet." As we noted in Chapter 11 the hard outer coating or cuticle of these animals is composed of a fibrous material called chitin which resembles the cellulose of plants. It may be thought of as a tough dry kind of mucus. Chitin is stiffened with an even tougher horny material called *sclerotin*, which is made up of proteins tanned like leather. Some arthropods such as lobsters also harden their cuticles by infiltrating them with lime in the form of the mineral calcite. But whenever real hardness is needed, such as in their jaws and the tips of the claws, they use an extra coating of sclerotin. When sclerotin is fully hardened, it is harder than calcite. The cuticle of the arthropods is more than a shell; it is a true skeleton. The muscles attached inside it move the body parts the way our muscles move our legs and hands, so that a crab can walk or a lobster can crush a clam in its big claw.

These jointed legs and tough exoskeletons enabled ancient marine arthropods to be fast-moving hunters, just like many of their modern descendants. Their exoskeleton provided them with a formidable defensive armor. It also permitted immense muscular agility that annelids simply did not have. Indeed rapid locomotion in the world of today is almost always associated with the possession of hard skeletons like those found in arthropods and vertebrates and conspicuously absent in annelids. It seems likely that an exoskeleton and jointed appendages have been the secret of the arthropods' continued success. Since exoskeletons cannot grow, they secrete new exoskeletons regularly and shed the old one. Other features include distinctive *hinged jaws* that open from side to side, instead of up and down, as our own and other vertebrate jaws do.

Two major annelid features that arthropods preserved are *segmentation* and an *annelidlike nervous system*. Although segmentation in an insect may not seem as obvious to an untrained eye as in an earthworm, in most arthropods at least one part of the body is still distinctly segmented. In a lobster it is the tail, and in a mosquito it is the abdomen. On closer examination one can see that other parts which once might have been free-moving have become fused together. Thus three fused segments are usually visible in the midsection or thorax of an insect.

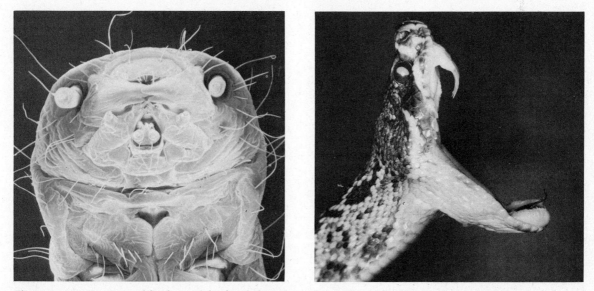

Figure 24-41 Insect jaws like those of this beetle larva open from side to side whereas the jaws of this rattlesnake and other vertebrates open up and down.

The nervous system also still shows many annelid traits (compare Figs. 24-42 and 13-31 with 24-29). There is a dorsal brain and a ventral nerve cord. An addition to the annelid nerve pattern, however, is the development of more complex and varied sense organs, such as compound eyes. These are necessary for animals that live the sort of active life characteristic of most arthropods.

Some evolutionary trends in the arthropods

It is generally believed that arthropods evolved from a polychaete annelid or from the ancestor of the polychaetes. When arthropods first appeared in Cambrian seas, they probably had long wormlike bodies composed of many segments, each with a pair of legs. During the succeeding 500 million years of their evolution, this ground plan has been modified by different groups. Three trends are conspicuous: (1) a reduction in number of segments and a combining of segments into groups such as head, thorax, and abdomen; (2) a concentration of nervous tissue and sense organs in or near the head of the animal; and (3) a specialization of certain legs for functions besides locomotion and the loss of legs from many other segments. The claws of a lobster, the fangs of a spider, and the mouth parts of a grasshopper—all these are modified legs. The immense diversity that arthropods have attained probably depended as much on the phenomenal versatility of their appendages as on anything else.

These three trends are evident to varying degrees in the eight classes of arthropods we shall mention. All of these classes may easily be recognized by the kind of appendages they have and their location on the body.

Two classes are completely extinct. One is the class **Trilobita,** made up of the fossil trilobites that were once common in Cambrian and Ordovician times (Fig. 22-5). They had many segments, every segment bore a pair of legs, and all of the legs were nearly identical. Unlike all other arthropods they had no appendages modified as mouth parts. The second extinct class is the **Eurypterida,** a group of ancient water scorpions we shall mention later along with a small class of living arthropods, the **Xiphosura** or horseshoe crabs.

Two other classes are very common, but do not warrant lengthy discussion. They are the **Chilopoda** and **Diplopoda**—centipedes and millipedes, respectively. These animals have many pairs of legs, but not "hundreds" or "thousands" as their common names imply. Although these segmented, wormlike animals look alike superficially, there are several fundamental differences between them. Centipedes have one pair of legs on a segment, but millipedes have two pairs. (Actually each segment in a millipede is really *two* segments fused together.) Differences in the location of sex organs and in diet (centipedes are carnivorous, whereas millipedes are herbivorous) also set them apart.

The three major classes within the phylum are the **Crustacea, Arachnida,** and **Insecta.**

Crabs, lobsters, and shrimps— Class Crustacea

Crustaceans are a huge and successful group with more than 28,000 members. They have undergone extensive diversification during their evolu-

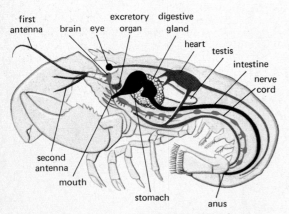

Figure 24-42 Anatomy of a lobster. Note the various specialized appendages. The anterior appendages include antennae and different kinds of legs. The abdominal appendages are swimming flaps. Also note the segmented ventral nerve cord and the segmented abdomen. The head and thorax have fused into a cephalothorax.

tion. The most familiar ones are favored delicacies from the sea — crabs, lobsters (and their freshwater relatives, crayfish), prawns, and shrimps. Less readily apparent but much more plentiful crustaceans, some microscopic and some nearly half an inch long, abound in the plankton of the oceans and fresh water. The class also includes barnacles — the stationary, hard-shelled animals that attach themselves to ships' hulls and to pilings under water — some bizarre parasites, and a small number of land animals such as pill bugs. Their body is usually composed of two main parts,

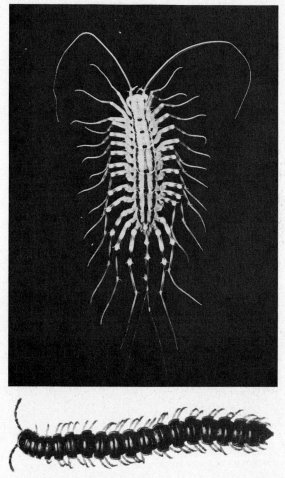

Figure 24-43   The centipede (above), *Scutigera*, which is a chilopod, has one pair of legs per segment. The millipede (below), which is a diplopod, has two pair of legs per segment.

an anterior *cephalothorax* and a posterior abdomen (Fig. 24-42). Each of these parts is composed of several segments. Simple crustaceans like the brine shrimp *Artemia* which are thought to resemble primitive crustaceans may have up to 40 segments, whereas so-called higher crustaceans, such as crabs, lobsters, or crayfish, have only half that number of segments. Crustaceans generally have a pair of jointed appendages in each segment. Unlike any other arthropods they have two pairs of antennae.

Crustaceans usually hatch from their eggs as a larva called a *nauplius*. This larva molts many times and may undergo striking changes in form when transforming or **metamorphosing** into an adult. Figure 24-45 illustrates some of these transformations for five species of crustaceans. Many are striking, but certainly the most bizarre are the changes occurring in certain barnacles that are parasitic in crabs and lobsters. One of these parasitic barnacles, called *Sacculina*, is probably the most highly modified parasite in the animal kingdom (Fig. 24-45E).

The nauplius larva of *Sacculina* when hatched looks like a typical crustacean. It swims about, molts into another larval stage called a *cypris*, and soon attaches itself to the skin of a crab by means of a suction cup on one of its antennae. When firmly secured it shakes about violently and rids itself of its thorax along with the legs, abdomen, and most of the internal organs except for its reproductive organs. The epidermis that remains with the parasite secretes a new cuticle and the larva is now transformed into a veritable hypodermic syringe with a long needle. The larval *Sacculina* expands its body and jabs the needle into the crab. The larva then invades the crab through the needle and finds its way to the intestine. At this point *Sacculina* is a small sac of epidermal cells enclosing a mass of reproductive cells. This mass of cells grows like a fungus and sends rootlike branches throughout the crab, by means of which the parasite feeds. A typical crustacean larva has metamorphosed into an animated fungus which feeds on the viscera of its host! When these bizarre events were first described at the turn of this century, they prompted the British zoologist Walter Garstang to offer the following verse.

Figure 24-44   Various crustaceans. A. Brine shrimps, *Artemia salina*, are primitive crustaceans which live in salty lakes. They swim upside down. B. A crab, *Partunus depurator*, in the process of molting. The old cuticle is above and the newly molted "soft-shell" crab is emerging below. C. *Caprella aequilibra* is a delicate crustacean commonly called a "skeleton shrimp" which lives among seaweeds in the tidal zone. It waits in this position using sense organs in its antennae to detect passing prey. Its claws are held ready to swoop down on any unfortunate creature that comes within range. D. The pill bug, *Porcellionides*, is one of a very few terrestrial crustaceans. It can roll up into a ball like an armadillo to avoid being seized by predators.

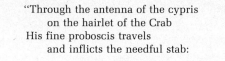

"Through the antenna of the cypris
  on the hairlet of the Crab
His fine proboscis travels
  and inflicts the needful stab:

Then, gathering all the salvage
  that he's rescued from the rout,
He slips along the tunnel
  of his own projecting snout."

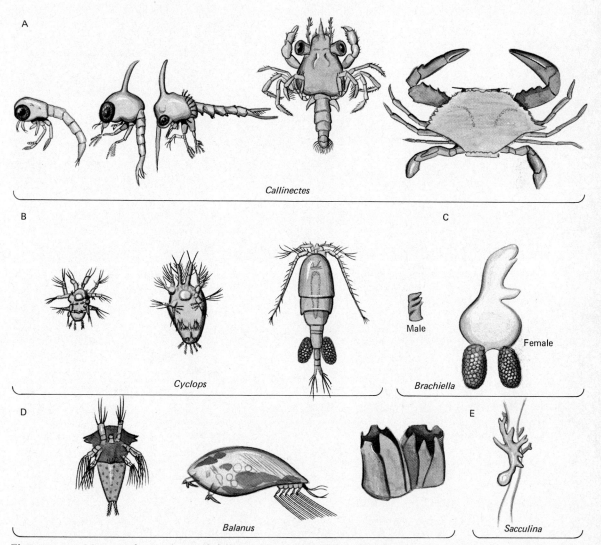

Figure 24-45 Metamorphosis of several different crustaceans. Only a few stages are shown for each organism (there may be a dozen). Molting occurs between each of the stages. A. Stages in the metamorphosis of the common blue crab, *Callinectes*. B. Stages in the metamorphosis of a free-living copepod, *Cyclops*. C. The early stages of the parasitic copepod *Brachiella* resemble those of *Cyclops* but the adult stage illustrated here looks nothing like an adult *Cyclops*. D. Stages in the metamorphosis of the common barnacle, *Balanus*. E. The early stages of the parasitic barnacle *Sacculina* resemble those of *Balanus*, but the adult stage illustrated here attached to the intestine of a crab is an amorphous fungus-like creature which sends root-like processes throughout the crab. These examples demonstrate the virtually unlimited potentialities of metamorphosis.

Most crustaceans have more conventional feeding habits than *Sacculina*. For instance, *Artemia* swims on its back and uses its limb movements not only to propel it forward but also to collect small food particles such as bacteria and diatoms which filter through bristles on its legs. *Artemia* looks very much like fossil crustaceans from the Devonian period, and this method of filter feeding is thought to resemble the way in which they fed. Many crustaceans still feed like *Artemia*. Others, like crabs and lobsters, are carnivorous and capture their prey.

One of the great interests of crustaceans to man is their role in the great food webs of the oceans and fresh water. We have already remarked that tiny crustaceans called *copepods* (*Calanus finmarchicus* is an example) are probably the most abundant animals in the world, both in numbers of individuals and of the fraction of the total animal biomass. Countless billions of these minute animals make up most of the *zooplankton*. They feed on the world's largest crop of green plants, the green algae, as well as on flagellates, diatoms, and other one-celled plants of the *phytoplankton*. Zooplankton is in turn eaten by larger animals. Whalebone whales like the blue whales, which are probably the largest animals that ever lived, feed exclusively on tiny crustaceans. They are also a major part of the food of some of the most abundant fishes such as sardines, herring, menhaden, and mackerel. These tiny creatures are the world's largest stock of living animal protein. Today a large industry is devoted to catching fish such as the menhaden and turning them into fish meal, which is used as a supplement in swine and chicken feed. Ultimately man eats 1/10,000th of the original protein in the copepods. In recent years scientists have endeavored to learn how they might harvest and process both zooplankton and phytoplankton to provide food directly for man, instead of indirectly through fish, pigs, and chickens.

Most crustaceans have remained aquatic, but in one group, the *isopods,* there are a few truly terrestrial forms. The most familiar are the armadillo-like pill bugs or wood lice which are common in damp places (Fig. 24-44). Interestingly, although they are perfectly good crustaceans and closely resemble their marine relatives, they have evolved a respiratory system of hollow

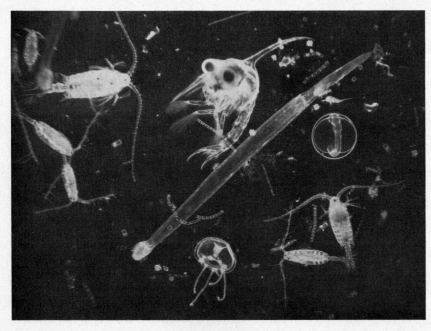

Figure 24-46 Zooplankton. Various copepods are evident along with a larva of a crab (with the two large eyes). Diagonally across is an arrow-worm, *Sagitta*, a representative of a small phylum called the Chaetognatha. A developing fish egg is just below the arrow-worm at 3 o'clock, while at 6 o'clock is a tiny medusa. Also visible are chains of diatoms, some of the "grasses" of the ocean upon which all these animals ultimately depend.

air-filled tracheae which is almost identical to the tracheal system of insects (Chapter 12). This is one of the best examples we have of independent or parallel evolution of similar structures (remember the eye of the squid and the human eye).

The largest present-day crustaceans are the lobsters and crabs. Lobsters, such as *Homarus americanus*, the clawed lobster of the Northeast, may weigh more than 50 pounds. Japanese spider crabs may be 9 feet across. These gigantic crustaceans are among the largest arthropods that ever lived. Their exoskeletons are heavily impregnated with lime in the form of calcite that makes them strong enough to support a large body size and provide attachments for powerful muscles. They apparently continue to molt throughout life. Undoubtedly the biggest ones are still out there somewhere.

Crustaceans have proved especially interesting in studies of nervous systems and muscles. Crustacean nerves have turned out to be ideal objects for examining the nature of inhibitory nerve fibers (mentioned in Chapter 13) that inhibit muscles from contracting. The giant muscle cells of barnacles are being used to investigate the role of calcium in the contraction of muscle cells. The hormonal control of crustacean molting and their physiological clocks are also being actively studied.

Spiders and their kin—Class Arachnida

Spiders and their relatives—mites, ticks, and scorpions—are widespread, diverse, and economically important groups of animals. More than 60,000 species have been named, and this probably represents only a small fraction of the species that will ultimately be identified. They range in size from microscopic mites a fraction of a millimeter long to the giant scorpions of Africa and Australia which may be 10 inches long. They have the dubious honor of being the most feared of all of the arthropods. This is mostly unwarranted, and useful arachnids far outnumber objectionable ones. Most spiders are helpful to man because they kill insects. The only two dangerous spiders found in the United States are the black widow and the brown recluse (Fig.

24-48). One need not dance the tarantella to cure the bite of the large false tarantulas of the Southwestern United States, because they are not poisonous to man. Daddy longlegs or harvestmen are completely harmless. There is some justification, however, in avoiding scorpions because of their painful stings and mites and ticks because of the diseases they may pass on to us.

Arachnids are usually called insects by most people, but they are not. Insects have three pairs of legs and have antennae. Adult spiders and their relatives have four pairs of legs all arising from the fused anterior segments or cephalothorax and have no antennae. They also possess two pairs of special appendages near the mouth called **chelicerae** and **pedipalps.** In most arachnids the chelicerae are used for holding prey. But in spiders they are used as fangs for injecting poison into the prey. The pedipalps are used by most arachnids to sieze and rip apart the prey. In scorpions, for instance, they are powerful grasping claws. But here again the spiders use them for a different purpose: male spiders use the pedipalps to transfer sperm into the females.

Most arachnids are carnivorous and tear apart their victims with their chelicerae and pedipalps. They usually secrete powerful digestive enzymes over the torn tissues of the prey which they hold and feed on the partially digested soup. As we mentioned in Chapter 12 many spiders actually inject digestive enzymes into their victim's body and suck up the partially digested fluid.

Spiders. The two largest orders of arachnids are the spiders and the mites and ticks. About 34,000 species of spiders have been described. They are all carnivorous and many have excellent sense organs. Almost all have eight simple eyes. In some spiders, such as wolf spiders and jumping spiders, vision is extremely acute.

One of the conspicuous features of spiders is silk production. Spiders have three pairs of spinnerets, which are tiny spigots at the tip of the abdomen through which silk is extruded. It comes out as a liquid but hardens immediately as it is pulled out by the legs. Spiders use their silk in many ways. Almost all of them continually lay a drag line which may serve as a safety line or to retrace their path. Many spiders use their silk to build webs or snares to capture prey. Others use silk to line their burrows and some make a silken nursery for their spiderlings. They often spin a molting pad to which they cling during molting. Some spiders use silk for balloon-

Figure 24-47   Various arachnids. A. This female black scorpion, *Centruroides gracilis*, carries her young on her back for several days after they are born. The young, incidentally, are produced viviparously. Although scorpions have a venom-bearing tail tip, they rarely use it except in defense or to subdue prey. B. The whip scorpion, *Mastigoproctus giganteus*, or "Vinegaroon" of Arizona gets its nickname from its habit of expelling a stream of acetic acid from the base of its tail. It looks fearsome but is harmless. C. Ticks like the dog tick, *Dermacentor variabilis*, shown here can transmit serious diseases such as Rocky Mountain spotted fever. D. A harvestman or "daddy long legs" like *Phalangium opilio* feeds on small insects. A tarantula spider is illustrated in Figure 24-60.

ing. They climb onto branches or fence posts and release silk threads which catch the wind. The tiny spiders are lifted off their perch and float to new places. In autumn the air may be filled with these ballooning threads called gossamer. Although spider silk is the strongest natural fiber known, it is not used commercially because the predatory habits of spiders make it difficult to rear them in large numbers. However, South Sea islanders twist the huge orb webs of certain spiders and use them to make fish nets.

It is from their ability to spin silk that the arachnids get their name. Greek mythology tells us about a Lydian girl named Arachne, who was such a splendid weaver that she impudently challenged the goddess Athena to a weaving contest. Arachne produced a faultless piece of cloth. This so enraged Athena that she tore the work to bits. Arachne in despair hung herself. The goddess loosened the rope and saved her life. But she

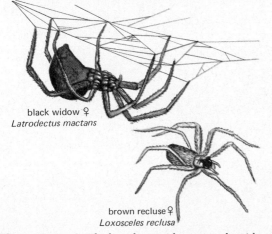

black widow ♀
Latrodectus mactans

brown recluse ♀
Loxosceles reclusa

Figure 24-48 A. Black widow spiders are web spiders and the female may bite if molested. The bite may go unnoticed but later severe abdominal pain, muscle pain and sweating occur. Usually no pain or swelling occurs at the site of the bite. Various widows are found throughout the United States and southern Canada. B. Brown recluse spiders commonly live in houses on the floor or behind furniture. After a bite, a red zone appears around the bite and may result in a wound that takes months to heal. Since there is no good first aid treatment for any poisonous spider bite, it is wise to consult a physician as soon as any signs of illness from a poisonous spider appear.

transformed Arachne into a spider and made her noose into a web. To insure that this lesson was not lost on future generations, Athena commanded that Arachne and her descendants weave forever with silk from their own bodies. Although spiders are the best known weavers among the arachnids, some mites also spin silk. Silk is also produced by centipedes and millipedes, and, of course, by the caterpillars of many moths.

Most of us are familiar with the ability of spiders to walk upside down on the ceiling and up a pane of smooth glass. To do this they use a brushlike tuft of flattened hairs which lies between the claws at the tips of their legs (Fig. 24-49). The brush adheres to the film of water which covers most surfaces and this permits the spider to walk on smooth surfaces.

Mites. Mites are found throughout the world from polar regions to the deserts and are the most numerous of spiderlike animals. They are also the most important economically of all arachnids, and are major pests of crops and carriers of dis-

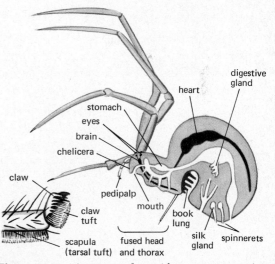

Figure 24-49 Anatomy of a spider, a representative arachnid. Note especially the fused head and thorax, four pairs of walking legs, the chelicerae and the pedipalps. Spiders have tracheae like those of insects and also special breathing organs called book lungs. Their legs often have a brush-like tuft of flattened hairs which enables them to walk upside-down on smooth surfaces. The foot is shown enlarged at the left.

ease such as scrub typhus. They resemble tiny spiders, but the body is fused into one piece with no separation between head and abdomen (Fig. 24-47). Most mites are terrestrial and are especially abundant in soil, humus, and rotten wood. In a famous study performed in England in 1943 it was shown that a single acre of farmland contained nearly 500 million mites. Some are also found in fresh water, a few live in the oceans, and some live in hot springs.

Most mites are carnivores and capture their prey, but some are scavengers and eat almost anything that once was alive. Numerous mites are parasites of animals and plants. The mange mites of dogs and the seven-year itch mite of man which tunnels into the skin are well-known pests. So are the red spider mites of citrus trees. Less well known are some parasitic mites which have unusual habitats. Some are found only in monkey lungs, or in bird nostrils, or under bat wings, or in the tracheal system of bees. One species lives in the ears of certain moths. Interestingly, they never infect both ears of the moth, so that the moth is still able to hear the supersonic hunting sounds of bats and drop to safety.

Mites are almost always small (0.5 to 3 millimeters), but one group, the ticks, includes some large forms which may reach 1 inch long after feeding. Ticks are all external parasites of mammals, birds, and reptiles. They attach themselves to their host and feed on blood. In man they are known to transmit Rocky Mountain spotted fever (which also occurs in New Jersey and Massachusetts), tularemia, and other diseases.

Ancient relatives. Interesting relatives of the arachnids are the horseshoe crabs (**Class Xiphosura**), common inhabitants of the Atlantic beaches of North America. They are the last representatives of an ancient class of arthropods which swam in Ordovician seas. If you turn a horseshoe crab over and look among its five pairs of legs, you will find its mouth. The animal has no jaws; it chews mollusks, worms, and bottom-dwelling algae with the bases of the legs. It chews by walking and vice versa. Today's horseshoe crabs appear identical to fossil horseshoe crabs 150 million years old and similar to horseshoe crabs 500 million years old.

Still older relatives of the arachnids were giant water scorpions called eurypterids (**Class**

Figure 24-50 The horseshoe crab, *Limulus polyphemus* (left), is one of the last relics of an ancient group of arthropods. It lives along the Atlantic Coast hunting for bivalves and worms in the sand and mud. Eurypterids (above), which are also known as water scorpions, disappeared about 250 million years ago. This restoration shows a large Silurian eurypterid, *Pterygotus*, which reached a length of more than 8 feet, and was probably the largest arthropod that ever lived.

Eurypterida) which lived at the same time as the trilobites but disappeared about 250 million years ago. They were the largest arthropods known, some exceeding 8 feet in length. Some eurypterids invaded freshwater, and fossil evidence indicates that some may have become terrestrial. In any event true terrestrial scorpions related to today's scorpions, and probably descendants of eurypterids, date back to the Silurian and are the oldest known arthropods which definitely lived on land.

Insects—Class Insecta

Insects make up nearly three-fourths of all of the animal species. They occupy almost every imaginable habitat on land and in fresh water. No other group of animals, plants, or protists can begin to compare with them in diversity. If you examine any particular countryside, you will discover a few species of insects with a very large number of individuals and many more species that are rare. By patiently examining many samples of soil and vegetation you can estimate how many species there are and how many individuals in each species. If you make a few assumptions, you can estimate how many insects there are in the whole world and how many different species there are. The British zoologist C. B. Williams made such calculations and estimated that at any moment the insect population of the world is 10^{18}, that is, a million million million, and that there are about 3 million different species of insects. In other words, there are roughly 300 million insects for every human being alive today. If biological success is judged by numbers, then insects are the most successful group of animals that has ever lived. Their only "failure" has been an inability to adapt to the oceans. Very few insect species live out their entire lives in the sea. Perhaps this was because they arose as terrestrial animals, and when they began their adaptive radiation they found the oceans already too well exploited by crustaceans. Although excluded from the oceans, some have adapted to fresh water, at least in the larval stages.

The most primitive insects known are found as fossils in the rocks of the Devonian period, some 350 million years ago. Soon after they first appeared, they became one of the dominant groups of animals. They underwent a great period of diversification 200 million years ago, in the Triassic period when the flowering plants arose, and this process continues today.

Insects evolved various adaptations to life on land, but the most important was a hard waterproof exoskeleton that protects the body from drying out. This waterproofing property resides in a thin waxy layer of the cuticle. The wax molecules are packed so tightly together in this layer that the molecules of water cannot escape between them. The insect protects this waxy layer from damage with a coat of varnish. This waterproofing is what makes insects truly terrestrial and it is absolutely crucial for their survival. It has been a common practice in North Africa since Roman times to mix a little fine road dust with stored grain to keep it free from weevils. Apparently the dust scratches the waxy layer of the cuticle, destroys the waterproofing barrier, and the insects die.

Another key adaptation to life on land is the tracheal breathing system. Tracheal breathing was also evolved independently by terrestrial isopods and by some arachnids. These air-filled tubes carry air directly to the body cells. Tracheal breathing is ideal for life on land as long as the animal remains small. Indeed all insects are small. The smallest are parasitic wasps about 0.2 millimeter long which develop inside the eggs of other tiny insects. The largest living insect is the 6-inch long giant rhinoceros beetle with the wonderful name *Dynastes hercules*. The largest fossil insects were giant dragonflies with bodies about 1 foot long and 2-foot wing spreads.

Why are all insects small? Part of the answer is to be found in their exoskeletons. The total weight a given kind of skeleton can support has a limit. Internal skeletons like those of vertebrates can support much more weight than external skeletons like those of arthropods. As a result the only place large arthropods (lobsters and eurypterids, for example) have ever been found is in the water, where the buoyancy of the water supports them. Because they have endoskeletons, large vertebrates could live on land, which is one of the big advantages that vertebrates have over insects. However, this still does not explain why

A

B

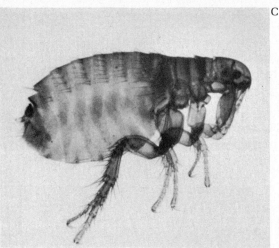

C

D

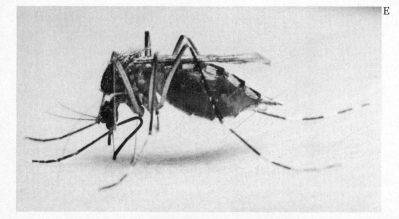

E

Figure 24-51 A variety of insects. A. The giant rhinoceros beetle, *Dynastes hercules,* is the largest living insect. B. The largest insects that ever lived were probably giant dragonflies like this one. C. Adult fleas feed on the blood of birds and mammals. They transmit numerous diseases such as plague (caused by a bacillus) and typhus (caused by a *Rickettsia*). D. The shield bug, *Acanthosoma haemorrhoidale,* feeds on juices of plants. E. An adult mosquito sucking blood.

more insects are not as big as rhinoceros beetles. Apparently being quite small has been a key to insect success. The reason is not hard to find. Small size opens up an entire world to insects that is not open to larger animals such as vertebrates. To develop inside the egg of another insect, to complete its growth within a tiny seed, to tunnel between the two walls of a living leaf are all possible only for a very small animal.

Smallness has had manifold consequences for insects. For example, it has limited their ability to evolve large complex nervous systems like those of vertebrates. After all, only so much can be crammed into a small package. Instead, as mentioned in Chapter 14, insects have developed miniaturized systems of programmed, or instinctive, behavior. Those appear to require fewer nerve cells and less space than systems which permit learning and complex behavior of the kind seen in higher vertebrates.

Although they are a diverse group, all insects share certain key features and probably had a common origin from some millipedelike ancestor. The insect body has three major parts: head, thorax, and abdomen. The head segments are fused firmly together, and in adults most of their boundaries cannot be distinguished. The head has one pair of antennae, one pair of compound eyes, and numerous other sensory receptors. It has three pairs of mouth parts which were derived from ancestral legs. These have been modified in various insects to meet special feeding requirements. Figure 24-52 illustrates the mouth parts of a grasshopper which are adapted

for chewing and of a moth whose mouth parts are suitable for sucking. The thorax is made up of three segments with a pair of legs attached to each, and usually two pairs of wings. The abdomen has 9 to 12 segments, usually without appendages. Sometimes its tip has modified appendages for mating and manipulating the eggs. We have already discussed numerous other features of insects in Chapters 11 to 14.

Wingless insects—Subclass Apterygota. There are about 28 orders of living insects. These are

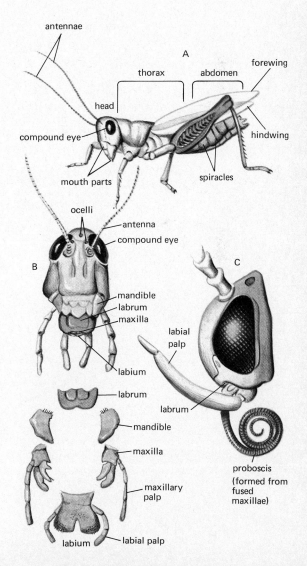

Figure 24-52 (Right) A. External anatomy of an insect. Note the separate head, thorax and abdomen and the three pairs of legs, one pair on each segment of the thorax. There are two pairs of wings. In this grasshopper, the forewings are hardened covers which protect the hindwings which are used for flying. B. Chewing mouth parts of a grasshopper. Top: front view of head showing mouth part. Bottom: the mouth parts have been removed from the head. The pair of *mandibles*, pair of *maxillae* and the lower lip or *labium* are derived from ancestral legs. Several of the segments of the legs from which they are derived are still evident on the maxilla and labium. The upper lip or *labrum* may also be derived from ancestral legs. C. Sucking mouth parts of a moth. The proboscis is formed from the fusion of the pair of maxillae. The mandibles are absent.

conveniently divided into two main subclasses. The first is represented by four orders of wingless insects and is called the **Apterygota,** meaning "without wings." The most familiar are the silverfish or firebrats (Fig. 24-54). These insects have certain primitive features and they have clearly come from wingless ancestors. When a silverfish hatches from its egg, it looks like a tiny adult, except for its gonads which mature later. After it matures, it continues to molt and to grow.

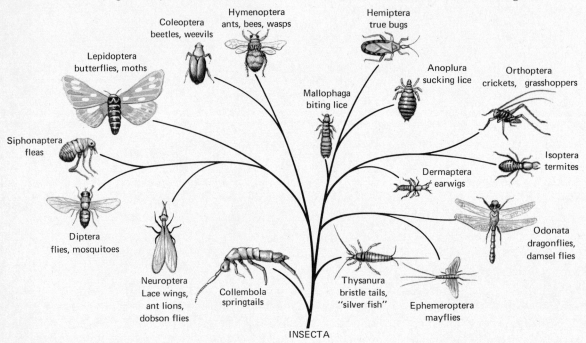

Figure 24-53   Representatives of some of the main orders of insects.

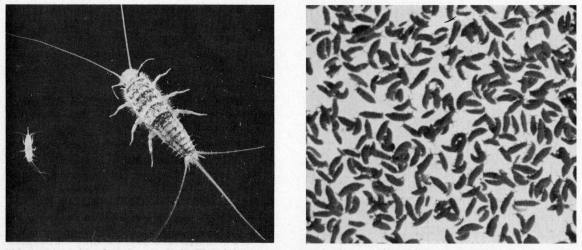

Figure 24-54   Wingless insects. (Left) Silverfish like *Lepisma saccharina* molt throughout adult life. (Right) Springtails are the most abundant insects. Rarely seen because of their small size, they live in the soil.

Another group of these primitive insects is less well known, but merits mention because they are immensely abundant. These are the springtails. These tiny creatures are found everywhere. The reason most of us have never encountered them is that they are small—a few millimeters long—and most of them live beneath our feet. A single square yard of Iowa farm soil may contain as many as 100 million springtails, 400 billion springtails per acre. Except for mites they are the most abundant of all land arthropods. This is true in both temperate and tropical climates and polar climates. Even on the Antarctic continent, wherever there is bare soil, springtails occur in abundance, feeding on fungi, bacteria, and decaying plant material. All told there are about 6500 species in the subclass Apterygota. There is still debate whether all of these are really insects. Perhaps some of them independently evolved six-leggedness from some many-legged ancestor, quite separately from other insects.

Winged insects—Subclass Pterygota. The second subclass is represented by 24 orders of winged insects and is called the **Pterygota,** meaning "winged." A few parasitic orders such as lice and fleas have become secondarily wingless, but these almost certainly had winged ancestors. The Pterygota differ from the Apterygota not only in possessing wings but also because they lose the ability to molt when they become adults. There are more than 800,000 different species of winged insects of which nearly half are beetles. The fact that species of winged insects outnumber wingless insects 100 : 1 indicates that the ability to fly had immense adaptive value. The success of insects as a group may result from various factors, but the evolution of flight certainly gave insects a big advantage over terrestrial invertebrates. They could disperse their young more widely, escape from predators and use new sources of food. Reptiles, birds, and mammals evolved flight at a later time, but insects were the first animals to conquer the air. Winged insects actually fall into two groups that differ in the way they develop and how their wings form.

The first group of winged insects contains 15 orders and includes such common insects as grasshoppers, cockroaches, aphids, milkweed bugs, and dragonflies. In these insects when the larva hatches from the egg it looks very much like the adult except that instead of wings it has two small projections on the outside where the future wings will be. For this reason these insects are called **Exopterygota,** which means "outside wings." In addition to lacking wings the larva also lacks reproductive organs. Also its cuticle may have a different pattern from the adult. It has antennae and compound eyes and usually lives a life like that of the adult. A young grasshopper or aphid feeds on the same food as the adult and is built on the same plan. At the end of larval life the wings and reproductive organs begin to develop and the immature grasshopper molts into a mature adult grasshopper with fully formed wings and reproductive organs. The proc-

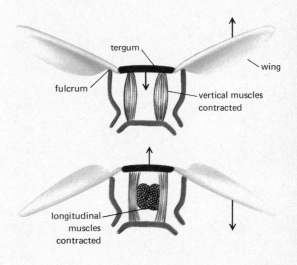

Figure 24-55 How insect wings work. Unlike birds, insects do not have flight muscles attached to their wings. Instead the wings are attached to the body wall of the thorax over a fulcrum in such a way that slight changes in the shape of the thorax causes the wings to move up and down. The contraction of vertical muscles pulls down the upper surface of the thorax called the *tergum*. This raises the wings. Contraction of longitudinal muscles causes the tergum to bulge upward and the wings are pulled downward. The movements of the thorax are extremely small. However, as the diagram shows, the length of the lever on the two sides of the fulcrum is very different and as a result the tips of the wings move a much greater distance.

ess of transforming from a larva to an adult is called **metamorphosis.** In exopterygote insects like grasshoppers, where the transformation is rather modest, we call the process gradual or **incomplete metamorphosis.** It contrasts sharply with the complete metamorphosis of other insects which are considered next.

The second group of winged insects contains nine orders and includes beetles, flies, butterflies and moths, bees and wasps, and fleas. When a larva of one of these insects hatches from its egg, it does not resemble the adult. Think for a moment of a fly larva or maggot and of an adult fly, or of a caterpillar and of a moth. There is no outward sign of the wings on the larva, because the wings develop inside the larva and are quite invisible. This gives the name **Endopterygota,** or "inside wings," to this group of insects. The larvae of endopterygote insects do not have compound eyes or antennae. They are often legless like fly larvae or bee larvae and have a completely different way of life from the adult. For example, caterpillars have chewing mouth parts, whereas adult moths have sucking mouth parts. We call the larval-pupal-adult transformation of endopterygote insects **complete metamorphosis,** because it involves many more changes than does the incomplete metamorphosis of exopterygote insects such as grasshoppers.

Usually, complete metamorphosis occurs in two stages. For instance, when the larva of the Cecropia silkmoth hatches from the egg, it begins to feed voraciously (Fig. 24-57). After a time it molts into a bigger caterpillar. It continues to feed and molt and increases in size about 2000 times. Finally, after four such larval molts, the larva undergoes dramatic internal changes and at the next molt it molts into a *pupa*. The pupa looks completely different from the larva. The transformation of the larva to the pupa is the first step in complete metamorphosis. Within the pupa dramatic developmental changes also occur, and at the next molt the pupa molts into an adult. This is the second step in complete metamorphosis. This process of complete metamorphosis occurs in all endopterygote insects. The number of molts before the adult stage ranges from 3 or 4 to more than 30 in different orders of insects.

Interestingly, there are ten times as many species of insects with complete metamorphosis as species with incomplete metamorphosis. Apparently complete metamorphosis has adaptive significance. This adaptive significance is not hard to identify. The larvae of most insects live in a completely different environment from the adult. A fly larva lives in decaying food and is almost aquatic. An adult fly is a flying organism, lives in a different environment, and eats dif-

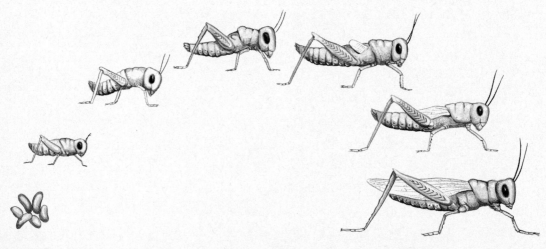

Figure 24-56   Incomplete metamorphosis of a grasshopper. The larval stages (called instars) resemble one another closely. The adult differs from the larvae in possessing wings and functional reproductive organs. There is no pupal stage and the last instar larva transforms directly into an adult.

ferent kinds of food. The ability to exploit two different environments was a major factor favoring the evolution of complete metamorphosis in insects because larva and adult do not compete for food or other resources.

Insect hormones. The periodic molting of immature insects is controlled by hormones. In the molting process, which was discussed in Chapter 11, the epidermal cells of an immature insect detach from the old cuticle and divide so that the

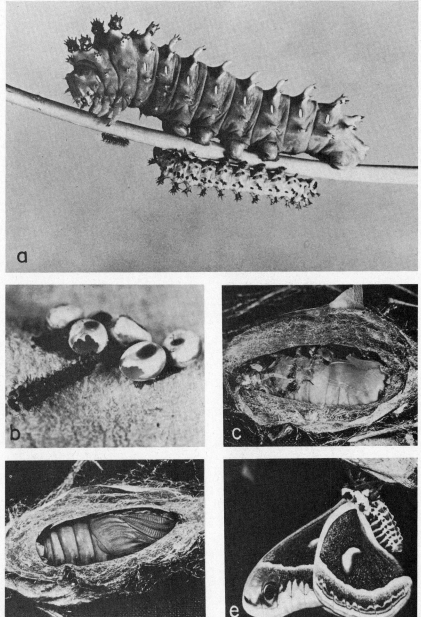

Figure 24-57 Complete metamorphosis of the Cecropia silkworm, *Hyalophora cecropia*. A. First, third, and fifth instar larvae. B. First instar larva hatching from egg. C. Larval-pupal molt within cocoon. D. Pupa within cocoon. E. Adult male of moth.

insect has a larger skin. These epidermal cells then secrete a new and enlarged cuticle and the insect sheds its old cuticle.

Molting is controlled by two groups of hormones. One hormone is produced by secretory cells in the insect's brain. This *brain hormone*, as it is called, activates endocrine glands in the thorax of the insect, the *prothoracic glands*, and causes them to secrete a molting hormone called *ecdysone*. Ecdysone acts on the epidermal cells of the insect and causes them to divide, to deposit a new cuticle, and to molt. This dual control of molting by brain hormone and ecdysone occurs in all insects that have been studied and probably occurs in other arthropods as well (see Fig. 24-60).

Metamorphosis is controlled by a third hormone, the *juvenile hormone*, which is secreted by a pair of endocrine glands, the *corpora allata*, located near the brain of the insect. Nearly 40 years ago Sir Vincent Wigglesworth showed that this hormone promotes larval development but prevents metamorphosis. The presence of juvenile hormone in the immature insect ensures that when the larva molts it will retain its larval characters and not differentiate into an adult. The juvenile hormone is a remarkable agent which permits growth but prevents maturation.

When larval cells are stimulated to grow and molt by the prothoracic gland hormone, ecdysone, the presence of juvenile hormone causes them to use their synthetic machinery to secrete larval cuticle. In the absence of juvenile hormone, the cells of exopterygote larvae secrete an adult cuticle whereas the cells of endopterygote insects secrete a pupal cuticle and then an adult cuticle. In the normal life cycle of an insect the corpora allata cease to secrete juvenile hormone at the end of larval life and metamorphosis occurs.

The following picture of insect molting and metamorphosis emerges. Molting is controlled by regulating the release of brain hormone. Metamorphosis is controlled by regulating the release of juvenile hormone. In simplest terms, in some way ecdysone stimulates the synthetic activity necessary for growth and molting. Juvenile hormone influences the kind of synthetic activities that occur. These various events are summarized schematically in Figure 24-58.

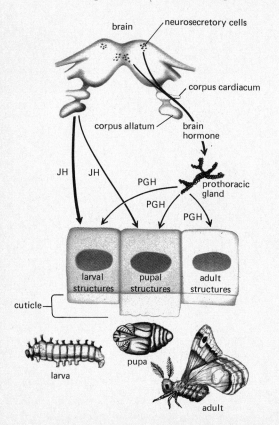

Figure 24-58   Endocrine control systems of insect. This diagram outlines the principal endocrine organs of the Cecropia silkworm. The scheme appears to hold for most other insects as well. The Cecropia larva molts four times. These larval molts appear to be initiated by a hormone from the brain, which stimulates the prothoracic glands to secrete the prothoracic gland hormone, ecdysone. At the same time other endocrine glands, the corpora allata, secrete juvenile hormone, which favors larval syntheses. As a result, when the larva molts in response to ecdysone, it molts into a larva. At the end of larval life the corpora allata cease secreting, and the mature larva is left with a low concentration of juvenile hormone. At the next molt, depending on the time and rate at which ecdysone is released and on the amount of juvenile hormone still remaining, the epidermal cells secrete pupal or adult cuticle. Also, in response to these hormonal conditions, other tissues within the insect either break down or transform into pupal or adult structures.

Insect societies—Chemical communication. Despite their small size, insects have reached a peak of evolutionary development in one area in which they are exceeded only by the higher vertebrates. This is in the formation of societies. The distinguishing feature of a true society is a division of labor. Ants and termites all live as members of a society, with each insect doing a specific job, such as laying eggs, collecting food, and caring for the young. Many, but not all, species of bees and wasps form societies too. This social structure is necessary for the perpetuation of the species. If an insect from one of the social species becomes permanently separated from its society, it is almost certainly doomed to die.

Unlike most human societies all insect societies are overgrown families. The 70,000 worker bees in a healthy hive are all the offspring of a single queen. This is also true of the half million termites that form a single colony and live in one nest. Even the huge marching columns of millions of driver ants that occur in the tropics are all the progeny of one female.

The key to the success of any society is that the members of the community must do the necessary work at the right time. To insure that this happens social insects constantly communicate with one another. Most of this communication is done by exchanging food materials and various chemicals. A striking example of chemical communication is the use of "queen substance" by bees; this substance is secreted by queen bees. Worker bees continuously lick the queen and pass this chemical to other workers by licking and feeding each other. Soon it is passed to every worker bee in the hive. When the old queen cannot secrete enough of this material or dies, the worker bees transfer one of her most recent eggs to a special compartment in the hive called a queen cell. They feed the larva a special diet—rich in a nutritious food called "royal jelly" —which causes the larva to develop into a new queen. Thus as soon as "queen substance" disappears from a hive, worker bees make a new queen. "Queen substance" serves as a means of chemical communication among the thousands of workers in the hive; its message is "the queen is alive and well." Many other chemicals are secreted by bees and other animals which influence the behavior of other members of the same species. Such chemicals are called *pheromones*. Ants use pheromones to blaze trails to

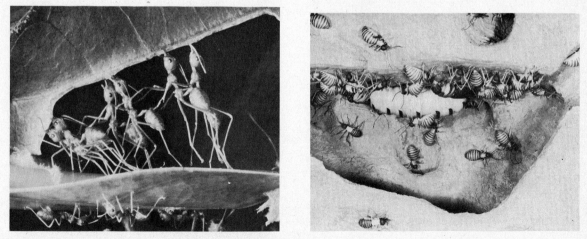

Figure 24-59 Social insects. Left: Worker ants building a nest. This species of Indian ant, *Oecophylla smaragdina*, builds its nest in trees in an unusual way. A group of workers pull the edges of two leaves together. Thereupon, other workers carry in ant larvae which produce silk threads. The workers use this silk to sew the leaves together. Right: This model of the royal chamber in the nest of a termite, *Constrictotermes cavifrons*, reveals a gigantic queen termite with her head toward the right. The king is just below her. A few special soldiers with heads like squirt guns are at the left. The rest of the individuals are workers. The queen may live for years producing a steady supply of eggs, thus maintaining the population of the colony at an efficient size.

food supplies. They also secrete pheromones which enable them to recognize members of their own family. Other animals including vertebrates use pheromones as sex attractants, to mark off territories, and in other ways.

Insects and man. Insects have never conceded that man is the "lord of creation," and man and insects are often in conflict. Aphids, corn borers, weevils, bugs, and sawflies eat the crops man grows for food and fiber, eat the grain he stores, and destroy millions of forest trees. The damage they cause is staggering. Despite the extensive use of insecticides, each year in the United States alone, insects account for crop and livestock losses of more than 5 billion dollars, enough to feed millions of people. In addition to their destructive effects on agriculture, certain insects, such as mosquitoes, carry disease organisms like malaria parasites, encephalitis viruses, and the nematode worms that cause diseases like elephantiasis.

Yet man cannot simply destroy insects *en masse*, because our lives depend on insects. Many of our crops and ornamental plants depend on insect pollination. The lives of many fishes and birds depend on insects. Also insects are their own worst enemies. The major natural enemies of the insect pests of our crops are usually other insects. The ruthless activities of these insect-eating insects keep most insect pests in check.

Consider, for example, the cottony cushion-scale insect which was a major pest of citrus trees in California. This insect almost destroyed the citrus groves. Its onslaught was stopped by an entomologist, who brought in a species of a ladybird beetle from Australia. Both larvae and adults of this ladybird beetle specialized in eating the scale insects. Within two years the citrus industry was saved and the scale insects had almost disappeared.

This, however, is only half the story. Later, grove owners began to spray their trees with insecticides to kill aphids. They also killed the ladybird beetles and cottony cushion scale broke out again. Clearly some sensible balance between the use of insecticides and natural controls must be reached.

Are insecticides necessary? The answer is Yes. There is no known substitute for the use of insecticides. If we stopped using them tomorrow, it is estimated that within two years the world's food production would be reduced by 25 to 50 percent and mass starvation would ensue. Without insecticides malaria would again become a scourge in many places from which it has been eradicated. We must develop insecticides which control dangerous insects without destroying useful insects; and insecticides which do not accumulate in the environment and thereby kill other animals.

Insects are certainly the most widely used experimental animals, with the exception only of mammals. The many roles of fruitflies in genetics and in developmental biology have already been mentioned. Their large chromosomes make visible some of the features of hereditary material; their short life cycle permits quick investigation of genetic changes over many generations. The nervous control of muscles is being studied in grasshoppers; the mode of action of hormones is being examined in caterpillars; the chemistry of vision is being examined in flies. These are only a few of the areas in which insects are used to answer basic biological questions.

Kinships among the arthropods

As already discussed, there is good evidence that the arthropods evolved from annelidlike animals. However, the kinships among the different classes of arthropods are not clear. Recently it was shown that molting in all six classes of living arthropods is controlled by exactly the same

Figure 24-60 Molting in the tarantula spider, *Dugesiella hentzi*, induced by an injection of an insect molting hormone, a steroid called beta ecdysone. This same hormone causes molting in many different groups of arthropods. The cast skin is on the left.

molting hormone. This result provides convincing evidence that all living arthropods descended from some ancestral arthropod that lived in Precambrian times. Many authorities believe that this ancestral arthropod gave rise to three major lines of arthropod evolution: one led to the trilobites, the second led to the arachnids and horseshoe crabs, and the third to the crustaceans, centipedes, millipedes, and insects.

A POSSIBLE CROSSROAD— LOPHOPHORATE PHYLA

There is a small group of strange phyla that have in common the possession of a *lophophore*, a horseshoe-shaped extension of the body surrounding the mouth which bears tentacles used for feeding. All of them are aquatic, and most are marine. The adults are usually permanently attached to the ocean bottom, to rocks, or to seaweed. They are all protostomes, have a coelom, and are unsegmented. The two best known representatives are *Ectoprocta* and *Brachiopoda*.

Ectoprocts (often called bryozoans or moss animals) are tiny creatures, usually less than a millimeter long, which are enclosed in a tough case. They are commonly found in colonies encrusting shells, seaweed or rocks. Their case encloses them like a cup, open at the lophophore end. They do not capture food directly with their tentacles. Instead cilia on the tentacles create a current that carries microscopic food particles into the mouth. Brachiopods are shelled animals that look somewhat like clams. However the resemblance is superficial. In clams the shells are on the right and left sides of the body, whereas in brachiopods they are on the top and bottom of the animal. There are many internal differences as well. Brachiopods are often called lampshells because their shells look like an old Roman oil lamp. Their name, "Brachiopoda," means "arm-footed." When brachiopods were first discovered, the lophophores which carried the tentacles were thought to be arms that had something to do with walking. This turned out to be a mistake but the the name has stuck anyway. Brachiopods feed the same way as do ectoprocts: they are attached to the bottom and use cilia on the tentacles to set up

feeding and respiratory currents.

Brachiopods are of special interest to geologists because they have left an extensive fossil record. There are fewer than 300 living species, but more than 30,000 fossil species have already been described. They are of special interest to us because they have some characteristics which suggest an affinity to the deuterostomes. For example, they have radial cleavage and form their mesoderm and coelom by pinching off pouches from the embryonic gut wall and, in this respect, resemble deuterostomes like the starfish and chordates. However, their mouth is derived from the blastopore and they have a trochophore-like larva. Thus they are perfectly respectable protostomes like the polychaete annelids. These facts suggest that lophophorates may represent the point at which the two major lines of the animal kingdom diverged.

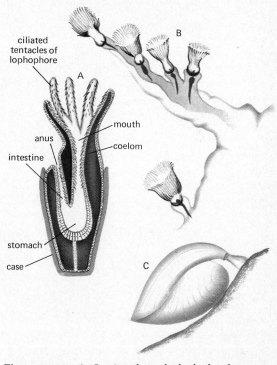

Figure 24-61   A. Section through the body of an ectoproct showing the lophophore with its tentacles and a U-shaped digestive tract. B. Part of a colony of ectoprocts. C. A brachiopod attached to a rock by means of a stalk.

STARFISHES, SEA URCHINS, AND SEA LILIES—PHYLUM ECHINODERMATA

Starfish and sea urchins symbolize the sea. And rightfully so, for that is the only place they are found. More than 550 million years ago they evolved the practice of circulating seawater through their body. Since their blood and the ocean are one and the same, they have never been able to invade fresh water. Yet for all this they have been rather successful. About 6500 living species of echinoderms have been described and more than 20,000 are known as fossils. They are largely bottom dwellers and are abundant in all the oceans of the world. They are found in all latitudes, on all types of bottoms, and at all depths from the intertidal zone to the deeps. Almost every underwater photograph of a stretch of ocean bottom reveals a carpet of echinoderms. They play an important role in marine ecology and consume great quantities of plankton and of dead organisms. Predatory forms feed on many animals, especially mollusks. Interestingly, no echinoderms have become parasites.

The echinoderms are a major phylum, and thousands of scientific articles have been written about them. Yet they remain a puzzle to zoologists. We still do not understand their evolutionary origins, although they left a vast array of fossils. We continue to be mystified by the functioning of some of their parts. And no one knows the adaptive significance of their most characteristic feature—five-sidedness. Echinoderms have maintained five-sidedness for about 500 million years, so it must be advantageous in some way.

The five classes of living echinoderms look quite different, but some common anatomical features demonstrate relationships among them. Starfishes make up one class and brittle stars another class; these animals are flat and usually have five radiating arms. Sea urchins are usually spiny balls. Sea lilies or crinoids also have five branching arms, but these arise from a stalk that is usually attached to the ocean bottom. Sea cucumbers, which make up the fifth class, look like crawling cucumbers with tentacles surrounding a mouth at one end. They include the largest echinoderms—some get to be 6 feet long.

Features shared by all adult echinoderms include radial symmetry and five-sidedness; they all have five similar parts projecting from a central axis. However, as larvae all echinoderms are bilaterally symmetrical. Only as they develop into adults does their typical five-rayed form appear. This five-rayed form is obvious in a starfish, and upon close examination is also evident in all the other forms. A study of echinoderm embryology shows that they are true deuterostomes that form their coelom in the same way chordates do. They are not segmented. Most of their organ systems are well developed except for the excretory system and circulatory system. Their nervous system, although extensive, does not include a brain. However, it does permit central coordination. The "righting" movements which are easily investigated by anyone with a healthy starfish must be centrally controlled.

Almost all echinoderms have bony plates in their bodies that serve as internal skeletons. The skeleton is covered with numerous bumps and knobs. Over these is stretched the animal's thin skin. The name echinoderm means "spiny skin" in Greek, and anyone who has picked up a warty starfish or a pin cushionlike sea urchin will agree that the name fits. This skeleton has had immense adaptive significance for the echinoderms. It is exceedingly tough and provides such extraordinary protection that they have few predators. After all, who could eat a starfish? This absence of most predators may have removed one of the major selective factors that ordinarily acts on animals in evolution. Perhaps this is one of the reasons that echinoderms have made little evolutionary progress in the past 250 million years, and contemporary forms closely resemble their late Paleozoic ancestors.

The most distinctive feature of the echinoderms is a **water-vascular** system. This consists of a system of water-filled tubes called canals which connect directly with the sea water by a tiny perforated *sieve plate* on the surface of the animal. Ciliary action draws water through the sieve plate down a tube into a *ring canal* which encircles the stomach (Fig. 24-63). *Radial canals* branch off the ring canal, run down each arm and connect with hollow *tube feet*. These are soft suction cups, controlled by hydraulic pressure, which enable

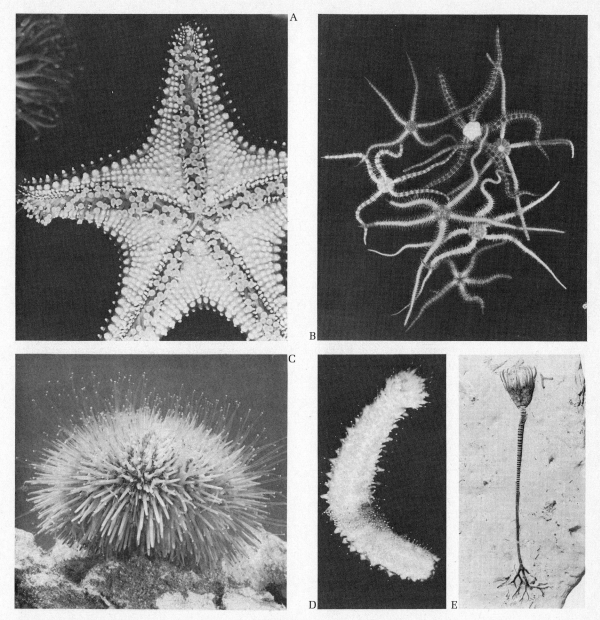

Figure 24-62 Five classes of echinoderms. A. The starfish, *Oreaster reticulatus,* turned upside-down to reveal its tube feet. B. Brittle stars, *Ophiothrix fragilis,* climbing up the side of an aquarium. C. Sea urchin, *Echinus sp.,* showing the tube feet extended well beyond the spines. By grasping a solid surface with the tube feet and then contracting them the animal moves along. Its mouth, which is on its underside, is armed with five pointed teeth that grind the animals and seaweeds upon which the urchin feeds. D. This sea cucumber, *Holothuria forskali,* crawls about over the bottom with its tube feet and stuffs mud into its mouth with its tentacles. It digests particles of food in the mud. E. A fossil crinoid or sea lily, *Eucalytocrinus crassus,* from the Silurian. Crinoids belong to the oldest and most primitive of the living classes of echinoderms. They were believed to be extinct until living specimens were discovered in 1873.

the animal to walk or grip prey. The hundreds of small tube feet of a starfish enable it to hold tightly to the shell and pull open the valves of an oyster or clam on which it feeds. In sea cucumbers the tube feet around the mouth are long tentacles. They are covered with a thick layer of mucus and are used in food capture. The water-vascular system with its tube feet is almost certainly an old invention of the echinoderms. Fossil echinoderms from the early Cambrian period had a pore plate, indicating that they too used a water-vascular system and had tube feet like contemporary echinoderms.

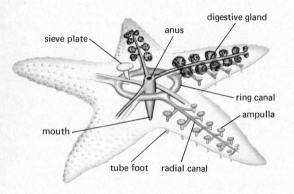

Figure 24-63 The water-vascular and digestive systems of a starfish. The water-vascular system (colored brown) is the starfish's means of locomotion and provides for the clinging and pulling activities of the tube feet. Sea water enters the sieve plate and is drawn by ciliary action down a tube into the ring canal. From there the sea water goes down the radial canal of each arm to the tube feet. Each tube foot is a hollow cylinder and has a soft suction cup at its tip. It connects with a muscular sac called an ampulla. Contraction of the ampulla forces the sea water in it into the tube foot. Backflow is prevented by a valve between the ampulla and the radial canal. By this means the tube feet can be controlled. The digestive system opens to the outside via a ventral mouth and a dorsal anus. Branches of the gut extend into the arms and are furnished with digestive glands.

In addition to a water-vascular system, echinoderms have a spacious coelom in which all of their internal organs are suspended. Like the water-vascular system, the coelom is filled with a liquid similar to seawater. In fact individual organs removed from a starfish and placed in plain sea water will live for many hours or even days. The seawater-filled coelom is important for both respiration and circulation. The bony plates of the skeleton have spaces between them. Through these plates thin fingerlike sacs poke out into the surrounding ocean. These sacs are in direct contact with the coelom on the inside and are covered with skin. They are the echinoderm's *gills*; oxygen diffuses into them from the seawater and then passes into the fluid in the coelom. Echinoderms have no heart to circulate the fluid in the coelom throughout their body, but the coelom is lined with cilia which stir the fluid slowly.

This combined respiratory and circulatory system works, and echinoderms are by no means at the verge of extinction. But they are limited to a fairly sluggish existence. A starfish does not suddenly scoot away when you try to grab it. Indeed all echinoderms are extremely slow movers. Sea lilies are actually anchored in place. They feed on plankton and debris that they capture with their mucus-covered tube feet. The other echinoderms move about, but slowly. Brittle stars use their arms to swim or crawl with and to capture small organisms or particles of food. Sea urchins grip rocks with their hydraulic tube feet and shove themselves slowly along with the five large teeth which surround their mouth. At the same time their teeth scrape algae and barnacles off the rocks for food. Some sea cucumbers use their tube feet to crawl with. Others squeeze themselves into the mud or in cracks in rocks by contracting longitudinal and circular muscle layers of the body wall, just like an earthworm. And, like earthworms, these burrowing sea cucumbers often eat mud. Other sea cucumbers use large mucus-covered tube feet as tentacles which sweep food into their mouths.

Starfish both move and feed by means of their tube feet. Most are predators and attack mollusks and other organisms, including other species of starfish. They create special problems for shell fishermen because of their ability to open the shells of oysters and clams and devour them. If you have ever tried, barehanded, to open a live oyster or clam, you may wonder how it can be done by a starfish which is not much bigger than the clam it attacks. The starfish climbs up on the shell and puts its mouth over the place where the two valves normally open. It attaches almost every tube foot to the two valves and pulls the valves slightly apart. Even a fraction of a millimeter is enough. The starfish everts its stomach through its mouth and slips it through this narrow slot. That is the beginning of the end of the clam. Soon the starfish's stomach has digested the clam deep in its own shell. Starfish are such a menace that fishermen attempt to remove them from oyster and clam beds by dragging a large mop over the bottom which the starfish grasp or become entangled in. They used to cut the starfish into pieces and throw them back into the sea. This proved of little use. Starfish have remarkable properties of regeneration and most of the pieces regenerated a new starfish! Nowadays starfish are destroyed.

We have already mentioned the huge venomous crown-of-thorns starfish, *Acanthaster planci*, which is undergoing massive population explosions throughout the Pacific Ocean. *Acanthaster* has a voracious appetite for coral. It everts its stomach through its mouth and spreads the stomach over the coral. It then digests the soft tissues of the coral and moves on, leaving behind a dead white skeleton, which is quickly overgrown with algae. The starfish's appetite seems to be stimulated by substances released by the coral, and they lay waste the coral reef at a rapid rate. No one knows why this starfish plague has started, nor does anyone yet know how to stop it. Control depends on limiting the migration of starfish into fresh areas. Sand is a poor surface for the gripping by the tube feet of the starfish, and some efforts are being made to produce sandy barriers to stem their advance.

Who were the ancestors of the echinoderms? Because of the structure and bilateral symmetry of their larvae, most zoologists believe that the ancester of the echinoderms was a bilaterally symmetrical ciliated animal. It may have diverged from the protostome line at about the same time that the brachiopods and other lophophorate phyla did. Since echinoderms appear to have arisen from a bilaterally symmetrical ancestor, it is worth asking why they now have radial symmetry. Radial symmetry is a useful adaptation for an animal that is attached to the bottom, but the only living echinoderms that are attached to the bottom are the sea lilies. Starfish, brittle stars, and sea urchins move about freely, but they are still radial. A good guess is that the bilaterally symmetrical ancestor of the echinoderms became anchored to the bottom, where it took up a sedentary mode of life and gradually evolved a more adaptive five-sided radial symmetry. Later in evolution, after they had become radial, some of these sedentary echinoderms became detached from the bottom and again took up a free-moving life. Most retained their radial symmetry and are today's starfish, brittle stars, and sea urchins. One group, the sea cucumbers, began to shift back to bilateral symmetry as an adaptation to a mobile life. Only the sea lilies remained attached. These speculations find some support in the fossil record.

Echinoderms have been widely used in research for more than 70 years, Sea urchin eggs are now employed the world over in studies of gene function in early development. This kind of chemical work requires large numbers of embryos at specific stages, and is easily solved by using sea urchin gametes. Sperm and eggs are mixed together in a dish to obtain tens of thousands of fertilized eggs, all cleaving and developing at the same rate. Starfish have also been the subject of much scientific investigation because of their ability to regenerate lost arms and their unusual ability to turn their stomachs inside out when they eat bivalve mollusks.

CHORDATE CHARACTERISTICS

We finally come to our own phylum, the **chordates.** This phylum gets its name from one of its three distinctive features. All chordates at some time during their lifetimes possess a flexible supporting rod along the back, called the **notochord,** which stiffens the animal's body. In most chordates the notochord is lost in the adults, and is replaced by a series of jointed *vertebrae,* the **backbone,** which develops around the notochord. Animals with backbones are called **vertebrates;** they comprise one of the three subphyla of the phylum Chordata. Along with the appearance of the vertebrae, the rest of the bony or cartilaginous endoskeleton appears.

The second important characteristic unique to chordates is the single *hollow tubelike nerve cord* that develops along the back of the embryo. During development this hollow nerve cord becomes the central nervous system, and in most members of the group the front end enlarges greatly to form the brain. It is worth noting that the main nerve trunk of chordates runs beneath the dorsal surface of the animal above the notochord. In contrast, the main nerve trunk of all of the major invertebrate phyla—annelids, arthropods and mollusks—is always near the ventral surface.

A third characteristic chordate feature is a series of openings in the wall of the pharynx called **gill slits.** Their adaptive significance to the earliest chordates is not known, but they proved to be a tremendous invention. In some aquatic forms they serve as strainers for filter feeding and respiration. In other forms they function mainly in respiration and provide the passageways through which water passes from the gills. In chordates adapted to land life, the gill slits and the part of the skeleton which supports them may appear only in the embryo and may later be altered to serve some purpose other than respiration. In reptiles, birds, and mammals, for instance, the embryonic gill slits disappear and some of their supporting tissues become part of

the jaws, tongue, the larynx, facial muscles, and inner ear as well as the thyroid and parathyroid glands.

All chordates are bilaterally symmetrical and most are segmented, at least to a certain extent. Segmentation is most clearly seen in the backbone of adult vertebrates. Most chordates also possess a well-developed muscular tail that extends beyond the anus, and a well-developed coelom which provides a fluid-cushioned chamber for many of the internal organs.

Chordates are customarily divided into three subphyla. Before examining our own subphylum, the Vertebrata, let us look briefly at the other two subphyla, the Urochordata and Cephalochordata.

Sea squirts and amphioxus— Invertebrate chordates

Familiar as you are with many of the chordates, you would probably not identify the *tunicates* or *sea squirts* as chordates. These animals compose the subphylum **Urochordata.** The most familiar adult tunicates are somewhat barrel-shaped animals and are usually attached by one end to rocks, seaweed, or pilings. They range in size from a pea to a large potato. They are a widespread, abundant group of marine animals and more than 2000 species have been described. They are covered with a characteristic cellulose *tunic,* one of the few places cellulose occurs in the animal kingdom. The only chordate feature of the adult tunicate is the presence of pharyngeal gill slits. Cilia on the gill slits create a current of water which is filtered by the gills for food and for respiration. This kind of behavior resembles the filter feeding of bivalve mollusks.

Larval tunicates look very different from the adults. The larvae are motile and resemble tiny tadpoles. In addition to gill slits they have a dorsal nerve tube. The notochord, however, is in their tails. (Urochordata means "tail-chorded".) Thus they are perfectly good chordates. Larval tunicates finally settle down and gradually transform into barrel-shaped adults. The transformation of the larva to the adult is called *tunicate metamorphosis.* It involves just as many changes in shape and structure as does insect or amphibian metamorphosis. Indeed the adult tuni-

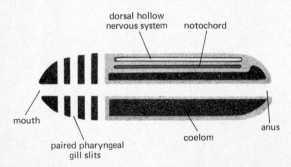

dorsal hollow
nervous system   notochord

mouth

paired pharyngeal
gill slits

coelom

anus

Figure 24-64   Basic design of the phylum Chordata.

cate entirely loses its notochord and most of its nerve cord.

Tunicates have been studied extensively by biologists, largely because of their relationship to other chordates. One of the most striking exam-

ples of this relationship is the fact that tunicates have cells in their pharynx which synthesize thyroid hormone in the same way that cells of the thyroid gland do in vertebrates. In tunicates these hormone-secreting cells are found in a ciliated

A

B

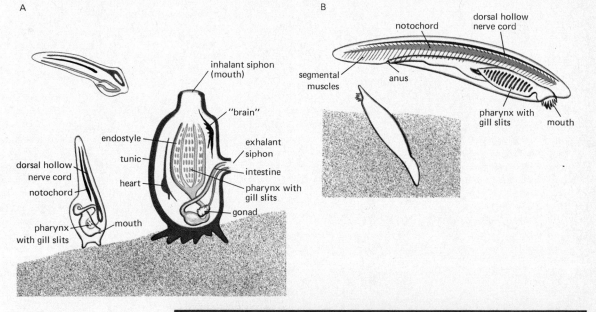

Figure 24-65 Invertebrate chordates. A. Tunicate structure and metamorphosis. Note the notochord in the larval stages. The photograph at right shows adults of a common tunicate, *Molgula manhattensis*. B. *Amphioxus*. The key chordate features of amphioxus are evident. The animal lives a rather sedentary life as a filter feeder, half buried in the sand.

groove in the pharynx called the *endostyle* which produces the mucus the animal uses in feeding. Hence the endostyle is thought to resemble the ancestor of the thyroid gland of the vertebrates. A similar structure is also found in the cephalo-chordates.

The second subphylum of chordates is the **Cephalochordata.** As their name indicates ("head-chorded") the notochord of these animals reaches to the tip of the head. They are also called *lancelets* and the most common form is *amphioxus.* These marine animals, which grow to 2 to 3 inches, also show characteristics in common with other invertebrates. One interesting characteristic is the nerve-muscle association. In common with a number of invertebrates (nematodes for example) long muscle processes extend from the periphery to make connection with the dorsal nerve cord, rather than the more normal arrangement in chordates where long nervous processes extend from the dorsal nerve cord to connect with muscles.

Although not common in our own coastal waters, in some areas off the China coast near Amoy amphioxus is unusually abundant and has been harvested by the ton for food. Unlike the Urochordates they keep their notochord, dorsal hollow nerve cord, and gill slits throughout their entire life cycle. In addition they have segmented muscles, a feature which emphasizes their close relationship to the vertebrates. They are fishlike in shape and can swim, but despite this they live much of their lives buried up to the head in sand. Like tunicates they use the cilia on their gill slits to create a current of water which enters their mouth. This water is strained over their gills to obtain food. Their main interest biologically is their close resemblance to the vertebrates.

The origin of chordates and vertebrates

The origin of chordates has long been the subject of speculation. The evidence is not complete, but several facts are clear. Curious as it may seem, the starfishes and sea cucumbers are among the closest living relatives to members of our own phylum. Admittedly they are unlikely-looking "blood relations." But from our earlier discussion of the differences between the protostomes and

deuterostomes it is clear that the echinoderms and the chordates have several important features in common that set them apart from all of the protostome phyla. Similarities shared by these animals are (1) the process by which the anus forms from the embryonic blastopore, (2) radial cleavage, (3) indeterminate cleavage, and (4) a mesoderm which forms from pouches of the gut. Further evidence for the common ancestry of echinoderms and chordates is shown by a group of wormlike marine animals, called **hemichordates,** which share features with both echinoderms and chordates and may have descended

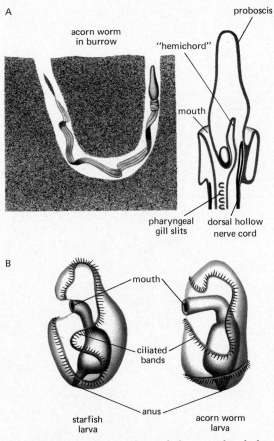

Figure 24-66 A. *Balanoglossus* belongs to the phylum Hemichordata. Here we see one in its burrow. It has several chordate features but lacks a notochord. The "hemichord" is not homologous with a notochord. B. Larva of a starfish (left) and of an acorn worm (right). Note the striking similarities.

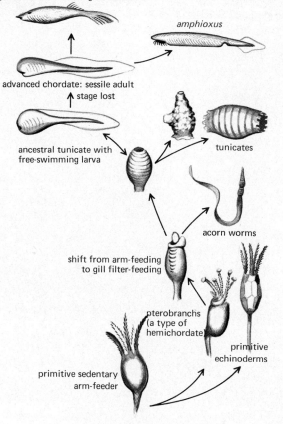

primitive filter-feeding vertebrate

amphioxus

advanced chordate: sessile adult stage lost

ancestral tunicate with free-swimming larva

tunicates

shift from arm-feeding to gill filter-feeding

acorn worms

pterobranchs (a type of hemichordate)

primitive echinoderms

primitive sedentary arm-feeder

Figure 24-67 One hypothesis of the origin of vertebrates. Central figures in this scheme are a group of tiny, rare marine animals called *pterobranchs*, related to acorn worms. Pterobranchs have a pair of gill slits but feed like ectoprocts by means of a lophophore composed of tentacles around the mouth. These tentacles, like those of ectoprocts, are richly supplied with cilia which set up water currents bringing food particles to the mouth. Echinoderms may have evolved from animals similar to pterobranchs. Some descendants evolved a gill-feeding system instead of a tentacle-feeding system. One group evolved into acorn worms. Another group developed the gill system extensively and became tunicates. Some sedentary tunicates evolved a free-swimming larva with a notochord and a nerve cord whose main functions were dispersal and locating favorable spots for the adult to settle down. Later, the old sedentary adult stage was abandoned while the larval form continued to advance.

from the common ancestor of both groups.

The most common hemichordates are called *acorn worms*. They live in U-shaped burrows in the mud or sand in shallow water. Most are only a few inches long, but a gigantic Brazilian acorn worm called *Balanoglossus gigas* is more than 4 feet long. The body possesses a large acorn shaped probiscis for burrowing, a collar, and a long wormlike trunk. Acorn worms have one distinctive chordate feature—gill slits in the wall of the pharynx. They pump water into the pharynx and out through the gill slits to obtain oxygen and unload carbon dioxide. This suggests a relationship to the chordates. However, unlike all true chordates, hemichordates lack a dorsal hollow nerve and a notochord. They have a structure resembling an incomplete notochord and their name. *Hemichordata*, means "half-chorded." However, recent studies indicate that this incomplete "notochord" is not a notochord at all and has a different embryonic origin and a different function. Hence, although hemichordates are related to chordates, they must be placed in a separate phylum.

In addition to being related to chordates, hemichordates are also closely related to echinoderms and have an almost identical embryonic development. Furthermore, the ciliated larvae of some hemichordates closely resemble the ciliated larvae of starfish. Because the hemichordates share important features with both echinoderms and chordates, most biologists believe that they originated from the ancestral organisms which give rise to both echinoderms and chordates.

What about the origin of the vertebrates? When the chordate features of tunicates were first described in 1866, Charles Darwin remarked that "we have at last gained a clue to the source whence the Vertebrata were derived." To this day the larvae of tunicates have been central to all discussions of chordate evolution. Many biologists accept the theory that tunicates and vertebrates descended from a common free-swimming ancestor that resembled a modern tunicate larva. One group of these ancestral tunicate larva-like creatures is supposed to have settled down on the bottom and then gradually evolved the barrel-shaped adult stage which is well adapted to a sedentary filter-feeding life.

While leading such a sessile mode of life permanently anchored to the bottom, there was no selective pressure upon the animal to evolve an elaborate skeleton or nervous system. A second group of these ancestral tunicate larva-like animals is supposed to have continued their free-swimming life and evolved into the vertebrates. Another and more popular theory suggests that the ancestor of the vertebrates was in fact a sedentary tunicate (Fig. 24-67). The larval form which originally evolved as a means of dispersal is thought to have developed the ability to reproduce, thus bypassing the sedentary adult stage of the life cycle. This latter theory is supported by the fact that some cephalochordates and some tunicates reproduce as larvae.

Animals with backbones— Subphylum Vertebrata

However the vertebrates originated, they have shown a remarkable potential for evolutionary change, and they are overwhelmingly more successful than the urochordates and cephalochordates. The unique characteristic of the vertebrates, which sets them apart from all other animals, is their **backbone.** It is composed of interlocking pieces of living tissue, either bone or cartilage. When it first evolved in ancient fish, it provided a rigid but flexible attachment for muscles to pull upon and so permitted powerful side-to-side bending movements of the whole body which enabled the animal to move through the water much more rapidly than a flatworm or a

polychaete. It also permitted other adaptations to evolve. The backbone supports the body, allowing the animals to reach large sizes. It protects the nerve cord and the large brain case. Since it is flexible, a backbone allows animals great freedom of motion, whether the animal is a fish swimming through the water, a snake wriggling over the ground, or a deer running at great speed.

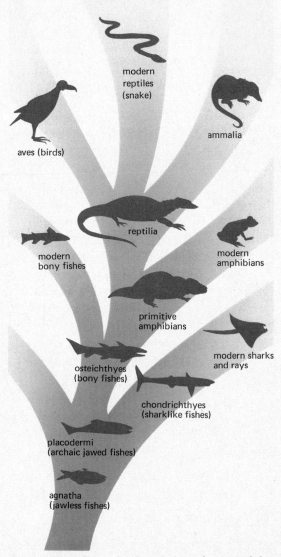

Figure 24-69   The eight classes of vertebrates and their relationships.

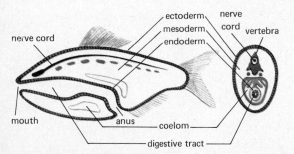

Figure 24-68 Vertebrate body plan. Compare with annelid body plan (Fig. 24-29).

Vertebrates are known as fossils from the Ordovician period, more than 500 million years ago. Because their bones are highly mineralized and resist decay, their skeletons form excellent fossils, and as we discussed in Chapter 22, we have been able to trace the path of vertebrate evolution with much more precision than that of most other groups of organisms.

Vertebrates are one of the most successful of all animal groups and are found nearly everywhere that life can exist. About 43,000 living species have been described. They have overcome the limitation in size that other terrestrial groups, such as the insects, have not. The largest animals on earth are vertebrates—whales that may be more than 100 feet long and weigh 150 tons. The largest animals ever to roam the land, the giant dinosaurs of the Mesozoic era, were also vertebrates. Two classes of vertebrates have evolved more efficient methods of temperature control than any other animal, allowing them to remain active in environmental extremes of heat and cold. The nervous systems of vertebrates have become the most complex known. Vertebrates can receive and respond to a greater variety of stimuli than any other group of organisms. To handle the input of information and process it, vertebrates have evolved complex brains, including the most complicated brains known, those of primates, the vertebrate order which includes apes and man.

The subphylum Vertebrata consists of eight classes. Four make up the superclass **Pisces,** the fish, which in simplest terms comprises the finned, gill-breathing aquatic vertebrates. The other four classes make up the superclass **Tetrapoda,** which includes the four-limbed, lung-breathing, terrestrial vertebrates.

Fishes—Superclass Pisces

The oldest vertebrates of which we have fossils are a group of jawless fishes called the **Agnatha** (which means lacking jaws). These ancient fishes had a perfectly good skeleton but lacked jaws and paired fins. They were probably filter feeders like amphioxus and the tunicates and strained the food material from mud and water flowing through their gills. When they first appeared 500 million years ago in the Ordovician period, they were heavily armored with bony plates, probably to protect them from vicious

Figure 24-70 A lamprey (below). Note the pharyngeal gill slits. On the right is an enlargement of its mouth which has no jaws.

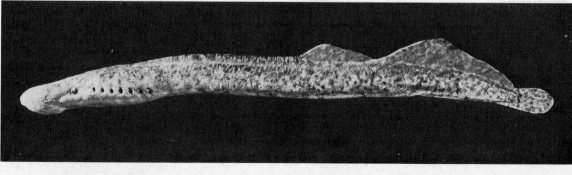

predators like the gigantic water scorpions which were abundant then. The modern descendants of these ancient armored fishes are the lampreys and hagfishes. These animals have elongated cylindrical bodies without armored plates or scales. In their larval stages they filter food through their gill slits like amphioxus. However, as adults these animals are not filter feeders. Instead they attach themselves to other fishes by a circular sucker around their mouth, rasp a hole in the fish with their tongue, and suck out blood or tissues. They still filter water through their gill slits, but only for respiration.

The *placoderms* (class **Placodermi**) were armored fishes with bony skeletons that evolved from agnaths. They were the first vertebrates to have **jaws.** Their jaws appear to have evolved from skeletal supports of the gills. It is worth noting that hinged jaws evolved independently in both the arthropods and the vertebrates. However, although they serve the same function, they are analogous and not homologous structures. The jaws of arthropods arose from ancestral legs, whereas the jaws of vertebrates evolved from the skeletal elements of the gills.

It is impossible to overestimate the adaptive significance of hinged jaws. Such jaws enabled the ancestral vertebrates to leave the bottom where small particulate food debris accumulates and to hunt and devour large prey. This changed their whole way of life and they evolved various adaptations for an active life, most notably paired fins. Placoderms gave rise to the two major classes of modern fish, one characterized by a skeleton made of cartilage, the other by a skeleton made of bone.

The **cartilaginous** fishes (class **Chondrich-thyes**) are represented today by the sharks, rays, and skates. Although they do not now have bony skeletons, their placoderm ancestors did, and the loss of bone and its replacement by cartilage are later developments. The bony toothlike structures in the skin called "*denticles*" that give shark skin its sandpaper-like texture are all that remains of their bones. They all have paired **pectoral** and **pelvic fins,** the precursors of the legs of tetrapods. This class probably was originally adapted to living in the ocean, and only a few forms have ever made the transition to a freshwater environment. They prevent water loss to the ocean by reabsorbing most of the urea they form until the concentration of their body fluids is similar to that of seawater.

Most cartilaginous fishes are predators. Rays and skates prey largely on mollusks found on the ocean floor and their flattened body is an adaptation for bottom dwelling. Sharks, in contrast, are streamlined and well adapted for their life as roving hunters. Interestingly, the largest cartilaginous fishes, indeed of all fishes, the basking and whale sharks, have forsaken a predatory life and returned to the ancient filter feeding habit.

The **bony fishes** (class **Osteichythes**) include the sardines, tunas, trout, indeed most of the fish we are familiar with as well as unusual fish like sea horses. They have a skeleton of true bone, scales, and two pairs of lateral fins. They probably originated in fresh water 400 million years ago in the Devonian period but spread to the sea early in their history. Soon after they evolved they became the dominant vertebrates in both fresh and salt water, a position they retain to this day. They live in every aquatic habitat from the surface of the sea to lightless oceanic depths where

Figure 24-71 Model of an extinct placoderm. The hinged jaws are noticeable.

pressures are in many thousands of pounds per square inch. They abound in the warm waters of the tropics and in polar seas. In both numbers of species and of individuals, they exceed all the other living classes of fish. Nearly 20,000 species are known and probably an equal number remains to be discovered. As a group they are at the height of their evolution and show no sign of being displaced. They range in size from guppies less than 1 inch long to gigantic ocean sunfish which may be more than 20 feet long. They feed on virtually every kind of food found in the water.

The primitive bony fishes had, in addition to gills, *lungs* which they used to increase gas exchange when the water was low in oxygen. Most modern body fishes use only their gills for breathing. The lungs have evolved into a **swim bladder** which regulates the density of the fish and helps maintain buoyancy. A few lone descendants of these early forms still retain both gills and lungs. One group, called the **lungfishes,** is represented by a handful of species. Some lungfish are so dependent on their lungs for breathing air that they will actually drown if their nostrils

A

B

Figure 24-72 Some modern cartilaginous fishes. A. A sting ray. This attractive creature uses its large pectoral fins to swim through water. The gill slits are evident on the underside. B. The whale shark, *Rhinodon,* is the largest of the cartilaginous fishes. C. The rough skin on the head of this sand shark (*Carcharias taurus*) reveals the denticles, the last remnants of bones.

C

Figure 24-73   A variety of bony fishes. A. Salmon (*Salmo salar*). B. Seahorse (*Hippocampus*) can hold on to seaweed with its unique prehensile tail. The males, like this one, incubate the fertilized eggs in an abdominal pouch until they hatch. The photo shows a male and his newly released offspring. C. Northern pufferfish (*Sphaeroides maculatus*). D. The African lungfish (*Protopterus*) lives in sluggish or stagnant waters and feeds on bottom fishes. In the dry season the lungfish wriggles into the mud which eventually hardens around it. Its mucous glands secrete a waterproof layer around its body which prevents it from drying out. Its mouth is left uncovered and it continues to take a few breaths every hour.

are kept under water. Another descendant of the primitive bony fishes is the coelacanth, *Latimeria*, described earlier, a bony fish that has both lungs and lobelike fins. Ancestors of *Latimeria* gave rise to the amphibians whose evolution opened up the land for the rest of the tetrapods.

As a group, the bony fishes are important to man and more than 100 billion pounds of fish are harvested each year as a source of protein for the world's population. As man's understanding of marine biology increases it is hoped that fish will be used increasingly as food. Unfortunately, over-fishing and environmental deterioration are destroying many of the world's great fisheries. It is no longer economical to fish for haddock off New England or for sardines or anchovies off California. In recent years the occurrence of mercury in certain fish such as swordfish and tuna has restricted their use as food in several countries, and is encouraging the development of international laws concerning the pollution of the ocean. The ocean is not as limitless as we once imagined, and man's activities are changing its composition.

Bony fishes present biologists with some of their most intriguing problems. For example, how do salmon "smell" their way home over thousands of miles from the middle of the Pacific Ocean to the particular mountain stream in North America or Asia where they were born? What navigational aids (sun, moon, stars, currents) do they use on their journey?

Newts and frogs—Class Amphibia

Although the amphibians were the first vertebrates to come ashore 350 million years ago, they were not fully adapted to a totally dry environment. Their living descendants still show a primitive dependence on water. Their name in fact means "a double life" and most adult amphibians return to fresh water to reproduce. In most cases the eggs hatch into larvae that are totally aquatic, and live almost as fish (they have gills, for instance). However, some amphibians have evolved in which the larval stage is completed inside the egg, and these species are independent of water for reproduction, although they need a moist environment.

The transformation of the aquatic larva to the terrestrial adult is called **amphibian metamorphosis.** It is accompanied by numerous changes necessary for the transition to terrestrial life, such as the replacement of gills by lungs and the development of legs. Most amphibians have two pairs of legs, modifications of the paired pectoral and pelvic fins of their lobe-finned fish ancestors, and lungs which they also inherited from their fish ancestors. The development of legs was certainly a crucial adaptation for terrestrial life.

Another less conspicuous but equally important adaptation was the development of **mucus-secreting glands** in the skin. Mucus enabled amphibians to remain out of water for sustained periods, yet not dry out. One can imagine an ancient Devonian summer in which many streams and ponds dried up. All the fish in a pond would die, but the amphibian with his newly developed legs and mucus-covered skin could crawl away to another stream or pond to continue his life. This ability to remain on land proved to have great selective advantage, not merely because it enabled amphibians to survive drought but because it opened up for them a whole new environment with numerous new food sources. It also freed them from aquatic predators.

Mucus-covered skin also proved important to amphibians for breathing purposes. Most adult amphibians have poorly developed lungs which are not well ventilated. They breathe mainly through their skin which has many fine blood vessels. The mucus glands help to keep the skin moist so that oxygen can diffuse in. When the skin dries out, oxygen ceases to diffuse in. Although their mucus-covered skin resists drying, it is not waterproof like the skin of an insect or a reptile. Hence most amphibians live in damp places or fresh water. A few occur in brackish water but none are marine. Some frogs and toads live in deserts such as the American Southwest. They hide underground during dry periods and come out at night to feed.

Because they have no mechanisms for regulating their body temperature amphibians are found mainly in tropical and temperate regions, but a few frogs range into the Arctic Circle. In winter many amphibians hibernate in deep lakes or streams that do not freeze, or enter crevices or

A

B

C

D

Figure 24-74 Modern amphibians. A. The common
newt, *Diemictylus viridescens*. B. A frog which has
captured a grasshopper with its tongue. C. The tiger
salamander or axolotl, *Ambystoma tigrinum*. D. Tadpoles of a bullfrog.

animal burrows and get below the frost line.

Many modern terrestrial amphibians eat insects which they capture with their unique tongue. It is long and attached anteriorly so that is can be flicked out a considerable distance (man's tongue is attached posteriorly). The tip is covered with a sticky substance to which insect prey adheres. Aquatic larvae of toads and frogs feed mainly on algae.

The 2000 living species of amphibians fall into two major orders. One order is the **Caudata** (caudal means tail or posterior) which includes the salamanders and newts, that have tails as adults. Examples of caudates are tiny creatures like the red eft, the "siren" of Florida which attains a length of 2 feet, as well as exotic 5-foot-long giants like the giant salamander of Japan, *Andrias japonicus*. The adults of one group of salamanders, the plethodontids, have achieved some notoriety because they lack both gills and lungs and breathe entirely through the skin. The other order of living amphibians, **Anura** (meaning no tail), consists of frogs and toads. These lose their tails during metamorphosis. Most of these animals are only a few inches long, but a few like the giant frogs of Lake Titicaca in the Andes and the giant toads of Brazil get to be 2 feet long.

Both groups have been widely used by biologists as experimental animals. Frog embryos have played a central role in experimental embryology and the clawed toad *Xenopus* is used extensively in analyzing the control of developmental processes (see Chapter 8). Newts and salamanders have received attention because, unlike most higher vertebrates, they have the ability to regenerate lost limbs and tails.

For instance, if you amputate the leg of a mouse, the mouse responds by healing the wound. But if you amputate the leg of a newt, the newt responds by making a new leg. After the newt's leg is amputated, the wound becomes covered by migrating epidermal cells. The other tissues in the vicinity of the wound—muscle, bone and connective tissue—form a mound of *"dedifferentiated cells"* beneath the epidermis. This mound of cells differentiates to form a new limb and replaces the one that was lost. There is good evidence that the presence of nervous tissue is necessary for the process of regeneration. Recently some regeneration (although not a normal limb) has been induced in the normally nonregenerating limbs of lizards and opossums by increasing the nerve supply to the stump.

Turtles, snakes, lizards, and crocodiles— Class Reptilia

Reptiles were the first true land vertebrates. When they first appeared some 300 million years ago, they quickly displaced most of the amphibians. There were many reasons for this, but probably the most significant was the evolution of the "land egg" or **amniotic egg.** Amniotic eggs are covered with a firm leathery or limey shell which keeps them from drying out. Within the shell is a large supply of yolk and water so that the embryo can develop within its own protected environment, independent of the outside world for food or water. This amniotic egg freed reptiles from returning to water to breed. Because most reptiles are wholly terrestrial, fertilization must occur within the body of the female and not in the water after the simultaneous release of eggs and sperm as in most fishes and amphibians. Reptiles thus became the first vertebrates to copulate.

Reptiles also evolved many other improvements that made them better fit for life on land than amphibians. Reptiles have strong rib muscles that enable them to ventilate their well-developed lungs more effectively than amphibians. No longer dependent on their skin for respiration, they waterproofed it and covered it with horny scales which prevent desiccation. As we saw in Chapter 12, they also show a number of improvements in their circulatory system, which prevent freshly oxygenated blood from the lungs from mixing with the nonoxygenated blood from the rest of the body.

Like amphibians, reptiles have no internal mechanisms of control of body temperature and they are markedly influenced by the temperature of their environment. For this reason most reptiles, both species and individuals, live in tropical and subtropical areas. Their numbers decrease rapidly in high altitudes and toward the poles. Thus 69 different kinds of reptiles are found in Louisiana but northern Alberta has only one, a garter snake.

Figure 24-75 Modern reptiles. A. A fox snake breaking out of its egg. B. The copperhead, *Agkistrodon contortrix*, a common venomous snake. More people are bitten by copperheads in the United States than by any other poisonous snake. A sensitive pit between the nostril and the eyes acts as a heat receptor and detects warm-blooded prey. C. The Komodo dragon lizard of Indonesia (*Varanus komodensis*), the largest of all living lizards, may exceed 10 feet in length. D. Two American alligators, *Alligator mississippiensis*. E. Musk turtles (*Armomochelys odoratus*). F. The tuatara of New Zealand, *Sphenodon punctatum*, is the sole living representative of an almost extinct order of reptiles.

Soon after reptiles appeared they invaded many different land habitats and some even returned to the sea. Altogether about 15 orders of reptiles evolved, of which only 4 survive today, and one of these is nearly extinct. The completely extinct orders included such forms as the marine ichthyosaurs and plesiosaurs, the flying reptiles called pterosaurs and the ornithischian (birdlike) and saurischian (reptilelike) dinosaurs. Included in this number were the largest land animals of all time, *Brontosaurus* and *Brachiosaurus*, which weighed more than 30 tons. One extinct order of reptiles, the *therapsids*, gave rise to the mammals and another, the *thecodonts*, gave rise to several other orders of reptiles and to the birds.

Three of the four orders of living reptiles are still quite abundant and include some 8000 species. Turtles, tortoises, and terrapins make up the order **Chelonia.** Here the keratinized scales have been modified to make the carapace, the "shell," and the teeth have been replaced by a horny beak. These reptiles range in size from tiny sliders which weigh only a few ounces to huge leatherback sea turtles which may weigh more than 1500 pounds. Some members of this order are renowned for their migratory ability, and they travel distances of over 2000 miles in the ocean to their breeding grounds. Crocodiles and alligators are in the order **Crocodilia** and include in their number the most massive modern reptile, the African crocodile which grows to a length of 16 feet. It feeds on fishes, other reptiles including its own young, and mammals sometimes including man.

The third prominent group is the order **Squamata** containing the lizards and snakes. They are the most plentiful living reptiles (5700 species) and occur everywhere from the desert to the open oceans. Although a legless undulating snake looks very different from a running lizard, the two are closely related. Snakes have evolved from ancient lizards. Interestingly the lack of legs has not slowed down snakes at all and some can move over the ground almost as fast as a man. They range in size from a few inches to 30 feet long. Giant Asian pythons can weigh more than 300 pounds. Several groups of snakes secrete a poisonous saliva which they inject with their teeth, "fangs," which are hollow like hypodermic needles.

Lizards have kept the body plan of a typical land vertebrate. They have legs with five toes on each foot, and they probably move in the same way as did the earliest reptiles. Most lizards are less than a foot long, but the Komodo "dragon" of Indonesia reaches a length of over 10 feet and weighs up to 300 pounds. Lizards are well known for their ability to discard their tails as a defense mechanism and to subsequently regenerate new ones. The exact spot where the tail can break off is preset by a crack in a vertebra, and the muscles nearby come apart neatly. The broken-off or *autotomized* section will often undergo violent contractions, distracting the predator, while the rest of the lizard slips away unnoticed.

The final order of living reptiles is almost extinct and is represented by a few individuals of one species, the **tuatara,** which live only on twenty tiny inaccessible islands off the New Zealand coast. It has many characteristics such as the absence of a male copulating organ which set it apart from other reptiles. Tuataras transfer sperm by bringing the genital openings into contact, as birds do, and probably as ancestral reptiles did. Tuataras can regenerate limbs, indicating that this ability is an ancient attribute of reptiles.

Unlike bony fishes, reptiles appear to be "past their prime" and now play only a modest role in the balance of nature. However, they are not without economic importance, often to their detriment. The carapace of the hawksbill turtle is used for tortoise-shell ornaments, the skin of crocodiles, snakes, and lizards is used in the leather industry, and thousands of marine and river turtles are captured for food when they come ashore to lay their eggs. In the absence of active conservation measures it seems likely that the long evolutionary history of some of these animals will soon be concluded.

Birds—Class Aves

Birds exceed all other vertebrates, except bony fish, in number of species and are found almost everywhere throughout the world from the slopes of the Himalayas to Antarctic icecaps to the roof-

Figure 24-76 Various birds. A. Common tern (*Sterna hirundo*) illustrating the use of the wings and feathers of a bird in flight. The wings propel the bird upward and forward. The feathers may be spread on both the wings and the tail to permit turning in flight. B. A great horned owl (*Bubo virginianus*). C. The kiwi, a flightless bird of New Zealand, leaving its burrow at dusk to forage. Unlike other birds, its nostrils open at the tip of the bill and enable it to detect insect larvae and earthworms with its keen sense of smell. D. Two gulls (*Larus glaucescens*) quarrelling over territory.

tops of our cities. They live in deserts where there is no water and over the oceans where there are no landing sites. Some species of terns live at sea their whole lives except when nesting. They are unable to land in water as their feathers are not waterproofed. The 8600 living species of birds have exploited virtually every available niche in the world, fishing, grazing, fly catching, hunting. They range from 8 foot ostriches which weigh 300 pounds to tiny hummingbirds which weigh less than a quarter ounce. The heaviest birds of all, *Aepyornis*, the elephant birds of Madagascar which became extinct within the last 1000 years, are thought to have weighed as much as half a ton. The tallest birds were the 13 foot *moas* of New Zealand which became extinct with the coming of the Maoris.

Usually we think of birds as flying animals, but their ability to fly is not their most important distinguishing characteristic. The unique feature of birds is that they all have feathers on their bodies. Feathers apparently evolved from the scales of the ancient reptiles which gave rise to the birds and their development during embryonic life resembles the development of scales. Birds still show their reptilian heritage in the scales which cover their legs and in their protected amniotic eggs. Also, although birds are toothless, when they are born they have an egg tooth on the upper jaw with which they break out of their shells. Snakes and lizards have a similar egg tooth and like birds they lose it soon after birth. These many reptilian features have led most zoologists to regard birds as "glorified" reptiles.

The oldest bird known, *Archaeopteryx*, appeared about 140 million years ago. Unlike modern birds *Archaeopteryx* had three clawed wing-fingers, a lizardlike jaw full of teeth, and a large jointed tail (Fig. 22-14). Most modern birds no longer have three clawed wing fingers, but their wings do end in three fused fingers, and their legs are covered with scales. Jointed tails are also missing from modern birds and the tooth-filled jaw of *Archaeopteryx* has been replaced by a toothless beak.

As birds evolved the ability to fly, they also evolved other adaptations for flight, such as good vision, quick reactions, sustained activity, large brains, superb balance and excellent reflexes.

They also evolved an improved circulatory system with complete separation of oxygenated and nonoxygenated blood. In addition, they share with mammals the capacity for maintaining their body temperature at a high and more or less constant level, a necessary adaptation for maintaining the high metabolic rate associated with sustained flight. Their feathers play an important role in insulating them from heat loss. A sparrow, for example, is covered with about 3500 feathers in winter and can easily keep its body temperature at 106°F even when the temperature drops below freezing.

As we saw in Chapter 11, the skeleton of birds is also adapted for flight. Their bones are hollow and air-filled. Also, most birds have evolved an enormous breastbone to which are attached the powerful chest muscles that move the wings (the white meat of a chicken). *Archeopteryx* did not have hollow bones and had a small breastbone indicating that it was a weak flier and probably did considerable gliding.

The young of most birds are blind, naked and feeble when they first hatch and cannot maintain their warm-bloodedness. They emerge from the egg in a relatively undeveloped condition and require constant parental care and feeding until they grow feathers, gain strength and can leave the nest. Birds of this type are called **altricial.** Because of this style of life most birds, unlike most reptiles and amphibians, have evolved extensive patterns of behavior related to parental care and nest building. Interestingly the closest bird ancestors among living reptiles, the crocodiles, also show well-developed parental care.

Certain birds are able to fend for themselves soon after they hatch. They are fully clothed and bright-eyed. As soon as their down dries they are able to run after their parents and peck at things. Chickens, ducks and geese fall into this category and are called **precocial** birds. As you might expect the eggs of precocial birds are usually larger than the eggs of altricial birds and their incubation time is longer. This enables a precocial chick to complete much more of its development in the egg before hatching than an altricial chick.

Some birds are **"determinate"** *layers.* A sandpiper, for instance, will lay four eggs and no more. If one egg is removed, it does not lay another one. It always stops after it has laid four eggs. Other birds are **"indeterminate"** *layers* and will keep on laying eggs if eggs are taken from the nest. Apparently they feel the proper number of eggs in their nest before they stop laying. Ducks and chickens fall into this category. One duck laid 363 eggs in 365 days! Most birds do not start to incubate their eggs until they have either laid the right number of eggs or have the sensation of sitting on the right number.

The number of eggs that different species of birds normally lay in a clutch varies. An albatross will lay one egg each year and a hummingbird only two. In contrast a wren will rear two broods of six in each summer and a pheasant may lay a dozen or more in a single clutch. Each species of bird has evolved a strategy to get as many young as possible into the next breeding population. Some birds, like the albatross, produce a few young but put much effort into each chick to insure survival. Others like pheasants produce many young but put little effort into each chick, and only a few survive to reproduce.

In some birds, the number of pairs which will breed decreases when the population density gets too high and food gets too scarce. They will also adjust the number of eggs in a clutch to conform to the carrying capacity of the environment. Apparently, successful species of birds practice family planning.

There are roughly 27 orders of birds. None has over a few hundred species in it except the order *Passeriformes.* It contains more than 5000 species, most of which have been described and classified. It includes the familiar "songbirds," such as sparrows, blackbirds, crows, larks, swallows, and thrushes.

Probably because of their ability to fly, birds have been worshipped and admired by many ancient and modern societies. The Hittites and the Babylonians built temples to eagles and the constant companion of Athena, the Greek goddess of wisdom, was the wise old owl. The national emblem of both Americans and Albanians is the eagle. But the most important of all birds from a human point of view is the red jungle fowl, *Gallus gallus.* This single species of pheasant has been domesticated for more than 5000 years and has given rise to the dozens of varieties of domestic chickens which cover the earth by the billions, and are without question the most numerous bird species on earth. In the

Figure 24-77 Altricial vs. precocial birds. Altricial birds like the cormorants (left) are born blind and feeble and remain in their nests, dependent on the care of their parents for days and weeks until they acquire feathers and become active. Precocial birds like the gull (right) are covered with down and have their eyes open when they hatch. They are active at the outset and promptly leave the nest after hatching.

United States alone nearly two billion chickens are raised yearly as broilers and more than 60 billion eggs are harvested. Many other birds have been associated with man in diverse ways.

Today the future of many groups of birds is precarious. Nearly 80 species have become extinct in the last three centuries, and 100 more are seriously endangered. In a few cases extinction appears to be a natural event and not due to man. For instance, the California condor which is now reduced to a handful of individuals may simply have reached the end of its evolutionary road. A relic of a past time when large mammals were abundant, this giant scavenger may no longer be able to earn its livelihood. However, most of the birds which have become extinct or are on the edge of extinction have been destroyed by man. Some, such as the passenger pigeon, were hunted to extinction. Others, such as the whooping crane and the ivory-billed woodpecker, have had most of their habitats destroyed, and are reduced to

fewer than a hundred individuals. Some, such as the brown pelican, the osprey, and the bald eagle, are dying out because insecticides like DDT become concentrated in the tissues of the fish they eat. Living with man remains the birds' greatest problem. It is to be hoped that the prospect of a world without songbirds and eagles will encourage their preservation.

Except for mammals and insects, birds are certainly the most widely used experimental animals. A good deal of what we know about embryology has come from the study of chick embryos. Birds are also used extensively in studies of behavior, including their ability to migrate long distances. The Arctic tern is the greatest traveler of all and makes a 10,000 mile journey each summer from northern Canada and Alaska to the seas near Antarctica. Recently evidence has accumulated that many birds chart their courses by the sun and by the stars and some may actually have a built-in magnetic compass!

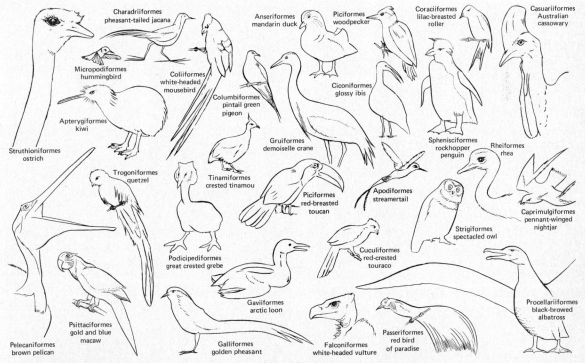

Figure 24-78 One representative of each the 27 orders of birds.

Porpoises, pangolins, and people— Class Mammalia

Fish-shaped porpoises, leather-winged bats, massive elephants, scaly pangolins, spiny porcupines; these and many others are the dominant vertebrates in the world, the mammals. Members of this large and diverse class, which includes about 5000 living species, occupy an immense number of niches on land, water, air, and underground in both frigid and tropical climates. They range in size from tiny shrews weighing a tenth of an ounce to the 100 foot long great blue whale weighing 130 tons. They include grazers, foragers, carnivores, nectar eaters, bloodsuckers, and the most efficient predator that evolution has yet produced—man. Yet for all this diversity in appearance and way of life, the mammals share a remarkably similar body plan and a number of unique features.

All mammals secrete milk and nurse their young by means of **mammary glands** (*"mamma"* means breast in Latin). Another unique feature of all mammals is an insulating layer of body hair which aids in temperature regulation. These hairs are surrounded by glands that produce an oily secretion that makes them waterproof and protects the mammal from both cold and rain. Another typical characteristic is their warm-bloodedness. Like birds they can maintain their body temperature at a high and more or less constant level. The methods used to accomplish this were discussed in Chapter 12, and include their hair, sweat glands, an internal thermostat and several other adaptations. This marvelous ability has enabled mammals to lead active lives in hostile climates. Arctic foxes move about at temperatures below −80°F; man can work in temperatures above 120°F as long as he has adequate water and is able to sweat.

Mammals also have a number of other features which enable them to maintain the high metabolism necessary for warm-bloodedness. Like birds, they completely separate oxygenated and nonoxygenated blood by a four-chambered heart; blood in half of the heart goes from the body to the lungs, in the other half from the lungs to the body. They also have a muscular wall or diaphragm separating the chest cavity from the abdominal cavity, which is an important improvement in respiration. In all but three species of mammals the embryo develops within the female's body. Mammals have a complex nervous system featuring a large brain, in which the cerebral cortex in particular is much larger than in reptiles. This improved brain is the most important single factor that gives mammals their superiority over other animals. They have very effective sense organs. Mammalian behavior is more easily modified by experience than is that of other animals. Their other body systems are also highly developed.

A distinctive feature which sets mammals apart from all other animals is their long period of **childhood.** When mammals are first born they are not ready to face the world alone, the way snakes are, for example. They depend upon parental care for a long period. Even those birds like the albatross, that care for their young for an extended period, cannot compete with the lioness in the care she lavishes on her cubs. Childhood for many mammals involves long periods of association with parents during which they learn by observing, and by being persuaded and coaxed. It reaches its extreme, of course, in humans, where childhood lasts at least a decade.

These mammalian features gradually evolved in the therapsid reptiles from which mammals arose, and were probably well developed 160 million years ago at the beginnings of the age of dinosaurs. Yet, as we noted earlier, the mammals were crowded out by the dinosaurs for about 100 million years. When the dinosaurs disappeared some 60 odd million years ago, the mammals took their place, underwent adaptive radiation throughout the world, and became the dominant vertebrates.

Platypuses and their kin—Egg-laying mammals. Living mammals are placed into one of three major groups, depending on the way the animal completes its embryonic development. One small group of mammals, called **monotremes,** includes only three species, the duckbill platypus and two kinds of echidnas or spiny anteaters (Fig. 22-16). All three are found only in Australia and New Zealand. These bizarre creatures differ from all other mammals because the female lays eggs, but when the young hatch she feeds them with her milk. She has no breasts or nipples; instead the milk seeps from small openings in her abdomen and is lapped up by the young. Other mammalian features are the presence of hair and warm-bloodedness. However, they also have a number of reptilian features; for instance, their skeletons have reptilian shoulder girdles, and like reptiles, they have only one external opening for eliminating all body wastes as well as for reproduction. Most zoologists believe that monotremes separated early in evolution from the main mammalian line and do not belong to a group which was ancestral to the other mammals.

Kangaroos and opossums—Pouched mammals. More plentiful than the monotremes are the **marsupials** or pouched mammals. The marsupial embryo stays in the uterus for only a short time. When the young are born they are little more than embryos. They crawl from the vagina into a pouch called a *marsupium* on the mother's abdomen. The mother may lick a pathway through her fur to smooth their way. Once they are inside the pouch the babies seek out a nipple which they grasp in their mouth and hang on to for weeks as they complete their development. The pouch serves the same function as the incubator in which we rear premature babies.

Marsupials have several other features which set them apart from the placental mammals. Their brains have a number of reptilian characteristics and their skeletons have two bones called marsupial bones attached to the pelvis. Apparently marsupials split off early from the main stem of mammalian evolution.

Most marsupials live in Australia where they have adapted to almost all the terrestrial environments that are occupied by different kinds of placental mammals in other parts of the world (Fig. 24-80). In addition to familiar kangaroos and koala bears, there are marsupial "wolves," "mice," "cats," and "anteaters." There is a marsupial "mole" which in almost every detail is a duplicate of the golden mole of South Africa, a true placental mammal. Phalangers look like marsupial "squirrels" and wombats look like huge marsupial "groundhogs."

The fact that most modern marsupials live only in Australia is explained by the geological history of that isolated island continent. Marsupials entered Australia before it became isolated as an island some 130 million years ago, and until man arrived they were the only mammals there (except for a few bats and rats which migrated from Indonesia). They had the whole continent to themselves, free of the flesh-eating placental mammals which were developing on other continents. These circumstances enabled marsupials to undergo adaptive radiation into almost every available niche. Marsupials also developed in South America, but only a few remain there today. They disappeared as a dominant group when the land bridge between North and South America was formed and the placental mammals from North America moved southwards. The most common of these marsupials are the opossums which have been expanding from their original home in South America. Opossums can now be found in the southern and middle regions of North America.

The future of marsupials as a group is precarious because their original development depended on their isolation from more efficient placental mammals with which they are unable to compete. Since man arrived in Australia, most marsupials have started on the road to extinction. Dogs, cats, and man have found them easy prey, and the introduction of animals like the rabbit and sheep has deprived them of their habitat and food supply. Except for a few such as the opossums, most marsupials appear doomed to be tomorrow's fossils.

Dominant land animals—Placental mammals. The most plentiful of all mammals are the placentals. Unlike the embryos of marsupials the embryos of placental mammals are nourished within the mother's body by a special organ, the **placenta,** and are not born until development is well advanced. Placenta means a "flat cake," which is the shape of this organ in humans. As we learned in Chapter 7 it is a development of the membranes which in the reptilian ancestors of the mammals enveloped the embryo within the eggs. In placental mammals these membranes have become connected to the wall of the uterus so that the embryo can obtain nourishment directly from the body of the mother. This adaptation freed the embryo from dependence on the limited food supply of the egg. It enabled the young animal to remain within the protective environment provided by the mother's body for a much longer period until its complex brain and other systems have had a chance to mature.

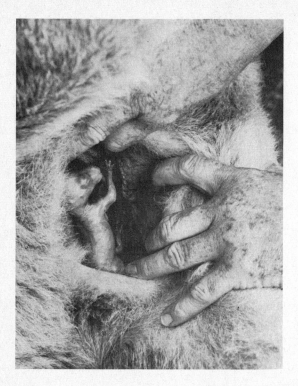

Figure 24-79 Marsupials. An infant kangaroo (upper left) attached to a nipple in its mother's pouch. The gestation period in marsupials is extremely short and an infant marsupial is born in a very immature state. It continues its further development in the protected pouch. A kangaroo with a well-grown infant called a joey in its pouch (upper right). The Virginia opossum (left) is an extremely common marsupial in the United States. It lives mainly in trees and is most active at night. When its young are first born they are barely a half inch long and are kept in the mother's pouch. Later on they are carried on the mother's back.

The sixteen orders of living placental mammals are listed below and illustrated in Figure 24-82.

Artiodactyla (even-toed): even-toed hoofed animals—reindeer, camels, giraffes, cattle, sheep, pigs, hippopotamuses

Carnivora (meat-eaters): bears, raccoons, skunks, wolves, seals, cats, weasels, otters

Cetacea (whale-like): whales, porpoises, dolphins

Chiroptera (hand wings): bats

Dermoptera (skin wings): flying lemurs

Edentata (toothless): armadillos, sloths and anteaters

Hyracoidea (hyrax-like): hyrax

Insectivora (insect eaters): shrews and moles

Lagomorpha (hare-shaped): rabbits, pikas

Perissodactyla (odd-toed): odd-toed hoofed animals—horses, rhinoceroses, tapirs

Pholidota (covered with scales): pangolins

Primates (the top): tree shrews, lemurs, monkeys, apes, men

Proboscidea (with a trunk): elephants

Rodentia (gnawers): rats, porcupines, beavers, capybara

Sirenia (siren-like): sea cows or manatees

Tubulidentata (little pipe teeth): aardvarks

By far the most plentiful of living species of placental mammals are those among the order

wolf *(Canis)*

tasmanian wolf *(Thylacinus)*

ocelot *(Felis)*

native cat *(Dasyurus)*

flying squirrel *(Glaucomys)*

flying phalanger *(Petaurus)*

ground hog *(Marmota)*

wombat *(Phascolomys)*

anteater *(Myrmecophaga)*

anteater *(Myrmecobius)*

mole *(Talpa)*

mole *(Notoryctes)*

mouse *(Mus)*

mouse *(Dasycercus)*

Figure 24-80 Marsupials in Australia have adapted to various environments occupied by placental mammals in the rest of the world. This figure illustrates a number of examples of this convergent evolution.

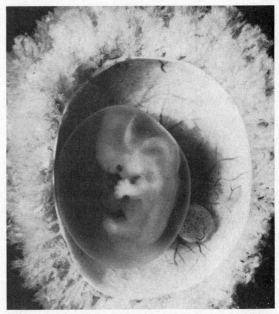
Figure 24-81 Human embryo at 39-40 days showing its relations to its membranes. (× 1.5).

Rodentia. Rats, mice, squirrels, and the many other rodents swarm almost everywhere. Some like the beaver are amphibious, although none are completely aquatic. Two orders of placental mammals have taken permanently to the water again (these include whales and sea cows) and two to the air (these include bats and flying lemurs). Even within individual orders there has been extensive adaptive radiation, and the order Carnivora includes such eminently terrestrial animals as lions along with such eminently aquatic animals as seals.

All of these orders descended from the first obscure and probably secretive little placental

Figure 24-82   16 orders of living placental mammals.

mammals that appeared some 160 million years ago. An analysis of the structures of members of various orders, along with fossil evidence, reveals some unexpected relationships among the orders. Whales and carnivores appear related, as do elephants and sea cows. The closest relatives of our own order, the Primates, are the shrews and moles. In the final chapter we consider the distinctive features of the Primates and their evolution.

The importance of other placental mammals to man, both now and in history, has been immense. Man's rise to dominance in life on earth has been closely interwoven with lives of other placental mammals. Even in today's highly mechanized society, man still depends on placental mammals for food, drink, fiber, and to some extent for transport and power. Also, because of their kinship, other mammals respond much as man does to disease, stress, and accident. For this reason they are more widely used than any other experimental animal in man's efforts to understand problems of life, health, and disease. For many years mice, rats, rabbits, cats, and dogs held the center of the experimental stage. In recent years, however, monkeys, kangaroo rats, dolphins, opossums, and many far more exotic mammals have become objects of intensive study by biologists and medical scientists.

There is much to be said in favor of this broader experimental look at diverse mammals. For above all man is a mammal and he faces the same problems of food, shelter, reproduction and social behavior that other mammals do. Mammals have solved these complex problems of existence in diverse ways and man can learn much from them.

SOME AFTERTHOUGHTS ON KINSHIPS AND SUCCESS

We have completed our survey of the animal kingdom and have a picture of both the diversity and kinships of animals. It is important at this point that we avoid falling into a common trap of thinking of animal phyla as an evolutionary series. It is easy to imagine that you are studying phylogeny and evolution when examining a flatworm, an earthworm, and a grasshopper in the laboratory. But reflect a moment. All of these animals are living today and none can be the ancestor of any other. Each is well adapted to a particular environment and way of life. What is true, as pointed out in this chapter, is that insects descended ultimately from annelids, these in turn from flatworms. The modern representatives of these groups share a heritage. However, over millions of years they have diverged from their ancestral forms, and their present diversity reflects their adaptation to their environment over hundreds of millions of years. Each living animal is thus the sum total of adaptive changes preserved by natural selection over millions of years.

A second point we should bear in mind concerns the word "success." We have used the term "successful organisms" several times in this chapter without defining it. To define it would require a lengthy discussion, which is not appropriate at this time. However, in a common-sense way the meaning is clear. Insects, bony fishes and placental mammals are obviously successful groups today, whereas brachiopods, echinoderms, and amphibians are far less successful. Yet in a sense all living organisms are profoundly successful. For they have survived the merciless filter of natural selection for more than 2 billion years. To be sure, brachiopods, horseshoe crabs, tuataras, and platypuses are no longer as diverse as other groups and occupy only a few of the many possible ecological niches, but their continued existence testifies to their success. For the object of the game of nature is not to win, but to keep on playing.

SUGGESTED READING LIST

Alvarado, C. A., and L. J. Bruce-Schwatt, "Malaria," *Scientific American,* May, 1962.

Applegate, V. C., and J. W. Moffett, "The Sea Lamprey," *Scientific American,* April, 1955.

Barnes, R. D., *Invertebrate Zoology* (second edition). W. B. Saunders, Philadelphia, Pennsylvania, 1968.

Barrington, E. J. W., *The Biology of Hemichordata and Protochordata.* W. H. Freeman, San Francisco, 1965.

Berrill, N. J., *The Origin of Vertebrates.* Oxford University Press, London, 1955.

Borror, D. J., and D. M. DeLong, *An Introduction to the Study of Insects* (third edition). Holt, Rinehart and Winston, New York, 1971.

Buchsbaum, R., *Animals without Backbones* (rev. ed.). The University of Chicago Press, Chicago, Illinois, 1948.

———, and L. J. Milne, *The Lower Animals. Living Invertebrates of the World.* Doubleday and Company, Inc., Garden City, New York, 1962.

Burnett, A. L., and T. Eisner, *Animal Adaptation.* Holt, Rinehart and Winston, New York, 1964.

Carr, A., and The Editors of *Life, The Reptiles* (Life Nature Library). Time-Life, New York, 1963.

Carrington, R., *The Mammals* (Life Nature Library). Time-Life, New York, 1963.

Chandler, A. C., and C. P. Read, *Introduction to Parasitology* (tenth edition). John Wiley and Sons, New York, 1961.

Cochran, D. M., and C. J. Goin, *The New Field Book of Reptiles and Amphibians.* G. P. Putnam's Sons, New York, 1970.

Colbert, E. H., "The Ancestors of Mammals," *Scientific American*, March, 1949 (Offprint 806).

———, *Evolution of the Vertebrates* (second edition). John Wiley and Sons, New York, 1969.

Deevey, E. S., Jr., "The End of the Moas," *Scientific American*, February, 1954.

Evans, H. E., *Life on a Little Known Planet*, Dutton, New York, 1968.

Fingerman, M., *Animal Diversity.* Holt, Rinehart and Winston, New York, 1969.

Gardiner, M. S., *The Biology of Invertebrates*, McGraw-Hill Book Company, New York, 1972.

Garstang, W., *Larval Forms and Other Zoological Verses.* Basil Blackwell, Oxford, 1962.

Glaessner, M. F., "Pre-Cambrian Animals," *Scientific American*, March, 1961 (Offprint 837).

Hanson, E. D., *Animal Diversity* (second edition). Prentice-Hall, Englewood Cliffs, New Jersey, 1964.

Herald, E. S., *Living Fishes of the World.* Doubleday and Company, Inc., Garden City, New York, 1961.

Holldobler, B., "Communication between Ants and Their Guests," *Scientific American*, March, 1971 (Offprint 1218).

Hyman, L., *The Invertebrates.* Vols. 1 through 6. McGraw-Hill Book Company, New York, 1940 through 1967.

Imms, A. D., *A General Textbook of Entomology* (ninth edition) (rev. by O. W. Richards and R. G. Davies). Methuen and Company, Ltd., London, 1964.

Larousse Encyclopedia of Animal Life. McGraw-Hill Book Company, New York, 1971.

MacGinitie, G. E., and N. MacGinitie, *Natural History of Marine Animals* (second edition). McGraw-Hill Book Company, New York, 1968.

Marshall, N. B., *The Life of Fishes.* The World Publishing Company, Cleveland, Ohio, 1966.

Miner, R. W., *Field Book of Seashore Life.* G. P. Putnam's Sons, New York, 1950.

Newell, N. D., "The Evolution of Reefs," *Scientific American*, June, 1972.

Noble, G. S., *Biology of Amphibia.* McGraw-Hill Book Company, New York, 1931. Reprinted by Dover Publications, 1955.

Orr, R. T., *Vertebrate Biology* (third edition). W. B. Saunders Co., Philadelphia, 1971.

Pennak, R. W., *Fresh-water Invertebrates of the United States.* Ronald Press. New York, 1953.

Peterson, R. T., *The Birds* (Life Nature Library). Time-Life, New York, 1963.

Porter, K. P., *Herpetology.* W. B. Saunders Company, New York, 1972.

Romer, A. S., *The Vertebrate Story* (fourth edition). The University of Chicago Press, Chicago, Illinois, 1959.

———, *The Vertebrate Body* (fourth edition). W. B. Saunders Company, Philadelphia, 1970.

Russell-Hunter, W. D., *A Biology of Lower Invertebrates.* The Macmillan Company, London, 1968.

———, *A Biology of Higher Invertebrates.* The Macmillan Company, London, 1969.

Schneiderman, H. A., and L. I. Gilbert, "Control of Growth and Development in Insects," *Science*, Vol. 143, 325-333, 1964.

Todd, J. H., "The Chemical Languages of Fishes," *Scientific American*, May, 1971 (Offprint 1222).

Vaughan, T. A., *Mammalogy.* W. B. Saunders Company, Philadelphia, 1972.

Wells, M., *Lower Animals.* World University Library series, McGraw-Hill Book Company, New York, 1968.

Welty, J. E., *The Life of Birds.* W. B. Saunders Company, Philadelphia, 1962.

Wigglesworth, V. B., "Metamorphosis, Polymorphism, Differentiation," *Scientific American*, February, 1959.

———, *The Life of Insects.* World Publishing Company, Cleveland, Ohio, 1964.

Williams, C. M., "Third-Generation Pesticides," *Scientific American*, July, 1967 (Offprint 1078).

Wilson, E. O., "Animal Communication," *Scientific American*, September, 1972.

Young, J. Z., *The Life of Vertebrates* (second edition). Oxford University Press, New York, 1962.

Ziswiler, V., *Extinct and Vanishing Animals* (The Heidelberg Science Library), Volume 2. Springer-Verlag, New York, 1967.

CHAPTER 25

Mankind Evolving

Modern man is a newcomer on earth. If we plotted the roughly 4.5 billion-year history of our planet on a 12-month calendar, man as we know him would only appear four or five minutes before midnight on December 31. The earliest manlike ape, with characteristics more human than apelike, would come into view at 8 P.M. the same day. Fossils of men much like ourselves are known from the time between 35,000 and 40,000 years ago. And primitive men lived as long ago as 10 to 15 million years. These timespans are so great in relation to the lifespan of an individual man that they seem to go back to great antiquity. It is only by comparing them with the vastly greater antiquity of other living forms, and with the age of the earth itself, that we get a true perspective (Fig. 25-1).

Nevertheless man shares in this antiquity. As a mammal his lineage can be traced back to the first mammal-like reptiles of the late Paleozoic era, approximately a quarter of a billion years ago, and, of course, we can carry it much farther back past that landmark into the origins of life itself. His primate lineage can be followed from primitive Paleocene insectivores, small insect-eating, ground-dwelling mammals, who were ancestors of the whole order, through the *prosimians* (pre-monkeys) and then through the evolution of monkeys and apes. The course of primate evolution culminating in man is the subject of this chapter (Fig. 25-2).

geologic era	millions of years ago	geologic period	first appearance of the indicated life form
Cenozoic	0.01	Recent	man
	2.5	Pleistocene	chimpanzees, gorillas, australopithecines
	7	Pliocene	apes
	26	Miocene	
	38	Oligocene	monkeys
	54	Eocene	whales, porpoises
	65	Paleocene	primates
Mesozoic	136	Cretaceous	mammals
	190	Jurassic	flowering plants
	225	Triassic	birds
Paleozoic	280	Permian	gymnosperms
	325	Pennsylvanian	reptiles
	345	Mississippian	amphibians
	395	Devonian	insects
	430	Silurian	land plants
	500	Ordovician	fishes algae
	570	Cambrian	all major invertebrate phyla
Precambrian	2700		oldest known green plants—algae
	3100		oldest known bacteria
	3400		origin of life
	4600		formation of the earth's crust

Figure 25-1 Summary of life on Earth.

MAN AS AN ANIMAL

Throughout this book we have mentioned many parallels between man and the other animals. They offer important evidence to back up the fossil record. Together the living evidence and the fossil record provide nearly incontrovertible proof of human evolution. We say "nearly," because this proof can never be absolute; there were no witnesses to the events. And yet the records left in man's body and in the earth's rocks verify his descent from ancestors who were not men. The records are such that there can be no reasonable doubt that evolution occurred.

Some of the parallels between man and other animals are so obvious that they have been acknowledged for hundreds of years—even when their significance was unrecognized. Humans breathe, eat and digest food, excrete bodily wastes, move about, and reproduce their own kind; and so do other animals, in ways that are similar or even identical. The human body has the same general plan as that of the other vertebrates, and the similarities grow greater the higher we go on the evolutionary scale.

Other parallels were unsuspected until modern techniques of biochemistry, genetics, and embryology had been developed. As we have learned these have added many new examples of similarity, not only to animals but to the entire living world. The processes by which human bodies produce, store, and utilize energy are found in microorganisms, plants, and animals. The enzymes that make these processes possible are similar in species as diverse as bacteria, yeasts, green plants, and man. The genetic code uses the same "four letter language" throughout nature. The fundamental patterns of development observed in many animals are also seen in the development of the human embryo.

When we reach the primates, the parallels with man become even closer. There are many common mammalian traits, such as hair, mammary glands, and internal temperature control. As is typical of all placental mammals, the young develop within the mother's body. There are also more specialized primate traits, including two critically important ones.

1. **Grasping hands.** Even the earliest fossil primates had fingers capable of gripping and holding. These animals lived in trees, and being able to grip branches gave them an advantage over other animals that could hold on only by digging their claws into the bark. Long fingers and an opposable thumb, with which they could hold things against the fingers, led to the great manual skill of the higher primates, including man (Fig. 25-3).

2. **Binocular** and **stereoscopic vision.** Life in the trees demanded accurate depth perception if an animal was to walk a narrow branch or jump to a distant tree trunk. Gradually the primates evolved flatter faces, shaped so that both eyes faced forward and could focus on

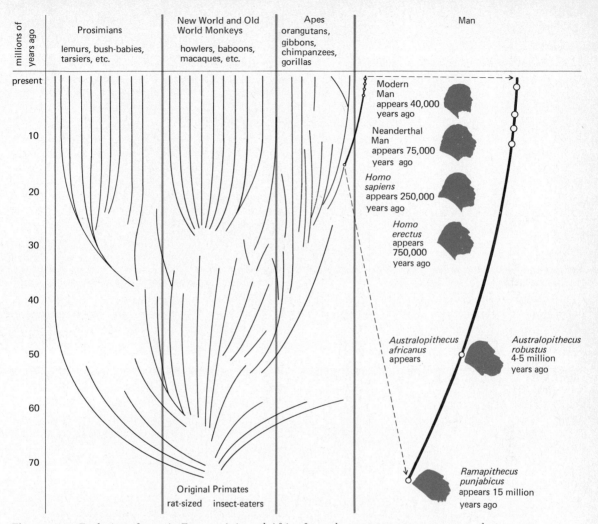

Figure 25-2 Evolution of man in Europe, Asia and Africa from about 70,000,000 years ago to the present.

one object together (binocular or two-eyed vision). Later they evolved eyes located on the head in such a way that the image from the right eye and the image from the left fused in the brain to produce stereoscopic vision or depth perception, the ability to see in three dimensions.

These two features led to the development of coordination among mind, eyes, and hands. Primates' brains became larger as time passed, permitting them to perceive and process visual information. Eventually one group of apes began

to walk upright. This freed their hands so that they could use tools. Primates of all sorts used their hands to capture and hold food, but only man has become a skilled user of tools. This led to a further expansion of the brain and to increased mental capacity, until man became more intelligent than any of his ancestors. Although it has not been proven that a larger brain necessarily means greater intelligence, there is evidence that animals with greater capacity for learning and remembering specific tasks in laboratory experiments are those with larger brains.

PROSIMIANS—THE FIRST PRIMATES

The first primates began evolving in the Paleocene epoch, the first subdivision of the Cenozoic era, about 60 to 70 million years ago. In the Eocene the early primates were plentiful in North America, Europe, and Asia. Closely related forms are now found only in Asia and Africa, particularly on a few tropical islands.

One of the more primitive forms living today is the *tarsier*, found only in the Philippine Islands and Indonesia. About the size of a rat, it has many obvious primate characteristics. Its hands and feet have long, slender bones, which give it a good grip as it leaps about the tree branches, and its eyes face forward (Fig. 25-4). The tarsier may be a distant and much modified descendant of early prosimian primates that led from lemurs to monkeys, apes, and man.

Lemurs are another group of prosimians, small tree-dwelling animals, now confined to Madagascar and a few nearby islands (Fig. 25-5). Members of one family occasionally walk on their hind feet during brief sojourns on the ground; however, they do not have a true upright posture. The lemurs on Madagascar have expanded into several genera. Except for men, no other primates live on the island.

"IMPROVED" PRIMATES— THE MONKEYS

Two groups of closely related prosimians gave rise to the two different types of monkeys on the earth today. Both developed 35 to 55 million years ago in the Eocene or early Oligocene epoch in the Cenozoic era, but evolved on different continents. One group includes the *New World monkeys* of Central and South America. Most of them have long prehensile tails which can be used to grasp objects. These animals often use their tails like fifth hands. The tail is strong enough to hold the animal when it is hanging from a branch and skilled enough to pick up a small piece of food (Fig. 25-6). Monkeys with such prehensile tails include the familiar spider monkey. Squirrel monkeys are among the New World forms that do not have prehensile tails. Their tails are long, however, and are used for balance.

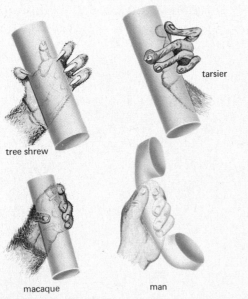

tarsier

tree shrew

macaque   man

Figure 25-3 Primate grips from claws to human grip with fully opposable thumb.

Figure 25-4 A tarsier, showing grasping hands and feet and binocular vision.

Old World monkeys are now found in Asia, Africa, and Europe, principally in the tropics. The vast majority of them, including the mangabeys and langurs, are arboreal. The ones most familiar to us, however, are terrestrial, the macaques (including Rhesus monkeys) and baboons (Fig. 25-7) Macaques have been widely used in medical research. An Oligocene fossil found in Egypt may be a transitional form between the earlier tarsierlike primates and the ancestors of modern Old world monkeys.

Despite the fact that New and Old World monkeys have lived apart for millions of years, there has been much *parallel evolution* and they resemble each other in certain important details. In general, parallelism in the monkeys has resulted in the same kinds of improvements on the basic prosimian model.

The monkeys' brains are larger than those of their prosimian ancestors, and by and large the monkeys are probably more intelligent. As they became better adapted to life in the trees, both the Old and New World species developed better eyesight and hands with good grips. Partly because of their better vision, monkeys are adept with their hands.

Life in the trees puts a premium on eyesight, whereas smell is less important than for an animal living on the ground. Modern monkeys have accordingly lost some acuteness in the sense of smell that earlier forms probably had. Their noses and the olfactory lobes of their brains are smaller and their faces are therefore usually flatter than those of the prosimians (Fig. 25-8).

ANTHROPOID APES

At nearly the same time the Old World monkeys were evolving during the Oligocene, apes were beginning to appear. It is difficult to differentiate between some of the early monkey fossils and those of the early apes, for they shared features that were not found among the prosimians. Apparently they either came from the same ancestral stock or the apes were an offshoot of the primitive Old World stock.

Today's apes live in central Africa (*chimpanzee* and *gorilla*) and in southeast Asia (*orangutan* and *gibbon*). The apes represent four very different animals, but despite their differences

Figure 25-5 A lemur on Madagascar.

Figure 25-6 Monkey using his prehensile tail.

Figure 25-7 Baboons moving in defense formation, adult males and older male juveniles guarding troop nucleus including mothers and infants. Early man probably moved in similar formations.

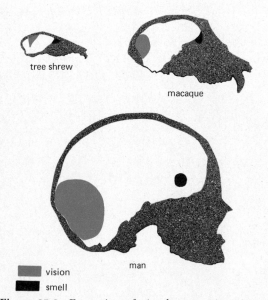

vision

smell

man

Figure 25-8 Expansion of visual centers in primate evolution: decline of old "smell" brain.

they all illustrate a similar potential for the evolution inherent in their stock. In some of the ways they have developed one can see why they are called the *anthropoid,* or "manlike apes." With the exception of the gibbon, whose intelligence is about on the level of the monkey, apes have much larger brains than monkeys. They are also not as confined to trees. Orangutans are awkward on the ground, but the other three forms can walk there comfortably and gorillas spend most of their time on the ground. Although gibbons are famous for their aerial grace in the trees (Fig. 25-9), they also walk upright with greater frequency than the others, maintaining their balance with their arms. Gorillas and chimpanzees can assume a partially upright stance, but generally walk with the help of their hands, supporting themselves on their knuckles.

Apes are not as skillful with their hands as humans, but they have progressed beyond the monkeys in that direction. Chimpanzees, in fact, are very primitive toolmakers and users (Fig. 25-10). They use sticks for digging, stones for cracking nuts, and leaves for cleaning themselves. When they cannot find suitable twigs or vines with which to probe for insects, they will make them by removing leaves from vines. The young learn to make these tools by observing their elders and then practicing to gain proficiency.

The animals carry their tools with them and store them in their nests.

We must emphasize however, that man did not evolve from apes like those living today. He shared some common ancestor with them. The apes evolved in directions suited for their way of life and man evolved in different directions. Some clues to the nature of their common ancestor may be found in the Miocene fossil known as *Proconsul* (or *Dryopithecus*) which was unearthed in Africa near Lake Victoria. The teeth and skull are apelike, but in some other features *Proconsul* resembles monkeys. It had not developed the elongated arm structure found in later apes, which allows them to *brachiate* or swing through the branches. Unlike monkeys, who walk on top of branches, the apes swing beneath them. *Proconsul* may be a relative of the animal that eventually gave rise to the two highest primate forms. From this stock there arose both the apes of today and the ancestors of man.

Proconsul is linked to the earliest distinguishable apelike men through the transitional form *Ramapithecus*. Only a few fossil fragments have been found in Africa and India; these specimens are about 10 to 15 million years old, "give or take a few million."

THE FIRST TOOLMAKER

There is a gap of 8 to 10 million years between *Ramapithecus* and the first primate that might be considered a member of the hominid family, the family of man. But no one seriously questions that it represents an important and very likely direct ancestor of modern man. This significant group of fossils is called the *Australopithecines*, literally "southern monkeys," although they were not exclusively southern and were not monkeys! Most specialists agree that there were two groups, the more primitive *Australopithecus robustus* and the more manlike *Australopithecus africanus* (or *Homo africanus*).

The first specimen was discovered near the edge of the Kalahari Desert in South Africa in 1924. Since then dozens of fossils have been found (Fig. 25-11), especially in East Africa, and together they tell a fascinating story. The earliest of these ape-men lived roughly 4 million years ago. The most recent are about 1 million years old, and by that time they had changed so much that they are not easily distinguished from forms considered to represent the first true men. In fact some consider them genuine, if very primitive, humans.

Figure 25-9   White-handed gibbon with young clinging to her.

Figure 25-10   Prehuman tool-user: chimpanzee probing for termites.

Figure 25-11   (Left) Looking down 300 feet into Olduvai Gorge in East Africa, where pre-men lived twenty thousand centuries ago. (Right) Louis and Mary Leakey examining early Olduvai living floor.

The earliest *Homo africanus* did not have unusually large brains. They probably averaged between 450 and 550 cubic centimeters, about the same size as those of modern gorillas (Fig. 25-12). However, their body size was one-third or less than that of the gorilla. The animals' heads were set more upright on their backbones than those of the apes. They walked in an upright position, according to the evidence from pelvic, thigh, and foot bones found in the fossil record. Thus their hands were free to be used while walking. Among many of their fossils are found the first stone tools that show signs of having been shaped by their users. The edges have been chipped slightly to provide a cutting edge. Were they used to dig for roots, kill animals, or crack open animal bones to get at the marrow or brain inside? Many cracked bones and punctured animal skulls are among *Homo africanus* relics.

As we have remarked two separate forms of this ape-man are known. One was a small creature, perhaps only 4 feet tall and under 100 pounds in weight. The other one was roughly twice as large, with the largest molar teeth found among the hominids. This larger form may have lived at a later time than the smaller one. Tools are associated with both of them, but paleontologists are still arguing about whether both

forms were making the tools. Most agree that at least one and perhaps both forms belong in the ancestry of modern man. If we accept the hypothesis that the use of tools encourages the evolutionary development of larger brains and increased mental capacity, then we can understand the process by which *Homo africanus* went on to the next stage in human evolution.

All of the earliest fossils accepted by paleontologists as truly *Homo africanus* have so far been discovered in South and East Africa. Thus the dark continent is regarded by most specialists as the cradle of humanity, the place where man arose. Later *Homo africanus* had spread to the rest of the tropics and warm temperate areas of the Old World.

EARLY WANDERERS

From the *Australopithecines* the next stage, *Homo erectus*, developed. Fossil remains of this stage were found in Asia and North and South Africa, suggesting that he was a wanderer. This stage used to be called *Pithecanthropus* (meaning "apeman") or Java man, and *Sinanthropus* (meaning "China man") or Peking man, since the latter was unearthed near Peking, China. Most of them

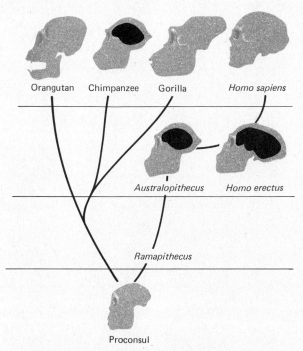

Figure 25-12 The ascent of man and the apes from a common ancestor, Proconsul (above). The average cranial capacities are compared for chimpanzee (398.3 cc), *Australopithecus* (507.9 cc) and *Homo erectus* (973.7 cc).

are now placed in the same species, *Homo erectus*.

Homo erectus lived in the middle of the Pleistocene epoch of the Cenozoic era, approximately 500,000 to 750,000 years ago. His brain capacity had increased so that it approached the size of that of modern man. The Java men's skulls indicate that their average cranial capacity was about 900 cubic centimeters (about a quart), while the China men had brains between 900 and 1200 cubic centimeters. These figures are within the lower parts of the range of variation of human brain size today.

Homo erectus had other human characteristics. He walked erectly and his thigh bones were virtually the same as ours. His skull, on the other hand, still had apelike features, such as a slanting forehead, a giant bony ridge over the eyes, and no chin (Figs. 25-12 and 25-13).

These men made primitive stone tools for chopping and cutting, distinctly advanced over those of the *Australopithecines*. Many believe they knew fire; carbon has been found in the caves where they presumably lived. *Homo erectus* probably lived in small bands, and these bands may even have cooperated in hunting large animals.

THE MAN WHO BECAME EXTINCT

The last 2 to 2½ million years of the earth's history, the epoch which we call the Pleistocene, produced many climactic changes. Cold spread five times from the North and where cold spread, snow accumulated to glacial proportions. Thus Arctic weather reached Illinois, the Black Sea, and northern China. *Homo erectus* may have appeared in the warm period between the first and second of these.

The most extensive glacial episode was the last, beginning roughly 75,000 years ago and ending about 10,000 years ago. In Europe there was a solid sheet of ice covering northern England, Scandinavia, and northern Germany. Many European mountains spawned glaciers, and most of the land was like the treeless Arctic tundra of today. Before this time there were already men approaching the status of modern man, *Homo sapiens* ("knowing man"). They are regarded as early members of our species, and a few of their fossil remains have been found in Germany, England, and France.

During the last glacial episode, however, these nearly modern men disappeared from Europe. After that the predominant fossil forms are those of Neanderthal man, *Homo sapiens neanderthalensis.* Neanderthal is perhaps the best known of fossil men, although his intelligence and temperament have been maligned in

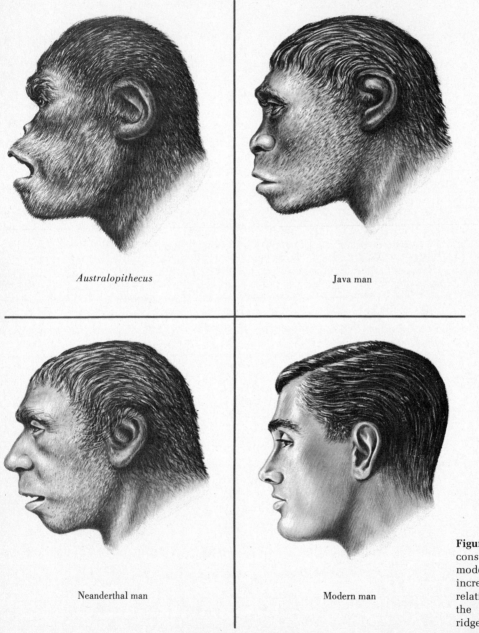

Australopithecus

Java man

Neanderthal man

Modern man

Figure 25-13 Artist's reconstruction of early and modern man. Note the increasing cranial size relative to the face and the decrease in brow ridges.

many popularizations. The first fossil was located at Neanderthal, Germany in 1856. Since then other remains have been found in many parts of Europe and the Near East.

Neanderthal, who first lived about 75,000 years ago, was probably not the brute he has been made out to be by some writers. Neanderthal men did have large, overhanging brows and receding chins, and large bones for their short size. But their brains were fully as large as ours today. They were good hunters and knew how to dress skins, which they probably wore as clothing. They buried their dead, sometimes with ornaments, food, and weapons as if for a journey to an afterlife. They even showed signs of esthetic appreciation, scratching designs on bones and polishing small, decorative objects.

Neanderthal coexisted with Cro-Magnon man for thousands of years in western Europe. Taller than Neanderthalers by as much as a foot (some were almost 6 feet in height), they were recognizably ancestral to peoples living today in Sweden, Spain, and Algeria. Other fossil remains in Europe from the same time have anatomical features resembling those of Negroes and Eskimos. Man has never been uniform in his characteristics, and he was not then.

The Cro-Magnon people were better toolmakers than their predecessors. They made spears and knives (Fig. 25-14); they trapped small animals and caught larger ones in pits. The esthetic tendency just beginning in Neanderthal bloomed among the Cro-Magnons, who made small statuettes of stone, bone, ivory, or fired clay and ground bone. They are most famous, however, for their magnificent cave paintings (Fig. 25-15). The paintings are not found in caves where men lived, which were usually under open, overhanging walls, but in true caves underground. These caves were probably shrines with mystic or religious significance. Their walls are decorated with pictures of the animals Cro-Magnon man hunted, such as deer, bison, horses, cave bears, and woolly mammoths. There are also pictures that may represent magicians or sorcerers, perhaps figures they believed helped them in the hunt.

While the last glacier was still covering much of Europe, 35,000 to 40,000 years ago, the Neanderthal line disappeared. The fossils disappear surprisingly quickly. For some reason we do not as yet understand the Neanderthals appeared to become extinct, at least in western Europe. Was Cro-Magnon man responsible for the extinction

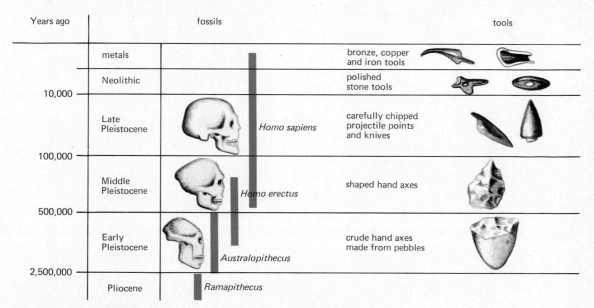

Figure 25-14 The rise of toolmaking.

of the Neanderthals? Were the Neanderthalers incorporated into the Cro-Magnon group by interbreeding (hybridization)? Is a combination of intergroup hybridization and warfare the answer? We cannot say.

In no more than 2 million years the apelike *Homo africanus* was gradually replaced by more intelligent, more skillful peoples who laid the groundwork for the world we live in today. By the end of the Pleistocene epoch, about 10,000 years ago, they ranged far and wide. They lived in Africa, Asia, Australia, and America.

We have described some of the landmarks of human evolution, but we should not leave the impression that we know the exact course of evolution from the apes to man. Many of the intermediate stages are hazy or unclear. But there is almost general agreement on one conclusion: Man did not evolve as the result of a straight-line sequence represented by the fossil men. There does seem to be a series of fossils in which we can trace some of the transitional stages in the evolutionary pattern leading to modern man. This series would include *Australopithecus* or *Homo africanus, Homo erectus, Homo sapiens sapiens*. But it would appear that some of these species lived together in the same area at the same time. Some groups disappeared from the scene and reappeared again, in the same location or elsewhere. As yet we cannot tell just which forms gave rise to the others. We have enough of the landmarks to know that human evolution followed a broad general path, but we cannot plot the exact points on the path. We are hopeful that new fossil finds and new, more exact ways of dating them will enable us to emerge with a more complete picture sometime in the future.

THE RACES OF MODERN MAN

When we apply the same taxonomic criteria to man that we apply to other species, we recognize only one human species. It is a single species because members of the different races can and do interbreed. However, there are more than two dozen racial varieties, or races, of man. *A race is a breeding population characterized by certain gene frequencies that are different from other breeding populations of the same species.*

Most classifications recognize three large groups, the Mongoloid, the Caucasoid, and the Negroid. In Goldsby's book about racial differences and their significance, *Race and Races*, he expands the list, as other authors have, to include the Australoid (the aboriginal populations of Australia); the Capoid, including the African Bushmen; the American Indian; the populations of the Indian subcontinent, including two subgroups, Pakistan and northern India, and southern India and Ceylon; and Melanesians, Micronesians, and Polynesians.

Within any race there is enormous variation, and the lines between races are blurred.

Figure 25-15   Paleolithic cave painting from Altamira, Spain. (Left) Large hind. (Right) Stationary bison.

Figure 25-16 The principal races of man: A. *Mongoloid* (A Korean woman and child); B. *Caucasoid* (An American biologist); C. *Negroid* (A tribal elder of Tanganyika); D. *Peoples of the Indian subcontinent* (East Pakistan refugees in West Bengal); E. *Capoid* (A Kalahari Bushman of South Africa); F. *Australoid* (Pintupi novices in the Warburton Range of Australia); G. *Melanesians, Polynesians, Negritos* (A Negrito man of the Phillipines).

The movement of men, accented by wars and improved travel, is breaking down the distinctions among races. There are no "pure" races of men.

We hardly need describe the races in detail. The Mongoloids ("Yellow Man") have skin colors ranging from "off-white" through "yellow" to brown. Their hair tends to be black and heavy, and grows long. Heavy beards and generous body hair are rarely seen. Faces tend to be flattened. Some anthropologists recognize five subgroups: North Chinese, Classic Mongoloid (Siberia, Mongolia, Japan), Southwest Asian (Viet Nam, Indonesia), Tibetan, and a subgroup which we have listed as a race, the American Indian. The latter classification stresses the origin of American Indians from Asian stock. Similarly, the Caucasoids have four subgroups—Northwest European, Northeast European, Alpine, and Mediterranean; and the Negroids at least three—North American Black, South American Black, and Sub-Saharan African, including Bantus, Hottentots, and Forest Pygmies.

THE ORIGIN OF THE RACES

Although we have an intense desire to understand how races came to exist, we can only theorize. The fossil record is not helpful. The student already has the background to understand the problem, however, within the framework of population genetics and evolution. The question of race formation and maintenance requires an understanding of what is necessary to establish and maintain differential gene frequencies in different populations. The features associated with race represent an adaptation to the environment made at some stage during evolution. Color, hair structure, blood chemistry can all be viewed as adaptations, proving the capacity of man, like other organisms, to respond to his environment. Genetic isolation and the pressure of a selective environment force a shift toward a higher frequency of adaptive genes.

IS MAN STILL EVOLVING?

Men today share the genetic heritage of the African ape-men. We probably have traits inherited from most of the fossil men who preceded us, from *Homo erectus*, even from Neanderthal man. Working with this raw material, nature produced modern man. But if our ancestors were so mutable that we could evolve from them, what is likely to happen to our kind? Will we in turn evolve, over thousands or millions of years, into another kind of man?

Asked these questions biologists would be sure to differ in their answers. Some would hold there is evidence that the physical evolution of man has changed direction—that it is no longer producing progress but deterioration. Others would contend that natural selection is still a positive force in determining the future of mankind. After all, they point out, many children with physical defects too serious to be compatible with life still die as infants or do not live to reproduce. There may be a difference of degree; the child with diabetes mellitus will now live because he can be given insulin, the child with a heart defect may have a long and productive life after heart surgery. But these human beings may offer as much to society as genetically more perfect individuals. Who is to say whether the diabetic Nobel Prize winner or Wimbledon finalist damages the genetic heritage of mankind? He may offer other genetic traits that could be considered more valuable.

Those who argue that natural selection as experienced by wild animals or primitive men or even men of a few thousand years ago is a thing of the past are right—up to a point. It has been relaxed in some respects. During the plague epidemics of the fourteenth century millions of people died. Other infectious diseases have had similar if not quite so drastic effects on the human population. Obviously individuals with greater resistance to these infections had a better chance of surviving. When there was a genetic tendency toward resistance it was selected; individuals with this tendency were more likely to survive an epidemic and reproduce.

This idea of selection, however, gives the impression that evolution always leads to improvement, to a better chance for a healthier, longer, more productive life. Selection does appear at times to be purposeful, producing genetic changes that increase the likelihood of survival and reproduction. But it is completely without plan. What may be suited at one time may be a failure at a later time, and yet the genetic heritage will provide for the failure as well as the success. There are so many examples of this that the Nobel laureate, Peter B. Medawar, was moved to write in his book on *The Future of Man*, "It is a profound truth . . . that nature does *not* know best; that genetical evolution . . . is a story of waste, makeshift, compromise, and blunder."

As an example of this, we might mention the distress human mothers go through at the time of childbirth. Birth is obviously an important event in terms of biological fitness. It must be successful millions upon millions of times if a species is to survive. Apparently no other female mammals suffer much pain when giving birth, but women do. Furthermore, the act of birth may lead to accidents and infections that may be fatal to mother, child, or both. This indeed seems like a form of waste in natural selection. The problems of childbirth are probably some of the consequences of the change from the horizontal to the vertical way of life. Walking upright has freed man's hands and perhaps led to his great mental growth; but it has left part of the body less than ideally suited to perform its functions.

To this process of makeshift, but at times surprisingly productive selection, nature has recently added a new development. Biological evolution has taken on a new dimension—cultural evolution. In addition to the genetic heritage every child receives from his parents, he also draws upon a cultural heritage. After he is born he will learn the language, customs, beliefs, and ways of doing things that have been passed down to his people from their ancestors, or even from those unrelated to him. This permits nongenetic evolution, that is, social and cultural evolution. This new sort of evolution, however, is still part of the biological evolution we have talked of throughout this book. It depends on the biological potential of man, the potential to understand, to record, and to transmit information. It depends on the development of an organ that can learn from the environment instead of responding blindly to it, that can be the basis for wise instead of makeshift action. This organ, of course, is the human brain.

We need the genetic equipment that provides the potential for culture. Is selection occurring to increase the cultural tendencies of modern man? The environment now shaping the future of man is much more a social environment than a physical one. If the pressures of urban living persist for many generations, perhaps man will evolve into a creature capable of surviving heart attacks and peptic ulcers as he managed to survive infectious diseases. If he does not, however, cultural evolution may succeed where biological chance does not. We and our children are already very dependent on modern science and technology, including medicine; we can only anticipate that our descendants will be even more so. It is probably too late to turn back. The solution to genetic problems may lie less in remaking our genetic heritage than in utilizing it even further to increase technological advances. In the past organisms have had to wait for genetic changes to occur at random. Today we can direct evolution by shaping it with our cultural experience.

Cultural evolution has provided man with the potential for acquiring ethics and values. They are not genetic traits, any more than the ability to read or to speak French, but they can be learned. And they can help us utilize our experiences better, more productively, more humanely, in directing human evolution.

We cannot predict the long-range future of man any more than a sapient visitor from another planet could have predicted the course of evolution if he had arrived on earth in Pre-Cambrian, Devonian, or even early Pleistocene times. We do know man has the biological as well as cultural potential for change through evolution, and that he now exerts greater control over this evolution than ever before because of his greater influence on the environment. He has the power to move mountains with his machines, when he used to have to be content with being able to dig roots up with a rock he held in his bare hand. He can plant entire valleys with fruit and grain, when he used to have pick berries from bushes he stumbled across. He is more nearly a master of his environment than ever before, although he must not delude himself that he is actually its master.

What man will do with his potential for shaping evolution still depends on the environment. The environment is now more of a social one than the physical one with which he used to have to contend. That he can help shape both these environments is one of man's greatest advantages, his even greater responsibility.

Just as biological evolution was not always good, or beneficial, so social evolution may not always produce results that are for the best. The management of social and cultural change may in the long run prove man's most demanding problem. It is up to him to use his abilities—to learn and to bring about change—in ways that will bring the greatest benefit to life on earth.

SUGGESTED READING LIST

Braidwood, R. J., *Prehistoric Man*, seventh edition. Scott Foresman, Chicago, 1967.

Dobzhansky, T., *Mankind Evolving*. Yale University Press, New Haven, Conn., 1962.

Garn, S. M., *Human Races*. C. C. Thomas, Springfield, Ill., 1965.

Goldsby, R. A., *Race and Races*. Macmillan, New York, 1971.

Howells, W., *Mankind in the Making*, rev. edition. Doubleday, Garden City, New York, 1967.

Lerner, I. M., *Heredity, Evolution and Society*. Freeman, San Francisco, 1968.

Morris, D., *The Naked Ape*. McGraw Hill, New York, 1968.

Pfeiffer, J. E., *The Emergence of Man*. Harper and Row, New York, 1969.

Reed, T. E., "Caucasian genes in American Negroes," *Science*, Vol. 165, p. 762 (1969).

Washburn, S. L., and P. C. Jay (eds.). *Perspectives on Human Evolution*, Vol. 1. Holt, Rinehart and Winston, New York, 1968.

GLOSSARY

Å: *see* angstrom.

a- or **an-** [Gk. *an-*, not]: Prefix, negates the succeeding part of the word.

ab- [L. away from, off]: Prefix meaning "away from" or "off."

abscisic acid [L. *abscissus*, cut off]: A plant hormone which promotes abscission and is involved in dormancy.

abscission layer: A layer of specialized thin-walled cells, loosely joined together, which forms at the base of the stalk of a leaf, fruit or flower, weakening the base and finally allowing the leaf or other structure to fall.

absolute temperature: Temperature on the absolute or Kelvin scale, in which absolute zero (0°K) is equivalent to $-273.1°C$ or $-459°F$, the temperature at which all molecular motion ceases.

absorption [L. *ab*, away, + *sorbere*, to suck in]: The taking up of a substance, as by the skin or the lining of the gut.

absorption spectrum: A graph showing the amount of light energy of specific wavelengths that is absorbed by a substance.

acetylcholine: A chemical transmitter of nerve impulses across synapses.

acid [L. *acidus*, sour]: A substance that dissociates in water, releasing hydrogen ions (H+); having a pH less than 7; a proton donor.

action potential: *See* potential.

action spectrum: A graph showing the degree of physiological response elicited by light energy of specific wavelengths.

active site: A region on the surface of an enzyme molecule where the substrate attaches during the catalytic reaction carried out by the enzyme.

active transport: The energy-requiring process by which a substance is transported across a cell membrane against a concentration gradient.

ad- [L. toward, to]: Prefix meaning "toward" or "to."

adaptation [L. *adaptare*, to fit]: Characteristics of an organism that better enable it to survive and reproduce in its particular environment; the adjustment of populations of organisms to their environment by means of natural selection.

adaptive radiation: The evolution of several

species from an ancestral species through the colonization of diverse habitats. Examples are the marsupials of Australia or Darwin's finches.

adenine: A major component of DNA and RNA; the nitrogenous base in ADP and ATP.

adenosine triphosphate (ATP): The primary source of chemical energy for physiological processes such as muscle contraction. ATP loses one phosphate through hydrolysis, becoming ADP and releasing usable energy.

adhesion [L. *adhaerere*, to stick to]: A sticking of one substance to another.

adrenal gland: A vertebrate endocrine gland located near the kidneys. The outer layer, or cortex, secretes various steroid hormones, whereas the inner part, or medulla, secretes adrenalin.

adrenalin: A hormone produced by the medulla of the adrenal gland which causes increases in blood pressure, blood sugar, and rate of heartbeat. By so doing it prepares an animal for "fight or flight."

adsorb [L. *sorbere*, to suck up]: To hold on a surface.

adventitious [L. *adventicus*, not properly belónging to]: In biology, applied to structures which arise from an unusual part of the body, such as roots growing out of stems.

aerobic [Gk. *aer*, air, + *bios*, life]: Requiring free oxygen for respiration.

afferent [L. *ad*, to + *ferre*, to bear]: Bringing inward toward a central part, such as blood vessels and nerves which conduct toward an organ.

agar [Malay]: A gelatinous material derived from red algae which is used to solidify culture media.

agglutinate: To join together. To clump or aggregate suspended particles, cells or molecules.

alcohol [Arabic]: Any of a class of organic compounds in which one or more OH groups are attached to a carbon backbone.

aldosterone: A hormone secreted by the adrenal cortex which stimulates the kidney to reabsorb sodium.

all-, allo- [Gk. *allos*, other]: Prefix meaning "other" or "different."

alga, *pl.* **algae:** Photosynthetic eucaryotes lacking vascular tissue and multicellular sex organs. Largely aquatic. Some are included in the kingdom Protista whereas others are included in the kingdom Plantae. Blue-green algae are photosynthetic procaryotes.

allantois [Gk. *allas*, sausage + *eidos*, form]: A pouch protruding from the posterior end of the embryonic gut. It serves as an embryonic urinary bladder in birds and reptiles, and is modified to carry blood vessels in the mammalian placenta.

allele [Gk. *allelon*, of one another]: One of the alternate forms a gene may take at a specific site (locus) on a chromosome.

all-or-none: A reaction in which there is no intermediate response; instead, the response is either maximal or nonexistent. The conduction of a nerve impulse is an all-or-none reaction.

allosteric: Refers to an enzyme that has two separate sites of action. One site is the *active site* which has the catalytic function and combines with the substrate. The second site is the *regulatory site* which interacts with a different substance to reduce the catalytic activity of the enzyme.

alternation of generations: A reproductive cycle characteristic of green plants in which a haploid (1*n*) phase, the gametophyte, gives rise to gametes. The gametes fuse to form a zygote which germinates to produce a diploid (2*n*) phase, the sporophyte. Spores produced by meiotic division from the sporophyte give rise to new gametophytes, completing the cycle. Since both the fusion of gametes in the haploid gametophyte generation and meiosis in the diploid sporophyte generation are essential features of sexual reproduction, both gametophyte and sporophyte are sexual generations.

alveolus, *pl.* **alveoli** [L. dim. of *alveus*, tub, cavity, hollow]: The microscopic air sacs which are the sites of gas exchange in the lungs of vertebrates.

amino acids [Gk. *Ammon*, the Egyptian god near one of whose temples ammonium salts were first prepared from dung]: The twenty-some organic acids that are joined end to end to make protein molecules. They contain nitrogen in the form of an amino group ($-NH_2$).

amnion [Gk. dim. of *amnos*, lamb]: A sac filled with fluid which encloses the embryo in birds, reptiles, and mammals.

amniotic egg: An egg of a vertebrate which is protected by a shell and requires only oxygen from its environment; such as the eggs of reptiles, birds and primitive mammals.

amoeboid [Gk. *amoibe*, change]: Amoebalike in the tendency to change form by cytoplasmic flow.

amphi- [Gk. *amphi*, on both sides]: Prefix meaning "on both sides," "of both kinds," or "both."

amylase [L. *amylum* starch]. An enzyme that hydrolyzes starch into sugars.

anaerobic: Able to live without oxygen.

analogous [Gk. *analogos*, proportionate]: In biology, refers to structures which are similar in function but not in origin or development, such as the wing of a bird and the wing of an insect.

anaphase [Gk. *ana*, up, + *phasis*, form]: The

stage in mitosis in which sister chromatids separate from each other and move along the spindle to opposite poles of the cell.

angiosperm [Gk. *angion*, vessel, +*sperma*, seed]: Literally a seed in a container. A group of vascular plants with flowers, and seeds enclosed in an ovary.

angstrom (Å) [after A. J. Ångström, Swedish physicist, 1814–1874]: A unit of length equal to one hundred-millionth of a centimeter or one tenth of a nanometer.

Animalia: The animal kingdom. Eucaryotic multicellular heterotrophs which move about by means of contractile fibrils.

anion [Gk. *ana*, upward + *ion*, going]: An ion (a charged atom) which carries a negative charge.

annelid [L. *anellus*, small ring]: A member of the phylum composed of ringed or segmented worms, such as earthworms and leeches.

annual: A plant which completes its life cycle in one growing season.

anther [Gk. *anthos*, flower]: The part of the stamen where pollen develops.

antibiotic [Gk. *anti*, against or opposite, + *biotikos*, pertaining to life]: A substance formed by one organism which inhibits the growth of other organisms. Antibiotics are usually made by microorganisms and act on other microorganisms.

anticodon: The set of three adjacent nucleotides on a tRNA molecule, which is complementary to another set of three adjacent nucleotides (the codon) on a molecule of mRNA. The matching of the codon and anticodon ensures that the amino acid is correctly positioned in a particular polypeptide.

antidiuretic hormone (ADH): A hormone secreted by the hypothalamus that makes the collecting tubules of the kidney more permeable to water. This promotes water reabsorption and inhibits urine secretion.

antigen [Gk. *anti*, against + *genesis*, formation]: A foreign substance, usually a protein or polysaccharide, that when introduced into the body stimulates the organism to form antibodies which can combine with it.

antimetabolites [Gr. *anti*, against + *metabalein*, to turn about, alter, change]: Substances bearing a close structural resemblance to ones needed for normal physiological functioning; they may replace or interfere with the utilization of the essential metabolites.

anus [L. *annus*, ring]. The posterior opening of the digestive tract through which undigested food is eliminated.

aorta [Gk. *airein*, to lift, heave]: The largest artery in the circulatory system.

apical: Refers to the apex, or tip of a shoot or root, as in apical meristem.

apical dominance: The inhibition of growth in lateral buds near the shoot apex, through the action of hormones secreted by the terminal bud.

apical meristem: The growing point at the tip of a root or stem, where cell divisions may continue indefinitely. It is composed largely of undifferentiated, embryonic cells.

appendage: An extension of the body, such as a limb.

arachnid [Gk. *Arachne*, a Lydian girl transformed into a spider for challenging Athena to a contest in weaving]. A member of the class of arthropods which includes spiders, scorpions, ticks, and mites.

arch-, archeo- [Gk. *archein*, to begin]: Prefix meaning "first," "earliest," "main."

archenteron [Gk. *arche*, beginning + *enteron*, intestine]: The central cavity of the early embryo. The archenteron becomes the digestive cavity in the adult.

arteriole: A small artery.

artery: A blood vessel which conducts oxygenated blood away from the heart. In vertebrates arteries usually have thick, muscular, elastic walls.

arthropod [Gk. *arthron*, joint + *pous*, foot]: A member of the phylum that includes animals such as insects, crustaceans and spiders, which have jointed legs and a tough exoskeleton.

artifact [L. *ars*, art + *facere*, to make]: A feature which is man-made, and does not occur naturally. For example, artifacts are sometimes introduced when tissue is prepared for light or electron microscopy.

ascus [Gk. *askos* bag]: An elongated sac at the end of a hypha which contains the spores in ascomycete fungi. These spores are called ascospores.

asexual reproduction: Reproduction, such as fission or budding, which does not require the union of sex cells or gametes. Botanists often use the term vegetative reproduction to refer to asexual reproduction in plants.

ATP: *See* adenosine triphosphate.

atom [Gk. *atomos*, indivisible]: The smallest unit of an element which retains the chemical properties of the element.

atomic number: The number of protons in the nucleus of an atom.

atrioventricular (A-V) node: A small specialized group of muscle cells in the heart of higher vertebrates, whose function is to transmit the heartbeat from the atrium to the ventricles.

atrium, *pl.* **atria** [Gk. *atrium*, room]: A chamber of the heart which receives blood from a vein and passes it to a ventricle.

auto- [fr. Gk. *autos*, self, same]: Prefix meaning "self" or "same."

autonomic nervous system: A special system of

motor nerves and ganglia in vertebrates which is not under voluntary control and which innervates smooth muscle, the heart, glands, and visceral organs. It is subdivided into the sympathetic and parasympathetic nervous systems.

autoradiography: A process which reveals the localization of radioactive compounds in cells and tissues. The radioactive specimen is either sectioned or squashed; it is placed on a glass slide and dipped in a photographic emulsion. After photographic development, silver grains are seen over areas containing the radioactive material. For instance, cells grown in H^3-thymidine (a radioactive precursor to DNA) would show silver grains clustered over the nucleus.

autosome [Gk. *soma*, body]: Any chromosome which does not carry the genes for "maleness" or "femaleness," i.e., which is not a sex chromosome.

autotroph [Gk. *autos*, self + *trophos*, feeder]: An organism which synthesizes all the organic compounds it requires from inorganic substances. Compare with heterotroph.

auxin [Gk. *auxein*, to increase]: A growth-regulating plant hormone, synthesized in meristematic tissue, which affects several processes including cell elongation.

axon [Gk. *axon*, an axle or axis]: An extension of a nerve cell that conducts impulses away from the nerve cell body; a nerve fiber.

bacillus [L. *baculum*, little rod]: A rod-shaped bacterium.

back (test) cross: A cross to determine whether an organism with a dominant phenotype is heterozygous or homozygous, by crossing the dominant with a homozygous recessive.

bacteriophage [Gk. *bakterion*, little rod, + *phagein*, to eat]: A virus that infects bacteria. Sometimes called "phage."

bacterium, *pl.* **bacteria** [Gk. dim. of *bactron*, staff]: A microscopic, unicellular procaryote lacking chlorophyll *a*.

bark: In the stems of woody plants, all of the tissue outside of the vascular cambium.

basal: At, near, or toward the base, or the point of attachment of a structure such as a limb.

basal body: A subcellular structure which resembles a centriole. It is found at the base of a cilium or flagellum and is probably involved in its movement. Basal bodies are found only in eucaryotes.

basal granule: A structure inside the cell membrane of procaryotes which gives rise to a procaryotic flagellum.

base: A substance that dissociates in water, releasing hydroxyl ions (OH^-); having a pH greater than 7; a proton acceptor.

base pairs and **base pairing:** The pairing of guanine with cytosine, and adenine with thymine (or uracil in RNA), which is the origin of specificity in the DNA-DNA double helix, in the synthesis of RNA from DNA, and in the synthesis of proteins.

basidium, *pl.* **basidia:** The spore-bearing structure of basidiomycetes (club fungi).

bi- [L. *bis*, twice, double, two]: Prefix meaning "two" or "twice."

bilaterally symmetrical: An arrangement of the body in which the left half of an organism is the mirror image of the right half. As in insects and vertebrates.

bile: A secretion of the liver which is stored in the gall bladder and used to emulsify fats in the small intestine.

binomial nomenclature: System of naming organisms by the combination of the names of genus and species.

bio- [Gk. *bios*, life]: Prefix meaning "life" or "living."

biogeochemical cycle: The movement of inorganic substances throughout the biological and geological components of an ecosystem. It encompasses biological components, that is, the producers, consumers, and decomposers of organic matter as well as geological components, such as the atmosphere, soil, and water systems.

biomass: The total weight or volume of all the organisms, both plant and animal, in a given area.

biome: A group of large natural communities characterized by distinctive climate and vegetation, such as the grassland biome or the tropical rain forest biome. All tropical rain forests collectively form the tropical rain forest biome.

biosphere: The total life-support system of the earth run by the energy of the sun, the zone, beginning at the surface of the earth and extending into the atmosphere, which is occupied by living organisms.

biotic: Pertaining to life.

bivalent [L. *bis*, twice, + *valens*, having power]: A pair of homologous chromosomes that have synapsed during the first meiotic division.

blasto- [Gk. *blastos*, bud]: Prefix meaning "embryo," "embryonic."

blastocoel [Gk. *koilos* hollow]: The hollow interior of a blastula.

blastopore [Gk. *poros*, passage]: In the gastrula, the opening connecting the cavity of the archenteron to the exterior.

blastula: An early embryonic stage in animals, which is a hollow or solid sphere of cells. The blastula precedes the appearance of the three principal tissue layers.

Bowman's capsule: A cup-shaped structure at one end of a kidney tubule where renal filtrate is filtered from the blood. Bowman's capsule encloses the glomerulus.

brain stem: The portion of the brain including the midbrain, the pons, and the medulla.

bronchus, *pl.* **bronchi** [Gk. *bronchos,* windpipe]: One of a pair of respiratory tubes branching into either lung at the lower end of the trachea; it subdivides into progressively finer passageways, called bronchioles, which terminate in the alveoli.

bryophyte [Gk. *bryon,* moss, + *phyton,* plant]: A division composed of the mosses and liverworts, that is, nonvascular, terrestrial green plants.

budding: A method of asexual reproduction in which a new organism forms from a small protuberance (bud) on the parent body.

buffered: Refers to a solution that contains a compound or series of compounds which are able to combine with H^+ ions or OH^- ions, and thus prevent pH changes from occurring in the solution.

caecum [L. *caecus* blind]: A blind sac in the digestive tract.

calorie: A unit of heat; the amount of heat needed to raise the temperature of one gram of water 1°C; a dietary calorie consists of 1,000 of these units.

cambium [L. *cambiare,* to exchange]: A layer of meristematic tissue in the stems and roots of many vascular plants which increases the diameter of the stem or root by cell division. See cork cambium and vascular cambium.

capillary [L. *capillus,* hair]: A thin-walled blood vessel just large enough to permit the passage of blood cells. Capillaries connect arteries and veins and are the site of exchange of substances between blood and the tissues.

capillary action: The movement of water upward, against the action of gravity, aided by cohesion among water molecules and adhesion between water and an alien surface, such as the inside of a thin glass tube.

carbohydrate [L. *carbo,* ember, + *hydro,* water]: A molecule composed of carbon, hydrogen, and oxygen in a ratio of 1 carbon : 2 hydrogen : 1 oxygen. This class of compounds includes sugar, starch, cellulose and glycogen.

carboxyl group: A carbon, hydrogen, oxygen group, -COOH, characteristic of organic acids.

cardiac [Gk. *kardia,* heart]. Referring to the heart.

carnivorous [L. *caro,* flesh + *vorare,* to devour]: Refers to "meat-eating" organisms, that is, animals (tigers) and plants (Venus' flytraps) that obtain nourishment by eating animals; opposed to herbivorous, or plant (herb) -eating organisms.

carotenoid: A class of yellow, orange, or red plant pigments that are fat-soluble; includes carotenes and xanthophylls.

carpel [Gk. *karpos,* fruit]: A leaflike female reproductive organ of a flower which bears one or more ovules and is often divided into ovary, style and stigma. Fused carpels form a pistil. Carpels are the sporophylls which produce the megaspores.

carrying capacity: The largest population that a given environment can support over a long period.

cartilage: A specialized connective tissue characterized by a dense, flexible intercellular matrix.

Casparian strip [after Robert Caspary, 19th century German botanist]: A thickened waxy strip that seals the radial and transverse walls of the endodermis, thus limiting the diffusion of dissolved materials into the vascular tissue of the root.

catalysis [Gk. *katalyein* to dissolve]. A process in which a chemical reaction is accelerated by an agent, the catalyst, which is not itself permanently changed by the reaction.

cation [Gk. *kata,* downward, + *ienai,* to go]: A positively charged atom.

cell [L. *cella,* small room]: The fundamental structural and functional subunit of a living system, composed of a plasma membrane, a nucleus with or without a nuclear membrane, organelles and cytoplasm. Some cells secrete a wall outside the plasma membrane. Many cells are able to duplicate themselves.

cell theory: The theory proposed in the mid-nineteenth century that predicted that all organisms are composed of similar units (called cells), each of which lives a partly independent existence.

cell division: See cytokinesis.

cell membrane: See plasma membrane.

cell plate: The structure which forms in most dividing plant cells at the equator of the mitotic spindle during early telophase, and serves to separate daughter cells. The cell plate is the precursor of the middle lamella.

cellulose: The major component of plant cell walls; an insoluble, high molecular weight polymer of glucose.

cell wall: The outermost layer of plant cells, some protists and all procaryotes. Its rigidity protects the cell and maintains the structure of the organism.

central nervous system: The portion of the nervous system that contains most of the nerve cell bodies and to some degree controls the rest of the nervous system. In the vertebrates, the brain and the spinal cord.

centriole [Gk. *kentron,* center]: A cytoplasmic organelle found in animal cells and in flagellated cells of plants and protists. It divides before mitosis and is involved in the organization of the spindle apparatus.

centromere [Gk. *kentron*, center, + *meros*, a part]: The structure on the chromosome where the spindle fiber attaches during meiosis and mitosis. Sometimes called the kinetochore.

cerebellum [L. dim. of *cerebrum*, brain]: An enlarged part of the hindbrain of vertebrates that coordinates muscular movements.

cerebral cortex: The "gray matter," or outer layer of nerve cells of the forebrain of vertebrates. It is well developed in mammals, where it integrates conscious sensations and voluntary motor activity.

cerebrum: An enlarged part of the forebrain of vertebrates. It integrates sensory information and voluntary functions, and is the source of memory, learning and conscious sensations.

character displacement or **character divergence:** The rapid evolution, in species living in the same area, of characters that minimize competition and of characters that minimize hybridization among them.

chelicera [Gr., *chele*, claw + *keras*, horn]: A pair of pincerlike appendages of the head found in spiders, scorpions, and other arachnids.

chitin [Gk. *chiton*, a coat of mail]: A tough nitrogen-containing polysaccharide that is an important component of the cuticle of arthropods and the cells walls of some fungi.

chlorinated hydrocarbons: Hydrocarbon molecules in which some of the hydrogen atoms have been replaced by chlorine atoms. DDT and Aldrin are well-known chlorinated hydrocarbons. Many chlorinated hydrocarbons like DDT persist in ecosystems for long periods of time before they are significantly degraded.

chlorophyll [Gk. *chloros*, green, + *phyllon*, leaf]: A photosynthetic pigment found in green plants and responsible for trapping energy from light, the first step in the conversion of light energy to chemical energy. There are several chlorophylls (*a*, *b*, *c*, etc.) with slightly different structures.

chloroplast: A chlorophyll-containing organelle, bounded by a double membrane, present in the cytoplasm of green photosynthetic eucaryotes; the site of photosynthesis.

cholinesterase: An enzyme that breaks down acetylcholine at synapses.

chordate: A member of the phylum Chordata. At some time during their life cycle chordates possess a dorsal hollow nerve cord, a notochord, and pharyngeal gill slits. Includes vertebrates, tunicates and amphioxus.

chorion: The membrane in birds and reptiles which forms an outer cover around the embryo, and in mammals participates in the formation of the placenta.

chroma [Gk. *chroma*, color]: Prefix meaning "color" or "pigment."

chromatid: One strand of a duplicated chromosome. Two chromatids, or daughter strands, are held together by a centromere.

chromatin: The deeply staining DNA-RNA protein complex of which chromosomes are composed.

chromatography [Gk. *chroma*, color, + *graphes*, writing]: A process of separating compounds in a mixture, which depends on differences in the affinities of the compounds for a medium.

chromosome [Gk. *chroma*, color, + *soma*, body]: Rod-shaped bodies in the nucleus of eucaryotic cells which are usually visible only during mitosis and meiosis. They are composed of chromatin and bear the genes in a linear order. In procaryotes, the chromosome is a circular molecule of DNA.

cilium, *pl.* **cilia** [L. *cilium*, eyelid]: A small hairlike locomotory organ found on the surface of some cells in large numbers. Each cilium is composed of two inner fibrils surrounded by nine pairs of outer fibrils, designated as a $(9 \times 2) + 2$ arrangement.

circadian rhythms [L. *circa*, about, + *dies*, day]: Rhythmic cycles of growth and activity which have a period of about 24 hours.

cistron: A linear sequence of nucleotides in DNA or RNA which specifies the sequence of amino acids in a polypeptide; a unit of genetic information.

class: A taxonomic category between phylum and order consisting of one or more orders.

cleavage: Cell division in the zygote.

climax community [Gk. *klima* ladder]: A community at the end of or at a stable point in a successional series, that is, a community is at equilibrium with the existing environment. Each climax community is composed of characteristic groups of species of animals and plants.

clone [Gk. *klon*, twig]: A population which has descended from a single ancestor through mitotic division only. A clone may be composed of cells, organs, or organisms.

co- [L. *com*, with, together]: Prefix meaning "with" or "together."

coccus, *pl.* **cocci** [Gk. *kokkos*, a berry]: A spherical bacterium.

codon: A set of three adjacent nucleotides on an mRNA molecule. The codon is the code word for an amino acid and is recognized by an anticodon on a tRNA molecule, bearing the amino acid specified by the codon.

coel-, -coel [Gk. *koilos*, hollow]: Prefix and suffix meaning "hollow cavity," "chamber."

coelenterate [Gk. *koilos*, hollow, + *enteron*, intestine]: A member of the phylum Coelenterata which includes hydras, jellyfish, sea anemones and corals. Coelenterates possess a single alimentary opening which serves as both mouth and anus. Their digestive system thus consists of a blind sac or coelenteron. They also possess characteristic stinging cells called nematocysts.

coelom [Gk. *koilos*, a hollow]: A body cavity between the gut and the body wall that is formed between layers of mesoderm.

coenocytic [Gk. *koinos* common]: Refers to a cell having more than one nucleus.

coenzyme: An organic molecule such as ATP or NAD which acts as an accessory in enzyme-catalyzed reactions. Coenzymes often donate or accept substances involved in the reaction. Unlike prosthetic groups, coenzymes are only loosely bound to enzymes.

cohesion [L. *cohaerere*, to stick together]: The holding together of like molecules.

colloid: A suspension of particles that do not settle out.

colon: The large intestine.

colony: A group of like organisms living together.

commensalism [L., *cum* together + *mensa* table]: A symbiotic relationship between two species in which one is benefited and the second is neither harmed nor benefited by the coexistence. For example, the tiny crabs that inhabit the tube of the polychaete worm *Chaetopterus*.

community: In ecology, all of the populations inhabiting a common environment and interacting with one another.

companion cell: A small narrow cell associated with the sieve elements in the phloem of angiosperms. The companion cell contains a nucleus and numerous organelles which are lacking in the sieve element. The sieve element is thought to depend upon the nucleus and organelles of the companion cell.

competition: In ecology, the utilization of a limited resource by two or more organisms or groups of organisms results in competition.

condensation reaction: A reaction joining two compounds with the resultant formation of water.

cone: In animals, the conical, light-sensitive nerve cell of the vertebrate retina involved in color vision and the perception of detail. In plants, the reproductive structure of the gymnosperms.

congenital: Pertaining to any condition present at birth.

conidium, *pl.* **conidia** [Gk. *konis*, dust]: A fungal spore characteristic of *Ascomycetes* which is produced asexually, that is, not within a sporangium; usually multinucleate.

conifer [Gk. *konos*, cone + L. *ferre*, to carry]: A tree which bears cones, such as a pine or spruce.

conjugation [L. *jugare*, to join, marry]: In bacteria, protozoans, certain algae and fungi, mating that occurs through the formation of a cytoplasmic bridge between two individuals, through which genetic material passes.

connective tissue: A type of animal tissue which generally surrounds and supports other tissues and organs. The cells of connective tissue secrete and are embedded in an intercellular matrix.

contractile vacuole: A vacuole, common in protozoans, that ejects material from the cell by visible contractions; involved in osmoregulation.

convergent evolution [L. *convergere*, to turn together]: Independent evolution of similar structures in groups of unrelated organisms that inhabit similar environments.

cork cambium: A lateral, cork-producing meristem in woody and some herbaceous plants. Cork cells are dead at maturity.

cornea [L. *corneus*, horny]: The transparent connective tissue which covers the iris and pupil of the vertebrate eye.

corpus luteum [L. *corpus*, body; *luteum*, yellow]: A yellowish structure in the ovary that develops from the follicle after the egg is released. It secretes progesterone, which stimulates growth of the uterine lining.

cortex [L. *cortex*, bark]: In stems and roots, the tissue lying between the central cylinder of vascular tissue and the epidermis. In animals, the outer tissue layer of an organ, such as the cerebral cortex.

cotyledon [Gk. *kotyledon*, cup-shaped, hollow]: Leaflike structure found inside the seed and after germination attached to the young shoot. It serves as a storage organ. Dicots have two cotyledons; monocots have one.

covalent bond: A chemical bond between atoms that results from sharing a pair of electrons.

cristae [L. *crista*, crest]: Folds in the inner membrane of a mitochondrion.

crop: A thin-walled and expanded portion of the digestive tract, primarily for food storage.

cross section: See sectioning.

crossing-over: During meiosis, exchange of pieces of chromatids between synapsed homologous chromosomes.

crustacean: A member of the class of arthropods which includes lobsters, crabs, pillbugs, and barnacles. The body of the adult is usually composed of cephalothorax and an abdomen. Crustaceans usually have a pair of jointed appendages in each segment and, unlike any other arthropods, they have two pair of antennae.

cryptic coloration: Refers to animals whose color and pattern blend with their environmental background so that they are inconspicuous to predators.

cuticle [L. *cuticula*, dim. of *cutis*, skin]: In plants, the waxy layer on the surface of epidermal cell walls exposed to air. In animals, such as arthropods, the outermost layer secreted by epidermal cells. Arthropod cuticle is composed largely of chitin and a horny protein called sclerotin and is covered with wax. In both plants and insects the cuticle prevents the loss of water from the external surface.

cybernetic [Gk. *kybernetes* pilot or governor]: Comparative study of the automatic control system formed by the nervous system and brain and by mechanical-electrical communication systems.

cyclic adenosine monophosphate (cyclic AMP or cAMP): Derivative of ATP that is of major importance in cellular control systems. Cyclic AMP is involved in neural transmission and is known to mediate the action of hormones on cells. It is often called a "second messenger."

cyst [Gk., *kystis* bladder]: A resistant protective covering formed about certain protozoans or other small organisms during unfavorable conditions or during reproduction; a small sac.

-cyte, cyto- [Gk. *kytos*, container, vessel]: Suffix or prefix meaning "pertaining to cell."

cytochrome: One of several iron-containing pigments involved in oxidative phosphorylation and photophosphorylation.

cytokinesis [Gk. *kinesis*, motion]: Cell division, or the division of the cytoplasm of a cell, and the formation of two distinct daughter cells.

cytokinins [Gk. *kytos*, vessel, + *kinesis*, motion]: A class of plant hormones that promote cell division.

cytoplasm [Gk. *kytos*, container, + *plasma*, form, mold]: The contents of the cell, excluding the nucleus.

cytosine: A pyrimidine base present in DNA and RNA, and thus one of the "letters" used in the genetic code.

daughter cells: The two cells formed by the mitotic division of one cell.

de- [L. *de-*, away from, down, off]: Prefix meaning "away from," "down," or "off."

deciduous [L. *decidere*, to fall off]: Shedding leaves at a certain season.

decomposers: Organisms such as fungi and bacteria that break down complex organic molecules, allowing the components of these molecules to recycle in the ecosystem.

demography [Gk., *demos* people + *graphos* writing]: The statistical study of human populations, especially with reference to size and density, geographical distribution, and vital statistics (age, sex, and the like).

dendrite [Gk. *dendron*, tree]: An extension of a neuron that conducts impulses toward the nerve cell body. Dendrites can be stimulated by transmitter substances but do not secrete them.

denitrification: A process in which denitrifying bacteria release nitrogen from the soil.

deoxyribonucleic acid (DNA): The nucleic acid that comprises the genetic material. In the cell two DNA strands are wound around each other in the form of a double helix. The genetic information contained in a DNA molecule is coded in specific sequences of four different nucleotides. DNA serves as a template for RNA synthesis.

depolarization: Reduced polarization (electrical potential or voltage difference) on two sides of a plasma membrane resulting from a change in the permeability of the membrane to ions. Stimulation of an axon causes depolarization at the point of stimulation.

-derm [Gk. *derma* skin]: Suffix meaning "skin," "covering," or "tissue layer."

desmosomes: A type of junction. Desmosomes are localized modifications of the plasma membrane that bind together the cells of multicellular animals like drops of glue.

determination: The process whereby a cell of an embryo becomes restricted to one particular pathway. This may occur before the cell shows visible differentiation. Thus a cell may be determined to become a muscle cell long before it can be recognized as a muscle cell.

deuterium: Heavy hydrogen, that is a hydrogen atom which possesses a neutron as well as a proton. A normal hydrogen nucleus has only a proton.

deuterostomes [Gk. *deuteros*, late, + *stoma*, mouth]: A large group of phyla including the echinoderms and chordates in which the embryonic blastopore becomes the anus and the mouth is a new opening.

di- [L. *di-*, two]: Prefix meaning "two."

diaphragm [Gk. *diaphrassein*, to barricade]: A sheet of muscle which separates the thoracic and abdominal cavities of reptiles, birds, and mammals, and is involved in breathing.

dicot or dicotyledon: A member of one of the two subclasses of angiosperms in which the seeds have two cotyledons.

differentiation: The process of cellular specialization which occurs during the course of development.

diffusion [L. *diffundere*, to pour out]: The movement of molecules through a medium as a consequence of their random motion. Diffusion results in a uniform distribution of dissolved or suspended particles, since the overall movement of particles is always toward a region of lower concentration.

digestion: The process of converting complex food molecules by enzymatic action into simple soluble forms that can be used by cells.

diploid: Having two of each type of chromosome.

The diploid (2n) number of chromosomes is characteristic of the somatic cells of animals and of the sporophyte generation of plants.

disaccharide: A sugar such as sucrose, lactose or maltose, which is composed of two monosaccharide units.

distal: Located away from some reference point (usually the main part of the body). Opposite to proximal.

disulfide bond: A covalent bond formed between the sulfur atoms of two molecules.

diverticulum [L. *devertere*, to turn aside]: A blind sac, such as the caecum, branching off a passageway or cavity.

division: The term used by most botanists to denote a phylum of plants. See "phylum."

DNA: *See* deoxyribonucleic acid.

dominant: An allele for which the phenotype is expressed to the exclusion of the phenotype of its allele. The opposite of recessive.

dormancy: In plants, an inactive period in plant organs such as bulbs, seeds, and buds. Dormancy continues until certain requirements (e.g., temperature, day length, moisture or time) are fulfilled. In animals, a prolonged period of inactivity. Many sorts of dormancy occur, for example hibernation.

dorsal [L. *dorsum*, back]: Referring to the upper side or to the back surface of an animal; the opposite of ventral.

dorsal root ganglia: Clusters containing the cell bodies of sensory neurons which lie just outside the spinal cord.

double fertilization: In angiosperms, the simultaneous fusion of one sperm with the egg nucleus to form a 2n zygote and a second sperm with the two polar nuclei, to form a 3n endosperm nucleus.

duodenum [L. *duodeni*, twelve each—from its length, about 12 fingers' breadth]: The part of the small intestine just below the stomach, into which ducts from the pancreas and gall bladder empty digestive enzymes and emulsifying agents.

echinoderm [Gk. *echinos*, spiny, bristly, + *derma*, skin]: A member of the phylum Echinodermata which includes radially symmetrical marine animals such as starfish, sea urchins and sand dollars.

ecological niche: The role played by a species in the physical and biological interactions of the ecosystem of which it is a part. An ecological niche is *not* a place; it is a role.

ecological succession: The process of change in the species composition of a community through several temporary stages, finally arriving at an equilibrium with the existing environmental conditions (a climax community).

ecology [Gk. *oikos*, house, + *logos*, a word]: The study of ecosystems, the relationships among organisms and of organisms to their environment.

ecosystem: All the organisms in a community plus the nonliving environment with which they interact.

ectoderm: In an animal embryo, the outermost tissue layer; also tissue derived from embryonic ectoderm in later development, such as skin and nervous tissue.

effector: The parts of the body by which an organism does something; the means by which it reacts to stimuli; such as cilia, muscles, glands.

efferent [L. *ex*, out + *ferre*, to bear]: Transporting away from an organ; applied to blood vessels and nerves.

electrolyte [Gk. *elektron*, amber, + *lytos*, soluble]: A substance which dissociates in solution into ions, so that an electric current can flow through the solution.

electron: One of the particles which make up an atom. Electrons surround the nucleus, and each carries a negative charge of the same magnitude as the positive charge on a proton. The electron, however, has much less mass than a proton.

electron transport system: System of enzymes which transfer electrons from foodstuff molecules to oxygen.

element: A substance made up of only one kind of atom. The atoms of each element all have the same number of protons in the atomic nucleus and the same number of electrons circling in the orbits. There are about 100 distinct elements which alone or in combination make up all materials of the universe.

embryo [Gk. *en*, in, + *bryein*, to swell]: An organism undergoing development, from the time it is a fertilized egg to the time it emerges from the egg, seed, or body of its mother.

embryo sac: The female gametophyte of an angiosperm. At maturity it usually consists of the egg, two synergids, two polar nuclei, and three antipodal cells.

embryonic induction: The process in which one group of embryonic cells influences the development of another group of cells with which they come into contact. For example, the induction of neural tube in overlying ectoderm by underlying mesoderm.

endemic [Gk., *en*, in, + *demos*, people]: Restricted to or native to a particular area or region.

endocrine glands [Gk. *krinein*, to separate]: Ductless glands, such as the pituitary or thyroid, which secrete hormones into the circulatory system.

endoderm [Gk. *endon*, within, + *derma*, skin]: In an animal embryo, the innermost tissue layer which forms the primitive gut.

endodermis [Gk. *endon*, within, + *derma*, skin]: A single layer of cells usually found in the root, separating the cortex from the vascular tissue. The endodermis limits diffusion of materials into the vascular tissue by means of the Casparian strip.

endometrium: The lining of the uterus that thickens in response to progesterone secretion during ovulation and sloughs off during menstruation.

endoplasmic reticulum: An intricate, three-dimensional network of double membranes which divide the cytoplasm into compartments in many cells.

endoskeleton: An internal skeleton, such as that of man.

endosperm [Gk. *endon*, within, + *sperma*, seed]: A nutritive material surrounding the embryo in the seeds of flowering plants. Usually a triploid (3n) tissue, the endosperm develops from the union of a sperm nucleus with the polar nuclei of the embryo sac. It is digested by the growing sporophyte during seed maturation or after seed germination.

enthalpy: Energy in a system that is not available to perform work. As the free energy of a system decreases, its enthalpy tends to increase.

entropy: A measure of the randomness or state of disorder of a system. A measure of the energy in a system that is unavailable to do work.

enzyme [Gk. *en*, in, + *zyme*, leaven]: A protein synthesized by cells, which acts as a catalyst, that is, accelerates a specific chemical reaction, without itself being permanently changed.

epi- [Gk. *epi*, upon] Prefix meaning "upon."

epidermis [Gk. *epi*, upon, + *dermis*, skin]: The outermost tissue of an organism. One cell layer thick in plants and in insects; several layers thick in vertebrates.

epithelial cell [Gk. *epi*, upon, + *thele*, nipple]: A type of cell found lining or covering an organ or structure. Epithelial cells usually have little intercellular material between them.

equilibrium [L. *aequus*, equal, + *libra*, balance]: A condition in which opposing forces exactly counteract each other. A state of balance.

erythrocyte [Gk. *erythros*, red, + *kytos*, vessel]: A red blood cell of a vertebrate, the carrier of oxygen and carbon dioxide through the circulatory system by means of hemoglobin.

esophagus: The portion of the digestive tract connecting the throat or pharynx with the stomach.

estrogen: One of a group of vertebrate female hormones. It is produced by the ovarian follicle and stimulates the development and maintenance of female secondary sex characteristics.

estrus [Gk. *oistros*, frenzy]: The periodic sexual receptivity of female mammals (higher primates excluded), marked by intense sexual urge.

ethylene: A potent plant hormone often present as a gas and noted for stimulating the ripening of fruit.

eucaryote [Gk. *eu*, good, + *karyon*, nut, kernel]: An organism whose cells contain a membrane-bound nucleus, and other membrane-bound organelles such as mitochondria, as opposed to a procaryote whose cells lack membrane-bound organelles.

excretion: The process through which metabolic wastes and excess water are released from the body in the form of urine and feces.

exoskeleton: An external skeleton like that of arthropods.

F_1: The first filial generation, or the first generation arising from a cross between two parents. F_2 and F_3 are the second and third generations.

Fallopian tubes: The tubes in female mammals through which the egg passes from the ovary to the uterus.

family: A taxonomic category between order and genus consisting of one or more genera.

fats: Compounds composed of carbon, oxygen and hydrogen, but with a much lower ratio of oxygen to carbon than in carbohydrates. Liquid fats are called oils.

fauna: The animal life of a given area or period.

feces: Indigestible wastes discharged from the digestive tract.

feedback: A circular system of control in a physiological or biochemical process, so that the control mechanism is itself regulated by the event it effects.

fermentation: A biochemical process which uses the first part of the glycolytic pathway in the absence of oxygen to convert a carbohydrate to lactic acid, alcohol, or a related compound.

fertilization: Fusion of an egg cell with a sperm cell, or more specifically, the fusion of their nuclei, to form a diploid zygote.

fiber: In vascular plants, an elongated, thick-walled, supporting cell, tapering at both ends, which is usually dead at maturity. In animals, a thin extension of a cell (as a nerve fiber) or a thin cell (as a muscle fiber).

fibrinogen: The blood plasma protein which upon conversion to fibrin forms blood clots.

filament: The stalk of the stamen at the end of which the anther is attached.

filter feeder: An animal that feeds by sweeping water through parts of the body and trapping food particles, for example, whales and clams.

fission [L. *fissus*, split]: The process of asexual reproduction in an organism by division into two or more nearly equal parts.

fixation: 1) Process of treating living tissue for microscopic examination. 2) Conversion of a substance into a form in which it is more usable biologically. For example, the conversion of CO_2 into carbohydrate by photosynthetic plants or the incorporation of N_2 into more complex molecules by nitrogen-fixing bacteria.

flora: The plant life of a given area or period.

flagellum, *pl.* **flagella** [L. *flagellum*, whip]: A fine threadlike structure on the surface of some eucaryotic cells with the same $(9 \times 2) + 2$ internal structure as a cilium, but longer. Bacterial flagella are completely different structures and have a much simpler organization.

flower: The male and/or female reproductive structure of angiosperms.

food chain: The sequence of energy transfers in a community from plant to animal to another animal. A food chain typically begins with producers (green plants) which are eaten by consumers (herbivores), which are in turn eaten by other consumers (carnivores). The largest carnivores are at the top. There are seldom more than five or six links in a food chain.

forebrain: One of the three subdivisions of the vertebrate brain. It consists of the cerebrum and interior structures such as the thalamus, hypothalamus, and limbic system.

fossil [L. *fossilis*, dug up]: The remains of an organism which have become preserved by infiltration with durable materials; also imprints from those remains, such as tracks.

free energy: The energy in a system, energy available to perform work.

fruit: The ripened plant ovary (or group of ovaries) which contains the seeds.

fungi: Heterotrophs which live chiefly on decaying matter. The Fungi are considered a separate kingdom of eucaryotes. Most fungi take the form of filaments (hyphae), which are linked together in a net-like structure (mycelium). Mushrooms and molds are examples of fungi.

gall bladder: A small sac attached to the liver for the storage of bile. A bile duct leads from the gall bladder to the duodenum.

gamete [Gk. *gamete*, wife, and *gametes*, husband]: A mature haploid (n) reproductive cell. The nuclei of two gametes of opposite sex fuse to form a zygote (2n), which then develops into a new organism.

gametophyte: The haploid (n) generation in plants having alternation of generations. The gametophyte produces gametes.

ganglion, *pl.* **ganglia** [Gk. tumor]: A mass of interconnecting nerve cell bodies.

gastr-, gastro- [Gk. *gaster*, belly]: Prefix meaning "stomach, ventral, like the stomach."

gastrocoel: The cavity of the archenteron or primitive digestive tract.

gastrodermis: The tissue lining the gut cavity that is responsible for digestion and absorption.

gastrovascular cavity: A primitive digestive cavity present in animals like flatworms which lack a circulatory system. The gastrovascular cavity allows nutrients to pass in and out of all parts of the organism; it has only one opening to the outside.

gastrulation: Process of germ layer formation in embryonic development.

-gen, -geny [Gk. *genos*, birth, race]: Suffix meaning "production, producing, generation."

gene [Gk. *genos*, birth, race]: 1) The basic unit of genetic information. Genes are located in a definite position (locus) on a particular chromosome. They are replicated when DNA is replicated, and are transmitted to the progeny via the chromosomes carried in the gametes. By interacting with the internal and external environments, genes control the development of a trait. 2) The sequence of nucleotides in a DNA molecule which specifies the complete sequence of amino acids in a particular polypeptide chain.

gene amplification: The replication of extra copies of genes in a cell. In the oöcytes of several animals, the genes for ribosomal RNA are replicated many times early in oögenesis and provide templates for massive ribosomal RNA synthesis.

gene pool: All of the genes and all of the forms of each gene (alleles) in a population.

generator potential: *See* potential.

genetic isolation: The blockage of genetic exchange between populations due to geographical, behavioral, morphological or physiological isolation.

genotype: The particular combination of genes carried in the cells of an individual. The sum total of all genes present in one individual, latent or expressed. Contrasts with phenotype.

genus, *pl.* **genera:** The taxonomic category between family and species composed of one or more species.

geotropism [Gk. *ge*, earth, + *tropos*, turn, direction]: A growth response which is oriented by a gravitational stimulus. For example, regardless of the orientation in which a plant is placed the stem generally grows up and the roots down.

germ cell: A sexual reproductive cell, an egg or a sperm.

germination [L. *germinare*, to sprout]: Resumption of growth by a bud, seed, spore or other structure after a period of dormancy.

gibberellins [fr. *Gibberella*, genus of fungi]: A class of plant growth hormones. Gibberellins are known to stimulate stem growth and play a role in seed germination.

gill: In animals, a thin-walled projection of the body surface, specialized for gas exchange. In fungi, the plates on the underside of the cap of a club fungus such as a mushroom.

gill slits: A series of openings in the wall of the pharynx of all embryonic and some adult chordates.

girdling: Removing a ring of bark from a woody stem down to the cambium.

gizzard: A chamber in the digestive tract, specialized for grinding food.

glial cell [Gr. *glia*, glue]: One of the small but numerous cells that fill the spaces between nerve cells and are thought to function in their support, nourishment, and in other ways.

globulin [L., *globulus* globule]: One of a class of proteins in the blood plasma of vertebrates. Some globulins function as antibodies.

glomerulus [L. *glomus* ball]: A cluster of capillaries at the proximal end of a kidney tubule which is enclosed by the cup of Bowman's capsule. Protein-free fluid from the blood is filtered under pressure through the walls of the glomerular capillaries into the kidney tubule.

glucose [Gk. *glykys* sweet]: A six-carbon sugar ($C_6H_{12}O_6$), the principal source of metabolic energy for most cells.

glycogen: The principal storage carbohydrate of most animals and fungi; a polymer of glucose. Glycogen is converted into glucose by hydrolysis.

glycolysis [Gk. *glykys*, sweet, + *lysis*, solution]: The anaerobic process which converts glucose to pyruvic acid. Glycolysis takes place in the cytoplasm and liberates only a small amount of useful energy, compared to the energy obtained through the Krebs cycle.

Golgi system [after Camillo Golgi, 19th century Italian anatomist]: An organelle found in the cytoplasm of eucaryotic cells. The Golgi system is composed of layers of membranous sacs, and is involved in the packaging and storage of cell secretory products.

gonad [Gk. *gone*, seed]: The organ which produces gametes in multicellular animals; the ovary in females and the testes in males.

gonadotropins: Hormones which stimulate the gonads. In vertebrates they are secreted by the pituitary.

granum, *pl.* **grana**]L. *granum* grain]: The stacked membranous sacs or thylakoids within the chloroplast where the light reactions of photosynthesis occur.

growth: An increase in size and total mass.

guanine: One of the nucleotides, a purine, present in DNA and RNA, and thus one of the "letters" of the genetic code.

guard cells: Paired epidermal cells surrounding the stomates (openings in the surface of the leaf). The expanding and contracting of guard cells controls the size of the opening of the stomates, and thus the passage of gases in and out of the leaf

gut: The digestive system.

gymnosperm [Gk. *gymnos*, naked, + *sperma*, seed]: A member of a class of seed plants whose seeds are borne on cone scales rather than enclosed in an ovary; for example, the conifers.

habit [L. *habitus*, to have]: In biology, the characteristic appearance or mode of growth of an organism.

habitat [L. *habitare*]: The natural environment of an organism; the place where it is normally found.

haploid [Gk. *haploos*, single]: Having a single set of chromosomes ($1n$), as in gametes, in the gametophyte generation in plants, and in most spores. Contrasts with diploid ($2n$).

hemoglobin [Gk. *haima*, blood]: A red, iron-containing pigment in the blood of vertebrates, which functions in oxygen transport.

herbivorous [L. *herba*, grass, + *vorare*, to devour]: Refers to an animal that eats plants.

Hardy-Weinberg law of genetic equilibrium: A mathematical representation of the equilibrium between gene frequencies, that is established in a popluation through random mating. The relative frequencies of the members of a pair of allelic genes, p and q, in a population are described by the expansion of the binomial equation $p^2 + 2pg + q^2$. The law demonstrates that the frequencies of alleles in a gene pool are not changed by the process of sexual recombination unless nonrandom forces such as selection alter the gene frequencies.

herbaceous: Refers to nonwoody plants, that is, herbs.

hermaphrodite [Gk. *Hermes* + *Aphrodite*, a mythical god and goddess, whence *hermaphroditos*, a person having the attributes of both sexes]: An organism which has both male and female reproductive organs.

hetero- [Gk. *heteros*, different, other]: Prefix meaning different or other.

heterocaryon or **heterokaryon** [Gk. *heteros*, other, + *karyon*, nut, kernel]: A cell in which two genetically different types of nuclei exist. Occurs naturally in many fungi for a considerable part of the life history and at the moment of fertilization in most organisms. May be produced experimentally in animal cells by cell fusion.

heterotroph [Gk. *heteros*, other, + *trophos*, feeder]: An organism which is incapable of synthesizing its own food (organic compounds) from inorganic substances, and must therefore feed on complex organic food materials that originate in other plants and animals. Contrasts with autotrophs.

heterozygous [Gk. *heteros*, other, + *zygous* yoke]: Having two different alleles of a given gene at the same locus on homologous chromosomes.

hibernation: A period of dormancy characterized by slow metabolism, inactivity, and in some cases low body temperature. Hibernation ends when the organism is exposed to favorable environmental conditions.

hindbrain: One of the three subdivisions of the vertebrate brain. It includes the medulla, pons, and cerebellum.

hippocampus [Gk., *hippokampos* sea horse]: Part of the limbic system of the forebrain associated with long term memory.

hist- [Gk. *histos*, web]: Prefix meaning "tissue."

histones: Basic proteins that are synthesized in the cytoplasm and are closely associated with the DNA of the chromosomes of eucaryotic cells.

holdfast: The basal part of an algal plant that attaches it to a solid object.

homeo-, homo-, homolo- [Gk. *homos*, similar, same]: Prefix meaning "same" or "similar."

homeostasis [Gk. *homos*, same, + *stasis*, standing]: The maintenance of an equilibrium in the internal physiological environment of an organism; also a steady-state relationship between an organism or a population and its ecosystem.

homeotherms: "Warm-blooded" animals, such as mammals and birds, that maintain a constant body temperature independent of their environment.

hominid [L. *homo*, man]: Any of a family of upright primates (Family Hominidae) which includes modern man, his immediate ancestors and related forms.

homologous [Gk. *homologia*, agreement]: *Pertaining to chromosomes:* Bearing genes for the same traits. Homologous chromosomes synapse (associate in pairs) at the beginning of meiosis. Each member of the pair is derived from a different parent. *Pertaining to evolution:* Homologous structures in different species have similar patterns of basic development and form because of a common ancestor and a similar genetic makeup. Homologous structures need not perform the same function, as for example, the forelimb of a mammal and the wing of a bird.

homozygous: Having identical alleles of a given gene at the same locus on homologous chromosomes.

hormone [Gk. *hormaein*, to excite]: A substance produced by one tissue and transported to a distant target tissue where it produces a specific effect. Most hormones are active in extremely small amounts and act as chemical messengers.

host: The organism on or in which a parasite lives and depends.

humus: Organic materials in the soil that are being decomposed (see decomposers).

hybrid: *In genetics,* the offspring of two parents that have different genes. *In evolutionary biology,* a cross between two different species such as a horse and a jackass. *In cell biology,* a heterocaryon. *In molecular biology,* a complex molecule formed when a molecule of single-stranded DNA is experimentally bound to a complementary strand of DNA or RNA.

hydro- [Gk. *hydor* water]: Prefix meaning "water;" "fluid;" or "hydrogen."

hydrocarbon: Organic compounds composed only of hydrogen and carbon atoms.

hydrogen bond: A weak bond formed between two molecules when a hydrogen atom in one molecule is shared by an atom, usually oxygen or nitrogen, of another molecule. Hydrogen bonds are of major importance in the structure of nucleic acids and proteins.

hydrolysis [Gk. *hydor*, water, + *lysis*, a loosening]: The breaking of a chemical bond (and thus the splitting of a molecule) by the addition of H^+ and OH^- ions.

hydrostatic or **hydraulic skeleton:** The use of the body fluid entrapped between body walls for support and movement. For example, the fluid that fills an earthworm enables it to hold its shape and transmit pressure from one part of the worm's body to another.

hydrostatic pressure: The pressure exerted against an object by a liquid at rest.

hydroxyl ion: The OH^- ion.

hyper- [Gr. above, over]: Prefix meaning "over" or "above."

hypertonic [Gk. *hyper*, above, + *tonos*, tension]: Refers to a solution having a higher concentration of dissolved material (solutes), relative to another solution on the opposite side of a semipermeable membrane. The solvent (usually water) will flow, by osmosis, across the membrane from the less concentrated solution into the hypertonic solution.

hypertrophy [Gk. *trophe* food]: Excessive growth, abnormal enlargement.

hyphae [Gk. *hyphe*, web]: One of the tubular fungal filaments that comprise a mycelium.

hypo- [Gk., less than]: Prefix meaning "under," "lower" or "less."

hypocotyl: The part of the axis of a plant embryo below the point of attachment of the cotyledons; forms the root.

hypothalamus [Gk. *thalamos*, inner room]: The part of the posterior portion of the vertebrate forebrain below the cerebral hemispheres, where centers that control body temperature, eating, and antidiuretic hormone are located. The hypothalamus controls the nervous mechanisms that are involved in holding the internal environment constant.

hypothesis [Gk. *hypo*, under, + *tithenai*, to put]: A supposition assumed as a basis of reasoning which can be tested by further controlled experiments. When hypotheses receive extensive experimental support, they become theories.

hypotonic [Gk. *hypo*, under]: Refers to a solution having a lower concentration of dissolved material relative to another solution on the opposite side of a semipermeable membrane. The solvent (usually water) will flow, by osmosis, across the membrane and out of the hypotonic solution into the solution with a higher concentration of solute.

ileum: The lower portion of the small intestine, where most molecules of digested food are absorbed through the intestinal walls.

indoleacetic acid (IAA): A plant growth hormone in the class called auxins. IAA is principally produced in the apical meristem.

inducer: In genetics, a substance which stimulates the transcription of a particular set of genes. In embryology, a substance or tissue which promotes the differentiation or development of a particular structure or group of cells.

inorganic compound: A chemical compound not having a carbon skeleton.

Insecta: A class of arthropods in which the adults have a body composed of a head, thorax and abdomen, three pairs of legs, one pair of antennae, and one pair of compound eyes.

in situ [L., in place]: In its original or natural position.

instinctive or **innate behavior:** An inherited pattern of behavior more complex than a reflex. For example, cocoon spinning by a caterpillar.

integument [L. *in*, on, + *tegere*, to cover]: In animals, skin, the covering of the body. In plants, the outer layer of the ovule which becomes the seed coat.

inter- [L., between]: Prefix meaning "between" or "in the midst of." For example, *intercellular*, between two or more different cells.

interneuron: A nerve cell that acts as an intermediary between other nerve cells. Interneurons process sensory input and finally send out impulses to motor neurons.

interphase: The period in a cell cycle when a cell is not dividing.

intra [L., within]: Prefix meaning within; for example, *intracellular*, within cells.

invaginated [L. *vagina*, sheath]: Protruded inward or folded.

invertebrate [L. *vertebra*, joint]: An animal without a backbone.

in vitro [L., in glass]: Not in the living organism; in culture or in the test tube (used as an adverb).

in vivo [L., in the living]: In a living organism (used as an adverb).

ion: An atom that has become charged through the loss or gain of one or more electrons. This process is called ionization.

iso- [Gk. *isos*, equal]: Prefix meaning "equal" or "uniform." Similar to "homo-."

isolating mechanism: Anything which prevents interbreeding among groups of organisms, intrinsically (behavioral or physiological barriers) or extrinsically (geographical barriers).

iris: The muscular, pigmented structure in the vertebrate eye which controls the size of the pupil.

isomer [Gk. *isos*, equal, + *meros*, part]: One of several compounds which differ in the structural arrangement of their atoms, but have identical chemical formulas. For example, glucose and fructose ($C_6H_{12}O_6$).

isotonic: Having the same osmotic concentration as the blood.

isomotic [Gk. *isos*, equal]: Refers to a condition in which two solutions on opposite sides of a semipermeable membrane contain the same concentration of dissolved substances.

isotope [Gk. *topos*, place]: An atom differing from another atom of the same element in the number of neutrons in its nucleus.

junctions: Modifications of the plasma membrane that bind the cells of multicellular organisms together and may function in cell to cell communication.

karyotpye [Gk. *karyon*, nut]: The number, size, and shape of the chromosomes in a cell.

keratin [Gk. *keratos*, horn]: A fibrous, water-insoluble protein found in the epidermis of vertebrates and in nails, feathers, hair, horn, and the like.

kidney: The vertebrate organ which controls the water and salt concentration of the body and selectively excretes waste in the form of urine.

kinetin [Gk. *kinetikos*, causing motion]: A purine which acts as a cytokinin (a class of plant hormones), but probably does not occur in nature. It influences the rate of cell dividios in experimental plant systems.

kingdom: The largest taxonomic division. The five kingdoms are: Monera, Protista, Fungi, Plantae, and Animalia.

Krebs cycle: A cyclical series of reactions in which the products of glycolysis are further broken down, releasing energy in the form of electrons which pass through the electron transport chain, allowing ATP to be synthesized from ADP.

lamella [L. *lamella*, thin metal plate]: A layer of cellular membranes such as the photosynthetic membranes of the chloroplast; a thin platelike structure.

larva [L. *larva*, ghost]: An immature stage in the development of some animals which is morphologically very different from the adult. For example, tadpoles and caterpillars are larvae.

lateral: Referring to the side.

lateral meristems: Meristems that give rise to secondary tissue and increase the diameter of a stem or root. For example, the vascular cambium and the cork cambium.

learning: The process through which experience modifies behavior.

lens: A structure in the eye, located behind the pupil, which focuses an image on the retina.

lichen: An obligatory symbiosis between an alga and a fungus, usually an ascomycete.

ligament [L. *ligare*, to bind]: A structure, composed of connective tissue, which binds two bones at a joint.

lignin [L. *lignum* wood]: An organic compound that is synthesized by xylem vessels, and which impregnates their cellulose walls, giving wood its rigidity.

limbic system: An integrated network of neurons in the forebrain which includes the hippocampus, the septal area, the amygdala, parts of the reticular formation, and the hypothalamus. Important in emotion, motivation, and the activities of visceral organs.

linkage: In genetics, the tendency of certain genes to be inherited together because they are located on the same chromosome.

lipase: An enzyme which breaks down fats.

lipid [Gk. *lipos*, fat]: A fatty substance, such as fats, oils, phosopholipids, steroids, waxes and carotenes. Lipids are insoluble in water but soluble in ethers and alcohols.

liver: A large glandular organ in vertebrates which converts blood sugars to glycogen, secretes bile, and detoxifies substances from the blood.

locus, *pl.* **loci** [L. *locus*, place]: The location of a particular gene on a chromosome.

long-day plant: A plant that is stimulated to flower by periods of darkness less than a critical length.

loop of Henle [After Friedrich Henle, German anatomist]: A sharp U-shaped bend in the kidney tubule prominent in mammalian kidneys and important for the production of a concentrated urine.

lophophore: A horseshoe-shaped extension of the body surrounding the mouth which bears tentacles used for feeding. Found in ectoprocts, brachiopods and a few other groups of organisms.

lumen [L. light, opening]: The cavity or space within a tube or sac.

lung: An inward folding of tissue through which gas exchange occurs between the environment and the blood of the animal.

lymph [L. *lympha*, water]: The fluid, derived from the blood plasma, which passes through capillary walls and fills intercellular spaces and special lymph vessels. Lymph is rich in white blood cells.

-lysis [Gk. *lysis*, a loosening]: Suffix meaning "loosening, disintegration, decomposition."

lysogenic bacteria [Gk. *lysis*, a loosening]: Bacteria carrying viral DNA incorporated into their genome. Under certain conditions the viral DNA is activated and active infection of the bacterium is resumed, resulting in the production of new viral particles (bacteriophages) and lysis of the bacterial host.

lysosome: A cell organelle bounded by a single membrane which contains and stores various hydrolytic or degradative enzymes. Lysosomes are the "recycling centers" of the cell.

macro- [Gk. *macros* large]: Prefix meaning "large."

macromolecule [Gk. *makros*, large]: A molecule such as a protein, nucleic acid or polysaccharide which has very high molecular weight; a complex of these molecules.

macronucleus: The single large nucleus in a ciliate protozoan which controls the metabolism of the cell.

macronutrient: One of the mineral nutrients required in substantial quantity for plant growth. The macronutrients include potassium, phosphorus, calcium, sulfur, magnesium, and nitrogen.

Malpighian tubule [After the 17th century Italian anatomist, Marcello Malpighi]: A diverticulum of the digestive tract in insects and some other arthropods which is involved in excretion.

mandible [L., *mandibula* jaw]: The lower jaw of a vertebrate; either jaw of an arthropod.

mantle: A heavy fold of tissue which covers the body of mollusks and which usually secretes a hard shell.

medulla [L. *medulla*, the innermost part]: The inner part of an organ, such as the *adrenal medulla*; the posterior part of the vertebrate hindbrain that connects with the spinal cord.

medusa [after Medusa, mythological monster with snakes in place of hair]: The "jellyfish" or free-swimming stage of coelenterates.

mega- [Gk. *megos*, great, large]: Prefix meaning "large."

megaspore: In plants with two kinds of spores such as angiosperms, the haploid spore which develops into a female gametophyte.

megaspore mother cell: The diploid (2n) plant cell that gives rise to four megaspores through meiosis.

meiosis [Gk. *meioun*, to make smaller]: Cell division that results in a reduction of the chromosome number from diploid (2n) haploid (n) Segregation and reassortment of genes occurs during meiosis. Gametes or spores are produced by meiosis.

Mendel's first law: The law of genetic segregation. Different forms (alleles) of the same factor (gene) are separated during meiosis. Only one is carried by a particular gamete.

Mendel's second law: The law of independent assortment. The pattern of inheritance of a given pair of genetic traits is not related to the inheritance of other traits. This law was later modified by the discovery of linkage.

menstrual cycle [L. *mensis*, month]: Cyclical changes in the uterine lining of some primates, characterized by the periodic discharge of blood and disintegrated uterine lining through the vagina.

meristem [Gk. *merizein*, to divide]: Undifferentiated embryonic tissue in plants in which cells divide mitotically. There are lateral maristems, apical meristems, and meristems in flowers, fruits, and leaves.

mesenchyme [Gk., *mesos* middle + *enchyme* an infusion]: A meshwork of loosely connected, mesodermal cells, found in the embryos of vertebrates and the adults of some invertebrates.

meso- [Gk. *mesos*, middle]: Prefix meaning "middle."

mesoderm: The embryonic tissue layer between the ectoderm and the endoderm. It gives rise to the circulatory system, excretory system, most of the reproductive system, the musculature, and the skeleton.

mesoglea: The jellylike layer between the outer epidermis and the inner gastrodermis of coelenterates. It contains some wandering amoeba-like cells and some connective tissue cells and is probably comparable to, but less developed than, the mesoderm of higher animals.

mesophyll [Gk. *phyllon*, leaf]: The middle tissue layer of a leaf; composed of parenchyma tissue.

messenger RNA (mRNA): The ribonucleic acid synthesized in the nucleus using a portion of nuclear DNA as a template. The mRNA carries genetic information into the cytoplasm, where the mRNA is used as a template for the synthesis of proteins on ribosomes.

meta- [Gk. *meta*, after, among, with]: Prefix meaning "posterior," "later" or "change in."

metabolism [Gk. *metabole*, change]: The sum of all chemical reactions taking place in a cell or an organism.

metamorphosis [Gk. *metamorphoun*, to transform]: In biology, the transformation of a larva to an adult, such as that of a caterpillar into a moth, or a tadpole into a frog.

metaphase [Gk. *meta*, middle, + *phasis*, form]: The stage in mitosis in which the chromosomes are aligned midway between the poles of the spindle.

micro- [Gk. *mikros*, small]: Prefix meaning small.

microfilaments: Thin (40–50Å in diameter) elongated structures that occur almost universally in eucaryotic cells.

micronucleus: A small nucleus in a ciliate protozoan that is involved mainly with reproduction and gives rise to the macronucleus.

micronutrient: A mineral required in very small amounts for plant growth. For example, chlorine, iron, boron, zinc, molybdenum, manganese, and copper.

microspore: In plants, with two kinds of spores, a spore that will germinate into a male gametophyte.

microtubules: Elongated rods about 200–300Å in diameter found universally in eucaryotic cells. They appear to play a role in movement and are the subunits of which cilia and spindle fibers are composed.

microvillus, *pl.* **microvilli:** A tiny outfolding of the plasma membrane of a cell. Microvilli increase the surface area and absorptive capacity of a cell.

midbrain: One of the three subdivisions of the vertebrate brain. Originally functioned as receiving center for nerve fibers coming into the eye and in other sensory functions. In higher vertebrates, particularly mammals, most of its functions have been taken over by the forebrain.

middle lamella: A layer between adjacent plant cell walls, derived from the cell plate and rich in pectins.

mineral [Medieval L., *minera*, ore, mine]: In biology, a naturally occurring, inorganic substance.

mitochondrion, *pl.* **mitochondria** [Gk. *mitos*, thread, + *chondrion*, small grain]: An organelle found in eucaryotic cells, bounded by a double membrane, containing its own DNA and protein-synthesizing machinery. Mitochondria are the site of the Krebs Cycle and electron transport and are the major source of ATP in nonphotosynthetic eucaryotic cells.

mitosis [Gk. *mitos*, thread]: A process in which the chromosomes which have replicated during interphrase divide longitudinally. The daughter chromosomes (chromatids) then separate and form two genetically identical nuclei. Mitosis is followed by cell division, resulting in two daughter cells, each containing identical nuclei.

molecular weight: The relative weight of a molecule when the weight of the carbon atom is taken as 12; the sum of the relative weights of the atoms in a molecule.

molecule [L. *moles*, mass]: Smallest particle of a compound having the composition and properties of a larger part of the substance; consists of two or more atoms.

mollusk: A member of the phylum Mollusca; includes organisms such as snails, bivalves (clams, etc.), chitons, and octopods; usually possesses a shell.

molting [L. *mutare*, to change]: The shedding and replacement of an outer covering such as exoskeleton, hair or feathers.

Monera: The taxonomic kingdom including all procaryotes, that is, the bacteria and blue-green algae.

mono- [Gk. *monos*, single]: Prefix meaning "single" or "one."

monocotyledon [Gk. *monos*, single, + *kotyledon*, a cup-shaped hollow]: A plant having one cotyledon; one of the two subclasses of angiosperms; abbreviated as monocot. The monocots include lilies, grasses, palms, and orchids.

monophyletic: A phylogenetic tree in which the organisms are all descended from one ancestral founder species.

monosaccharide [Gk. *monos*, single, + *sakcharon*, sugar]: A simple sugar, such as a five- or six-carbon sugar.

-morph, morph- [Gk. *morphe*, form]: Suffix or prefix meaning "form" or "structure."

morphogenesis [Gk. *morphe*, form, + *genesis*, birth]: The organization of various body parts into the proper form or shape.

morphology [Gk. *morphe*, form, + *logos*, discourse]: The form and structure of organisms or parts of organisms; the study of form.

motivation: A change in the internal state of an animal that leads to a particular behavior pattern.

motor neuron: A neuron conducting impulses from the central nervous system to an effector, such as a muscle (*see* reflex arc.).

mucus: A slimy protective substance secreted by mucous cells. It consists of proteins and polysaccharides.

muscle: A contractile tissue of animals.

mutant [L. *mutare*, to change]: A mutated gene or an organism carrying a mutated gene.

mutation: A stable, inheritable change in a gene.

mutualism: A living together or symbiosis in which both parties benefit.

myc-, myco- [Gk. *mykes*, fungus]: Prefix meaning "relating to fungi."

mycelium [Gk. *mykes*, fungus]: A mass of hyphae that form the body of a fungus.

mycorrhiza [Gk. *mykes*, fungus, + *rhiza*, root]:

A symbiotic relationship between the hyphae of certain fungi and the roots of a vascular plant.

myelin sheath: The plasma membranes of the sheath cells (Schwann cells) which are wound many times around the axons of many nerve cells in vertebrates.

myo- [Gk. *mys*, mouse, muscle]: Prefix meaning "muscle."

NADP: *See* nicotinamide adenine dinucleotide phosphate.

nanometer (nm): A unit of length equal to one billionth of a meter, 10^{-9} meter (0.000025 inch); formerly termed millimicron.

natural selection: The fundamental mechanism of evolution. The individuals that are best suited to given environmental conditions will be most likely to survive and reproduce. As a result of natural selection there is an increase in the frequency of some genes or gene combinations and a decrease in the frequency of others.

nematocyst [Gk. *nema*, thread, + *kystis*, bag]: A cell in the coelenterates, specialized for stinging, which ejects a threadlike structure. Used for defense, capture of prey, and anchorage.

nematode: A roundworm; a member of the class Nematoda in the phylum Aschelminthes. They have a three-layered body and a pseudocoel instead of a true coelom. Includes trichina worms and pinworms.

nemertine [Gk. *Nemertes*, one of the sea nymphs or Nereids who attended Poseidon]: A member of the phylum Nemertinea. The nemertines are soft-bodied, unsegmented worms, commonly called ribbon worms. They appear to be the simplest organisms with both a mouth and an anus, and a circulatory system.

neolithic [Gk. *neos*, new, + *lithos*, stone]: Relating to the latest period of the Stone Age characterized by polished stone implements.

nephron [Gk. *nephros*, kidney]: The functional unit of the kidney consisting of Bowman's capsule, convoluted tubule and loop of Henle. There are about one million nephrons in a human kidney.

nerve [L. *nervus*, sinew, nerve]: A bundle of axons bound together. There are no synapses in nerves, just axons.

nerve net: A net-like intermingling of axons and dendrites, without central control, as in the nervous system of coelenterates.

neural tube: The embryonic tube that forms the brain and spinal cord of a vertebrate.

neuron [Gk. *neuron*, sinew, nerve]: A nerve cell, the basic unit of the nervous system; includes the cell body, axons, and dendrites.

neutron [L. *neuter*, neither]: A subatomic particle that is electrically neutral and of slightly greater mass than a proton. Neutrons are present in all atomic nuclei except hydrogen, in which the nucleus is composed of a single proton.

nicotinamide adenine dinucleotide phosphate (NADP): A coenzyme and electron acceptor.

nondisjunction: A failure of homologous chromosomes to separate during meiosis, resulting in an abnormal number of chromosomes in the gametes.

notochord [Gk. *notos*, back, + *chorde*, string]: A longitudinal, rod-shaped structure that serves as a skeletal axis in the embryos of all chordates. In the vertebrates, it is replaced by the vertebral column as the organism develops.

nuclear envelope: The double membrane which separates the nucleus from the cytoplasm.

nucleic acid: An organic acid composed of nucleotides joined together as a long polymer; DNA or RNA; genetic information is stored and replicated in nucleic acids.

nucleolus, *pl.* **nucleoli** [L. *nucleolus*, a small kernel]: A spherical body composed of RNA and protein present in the nucleus of eucaryotic cells; the site of ribosomal RNA synthesis.

nucleotide: A molecule composed of a phosphate group, a five-carbon sugar (dexoyribose or ribose) and a nitrogenous base (a purine or a pyrimidine); a subunit of a nucleic acid.

nucleus, *pl.* **nuclei** [L. *nucleus*, a kernel]: In biology, the chromosome-containing body of the eucaryotic cell. It is bounded by a double membrane, the nuclear envelope; in physics, the central part of an atom.

-oid [Gk. like, resembling]: Suffix meaning "similar to" or "like."

omnivorous [L. *omnis*, all + *vorare*, to devour]: Refers to an organism that eats both plants and animals.

ontogeny [Gk. *on*, being, + *genesis*, origin]: The complete developmental history of an individual organism from zygote to maturity.

oo- [Gk. *oion*, egg]: Prefix meaning "egg."

oocyte [Gk. *oion*, egg, + *kytos*, vessel]: One of the cells which gives rise by meiosis to an egg (ovum).

oogenesis: The process of egg formation from sex cells in plants and animals.

oogonium, *pl.* **oogonia:** A cell in the ovary from which the egg arises. An oogonium multiplies by mitosis to form oocytes.

operator gene: The gene in an operon which is repressed by a repressor protein, and thereby prevents transcription of that operon.

operon: A cluster of adjacent genes whose activities are coordinated by the operator gene and thus under the control of a single repressor.

order: The taxonomic category between family and class consisting of one or more families.

organ [Gk. *organon*, tool]: A group of tissues which function together to perform one or more collective tasks; for example, heart, lung, leaf or root.

organelle: Any specialized structure within the cytoplasm of a cell, such as, ribosomes, mitochondria, chloroplasts, or the Golgi complex.

organic compound: A chemical compound based on a skeleton of carbon atoms.

organism: An individual living creature.

orientation: The spatial disposition of an organism or organ relative to a stimulus or reference point; for example, the orientation of a plant stem relative to a light source.

osmoregulation: Maintenance of a relatively constant osmotic concentration in body fluids, despite osmotic changes in the external medium.

osmosis [Gk. *osmos*, impulse or thrust]: The movement of water across a membrane that is permeable to water but not to certain substances (solutes), such as sugar, dissolved in the water. Given two solutions separated by such a semipermeable membrane, water will flow from the solution containing less dissolved material (solutes) into the solution containing a higher concentration of solutes.

ovary [L. *ovum*, egg]: The organ that produces eggs in female animals. In plants, the basal portion of the carpel, which will become the fruit after fertilization.

oviduct [L. *ovum*, egg, + *duccre*, to lead]: A duct which transports eggs from the ovary to the uterus, or to the outside.

ovum, *pl.* **ova:** The female gamete, the egg cell.

ovulation: The release of an egg from an ovary.

ovule: Sporangia attached to the wall of the ovary in seed plants which produce megaspores. Each ovule is composed of an integument, sporangium and female gametophyte. It develops into a seed after fertilization.

oxidation: The process in which an atom loses an electron. Reduction, the gain of an electron, takes place simultaneously as the electron lost by one atom is accepted by another. In biological systems, oxidation generally involves the removal of hydrogen (removal of an electron plus a proton). Oxidation–reduction systems are an important means of transferring energy within cells.

pacemaker: A knot of muscular tissue located where the superior vena cava enters the right atrium, which initiates the heartbeat. Commonly called the sino-atrial node or the S-A node.

paleolithic [Gk. *palai*, long ago]: Relating to the earliest part of the Stone Age characterized by rough or chipped stone implements.

pancreas [Gk. *pan*, all, + *kreas*, meat, flesh]: In

vertebrates, a gland that secretes various digestive enzymes and the hormones, glucagon and insulin.

para- [Gk., at the side of, beside]: Prefix meaning "beside."

parasite [Gk. *para*, beside, + *sitos*, food]: An organism which lives in or on an organism of a different species for at least part of its life cycle and benefits at the expense of its host.

parasympathetic nervous system: One of the two divisions of the autonomic nervous system. Parasympathetic fibers originate in the brain and the pelvic region of the spinal cord, and innervate the internal organs. It is the parasympathetic system, for example, which stimulates digestion. Acetylcholine is usually released by parasympathetic nerves at synapses with the tissues they innervate.

parenchyma [Gk. *para*, beside, +*en*, in, +*chein*, to pour]: *In plants*, a supporting tissue composed of an unspecialized type of cell. Parenchyma cells are living, thin-walled, loosely packed and spherical in shape, with large vacuoles; often a storage or photosynthetic tissue. *In animals*, a loose connective tissue.

parthenogenesis [Gk. *parthenos*, virgin]: A process in which offspring are produced without fertilization.

pectin: A complex plant polysaccharide present in the cell wall and middle lamella.

pectoral [L. *pectoralis*, pertaining to the breast]: Pertaining to the upper thoracic region or breast of vertebrates.

penis: The male organ through which sperm is deposited in the female reproductive tract during copulation. In vertebrates it also functions as a passage for the excretion of urine.

peptide bond: The chemical bond which joins adjacent amino acids in a polypeptide. It results from a condensation reaction between the amino group of one amino acid and the acidic group of the other.

perennial [L. *per*, through, + *annus*, year]: A plant which lives and reproduces through two or more growing seasons.

peri- [Gk., around]: Prefix meaning "surrounding" or "around."

pericycle [Gk. *peri*, around, + *kyklos*, circle]: A cell layer in the root, between the endodermis (on the outside) and phloem (on the inside), which gives rise to lateral roots.

peristalsis: Waves of muscle contraction in the digestive tract of vertebrates, that mix food matter and push it down the tract.

permeable [L. *permeare*, to pass through]: Ordinarily applied to membranes through which substances may pass through by diffusion.

petal [Gk. *petalon*, a leaf]: A flower part; modified leaves, often brightly covered, that advertise the flower to prospective pollinators.

pH: Symbol denoting the concentration of hydrogen ions (H^+) in a solution. pH = $-\log (H^+)$. Therefore, the greater the hydrogen ion concentration, the more acid the solution, and the lower the pH value. pH values vary from 0 (acid) to 14 (basic alkaline). pH 7 is neutral.

phage: See bacteriophage.

phagocytosis [Gk. *phagein*, to eat, + *kytos*, container]: The active engulfing of particles by cells; usually involves isolation of the particle in a vacuole and its digestion.

pharynx [Gk. *pharynx*, throat]: The throat, that is, the connection between the oral cavity and the esophagus.

phenotype [Gk. *phainein*, to show]: The physical characteristics of an organism; the product of interactions between the genotype (the genetic information carried by the organism) and the environment.

pheromone [Gk. *pherein*, to carry, + hormone]: A substance secreted by an organism, which affects the behavior or physiology of other organisms of the same species.

phloem [Gk. *phloos*, bark]: A vascular tissue in plants which conducts the products of photosynthesis throughout the plant. Unlike the xylem, the phloem consists of living cells.

-phore [Gk. *pherein*, to carry]: A suffix meaning "carrier."

phospholipid: A molecule composed of two fatty acids linked to a glycerol molecule by a molecule of phosphoric acid.

phosphorylation: The addition of a phosphate group to a molecule, such as the phosphorylation of ADP to ATP.

photo-, -photic [Gk. *photos*, light]: Prefix or suffix meaning "light."

photoperiodism: A response by an organism to the duration and timing of the light and dark conditions.

photosynthesis [Gk. *photos*, light, +*syn*, together, + *tithenai*, to place]: The synthesis of carbohydrate from carbon dioxide and water, using the radiant energy of light.

phototropism [Gk. *photos*, light, + *tropos*, turn, direction]: A growth response which is oriented by a light stimulus, for example, the turning of a leaf toward a light source.

phyllo-, phyll- [Gk. *phyllon*, leaf]: Prefix or suffix meaning "leaf."

phylum [Gk. *phylon*, race, tribe]: The taxonomic category between class and kingdom, comprised of one or more classes.

phylogenetic tree: A diagram of the evolutionary

relationships among a group of organisms. A family tree of a group of organisms.

phylogeny [Gk. *phyle*, tribe]: Evolutionary history of an organism; kinship among organisms.

phyto-, -phyte [Gk. *phyton*, plant]: Prefix or suffix meaning "plant."

phytochrome: A growth-regulating plant pigment, sensitive to red and far-red light stimuli. It is the photoreceptor for many developmental events, such as flowering and seed germination.

phytoplankton: Photosynthetic plankton.

pigment: A light-absorbing molecule.

pinocytosis [Gk. *pinein*, to drink]: The process by which cells engulf small droplets of liquid.

pistil: A floral organ consisting of the ovary, style, and stigma; formed from one or more fused carpels.

pith: The core of the vascular cylinder in the stem and some roots; usually composed of parenchyma cells.

pituitary [L. *pituita*, phlegm]: A vertebrate endocrine gland located near the brain. It secretes hormones that regulate other endocrine glands and is often called the master gland.

pK: Provides a measure of the strength of an acid; the lower the number, the stronger the acid. At a pH numerically equivalent to the pK, half of the groups are dissociated and half are not.

placenta [Gk. *plax*, flat surface]: A flat platelike structure formed partly of the inner lining of the uterus and partly from embryonic tissue that connects the mother and the embryo in most mammals. In the placenta substances are exchanged between the blood of the mother and that of the embryo. Placental mammals include all living mammals except monotremes and marsupials.

planarians: A group of flatworms in the class Turbellaria. They have part of their body covered with cilia and have an extensible pharynx in the middle of their underside with which they suck in food.

plankton [Gk. *planktos*, wandering]: Free-floating aquatic organisms, both plant, animal, and protist, largely microscopic.

Plantae: The plant kingdom, comprising all multicellular eucaryotic photosynthetic organisms and some closely related unicellular forms.

planula [L. dim. of *planus*, flat]: A ciliated larva found in many coelenterates.

plasm-, plasmo-, plast [Gk. *plasma*, something formed or molded]: Prefix or suffix meaning "formed" or "molded"; for example, chloroplast, "something green-formed" or protoplasm, "something first-molded."

plasma [Gk. *plasma*, form]: The fluid, noncellular component of the blood. It is composed of dissolved salts and proteins.

plasma membrane: The cell membrane or the outermost semipermeable membrane which encloses the protoplasm of all cells. Also called the plasmalemma. It may be arranged as a "sandwich" in which the two outer layers are composed of protein molecules and the middle is a double layer of lipid molecules.

plasmodesmata [Gk. *plassein*, to mold, + *desmos*, bond]: Threads of cytoplasm bounded by plasma membrane which pass through plant cell walls, thus connecting adjacent cells.

plasmodium: A multinucleate mass of protoplasm like that found in myxomycete slime molds.

plasmolysis [Gk. *plasma*, form, + *lysis*, a loosening]: Shrinkage away from the wall of the protoplasm of a plant cell, due to loss of water through osmosis.

plastid [Gk. *plastos*, formed]: A relatively large plant cell organelle bounded by a double membrane that functions in photosynthesis and/or storage of the products of photosynthesis.

platelet: Very small particle found in mammalian blood that is involved in clotting.

Platyhelminthes: The phylum composed of the flatworms.

poikilotherm [Gk. *poikilos*, various, + *therme*, heat]: A "cold-blooded" animal; that is, an animal whose body temperature fluctuates with environmental temperature changes. An animal incapable of precise self-regulation of body temperature.

polar body: A cell formed during the meiotic divisions that produce egg cells. It has a nucleus but contains almost no cytoplasm.

polar molecule: A molecule on which the electrical charge is not evenly distributed. Polar molecules like amino acids tend to dissolve in water, whereas nonpolar molecules like hydrocarbons tend to dissolve in lipid solvents like alcohols or ethers.

polar nuclei: Two nuclei, each derived from one end (pole) of the embryo sac, which move to the center of the embryo sac. They fuse with a sperm nucleus to form the primary ($3n$) endosperm nucleus.

polarity: The existence of opposite or contrasted properties in opposite parts of a substance or structure. In animal embryos, the presence of a head end and a tail end illustrates polarity. In chemistry, the unequal distribution of electrical charges on a molecule, such as a water molecule, reflects polarity.

pollen [L. *pollen*, fine dust]: The male gametophytes of seed plants at the stage when they are shed from the sporophyte.

pollen tube: A tubular outgrowth of the germinated pollen grain. It provides a passage for the male gametes into the ovule.

pollination: The transfer of pollen from the male reproductive organ or anther to the receptive surface of the female reproductive structures, usually the stigma.

poly- [Gk. *poly*, many]: Prefix meaning "many."

polygenic inheritance: Inheritance in which interaction among several genes determines a given trait, such as weight or height in man, or milk production in cows.

polymer [Gk. *meros*, part]: A macromolecule composed of many small molecules called monomers, linked together end-to-end in a chain.

polymorphism [Gk. *polys*, many, + *morphe*, form]: The simultaneous occurrence of several distinct forms of a species in a population, such as the polyp and medusa stages of many coelenterates, or workers and queens among ants, bees and termites.

polyp [Gk. *polypous*, many-footed]: A stage in the life cycle of a coelenterate. In contrast to the medusa, the polyp is usually non-motile.

polypeptide: A chain of amino acids linked together by peptide bonds.

polyploidy [Gk. *poly*, many, + *ploos*, fold or times]: The occurrence of more than two sets of chromosomes in a cell; for example, 3n or 4n.

polysaccharide: A carbohydrate composed of simple sugars (monosaccharides) linked together in a chain. Starch, cellulose, and glycogen are polysaccharides.

polysome: Polyribosome; cluster or row of several ribosomes attached to a single mRNA molecule and found within the cytoplasm of a cell.

population: An ecological term, meaning a group of organisms of the same species that interbreed in a defined geographical area.

posterior: Toward the hind end.

potential: Shorthand for *electrical potential* difference, the difference in the charge of one point or object relative to another point or object. *Resting potential:* a fairly steady electrical potential difference across the membrane of a cell, particularly of a nerve cell that is not firing or a muscle cell that is relaxed. *Action potential:* an abrupt, all-or-none change in the potential difference across the membrane of a nerve or muscle cell, which moves along the cell in a wave-like manner, constituting a nerve impulse or a muscle action potential. *Generator potential:* a graded change in potential difference that occurs across the membrane of a sensory cell when it is stimulated. If the generator potential reaches a certain threshold, an action potential is triggered in the associated neuron.

potential energy: The energy that an object may derive from its position relative to its surroundings.

preadaptation: A structure or function which an organism comes to use for new purposes during the course of evolution.

primary consumers: Animals that eat plants directly. The primary consumers in an ecosystem may be eaten by secondary consumers or carnivores.

primary growth: Plant growth that originates in apical meristems and results in increased length; as opposed to secondary growth, which results in increased girth and originates in lateral meristems.

primary producers: Usually, the green plants in an ecosystem which convert energy from the sun and manufacture the food substances that supply the rest of the system.

primary structure: The linear sequence of amino acids in a protein. *See* secondary and tertiary structure.

primary succession: Ecological succession that occurs in a region that was previously devoid of vegetation.

primate: A member of the order Primates which includes monkeys, apes, and man.

primordium [L. *primus*, first, + *ordiri*, to begin]: The earliest stage in the development of an organ; for example, a leaf primordium in the shoot apex.

pro- [L. before]: Prefix meaning "before."

procaryotes [Gk. *pro*, before, + *karyon*, nut, kernel]: Organisms which lack membrane-bound nuclei, mitochondria, Golgi complex, and chloroplasts; bacteria and blue-green algae.

progesterone: A steroid vertebrate sex hormone which helps prepare the uterus for implantation of the egg; it is produced by the corpus luteum.

prophase [Gk. *pro*, before, + *phasis*, form]: The first stage in mitosis. During prophase threads of chromatin (already duplicated) condense into distinct chromosomes and move toward the equator of the spindle.

prosimians [L., *pro*, before, + *simia*, an ape]: A primitive living primate or an early ancestral primate.

prostate gland [Gk. *prostates*, one who stands before]: A male gland found at the base of the urethra, where it is joined by the vasa deferentia. The prostate gland secretes a large portion of the seminal fluid.

prosthetic group: An organic compound that is permanently attached to an enzyme and is necessary for enzyme activity. The heme group in hemoglobin is a prosthetic group.

protein [Gk. *proteios*, primary]: An organic compound composed of many amino acids linked end to end by peptide bonds.

Protista: A taxonomic kingdom including unicellular, eucaryotic organisms, and colonies of similar eucaryotic cells. Includes protozoans and many algae.

proto- [Gk, *protos*, first]: Prefix meaning "first"; for example, Protozoa, "first animals."

proton: A subatomic particle which carries a

single positive charge, equal in magnitude to the charge of an electron. Protons have a mass of 1 and are found in the atomic nucleus.

protoplasm: The living material of all cells.

protostomes [Gk. *protos*, first, + *stoma*, mouth]: A large group of phyla embracing the nemertines, nematodes, annelids, mollusks, and arthropods, in which the embryonic blastopore becomes the mouth and the anus is a new opening.

protozoan [Gk. *protos*, first, + *zoe*, life]: A member of the phylum composed of microscopic, motile, unicellular, eucaryotic organisms. They are heterotrophic and feed mainly by ingestion.

proximal [L. *proximus*, near]: Located near some reference point, often the main part of the body; opposite of distal.

pseudo- [Gk. *pseudes*, false, temporary]: Prefix meaning "false."

pseudocoel: A false coelom. A body cavity that is lined partly by mesoderm and partly by the endoderm of the gut; characteristic of nematodes and other aschelminths.

pseudopod [L. *podium*, foot]: A temporary protrusion of the cytoplasm of a cell capable of amoeboid movement; functions in feeding and in locomotion.

pulmonary artery [L. *pulmonis*, lung]: The blood vessel in vertebrates which carries blood from the heart to the lungs.

pulmonary vein: The blood vessel in the vertebrates which carries oxygenated blood from the lungs to the heart.

Punnett Square: A checkerboard diagram used to predict the distribution of genes arising from a cross where the genotype is known.

pupa [L. *pupa*, doll]: A developmental stage between the larval stage and the adult stage of insects with complete metamorphosis.

pure cultures: A population of cells known to have come from a single cell.

purine: One of the nitrogenous bases in nucleic acids, ATP, NADP and other biologically active substances, with a double ring of carbon and nitrogen atoms; such as adenine or guanine.

pyloric sphincter: A muscular valve located at the junction between the stomach and the duodenum, which controls the entry of food into the duodenum.

pyrimidine: One of the nitrogenous bases in nucleic acids, consisting of a single ring of carbon and nitrogen atoms; such as cytosine, thymine, or uracil.

quaternary structure: A complex protein structure arising from the association of several folded protein chains.

race: A subspecies.

radial symmetry: Arrangement of body parts in a circular fashion around a central point. A radially symmetrical organism can be divided into two equal parts by any plane that passes through the axis (center) of symmetry. Coelenterates and starfish are radially symmetrical.

radioactivity: The emission of radiation displayed by certain elements as a result of changes in the nuclei of their atoms. The introduction of a radioactive atom into a molecule makes it possible to follow more easily the metabolic transformations of that molecule. Such molecules are said to be radioactively labeled.

radula: A ribbonlike rasping device which is used for feeding located just inside the mouth of most mollusks.

receptacle: The part of the stem where the floral organs form.

receptor: Specialized cells, often not nerve cells, that make close contact with sensory neurons. They respond to internal or external stimuli and send a nervous signal to the animal. They may be single cells or groups of cells (sense organs).

recessive allele [L. *recedere*, to recede]: An allele which is not expressed when paired with a different (dominant) allele of the same gene. A recessive allele is usually expressed only when both homologous chromosomes carry the recessive allele. Heterozygotes involving recessives are not distinguishable phenotypically from dominant homozygotes.

recombination: The occurrence of gene combinations in the progeny that were not present in the parents. Recombination results from independent segregation and assortment of genes on separate chromosomes during meiosis.

rectum [L. *rectus* straight]: The short section at the end of the large intestine where wastes are stored prior to elimination.

reduction [L. *reducere*, to lead back]: The process through which an atom gains an electron: *see* oxidation.

reflex [L. *reflectere*, to bend back]: The basic unit of action of the nervous system that produces a simple behavioral response. The simplest reflexes, like the knee-jerk reflex, involve a receptor, which stimulates a sensory neuron; the sensory neuron then conducts an impulse to a motor neuron in the spinal cord, which conducts an impulse to a muscle. Most reflexes are more complicated and involve a sensory neuron, one or more interneurons, and one or more motor neurons. The complete nervous pathway involved in a reflex is often called a reflex arc.

regulator gene: A gene which makes a repressor molecule which combines with an operator gene and prevents transcription of the structural genes in an operon. The action of the repressor may be inhibited by an inducer molecule.

renal [L. *renes*, kidneys]: Referring to the kidney.

replicate: A verb meaning to make an exact copy of something, or a noun referring to the copy itself. For example, DNA is replicated in the nucleus. The new DNA is a replicate of the original.

repressor: The molecule encoded by the regulator gene, that binds to the operator and represses transcription of the structural genes of the operon.

respiration [L. *respirare*, to breath]: In cells, the release of chemical energy from fuel molecules through controlled oxidation. In organisms, the uptake of oxygen and release of carbon dioxide.

resting potential: *See* potential.

reticular formation: A core of fibers and neurons that runs through the medulla, the midbrain and part of the forebrain. It monitors incoming stimuli, analyzes them and sends appropriate arousal signals to other areas of the brain. It is important in both consciousness and selective attention.

retina [L. *rete*, a net]: The layers of neurons and light-sensitive receptor cells (rods and cones) lining the inner surface of the eyeball (opposite the lens). The retina receives the image formed by the lens, translates it into nerve impulses, and sends the impulses to the brain via the optic nerve.

Rh factor: A type of antigen on red blood cells.

rhizoid [Gk. *rhiza*, root]: A root-like structure found in fungi, algae, liverworts, and mosses, that absorbs water and nutrients. True roots only occur in the vascular plants.

ribonucleic acid (RNA): A nucleic acid containing the five-carbon sugar, ribose. RNA is synthesized as a complement to a DNA molecule, using the DNA as a template. RNA carries genetic information from the nucleus to the cytoplasm (messenger or mRNA), it acts as an adaptor molecule in protein synthesis (transfer or tRNA), and it is a major structural component of ribosomes (ribosomal or rRNA). RNA replaces DNA as the genetic material in many viruses.

ribosome: A small organelle composed of protein and RNA which is the site of protein synthesis.

rod [Anglo-Saxon, *rodd*]: A type of light-sensitive receptor cell in the retina of the vertebrate eye. In contrast to cones, rods are particularly sensitive to dim light and permit night vision.

root [Anglo-Saxon, *rot*]: A geotropically sensitive plant organ that possesses an apical meristem at its tip; is usually highly branched, and normally grows below the ground. Certain regions are covered with root hairs. Roots anchor the plant in the soil, and take up and conduct water and minerals.

root cap: A sheath-like mass of specialized cells covering the tip of a root. The root cap is produced by the root apical meristem which it protects as the root grows through the soil. Gravity perception is thought to take place in the root cap.

root hairs: Tubular outgrowths of epidermal cells of a root that appear just above the apical meristem. They greatly increase the surface area of the root, aiding water uptake.

salt: A class of compounds formed from the ionic reaction between an acid and a base, for example, NaCl (table salt).

sap: An aqueous solution of minerals, sugars, and other organic substances found in the xylem.

savanna: A tropical grassland, transitional between desert and tropical rain forest. Savanna is characterized by scattered trees, perennial herbs and grasses and seasonal drought.

sclerenchyma [Gk. *skleros*, hard, + L. *enchyma*, infusion]: A supportive tissue of plants characterized by cells with thick, lignified secondary walls, often dead at maturity. Fibers like those found in flax as well as shells of nuts are sclerenchyma tissue.

sclerotin [Gk., *skleros* hard]: Horny proteins which arthropods incorporate with chitin to stiffen their skeletons.

secondary growth: Growth that occurs in lateral meristems, the vascular cambium and cork cambium, and increases the diameter of a vascular plant.

secondary structure: The helical structure assumed by parts of a protein molecule as a result of hydrogen bonding among amino acids in the protein.

secondary succession: Ecological succession that occurs in a region already occupied by plants.

secondary tissues: The tissues which arise from the vascular cambium and cork cambium; for example, secondary phloem, secondary xylem, and cork.

secondary wall: A thick layer of cell wall deposited by the cell inside of the primary cell wall. Often impregnated with lignin.

sectioning: Cutting a specimen in thin slices perpendicular to the long axis (cross or transverse sectioning), or parallel to the long axis (longitudinal sectioning). A sagittal section is a longitudinal section along the midline of a bilaterally symmetrical animal.

seed: A complex organ that protects the embryo during dispersal from the parent plant. It forms from the ovule following fertilization. In the angiosperms the seed is composed of a seed coat and an embryo; many angiosperm seeds also have an endosperm ($3n$), which serves as a storage tissue for the embryo. In conifers, the seed is composed of a seed coat, embryo and storage tissue ($1n$) of the female gametophyte.

segmentation: The subdivision of an organism

or parts of an organism into more or less equivalent serially arranged units. As in earthworms, insects, and the vertebrate spinal cord.

segregation: The separation of homologous chromosomes (and thus alleles of the same gene) during meiosis.

selection: The nonrandom reproduction of certain genotypes in a population, resulting in an increased frequency of some genes, and a decreased frequency of others. *Selection pressure* is force for this genetic change resulting from natural selection.

selectively, differentially, or **semipermeable membrane:** A membrane that allows water to pass, but not certain substances (solutes) dissolved in the water.

seminal vesicle [L. *semen*, seed]: A storage organ for sperm in the male reproductive system.

semipermeable: *See* selectively permeable membrane.

sensation: The brain's interpretation of certain incoming stimuli.

sensory neuron: A neuron that conducts impulses from a receptor cell to the central nervous system.

sepal [Medieval Latin *sepalum*, a covering]: The outermost parts of a flower which look most like ordinary green leaves.

septum [L. *septum*, fence]: A wall or partition. *Septate* means divided by crosswalls into compartments, as in a fungal filament.

serotonin: A transmitter substance, which transmits nervous impulses across certain synaptic junctions in the vertebrate brain.

sessile [L. *sessilis*, fit for sitting]: Of animals, fixed in place or sedentary.

seta, *pl.* **setae** [L. *seta*, bristle]: A bristle or slender, stiff bristle-like structure which occurs, for example, in earthworms and polychaetes.

sex-linked characteristic: A genetic characteristic, such as color blindness, governed by a gene located on a sex chromosome (X or Y).

sexual reproduction: A life cycle involving meiosis and recombination and the fusion of gametes.

sheath cell or **Schwann cell:** A specialized cell wrapped around one or more axons that forms the myelin sheath.

shoot: The organs of a vascular plant above the soil — the stem, leaves and flowers.

short-day plant: A plant that flowers after exposure to periods of darkness greater than a critical length.

sieve element: A component of a sieve tube; a cell specialized to conduct sugars and other organic solutes in the phloem of the angiosperms. A mature sieve element lacks organelles and a nucleus, and depends upon close association with its companion cell.

sieve plate: The perforated plate separating sieve elements in a sieve tube of the phloem. Strands of protoplasm pass through the sieve plate and join the protoplasm of adjacent sieve elements.

sieve tube: A vertical column of sugar-conducting cells (sieve elements) in the phloem of angiosperms.

sino-atrial node (S-A node): The pacemaker of the vertebrate heart. A tiny mass of muscular tissue in the right atrium that contracts rhythmically and initiates each heartbeat.

skeleton: The hardened framework of an animal body which may serve for support, to protect soft parts or to attach muscles. It may be internal (as in man) or external (as in insects) and either solid (as in corals) or jointed.

skin: The boundary between the tissues and the external environment; the integument.

sodium-potassium exchange pump: An active transport mechanism which pumps out sodium ions and enables a cell to maintain an excess of negative ions inside itself.

solution [L. *solutio*, loosening]: A mixture in which the molecules of a dissolved substance such as sugar (called a solute) are dispersed among the molecules of a liquid such as water (called a solvent).

-soma, somat-, -some [Gk. *soma*, body]: Prefix or suffix meaning "body," "entity."

somatic [Gk., *soma* body]: Pertaining to the body. *Somatic cells* are the differentiated cells that compose the body tissues of a plant or animal, as opposed to germ cells, or gametes. *Somatic nerves* are part of the nervous system that is potentially under voluntary control, in contrast to autonomic nerves.

specialized: Having special adaptations to particular environmental or physiological conditions. A cell such as muscle is specialized to perform a certain function. An organism is specialized to inhabit a certain environment.

speciation: The process by which new species evolve from an interbreeding population. Groups of individuals become sufficiently different in their genetic composition to be able to share a common environment without interbreeding.

species [L. *species*, kind, sort]: A reproductively isolated group of plants or animals that are capable of interbreeding. Species are denoted by binomial names written in italics.

sperm [Gk. *sperma*, seed]: The male gamete. Sperm are usually motile and smaller than the egg (the female gamete).

spermatids [Gk. *sperma*, seed]: A haploid, undifferentiated precursor of a sperm cell. Four spermatids are produced when a spermatocyte undergoes meiosis.

spermatocyte: The diploid cell which gives rise through meiosis to four spermatids.

spermatogonium, *pl.* **spermatogonia:** A diploid cell found in the testes. It divides mitotically to form spermatocytes, which in turn undergo meiotic division to become spermatids, and finally sperm.

spinal cord: A part of the central nervous system of vertebrates. The spinal cord is a rope of nerve fibers, nerve cells and synapses, which runs from the brain down the vertebral column.

spindle: A structure composed of microtubules which appears during mitosis. It functions in the migration of chromosomes to opposite poles of the cell during anaphase (*see* spindle fibers).

spindle fibers: Bundles of microtubules that appear in dividing cells and extend from the centromeres of the chromosomes to the poles of the spindle, and from pole to pole.

sponge: A member of the phylum Porifera. Mostly marine animals composed of three layers perforated by pores, enclosing a central cavity.

sporangium, *pl.* **sporangia:** A structure in plants that produces spores.

spore: A reproductive cell capable of developing into a complete organism asexually, that is, without fusing with another cell (such as in the sexual fusion of gametes).

sporophyll [Gk. *phyllon*, leaf]: A modified leaf that bears sporangia. A term applied to the fertile fronds of ferns, and to carpels and stamens of angiosperms.

sporophyte: The diploid (2n), spore-producing stage in the life cycle of a plant that has alternation of generations.

stamen [L. *stamen*, thread]: The pollen-producing organ of a flower, consisting of an anther and a filament.

starch [Middle English *sterchen*, to stiffen]: A carbohydrate composed of hundreds of glucose subunits linked end to end in branched chains. Starch is the chief food-storage substance in plants; it is insoluble. Specific enzymes degrade starch into glucose units.

steady state: A condition which does not fluctuate or change, in which input of energy and matter is balanced by output.

stem: The part of the axis of a vascular plant that is above ground. Also refers to anatomically similar structures such as rhizomes which are below ground; for example, a potato is an underground stem.

stereo- [Gk. *stereos*, solid]: Prefix meaning "solid" or "three-dimensional."

steroid: A class of biologically active organic compounds based on a skeleton of four interlocking rings of carbon atoms. The vertebrate sex hormone, estrogen, and the arthropod molting hormone, ecdysone, are steroids.

stigma: The part of the carpel where pollen grains adhere and germinate.

stimulus: Any chemical or physical change that is detected by a receptor.

stomach: In vertebrates, an expanded portion of the digestive tract just beyond the esophagus. In vertebrates, an analogous cavity.

stomate [Gk. *stoma*, mouth]: Microscopic openings in the epidermis of plant leaves and stems. The size of the opening is controlled by two guard cells which border the stomate. Stomates allow gases to diffuse in and out of the plant.

structural gene: One of the genes of an operon that codes for a protein that has physiological or structural significance, in contrast to genes that have only a regulatory role.

style [Gk. *stylos*, column]: A stalk which separates the stigma from the ovary. The pollen tube grows down through the tissue of the style before it reaches the ovary.

sub-, sus- [L. under, below]: Prefix meaning "under" or "below"; for example, subcutaneous, under the skin.

subspecies: A genetically distinct subdivision of a species. Commonly a geographic subdivision.

substrate [L. *substratus*, strewn under]: The molecule on which an enzyme acts; the object or material on which an organism dwells, such as rocks or soil.

succession: Gradual changes in the kinds of organisms in an ecosystem, beginning with colonization and ending with a climax community.

succulent: A plant whose fleshy leaves and stems are capable of storing large amounts of water.

sucrose: A disaccharide composed of a glucose and a fructose molecule. Sugar produced by photosynthesis is generally transported within plants in the form of sucrose.

surface tension: The cohesion of molecules on the free surface of a liquid as a result of an unbalanced distribution of intermolecular forces on the molecules on the surface.

suspension: A heterogeneous mixture which depends on agitation to keep the particulate components from settling to the bottom.

sym-, syn [Gk. *syn*, together]: Prefix meaning "together."

symbiosis [Gk. *syn*, with, + *bios*, life]: The living together of two dissimilar organisms, in an intimate relationship, so that both organisms benefit from the relationship (mutualism), or one benefits at the expense of the other (parasitism).

sympathetic nervous system: One of the two divisions of the autonomic nervous system, with centers located in the mid-portion of the spinal cord. It employs the transmitter chemical noradrenalin and the hormone adrenalin to mobilize energy resources, and prepare the animal for "fight or flight." It generally counteracts the parasympathetic nervous system.

synapse [Gk. *synapsis*, union]: The junction between neurons or between a neuron and an effector or receptor cell, across which nerve impulses are transmitted.

synapsis: The pairing that takes place between homologous chromosomes during the first meiotic division. Crossing-over takes place during synapsis.

syncytium: A multinucleate structure formed by the fusion of the cytoplasm of two or more cells and by the loss of cell membranes; for example, the fusion of embryonic myoblasts to form a skeletal muscle cell.

syngamy: The fusion of gametes during sexual reproduction.

synthesis: The construction of more complicated molecules from simple molecules or atoms.

system: Any piece or part of the world that we isolate conceptually from its surroundings for experimental manipulation and study.

systematics: The study of plant and animal diversity and of the relationships among groups of organisms.

taiga: A wet and cold biome of coniferous forests found in Canada, Northern Europe, and Siberia.

taproot: A thick, tapering main root which gives rise to smaller, lateral branches.

taxonomy [Gk. *taxis*, arrangement, + *nomos*, law]: The classification or cataloging of organisms.

telophase [Gk. *telos*, end, + *phasis*, form]: The last stage in mitosis and meiosis, in which chromosomes are reorganized into two new nuclei.

temperate phage: A virus that is capable of incorporating its DNA into the genome of a host bacterium; it may remain in this latent or "prophage" stage through many generations of bacteria, since the DNA of the prophage is replicated along with that of the bacterium.

template: A structure which serves as a mold or pattern for the construction of another, complementary structure. DNA serves as a template for the replication of a complementary DNA or RNA molecule.

tendon [L. *tendere*, to stretch]: A cord-like structure composed of connective tissue which anchors a muscle to a bone.

tertiary structure: The structure given to a helical protein by specific bending at certain sites.

test cross: *See* back cross.

testis, *pl.* **testes** [L. *testis*, witness]: The male sex organ which produces gametes and may produce male sex hormones.

testosterone: A male sex hormone of the higher vertebrates. It is secreted by the testes, and stimulates the production of sperm, as well as the development and maintenance of male sex characteristics.

tetrad: In plants, the four spores formed by meiosis from a single spore mother cell.

thalamus [Gk., *thalamos* inner chamber]: A cluster of neurons and glial cells in the back of the forebrain and just above the midbrain. Almost all sensory impulses are integrated in the thalamus before they are relayed to the cerebral cortex.

thallophyte: A term sometimes used to designate fungi and algae collectively.

theory [Gk. *theorein*, to look at]: A hypothesis supported by a large body of observations and experiments.

thermodynamics [Gk. *therme*, heat, + *dynamis*, power]: The study of energy relationships. The first law of thermodynamics states that the total energy of the universe remains constant; energy is neither created nor destroyed. The second law states that the degree of randomness, or entropy, always tends to increase and the energy available to do work decreases.

thylakoid: A membranous sac in the chloroplast. The grana are composed of stacks of thylakoids.

thymine: The pyrimidine subunit of DNA, not found in RNA. Thymine is replaced by uracil in RNA.

thyroid [Gk. *thyreoeides*, shield-shaped]: A vertebrate endocrine gland located in the neck region; it secretes the iodine-containing hormone, thyroxin, which stimulates metabolic rate.

tissue [L. *texere*, to weave]: A group of cells, usually similar in both structure and function, that are usually bound together by intercellular material.

tissue culture: A method that allows fragments of organisms or single cells to be grown *in vitro* (in glass), that is, apart from the organism.

trace element: Same as micronutrient element.

trachea [Gk. *tracheia*, rough[: A breathing tube, such as the windpipe of mammals or the tubes composing the breathing system of insects.

tracheid [Gk. *tracheia*, rough]: A cell type found in xylem that is specialized to conduct water and give the stem rigidity. Tracheids are elongated and taper at the ends; they have thick walls without true perforations. They are found in most vascular plants and are dead at maturity.

tracheoles: The fine tubular endings of the tracheal system of arthropods. Tracheoles are the principal site of gas exchange between the tissues of the animal and the air in the tracheal system.

tracheophyte [Gk. *tracheia*, rough, + *phyton*, plant]: A plant having xylem and phloem tissue; that is, a vascular plant.

trans- [L. across, beyond]: Prefix meaning "across" or "beyond."

transcription: The biosynthetic process by which the base sequence of chromosomal DNA is transferred to messenger RNA, which forms a complementary copy.

transduction [L. *ducere*, to lead]: The transfer of DNA from one bacterium to another by a temperate bacteriophage.

transfer RNA (tRNA): A small RNA molecule about 80 nucleotides long that acts as a carrier of specific amino acids during protein synthesis. A molecule of tRNA which is specific for a particular amino acid binds to that amino acid. The sequence of three nucleotides that make up the anticodon on the tRNA recognize and bind to the codon on the messenger RNA, which is complexed with a ribosome on which a polypeptide chain is developing. The amino acid is thus guided into the proper position on the ribosome-mRNA complex to be enzymatically added to the polypeptide chain. There is at least one tRNA for each kind of amino acid.

transformation: The incorporation by a cell of DNA from another cell into its genome.

translation: The biosynthetic process by which amino acid sequences in proteins are determined by base sequences in the mRNA template on the ribosome; the translation of an mRNA into a protein product.

translocation: In genetics, the interchange of segments between nonhomologous chromosomes. In plant physiology, the transport of dissolved foods from one place to another within the plant, largely through the phloem.

transmitter substance: A chemical secreted by an axon at a synapse which affects the membrane permeability of the postsynaptic cell. Acetylcholine and noradrenalin are transmitter substances.

transpiration [Fr. *transpirer*, to perspire]: The evaporation and loss of water vapor from a plant through the stomates.

tritium: A radioactive isotope of hydrogen, the nucleus of which contains two extra neutrons and one proton, instead of the usual single proton.

trochophore [Gk. *trochos*, wheel, + *phoros*, bearing]: A distinctively shaped ciliated larva found in some annelids and mollusks.

-trophic [Gk. *trophe*, food]: Suffix or prefix meaning "feeding" or "feeder." For example, autotrophic, "self-nourishing."

trophic level: A step in the movement of energy through an ecosystem. A link in a food chain.

tundra: A subarctic biome characterized by a ground layer of grasses, mosses, and lichens, and a few perennials. In tundra the ground is permanently frozen a few inches below the surface (permafrost).

turgor pressure [L. *turgere*, to be swollen]: The pressure exerted against the cell wall or membrane by the cytoplasm of a hypertonic cell, that is, a cell that is osmotically taking up water.

unicellular: Composed of a single cell.

uracil: A pyrimidine base found in RNA. Uracil is replaced by thymine in DNA.

urea [Gk. *ouron*, urine]: The principal form of nitrogen excretion in most mammals, formed in the liver from ammonia and CO_2 and excreted by the kidneys. Urea is water-soluble and relatively nontoxic.

ureter [Gk. *ourein*, to urinate]: The tube which conducts urine from the kidney to the bladder in amphibians and mammals, and from the kidney to the cloaca in birds and reptiles.

urethra: The tube which conducts urine from the bladder to the exterior in mammals.

uric acid: The principal form of nitrogen excretion in reptiles, birds, and insects. Uric acid is almost insoluble in water.

uterus [L. *uterus*, womb]: The womb; the muscular chamber in the mammalian, female reproductive tract where most of the development of the embryo takes place.

vacuole [L. *vacuus*, empty]: A membrane-bound chamber or vesicle within a cell, filled with water and various crystalized or dissolved substances.

vagina [L. *vagina*, sheath]: The most exterior part of the mammalian female reproductive tract, into which the penis is inserted during copulation.

vagus nerve [L. *vagus*, wandering]: One of a pair of autonomic nerves that conducts impulses from the medulla of the vertebrate brain to visceral organs.

variation: Differences that exist among the members of a species.

vascular: A term referring to a fluid-conducting tube or system of tubes.

vascular cambium: A meristematic cylinder, one cell thick, which produces secondary xylem on the inside and secondary phloem on the outside.

vascular tissue [L. *vasculum*, small vessel]: Tissue specialized to conduct substances from one part of the organism to another; for example, xylem and phloem in plants, and tissues of the circulatory system in animals.

vas deferens: In mammals, the tube that carries sperm from the testes to the urethra.

vaso- [L. *vas*, vessel]: Prefix meaning "blood vessel."

vegetative: Of reproduction, asexual. Of plant

cells and organs, not specialized for reproduction.

vein [L. *vena,* blood vessel]: In animals, a blood vessel carrying blood from the tissues and organs to the heart. In plants, a vascular bundle in a leaf.

vena cava: A large vein leading directly into the right atrium of the heart, composed of two branches, the superior and the inferior vena cava.

ventral [L. *venter,* belly]: Referring to the front surface of an animal that keeps itself erect and to the under-surface of an animal that moves on all fours or creeps, in contrast to dorsal which refers to the back.

ventricle [L. *ventriculus,* the stomach]: A chamber of the heart which receives blood from an atrium and pumps it out of the heart into an artery.

vesicle [L. *vesicula,* small bladder]: A small sac or bladder surrounded by a membrane. Vesicles may form within cells as in pinocytosis, or may be multicellular as in the seminal vesicles which store sperm in animals.

vessel: In flowering plants, a nonliving tube of cells (vessel elements placed end to end) in the xylem, which conducts water and minerals upward from the roots.

vessel element: One of the dead cells composing a vessel.

villi [L. *villus,* a tuft of hair]: Small projections in the lining of the vertebrate small intestine, which greatly increase the absorptive area.

virus [L. *virus,* poison, slimy liquid]: A parasitic, noncellular particle composed of a DNA or RNA core usually surrounded by a protein coat. Viruses replicate by using the protein-synthesizing machinery of the host cell. They exhibit some features of living organisms, including the ability to mutate and evolve.

vitamin [L. *vita,* life]: An organic compound required by an organism in small quantities, but not synthesized by that organism in adequate amounts. For this reason vitamins must be present in the diet.

-vorous [L. *vorare,* to devour]: Suffix meaning "feeding on." For example, "herbivorous," herb feeding..

wood: The xylem or plant vascular tissue that has lignified secondary walls.

xylem [Gk. *xylon,* wood]: The vascular tissue in plants which conducts water and minerals upward from the roots, and also gives the stem rigidity. Mature xylem tissue is nonliving, and is composed of several cell types: tracheids or vessel elements, parenchyma cells, and fibers. The wood in trees and shrubs consists of xylem.

yolk sac: The energy source stored in the egg cell of animals to nourish the developing embryo.

zoo- [Gk. *zoion,* animal]: Prefix meaning "animal" or "motile."

zoospore: A motile spore found among algae and fungi.

zygote [Gk. *zygotos,* joined together]: The diploid (2n) cell, such as the fertilized egg which results from the fusion of gametes.

APPENDICES

APPENDIX A – A CLASSIFICATION OF ORGANISMS

The system of classification presented here is one of several in current use. It follows the outline used in Chapters 9, 21, 23, and 24. Other classifications recognize more or fewer divisions and phyla and may divide or combine classes in slightly different ways, but most parts of the classification outlined here are accepted by most biologists. The principal taxonomic groups are kingdom, phylum, class, order, family, genus, and species. The term "division" is commonly used by botanists, instead of the term "phylum," to denote major groups of plants. In this summary classification the term phylum has been used throughout for simplicity. Either term is acceptable.

The basic separation into five kingdoms follows the suggestion of Whittaker. Almost all of the phyla and most of the classes are listed here. Certain orders discussed extensively in this book are also included. A few extinct groups of special evolutionary importance (such as seed ferns and armored fishes) are included, but for the most part only groups with living members are listed. A few of the common genera (or genera illustrated in this book) are cited along with each group. An estimate of the number of living species of each group that has been described and named is given in parentheses. For many groups there are many additional species as yet undescribed.

Only the briefest description is given of groups that are described in the text. However several groups that were not discussed in the text but which may be encountered in general reading or in the laboratory (such as rotifers and plasmodial slime molds) are described sufficiently so that their relationships to other groups of organisms is clear.

KINGDOM MONERA

Procaryotic cells which lack a nuclear membrane, plastids, mitochondria, and $(9 \times 2) + 2$ flagella. Unicellular, may form filaments, sheets or other structures. They feed mainly by absorption but some groups are chemosynthetic or photosynthetic.

PHYLUM SCHIZOMYCETES (1600). Bacteria.

Class Eubacteria. True bacteria. *Escherichia, Salmonella, Staphylococcus, Streptococcus, Bacillus.*

Class Myxobacteria. Unicellular organisms with thin flexible cell walls and the ability to glide. *Myxococcus.*

Class Spirochaetes. Unicellular organisms with a helical shape and two tufts of special flagella. Includes causative organisms of syphilis. *Treponema. Leptospira.*

Class Rickettsiae. Small bacteria-like organisms. Occur widely as parasites of arthropods. Includes causative agent of Rocky Mountain fever. *Rickettsia.*

PHYLUM CYANOPHYTA (200). Blue-green algae. Photosynthetic procaryotes which contain chlorophyll *a. Anabaena, Nostoc, Gloeocapsa.*

KINGDOM PROTISTA

Eucaryotic unicellular organisms. Cells are sometimes aggregated together to form filaments, sheets or other structures. Includes both photosynthetic organisms (several phyla of algae) and organisms which feed by ingestion and absorption (protozoans). Certain fungi-like organisms (slime molds) are also included in this kingdom.

PHYLUM CHRYSOPHYTA. Diatoms and golden algae. Autotrophic organisms that contain chlorophylls *a* and *c* and the pigment fucoxanthin. Cell walls composed mainly of pectin and sometimes impregnated with silica.

Class Bacillariophyceae (6000+). Diatoms. Possess two shells which fit together like a box and its lid. *Chaetoceras.*

Class Chrysophyceae (1000). Golden algae. A varied group of organisms some of which are nonmotile whereas others move by means of flagella or by amoeboid motion. Many have cell walls with silica scales whereas others look like amoebae with chloroplasts. *Distephanus.*

PHYLUM PYRROPHYTA. Golden-brown algae. Autotrophic organisms that contain chlorophylls *a* and *c*. Cell walls contain cellulose. The most important class is composed of the dinoflagellates.

Class Dinophyceae (1000). Dinoflagellates. Possess two grooves at right angles with a flagellum extending from each groove.

PHYLUM EUGLENOPHYTA (450). The euglenas. Autotrophic (or very similar heterotrophic) organisms which possess chlorophylls *a* and *b*. Their cell wall or pellicle is flexible and composed mainly of proteins. Most have a contractile vacuole and a single apical flagellum. *Euglena, Astasia.*

PHYLUM GYMNOMYCOTA. The slime molds. Amoebalike organisms usually lacking a cell wall, that feed heterotrophically, mainly by ingestion. At some stage in their development they form spores in a sporangium.

Class Acrasiomycetes (26). Cellular slime molds. Separate amoebas eventually swarm together to form a mass within which the individual amoebas retain their identity. Finally differentiates into a sporangium within which are spores. *Dictyostelium.*

Class Myxomycetes (450). Plasmodial slime molds. These slime molds form a multinucleate plasmodium which slowly creeps along like a giant multinucleate amoeba, often several feet in diameter. Finally it ceases crawling and differentiates into a number of sporangia, each of which forms many spores. *Physarum.*

Class Protostelidomycetes (12+). A new order of slime molds first described in 1966 which appear to have features of both the cellular and the plasmodial slime molds. It may feed as a uninucleate amoeba or as a multinucleate plasmodium. *Schizoplasmodium.*

Class Labyrinthulomycetes. Cell net slime molds. A strange group of organisms composed of amoebalike cells which do not ingest food but absorb food as do the fungi. Unlike any other fungi or slime molds the cells possess a photosensitive eyespot. *Labyrinthula.*

PHYLUM PROTOZOA (50,000). Unicellular heterotrophs.

Subphylum Sarcomastigophora

Class Flagellata (or Mastigophora). Flagellates. Move by means of flagella. *Trypanosoma, Calonympha, Trichomonas.*

Class Sarcodina. Protozoans which move by means of pseudopods. Includes amoebas, radiolarians and forams. *Amoeba, Difflugia, Globigerina, Acanthometra.*

Subphylum Sporozoa

Class Sporozoa. Parasitic protozoans that form spores. Usually lack organs of motility during most of their life cycle. Includes malarial organisms. *Plasmodium.*

Subphylum Ciliophora

Class Ciliata (or Ciliatea). Have two kinds of nuclei; locomotion by cilia. Includes ciliates and suctorians. *Paramecium, Vorticella, Stentor, Ephelota.*

KINGDOM FUNGI

Eucaryotic unicellular or multinucleate organisms in which the nuclei occur in interconnecting, branching tubular filaments called hyphae. They are heterotrophic and feed by absorption.

PHYLUM EUMYCOTA (or MYCOTA).

Class Oomycetes (several hundred). Water molds and downy mildews, late blight fungus of potatoes. Mainly aquatic fungi with cell walls containing cellulose, Have motile flagellated spores. *Saprolegnia, Phytophthora.*

Class Zygomycetes (several hundred). Bread molds and similar organisms. Terrestrial fungi in which hyphae do not have crosswalls. Cell walls contain chitin. *Rhizopus.*

Class Ascomycetes (30,000). Sac fungi. Includes yeasts, Dutch Elm disease fungus, truffles. Aquatic and terrestrial fungi. Hypae usually have perforated crosswalls and contain chitin. Form a characteristic reproductive cell, the ascus. Some are unicellular. *Neurospora, Aspergillus, Saccharomyces, Claviceps.*

Class Basidiomycetes (25,000). Club fungi. Includes edible mushrooms, toadstools, puffballs, rusts, Terrestrial fungi in which hyphae have perforated crosswalls and contain chitin. Form special reproductive cells called basidia which bear spores. *Agaricus, Amanita, Psilocybe, Puccinia.*

Fungi Imperfecti (25,000). Principally microscopic fungi with characteristics of Ascomycetes but in which the sexual stages have not been observed. *Penicillium.*

Lichens (17,000). Mostly Ascomycetes that have an obligatory symbiotic relationship with unicellular algae which live and multiply within their hyphae. *Parmelia, Cladonia.*

KINGDOM PLANTAE

Multicellular autotrophic eucaryotes and related unicellular forms. Mainly photosynthetic but a few have secondarily become heterotrophs. Contain chlorophyll *a* and carotenoids. Have cell walls that contain cellulose, are generally nonmotile, and attached. Alternating gametophyte and sporophyte phases.

PHYLUM RHODOPHYTA (4000). Red algae. Unicellular or multicellular plants, mainly marine, possessing chlorophyll *a* and phycobilins. No motile cells. Composed of densely-packed filaments held together in a gelatine-like matrix. *Ceramium, Chondrus, Porolithon.*

PHYLUM PHAEOPHYTA (1500). Brown algae. Multicellular marine plants possessing chlorophyll *a* and *c* and fucoxanthin. Also possess flagellated reproductive cells. *Fucus, Laminaria, Sargassum.*

PHYLUM CHLOROPHYTA (7000). Green algae. Unicellular or multicellular. Possess chlorophyll *a* and *b* and carotenoids. Also possess flagellated cells. Unlike other algae, their carbohydrate food reserve is starch like that found in higher plants. *Chlamydomonas, Ulothrix, Acetabularia, Volvox, Ulva, Oedogonium.*

PHYLUM BRYOPHYTA (23,600). Mosses, liverworts and hornworts. Multicellular plants possessing chlorophyll *a* and *b* and carotenoids and containing starch as do the green algae. Gametes produced within multicellular sex organs. Sperm have flagella. Lack well-formed conducting tisues. Gametophyte phase is dominant.

Class Hepaticae (9000). Liverworts. *Marchantia, Riccia.*

Class Anthocerotae (100). Hornworts. *Anthoceros.*

Class Musci (14,500). Mosses. *Physcomitrium, Polytrichum, Sphagnum.*

PHYLUM TRACHEOPHYTA. Vascular plants. Terrestrial plants with stems, leaves and roots and well-developed conducting tissue for the transport of water and products of photosynthesis. Sporophyte phase is dominant.

Subphylum Psilopsida. Leafless and rootless vascular plants. All thought to be extinct. *Rhynia.* Some living plants (*Psilotum, Tmesipteris*) long classified as psilopsids are now thought to be primitive ferns.

Subphylum Lycopsida (1200). Club mosses or ground pines. Simple conducting systems and small green leaves. *Lycopodium Selaginella.*

Subphylum Sphenopsida (40). Horsetails. Simple conducting systems, jointed stems, and reduced scalelike leaves. *Equisetum.*

Subphylum Pteropsida (11,000). Ferns. *Polystichum, Polypodium, Dryopteris, Cyathea.*

Subphylum Spermopsida. Seed plants.

Class Pteridospermae. Seed ferns. No living examples. *Medullosa.*

Class Cycadinae (100). Cycads. *Zamia.*

Class Ginkgoinae (1). Ginkgoes. *Gingko.*

Class Coniferinae (550). *Pinus, Taxus, Sequoia, Tsuga.*

Class Gnetinae (70). *Ephedra, Weltwitschia, Gnetum.*

Class Angiospermae (250,000). Flowering plants.

Subclass Dicotyledonae (190,000). Dicots. *Magnolia, Quercus, Rosa, Ranunculus, Aster, Euphorbia, Cereus, Taraxacum, Nicotiana, Cannabis, Digitalis.*

Subclass Monocotyledonae (60,000). Monocots. *Zea, Tulipia, Poa, Lilium, Yucca, Cattleya.*

KINGDOM ANIMALIA

Eucaryotic multicellular heterotrophic organisms whose principal mode of nutrition is by ingestion or occasionally by absorption. Usually lack the rigid cell walls characteristic of plants and are commonly motile.

Subkingdom Parazoa

PHYLUM PORIFERA (5000). Sponges. Radially symmetrical multicellular animals with no digestive cavity. Body perforated by many pores through which water containing food particles flows. No well-defined tissues or organs.

Class Calcarea. Marine forms found in shallow water with calcareous (chalky) spicules. *Leucosolenia, Sycon, Grantia.*

Class Hexactinellida. Deep water marine forms with spicules made of silica called glass sponges. *Euplectella.*

Class Desmospongia. Marine and freshwater sponges with skeletons composed of protein fibers, silica or both. *Spheciospongia, Spongia.*

Subkingdom Metazoa

PHYLUM COELENTERATA (or CNIDARIA) (9000). Radially symmetrical animals with a single alimentary opening which serves as both mouth and anus. They have tentacles around their mouth and possess cells which produce characteristic stinging organs called nematocysts. Body contains three layers of cells but the middle layer is not as well developed as the mesoderm of higher animals.

Class Hydrozoa (2,700). Hydra-like animals, either single or colonial. *Hydra, Obelia, Tubularia, Physalia.*

Class Scyphozoa (200). True jellyfish. *Cyanea, Aurelia, Chrysaora, Haliclystus.*

Class Anthozoa (6000). Corals and sea anemones. *Stoichactis, Aptasia, Metridium, Astrangia.*

PHYLUM CTENOPHORA (90). Comb jellies or sea walnuts. Free-swimming, almost spherical marine animals, with eight longitudinal rows of ciliary combs and usually with two tentacles.

Class Tentaculata. Ctenophores with tentacles. *Pleurobrachia, Mnemiopsis.*

Class Nuda. Ctenophores without tentacles. *Beroe.*

Section Protostomia

PHYLUM PLATYHELMINTHES (12,700). Flatworms. Bilaterally symmetrical animals with three tissue layers. Body dorsoventrally flattened. The digestive cavity, when present, has only a single opening. No coelom and no circulatory system.

Class Turbellaria (3000). Free-living flatworms, marine, freshwater, and a few terrestrial. Includes planarians, polyclads, and acoels. All ciliated and carnivorous. *Planaria, Dugesia.*

Class Trematoda (6250). Flukes. Parasitic flatworms with one or more suckers. *Schistosoma, Clonorchis, Fasciola.*

Class Cestoda (3400). Tapeworms. Parasitic flatworms with no digestive tract. Obtain nutrients by absorption through body wall. *Taenia.*

PHYLUM MESOZOA (50). Minute wormlike parasites of marine invertebrates with the simplest structure of any multicellular animals. They are solid, up to 7 mm long and composed of an outer layer of about 25 ciliated somatic cells and about a dozen other cells. Their kinship is unknown. They may be a group of degenerate flatworms or perhaps an offshoot of the early metazoans.

PHYLUM NEMERTINEA (or NEMERTINA, RHYNCHOCOELA) (750). Ribbon worms or proboscis worms. Simplest animals with a blood vascular system and a digestive tract with a mouth and an anus. Mostly marine, but a few freshwater and terrestrial. *Cerebratulus, Tubulanus.*

PHYLUM ASCHELMINTHES. All possess a body cavity partly lined with endoderm called a pseudocoel or false coelom. Complete digestive tract with mouth and anus. Covered with a tough cuticle. Few or no cilia on the surface.

Class Nematoda (10,000). Roundworms. The most numerous of the aschelminths. Elongated, cylindrical body, covered with a tough cuticle. Free-living in soil or water, or parasitic in plants or animals. *Ascaris, Trichinella, Enterobius.*

Class Nematomorpha (100). Hairworms or horsehair worms. Extremely thin, brown or black worms about a foot long but only a millimeter thick. Larvae are parasitic in insects whereas the adults are free-living. *Gordius.*

Class Rotifera (1500). Rotifers. Small, wormlike animals with a circle of cilia on the head which, when beating, looks like a rotating wheel. Abundant in fresh water but some are marine. *Asplanchna, Floscularia.*

Class Gastrotricha (175). Up to 0.6 mm long. Resemble rotifers but lack the crownlike circle of cilia. Both marine and fresh water. *Macrodasys, Chaetonotus.*

Class Kinorhyncha (60). Small marine animals up to 1 mm long with a segmented cuticle. Periodically molt. Live in mud and silt. *Echinoderes.*

PHYLUM ACANTHOCEPHALA (500). Spiny-headed worms. Wormlike animals with a psuedocoel. No digestive tract and food is absorbed through the body wall as in tapeworms. Larvae parasitic in arthropods, adults in vertebrates. *Acanthocephalus. Echinorhynchus.*

PHYLUM ENTOPROCTA (60). Minute stalked, sessile animals, may be solitary or colonial. Have a U-shaped digestive tract. Uncertain whether the body cavity is a pseudocoel and relation to other animals unclear. Mouth and anus surrounded by a circle of ciliated tentacles. Mostly marine. *Pedicellina.*

PHYLUM ANNELIDA (7000). Segmented worms with a thin cuticle, digestive tract with mouth and anus, true coelom lined with mesoderm, closed circulatory system, hydrostatic skeleton. Large longitudinal ventral nerve cord. Some have non-jointed appendages. Mostly free-living.

Class Polychaeta. Marine worms, either free-moving or live a sedentary life in a tube. Includes sandworms, clamworms, tubeworms. *Nereis, Chaetopterus, Sabella, Arenicola.*

Class Oligochaeta. Earthworms and many fresh-water annelids. Terrestrial or fresh water. *Lumbricus, Tubifex.*

Class Hirudinea. Leeches. *Hirudo, Macrobdella.*

PHYLUM PRIAPULIDA (8). Cucumber-shaped marine animals, several inches long, with a body covered with small spines and tubercles, which feed largely on polychaetes. They appear to have a true coelom. Their relationship to other groups of organisms is unknown. *Priapulus.*

PHYLUM SIPUNCULIDA (250). Slender marine worms that live in sand or mud. Usually 6 to 12 inches long. Possess a coelom. Lack segmentation and setae like that of annelids but have a number of annelid features including a trachophore larva. They probably diverged from the ancestors of the annelids before segmentation developed. *Golfingia ,Phascolosoma., Sipunculus.*

PHYLUM ECHIURIDA (60). Marine worms that resemble sipunculids in size and general habit. Possess a coelom. The adults lack segmentation but segmentation occurs during embryonic development. They have a trochophore larva and setae like polychaetes and probably diverged from the line leading to the polychaetes. *Urechis, Echiurus.*

PHYLUM MOLLUSCA (80,000). Unsegmented, soft-bodied animals, usually covered by a shell and with a ventral muscular foot.

Class Monoplacophora. Primitive marine mollusks with a single shell, 5 to 6 pairs of gills. Superficially resemble limpets. *Neopilina.*

Class Amphineura.
Subclass Polyplacophora. Chitons. Marine forms with a shell composed of eight plates. Head very much reduced. *Chaetopleura.*
Subclass Aplacophora. A small group of worm-like mollusks about an inch long. They have no shells as adults, but in early development they have shell plates like those of chitons. They appear to be derived from chitons. *Nomenia.*

Class Scaphopoda. Tusk or tooth shells. Marine forms living in sand or mud with tubular shells open at both ends. *Dentalium.*

Class Gastropoda. Snails, slugs, abalones, whelks, cones, pteropods. The largest group of mollusks. Have a single spiral shell or no shell. *Limax, Helix, Buccinum.*

Class Pelecypoda. Bivalve mollusks such as clams, mussels, scallops and oysters. Lack a head, filter feeders. *Mercenaria, Mytilus, Ostrea, Pecten, Anodonta.*

Class Cephalopoda. Octopods, squids, nautiloids. Marine molluscs with a distinct head, with 8 or 10 arms around their mouth. *Loligo, Octopus, Argo.*

PHYLUM ONYCHOPHORA (70). Onychophores or walking worms. Tropical animals structurally intermediate between annelids and arthropods. Soft segmented bodies, hydrostatic skeleton, an anterior pair of antennae and many pairs of unjointed legs. Possess a cuticle of chitin; molt. *Peripatus.*

PHYLUM ARTHROPODA (900,000+). Segmented animals with jointed appendages, a hard chitinous skin, with body divided into head, thorax, and abdomen. Molt.

Subphylum Trilobitomorpha.

Class Trilobita. Trilobites. Primitive marine arthropods, all extinct. Segmented body divided into three lobes by two longitudinal furrows. All segments except last had a pair of jointed appendages.

Subphylum Chelicerata. No antennae. The first pair of appendages called chelicerae, are in front of the mouth and function in feeding. Behind the mouth is another pair of appendages called pedipalps.

Class Eurypterida. Eurypterids or water scorpions, all extinct. *Pterygotus.*

Class Xiphosura (5). Horseshoe crabs. *Limulus.* Horseshoe crabs and eurypterids are sometimes grouped together in the class Merostomata.

Class Arachnida (60,000). No antennae. Chelicerae used as pinchers, pedipalps used as jaws. Four pairs of legs arise from cephalothorax. There are about 10 different orders of which the following are fairly common.
Order Araneae. Spiders. *Dugesiella.*
Order Acarina. Mites and ticks. *Tetranychus, Dermacentor.*
Order Opiliones. Harvestmen or daddy longlegs. *Phalangium.*
Order Scorpiones. Scorpions. *Centruroides.*
Order Uropygi. Whip scorpions or vinegaroons. *Mastigoproctus.*

Class Pycnogonida (440). Sea spiders or nobody crabs. Common, tiny marine animals with tiny bodies and long legs. Probably related to the chelicerates but their evolutionary affinities are uncertain. *Nymphon.*

Subphylum Mandibulata. Have antennae, maxillae and mandibles. Usually have compound eyes.

Class Crustaea (28,000). Crabs, lobsters, barnacles, copepods, brine shrimp, pill bugs. Usually aquatic. Have two pair of antennae. *Callinectes, Homarus, Balanus, Sacculina, Calanus, Daphnia, Artemia, Porcellionides, Armadillidium.*

Class Chilopoda (3000). Centipedes. Each body segment, except the head and tail, possesses a pair of legs. *Scutigera.*

Class Diplopoda (8000). Millipedes. Each external segment (actually two fused segments) bears two pair of legs. *Platyrachus, Julus.*

Class Insecta (700,000+). Insects. Body divided into a distinct head, a thorax and an abdomen. The head has four pairs of appendages; the thorax has three pairs of legs and usually two pairs of wings; the abdomen usually has no appendages.

Subclass Apterygota. Wingless insects in which the absence of wings is a primary condition.

Order Collembola. Springtails. Minute wingless insects usually less than 6 mm long. Abdomen with 6 or fewer segments. Possess a forked structure on the abdomen with which they jump. *Sminthurus.*

Order Protura. Proturans. Minute whitish insects with a cone-shaped head, about 1 mm long. Lack eyes, wings, cerci and antennae. Adult abdomen has 12 segments. Occur in soil and leaf mold. *Acrentulus.*

Order Thysanura. Silverfish, firebrats, bristletails, fast-running wingless insects, with two or three tail-like appendages at the tip of the abdomen. Abdomen has 11 segments. *Lepisma, Thermobia, Machilis.*

Order Diplura. Diplurans. Similar to bristletails but have only two tail-like appendages at the tip of the abdomen. Usually 6 mm long or less. Found in leaf mold and soil. Lack compound eyes. *Campodea.*

Subclass Pterygota. Winged insects, or, if lacking wings, the wingless condition is secondary.

Order Ephemeroptera. Mayflies. *Hexagenia, Ephemerella.*

Order Odonata. Dragonflies and damselflies. *Celithemis, Aeshna, Lestes.*

Order Orthoptera. Grasshoppers, katydids, crickets, roaches, mantids and walking sticks. *Melanoplus, Romalea, Pterophylla, Gryllus, Blatta, Periplaneta, Mantis, Megaphasma.*

Order Isoptera. Termites. *Reticulitermes, Constrictotermes.*

Order Dermaptera. Earwigs. *Forficula, Labia.*

Order Embioptera. Webspinners. *Anisembia, Oligembia, Oligotoma.*

Order Plecoptera. Stoneflies. *Isoperla, Nemoura, Taeniopteryx.*

Order Psocoptera. Booklice or psocids. *Liposcelis, Psocus, Trogium.*

Order Zoraptera. Zorapterans. *Zorotypus.*

Order Mallophaga. Chewing lice, bird lice. *Menopon, Bovicola, Chelopistes.*

Order Anoplura. Sucking lice, human body louse, crab louse. *Pediculus, Phthirus, Haematopinus.*

Order Thysanoptera. Thrips. *Taeniothrips, Selenothrips, Franklinella.*

Order Hemiptera. True bugs including milkweed bugs, bedbugs, backswimmers. *Oncopeltus, Cimex, Lygus, Rhodinus, Notonecta.*

Order Homoptera. Cicadas, leaf hoppers, aphids, scale insects. *Megacicada, Empoasca, Megoura, Aphis, Phylloxera.*

Order Neuroptera. Fishflies, dobsonflies, snakeflies, lacewings, and antlions. *Corydalus, Sialis, Agulla, Chrysopa, Dendroleon.*

Order Coleoptera. Beetles and weevils. *Cicindela, Scolytus, Photinus, Dytiscus, Tenebrio, Hercules, Dermestes, Leptinotarsa, Epilachna, Popillia, Anthonomus.*

Order Strepsiptera. Twisted wing parasites. *Stylops.*

Order Mecoptera. Scorpionflies. *Panorpa, Bittacus.*

Order Trichoptera. Caddisflies. *Hydropsyche, Rhyacophila, Limnephilus.*

Order Lepidoptera. Butterflies and moths. *Danaus, Pieris, Popilio, Heliothis, Vanessa, Pyrausta, Sphinx, Hyalophora, Porthetria, Malacasoma.*

Order Diptera. True flies, mosquitoes, midges, *Drosophila, Musca, Sarcophaga, Chironomus, Tipula, Tabanus, Aedes, Anopheles.*

Order Siphonaptera. Fleas. *Pulex, Xenopsylla.*

Order Hymenoptera. Wasps, bees, ants, saw-flies, ichneumons, chalcids. *Vespa, Cimbex, Apis, Bombus, Sphex, Habrobracon, Mormoniella, Cephus, Formica, Camponotus.*

Class Symphyla (120). Small arthropods between 2 and 10 mm long found in soil and leaf mold which superficially resemble centipedes. Adults have 12 pairs of legs. May be related to insects. *Scutigerella.*

Class Pauropoda (360). Small, soft-bodied grub-like animals found in soil and leaf mold. Adults have nine pairs of legs. *Pauropus.*

PHYLUM TARDIGRADA (180). Waterbears. Tiny animals about 0.5 mm long which are aquatic or live in water films on plants. Cylindrical bodies with four pairs of stubby legs, segmented nervous system. Secrete a cuticle without chitin and molt. Related to annelids and arthropods. *Echiniscus, Macrobiotus.*

PHYLUM PENTASTOMIDA (70). Tongue worms. Parasites of lungs and nasal passages of vertebrates. Usually several centimeters long. Chitinous cuticle, molting, nervous system like that of annelids and arthropods. Clearly related to arthropods. *Cephalobaena, Linguatula.*

PHYLUM PHORONIDA (15). Wormlike marine animals less than 8 inches long which secrete and live in a leathery tube. They have a U-shaped digestive tract and a lophophore, a horseshoe-shaped fold of the body wall that encircles the mouth and bears tentacles and which the animal uses for filter feeding. They also possess a true coelom. *Phoronis.*

PHYLUM ECTOPROCTA OR BRYOZOA (4000). Bryozoans, moss animals. Small colonial animals, largely marine, with a lophophore and a true coelom.

Class Gymnolaemata. *Bugula, Schizoporella.*

Class Phylactolaemata. *Plumatella.*

PHYLUM BRACHIOPODA (260). Lamp shells. Marine animals superficially resembling clams. Body attached by a stalk and enclosed within two dorsoventrally oriented shells. Feed by means of a lophophore and possess a true coelom.

Class Inarticulata. *Lingula.*
Class Articulata. *Terebratula.*

SECTION DEUTEROSTOMIA

PHYLUM ECHINODERMATA (5700). Marine animals, radially symmetrical and five-sided as adults, bilaterally symmetrical as larvae. Skin contains calcareous, plates bearing spines. Possess a water-vascular system and tube feet for locomotion and food collecting.

Class Crinoidea. Crinoids, sea lilies. *Ptilocrinus, Comactinia.*

Class Asteroidea. Starfish or sea stars. *Asterias, Oreaster, Ctenodiscus.*

Class Ophiuroidea. Brittle stars, basket stars, serpent stars. *Ophiura, Ophioderma, Asteronyx.*

Class Echinoidea. Sea urchins and sand dollars. *Arbacia, Strongylocentrotus, Echinarachnius.*

Class Holothuroidea. Sea cucumbers. *Thyone, Cucumaria, Leptosynapta.*

PHYLUM HEMICHORDATA (90).

Class Enteropneusta. Acorn worms. Marine animals with an anterior muscular proboscis connected by a collar region to a long wormlike body. Have gill slits. Larvae resemble echinoderm larvae. *Balanoglossus, Saccoglossus.*

Class Pterobranchia. Pterobranchs. Bottom dwelling marine animals which have arms bearing ciliated tentacles that are used in feeding. Most have a pair of gill slits. Thought to be similar to common ancestor of both echinoderms and acorn worms. *Cephalodiscus, Rhabdopleura.*

PHYLUM POGONOPHORA (80). Beard worms. Slender deep sea worms related to hemichordates. First discovered in 1900. Live in chitinous tubes and may be 40 cm long. Possess a coelom and long ciliated tentacles at the anterior end. Lack a mouth, digestive system and an anus. The tentacles are thought to form a cylindrical chamber where feeding and external digestion presumably take place. *Siboglinum, Spirobranchia, Lamellisabella.*

PHYLUM CHAETOGNATHA (50). Arrow worms. Transparent slender carnivorous animals common in marine plankton. Usually about 3 cm long and dart-shaped. Complete digestive tract. Bristles or hooks around mouth used in capturing prey. Have certain features of aschelminths. For example, they are covered with a thin cuticle. Also, the coelom is not lined with mesoderm. However their embryology has many deuterostome features. Apparently they diverged early from the deuterostome line and are not closely related to the other deuterostome groups.

PHYLUM CHORDATA (46,000). Chordates. At some stage of development or throughout life all chordates have a notochord, gill slits in the pharynx and a dorsal hollow nerve cord.

Subphylum Urochordata or Tunicata (2000). Tunicates.

Class Ascidiacea. Ascidians or sea squirts. *Ciona, Molgula, Botryllus.*

Class Thaliacea. Chain tunicates, salps. *Salpa, Doliolum, Pyrosoma.*

Class Larvacea. *Oikopleura, Appendicularia.*

Subphylum Cephalochordata (13). Lancelets, amphioxus. *Brachiostoma.*

Subphylum Vertebrata (44,000). Vertebrates.

Class Agnatha (50). Jawless fishes, without paired fins, including extinct armored ostracoderms and living lampreys. *Cephalaspis, Petromyzon, Myxine.*

Class Placodermi. Extinct armored fishes with jaws. *Dinichthys.*

Superclass Pisces

Class Chondrychthyes (700). Cartilaginous fishes including sharks and rays. *Squalus, Raja, Chimaera.*

Class Osteichthyes (20,000). Bony fishes.
Subclass Sarcopterygii
Order Crossopterygii. Lobe finned fishes. *Latimeria.*
Order Dipnoi. Lungfishes. *Protopterus, Lepidosiren.*
Subclass Actinopterygii. Ray-finned fishes. Including bowfins, bichirs, sturgeon, salmon. *Amia, Polypterus, Acipenser, Salmo, Perca, Hippocampus.*

Superclass Tetrapoda

Class Amphibia (2000).
Order Anura (Salientia). Frogs and toads. *Bufo, Hyala, Rana.*
Order Urodela (Caudata). Salamanders. *Diemictylus, Ambystoma, Necturus, Plethodon.*
Order Apoda (Gymnophiona). Coecilians. Limbless amphibians from the tropics. *Gymnopis, Typhlonectes.*

Class Reptilia (8000).
Order Chelonia. Turtles. *Chelydra, Kinosternon, Terrapene, Graptemys.*
Order Rhynchocephalia. Tuatara. *Sphenodon.*
Order Squamata. Lizards and snakes. *Anolis, Iguana, Sceloporus, Sauromalus, Heloderma, Varanus, Thamnophis, Agkistrodon, Crotalus.*
Order Crocodilia. Crocodiles, alligators, caimans, and gavials. *Crocodylus, Alligator, Gavialis.*

Class Aves (8600). Birds.
Order Sphenisciformes. Penguins. *Aptenodytes.*
Order Struthinoformes. Ostriches. *Struthio.*
Order Rheiformes. Rheas. *Rhea.*
Order Casuariiformes. Cassowaries and emus. *Cassuarius.*
Order Apterygiformes. Kiwis. *Apteryx.*
Order Tinamiformes. Tinamous. *Tinamus.*
Order Gaviiformes. Loons. *Gavia.*
Order Podicipediformes. Grebes. *Podiceps.*
Order Procellariiformes. Albatrosses and petrels. *Diomedea, Oceanodroma.*

Order Pelicaniformes. Pelicans, cormorants and gannets. *Pelicanus, Phalacrocorax, Morus.*
Order Cicioniiformes. Herons, storks, ibises, and flamingos. *Ardea, Egretta, Phoenicopterus.*
Order Anseriformes. Ducks, geese and swans. *Anas, Branta, Cygnus.*
Order Falconiformes. Hawks, falcons, eagles, and vultures. *Accipter, Buteo, Falco, Aquila, Cathartes.*
Order Galliformes. Grouse, quail, pheasants, chickens, turkeys. *Bonasa, Colinus, Gallus, Meleagris.*
Order Gruiformes. Cranes, rails, coots. *Grus, Rallus, Fulica.*
Order Charadriiformes. Shorebirds like sandpipers, also gulls, auks. *Calidris, Sterna, Larus.*
Order Columbiformes. Pigeons and doves. *Columba, Zenaidura.*
Order Psittaciformes. Parrots. *Rhynchopsitta.*
Order Cuculiformes. Cuckoos, roadrunner. *Coccyzus, Geococcyx.*
Order Strigiformes. Owls. *Tyto, Bubo.*
Order Caprimulgiformes. Goatsuckers, whippoorwill, nightjars. *Caprimulgus, Chordeiles.*
Order Apodiformes. Swifts and hummingbirds. *Chaetura, Archilochus.*
Order Coliiformes. Mousebirds or colies. *Colius.*
Order Trogoniformes. Trogons, quetzal. *Trogon, Pharomacrus.*
Order Coraciiformes. Kingfishers, hornbills. *Megaceryle.*
Order Piciformes. Woodpeckers, toucans. *Dendrocopos, Colaptes.*
Order Passeriformes. Perching birds, includes songbirds, flycatchers. *Corvus, Melospiza, Turdus.*

Class Mammalia (4500). Mammals.
Subclass Prototheria.
Order Monotremata. Egg-laying mammals. *Ornithorhynchus, Tachyglossus.*
Subclass Metatheria. Marsupials.
Order Marsupialia. Marsupials, *Didelphia, Notoryctes, Dasyurus, Macropus.*
Subclass Eutheria. Placental mammals.
Order Insectivora. Insectivores including moles, shrews, hedgehogs. *Scalopus, Sorex, Erinaceus.*

Order Dermoptera. Flying lemurs. *Galeopithecus*.

Order Chiroptera. Bats. *Myotis, Eptesicus, Desmodus*.

Order Primates. Lemurs, monkeys, apes, man. *Lemur, Tarsius, Macaca, Cebus, Hylobates, Gorilla, Pongo, Pan, Homo*.

Order Edentata. Anteaters, sloths, armadillos. *Myrmecophagus, Bradypus, Dasypus*.

Order Pholidota. Pangolins or scaly anteaters. *Manis*.

Order Lagomorpha. Pikas, rabbits, hares. *Ochotona, Sylvilagus, Oryctolagus, Lepus*.

Order Rodentia. Rodents or gnawing mammals including squirrels, rats, mice, beavers, porcupines. *Sciurus, Geomys, Dipodomys, Peromyscus, Rattus, Mus, Castor, Erethizon*.

Order Cetacea. Whales, dolphins, porpoises. *Balaena, Eschrichtius, Delphinus, Phocaena*.

Order Carnivora. Carnivores including dogs, cats, bears, mink, seals, walruses. *Canis, Vulpes, Felis, Ursus, Mustela, Phoca, Zalophus*.

Order Tubulidentata. Aardvarks. *Orycteropus*.

Order Proboscidea. Elephants. *Elephas, Loxodonta*.

Order Hyracoidea. Hyraxes or coneys. *Procavia*.

Order Sirenia. Manatees or sea cows. *Trichechus, Dugong*.

Order Perissodactyla. Odd-toed hoofed mammals including horses, tapirs, rhinoceroses. *Equus, Tapirus, Rhinoceros*.

Order Artiodactyla. Even-toed hoofed mammals including pigs, camels, deer, giraffes, pronghorned antelopes, cows, and sheep. *Sus, Camelus, Cervus, Alces, Giraffa, Antilocapra, Bos, Bison, Ovis*.

APPENDIX B—SOME IMPORTANT UNITS OF MEASUREMENT

gram (g)	0.035 ounce
kilogram (kg)	2.2 pounds
liter (l)	1.057 quarts
milliliter (ml)	0.034 fluid ounces
cubic centimeter (cm³)	0.061 cubic inches
millisecond (msec)	0.001 second
meter (m)	39.37 inches
centimeter (cm)	0.39 inch
millimeter (mm)	0.039 inch
micron (μ) or	
micrometer (μm)	one thousandth mm
nanometer (nm) or	
millimicron (mμ)	one millionth mm
angstrom (Å)	one ten millionth mm

Credits and Acknowledgments
(continued from copyright page)

Biology, **32**:193-208, 1967. Reproduced by permission of The Rockefeller University. 2-29A: L. K. Shumway and T. E. Weier, *American Journal of Botany,* **54**:773, 1967; 2-31: D. W. Fawcett; 2-32: A. H. Sparrow and R. F. Smith, Brookhaven National Laboratory.

Chapter 3. 3-1: Don W. Fawcett and Everett Anderson; 3-2: Saundra Villafane; 3-6B: J. Kezer; 3-16: From *Genetics,* 2/E, by Robert Paul Levine. Copyright © 1968 by Holt, Rinehart and Winston. Reprinted by permission of Holt, Rinehart and Winston; 3-12: Richard M. Tullar, from *Life: Conquest of Energy* by Richard M. Tullar. Copyright © 1972 by Holt, Rinehart and Winston. Reprinted by permission of Holt, Rinehart and Winston; 3-14: E. W. Sinnott, L. D. Dunn, and T. Dobzhansky, *Principles of Genetics,* 5/E, © 1958. With permission of McGraw-Hill Book Company; 3-17: Brenner, Horne, et al.; 3-18: Reprinted with permission of Macmillan Publishing Co., Inc. from *Genetics* by Monroe W. Strickberger. Copyright © 1968 by Monroe W. Strickberger.

Chapter 4. 4-7, -8, -10, -11B, -12, -13, -15, -16, -17, -18, -19, -20, -21: Loewy/Siekevitz; 4-14: S. Fleischer; 4-22: Loewy/Siekevitz, after Kendrew; 4-23: Loewy/Siekevitz, after Low and Esdall, 1965; 4-24: After M. O. Dayhoff and R. V. Eck, *Atlas of Protein Sequence and Structure,* **3**, National Biomedical, 1968.

Chapter 5. 5-3, -4, -5, -7: Loewy/Siekevitz; 5-6: From the *Atlas of Protein Sequence and Structure 1967-68,* Margaret O. Dayhoff and Richard V. Eck, National Biomedical Research Foundation, Washington, D.C., 1968. The drawing was made by Irving Geis based on his perspective painting of the molecule which appeared in *Scientific American,* November, 1966. The painting was made of an actual three-dimensional model assembled at the Royal Institution, London, by D. C. Phillips and his colleagues, based on their X-ray crystallography results; 5-8: G. E. Palade and R. Bruns; 5-9: G. E. Palade; 5-20: H. Fernandez-Moran, T. Oda, P. V. Blair and D. E. Green, *Journal of Cell Biology,* **22**:73, no. 1, 1964. Reproduced by permission of The Rockefeller University.

Chapter 6. 6-5: Loewy/Siekevtiz, after H. Taylor; 6-6, -8, -10, -11, -12, -14B, -15, -16, -17, -18, -19, -20: Loewy/Siekevitz; 6-13: H. E. Huxley; 6-14A: P. Siekevitz; 6-18: Jack Griffith; 6-21: H. Fernandez-Moran, University of Chicago.

Chapter 7. 7-1, 7-22B, C, D and E, -24: Carnegie Institution of Washington; 7-2: After G. B. Moment, *Journal of Experimental Zoology,* **117**:6, no. 1, 1951, The Wistar Institute Press; 7-3: From *Interacting Systems in Development,* 2/E, by James D. Ebert and Ian M. Sussex. Copyright © 1970 by Holt, Rinehart and Winston. Reprinted by permission of Holt, Rinehart and Winston; 7-5: From *Modern Embryology* by Charles W. Bodemer. Copyright © 1968 by Holt, Rinehart and Winston. Reprinted by permission of Holt, Rinehart and Winston; 7-6: Reproduced by courtesy of Professors Hamilton, Boyd and Mossman and Messrs Heffer & Sons, Cambridge, from *Human Embryology;* 7-7 top, 7-8: from original drawings by D. W. Bishop; 7-7 bottom: D. W. Bishop, from *Sex and Internal Secretions,* II, 2/E, W. C. Young, ed., © 1961, The Williams & Wilkins Co., Baltimore; 7-9: A. L. Colwin and L. H. Colwin, from *Cellular Membranes in Development* (the 22nd Symposium of the Society for the Study of Development and Growth), ed. by Michael Locke, © 1964 Academic Press; 7-10: A. L. Colwin and L. H. Colwin, *Journal of Biochemical and Biophysical Cytology,* **10**:233, 257, no. 2, 1961. Reproduced by permission of The Rockefeller University; 7-11, -20c: T. W. Torrey, *Morphogenesis of the Vertebrates,* © 1962, John Wiley & Sons, Inc. By permission; 7-12, -18: Bodemer; 7-14: B. I. Balinsky, *An Introduction to Embryology,* 2/E, © 1970, W. B. Saunders Company; 7-17: J. Holtfreter and V. Hamburger, from B. H. Willier, *Analysis of Development,* W. B. Saunders Company; 7-19: After Torrey; 7-20A and B: After Hamilton and Mossman; 7-21, -22A: A. T. Hertig, J. Rock, E. C. Adams, and W. J. Mulligan, *Contrib. Embryol, Carnegie Inst.,* **240**. By courtesy of the Carnegie Institution of Washington; 7-23: From F. C. Steward "The Control of Growth in Plant Cells." Copyright © 1963 by Scientific American, Inc. All rights reserved; 7-25a-i; From *Plant Morphology* by A. W. Haupt. Copyright © 1953 by McGraw-Hill Book Company, Inc. Used with permission of McGraw-Hill Book Company; 7-25j-m: M. Schaffner, *The Ohio Naturalist*, **vii**:6, 7, no. 1, 1906, The Ohio Journal of Science. **Chapter 8.** 8-1: After J. Brachet, *Biochemical Cytology*, Academic Press; 8-2: Ebert/Sussex, after Hammerling; 8-3: Ebert/Sussex, after H. Spemann; 8-4: R. Briggs and T. J. King, *Biological Specificity And Growth* (12th Symposium of the Society for the Study of Development and Growth), ed. by E. G. Butler, Princeton University Press; 8-5: F. C. Steward, et al., *Science*, **143**: 20-27, no. 3601, Jan. 1964. Copyright 1964 by the American Association for the Advancement of Science; 8-6: V. Vasil and A. C. Hildebrandt, *Science*, **150**:889-892, no. 3698, Nov. 1965. Copyright by the American Association for the Advancement of Science; 8-7 right: J. M. Whitten, "Coordinated Development in the Foot Pad of the Fly *Sarcophaga bullata* during Metamorphosis: Changing Puffing Patterns of the Giant Cell Chromosomes," *Chromosoma* (Berl.), **26**:215-244, 1969. Berlin-Heidelberg-New York: Springer; 8-8: U. Clever, "Genaktivitäten in den Riesenchromosomen von Chironomus Tentans und ihre Beziehungen zur Entwicklung. I. Geneaktivierungen durch Ecdyson," *Chromosoma* (Berl.), **12**:607-675, 1961. Berlin-Göttingen-Heidelberg: Springer; 8-9: O. L. Miller, Jr. and Barbara R. Beatty, Oak Ridge National Laboratory; 8-10: D. D. Brown and I. B. Dawid, *Science*, **160**:272-280, no. 3825, Apr 1968. Copyright 1968 by the American Association for the Advancement of Science; 8-12: From *Embryology*, rev. ed., by Lester G. Barth. Copyright © 1953, by Holt, Rinehart and Winston. Reprinted by permission of Holt, Rinehart and Winston; 8-13: I. B. Dawid and D. B. Wolstenholme; 8-14: A. J. Coulombre; 8-15, -16: I. R. Konigsberg, *Science*, **140**:1273-1284, no. 3573, June 1963. 8-17, -18, -19: From *Proceedings of The National Academy of Sciences*, **58**:344-351, 1967, by permission of the author, Beatrice Mintz; 8-22: M. S. Steinberg, *Cellular Membranes in Development*, **22**; 8-24A: Reproduced by permission of the National Research Council of Canada from the *Canadian Journal of Botany*, **47**:1367-1375, 1969; 8-24B: Reproduced by permission of the National Research Council of Canada from the *Canadian Journal of Botany*, **42**:1615-1628, 1964; 8-25,

-26: After J. W. Saunders, M. T. Gasseling, and L. C. Saunders, *Developmental Biology* (5th Symposium of the Society for the Study of Growth and Development), Academic Press. **Chapter 9.** 9-2: From *Microbial Life*, 2/E, by W. R. Sistrom. Copyright © 1969 by Holt, Rinehart and Winston. Reprinted by permission of Holt, Rinehart and Winston; 9-3: N. Lang, *Journal of Phycology*, **1**, 1965; 9-4, -5, -6, -7, -13, -14, -16, -17, -18: Sistrom; 9-15: E. Cota-Robles and M. D. Coffman, *Journal of Bacteriology*, **86**:267, 1963. **Chapter 10.** 10-1, -5, -6, -9, -10, -11, -12, -16, -19, -24: After *Botany*, 5/E, by Carl L. Wilson, Walter E. Loomis, and Taylor A. Steeves, Copyright © 1971 by Holt, Rinehart and Winston. Reprinted by permission of Holt, Rinehart and Winston; 10-2: Paul B. Green, from *The Living Plant*, 2/E, by Peter Martin Ray. Copyright © 1972 by Holt, Rinehart and Winston. Reprinted by permission of Holt, Rinehart and Winston; 10-3: Myron C. Ledbetter, from M. C. Ledbetter and R. K. Porter, *An Introduction to the Fine Structure of Plant Cells*, © 1970, Springer-Verlag; 10-4: From *The Plant World*, 5/E, by Harry J. Fuller et al. Copyright © 1972 by Holt, Rinehart and Winston. Reprinted by permission of Holt, Rinehart and Winston; 10-6B: J. Heslop-Harrison; 10-8: R. Anderson and J. Cronshaw, "Sieve-Plate Pores in Tobacco and Bean," *Planta* (Berl.) **91**:173-180, 1970. Berlin-Heidelberg-New York: Springer; 10-14, -35: Ray; 10-15: Reprinted by permission from *Botanical Microtechnique*, 3/E, by John E. Sass, © 1958 by The Iowa State University Press, Ames, Iowa; 10-18: Bob Broder; 10-20: Norman Hodgkin, UCI; 10-21: After data of R. Bohning and C. Burnside, *American Journal of Botany*, **43**:560, 1956; 10-22: Wilson, Loomis, Steeves; 10-29: M. H. Zimmerman, *Science*, **133**: cover, 76, no. 3446. Jan. 1961. Copyright 1961 by the American Association for the Advancement of Science; 10-36: S. H. Wittwer and M. J. Bukovac, *Economic Botany*, **23**:226, no. 3, 1958. **Chapter 11.** 11-4: Norman Hodgkin, UCI: 11-7: After Novikoff/Holtzman; 11-8A, -8E: Courtesy of the American Museum of Natural History; 11-8B: Lynwood M. Chase, National Audubon Society; 11-8C: © Douglas P. Wilson; 11-8D: United States Department of Agriculture; 11-17: Howard A. Schneiderman; 11-19: Robert D. Allen.

Chapter 12. 12-1A, and B: UNICEF; 12-2B, -3B: Treat Davidson, National Audubon Society; 12-3A, -24A: © Douglas P. Wilson; 12-3C: Roger Seapy, UCI; 12-3D: Frank Williamson, National Audubon Society; 12-3E: Ed Cesar, National Audubon Society; 12-4: Eric V. Gravé, Photo Researchers; 12-9 left and right: Jeanne M. Riddle; 12-17: The Cystic Fibrosis Foundation; 12-19A, -19B: American Cancer Society; 12-21: After Griffin/Novick; 12-22: Charles L. Trainor; 12-25: Pierre Lyonet, *Traite Anatomique de la Chenille qui Ronge le Bois de Saule*, La Haye, 1760; 12-32: Norman Hodgkin, UCI; 12-33: Clinton Van Zandt Hawn, Keith R. Porter, and the Rockefeller Institute for Medical Research; 12-37A, -37B: American Heart Association; 12-43: Knut Schmidt-Neilsen, Duke University; 12-48, -49 left: Karl H. Maslowsky, National Audubon Society; 12-49 right: Leonard Lee Rue III, National Audubon Society; 12-50: After C. M. Bogert, "How Reptiles Regulate Their Body Temperature." Copyright © 1959 by Scientific American, Inc. All rights reserved.

Chapter 13. 13-1A: John Parnavelas; 13-1B and C, -5: Camillo Golgi, 1886; 13-2A: J. David Robertson; 13-3: E. B. Lewis and Y. Y. Zeevi, from Everhart and Hayes, *The Scanning Electron Microscope;* 13-6: A. L. Hodgkin, *Journal of Physiology* (London), **131**: 1956; 13-16: After D. Marsland, *Principles of Modern Biology*, 4/E. Copyright © 1964 by Holt, Rinehart and Winston. Reprinted by permission of Holt, Rinehart and Winston; 13-17: Frederic A. Webster; 13-20A: After W. C. Van der Kloot, *Behavior.* Copyright © 1968 by Holt, Rinehart and Winston. Reprinted by permission of Holt, Rinehart and Winston; 13-20C: P. P. C. Graziadei; 13-28: After Peter N. Witt; 13-31: Lyonet; 13-32A and B, -33: Howard A. Schneiderman and John H. Postlethwait.

Chapter 14. 14-1A-G: Howard A. Schneiderman; 14-6: Nina Leen, *Life Magazine*, © Time, Inc.; 14-8: Wolfgang Köhler, *The Mentality of Apes*, Kegan Paul, Trench, Trubner and Company, Ltd., 1925; 14-9: Peter Bryant, UCI; 14-12: Patricia Caulfield; 14-13: John Theberge.

Chapter 15. 15-3, -4, -10: From *Essentials of Biology* by W. H. Johnson, L. E. DeLanney and T. A. Cole. Copyright © 1969 by Holt, Rinehart and Winston. Reprinted by permission of Holt, Rinehart and Winston; 15-5, -12: Ebert/Sussex; 15-8: Griffin/Novick.

Chapter 16. 16-1, -2, -3: Michigan Department of Conservation; 16-4: United States Department of Interior; 16-6: The Nitragin Company; 16-8: Roman Vishniac; 16-11: H. W. Kitchen, National Audubon Society; 16-12: Philip Hyde; 16-13: Walter Dawn; National Audubon Society; 16-14: United States Forest Service; 16-15: Hermann Postelthwaite, United States Department of Agriculture; 16-16: John H. Gerrard, National Audubon Society.

Chapter 17. 17-6: G. Ronald Austing, National Audubon Society.

Chapter 18. 18-1: John L. Shrader, EPA; 18-2: H. B. D. Kettlewell.

Chapter 19. 19-2: Rudolf Freund, Photo Researchers; 19-3: Courtesy of American Museum of Natural History; 19-4: Arthur Ambler, National Audubon Society; 19-5, -8: Tullar.

Chapter 20. 20-1: J. W. Schopf and E. S. Barghoorn, *Science*, **156**:508-512, no. 3774, April 1967. Copyright 1967 by the American Association for the Advancement of Science; 20-2, -3: J. Gross, *Cytodifferentiation and Macromolecular Synthesis* (21st Symposium of the Society for the Study of Development and Growth), ed. by Michael Locke, Academic Press.

Chapter 21. 21-2A top: Allan D. Cruickshank, National Audubon Society; 21-2A bottom: Robert C. Hermes, National Audubon Society; 21-2B: Howard A. Schneiderman; 21-4: Roberts Rugh; 21-5: After *Evolutionary Biology* by S. N. Salthe. Copyright © 1972 by Holt, Rinehart and Winston. Reprinted by permission of Holt, Rinehart and Winston; 21-6: After E. T. Bolton, Carnegie Institution of Washington, D.C. Yearbook, 1963; 21-7A: Ralph W. Chaney, Save the Redwoods League; 21-10: Robert J. Thornton, *New Illustration of the Sexual System of Carolus von Linnaeus*, Prints Division, New York Public Library; 21-11: Yvonne Freund.

Chapter 22. 22-1, -5, -10, -11, -12, -13 top, -14, -15: Courtesy the American Museum of Natural History; 22-3: Field and Wildlife Service; 22-6, -13B: Field Museum of Natural History; 22-8, -15: A. Lee McAlester, *The History of Life*, © 1968, pp. 78, 122. By permission of Prentice-Hall, Inc., Englewood Cliffs, New Jersey; 22-9: Yale Peabody Museum; 22-16 left: New York Zoological Society; 22-19 right: Arthur Ambler, National Audubon Society.

Chapter 23. 23-3A, B, and D, -23C, -24B and C, -29B, -31B, -35D: Walter Dawn; 23-3C: National research Council of Canada; 23-3E, -16D, -29C, -32B, -35A and B: Ross E. Hutchins; 23-5 left and right: Richard W. Greene; 23-6 top, -11A: © Douglas P. Wilson; 23-6 bottom: Norman Hodgkin; 23-8: Mark Littler; 23-9A: Verna R. Johnston, National Audubon Society; 23-9B, -16A: Russ Kinne, Photo Researchers, Inc.; 23-9C, -21 left, -22A and C, -25B, and D, -36 left: Hugh Spencer, National Audubon Society; 23-11B: W. R. Taylor, *Plants of Bikini*, © 1950 University of Michigan Press; 23-11D: B. Gantt and S. F. Conti, *Journal of Cell Biology*, **26**:367 No. 2, 1965. Reproduced by permission of The Rockefeller University; 23-12A: Richard Campbell and Howard A. Schneiderman; 23-12B: Charles E. Bracker; 23-15A: Lynwood M. Chace, National Audubon Society; 23-15B: Roman Vishniac; 23-16B: M. J. Baum; 23-16C: Jerry Cooke; 23-17: From *Fundamentals of Cytology* by L. W. Sharp. Copyright © 1943 by McGraw Book Company, Inc. Used with permission of McGraw-Hill Book Company. 23-19A: Howard A. Schneiderman; 23-19B: Turtox; 23-19C: Verne R. Rockcastle; 23-21 right: Danish Information Office; 23-22B: Grant Haist, National Audubon Society; 23-22D: C. G. Maxwell, National Audubon Society; 23-22E: National Park Service; 23-23A: Field Museum of Natural History; 23-23D: Ernest Ball; 23-24A: Dr. Wrice; 23-25A: Clemens Kalischer; 23-25C; -29A: Lola Beall Graham, National Audubon Society; 23-25E: Robert C. Hermes, National Audubon Society; 23-29D: Jesse Lunger, National Audubon Society; 23-31A: E. R. Degginger; 23-31C: Robert J. Rodin; 23-32A: F. E. Westlake, National Audubon Society; 23-32C: C. G. G. G. van Steenis; 23-32D: Australian News and Information Bureau; 23-35C: San Diego Zoo; 23-35E: H. E. Hinton; 23-36 right: Henry M. Mayer, National Audubon Society; 23-39: Courtesy the American Museum of Natural History. Line drawings: 23-2, -26: P. Raven and H. Curtis, *Biology of Plants*, Worth Publishers, Inc., 1970; 23-10: W. A. Jensen and F. B. Salisbury, *Botany: An Ecological Approach*, Wadsworth, also Raven and Curtis; 23-14, -30:

Jensen and Salisbury; 23-17: L. W. Sharp, *Fundamentals of Cytology*, McGraw-Hill, 1943; 23-28A: Jensen and Salisbury, from a sketch by Charles B. Beck; 23-28B: W. N. Stewart and T. Delevoryas, *Botanical Review*, **22 (1)**, 45-80, 1956; 23-34: Raven and Curtis, also Fuller, et. al.

Chapter 24. 24-2B: Martin Deckart, courtesy Carl Zeiss, Inc.; 24-6A. -40 left: Norman Hodgkin; 24-6B, -22A, -27, -30B, -44D, -76A: Walter Dawn; 24-11A, -22C: Western Marine Laboratory; 24-7, -23, -64: From *Principles of Zoology* by W. H. Johnson, L. E. De Lanney, E. C. Williams, and T. A. Cole. Copyright © 1969 by Holt, Rinehart and Winston. Reprinted by permission Holt, Rinehart and Winston; 24-9B: Tullar; 24-11B, -14, -15A and C, -16 left and right, -19, -22B, -35B and C, -36A, -37 right, -40 right, -44B and C, -46, -62B and D: © Douglas P. Wilson; 24-11C, -36B: Ralph Buchsbaum; 24-15B, -54 left: Roman Vishniac; 24-15D: Menachiem Rahat and Norman Hodgkin; 24-18A, -62A: Howard M. Lenhoff and William Diffenderfer; 24-18B, -36C, -50 left, -70 both, -75E and F: Courtesy the American Museum of Natural History: 24-22D: Luvenia C. Miller; 24-30A and C, -47C: John H. Gerrard, National Audubon Society; 24-31, -35A, -47B, -62C, -72A, -74B: Robert C. Hermes, National Audubon Society; 24-33A, -39B: H. B. Whittington, from B. Kummel, *History of the Earth, 2/E*, W. H. Freeman and Company, © 1970; 24-33B: M. F. Glaessner; 24-35F, -51A: Lynwood M. Chase, National Audubon Society; 24-37 bottom, -73B: Marine Studios, Marineland, Florida; 24-39A: Carolina Biological Supply Company; 24-41: William J. Jahoda, National Audubon Society; 24-43A and B: United States Department of Agriculture; 24-44A, -51C and E, -73D: Russ Kinne, Photo Researchers; 24-47A, -74C: New York Zoological Society; 24-47D: Jerome Wexler, National Audubon Society; 24-50 right, -59 right: Buffalo Museum of Science; 24-51B, -71: Field Museum of Natural History; 24-51D, -74D: Ross E. Hutchins; 24-57: Charles Walcott; 24-60: Howard A. Schneiderman; 24-72C, -73A: Mitchell Campbell, National Audubon Society; 24-73C: Arthur Ambler, National Audubon Society; 24-74A: Susan V. Bryant; 24-75A: N. E. Beck, Jr., National Audubon Society; 24-75C: Rhea Warren; 24-75D: Florida

Development Commission; 24-76C, -77 left and right: George Hunt, Jr.; 24-76B: Michigan Conservation Department; 24-76D: M. F. Soper, National Audubon Society; 24-79A: Standard Oil Company of New Jersey; 24-79B: Acme News-pictures, Inc.; 24-79C: Gordon S. Smith, National Audubon Society; 24-80: After A. E. Brehns, *Tierleben Allgemaine Kunde des Tierreichs*, 4/E. Leipzig und Wien Bibliographisches Institut, 1912; 24-81: Courtesy Carnegie Institution of Washington.

Chapter 25. 25-1, -14: Reprinted with permission of Macmillan Publishing Co., Inc. from *Race and Races* by Richard A. Goldsby. Copyright © 1971 by Richard A. Goldsby; 25-4: New York Zoological Society; 25-5: Jean Jacques Petter; 25-6, -15 left and right, -16E, F, G: Courtesy the American Museum of Natural History; 25-7: Courtesy Irven DeVore; 25-9: Arthur Ambler, National Audubon Society; 25-10: Courtesy of the National Geographic Society and Baron Hugo van Lawick; 25-11 right: © National Geographic Society; 25-12: By permission of the Trustees of the British Museum (Natural History); 25-13: From *Animal Diversity* by M. Fingerman. Copyright © 1969 by Holt, Rinehart and Winston. Reprinted by permission of Holt, Rinehart and Winston; 25-16A, C: United Nations; 25-16D: UNICEF.

Index*

*Definition on *italicized* page, illustration on **bold** page.

R

rabbits, 715, 717
raccoons, 715
races, *519*
    of modern man, 730, **731**, 732
    See also adaptive radiation
radial symmetry, 327–328
radiation, and mutations, 81–83, **84**
radioactive dating of fossils, 555
radiolarians, 632
radula, 660, 661
ragweed, 618
Ramapithecus, 725
Ramapithecus punjabicus, 721
Rana, 212–213, **213**
    pipiens, 213, **213**, 214
random segregation of alleles, 510
rats, 715–717
    behavior, 441–442, **442**
    ionic composition of body fluids
        of, 104 (table)
    and memory, 449
Rauwolfia, 623
rays, 700, **701**
reabsorption, in excretory sys-
    tems, 388–390
reasoning, 448
receptacle, 613
receptors, 416–421, **418, 419, 421,
    422, 423**
    and hormones, 464–466
recombination of genes, 63, 65–67,
    67, 75–77, 90, 510
rectum, 361, **457, 463**
red algae, 586–587, **587**
red blood cells, **31,** 366, 379–380,
    380
    and *Plasmodium,* 634
red cell ghosts, **31**
red tides, 584
Redi, Francesco, and spontaneous
    generation, 532
redwoods, 541, **542,** 600, **601,** 611
reflexes, 439
    spinal, 412–413
regeneration, 183, **183,** 184, 705
regulator genes, **172,** 173, **173**
reindeer, 715
relaxin, 460 (table), 461 (table)
replication, *49*
    bacteriophage, 89–90, **89**
    of DNA, 124, **124,** 160–163, **161,**
        161 (table), **162,** 162 (table),
        163, 268–270, **270,** 278–281,
        279
repressor protein, **172,** 173–174,
    173, 174, 270–272, **271**
reproduction, (*see* asexual repro-
    duction; human reproduc-
    tion; sexual reproduction)
reptiles
    evolution of, 562, 563
    excretion in, 391
    temperature regulation in, 397–
        398, **397**
respiration, 115, 263, *364–365*

bacterial, 265, 265 (table)
    in birds, **372**
    carbon dioxide in, 365–367
    in echinoderms, 692
    in fish, 370–373, **371**
    of glucose, *133,* 145–146, *148–149,*
        150, 151, 152–153, **152, 153,**
        265
    in insects, 373–374, **374**
    in man, 365–370, **365, 366, 368,
        369**
    and origin of life, 531
    in vertebrates, 370–373, **370, 371,
        372, 373**
retina, 198
revolutions
    biological, 10, 12
    cybernetic, 10, 11
    and evolution, 5, 7–15
    industrial, 5, 7
    and population growth, **6,** 12–13
    scientific, 7, 9–10, **11**
    urban, 7, **9**
Rh factor, 383
Rhesus monkeys, 723
rhinoceros beetle, 679, **680**
Rhinodon, 701
rhizoids, 598
rhizomes, 620
Rhizopus, 590, **591**
Rhodophyta, 586–587, **587**
Rhynchocoela, Appendix A
Rhynchosciara, chromosomes of,
    217
Rhynia Gwynne-Vaughni, 603
rib cage, 336, **336**
ribbon worms, 650–651, **650**
ribonucleic acid, (*see* RNA)
ribosomes, **26, 27,** *31, 33,* 83, **170**
    in procaryotes, 258
    and RNA, 163–166, **169,** 217, 219–
        223
rickets, **353**
Rickettsia, 680
Rickettsiae, Appendix A
rivers, as ecosystem, 489
RNA
    and cancer, 246–247
    and chromosome puffing, 217
    discovery of, 119
    and memory, 449
    messenger, 163–168, **169,** 170,
        172, 217, 221
    in protein synthesis, 163–168,
        165, 167, 168, 169, 170
    and ribosomes, 163–166, **169,** 217,
        219–223
    structure of, 120, **120, 121,** 124
    transfer, 165–168, **165, 167, 168,
        169,** 170
Rock, John, and oral contraceptive
    pills, 471
Rodentia, 715
root caps, 293, 295
root hairs, 307–309, **307**
root pressure, 312
roots

structure and growth of, 292–
    293, **292, 293, 294, 295**
    and water absorption, 306–308,
        307, 308
Rothschildia, 440, 547
Rotifera, Appendix A
roundworms, 653–654, **653, 654**
rubella, 246
rubidium, 100
rumen, 363
rusts, 592, 595

S

Sabella pavonia, **373**
sac fungi, 590–591, **591**
Sacculina, 671
Sagitta, **674**
salamanders, **704,** 705
    cells of, **32**
    embryo development in, 193–
        198, **193, 194, 195, 196, 197,**
        200, **212,** 212
    regeneration in, 184
    skin, 334
salivary glands, 358
    of *Drosophila,* chromosomes in,
        78, 79, 80, **80**
    of mice, development in, 231–
        232, **232**
Salmo salar, **702**
salmon, **702**
Salmonella typhimurium, 280
Salsola, 619
salt glands, 391
Samia cynthia, **440**
sand dunes, and succession proc-
    ess, 477–479, **477, 478, 479**
Sanger, Fred, and protein insulin,
    128, 167
sarcodines, 630–632, **631, 632**
Sarcomastigophora, Appendix A
sarcomeres, 342–343
Sarcophaga bullata, 216
Sarcopterygii, Appendix A
Sargassum, 584
saturation, 116
scallops, **664**
Scaphopoda, Appendix A
Schistosoma, 649
schistosomes, 649
schistosomiasis, 649–650
Schizomychetes, Appendix A
schizophrenia, concordance of
    identical twins with, 91–92,
    91 (table)
Schleiden, Matthias, and plant
    cells, 19
Schwann, Theodor, and animal
    cells, 19
scientific revolution, 7, 9–10, **11**
sclerotin, 340
scorpions, **676, 678**
Scutigera, **671**
Scyphozoa, (*see* jellyfishes)
sea anemones, **644,** 645
sea butterflies, **662**